Inter-Basin Water Transfer
Case Studies from Australia, United States, Canada, China and India

Since the Second World War increasing demands for irrigation, domestic and industrial water have generated a massive growth world-wide in the number of large water infrastructure projects. Many of these projects involved the transfer of water from basins considered to have surplus water to those where the demand for water has exceeded or is expected to exceed supplies. While these inter-basin water transfers have substantially contributed to the overall development of numerous countries, they also have caused environmental, social, cultural and economic problems.

Using the experience of inter-basin water transfer projects in Australia, United States, Canada, China and India this book examines case studies within the diverse geographical, climatic, economic, and policy regimes operating in these countries. The first part of the book is an overview of world challenges with respect to water resources and discusses the key issues in inter-basin water transfers. The second part examines the water resources of Australia, the driest inhabited continent. It describes the benefits and impacts of a number of inter-basin transfer projects developed or proposed in Australia. The third part explores inter-basin water transfer projects in the United States, Canada, China and India, examining their benefits and impacts within these nations' contrasting economies and governance systems. The fourth part consists of numerous appendices. The book concludes by highlighting the successes and failures of the case examined, and provides pointers for the future of inter-basin water transfer in meeting urgent and growing water demands. This comprehensive and well-illustrated text will be of great interest to professionals and researchers in the fields of hydrology, water resources, and to those engaged in environmental science, policy and regulation.

FEREIDOUN GHASSEMI is Visiting Fellow at the Centre for Resource and Environmental Studies, The Australian National University. He is a Fellow of the Modelling and Simulation Society of Australia and New Zealand and was recipient of the G. Burton Medal from the Hydrological Society of Canberra in 1995. Dr Ghassemi has more than 35 years of experience in various aspects of water resource research in Australia, France, Iran and Vietnam.

IAN WHITE is Professor of Water Resources at the Centre for Resource and Environmental Studies, The Australian National University. He is a Fellow of the American Geophysical Union and the Australian Academy of Technological Sciences and Engineering. Professor White was awarded a Centenary Medal for service to Australian society in environmental science and technology in 2003 and has twice (in 1994 and 1997) received the G. Burton Publication Medal from the Hydrological Society of Canberra. He has worked in water and land resources in Australia, the United States, Pacific small island nations, Vietnam, China and France.

INTERNATIONAL HYDROLOGY SERIES

The **International Hydrological Programme** (IHP) was established by the United Nations Educational, Scientific and Cultural Organization (UNESCO) in 1975 as the successor to the International Hydrological Decade. The long-term goal of the IHP is to advance our understanding of processes occurring in the water cycle and to integrate this knowledge into water resources management. The IHP is the only UN science and educational programme in the field of water resources, and one of its outputs has been a steady stream of technical and information documents aimed at water specialists and decision-makers.

The **International Hydrology Series** has been developed by the IHP in collaboration with Cambridge University Press as a major collection of research monographs, synthesis volumes and graduate texts on the subject of water. Authoritative and international in scope, the various books within the series all contribute to the aims of the IHP in improving scientific and technical knowledge of fresh-water processes, in providing research know-how and in stimulating the responsible management of water resources.

EDITORIAL ADVISORY BOARD

Secretary to the Advisory Board
Dr Michael Bonell *Division of Water Science, UNESCO, 1 rue Miollis, Paris 75732, France*

Members of the Advisory Board
Professor B. P. F. Braga Jr *Centro Technológica de Hindáulica, São Paulo, Brazil*
Professor G. Dagan *Faculty of Engineering. Tel Aviv University, Israel*
Dr J. Khouri *Water Resources Division, Arab Centre for Studies of Arid Zones and Dry Lands, Damascus, Syria*
Dr G. Leavesley *US Geological Survey, Water Resources Division, Denver Federal Center, Colorado, USA*
Dr E. Morris *Scott Polar Research Institute, Cambridge, UK*
Professor L. Oyebande *Department of Geography and Planning, University of Lagos, Nigeria*
Professor S. Sorooshian *Department of Civil and Environmental Engineering, University of California, Irvine, California, USA*
Professor K. Takeuchi *Department of Civil and Environmental Engineering, Yamanashi University, Japan*
Professor D. E. Walling *Department of Geography, University of Exeter, UK*
Professor I. White *Centre for Resource and Environmental Studies, Australian National University, Canberra, Australia*

TITLES IN PRINT IN THE SERIES

M. Bonnell, M. M. Hufschmidt and J. S. Gladwell *Hydrology and Water Management in the Humid Tropics: Hydrological Research Issues and Strategies for Water Management*
Z. W. Kundzewicz *New Uncertainty Concepts in Hydrology*
R. A. Feddes *Space and Time Scale Variability and Interdependencies in the Various Hydrological Processes*
J. Gibbert, J. Mathieu and F. Fournier *Groundwater and Surface Water Ecotones: Biological and Hydrological Interactions and Management Options*
G. Dagan and S. Neuman *Subsurface Flow and Transport: A Stochastic Approach*
J. C. van Dam *Impacts of Climate Change and Climate Variability on Hydrological Regimes*
J. J. Bogardi and Z. W. Kundzewicz *Risk, Reliability, Uncertainty and Robustness of Water Resources Systems*
G. Kaser and H. Osmaston *Tropical Glaciers*
I. A. Shiklomanov and John C. Rodda *World Water Resources at the Beginning of the Twenty-First Century*
A. S. Issar *Climate Changes during the Holocene and their Impact on Hydrological Systems*
M. Bonnell and L. A. Bruijnzeel *Forests, Water and People in the Humid Tropics: Past, Present and Future Hydrological Research for Integrated Land and Water Management*
F. Ghassemi and I. White *Inter-Basin Water Transfer: Case Studies from Australia, United States, Canada, China and India*

INTER-BASIN WATER TRANSFER:

Case Studies from Australia, United States, Canada, China and India

By:

Fereidoun Ghassemi and Ian White

CAMBRIDGE
UNIVERSITY PRESS

CAMBRIDGE UNIVERSITY PRESS
Cambridge, New York, Melbourne, Madrid, Cape Town, Singapore, São Paulo

Cambridge University Press
The Edinburgh Building, Cambridge CB2 2RU, UK

Published in the United States of America by Cambridge University Press, New York

www.cambridge.org
Information on this title: www.cambridge.org/9780521869690

© Cambridge University Press 2007

This publication is in copyright. Subject to statutory exception and to the provisions of relevant collective licensing agreements, no reproduction of any part may take place without the written permission of Cambridge University Press.

Printed in the United Kingdom at the University Press, Cambridge

A catalogue record for this publication is available from the British Library

Library of Congress Cataloging-in-Publication data

Ghassemi, F. (Fereidoun), 1940-
  Inter-basin water transfer : case studies from Australia, United States, Canada, China, and India / by Fereidoun Ghassemi and Ian White.
    p. cm. – (International hydrology series)
  Includes bibliographical references and index.
  ISBN-13: 978-0-521-86969-0 (hardback)
  ISBN-10: 0-521-86969-2 (hardback)
  1. Water transfer–Case studies. 2. Water-supply–Management–Case studies. 3. Water consumption–Forecasting–Case studies. I. White, Ian, 1943- II. Title. III. Series.

TC409.G49 2006
363.6'1–dc22

2006034149

ISBN-13: 978-0-521-86969-0 hardback
ISBN-10: 0-521-86969-2 hardback

Cambridge University Press has no responsibility for the persistence or accuracy of URLs for external or third-party internet websites referred to in this publication, and does not guarantee that any content on such websites is, or will remain, accurate or appropriate.

Dedication

This book is dedicated to the memory of Benedict (Ben) Chifley, the Post-War visionary Labor Prime Minister of Australia (July 1945 to December 1949) and founder of the Australian National University 60 years ago on the 1st August 1946 who understood the importance of water in Australia and had the courage and tenacity to act on that understanding.

Note
Throughout this book:

The Australian dollar* is represented by $
The US dollar is represented by US$, and
The Canadian dollar is represented by CAN$

* In February 1966 the Australian currency system was
converted from the British system of pounds to
Australian dollars, which were worth
half a pound.

Disclaimer

The authors, publisher and the Centre for Resource and Environmental Studies, the Australian National University, would like to advise that the information contained in this publication is based on scientific publications and research results. As such, this information may be incomplete or not suitable to be used in any specific situation. No reliance or actions should be made on that information without seeking prior expert advice.

The authors, publisher and the Centre for Resource and Environmental Studies, the Australian National University exclude all liability to any individual person, organisation, government department, research institution, and others for any consequences including but not limited to all losses, damages, costs, expenses and any other compensation, arising directly or indirectly from using this publication (in part or as a whole) and any information or material contained in it.

Contents

Foreword		page xv
Overview and Scope		xix
Acknowledgements		xxiii
List of Abbreviations		xxv

Part I   The Challenges — 1

1   World population and pressures on land, water and food resources — 3

1.1	Population	3
1.2	Dryland areas	4
1.3	Extent of human-induced land degradation	4
1.4	Water resources	8
1.5	Agricultural land use	13
1.6	Food and fibre production	15
1.7	Feeding the world population	16
1.8	World water and food to 2025	17
1.9	Challenge Program on water and food	18
1.10	Conclusions	19
	References	20

2   Issues in inter-basin water transfer — 22

2.1	Introduction	22
2.2	Knowledge requirements and inter-basin water transfer	23
2.3	Planning and public participation	27
2.4	Assessment of the impacts	28
2.5	Environmental flow requirements of rivers	31
2.6	Social and cultural issues	35
2.7	Economic appraisal	37
2.8	Water rights	38
2.9	Conflicts and their resolution	40
2.10	Integrated assessment and modelling	43
2.11	Conclusions	45
	References	45

Part II   Inter-basin Water Transfer in Australia — 49

3   Land and water resources of Australia — 51

3.1	Geography	51
3.2	Population	51
3.3	Climate	54
3.4	Climate change	56
3.5	Drought	59
3.6	Flood	60
3.7	Soil resources	61
3.8	Agricultural land use	64
3.9	Water resources	65
3.10	Environmental degradation	70
3.11	Management reforms and programmes	74
3.12	Estimates of future water requirements	79
3.13	National water initiative	84
3.14	Potential role of inter-basin water transfer	85
3.15	Conclusions	87
	References	87

4   The Snowy Mountains hydro-electric scheme — 91

4.1	Location	91
4.2	Hydrology	91
4.3	Decline in precipitation	91
4.4	Historical background	92
4.5	Snowy Mountains Act	95
4.6	Cost of the scheme	96
4.7	Technical features of the scheme	96
4.8	Water releases	97
4.9	Electricity production	97
4.10	Workforce	99
4.11	Environmental impacts of the scheme	101
4.12	Corporatisation of the scheme	101
4.13	The Snowy water inquiry	102
4.14	The environmental flow agreement	104
4.15	Precipitation enhancement project	104

5 Inter-basin water transfer from coastal basins of New South Wales 107

- 5.1 Introduction 107
- 5.2 Environmental problems of the North Coast river basins 107
- 5.3 Proposed diversion schemes 110
- 5.4 The scoping study 122
- 5.5 Clearance scheme and water supply of Adelaide 122
- 5.6 Conclusions 123
- References 123

6 The Bradfield and Reid schemes in Queensland 125

Section A: The Bradfield scheme 125

- 6.1 Introduction 125
- 6.2 Water availability 125
- 6.3 Outline of the Bradfield scheme 125
- 6.4 Costs and benefits of the scheme 126
- 6.5 The 1947 review of the scheme 126
- 6.6 The expanded Bradfield scheme 129
- 6.7 The 1982 review of the scheme 131
- 6.8 Bradfield scheme and water supply of Adelaide 134

Section B: The Reid scheme 135

- 6.9 Introduction 135
- 6.10 Description of the scheme 135
- 6.11 Cost of the scheme 137
- 6.12 Expected benefits of the scheme 137
- 6.13 Conclusions 137
- References 137

7 Three schemes for flooding Lake Eyre 139

- 7.1 Introduction 139
- 7.2 Characteristics of the Lake Eyre Basin 139
- 7.3 Port Augusta–Lake Eyre canal scheme 144
- 7.4 The Great Boomerang Scheme 147
- 7.5 Flooding Lake Eyre with waters of the Great Artesian Basin 149
- 7.6 Conclusions 149
- References 150

8 The Goldfields pipeline scheme of Western Australia 151

- 8.1 Introduction 151
- 8.2 Water shortage 151
- 8.3 Pipeline proposals 152
- 8.4 Conclusions 162
- References 164

9 Supplying Perth, Western Australia with water: the Kimberley pipeline scheme 165

- 9.1 Introduction 165
- 9.2 Water conservation strategy 165
- 9.3 Long-term water supply options for Perth 165
- 9.4 Perth's water supply options 167
- 9.5 Inter-basin water transfer from Kimberley 169
- 9.6 Bulk water transport by ship from Kimberley to Perth 177
- 9.7 Seawater desalination for Perth's water supply 177
- 9.8 Conclusions 178
- References 179

10 Other schemes in Australia 180

- 10.1 Introduction 180

Section A: River Murray pipelines in South Australia 180

- 10.2 Introduction 180
- 10.3 Morgan–Whyalla pipelines 181
- 10.4 Mannum–Adelaide pipeline 183
- 10.5 Swan Reach–Paskeville pipeline 183
- 10.6 Tailem Bend–Keith pipeline 183
- 10.7 Murray Bridge–Onkaparinga pipeline 183

Section B: Mareeba–Dimbulah irrigation scheme, Queensland 184

- 10.8 Introduction 184
- 10.9 History of the scheme 184
- 10.10 Agricultural development 186
- 10.11 Water allocation and water use 186
- 10.12 Power generation and town water supply 186
- 10.13 Water quality issues of the Tinaroo Falls Lake 187
- 10.14 Barron water resources plan 187
- 10.15 Possibilities for future expansion 188
- 10.16 Impacts of water resources development 188

Section C: Domestic and industrial water supply in North Queensland 188

- 10.17 Water supply from Eungella Dam 188
- 10.18 Pipelines for water supply of Townsville and Thuringowa 189

4.16 Conclusions 105
References 105

10.19	Pipeline to Bowen area	189

Section D: Water supply to the Broken Hill mines and township, New South Wales — 189

10.20	Introduction	189
10.21	Water supply	191
10.22	Conclusions	197
	References	198

Part III Inter-basin Water Transfer in Other Selected Countries — 199

11 Inter-basin water transfer in the United States of America — 201

Section A: Overview of geography, population, land and water — 201

11.1	Geography	201
11.2	Population	202
11.3	Precipitation and climate	202
11.4	Land use	203
11.5	Water resources	204
11.6	Flood	207
11.7	Drought	208
11.8	Climate change impacts	209
11.9	Water transfer projects in the United States	209
11.10	Ambitious plans for water transfer	211
11.11	Federal water plan for the west (water 2025)	212

Section B: Inter-basin water transfer in California — 215

11.12	Geography and population	215
11.13	Water supply and demand	215
11.14	Water transfer projects	217
11.15	Major management programs and strategies	229

Section C: Inter-basin water transfer from the Colorado River — 240

11.16	Colorado River Basin	240
11.17	Water transfer projects	243
11.18	Conclusions	257
	References	258

12 Inter-basin water transfer in Canada — 261

Section A: Overview of geography, population, land and water — 261

12.1	Geography	261
12.2	Population	261
12.3	Economy	262
12.4	Climate and precipitation	262
12.5	Land cover and use	263
12.6	Water resources	264
12.7	Flood	269
12.8	Drought	270
12.9	Hydro-power generation	270
12.10	Climate change impacts	270
12.11	Management of water resources	272

Section B: Inter-basin water transfer projects — 275

12.12	Introduction	275
12.13	Examples of water transfer projects	276
12.14	Great Lakes Basin diversions	281
12.15	Impacts of the diversion projects	281
12.16	Learning from Canadian experience	284
12.17	Large-scale water export proposals	284
12.18	Water export policy	290
12.19	Conclusions	292
	References	293

13 Inter-basin water transfer in China — 295

Section A: Overview of geography, population, land and water — 295

13.1	Geography	295
13.2	Population	296
13.3	Economy	296
13.4	Climate and precipitation	297
13.5	Land cover and use	297
13.6	Irrigation	298
13.7	Water resources	299
13.8	Flood	304
13.9	Drought	305
13.10	Climate change impacts	305
13.11	Sustainable water resources development	305
13.12	Water conservation	306

Section B: Inter-basin water transfer projects — 307

13.13	Introduction	307
13.14	South to North Water Transfer Project	307
13.15	Action plan for the North China Plain	314
13.16	Conclusions	316
	References	316

14 India: The National River-Linking Project — 319

Section A: Overview of geography, population, land and water — 319

14.1	Geography	319
14.2	Population	319

14.3	Economy	319		A.7	O'Connor, Charles Yelverton (1843–1902)	369
14.4	Climate and precipitation	320				
14.5	Irrigation	321				
14.6	Water resources	323				
14.7	Flood	326				

14 (continued)

14.3	Economy	319
14.4	Climate and precipitation	320
14.5	Irrigation	321
14.6	Water resources	323
14.7	Flood	326
14.8	Drought	326
14.9	Climate change impacts	326
14.10	Impacts of dam building	327
14.11	National water policy	329
14.12	Inter-state water disputes	330

Section B: The National River-Linking Project — 330

14.13	Introduction	330
14.14	Existing projects	330
14.15	River-linking proposals of the 1970s	331
14.16	The National River-Linking Project	331
14.17	Conclusions	342
	References	344

15 Inter-basin water transfer, successes, failures and the future — 345

15.1	Introduction	345
15.2	Benefits of inter-basin water transfer projects	346
15.3	Impacts of inter-basin water transfer projects	350
15.4	Mega-scale water transfer proposals	353
15.5	Necessary knowledge for inter-basin water transfer	353
15.6	Inter-basin water transfer, water conservation and new sources of supply	354
15.7	Inter-basin water transfer and cross jurisdictional agreements	355
15.8	Recommendations of the World Commission on Dams	356
15.9	Concluding comments	356

Part IV Appendices — 359

Appendix A Some of the Australian pioneers of inter-basin water transfer — 361

A.1	Bradfield, John Job Crew (1867–1943)	361
A.2	Chifley, Joseph Benedict "Ben" (1885–1951)	362
A.3	Forrest, Sir John (1847–1918)	364
A.4	Hudson, Sir William (1896–1978)	366
A.5	Idriess, Ion Llewellyn (1889–1979)	367
A.6	Menzies, Sir Robert Gordon (1894–1978)	368
A.7	O'Connor, Charles Yelverton (1843–1902)	369

Appendix B Construction timetable of the Snowy Mountains Hydro-electric Scheme — 371

Appendix C Details of diversion schemes from the Clarence River Basin — 374

C.2	Details of diversion schemes from the Macleay River Basin	376

Appendix D Chronological table of the most important events in the Goldfields Pipeline Scheme, Western Australia — 377

Appendix E Flooding of the Sahara depressions — 379

E.1	Introduction	379
E.2	Roudaire's expeditions	379
E.3	Commission of inquiry	380
E.4	Continuation of the inland sea affair (1882–1936)	381
E.5	Developments from 1957 to 1968	381
E.6	The joint Algeria and Tunisia project (1983–85)	382
	References	383

Appendix F The Ord River Irrigation Scheme — 384

F.1	Introduction	384
F.2	Hydrology and water quality of the Ord River	386
F.3	Economic evaluation of the scheme	386
F.4	Recent gross values of agricultural production	388
F.5	Hydro-power generation	388
F.6	Stage 2 of the scheme	388
	References	392

Appendix G The West Kimberley Irrigation Scheme — 393

G.1	Introduction	393
G.2	Groundwater allocation and stakeholders concerns	393
G.3	Cultural values of groundwater	395
G.4	Cotton research	395
G.5	Benefits of the WAI proposal	395
G.6	Progress of the feasibility study	396
G.7	Failure of the proposal	396
	References	396

Appendix H Some other water transfer schemes in Australia	**397**
H.1 Introduction	397
H.2 Shoalhaven Diversion Scheme	397
H.3 Thomson Diversion Scheme	403
H.4 Hydro-power generation in Tasmania	405
References	412

Appendix I Selected technical features of the Central Valley Project in California	**413**
Reference	414

Appendix J Selected technical features of the State Water Project in California	**415**
Reference	416

Appendix K Selected characteristics of some of the completed or proposed inter-basin water transfer projects in Australia, United States, Canada, China and India, in chronological order	**417**
Glossary	423
Index	429

Foreword

A fundamental problem that is facing the water profession at present is its inability to look to the future. An implicit assumption has been that future water availability, use and demand patterns will basically be similar to what have mostly been witnessed in the past, with perhaps only incremental changes. The water profession has been repeating ad nauseaum for the last four decades that "business as usual" is not an option but continues to behave as if there is no other option. The only difference that can be noted during the past decade is that the rhetoric of "business as usual" is not an option has intensified immensely, but it has not resulted in any perceptible change in terms of actions.

Based on the research carried out at the Third World Centre for Water Management, it can be said with considerable confidence that the world of water management will change more during the next 20 years, compared to the past 2000 years. The structures of water availability, use patterns and overall demands will change radically because of many factors, some known but the others mostly unknown. The factors that are mostly being ignored at present are likely to have increasingly more impacts on water-related issues during the coming decades. Among these factors are radically changing population dynamics (declining population in many countries, population stabilisation in other countries, increasing number of elderly people all over the world, and especially in China during the post-2025 period, etc.), concurrent urbanisation and ruralisation, globalisation and free trade in agricultural and industrial products, information and communication revolution, advances in technology (especially in areas like biotechnology and desalination), scramble for energy security by the major nations, and uncertainties associated with climate change. All of these will have major implications for water planning and management in the coming decades. Yet, none of these issues are being seriously considered at present.

These uncertainties are especially important for considering future major inter-basin water transfer (IBWT) projects. These projects often have gestation periods of 15 years or more. Thus, unlike in the past, when it was comparatively easy to predict future developments, and thus water requirements, the forecasting process will become exceedingly more complex in the coming years. If the future water demands cannot be predicted with any degree of certainty, it will not be an easy task to analyse the needs, desirability and cost-effectiveness of the proposed new IBWT projects.

Let us take only one example: the current on-going discussions under the Doha round of negotiations under the World Trade Organisation, and how this activity that is seemingly unrelated to water could have major implications in the future on the water sector. Irrespective of whatever may be the final results of the Doha round, it is now certain that agricultural subsidies and tariffs will be reduced quite significantly within the next 10 to 20 years. The only question is when and by how much. By 2020, only 14 years from now, we can say with certainty that we shall see considerable progress in terms of reduction in agricultural subsidies, even though we cannot say definitively when exactly this will occur, or by how much. Because of these important changes, the structure of agricultural production in numerous countries will change very substantially, along with their agricultural water requirements, which globally is the largest user of water at present.

When our Centre was requested to undertake an independent review of the Spanish Plan to transfer water from the Ebro River to the southern coastal areas of Spain, our conclusions were that if we consider the conditions that are likely to prevail during the post-2020 period, when the Plan may become operational, it may be difficult to justify even the existing agricultural water use patterns, let alone expect higher water uses. This is because the structure of water demand is likely to change radically in Western Europe because of new global agricultural trade agreements, changing socio-political considerations, and economic and technological developments. In addition, the officially estimated cost of delivering per cubic metre of the Ebro water to the Levante basins is nearly 50 percent higher than the current cost of desalination of sea water. Accordingly, even though the Spanish Parliament had earlier approved the Ebro water transfer, it later decided to cancel this plan.

The Ebro example, however, should not be construed to mean that in the future no inter-basin water transfer schemes will be necessary. Rather, each case must be carefully considered and analysed in terms of future water requirements and societal expectations when the projects are expected to be completed, and not on the basis of the prevailing conditions when the planning starts. The two sets of conditions are likely to be very different, a fact that has thus far been mostly ignored by the water professionals. If after objective analyses, it is considered that an IBWT project is necessary and can be *justified* on economic, social and environmental terms, its construction should proceed.

A major problem facing the developing world at present is the knee-jerk reactions of certain activist groups, primarily from the Western countries, that large scale water developments are no longer necessary, and that the water requirements of the future can be taken care of by small-scale projects like rainwater harvesting. It is difficult to have any sympathy with such a dogmatic view. First, large dams or small projects are not an either/or proposition. At a certain location and at a certain time, a large project may prove to be the best solution. Equally, at another place, a small project may be more appropriate. Many times, the two alternatives may even have to co-exist. An objective analysis of past water development projects from different parts of the world indicates that small can be beautiful, but it can also be ugly. Similarly, big can be magnificent, but it can also be a disaster. Each case must be judged by its site-specific conditions and its own merits. Dogmatic views are invariably wrong and socially unproductive on a long-term basis. For a heterogeneous and rapidly changing world, there is no other alternative but to consider plurality of paradigms. One size simply does not fit all.

In addition, a vast majority of water professionals and international institutions do not understand the water problems of developing countries, all of which are in tropic and semi-tropical climates with pronounced seasonality in precipitation patterns. This is in sharp contrast to developed countries, all of which (except Australia) are in temperate climates with a much more even distribution of precipitation within the year, and also between the years.

Let us take the case of India, much of which receives its annual rainfall in less than 100 hours (not necessarily consecutive). The main water problem of India thus is how to store this immense amount of rainfall over such a short period so that water is available for various uses throughout the year. For the large Indian cities, there is simply no other alternative but to build large dams so that water is available on a reliable basis throughout the year. In other parts of India, depending upon the local conditions, rainwater harvesting may prove to be the best solution. Thus, the main questions with large dams, which are invariably components of IBWT projects, is not whether they should be built, since there may not be any alternative to them under certain conditions, but to ensure that they are built and managed in a way that is economically efficient, socially desirable and environmentally acceptable.

Another important problem in the water resources area is the lack of reliable and basic information. For example, current estimates of global water withdrawal figures can at best be of very limited use. First, we do not even know with any degree of reliability how much water a major country like India or China withdraws, let alone many other smaller countries. Thus, one has no idea about the accuracy of the current global water abstraction and use estimates. Almost certainly, they are all wide of the mark.

Second, the quantity of water abstracted, even if this estimate was known reliably, is increasingly becoming less and less meaningful for planning and management purposes. Water is not like oil which can be used only once. Some have estimated that each drop of the Colorado River water is used several times. If the management practices can be improved, the extent of water reuse will increase very substantially. As water is reused more and more, both formally and informally, the information on how much water is being withdrawn becomes increasingly less and less relevant. Even for highly developed countries of Western Europe, or the United States, we have only very limited information as to the quantity of water that is being reused. For developing countries, we simply do not have any idea. All we can say is that the amount that is being reused is very high, and the extent of reuse is increasing everywhere.

In this context, a few comments on the World Commission on Dams are appropriate. Regrettably, the report of the Commission leaves much to be desired. Not surprisingly, one of its two god-fathers, the World Bank, did not endorse the report, and the major dam-building countries like China, India and Turkey, have very specifically rejected this report.

Some of its views are fundamentally erroneous. For example, the Commission has claimed that 40–80 million people have been displaced by large dams. No knowledgeable and objective expert will accept even the lower estimate of 40 million, which is wide of the mark. The total estimate is likely to be very significantly less. To claim that it could be as high as 80 million is patently ridiculous. The main problem with the so-called knowledge-base developed by the Commission is that it is full of chaff, but it may contain some wheat. However, absence of serious peer reviews of its case studies has meant that it is impossible to separate the wheat from the chaff.

The authors of this book, Fereidoun Ghassemi and Ian White, have done a remarkable job in assembling and analysing an immense amount of data on inter-basin water transfer projects from Australia, United States, Canada, China and India. Many of these data are not easily available. Some of the information like those on the Australian pioneers of IBWT is mostly unknown at present. Thus, the book should be of special significance to all the water professionals interested in IBWT. I am thus confident that the water profession will consider this book to be an important contribution to the literature.

Atizapan, Mexico
April 2006

Asit K. Biswas
President Third World
Centre for Water Management
and the 2006 Stockholm Water
Prize Laureate

Overview and Scope

Large water infrastructure projects were completed throughout the world during the twentieth century to meet the increasing demands of burgeoning populations for irrigation and domestic water supplies. These projects saw the construction of dams, reservoirs, pipelines, pumping stations, hydro-power plants and irrigation systems within river basins. In several countries, major and in some cases almost heroic, projects were undertaken to transfer water from basins considered to have surplus water to basins where water demand exceeded or was expected to exceed the available supply. This book compares the contexts and experiences in inter-basin water transfer in countries with widely different water needs, population pressures, economies and forms of government.

Most large water infrastructure and inter-basin water transfer projects in the past were the domain of engineers and government bureaucrats. Many were undertaken with minimal assessment of environmental or social impacts and with rudimentary and in some cases doubtful cost–benefit analyses. Community participation in such schemes was either nonexistent or token. While many have benefited from such schemes, there has often been marked inequity in the distribution of benefits. There have been significant social, economic and environmental impacts, with poor and indigenous communities frequently bearing a disproportionate share of the impacts. Globally, millions of people have been displaced by large water projects. The predicted performance of water projects and projected cost recovery and profitability has often proved illusory. Rivers and lakes have dried to a trickle, aquatic ecosystems and biodiversity have declined, and sediment delivery to floodplains has been reduced while expensive dams have silted up. As a result of these issues, the World Bank has been impelled to change its policy and currently demands detailed impact assessment of water resources development projects before approving their funding. Furthermore, the World Commission on Dams, following its extensive review of major water infrastructure projects, has recommended seven strategic priorities and related policy principles for making decisions on dam construction and inter-basin water transfer.

The proposal early in the twentieth century in the United States to build the Hetch Hetchy Aqueduct to meet San Francisco's increasing demands for freshwater was possibly the first inter-basin transfer scheme to face significant opposition because of perceived adverse environmental impacts. In 1913, that opposition failed to stop construction and the dispute over its impacts continues to the present. Communities, particularly in the developed world, have become increasingly vocal over proposed water projects, questioning needs, benefits, costs and impacts, demanding better information, protection of the environment and social and cultural values, and a voice in the decision process.

Proposals for the inter-basin transfer of water continue to evoke heated disputes because of disagreements over benefits, costs and impacts. For example, in the 2005 Western Australian State election, the US$9 billion inter-basin water transfer proposal from the Kimberley region in the north of the State to Perth in the south was a key and deciding election issue which resulted in the defeat of the opposition who supported the project. In these often lengthy disputes, limited use has been made of analyses of previous inter-basin transfer projects. The aim of this book is to present as dispassionate an account as possible of the history and technology of inter-basin water transfer under contrasting conditions in Australia, the western United Sates, Canada, China and India. These countries vary dramatically in climate, from the driest inhabited continent to one with the highest per capita annual quantity of freshwater; in political systems, from centrally planned to free market; and in different stages of economic development. Our goal in this wide-ranging analysis is to draw general lessons from the experiences of these widely diverse countries in inter-basin water transfer so that past mistakes will not be repeated.

In developed countries with relatively low rates of population increase, such as Australia, the United Sates and Canada, priorities have now moved from increasing water harvesting to meet untrammelled water demand to water conservation, especially through improvements in water use efficiency in all sectors of the economy and particularly in irrigation. Emphasis is being placed on water pricing and water trading and on the reuse of treated wastewater, conjunctive use of surface and groundwater, precipitation enhancement, rainwater harvesting, and to a lesser extent desalination. In developing countries, with rapidly expanding economies, increasing populations and urgent water demands, such as China and India, the imperative is to meet regional water and power needs. In such countries, and in areas that are expected to experience decreases in water availability due to global warming, inter-basin transfer of water remains attractive.

This book is divided into four parts. Part I overviews information about world water resources and summarises the key issues that have arisen in inter-basin water transfer and in large water infrastructure projects throughout the world. It provides a framework for examining inter-basin transfer proposals. Part II focuses on land and water scarcity issues, policy changes and the Australian experience in inter-basin transfer. Australia is undergoing the most profound changes in water policy and strategy since federation in 1901. These changes are based on the need for pricing mechanisms to reflect the true costs in supplying water, and the need to better balance water allocation between consumers and the environment. Part III examines selected inter-basin transfers in the United States, Canada, China and India. Finally, Part IV consists of numerous appendices.

In Part I, **Chapter 1** provides an overview of world challenges, which includes topics such as: population, land degradation, water resources and the extent of their developments, dams and transfer of water from one basin to another, climate change and its impacts on water resources, agriculture, and food production. Here, the limitations of the world's land and water resources, faced with an increasing population and prospects of global warming, are explored. **Chapter 2** describes major issues relevant to the inter-basin water transfer including topography, geology, hydrology, environmental considerations, land degradation, social and cultural issues, economic appraisal, and conflicts and their resolution. It concludes that inter-basin water transfer projects require detailed multidisciplinary investigations and an integrated approach in assessment of projects.

In Part II, **Chapter 3** provides an introduction to Australia's geography, population, climate, agriculture, water resources, and estimates of its future water requirements. This is a prelude to the following chapters on inter-basin water transfer in Australia. **Chapter 4** describes the Snowy Mountains Hydro-electric Scheme, its history, technical features, finance, and other related issues. **Chapter 5** describes numerous proposals developed for the inland transfer of water from coastal river basins of New South Wales, such as the Clarence, Macleay, Manning and Tuross and outlines the reasons for their rejection. **Chapter 6** details the Bradfield and the Reid schemes for inland diversion of coastal rivers of Queensland. **Chapter 7** describes three schemes for flooding of Lake Eyre, located at the centre of the continent, by diversion of surface water from coastal rivers of Queensland, by seawater from South Australia and by groundwater from the Great Artesian Basin. The idea was inspired by a similar proposal for flooding of the Sahara depressions in north Africa with Mediterranean Sea water under the erroneous assumption that this would change local rainfall and climate. **Chapter 8** examines the history and construction of the Goldfields Pipeline in Western Australia, the first major water transfer project in Australia, completed in 1903. **Chapter 9** examines the politically contentious proposals for water transfer from the Kimberley region in the north of Western Australia to Perth and Adelaide. **Chapter 10** covers a number of large to relatively small projects for domestic, irrigation and mining water supply in South Australia, Queensland and New South Wales.

In Part III, **Chapter 11** explores water transfer projects in the United States. It reviews water transfer projects in California, and from the Colorado River Basin to its neighbouring states. This chapter outlines policies developed by the Federal and State Governments for the better management of their currently developed resources in order to satisfy water requirements in the ensuing two or three decades without building new dams and initiating inter-basin water transfer projects. **Chapter 12** covers inter-basin water transfer projects in Canada, developed mainly for hydro-power generation rather than for irrigation or domestic water supply. **Chapter 13** examines the South to North Water Transfer Project in China planned to overcome serious water shortage and environmental degradation in the North China Plain. It also describes China's continuing dam construction and inter-basin water transfer projects, and its efforts to implement water conservation measures. In **Chapter 14**, India's response to its growing water demands, rapidly developing economy, and variable distribution of water are discussed. Its highly controversial planned National Rivers-Linking Project is considered. Finally, **Chapter 15** highlights successes, failures and provides pointers for the future of inter-basin water transfer projects.

International meetings over the past two decades have increasingly drawn attention to the shortfalls in good quality water for human needs, particularly in drier areas with high population growth rates, and to the environmental and ecological impacts of human activities and interventions in the hydrologic cycle on water systems. The United Nations General Assembly Millennium Declaration in 2000 resolved, "*to halve by the year 2015 the proportion of the world's population who are unable to reach or afford safe drinking water*" and "*to stop the unsustainable exploitation of water resources*". The Implementation Plan of the World Summit on Sustainable Development in Johannesburg in 2002 had as one of its aims to "*improve the efficient use of water resources and promote their allocation among competing uses in a way that gives priority to the satisfaction of basic human needs and balances the requirement of preserving or restoring ecosystems and their functions, in particular fragile environments, with human domestic, industrial and agricultural needs, including safeguarding drinking water quality*". These goals represent enormous tasks.

In the developed world, with more stable populations, emphasis is being placed on water conservation and reuse and on restoring or mitigating aquatic ecosystems impacted by water developments. In the developing world, rapidly increasing water demand requires new water infrastructure and perhaps inter-basin transfer projects to assist in alleviating poverty, and satisfying basic water, food and fibre demands. In numerous cases alternative options to inter-basin water transfer may exist. They need to be explored and implemented where possible. It is our hope that the material and analyses presented in this book will be useful to decision-makers, researchers, university students and general public in both developed and developing worlds in stimulating debate and informing decisions on new inter-basin water transfer proposals and in achieving negotiated outcomes with active participation of all stakeholders.

Fereidoun Ghassemi and Ian White

Acknowledgements

The authors would like to thank sincerely all those people who reviewed various chapters/sections of the book and made constructive comments or assisted us by providing information. These are:

A. Australia

Arthington, Angela (Prof.): Centre for Riverine Landscapes, Faculty of Environmental Sciences, Griffith University, Nathan, Brisbane, Queensland.

Ballard, Jeff (Mr): Infrastructure Engineer, NQ Water, Townsville, Queensland.

Barnes, Marilla (Ms): Corporate Communications, SA Water Corporation, Adelaide, South Australia.

Braaten, Robert (Mr): Water Management Division, DIPNR,[1] Sydney, New South Wales.

Chartres, Colin (Dr): Deputy Chief, CSIRO Land and Water, Canberra.

Close, Andrew (Mr): Manager, Water Resources Group, Murray–Darling Basin Commission, Canberra.

Commander, Philip (Mr): Department of Environment, Perth, Western Australia.

Crabb, Peter (Dr): Visiting Fellow, CRES,[2] ANU.[3]

Croke, Barry (Dr): Joint CRES and iCAM[4] Research Fellow at ANU.

Dovers, Stephen (Prof.): CRES, ANU.

Dunlop, Michael (Dr): Resource Futures Program, CSIRO Sustainable Ecosystems, Canberra.

Everson, Derek (Mr): Water Management Division, DIPNR, Sydney, New South Wales.

Fisher, Sarah (Ms): Senior Planning Engineer, Infrastructure Planning Branch, Water Corporation, Leederville, Western Australia.

Fitt, Gary P. (Dr): Chief Executive Officer, Australian Cotton Cooperative Research Centre, Narrabri, New South Wales.

Fitzgerald, Bruce (Mr): Water Management Division, DIPNR, Sydney, New South Wales.

Ghadiri, Hossein (Dr): Senior Lecturer, Centre for Riverine Landscapes, Faculty of Environmental Sciences, Griffith University, Nathan, Brisbane, Queensland.

Grafton, R. Quentin (Prof.): International and Development Economics, Asia Pacific School of Economics and Government, ANU.

Hamblin, Ann (Dr): Visiting Fellow, CRES, ANU.

Hazell, Donna (Dr): Post Doctoral Fellow, CRES, ANU.

Hughes, Robert (Mr): Manager System Control, SA Water Corporation, Adelaide, South Australia.

Jakeman, Anthony, J. (Prof.): Director, Integrated Catchment Assessment and Management (iCAM) Centre, ANU.

Johnson, Ken (Mr): School of Resources, Environment and Society, ANU.

Jotzo, Frank (Mr): PhD candidate, CRES, ANU.

Locher, Helen (Dr): Environmental Programs Manager, Hydro Tasmania, Hobart, Tasmania.

Logan, John (Mr): Chairman, Western Agricultural Industries Pty Limited, Neutral Bay, Sydney, NSW.

Magee, John (Dr): Australian Research Council Queen Elizabeth II Fellow, Department of Earth and Marine Sciences, Faculty of Science, ANU.

Martin, Gary (Mr): Manager Water Services, Bowen Shire Council, 67 Herbert Street, Bowen, Queensland.

McKenzie, Neil (Dr): Research Group Leader, CSIRO Land and Water, Canberra.

McLeod, Ivan (Dr): Project Manager, Western Agricultural Industries Pty Limited, Perth, Western Australia.

Meehan, David (Mr): Project Manager, Office of Major Projects, Department of Industry and Resources, Perth, Western Australia.

---

[1] Department of Infrastructure, Planing and Natural Resources.
[2] Centre for Resource and Environmental Studies.
[3] The Australian National University, Canberra, Australia.
[4] Integrated Catchment Assessment and Management Centre.

Neilson, Danielle (Ms): Marketing Services Officer, Snowy Hydro Limited, Cooma, New South Wales.

Nix, Henry (Emeritus Prof.): Visiting Fellow, CRES, ANU.

Ollier, Cliff (Prof.): School of Earth and Geographic Science, University of Western Australia, Nedlands, Western Australia.

Pagan, Adrian (Emeritus Prof.): Economics Program, Research School of Social Sciences, ANU.

Parsons, Andrew (Mr): Engineering and Projects, SA Water Corporation, Adelaide.

Perkins, Paul (Adjunct Prof.): CRES, ANU.

Ray, Binayak (Mr): Visiting Fellow, Department of Political and Social Change, Research School of Pacific and Asian Studies, ANU.

Rebello, Gerry (Mr): Water Management Division, DIPNR, Sydney, New South Wales.

Rose, Deborah (Dr): Senior Fellow, CRES, ANU.

Smith, David Ingle (Mr): Visiting Fellow, CRES, ANU.

Smith, Peter (Mr): Manager of the Utility Services, BHP Billiton Mitsubishi Alliance, Riverside Centre, Brisbane, Queensland.

Stein, Janet (Mrs): Research Officer and PhD Candidate, CRES, ANU.

Walkemeyer, Peter (Mr): Project Manager, Project Management Branch, Water Corporation, Leederville, Western Australia.

West, Adam (Mr): Water Planning Coordinator, Queensland Department of Natural Resources and Mines, Townsville.

White, Geoffrey B. (Mr): Chairman, White Industries Australia Limited, Suite 214, Harrington Street, Sydney, New South Wales.

B. Other countries

Alemi, Manucher (Dr): Office of Water Use Efficiency, Department of Water Resources, Sacramento, California, USA.

Day, J. Chadwick (Emeritus Prof.): School of Resource and Environmental Management, Simon Fraser University, Burnaby, British Columbia, V5A 1S6, Canada.

Flugel, Wolfgang-Albert (Prof.): Chair and Head, Department of Geoinformatics, Hydrology and Modelling, Friedrich-Schiller-University, Jena, Germany.

Fried, Jean (Prof.): Université Louis Pasteur, Strasbourg, France.

Howard, Ken (Prof.): Groundwater Research Group, Scarborough Campus, University of Toronto, Ontario, Canada.

Letolle, René (Prof.): Université Pierre et Marie Curie (Paris 6), Campus Jussieu, Paris, France.

Quinn, Frank (Dr): Formerly, Chief of Water Policy and Transboundary Issues, Environment Canada, Ottawa, Canada.

Renzetti, Steven (Prof.): Department of Economics, Brock University, Ontario, Canada.

Reynolds, Dean (Dr): Associate Land and Water Use Analyst, Department of Water Resources, Sacramento, California, USA.

Shao, Xuejun (Prof.): Department of Hydraulic Engineering, Tsinghua University, Beijing, China.

Shields, Tina (Ms): Assistant Manager, Water Department, Resource Planning and Management, Imperial Irrigation District, California, USA.

Storey, Brit Allan (Dr): Senior Historian, US Bureau of Reclamation, Denver, Colorado, USA.

Tharme, Rebecca (Ms): International Water Management Institute (IWMI), Colombo, Sri Lanka.

Wolfgang, Carolann (Dr): Geohydrologist, SAIC, 525 Anacapa Street, Santa Barbara, California, USA.

Our special thanks go to Professor Anthony Jakeman for his support of this project and Professor Angela Arthington for writing the section on "*Environmental Flow Requirements of Rivers*". We also thank Dr Anthony Scott for his valuable comments and copy-editing, Mr Clive Hilliker for graphics and Dr McComas Taylor for his valuable editorial advice.

List of Abbreviations

ABARE	Australian Bureau of Agricultural and Resource Economics	EC	Electrical Conductivity
ACT	Australian Capital Territory	EFA	Environmental Flow Assessment
ADR	Alternative Dispute Resolution	EFR	Environmental Flow Requirement
AHD	Australian Height Datum	EIS	Environmental Impact Statement
ALP	Australian Labor Party	EMBUD	East Bay Municipal Utility District
AMSL	Above Mean Sea Level	EPA	Environmental Protection Authority
ANF	Average Natural Flow	FAO	Food and Agriculture Organization of the United Nations
ASSOD	Assessment of the Status of Human-Induced Soil Degradation	Fry-Ark	Fryingpan-Arkansas
ATSIC	Aboriginal and Torres Strait Islander Commission	FSL	Full Supply Level
BBM	Building Block Methodology	GDP	Gross Domestic Product
BHP	Broken Hill Proprietary Company Limited	GEWEX	Global Energy and Water Cycle Experiment
BMA	BHP Billiton Mitsubishi Alliance	GIS	Geographic Information System
CALFED	CALiforniaFEDeral	GLASOD	Global Assessment of Soil Degradation
C-BT	Colorado-Big Thompson	GWh	Gigawatt hours
CIMIS	California Irrigation Management Information System	ha	Hectare ($10\,000\,m^2$)
CMG	Commander of order of St Michael and St George	HEC	Hydro-Electric Commission
CRC	Cooperative Research Centre	HRC	Healthy Rivers Commission
CSIRO	Commonwealth Scientific and Industrial Research Organisation	IDC	Infrastructure Development Corporation
CUP	Central Utah Project	IGBP	International Geosphere–Biosphere Programme
CUWCD	Central Utah Water Conservancy District	IID	Imperial Irrigation District
CVP	Central Valley Project	IPCC	Intergovernmental Panel on Climate Change
CVPIA	Central Valley Project Improvement Act	ISRIC	International Soil Reference and Information Centre
DIMIA	Department of Immigration and Multicultural and Indigenous Affairs	IWMI	International Water Management Institute
DLWC	Department of Land and Water Conservation (currently Department of Infrastructure, Planning and Natural Resources)	IWSS	Integrated Water Supply Scheme
		kW	Kilowatt
		kWh	Kilowatt hours
		$L\,h^{-1}\,d^{-1}$	Litre per head per day
		LPG	Liquefied petroleum gas
		m	Metre
		mm	Millimetre
		m^3	Cubic metre
DRIFT	Downstream Response to Imposed Flow Transformations	M	Million
DWR	Department of Water Resources	MDB	Murray–Darling Basin
		MDBC	Murray–Darling Basin Commission.
EA	Environmental Assessment	MDBMC	Murray–Darling Basin Ministerial Council

MDIA	Mareeba–Dimbulah Irrigation Area	**SNWTP**	South-to-North Water Transfer Project
MDWSS	Mareeba–Dimbulah Water Supply Scheme	**SOI**	Southern Oscillation Index
MoU	Memorandum of Understanding	**SWP**	State Water Project
Mt	Million tonnes	**SWRCB**	State Water Resources Control Board
MW	Megawatt	**SWUA**	Strawberry Water User's Association
MWD	Metropolitan Water District	**t**	Tonne (1000 kg)
NCWCD	Northern Colorado Water Conservancy District	**tpa**	Tonnes per annum
		TDS	Total Dissolved Solids
NGOs	Non Government Organisations	**UNEP**	United Nations Environment Programme
NSW	New South Wales	**UNESCO**	United Nations Educational Scientific and Cultural Organisation
NT	Northern Territory		
NWC	National Water Commission	**URL**	Uniform Resource Locater
NWDA	National Water Development Agency	**USBR**	United States Bureau of Reclamation
NWI	National Water Initiative	**USRS**	United States Reclamation Service[1]
ORIA	Ord River Irrigation Area	**VIC**	Victoria
ORIS	Ord River Irrigation Scheme	**WA**	Western Australia
OWUE	Office of Water Use Efficiency	**WAI**	Western Agricultural Industries Pty Limited
PCA	Permanent Court of Arbitration	**WCD**	World Commission on Dams
ppb	Part per billion	**WRC**	Water and Rivers Commission
ppm	Part per million	**yr**	Year
PRC	People's Republic of China		
QLD	Queensland		
R&D	Research and Development		
s	Second		
SA	South Australia		
SDCWA	San Diego County Water Authority		
SMHEA	Snowy Mountains Hydro-electric Authority		

[1] The United States Reclamation Service was established within the U.S. Geological Survey (USGS) in July 1902. Then, in 1907, the Secretary of Interior separated the Reclamation Service from the USGS and created an independent bureau within the Department of the Interior. In 1923 the agency was renamed the "United States Bureau of Reclamation" (http://www.usbr.gov/history/borhist.html visited in April 2005).

Part I
The Challenges

1 World Population and Pressures on Land, Water and Food Resources

1.1 POPULATION

The world population was about 200 million in the year 500 AD, 275 million in the year 1000, 450 million in 1500 and one billion around 1800 (Cohen, 1995, Appendix 2). While the world population took most of human history to reach one billion, subsequent additions came much faster: 130 years to reach 2 billion, 30 years to reach 3 billion, then 14, 13, and 12 years to reach 4, 5, and 6 billion respectively (Gilbert, 2001, p. 1). The high rate of population growth in recent decades has been the result of improvements in public health and sanitation that have reduced the mortality rate, particularly in the developing countries. The population growth rate peaked at 2.1 percent per year during the period 1965–70 (World Bank, 1992, pp. 25–26) and then started to decline to 1.7 percent over the period 1975–80 and 1.3 percent (or approximately 80 million a year) over the period 1995–2000. The fertility rate declined from 3.9 in 1975–80 to 2.7 for the period 1995–2000 (Table 1.1).

The world population increased from about 2.5 billion in 1950 to 6 billion in 2000, representing an increase of 2.4 times, and is expected to reach 7.8 billion by 2025 (Table 1.1). Most of this increase will take place in the developing world. It is expected that Asia will reach a population of 4.3 billion and Sub-Saharan Africa 1.1 billion by the year 2025. In the year 2000, 78 percent of the world's 6 billion people lived in the developing countries. By 2025 this is predicted to rise to 83 percent of a total population of 7.8 billion. Population growth increases the demand for food, fibre, goods and services, and many of the Earth's new citizens will not be offered the health and educational resources necessary to reach their potential (World Resources Institute, 1992, Chapter 6). Cities are drawing people into ever-increasing concentrations. Urban regions

Table 1.1. *Estimated and projected world population from 1950 to 2025*

Region	Population (million)			Average annual population change (%)		Fertility rate (%)	
	1950	2000	2025 (projected)	1975 to 1980	1995 to 2000	1975 to 1980	1995 to 2000
Asia (excluding Middle East)	1338	3420	4308	–	–	–	–
Europe	504	728	702	0.5	0.0	2.0	1.4
Middle East and North Africa	112	404	614	–	–	–	–
Sub-Saharan Africa	177	641	1095	–	–	–	–
North America[a]	172	310	364	0.9	0.8	1.8	1.9
Central America and Caribbean	54	173	236	–	–	–	–
South America	113	346	461	2.3	1.5	4.3	2.6
Oceania	13	30	40	1.1	1.3	2.8	2.4
World	2521	6055	7823	1.7	1.3	3.9	2.7
Developed countries	853	1306	1358	0.6	0.3	1.9	1.6
Developing countries	1668	4746	6459	2.1	1.6	4.7	3.0

Note: [a] Updated population data for the United States and Canada are provided in Chapters 11 and 12.
Source: World Resources Institute (2000, Data Table HD.1).

tend to offer more opportunities economically as well as better education and health resources. Although these regions occupy only 4 percent of the Earth's land area, they are home to nearly half the world's population. Densely populated cities, the so-called mega cities, are a major source of pollution, particularly of surface and groundwater.

One of the main characteristics of the world population is its ageing profile, which is unprecedented in the history of humanity (United Nations, 2002). During the twentieth century the proportion of older persons (60 years or more) continued to rise. It was 8 percent in 1950, 10 percent in 2000, and is projected to reach 21 percent in 2050. This is being accompanied by a decline in the proportion of the young under the age of 15. By 2050, it is expected that the number of older persons in the world will exceed the number of young for the first time in history. Population ageing has major repercussions for many aspects of human life. It has an impact on economic growth, savings, investment, consumption, labour markets, pensions and taxation. Population ageing affects health care, family composition, housing and migration, and can influence voting patterns and representation. The number of support persons aged 15–64 years per one older person aged 65 years or older, fell from 12 to 9 between 1950 and 2000. It is expected to fall to 4 by 2050.

Cohen (1995) provides an analysis of the upper limit of population that the Earth can sustain. Because of numerous ecological, social and technological constraints on the Earth's population, and different views on what is an acceptable standard of living for human beings, the carrying capacity of the Earth can be defined in many different ways. Cohen (1995, Appendix 3) gathered 66 estimates of how many people the Earth can support. These estimates range from less than one billion to more than 1000 billion. He demonstrated that one-quarter of them fall below 6.1 billion, half fall below 12 billion, and three-quarters fall below 30 billion. Others have estimated that the world population will stabilise at about 9.3 billion in the middle of the twenty-first century (UNESCO, 2003, p. 12), about 50 percent higher than the 2000 population of 6.1 billion (see Table 1.1).

1.2 DRYLAND AREAS

The extent of the world's dryland areas has been estimated by using an aridity index (Dregne et al., 1991). The index is expressed as the ratio of precipitation over potential evapotranspiration. The various categories of dryland have the following aridity index ranges: hyper-arid (<0.05); arid (0.05–0.20); semi-arid (0.21–0.50); dry sub-humid (0.51–0.65); moist sub-humid and humid (>0.65). With this method, the driest inhabited continent of the world is Australia where 75 percent of its area is dry (Table 1.2). It is followed by Africa and Asia. Drylands comprise about one-third of the areas of Europe, North America and South America. In total area, however, the largest drylands occur in Africa (1959 Mha), and Asia (1949 Mha) totalling about 64 percent of the world's drylands, whose area is about 6150 Mha, or 41 percent of the land area of the world. Of this nearly 978 Mha are hyper-arid deserts and 5172 Mha are arid, semi-arid and dry sub-humid. Figure 1.1 shows the distribution of the world's dryland areas.

1.3 EXTENT OF HUMAN-INDUCED LAND DEGRADATION

The International Soil Reference and Information Centre (ISRIC) published the results of a Global Assessment of Soil Degradation (GLASOD) in 1991. The assessment is based on the World Map of the Status of Human Induced Soil

Table 1.2. *World drylands (in million hectares)*

Dryland class	Africa	Asia	Australia	Europe	North America	South America	World total
Hyper-arid	672	227	0	0	3	26	978
Arid	504	626	303	11	82	45	1571
Semi-arid	514	693	309	105	419	265	2305
Dry sub-humid	269	353	51	184	232	207	1296
Total	1959	1949	663	300	736	543	6150
Percent of world total	32	32	11	5	12	8	100
Percent of continent	66	46	75	32	34	31	41

Source: Dregne et al. (1991, Table 1).

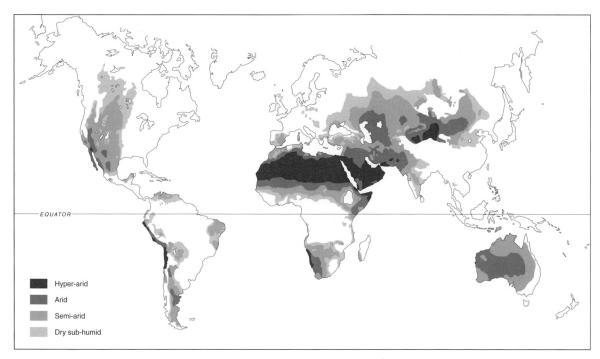

Figure 1.1 Drylands of the world (UNEP, 1992).

Degradation (Oldeman *et al.*, 1991a). The map, at a scale of 1:10 million, was prepared with financial support from the United Nations Environment Programme (UNEP) through a cooperative effort of about 250 soil scientists from international institutions throughout the world. Soil scientists were asked to only categorise soils degraded since the Second World War as a result of human intervention (World Resources Institute, 1992, pp. 111–118).

A primary objective for the creation of the soil degradation map was to generate awareness of the status of soil degradation in the mind of policy makers, and the general public (Oldeman *et al.*, 1991a). The GLASOD map covers 13 billion hectares of the land surface between 72° N and 57° S. Its results are alarming because, unlike other attempts to estimate land degradation, they do not include land degraded by ancient civilisations or even by colonial expansions, nor do they include land that is naturally barren.

GLASOD considered two categories of human-induced soil degradation processes. The first deals with soil degradation by displacement of soil material and the second with physical and chemical soil degradation. The two major types of soil degradation in the first category are erosion by water and wind. Water erosion includes loss of topsoil and terrain deformation. The most common forms are rill and gully erosion. Wind erosion includes loss of topsoil, terrain deformation and over blowing. Chemical deterioration is caused by a loss of nutrients and/or organic matter, salinisation, acidification and contamination by pollutants. Physical deterioration includes compaction, waterlogging and subsidence of organic soils caused by drainage and/or oxidation (Oldeman *et al.*, 1991a).

Globally, water erosion is by far the most important type of soil degradation, occurring in 1094 Mha or 56 percent of the total area affected by human-induced soil degradation (Table 1.3). The area affected by wind erosion is 548 Mha (28 percent); by chemical soil degradation, 239 Mha (12 percent); and by physical soil degradation, 83 Mha (4 percent).

Four degrees of soil degradation are recognised (Table 1.3). Light soil degradation, implying somewhat reduced productivity, which is manageable, by local farming systems, is identified for 38 percent of all degraded soils. A large percentage (46 percent) has a moderate soil degradation and greatly reduced productivity. Major improvements, often beyond the means of local farmers in developing countries, are required to restore productivity. Strongly degraded soils cover 296 Mha worldwide. These soils are no longer reclaimable at farm level and are virtually lost. Major engineering work or international assistance is required to restore these soils. Extremely degraded soils are considered to be beyond restoration. Their worldwide coverage is estimated to be around 9 Mha.

Table 1.3. *Global human-induced soil degradation*

Type	Light (Mha)	Moderate (Mha)	Strong (Mha)	Extreme (Mha)	Total (Mha)	(%)
Loss of topsoil	301.2	454.5	161.2	3.8	920.3	
Terrain deformation	42.0	72.2	56.0	2.8	173.3	
Water	343.2	526.7	217.2	6.6	1093.7	55.7
Loss of topsoil	230.5	213.5	9.4	0.9	454.2	
Terrain deformation	38.1	30.0	14.4	–	82.5	
Overblowing	–	10.1	0.5	1.0	11.6	
Wind	268.6	253.6	24.3	1.9	548.3	27.9
Loss of nutrients	52.4	63.1	19.8	–	135.3	
Salinisation	34.8	20.4	20.3	0.8	76.3	
Pollution	4.1	17.1	0.5	–	21.8	
Acidification	1.7	2.7	1.3	–	5.7	
Chemical	93.0	103.3	41.9	0.8	239.1	12.2
Compaction	34.8	22.1	11.3	–	68.2	
Waterlogging	6.0	3.7	0.8	–	10.5	
Subsidence of organic soils	3.4	1.0	0.2	–	4.6	
Physical	44.2	26.8	12.3	–	83.3	4.2
Total (Mha)	749.0	910.5	295.7	9.3	1964.4	
Total (percent)	38.1	46.1	15.1	0.5		100

Source: Oldeman *et al.* (1991b, Table 9).

Table 1.4. *Global extent of human-induced salinisation*

Continent	Light (Mha)	Moderate (Mha)	Strong (Mha)	Extreme (Mha)	Total (Mha)
Africa	4.7	7.7	2.4	–	14.8
Asia	26.8	8.5	17.0	0.4	52.7
South America	1.8	0.3	–	–	2.1
North and Central America	0.3	1.5	0.5	–	2.3
Europe	1.0	2.3	0.5	–	3.8
Australia	–	0.5	–	0.4	0.9
Total	34.6	20.8	20.4	0.8	76.6

Source: Oldeman *et al.* (1991b, Tables 2–8).

Five types of human intervention resulting in soil degradation were identified:

(1) degradation and removal of natural vegetation, 579 Mha;
(2) overgrazing of vegetation by livestock, 679 Mha;
(3) improper management of agricultural land, 552 Mha;
(4) overexploitation of vegetation cover for domestic use, 133 Mha; and
(5) industrial activities leading to chemical pollution, 23 Mha.

Table 1.4 shows that more than 76 Mha of the world's land is salt affected, out of which 52.7 Mha (69 percent) are in Asia, 14.8 Mha (19 percent) in Africa and 3.8 Mha (5 percent) in Europe. The four degrees of light, moderate, strong and extreme salt-affected land cover 34.6 Mha, 20.8 Mha, 20.4 Mha and 0.8 Mha respectively.

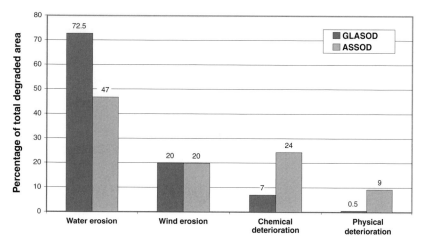

Figure 1.2 Distribution of main land degradation types in South and Southeast Asia as percentage of total degraded area assessed by GLASOD and ASSOD projects (van Lynden and Oldeman, 1997, Figure 1).

Because of funding problems, GLASOD has unfortunately not been refined or updated. Luckily, however, a number of other documents have been published regarding assessment of land degradation at the regional or national scale. These include the following publications:

- Acton and Gregorich (1995) describe the status of soil degradation in Canada, where almost all land suitable for crop production has been developed. Thus agricultural productivity must now be maintained through wise use of the existing resources, preserving both the area and quality of this land. The publication indicates that: (1) some Canadian agricultural soils are improving in health and becoming less susceptible to erosion and damage, mainly because of increased use of conservation farming methods; (2) this trend does not apply to all soils; (3) further maintenance and improvement of agricultural soil health depends on selecting appropriate land use and management practices; (4) a new government policy for soil conservation is needed, aimed at achieving sustainable agriculture and built on the understanding that agro-ecosystems are part of the broader environment; and (5) soil management programmes are best designed at the farm level, integrating management practices to suit specific, local soil needs.
- van Lynden and Oldeman (1997) describe the Assessment of the Status of Human-induced Soil Degradation in South and Southeast Asia (ASSOD). This study was commissioned by the UNEP and used a slightly modified GLASOD methodology. It covers the following 17 countries: Bangladesh, Bhutan, Cambodia, China, India, Indonesia, North Korea, South Korea, Laos, Malaysia, Myanmar, Nepal, Pakistan, Philippines, Sri Lanka, Thailand and Vietnam. Figure 1.2 compares results of GLASOD and ASSOD assessments. It indicates that GLASOD showed a high predominance of water erosion (72.5 percent of degraded lands), while the picture that emerged from ASSOD is more varied. Although water erosion remained a dominant feature in the ASSOD inventory (47 percent of degraded land), chemical and physical deteriorations were also prominent.
- Stolbovoi and Fischer (1998) describe the compilation of a new digital georeferenced database on human-induced soil degradation for Russia. The Russian territory covers 1710 Mha, which includes 131 Mha of cultivated land and 91 Mha of other agricultural lands. The extent of land degradation is estimated at 234 Mha, which includes: 58.3 Mha by compaction, 35.7 Mha by desertification, 25.8 Mha by water and wind erosion, 25.4 Mha by disturbance of soil organic horizons due to cutting and fire and 3.5 Mha by secondary salinisation.
- Hamblin (2001) describes the state of Australian lands as part of the Australia State of the Environment 2001 report for the period of 1995–2000. The publication covers various issues including: accelerated erosion and loss of surface soil, salinity and acidity, nutrient and carbon cycle issues, and land pollution. Soil acidification looms as a major soil degradation issue in Australia. Estimates indicate that 50 Mha and 23 Mha of Australia's agricultural zone are already experiencing impacts from soil acidity in surface and subsoil layers respectively (National Land & Water Resources Audit, 2001). It is estimated that in the absence of remedial lime application, which neutralises acidity, between 29 Mha

Table 1.5. *Global distribution of water*

Location	Volume (10^3 km^3)	Percentage of total volume in hydrosphere	Percentage of freshwater	Renewal period (years)
Ocean	1 338 000	96.5	–	2500
Groundwater (gravity and capillary)	23 400[a]	1.7	–	1400
Predominantly fresh groundwater	10 530	0.76	30.1	–
Soil moisture	16.5	0.001	0.05	1
Glaciers and permanent snow cover:	24 064	1.74	68.7	–
Antarctica	21 600	1.56	61.7	–
Greenland	2340	0.17	6.68	9700
Arctic islands	83.5	0.006	0.24	–
Mountainous regions	40.6	0.003	0.12	1600
Ground ice (permafrost)	300	0.022	0.86	10 000
Water in lakes:	176.4	0.013	–	17
Fresh	91.0	0.007	0.26	–
Salt	85.4	0.006	–	–
Marshes and swamps	11.5	0.0008	0.03	5
River water	2.12	0.0002	0.006	16 days
Biological water	1.12	0.0001	0.003	–
Water in the atmosphere	12.9	0.001	0.04	8 days
Total volume in the hydrosphere	1 386 000	100	–	–
Total freshwater	35 029.2	2.52	100	–

With the exception of the last column, data provided in this table have been previously published by Korzun *et al.* (1978).
Note: [a] Excluding groundwater in the Antarctica estimated at 2 million km^3, including predominantly freshwater of about 1 million km^3.
Source: Shiklomanov and Rodda (2003, Tables 1.8 and 1.14).

and 60 Mha will reach the limiting soil pH value of 4.8 within 10 years, and a further 14 Mha to 39 Mha will reach the pH value of 5.5, where growth of sensitive plant species is impaired.

1.4 WATER RESOURCES

Humans, and almost all other terrestrial life, depend on the availability of freshwater resources. However, the global distribution of water is highly uneven. Water is also limited by its accessibility and suitability. Of the Earth's total volume of about 1386 million km^3, some 96.5 percent is saline ocean water, unsuitable for human use (Table 1.5). Of the remaining 3.5 percent, 35 million km^3 is fresh, but 24 million km^3 is stored in ice sheets and glaciers,[1] and 10.5 million km^3 is groundwater resources. Freshwater in lakes totals 91 000 km^3 and rivers 2120 km^3.

The average annual precipitation on the Earth's surface is about 800 mm (Chow *et al.*, 1988, p. 71). However, the hydrological cycle distributes water unevenly around the globe, and the world can be divided into water surplus and water deficit regions. Water is in surplus when precipitation is high enough to satisfy the potential evapotranspiration demand of the vegetation cover. When precipitation is lower than potential demand, there is a water deficit. In general, most of Africa, much of the Middle East, the western United States, north-western Mexico, part of Chile and Argentina, and major parts of Australia are water deficit regions (World Resources Institute, 1986).

Globally, river run-off is one of the main sources of freshwater for human use. Through its continuous renewal by the hydrological cycle, river run-off represents the dynamic component of the Earth's total water resources, compared to the less mobile volume of water contained in lakes, groundwater reservoirs and glaciers (Shiklomanov, 1990). Table 1.6 shows the distribution of river run-off by continents. The average annual river run-off of the

---

[1] For details of freshwater reserves in glaciers and ice sheets see Shiklomanov and Rodda (2003, Table 1.9, p. 14). Wadhams (2000) provides an introduction to our modern knowledge of sea ice and icebergs, while Lewis *et al.* (2000) describe freshwater balance of the Arctic Ocean.

world is about 43 000 km³. Asia has the highest run-off (13 510 km³), followed by South America (12 030 km³), and North America (7870 km³).

La Rivière (1989) argues that about 9000 km³ of water are available for human exploitation worldwide, which is enough to sustain 20 billion people. Yet, because both the world's population and usable water are unevenly distributed, the local availability of water varies widely. Currently, much of the Middle East and North Africa, parts of Central America and many other countries are experiencing extreme scarcity of water due to increasing demands to satisfy their agricultural, industrial and domestic requirements.

Water resources of the world have been developed rapidly to satisfy demand. These developments included construction of large dams, and numerous inter-basin water transfer projects in all continents. During the twentieth century about 23 700 large dams higher than 15 m were constructed for town water supply, irrigation, flood control and hydro-power generation. However, this does not include a substantial number of large dams in China (Gleick, 2002, p. 301). Current estimates suggest that dams and diversion structures have affected some 60 percent of the world's rivers. The total investment in large dams is estimated at more than US$2000 billion, supplying some 30–40 percent of irrigated lands and generating 19 percent of the world electricity (World Commission on Dams, 2000, p. XXIX). Figure 1.3 shows that the peak in large dam construction occurred in the 1970s with construction of 5418 dams. Since then, dam construction has declined significantly and only 2069 dams were built in the 1990s. This has been due to opposition against dam construction for ecological, economical and social reasons (see section 2.1). The number of high dams worldwide is estimated at about 47 000, which includes 22 000 in China, 6575 in the United States of America, 4291 in India, 2675 in Japan, and 1196 in Spain (Gleick, 2002, pp. 291–295).

Table 1.7 shows the number of reservoirs (listed by continent) with storage capacities greater than 0.1 km³. The largest number is located in North America (915) followed by Asia (815) and Europe (576). In terms of total reservoir capacity, Asia has the greatest volume (1980.4 km³), followed by North America (1692.1 km³) and Africa (1000.7 km³). Further information on the distribution of dams, their dimensions and functions, is provided by the World Commission on Dams (2000, pp. 368–382).

1.4.1 Water Use

Global water withdrawal and consumption have been rapidly increasing due to the increasing world population

Table 1.6. *River run-off in various continents*

Continent	Annual run-off in:	
	(mm)	(km³)
Europe	274	2900
Asia	311	13 510
Africa	134	4047
North America	324	7870
South America	672	12 030
Australia and Oceania	268	2400
Total	–	42 757

Source: Shiklomanov and Rodda (2003, Table 10.1).

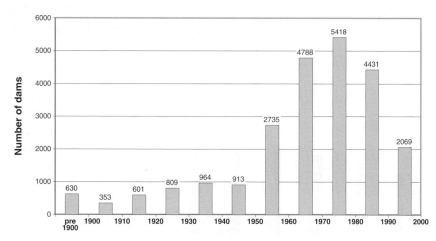

Figure 1.3 Number of large dams higher than 15 m commissioned in the twentieth century, listed by decades (World Commission on Dams, 2000, Figure V.3, and Gleick, 2002, Table 15).

and rising living standards. Table 1.8 lists the changes in world water withdrawal for the major sectors of the economy during the period of 1900–2000, and its projection to the year 2025. Global water withdrawal has increased by about seven-fold during the twentieth century compared to a four-fold increase in population from 1.5 billion to 6 billion.

Table 1.7. *Number of reservoirs with capacities of more than $0.1\,km^3$, by continent, for mid-1990s and their capacities*

Continent	Number of reservoirs	Volume of reservoirs (km^3)
Asia	815	1980.4
North America	915	1692.1
Africa	176	1000.7
Central and South America	265	971.5
Europe	576	645.0
Australia and New Zealand	89	94.8
Total	2836	6384.5

Source: Gleick (2000, Table 15).

By the year 2025 more than half of the $9000\,km^3$ of available water supply estimated by La Rivière (1989) will be in use. Agriculture is the largest consumer of water resources (Figure 1.4). Its share of total water use was about 91 percent in 1900, decreased to 66 percent by the year 2000 (although total volume increased) and is expected to decline to about 61 percent by 2025. Industry is the second largest water consumer and is followed by domestic water use. Excessive use of water for irrigation has led to waterlogging and salinisation (Ghassemi *et al.*, 1995), thereby accelerating land degradation and associated environmental problems.

Water use has not been efficient and there has been a significant difference between the annual volume of water withdrawn and consumed (Figure 1.5). In 1900, the ratio of water consumption to water withdrawal was about 71 percent. The gradual introduction of more efficient technologies, especially in the agricultural sector, resulted in this ratio declining to 66 percent in 1940 and 60 percent in year 2000. It is estimated that this trend will continue, and by 2025 the ratio will drop to 55 percent (Shiklomanov and Rodda, 2003, Chapter 11).

Table 1.8. *World water withdrawal by sectors of economic activity from 1900 to 2025 (in km^3)*

Water use	Year							
	1900	1950	1970	1980	1990	1995	2000	2025
Agriculture	525	1125	1834	2190	2412	2494	2571	3114
Industry	38	182	544	686	681	714	748	1105
Domestic	16	52	130	206	321	356	388	650
Reservoirs[a]	0.3	10	76	130	170	188	210	270
Total	579	1369	2584	3212	3584	3752	3917	5139

Note: [a] This is mainly because of evaporation.
Source: Shiklomanov and Rodda (2003, Table 11.3).

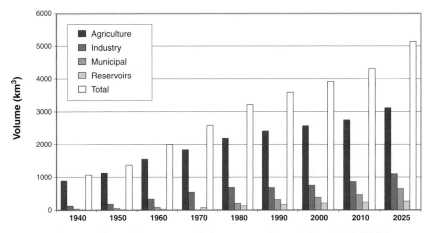

Figure 1.4 Water withdrawal for various sectors of the economy from 1940 to 2025 (Shiklomanov and Rodda, 2003, Table 11.3).

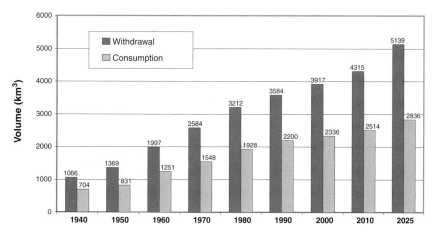

Figure 1.5 Water withdrawal and consumption from 1940 to 2025 (Shiklomanov and Rodda, 2003, Table 11.3).

Lack of maintenance of water delivery systems and overuse of water for domestic, commercial and industrial purposes, especially in developed countries, has caused a host of environmental and economic problems. Water losses, which in some cases amount to 70 percent of water delivered, put increasing and costly pressure on water works, which are struggling to meet the ever-increasing demand for water (UNEP, 1992, Chapter 5).

Another major concern is the overexploitation of groundwater mainly for irrigation and domestic use. This has led to the depletion of these resources in the arid and semi-arid areas of the world. Extraction of groundwater has also been responsible for the increased encroachment of seawaters into aquifers along coastal zones, land subsidence and decline of groundwater dependent ecosystems in many countries.

Adequate supply is not the only water problem facing many countries throughout the world. Water quality also is a cause of considerable worry. Concerns about water pollution have become widespread since the 1960s. At first, attention centred on surface water pollution from point sources (human, industrial and mining wastes). But more recently, pollution of groundwater and surface water from non-point sources (mainly agricultural pesticides, herbicides and fertilisers) has been found to be equally serious. It has been estimated that half of the population of the developing world is exposed to polluted sources of water that may increase the incidence of disease (UNESCO, 2003, p. 11). Key forms of pollution include: faecal coliforms, industrial organics, heavy metals, ammonia, nitrate, phosphate, pesticide residues, endocrine disruptors, salt and sediments. Further information on the state of the world's water resources is available in Gleick (1993).

1.4.2 Climate Change and Water Resources

There is strong evidence to suggest that the increasing atmospheric concentrations of pollutants such as carbon dioxide (CO_2), methane (CH_4), nitrous oxide (N_2O), and CFCs (chlorofluorocarbons) are causing an increase in air temperatures and changing the world's climate. Over the period 1950 to 1996 carbon dioxide emissions were highest in Europe, followed by North America and Asia (Table 1.9). Per capita CO_2 emissions in 1996 were highest in North America, followed by Oceania and Europe.

The atmospheric concentration of CO_2 has increased by 30 percent from 280 ppm at the end of the eighteenth century to 370 ppm in 2000. During the same period, the methane concentration increased from 750 ppb to 1750 ppb, and nitrous oxide from 270 ppb to 315 ppb (IPCC, 2001a, p. 36). Despite international efforts to reduce emissions of greenhouse gases, future changes in the world's climate are expected. The Intergovernmental Panel on Climate Change (IPCC) projected a global average warming of 1.4 to 5.8°C by 2100, relative to 1990 (IPCC, 2001a, p. 70). These rates of global warming are unprecedented over the past 10 000 years (UNESCO, 2003, p. 76). Among the predicted effects of global warming are rises in the ocean level, in the range of 0.1 m to 0.9 m (IPCC, 2001a, p. 74), and dramatic changes in climate. Global warming will impact on domestic, agriculture, natural ecosystems, forests, and the energy sector. In particular, global warming will impact on the hydrological cycle and associated water resources. Potential impacts include: changes in spatial and temporal rainfall patterns; changes in streamflow and groundwater recharge; changes in soil moisture; and an increase in magnitudes and frequencies of extreme events such as floods, droughts, typhoons and cyclones.

Table 1.9. *Regional carbon dioxide (CO_2) emissions in order of 1996 per capita emission*

Region	Carbon dioxide emissions (million tonnes)			Per capita emission (kg)
	1990	1996	1950 to 1996	1996
North America	5234	5710	200 969	19 074
Oceania	296	342	8583	11 842
Europe	–	6125	264 992	8414
Middle East and North Africa	1057	1408	24 847	3792
Central America and Caribbean	412	499	11 518	3078
Asia (excluding Middle East)	5194	7452	145 131	2296
South America	572	736	17 876	2260
Sub-Saharan Africa	467	520	12 928	894
World	22 361	23 882	718 514	4157

Source: World Resources Institute (2000, Data Table AC.1).

IPCC (2001b, Chapter 4) provides a comprehensive assessment of global climate change impacts on water resources. A brief description based on this publication follows.

The impacts of climate change on hydrology are usually estimated by linking global climate models, also known as general circulation models (GCMs), with hydrological models. The three key challenges are:

(1) constructing scenarios that are suitable for hydrological impact assessments;
(2) developing realistic hydrological models; and
(3) understanding the linkages and feedbacks between climate and hydrological systems.

The most important issue is the mismatch between global climate models (data generally provided on a monthly time step at a spatial resolution of tens of thousands of square kilometres) and catchment hydrological models which require data at daily time scales and at a resolution of perhaps a few square kilometres. Considerable effort has been expended on developing improved hydrological models for estimating the effects of climate change. Models have been developed to simulate water quantity and quality, with a focus on realistic representation of the physical processes involved. However, the greatest uncertainties arise from uncertainties in the climate change scenarios. Estimating the impacts on groundwater recharge and water quality are even less well understood.

The potential effects of climate change on some components of the water balance are:

Precipitation

Although there are large differences between the predictions of the various climate models, they all suggest that the greatest increases in precipitation will occur in high latitudes, some equatorial regions, and South East Asia, together with a general decrease in the sub-tropics. The predicted changes in seasonal precipitation are more variable. However, the frequency of heavy rainfall events is likely to increase with global warming. Increasing temperatures mean that a smaller proportion of precipitation may fall as snow, and in the areas where snowfall is currently marginal, snow may cease to occur. This has the potential to change the seasonality of run-off.

Soil Moisture

Global climate models have simulated soil moisture contents at a very coarse spatial resolution. Outputs of the models indicated that a rise in greenhouse gas concentrations is associated with reduced soil moisture in the northern hemisphere, mid-latitude summer. This is due to higher winter and spring evaporation, reduced snow cover, and lower rainfall inputs during summer. However, the local effect of climate change on soil moisture will vary not only with the degree of climate change but also with soil characteristics and location.

Groundwater

Unconfined aquifers are recharged via local rainfall. For confined aquifers recharge areas could be tens or hundreds of kilometres away. Quantification of recharge is complicated by the characteristics of the aquifers and overlaying deposits. As projected under most scenarios for mid-latitudes, increased winter rainfall is likely to result in increased groundwater recharge. In the arid and semi-arid environments, seasonal streamflows contribute to aquifer recharge, and changes in duration of flow of these streams will also impact on the recharge of aquifers.

A predicted sea level rise would have significant impacts on the livelihoods of inhabitants on low islands. It could also move the groundwater interface between freshwater and saline water further inland, impacting on the water supplies that are available on these small islands, and along coastal areas in general.

Streamflows

The predicted changes in streamflows are broadly similar to the changes in annual precipitation. Streamflows are expected to increase in high latitudes and many equatorial regions. However, the general increase in evaporation means that some areas with higher precipitation will actually experience a reduction in run-off (see IPCC, 2001b, Figure 4.1). Forecasts for changes in river flow for specific regions are:

- *Cold and cool temperate regions.* These areas are characterised by precipitation falling as snow during winter and include large parts of North America, northern and eastern Europe, most of Russia, part of China and much of Central Asia. In these regions a smaller proportion of precipitation during winter will fall as snow, providing more run-off in winter and less run-off during spring. In very cold areas such as Siberia and northern Russia, there would be little change in timing of streamflow because winter precipitation will continue to fall as snow.
- *Mild temperate regions.* Across most mid-latitude regions, winter run-off will increase due to increased rainfall, while in summer streamflows will be reduced because of a reduction in summer rainfall.
- *Humid tropical regions.* Run-off regimes in these regions are very much influenced by the timing and duration of the rainy season. Climate change therefore may affect streamflows, not only through a change in the magnitude of rainfall, but also through possible changes in the onset or duration of the rainy season, such as those caused by monsoons.
- *Arid and semi-arid regions.* Reduction of rainfall in these regions will lead to reduced streamflows.

Changes in Flood Frequency

Although a change in flood risk is frequently cited as one of the potential effects of climate change, relatively few studies have looked explicitly at possible changes in high flows. Global climate models currently cannot simulate with accuracy short-duration, high intensity, localised heavy rainfall, and a change in mean monthly rainfall may not be representative of a change in short-duration rainfall. However, a few studies have tried to estimate possible changes in flood frequencies, largely by assuming that changes in monthly rainfall also apply to flood-producing rainfall. In addition, some have examined the possible additional effect of changes in rainfall intensity.

Climate Change and Water Resources Management

In some countries, water managers have started to incorporate climate change in their planning.[2] However, Shiklomanov and Rodda (2003, Chapter 12) argue that the problem of providing reliable forecasts of global warming and its influence on water resources is one of the most difficult problems facing contemporary hydrology, because of the great sensitivity of the water system to even small changes in climate. They conclude that:

- Scenarios of possible changes in regional climate are at present extremely uncertain and do not agree with each other. This is particularly true for changes in precipitation, which is the key factor influencing water resources and water use. None of the existing climate scenarios provide a reliable basis for estimating changes in water resources at the regional and global scales.
- The slow rate of global warming compared with climate variations and the large uncertainty surrounding climate scenarios at the regional and global scales over the next 20 to 30 years, does not justify water resource managers investing large amounts of time planning for the possible impacts on streamflows and groundwater supplies.

1.5 AGRICULTURAL LAND USE

Like freshwater resources, the agricultural lands of the world are also limited and unevenly distributed. Some 3190 Mha of the world's land is potentially arable (Buringh, 1977). As shown in Table 1.10, 1497 Mha or 47 percent of these lands

---

[2] In the United States, the American Water Works Association urged water agencies to explore the vulnerability of their systems to plausible climate changes (AWWA, 1997). Also in the United Kingdom, water supply companies were required by regulators in 1997 to consider climate change in estimating their future resources and investment projections (Subak, 2000).

Table 1.10. *Land area, arable land, permanent crops and irrigated lands of each continent, in 2000*

Continent	Land area[a] (Mha)	Arable land (Mha)	Permanent crops (Mha)	Arable land and permanent crops (Mha)	Irrigated land (Mha)
Africa	2963	179.7	25.4	205.1	12.7
Asia	3098	486.3	59.6	545.9	190.1
North and Central America	2137	259.2	8.5	267.7	31.4
South America	1753	96.8	19.6	116.4	10.3
Europe	2261	289.8	16.9	306.7	24.5
Oceania	849	52.5	3.1	55.6	2.7
World total	13 061	1364.3	133.1	1497.4	271.7

Note: [a] Total area includes areas under inland water bodies.
Source: FAO (2002a, Tables 1 and 2).

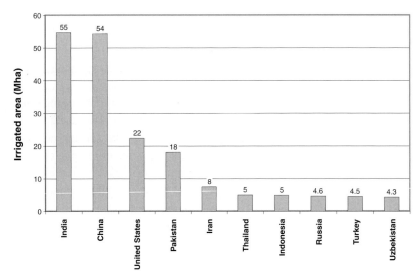

Figure 1.6 Ten countries with the largest area of irrigated lands, in the year 2000 (FAO, 2002a, Table 2).

are currently under cropping (arable land and permanent crops). However, it should be noted that the best agricultural lands have already been developed, particularly in the tropics and sub-tropics where land use pressure is greatest, and the remaining lands are of marginal quality or have unreliable rainfall.

In 1800, the area of agricultural land under irrigation was about 8 Mha (Table 1.11), and increased five-fold to 40 Mha by 1900. The area rapidly expanded through the twentieth century and reached 272 Mha in the year 2000. Asia has the largest area of irrigated land (190 Mha) while Oceania has only 2.7 Mha (Table 1.10). Figure 1.6 shows the ten countries with the largest areas of irrigated land.

Table 1.11. *Increase in world population and irrigated area*

Year	Population (billion)	Irrigated area (Mha)
1800	1.0	8[a]
1900	1.5	40[a]
1950	2.5	95[b]
1990	5.3	243[c]
2000	6.1	272[c]

Sources: [a] Framji *et al.* (1981); [b] Brown (1989); [c] FAO (2002a, p. 15).

As the world population continues to grow there will be increasing pressure on arable lands, especially in countries where it is in short supply. By 2025, it is predicted that in 14 countries (Japan, Egypt, China, Bangladesh, Israel, Vietnam, Kenya, Somalia, Tanzania, Nepal, Haiti, Yemen, Jordan and Saudi Arabia), there will be less than 0.07 ha of arable land per capita (World Resources Institute, 2000, p. 28).

1.6 FOOD AND FIBRE PRODUCTION

Globally, about 31 percent of agricultural lands are croplands and 69 percent are pasture, but actual proportions of each vary widely among countries and regions. For example: in south Asia croplands cover 95 percent while pasture is limited to 5 percent; in North America the distribution is about 50 percent each; and in Oceania some 90 percent is pasture and the remaining 10 percent is cropland (World Resources Institute, 2000, p. 57). While the total area of agricultural lands has increased worldwide over the past four decades, it has decreased in some industrialised countries.

Agricultural output and food production increased in both developed and developing countries over the past few decades. However, the annual rate of increase was higher in the developing countries than in the developed ones. Globally, agriculture produces enough food to provide every person on the planet with about 2760 calories each day, a level considered nutritionally adequate. But, this global average has little significance while many people living in developing countries suffer from inadequate nutrition. The average intake of calories varies from about 2190 in Africa (south of Sahara) to 3350 in western Europe, a difference of 1160 calories (Figure 1.7).

Buringh (1977) argued that taking into account the regional conditions of the soils, climate and farm management, enough food could be produced for a population of five to ten times the 1977 world population of about 4 billion, but this would require massive investment, substantial political will and would result in widespread removal of native vegetations.

Agriculture has been remarkably successful in keeping pace with the demand for food and fibre. Per capita food production was higher in 2000 than in 1970, even though the global population increased significantly over this period. Globally, agriculture is faced with an enormous challenge to continue meeting the needs of an increasing world population. Historically, agricultural output has increased mainly by bringing more land into production. But the amount of remaining land suitable for crop production is limited and there is growing competition for land from residential, industrial, commercial interests and for maintaining ecosystems (World Resources Institute, 2000, p. 53).

Intensification of agricultural production (obtaining more output from a given area of land) has become essential. In some regions, particularly in Asia and Egypt, farmers have intensified their production by growing two or three crops per year. This was possible by irrigating fields, using new crop varieties with shorter growth cycles and application of fertilisers. On non-irrigated lands, farmers have intensified their production mainly by abandoning or shortening fallow periods (World Resources Institute, 2000, pp. 53, 58).

The unprecedented scale of agricultural expansion and intensification has raised concerns about the sustainability

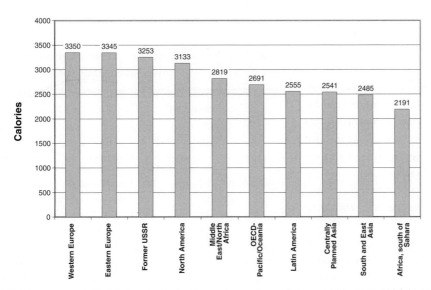

Figure 1.7 The average daily intake of calories for major regions of the world in 1989 (Gleick, 2000, Table 4.2).

of the agro-ecosystems and their long-term productive capacity. Stresses include increased erosion, soil nutrient depletion, salinisation, and waterlogging. There is also concern about the harmful effects of increased soil erosion on downstream fisheries, and the damage to both aquatic ecosystems and human health from fertilisers and pesticide residues. Soil degradation is widespread and severe enough to reduce productivity on about 16 percent of agricultural land, especially the croplands in Africa and Central America and pastures in Africa. Although new technologies may offset this decline in the foreseeable future, regional differences are likely to increase (World Resources Institute, 2000, p. 54).

In order to achieve the required increase in food production in the framework of sustainable rural development, the following issues must be considered (Schultz, 2001):

- Availability of water in space and time.
- Links between irrigation, drainage, flood protection, food security, protection of the environment, sustainable rural development and the livelihoods of rural communities.
- Socio-economic and ethical issues such as poverty alleviation, employment generation, and migration from rural to urban areas.
- The need for increased water withdrawals of 15–20 percent to meet the mismatch between demand and supply, in combination with water saving initiatives and improved efficiency in irrigation.
- The need for increased water storage volumes of 10 to 15 percent.
- Basin wide planning for integrated development and management.
- Options for inter-basin water transfers.
- Financing integrated water resources development and management, in particular the modernisation and replacement of existing infrastructure.

Increasing water use efficiency and water savings are particularly important issues, and there is a critical need to modernise irrigation and drainage facilities, especially in the developing countries. It should be noted that a significant part of the existing irrigation and drainage systems are more than 30 years old. Modernisation of infrastructure, including the related institutional reforms and cost recovery aspects, deserve major attention in the coming decades (Schultz, 2001).

Inter-basin water transfers can play a major role in reducing the risk of flooding in some basins while providing extra water in other areas where surface and groundwater resources are depleted. Increasingly, these depletions occur in the arid and semi-arid regions of the world. A combination of inter-basin water transfers, improved water use efficiency and an increased volume of water storages, can help meet the growing demand for water (Schultz, 2001).

1.7 FEEDING THE WORLD POPULATION

As mentioned in the previous section, the amount of food produced around the world is theoretically enough to feed all the Earth's population. However, a high percentage of people living in the developing world are faced with malnutrition because of their poverty and lack of adequate resources (water and land) for producing their own food, as well as other factors such as the impact of wars and disturbances. At the same time, a significant number of people in the developed world are overweight because they eat beyond their requirements (Table 1.12).

The number of hungry people in the developing countries was about 956 million in 1970, 919 million in 1980, and 816 million in 1990–92 (FAO, 2002b, Figure 3). In 1996, world leaders met in Rome at the World Food Summit and pledged to eradicate hunger. As a first step, they made a commitment to cut by half the number of undernourished people in developing countries by 2015, compared with the 1990–92 number (816 million). The latest data available indicate that the number of undernourished people in the developing countries only dropped slightly to 777 million in 1997–99 (Figure 1.8). If this slow trend continues, the number of undernourished people would be about 682 million by 2015, or 274 million above the target.

The average growth rate in world agricultural production (crop and livestock) over the period 1997–2001 was 1.7 percent, compared with 2.1 percent over the preceding

Table 1.12. *Share of children who are underweight and adults who are overweight in selected countries for the mid 1990s*

Country	Share of children underweight (%)	Country	Share of adult overweight (%)
Bangladesh	56	United States	55
India	53	Russian Federation	54
Ethiopia	48	United Kingdom	51
Viet Nam	40	Germany	50
Nigeria	39	Colombia	43
Indonesia	34	Brazil	31

Source: Gardner and Halweil (2000).

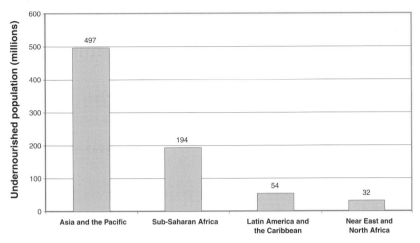

Figure 1.8 Regional distribution of undernourished people in the developing countries in 1997–99 (FAO, 2002b, Figure 1).

5 years and 2.5 percent in the 1980s (FAO, 2002b, p. 6). This trend towards lower growth rates in agricultural output is particularly evident in the developing countries.

The main cause of hunger is poverty, which limits people's access to food in the market or to land and other inputs (water, seed, and fertilisers) required to produce food. Also, many developing countries have turned to cash crops (such as coffee, bananas and flowers) for export, often at the expense of their own domestic food supplies. The main beneficiaries of these exports are large corporations, foreign investors and large landowners, but not the local communities (Gardner and Halweil, 2000).

The dependence of poor countries on foreign markets for staple foods leaves them vulnerable to price fluctuations and currency devaluations. The growing emphasis on free trade in agricultural products may bring additional vulnerability. Under the World Trade Organization agreements on agricultural products, farmers of the industrial nations are permitted to sell their subsidised surplus products cheaply to developing nations, forcing local farmers off the land (Gardner and Halweil, 2000).

Gleick (2000, pp. 63–92) has provided an analysis of the factors related to feeding the future world population. These include: total number of people to feed; basic and desired food requirements per person; preferences in type of food (grain, meat, etc.); availability of land; extent of land degradation; crop yields; cropping intensity; water requirements; irrigation efficiency; and the source of water to be used (rainfall, streamflow, water stored in lakes and reservoirs, groundwater and recycled wastewater). He concluded that the demands of the growing population will not be met by simply addressing this series of technical factors in isolation. Feeding the world population requires carefully planned integration of information about the various issues, effective and equitable international trade, exchange of technology and ideas, development of adequate food distribution systems, and suitable economic structures.

In the following two sections, the results of a recent study entitled *World Water and Food to 2025* and a short description of the *Challenge Program on Water and Food* are summarised.

1.8 WORLD WATER AND FOOD TO 2025

Using computer modelling techniques, Rosegrant *et al.* (2002) simulated feasible strategies that would enable the global community to avoid a water crisis, provide sufficient food, and protect the environment. They considered the year 1995 as the baseline and simulated three possible scenarios, described below, for global water and food production in 2025. Some of the 1995 data and the results of simulations are presented in Tables 1.13 and 1.14.

- **The businesses as usual scenario (BAU)** predicts the likely water use and food production based on the assumption that current trends for water management are broadly maintained. Under this scenario, water demand is projected to grow for irrigation, industrial and domestic uses (Table 1.13). The area of cereal cropping will expand and contribute to additional food production (Table 1.14).

- **The water crisis scenario (CRI)** examines the impact of a deterioration of current trends in water and food policy and investment. Under CRI, water consumption will be higher than under BAU (Table 1.13). A significant part

Table 1.13. *Total worldwide withdrawal and consumption of water, and irrigation, industrial and domestic water use for 1995 baseline and 2025 projections (in $10^9 m^3$)*

Baseline and various scenarios	Total Withdrawal	Total Consumption	Irrigation	Industrial	Domestic
1995 baseline	3906	1799	1436	159.5	169.3
2025 projection scenarios:					
BAU	4772	2081	1492	239.5	289.6
CRI	5231	2342	1745	319.6	222.5
SUS	3743	1673	1196	154.6	264.5

BAU, Business as usual scenario; CRI, Water crisis scenario; SUS, Sustainable water use scenario.
Source: Rosegrant *et al.* (2002, Tables 5.1, 5.2, 5.5 and 5.7).

Table 1.14. *Irrigated and rainfed cereal production for 1995 baseline and 2025 projections (in Mt)*

Baseline and various scenarios	Irrigated cereal	Rainfed cereal	Total cereal
1995 baseline	742.3	1033.3	1775.6
2025 projection scenarios:			
BAU	1161.0	1453.3	2614.3
CRI	1037.3	1327.7	2365.0
SUS	1108.5	1524.3	2632.8

BAU, Business as usual scenario; CRI, Water crisis scenario; SUS, Sustainable water use scenario.
Source: Rosegrant *et al.* (2002, Table 5.13).

of this increase is in the irrigation sector, mainly as a result of declining water use efficiency. Compared with BAU, CRI results in lower irrigated and rainfed cereal production (Table 1.14).

- **The sustainable water use scenario (SUS)** explores the potential for dramatically increasing water use efficiency and environmental water allocations. Under this scenario, the total water consumption will be lower than under the BAU scenario because of higher water prices and higher water use efficiency. This reduction in consumption frees water for environmental use. In 2025, SUS results in slightly higher cereal production compared with BAU (Table 1.14).

Rosegrant *et al.* (2002, p. 206), provide the following summary and conclusions for their study.

> A large part of the world is facing severe water scarcity. With a continued worsening of water supply and demand trends and water policy and investment performance, water scarcity could become a fully-fledged crisis with severe impacts on food production, health, nutrition, and the environment. But solutions to the potential water crisis are available, including increasing the supply of water for irrigation, domestic, and industrial purposes through highly selective investments in infrastructure. Even more important, however, are water conservation and water use efficiency improvements in existing irrigation and water supply systems through water management reform, policy reform, and investment in advanced technology and infrastructure, and improving crop productivity per unit of water and land through integrated efforts in water management and agricultural research and policy. These solutions are not easy, and will take time, political commitment and money. One thing is certain; the time to act on fundamental reform of the water sector is now.

1.9 CHALLENGE PROGRAM ON WATER AND FOOD

The Challenge Program (CP) on *Water and Food* was launched in November 2002 to assess innovative methods of Integrated Water Resources Management, with the aim of producing enough food for the world's growing population, while satisfying the requirements of other water users and also preserving environmental sustainability. Some 200 researchers representing 20 countries and 50 institutions have contributed to the development of this program. The following is a brief description of the CP on *Water and Food* based on the International Water Management Institute (2002).

The immediate objectives of the CP on *Water and Food* are:

(1) food security at household level;
(2) poverty alleviation through increased sustainable livelihoods in rural and urban areas;
(3) improved health through better nutrition, lower agriculture-related pollution and reduced water-related diseases; and

(4) environmental security through improved water quality as well as the maintenance of biodiversity and water-related ecosystem services.

The CP on *Water and Food* is structured in five key themes:

- **Theme 1: Crop water productivity improvement.** The objective is to increase crop water productivity such that food security can be ensured and farmers' livelihoods enhanced without increasing water diverted for agriculture over the amount diverted in the year 2000.
- **Theme 2: Multiple use of catchments.** The objective is to improve sustainable livelihoods for people who live in both upper catchments and downstream, through significant improvements in water productivity.
- **Theme 3: Aquatic ecosystems and fisheries.** The objective is to enhance food security and livelihoods by maintaining aquatic ecosystems and optimising fisheries.
- **Theme 4: Integrated basin water management systems.** The objective is to improve the productivity of water (in crop, livestock and fisheries production systems and ecosystem services) within the basin, by generating and applying knowledge on how to manage trade-offs and promote synergies, while maintaining or improving food security and environmental sustainability.
- **Theme 5: Global and national food and water systems.** The objective is to facilitate better policy-making and the implementation of necessary changes within the food and water system, at international, national and regional levels.

Although the initial intention was to focus activities of the CP on *Water and Food* on twelve Benchmark Basins representing large-scale real world problems, in January 2004 a decision was made to reduce the number of basins to nine. These basins and their catchment countries are:[3]

- Yellow River Basin (China), Mekong River Basin (China, Laos, Myanmar, Thailand, Cambodia and Vietnam), and Indo-Gangetic Basin (Pakistan, India, Nepal and Bangladesh) in Asia.
- Limpopo River Basin (Mozambique, Zimbabwe, Botswana and South Africa), Volta River Basin (Benin, Burkina Faso, Ivory Coast, Ghana, Mali and Togo) and Nile River Basin (Burundi, D.R. of Congo, Egypt, Eritrea, Ethiopia, Kenya, Rwanda, Sudan, Tanzania and Uganda) in Africa.
- Karkheh (Iran) in the Middle East.
- Sao Francisco Basin (Brazil), and Andean System of Basins (Bolivia, Colombia, Ecuador and Peru) in South America.

The CP on *Water and Food* proposed a minimum core budget of US$82 million for the first 5 year phase and is expected to attract a further US$50 million in matching funds. To have a significant impact, the expected duration of the CP on *Water and Food* will be considerably longer than the initial 5 year phase. Assuming that the first 5 years show sufficient progress, the duration of the program could be at least 10–15 years.

1.10 CONCLUSIONS

This chapter has outlined the limitations of the world's land and water resources for supporting an increasing population, which is expected to reach 9.3 billion by 2050. This is about 50 percent higher than the 2000 population of 6.1 billion. The per capita supply of both arable land and freshwater are declining. This has not only been caused by the increasing world population, but also by the degradation and pollution of the environment. Approximately 2 billion hectares of land throughout the world has been degraded by various processes such as salinisation, acidification and excessive application of agricultural chemicals (nutrients, herbicides and pesticides). Our planet is also faced with the prospect of global warming which inevitably will increase the demand for water. Millions of people in the third world countries are malnourished and remain in poverty because they do not have access to sufficient land and water. Their problems are exacerbated by global food trade and food distribution. At the same time, in developed countries excessive intake of food is causing various health problems.

Politicians and government bureaucrats must focus on policies and laws which address these urgent issues. Land and water resource managers and International Institutions must provide better management of the world's land and water resources to meet the growing demand for food while protecting the long-term sustainability of agricultural systems and the natural environment. Better management of water includes maximising the use of already developed

---

[3] It should be noted that up to 25 percent of the CP on *Water and Food* project investment can be made in non-benchmark basins. Therefore, the three basins initially considered (Amu Darya in Aral Basin, Euphrates in Iraq and Ulua in Honduras) which are not on the list of Benchmark Basins, plus other basins in developing countries, are eligible for consideration (Pamela George, Program Manager, Challenge Program on *Water and Food*, IWMI, Colombo, Sri Lanka, personal communication). Brief descriptions of the Benchmark Basins as well as more detailed descriptions of the previously mentioned five themes are available at http://www.waterforfood.org/docPresent/nairobi.asp (visited in June 2005).

resources without causing further degradation. This could be achieved by increasing water use efficiency in the irrigated agricultural sector, which is the largest water user throughout the world. Water use efficiency can also be improved in other sectors such as domestic, industrial and mining. Implementation of other measures, such as reuse of treated sewage and stormwater, water trading, and elimination of subsidies, are also important, as is judicious choice of new water sources.

References

Acton, D. F. and Gregorich, L. J. eds. (1995). *The Health of Our Soils: Towards Sustainable Agriculture in Canada*. Ottawa, Ontario: Centre for Land and Biological Resources Research, Research Branch, Agriculture and Agri-Food Canada.

AWWA (1997). Climate change and water resources. *Journal of the American Water Works Association* **98**: 107–110.

Brown, L. R. (1989). Reexamining the world food prospect. In *State of the World 1989*. A World Watch Institute report on progress toward a sustainable society. Washington D.C.: World Watch Institute. pp. 41–58.

Buringh, P. (1977). Food production potential of the world. In Radhe Sinha, ed. *The World Food Problem: Consensus and Conflict*. Oxford: Pergamon Press. pp. 477–485.

Chow, V. T., Maidment, D. R. and Mays, L. W. (1988). *Applied Hydrology*. New York: McGraw-Hill.

Cohen, J. E. (1995). *How Many People Can the Earth Support?* New York: W. W. Norton & Company.

Dregne, H., Kassas, M. and Razanov, B. (1991). A new assessment of the world status of desertification. United Nations Environment Program. *Desertification Control Bulletin* **20**: 6–18.

FAO (2002a). *FAO Production Yearbook 2001*. Rome: Food and Agriculture Organization of the United Nations.

FAO (2002b). *The State of Food and Agriculture*. Rome: Food and Agriculture Organization of the United Nations.

Framji, K. K., Garg, B. C. and Luthra, S. D. L. (1981). *Irrigation and Drainage in the World: A Global Review*. Third Edition. New Delhi: International Commission on Irrigation and Drainage. Volume I.

Gardner, G. and Halweil, B. (2000). Nourishing the underfed and overfed. In *State of the World 2000*. United Kingdom: Earthscan Publications Ltd. pp. 59–78.

Ghassemi, F., Jakeman, A. J. and Nix, H. A. (1995). *Salinisation of Land and Water Resources: Human Causes, Extent, Management and Case Studies*. Sydney: University of New South Wales Press.

Gilbert, G. (2001). *World Population: A Reference Handbook*. Santa Barbara, California: ABC-CLIO, Inc.

Gleick, P. H. ed. (1993). *Water in Crisis: A Guide to the World's Freshwater Resources*. New York: Oxford University Press.

Gleick, P. H. (2000). *The World's Water 2000–2001: The Biennial Report on Freshwater Resources*. Washington D.C.: Island Press.

Gleick, P. H. (2002). *The World's Water 2002–2003: The Biennial Report on Freshwater Resources*. Washington D.C.: Island Press.

Hamblin, A. (2001). *Land, Australia State of the Environment Report 2001*. Collingwood, Victoria: CSIRO Publishing.

International Water Management Institute (2002).[4] *CGIAR Challenge Program on Water and Food*. Colombo, Sri Lanka: IWMI.

IPCC (2001a). *Climate Change 2001: The Scientific Basis*. Contribution of Working Group I to the Third Assessment Report of the Intergovernmental Panel on Climate Change. Cambridge: Cambridge University Press.

IPCC (2001b). *Climate Change 2001: Impacts, Adaptation and Vulnerability*. Contribution of Working Group II to the Third Assessment Report of the Intergovernmental Panel on Climate Change. Cambridge: Cambridge University Press.

Korzun, V. I., Sokolov, A. A., Budyko, M. I., Voskresensky, K. P., Kalinin, G. P., Konoplyantsev, A. A., Korotkevich, E. S., Kuzin, P. S. and Lvovich, M. I. eds. (1978). *World Water Balance and Water Resources of the Earth*. Paris: UNESCO.

La Rivière, J. W. M. (1989). Threats to the world's water. *Scientific America*. September: 48–55.

Lewis, E. L., Jones, E. P., Lemke, P., Prowse, T. D. and Wadhams, P. (2000). *The Freshwater Budget of the Arctic Ocean*. NATO Science Series, Environmental Security, Volume 70. Dordrecht: Kluwer Academic Publishers.

National Land & Water Resources Audit (2001). *Australian Agriculture Assessment 2001*. Volume 1. Canberra: Natural Heritage Trust.

Oldeman, L. R. Hakkeling, R. T. A. and Sombroek, W. G. (1991a). *World Map of the Status of Human-Induced Soil Degradation: An Explanatory Note*. Second revised edition. Wageningen: International Soil Reference and Information Centre (ISRIC).

Oldeman, L. R., van Engelen, V. W. P. and Pulls, J. H. M. (1991b). The extent of human-induced soil degradation. In Oldeman, L. R., Hakkeling, R. T. A. and Sombroek, W. G. *World Map of the Status of Human-Induced Soil Degradation: An Explanatory Note*. Second revised edition. Wageningen: International Soil Reference and Information Centre (ISRIC). pp. 27–33.

Rosegrant, M. W., Cai, X. and Cline, S. A. (2002). *World Water and Food to 2025: Dealing With Scarcity*. Washington D.C.: International Food Policy Research Institute.

Schultz, B. (2001). Water for food and environmental security. In *Transbasin Water Transfers*. Proceedings of the 2001 USCID Water Management Conference, Denver, Colorado, 27–30 June 2001. Denver, Colorado: U.S. Committee on Irrigation and Drainage. pp. 1–14.

Shiklomanov, I. A. (1990). Global water resources. *Nature & Resources* **26**(3): 34–43.

Shiklomanov, I. A. and Rodda, J. C. eds. (2003). *World Water Resources at the Beginning of the Twenty-First Century*. Cambridge: Cambridge University Press.

Stolbovoi, V. and Fischer, G. (1998). A new digital georeferenced database of soil degradation in Russia. In Blume, H.-P., Eger, H., Fleischhauer, E., Hebel, A., Reij, C. and Steiner, K. G. eds. *Towards Sustainable Land Use: Furthering Cooperation Between People and Institutions*. Reiskirchen, Germany: Catena Verlag GMBH, Volume 1. pp. 143–152.

Subak, S. (2000). Climate change adaptation in the U.K. water industry: manager's perceptions of past variability and future scenarios. *Water Resources Management* **35**: 137–157.

UNEP (1992). *Saving Our Planet: Challenges and Hopes*. Nairobi: United Nations Environmental Programme.

UNESCO (2003). *Water for People Water for Life: The United Nations World Water Development Report*. Paris: UNESCO.

United Nations (2002). *World Population Ageing 1950–2050*. New York: Department of Economic and Social Affairs, Population Division.

van Lynden, G. W. J. and Oldeman, L. R. (1997). *The Assessment of the Status of Human-Induced Soil Degradation in South and Southeast Asia*. Wageningen: International Soil Reference and Information Centre (ISRIC).

Wadhams, P. (2000). *Ice in the Ocean*. Australia: Gordon and Breach Science Publishers.

World Bank (1992). *World Development Report 1992: Development and the Environment*. New York: Oxford University Press.

World Commission on Dams (2000). *Dams and Development: A New Framework for Decision-Making*. London: Earthscan Publications.

World Resources Institute (1986). *World Resources 1986*. A report by the World Resources Institute and the International Institute for Environment and Development. New York: Basic Books.

---

[4] This document is also available at the URL http://www.waterforfood.org (visited in June 2005).

World Resources Institute (1992). *World Resources 1992–93*. A Report by the World Resources Institute in collaboration with the United Nations Environmental Programme and the United Nations Development Programme. New York: Oxford University Press.

World Resources Institute (2000). *World Resources 2000–2001: People and Ecosystems, The Fraying Web of Life*. A collaborative product of the United Nations Development Programme, United Nations Environmental Programme, World Bank and World Resources Institute. Oxford, UK: Elsevier Science.

2 Issues in Inter-Basin Water Transfer

2.1 INTRODUCTION

A significant portion of the world's limited freshwater resources have already been developed to supply the increasing world population with water for drinking, sanitation, production of food and fibre, flood control and hydro-power generation. Developments have included inter-basin water transfer, construction of dams, canals, pipelines, pumping stations, and well fields for groundwater extraction. As the world approaches the limits of its finite freshwater resources, efforts are being increasingly directed towards their better management through improving water use efficiency, recycling of used water, improved laws and the establishment of a water market.

Globally, many water resources development projects have been criticised for their negative environmental, social and cultural impacts, as well as their poor economic performances (Petts, 1984; Goldsmith and Hildyard, 1984; Blackwelder and Carlson, 1986; Pearce, 1992). The World Bank also has been criticised for funding numerous large-scale water resources development projects around the world, resulting in environmental degradation, population displacements, and unrealised economic outcomes. Following these criticisms, the World Bank developed policies and guidelines for the environmental and social assessment of its various projects, including water resources development prior to any future funding (World Bank, 1991a,b,c).

The World Commission on Dams[1] (WCD) has undertaken the most recent critical review of the large dams around the world. It started its activities in May 1998 by:

(1) creating a 68 member stakeholder forum;
(2) drawing on the wider stakeholder community of experts and analysts in developing the WCD knowledge base;
(3) undertaking a programme of four regional consultations in different parts of the world with participation of 1400 individuals from 59 countries;
(4) analysing 900 submissions made to the Commission;
(5) initiating eight independent in-depth case studies and two country studies of India and China;
(6) undertaking 17 thematic reviews in five categories of social, environmental, economic, financial and institutional issues; and
(7) examining around 1000 dams with varying degrees of intensity, including comprehensive surveys of 125 dams.

The World Commission on Dams (2000) found that the unprecedented expansion of large dam building during the twentieth century had clearly benefited many people globally. However, this positive contribution has been marred in many cases by significant social, economic and environmental impacts, which are now unacceptable. Some of the identified problems include:

- A high number of displaced people, estimated at about 40 to 80 million.
- Lack of equity in the distribution of benefits.
- Extensive negative impacts on rivers, watersheds and aquatic ecosystems.
- Lack of development of alternative options for water and electricity supply such as reducing consumption, increasing water use efficiency, water recycling and development of local renewable energy resources (micro-hydro, solar, wind and biomass systems).
- Lack of integration between economic, social and environmental dimensions of developments.
- Lack of stakeholder participation and transparency in project developments.
- High degree of variability in delivering predicted water and electricity services, as well as social and economical benefits.

[1] The World Commission on Dams was born in April 1997 in Gland, Switzerland, out of a workshop sponsored jointly by the World Conservation Union and the World Bank to discuss the implications of the World Bank review of 50 dams funded by the Bank. Its establishment was formally announced in February 1998, and began its activities the following May (World Commission on Dams, 2000).

- Shortfalls in achieving physical targets, cost recovery and economic profitability.
- Distortion of decision-making processes due to corruption.
- Delays in the construction period and significant cost overruns.

In order to prevent these past mistakes in future water resources development, the World Commission on Dams (2000) recommended the following seven strategic priorities and related policy principles for decision-making:

- **Gaining public acceptance.** Acceptance of key decisions by all groups of affected people, particularly indigenous and tribal peoples, women and other vulnerable groups, is essential for equitable and sustainable water and energy resource developments.
- **Comprehensive options assessment.** Alternatives to dams should be explored. The selection should be based on comprehensive and participatory assessment of a full range of policy, institutional and technical options. In the assessment process, social and environmental aspects should have the same significance as economic factors.
- **Addressing existing dams.** Dams and the context in which they operate are not static over time. Therefore, opportunities exist to optimise benefits from many existing dams, and addressing outstanding social, economic and environmental issues. This can be achieved by numerous measures including changes in water use priorities, land use changes, changes in public policies, management and operation practices.
- **Sustaining rivers and livelihoods.** Rivers, watersheds and aquatic ecosystems are the basis for life and livelihood of local communities. Dams transform landscapes and create risks of irreversible impacts. Therefore, understanding, protecting and restoring ecosystems at river basin level is essential to foster equitable human development and welfare of all species. Avoiding impacts through good site selection and project design are crucial. Release of tailor-made environmental flows[2] can help maintain downstream ecosystems and the communities that depend on them.
- **Recognising entitlements and sharing benefits.** Negotiations with adversely affected people can result in mutually agreed and legally enforceable mitigation, resettlement and development provisions. These recognise entitlements that improve livelihoods and quality of life, so that affected people become beneficiaries of the project.
- **Ensuring compliance.** Compliance with applicable regulations and guidelines, and project-specific negotiated agreements, should be secured at all critical stages in project planning and implementation. A set of mutually reinforcing incentives and mechanisms is required for social, environmental and technical measures. These should involve an appropriate mix of regulatory and non-regulatory measures, incorporating incentives and sanctions.
- **Sharing rivers for peace, development and security.** Storage and diversion of water on transboundary rivers has been the source of considerable tension both between countries and within countries. As specific interventions for diverting water, dams require constructive cooperation. Consequently, the use and management of resources increasingly becomes the subject of agreement between states to promote mutual self-interest for regional cooperation and peaceful collaboration. This leads to a shift in focus from the narrow approach of allocating a finite resource, to the sharing of rivers and their associated benefits.

These strategic priorities provide a sound basis for considering and proposing an inter-basin water transfer project.

2.2 KNOWLEDGE REQUIREMENTS AND INTER-BASIN WATER TRANSFER

The development of water resources, including inter-basin water transfer, involves complex issues. It requires careful consideration of the costs and benefits, environmental issues, a long planning period, and painstaking negotiations between numerous stakeholders before decisions can be made.

Diversions of water from one river basin to another are multidisciplinary problems (Yevjevich, 2001). These are geomorphological, hydrological, water quality, water resources planning, and hydraulic engineering components, which are intimately connected with political, administrative, economic, environmental, cultural and legal components. Controversies that have surrounded water diversion schemes in the past have many issues in common. However, each case also has its own specific problems (Committee on Western Water Management, 1992).

It is now clear that to avoid controversies, policy makers and water resources managers of basins with apparent water

---

[2] See Section 2.5 in this chapter.

shortages should firstly exhaust all possibilities for better management of their own water by:

(1) eliminating losses in their current water supply network;
(2) increasing water use efficiencies;
(3) conjunctive use of surface and groundwater resources;
(4) increasing water prices to promote water use efficiency and shift water use from low value to higher value production systems;
(5) reclamation of wastewater in municipal areas;
(6) reviewing policy and regulations; and
(7) improving monitoring.

Exploration of the possibilities for building new dam(s) in their own basin(s) may show that this option is cheaper and less controversial than water transfer from other basin(s), particularly if they are located in another jurisdiction. River basins, which currently seem to have an excess of water for inter-basin transfer should be carefully assessed for their long-term potential water requirements for agricultural, industrial and domestic water supplies, and also for the potential effects of climate change. It is fundamentally important that assessments include a realistic appraisal of environmental water requirements before any decision on the export of excess water to other basins are made.

Both topography and geology of the two basins are fundamentally important in all inter-basin water transfer projects. These define location of reservoirs and determine largely the project costs. In this chapter, some of the issues related to inter-basin water transfer are discussed.

2.2.1 Topography

Topographic maps are indispensable for the investigation, design and construction of a diversion scheme. Existing topographic maps at suitable scale are essential for investigations of the catchments, reservoirs, dam sites and potential sources of construction materials. These maps are also useful for design of the significant elements of the scheme such as dams, tunnels and canals. Detailed topographic maps and profiles can be readily prepared using various techniques such as Airborne Laser Scanning, which has been in commercial operation since 1993. This can make over 50 000 measurements per second and cover over 50 km^2 per hour, to a typical vertical accuracy of 0.15 m or better (Jonas and Byrne, 2003).

2.2.2 Geology

The local geology critically determines suitable sites for dams, spillways, tunnels, aqueducts and, if required, sub-surface power stations associated with the inter-basin water transfer projects. Geological investigations are usually carried out in several phases. These are (Best, 1984):

- **Preliminary or reconnaissance investigations.** Alternative locations are considered for any dams. The objective is to provide sufficient data to select the best sites for dam construction.
- **Feasibility investigations.** Once the preferred sites have been selected, investigations are carried out to confirm the economic feasibility of the project. Alternative types of structures such as concrete arch, concrete gravity, gravity-arch, buttress, rockfill and earthfill, are evaluated at this stage. Investigations need to provide sufficient data and to allow comparative designs and cost estimates to be evaluated.
- **Design investigation.** Once the general type of engineering structure has been determined, more information is required for the detailed design of each structure. This is a critical stage, as all geological conditions to be encountered during construction need to be predicted and addressed in the design document.

Site investigations require detailed lithological and geological maps. Lithological maps represent the extent of various rock and sediment types (igneous, metamorphic and sedimentary rocks; unconsolidated fluvial sediments; glacial and aeolian deposits; and others). Geological maps demonstrate the extent of geological units, geological structures, and faulting systems. Aerial photographs and satellite images greatly facilitate mapping. Geological cross-sections, which display the shape and depth of the various geological units, provide a better understanding of the geology of the area. However, cross-sections based solely on surface geological data are of doubtful value and are not accurate in depth. Surface geology needs to be combined with bore records for accurate geological cross-sections.

Karstic or dissolution cavities in limestone and dolomite rocks present particular problems as they can cause severe leakage from the reservoir areas or from beneath the dam sites. If these cavities are extensive, grouting is often necessary (Walhstrom, 1974, Chapters 7 and 10).

The type of rocks and sediments at the dam sites, their geotechnical properties, and the depth to the bedrock are critical for the design and construction of dams (Walhstrom, 1974). Geophysical investigations using seismic refraction and resistivity methods (Parasnis, 1986; Milson, 1989), followed by sub-surface geophysical investigations and core drilling, provide useful information on the suitability of a particular site. This information also assists selection of the most appropriate type of dam. Geological factors also play a significant role in the construction of tunnels, for the choice of the best route, the cross-sectional design and the construction method, and in the design of underground power stations.

Seismic activity of the region is another important issue in the design and construction of dams, tunnels, canals, aqueducts, spillways and underground power stations (Jansen, 1988, Chapter 5: Seismology, and Chapter 8: Earthquake response analysis of embankments). Structures are required to resist earthquakes with the most probable intensity in the area of interest.

2.2.3 Hydrology

The hydrology and water quality of the river are critically important factors to consider when planning a new water diversion. The volume of available water for transfer is probably the most important factor determining the feasibility of a diversion. It is now unacceptable to divert a large proportion of one basin's run-off to another basin. Experience has shown that the maintenance of a reasonable flow downstream of the diversion point is critical to satisfy environmental and other long-term water requirements of the water-exporting basin. Environmental releases must consider the volume of water, its quality, and also its seasonal pattern, often designed to mimic the natural flow regime.

The important hydrological issues include:

(1) the long-term monthly and annual average rainfall and evaporation of the basin;
(2) river run-off;
(3) water quality issues such as salinity, turbidity and nutrient concentrations;
(4) potential impacts of climate change on medium to long-term run-off and water quality of the basin;
(5) current and projected withdrawals; and
(6) downstream ecosystem flow requirements.

Analysis of the hydrological data is undertaken by conventional methods coupled to modelling and simulation techniques. These analyses, combined with other factors such as topography and geology of the suitable sites, determine the possible capacity of the reservoirs. The capacity of the reservoirs and the downstream water requirements for both environmental and other uses then provides estimates of the volume of available water for diversion out of the basin.

The volume of water that can be diverted depends on the location of the diversion point. If this is in the upper part of the basin, river run-off may be small. However, upstream sites have the advantage of providing an opportunity for water diversion by gravity via canal, pipeline or tunnel. Locating the diversion point in the lower part of the basin may increase the potential for diverting larger volumes of water, but this could make the water transfer conduits longer and may also require expensive pumping. During investigations, different combinations of dams with various capacities at one or a number of locations are often considered, to select the most suitable alternative.

Computer models are in widespread use for assessment of streamflow and water quality issues, as well as simulation of changes in the catchment, such as future changes to land use that might affect water yield and quality. The selection of an appropriate model depends on the model assumptions, data availability and the application.

Hydrological models fall into three main categories depending on how the processes are represented (Wheater *et al.*, 1993; Letcher and Jakeman, 2002). These are:

- **Empirical or metric models** which are generally the simplest and are based primarily on analysis of observations. These models are based on the development of the unit hydrograph and many of them use stochastic techniques for the correlation of catchment data. The main features of this class of models are their high level of spatial and temporal aggregation and the use of a small number of variables.
- **Conceptual models** represent important catchment-scale hydrological processes and vary considerably in complexity. They usually incorporate some underlying physical mechanisms for describing water flow within their structure. They tend to lump representative processes over the scale at which outputs are simulated. Parameter values are obtained through a calibration process using observed data such as stream discharge and other measurements.
- **Physically based models** depend more upon the numerical solution of equations representing the physics governing water transport within the catchment. In theory, the parameters used are measurable within the catchment. However, in practice the large number of parameters involved and the heterogeneity of the catchment means that these parameters must often be calibrated against observed data.

Croke and Jakeman (2001) discuss the selection of hydrological models, and consider the size and other characteristics of the catchment, the intended use of the model, and the hydrological data that is available for model calibration. They also discuss factors affecting model performance such as rain gauge density, stream gauge rating quality, actual catchment response dynamics and data sampling frequencies.

In developed countries, stakeholders have become increasingly suspicious of the output of models because their assumptions and operation are often opaque. There is a clear need to specify transparently both in any model used to assess water resources.

Further detailed information regarding hydrological investigations for design and construction of dams is available in the United States Bureau of Reclamation (1987, Chapter 3: Flood Hydrology Studies), and Jansen (1988, Chapter 3: Hydrology).

Reservoir Sedimentation

Sediments transported by rivers are trapped when reservoirs are constructed. This reduces the effective storage capacity of the reservoirs and their longevity. Sediments also may cause dam safety problems by applying pressure on the dam wall and intake structures. Although dam designers provide a dead storage for trapped sediments, this storage may not be adequate if the annual volume of sediments exceeds the estimated design. Rates of sedimentation vary markedly in river catchments, depending on their geology, soils and the vegetation cover.

Globally, it is estimated that annually some 0.5 to 1 percent of the world reservoir volume is lost due to sedimentation (World Commission on Dams, 2000, p. 65). In the United States the rate of storage capacity loss is 0.22 percent per annum representing a volume of $2 \times 10^9 \, m^3$, and almost 25 percent of the sediments are estimated to originate from croplands (Gleick, 1993, p. 367). Mahmood (1987) estimated the worldwide replacement cost of the lost reservoir capacity at US$6 billion per annum, or about US$10 billion in 2005 values. It is therefore important to accurately estimate sediment loads of rivers discharging into reservoirs, and their distribution within the reservoirs. If sediment loads are expected to be high, adequate soil conservation measures and watershed management initiatives such as tree planting and land tracing should be envisaged. Mahmood (1987) has discussed global magnitude of reservoir sedimentation, erosion in drainage basins, sedimentation processes, methods for predicting reservoir sedimentation, and research needs. He also describes practical limitations of mitigation methods such as debris dams, sediment bypassing, sediment flushing, and sediment dredging. Annandale (1987) describes a wide range of technical issues related to reservoir sedimentation including sediment transport theories, practical techniques for estimating sediment yields, strategies to prevent reservoir sedimentation, and techniques for forecasting distribution of deposited sediments.

2.2.4 Hydrogeology

The fundamental issues in hydrogeology are described in various books including Nielsen (1991), Fetter (2001), and Schwartz and Zhang (2003). For investigations of inter-basin water transfer, it is important to understand hydrogeology of both the exporting and importing water basins. In the importing water basin detailed investigations of the geology, structure and extent of the aquifer systems, their piezometry, hydrodynamic parameters, depth to water-table, recharge and discharge, current and projected withdrawals, water balance, and water quality are important. This understanding assists the assessment of groundwater resources of the basin, interactions between surface water and groundwater systems, and allows the planning of effective conjunctive management of groundwater and surface water resources. The establishment of monitoring facilities and programmes to prevent watertable rise and salinisation from any expansion of irrigation activities, and/or depletion of groundwater resources from excessive extraction, is also an essential element.

Groundwater numerical models, which assess the impacts of rainfall, irrigation, cropping activity, pumping and land use changes on groundwater resources are increasingly important components of groundwater studies. McLaughlin *et al.* (1993) provide a review of groundwater flow and transport modelling. Other publications regarding simulation of groundwater flow and contaminant transport include McDonald and Harbaugh (1984, 1988), which describe the widely used groundwater flow model MODFLOW, Anderson and Woessner (1992), Fetter (1993), Zheng and Bennett (1995a), Spitz and Moreno (1996), and Holzbecher (2002). Relevant numerical contaminant models include SUTRA (Voss, 1984), SUTRA-GUI (Voss *et al.*, 1997), HST3D (Kipp, 1987) and MT3D (Zheng and Bennett, 1995b). MODFLOW has been steadily developed by the inclusion of various modules to make it more efficient and user friendly. MODFLOW-2000[3] (Harbaugh *et al.*, 2000; Hill *et al.*, 2000) was a major update of the model. Visual MODFLOW was first released in August 1994 and is used by consulting firms, research institutions and government agencies worldwide (Waterloo Hydrogeologic, 2000).

---

[3] For information regarding MODFLOW and its latest version visit: http://water.usgs.gov/software/modflow.html (visited in June 2005).

Visual MODFLOW Pro[4] is a fully integrated three-dimensional (3D) groundwater flow and contaminant transport model, which combines MODFLOW, MODPATH, MT3D, automatic model calibration and built-in 3D visualisation and animation. The innovative menu structure allows an easy dimensioning of the model domain and selected units, conveniently assigns model parameters and boundary conditions, runs model simulations, and visualises the results with line contours or colour shading. Similar to hydrological modelling, groundwater modelling needs to be transparent in terms of assumptions in model structure, uncertainties in data and model parameters, and simulated results.

2.3 PLANNING AND PUBLIC PARTICIPATION

Almost all of the inter-basin water transfer schemes now operating around the world were constructed when politicians and engineers had the authority to plan, finance and construct the schemes with little or no consultation with the public. Today, the public often plays a major role in the development of any new scheme, particularly in developed countries. In many countries, the public now participates in project development and assessment and without their approval the project may not proceed.

Planning for any proposed inter-basin transfer of water must be based on the principles of *Ecologically Sustainable Development* with the active involvement of stakeholders, such as citizens, indigenous peoples, farmers, tourist operators, industry groups, scientists, environmentalists, NGOs, and government officials at local, state and federal levels. Until the 1970s, governments and their consultant engineers dominated the planning of water resources development. Over the last 30 years, public participation in the planning process in developed countries has become the rule rather than the exception. Effective public participation in the planning process is democratic, clarifies issues, reduces misunderstandings, softens one-sided viewpoints and facilitates the resolution of differences (Schwass, 1985). Nancarrow (1994) provides an insight into Australian, Canadian and American experiences with public participation in water resources, and identifies some priority areas for increased public involvement.

Leading government water agencies have a major role in presenting new projects to the stakeholders, addressing their concerns in a revised proposal, and then reaching a final decision. During the planning process for inter-basin water transfer schemes, it is fundamentally important to investigate possibilities for alternative sources of water supply such as increasing water use efficiency, building new dams in the basin with the water shortage, desalination and reuse of treated wastewater. Environmental and social impacts of these alternatives require assessment and comparison with the option of inter-basin water transfer. In general, a decision to transfer large volumes of water from one basin to another will only succeed if the stakeholders are satisfied that the benefits are much greater than the benefits of any non-transfer alternatives.

2.3.1 Engineering Issues

The engineering problems of water diversion are likely to be among the simplest problems to resolve since the technology for designing and constructing hydraulic structures is well developed. In diversion projects, the following engineering issues are usually investigated:

- Diversion structures (dams), their type (concrete, earthfill, rockfill, etc.) storage capacities, full supply levels and the monthly and annual diversion volumes.
- Type of conveyance facilities (aqueduct, canal, tunnel or pipeline) between the exporting and importing basins, and their details (length, diameter, capacity and gradient). Pipelines for the transfer of water in the receiving basin, despite their cost, have benefits over transfer via rivers because of the minimisation of water loss by percolation and evaporation, and elimination of riverbank erosion.
- Pumping facilities (number of pumping stations, their capacities, pumping heads, and energy requirements) if water is to be lifted to higher elevation before being transferred to another basin.
- Opportunities for hydro-electric power station(s) if there are significant drops in level as the water is transferred into the receiving basin.

2.3.2 Role of Governments

Governments play a major role in providing reliable sources of water while this may include transferring water from one basin to another (Yevjevich, 2001). Inter-basin transfers must not only consider the interests of the country as a whole, but also the local interests in any river basin from which water is being diverted. Each region within a country requires a long-term plan for water demand and the associated water schemes. Regional plans often create

---

[4] http://www.waterloohydrogeologic.com/software/visual_modflow_pro/ (visited in June 2005).

conflicts between different water users within a region and between adjoining regions that wish to share water supplies. Negotiations to develop a mutual understanding and consensus between the different interest groups are often required. Forward thinking governments encourage regional plans and integrate them into state and countrywide strategies and into legislation. An additional key role of governments is in the initiation of impact assessment of proposed projects.

2.4 ASSESSMENT OF THE IMPACTS

Environmental assessment[5] (EA) is a process for identifying, predicting and evaluating the potential biological, physical, social and health effects of a proposed development action, and for communicating the findings in a way that encourages these concerns to be adequately addressed by stakeholders prior to any decision being made (Harrop and Nixon, 1999, p. 2).

The following sections describe the key environmental issues associated with the construction of dams, development of irrigation and drainage systems, and hydro-power generation, as components of inter-basin water transfer projects.

2.4.1 Dams and Reservoirs

Construction of large dams causes irreversible environmental changes over wide geographic areas (Dixon *et al.*, 1989; UNESCO, 1990; World Bank, 1991b, p. 32; World Commission on Dams, 2000, pp. 15–17). The area of influence of a dam project extends from the dam site and its reservoir to as far downstream as the estuary, coast and offshore zone. While there are direct environmental impacts associated with the construction of the dam, the greatest impacts result from the flooding of land above the dam site to form the reservoir, and alteration of the flow regime downstream. These effects have direct impacts on soils, vegetation, wildlife, fisheries and human populations in the area.

2.4.2 Irrigation and Drainage

Irrigation and drainage projects use water for agricultural production. The dominant irrigation method is surface irrigation (flood and furrow) in which water is distributed by gravity as overland flow. Other systems use sprinkler and drip (trickle) irrigation (Withers and Vipond, 1974; Rural Water Commission of Victoria, 1988; Christen and Hornbuckle, 2001). Drainage uses both surface and sub-surface methods to remove excess water from the root zone.

The potential negative impacts of most large irrigation projects include:

(1) waterlogging and salinisation;
(2) increased erosion;
(3) pollution of surface and groundwater from agricultural chemicals;
(4) deterioration of water quality, and increased nutrient levels in the irrigation and drainage water;
(5) proliferation of aquatic weeds;
(6) eutrophication in irrigation canals and downstream waterways;
(7) increased incidence of water-borne diseases such as malaria and schistosomiasis;
(8) changes in the lifestyle of the local population; and
(9) increases of agricultural pests and diseases resulting from the creation of a more humid microclimate (World Bank, 1991b, p. 94).

2.4.3 Hydro-electric Power Generation and Transmission

Hydro-electric power generation provides an alternative to the burning of fossil fuels or to nuclear power, which allows the power demand to be met without producing greenhouse gases, air emissions and radioactive waste. Hydro-electric power generation often benefits communities and industries well away from the dam. Potential negative environmental effects occur during the construction period and are: air and water pollution from construction and waste disposal; soil erosion; and destruction of vegetation (World Bank, 1991c, p. 69). The generated hydro-electric power (like the thermal-electric power) is transported to the population centres and industries via transmission lines, which can also have environmental impacts caused by their construction, operation and maintenance (World Bank, 1991c, pp. 25, 26).

---

[5] In different countries different terms have been used for Environmental Assessment. For example, in Australia the term environmental impact assessment (EIA) is frequently used. This is defined as a systematic process for the examination and evaluation of the environmental effects of proposed activities that are considered likely to affect the environment. The purpose of EIA is the production of an environmental impact statement (EIS), which is used to inform decision makers and the general public about the predicted outcomes (environmental, social and economic) of a proposal and what should be done to manage any impacts (Bates, 2002, p. 275).

2.4.4 Monitoring

Planning for water resources development, including inter-basin water transfer projects, requires provision of an array of monitoring facilities. The following sections discuss some of these requirements.

Monitoring of the Reservoirs and the Catchment

Factors that should be monitored in managing reservoirs and the catchment include (World Bank, 1991b, p. 36):

(1) climatic factors such as temperature, rainfall, evaporation and wind;
(2) volume of stored water in the reservoir;
(3) annual volume of sediment transported into the reservoir;
(4) water quality (such as salinity, TDS, pH, temperature, electrical conductivity, turbidity, dissolved oxygen, suspended solids, phosphates, and nitrates) at dam discharge and at various points along the river;
(5) hydrogen sulphide and methane generation behind the dam;
(6) limnological sampling of microflora, microfauna, aquatic weeds and benthic organisms;
(7) fisheries assessment surveys (species, population, etc.) in the river and reservoir;
(8) wildlife (species, distribution, numbers);
(9) vegetation changes (cover, species, composition, growth rate, biomass, etc.) in the upper watershed, reservoir zone, and downstream areas;
(10) increases in erosion in the catchment;
(11) impacts on wildlife species or plant communities of special ecological significance;
(12) public health and disease vectors;
(13) migration of people in and out of the area; and
(14) changes in economic and social status of resettled populations as well as people remaining in the river basin.

Monitoring of the Irrigation Systems

Factors that should be monitored in irrigation systems include (World Bank, 1991b, p. 97):

(1) climatic factors;
(2) stream discharge above and below the irrigation project at various points;
(3) nutrient content of discharge water;
(4) flow and water levels at critical points in the irrigation system;
(5) watertable depth in project area and downstream;
(6) water quality of project inflows and return flows;
(7) quality of groundwater in project area;
(8) soil physical and chemical properties in the irrigation area;
(9) crop yield per unit of land and water;
(10) erosion and sedimentation rates in project area;
(11) condition of distribution and drainage canals (siltation, presence of weed, condition of lining);
(12) incidence of disease and presence of disease vectors;
(13) health condition of project population;
(14) changes in wildlife population in the project area and on the floodplain downstream; and
(15) distribution of fish species and the population numbers.

2.4.5 Fluvial Geomorphology

The stability of the riverbed and bank needs to be investigated in any proposal for an inter-basin water transfer scheme. River flows have significant impacts on channel morphology through erosion, sediment transport and sediment deposition. Any changes to the flow regime caused by regulation or water diversion can have considerable impacts on the geomorphology of both exporting and importing rivers.

The particle size of transported sediment changes along the length of a river, from gross materials (boulders, cobbles and gravel) in the turbulent steep upper reaches, to sands and silt in the meandering low gradient lower reaches. Dams alter the flow regime (flood peaks and seasonal distribution of flow) thereby profoundly change the character of the rivers. They also disrupt the longitudinal continuity of the river system and interrupt the movement of sediment downstream. At the upstream end of the reservoir, all bed-load sediments and part of the suspended load are deposited in the form of a delta. Deltas can also develop where tributaries discharge into impoundments. Water released from the upper levels of a dam has little sediment but has the energy to erode the river channel and its banks downstream of the dam for some years, until a new equilibrium is reached (Kondolf, 1997).

Channel erosion below dams is frequently accompanied by a change in particle size on the bed, as gravel and finer materials are separated from the bed and transported downstream, leaving behind coarse materials such as large gravel, cobbles or boulders. The increase in particle size can threaten the success of spawning by salmonids (salmon and trout), which use gravels to incubate their eggs (Kondolf, 1997). Below dams, the bed may coarsen to such an extent that the fish can no longer move the gravel to make a pit for depositing eggs. To remedy this problem, gravels have been artificially added to 13 rivers in California since 1992, to enhance the supply of spawning gravel downstream of the dams.

Reduction in the frequency and duration of flows reduces their capacities to transport sediments further downstream. This can have implications for channel morphology, as sediment bars can develop near tributary junctions, or the channel mouth may become wider. The formation of bars at the tributary confluences may increase frequency of flooding, or in extreme cases, flows might be forced out of the channel by the blockage (Brizga, 1998).

The concepts of fluvial geomorphology that were developed for alluvial channels do not readily apply to bedrock channels. Tinkler and Wohl (1998, p. 1) define bedrock channels as "those reaches along which a substantial proportion of the boundary (≥ 50 percent) is exposed bedrock, or is covered by an alluvial veneer which is largely mobilised during high flows such that underlying bedrock geometry strongly influence patterns of flow hydraulics and sediment movement".

Brizga (1998) described methods for addressing environmental flow requirements of the fluvial geomorphology. The objectives of her study were:

(1) to review available techniques for assessing river flow as they relate to river channel morphology, habitat structure, substrata condition, flushing flows and estuaries;
(2) to assist in the selection of a "best practice" framework for the application of techniques to environmental flow assessment; and
(3) to provide research and development priorities for the refinement, development and integration of the techniques to facilitate their use in water allocation and water reform.

Further information about this topic is available in numerous publications, including: Olive et al. (1994), and Brizga and Finlayson (2000).

2.4.6 Land Degradation

Land degradation is defined as processes which lower the actual or potential capacity of the land to produce crops (Riquier, 1982). Soil erosion is one form of land degradation, which involves the removal and transport of sediments by water and wind. Soil structural decline, salinity, sodicity, acidity, leaching of nutrients and soil pollution are other forms of land degradation. Land degradation may have both on-site and off-site effects. On-site effects include loss of top soil, loss of plant nutrients, loss of soil organic matter, damage to soil structure, surface sealing and crusting, low infiltration, high run-off and low water storage, exposure of infertile sub-soil, and reduced crop yield. Off-site effects include pollution of surface water, siltation of rivers, dams and reservoirs, thus reducing their flood carrying capacities, eutrophication and algal bloom of rivers and lakes, and damage to in-stream ecosystems. In the following paragraphs only water erosion and soil salinisation, which are relevant to the development of irrigated agriculture associated with inter-basin water transfer projects, are discussed.

Water erosion occurs mainly on sloping lands. Soil is washed away by rain or irrigation water and is carried to waterways, thus lowering water quality through increased turbidity and nutrient build-up. Also, the impact of raindrops can cause water erosion. Soil erosion by water can be classified into splash erosion, sheet erosion, rill erosion and gully erosion (Morgan, 1986). The small channels, which are cut out by flowing water, are called rills. A number of rills may come together downslope to form a lesser number of larger rills. The term gully erosion is used to describe a deeply incised channel formed by run-off. Rose (1993) and Marshall et al. (1996, pp. 274–320) provide detailed descriptions of the different types of water erosion and the relevant models for estimation of soil loss.

Numerous methods and mathematical models have been developed as:

(1) tools for understanding erosion processes;
(2) tools for predicting where and when soil erosion is likely to occur; and
(3) predictive means for planning and management of soil erosion.

Ghadiri and Rose (1992) provide a critical review of the most widely used models, or those with potential of becoming planning and research tools in soil erosion, sediment transport, sediment deposition and the transport of pollutants such as nutrients (nitrogen and phosphorous), pesticides and salt. The models described include: USLE, SOILLOSS, RUSLE, GUEST, WEPP, CREAMS, GAMES, KYERMO, ANSWERS, PWM, PERFECT, NTRM, NPS and ARM.

The annual quantity of soil loss by water erosion can be quite significant. Hamblin (2001, Table 6) provided an estimate of the average annual rate of soil loss for a number of irrigated crops such as sugarcane 37.9 t/ha, nuts 16.6 t/ha, vegetables 12.9 t/ha, cotton 7.3 t/ha, oilseeds 4.6 t/ha, and rice 0.93 t/ha.

Soil salinisation is the process whereby the concentration of total dissolved salt increases due to natural processes (primary salinisation) or human-induced processes (secondary salinisation). Secondary salinisation occurs when the salt stored in the soil profile and/or groundwater is mobilised by extra water provided by human activities such as irrigation or land clearing. The extra water raises watertables or increases the pressure of confined aquifers, creating an upward leakage. When the watertable nears the soil surface, water is evaporated, depositing salts and causing land salinisation. Mobilised salt can also move laterally or vertically towards watercourses and increase their salinity. Ghassemi *et al.* (1995) provide a comprehensive description of salinisation processes, its management and the extent of human-induced salinisation in 11 countries (Argentina, Australia, China, former Soviet Union, Egypt, India, Iran, Pakistan, South Africa, Thailand and the United States), which have large tracts of secondary salinisation in their irrigated areas and dryland farms.

2.4.7 Soil Conservation

Numerous methods for protecting soils from erosion and degradation have been developed over the past 50 years. Physical soil conservation measures can be categorised as (Hudson, 1995, Chapter 10):

- **Soil management measures** intend to: (1) modify soil slope by bench terraces; (2) slowly reduce soil slope by progressive terracing; (3) contain erosion with low inputs by using ladder terrace and trash lines; and (4) contain erosion with minimal earth-moving on steep slopes by using step terraces, hillside ditches and intermittent terraces.
- **Water management measures** intend to improve rainfall absorption and retention capacity of soil by: (1) construction of conservation bench terraces and run-off level terraces; (2) catching and holding run-off by construction of absorption ridges; (3) absorption of some run-off with emergency overflow by construction of contour furrow and contour bund; (4) controlling unavoidable run-off by graded channel terraces; (5) controlling reduced run-off by ridging; and (6) reducing run-off velocity and promoting infiltration by minimum tillage, strip cropping, grass strips, and permeable barriers. Evidently, there will be situations where maximising infiltration is not practical or not desirable, such as the case of shallow soils, soils with poor drainage, or crops which suffer from restricted drainage in the root zone. However, in semi-arid regions, or regions with moderate rainfall and good soils, maximising rainfall retention is usually desirable.
- **Crop management measures** reduce soil erosion by: (1) reducing the velocity and sediment carrying capacity of run-off or overland flow on the slopes; (2) protecting the surface aggregates from direct impact by raindrops; (3) filtering out the suspended sediment from overland flow thus reducing its erosivity; (4) stabilising slopes through their root networks; and (5) enhancing soil aggregation by providing organic binding agents.

Other forms of soil conservation include erosion control by improved farming practices (Hudson, 1995, Chapter 11), control of gully erosion by vegetation and structures (Hudson, 1995, Chapter 12) and erosion control on non-arable lands (Hudson, 1995, Chapter 13). Soil conservation must not only be technically feasible, but socially acceptable and economically viable (Marshall *et al.*, 1996, p. 313). Sombatpanit *et al.* (1997) provide a wide ranging description of soil conservation extension measures in Australia, Bangladesh, China, India, Indonesia, Japan, Kenya, Malaysia, Nepal, Papua New Guinea, Philippines, Sri Lanka, Taiwan, Thailand, United States of America, Vietnam and Zimbabwe.

2.5 ENVIRONMENTAL FLOW REQUIREMENTS OF RIVERS

The environmental flow requirements of rivers and streams is defined as the quantities and seasonal patterns of flow needed to restore and maintain water-dependent ecosystems.

The growing recognition of the escalating hydrological alteration of rivers on a global scale, and the resultant environmental degradation, has led to the establishment of a field of research termed environmental flow assessment (EFA). Assessment addresses how much, and which specific temporal characteristics of the original flow regime of a river should continue to flow down the river, and onto its floodplains, in order to maintain specified features of the riverine ecosystem (Arthington *et al.*, 1992; Tharme and King, 1998; King *et al.*, 2002). An EFA produces descriptions of possible modified hydrological regimes for the river, and the environmental flow requirements (EFRs) or environmental water allocations. Each option is linked to a predetermined objective in terms of the ecosystem's future condition (Tharme, 2003). However, definition of conceptual frameworks and practical methods for assessing the water requirements of environmental systems remains a difficult and controversial task.

Tharme (1996, 2003) recognised four relatively discrete types of environmental flow methodology:

(1) *hydrological;*
(2) *hydraulic rating;*
(3) *habitat simulation;* and
(4) *holistic methods.*

The first three types correspond to those of Jowett (1997) thus: *hydrological methods* equates to historic flow methods, *hydraulic rating methods* to hydraulic methods and *habitat simulation methods* to habitat methods.

The *hydrological methods* represent the simplest set of techniques where hydrological data (as naturalised, historical monthly or average daily flow records) are analysed to derive standard flow indices, which then become the recommended environmental flows. Commonly, the EFR is represented as a proportion of flow (often termed the *minimum flow*, e.g. Q_{95} – the flow equalled or exceeded 95 percent of the time) intended to maintain river health, fisheries or other key ecological features, at some acceptable level. In a few instances, secondary criteria, in the form of catchment variables such as hydraulic, biological or geomorphological parameters, are also incorporated. Hydrological methods have been used mainly at the planning stage of water resource developments, or in situations where preliminary flow targets and exploratory water allocation trade-offs are required (Tharme, 1996; Arthington et al., 1998a; Tharme, 2003).

Hydraulic rating methods use changes in simple hydraulic variables of the river channel, such as wetted perimeter or maximum stream depth, usually measured across single, flow-limited river cross-sections. These are used as surrogates for habitat factors that are known, or assumed, to be limiting to target biota. Environmental flows are determined from a plot of the hydraulic variables against discharge, commonly by identifying curve breakpoints where significant percentage reductions in habitat quality occur with decreases in discharge. The assumption is that some threshold values of the selected hydraulic parameters at a particular level of altered flow will maintain aquatic biota and thus, ecosystem integrity. These relatively low-resolution hydraulic techniques have largely been superseded by more advanced habitat modelling tools, or assimilated into holistic methodologies (Tharme, 1996; Jowett, 1997; Arthington and Zalucki, 1998; Tharme, 2003). However, some of these approaches continue to be applied and evaluated, notably the Wetted Perimeter Method (Gippel and Stewardson, 1998).

Habitat simulation methods are based on the understanding of hydraulic conditions that meet specific biological requirements for target aquatic species. They make use of hydraulic habitat-discharge relationships, but provide more detailed, modelled analyses of both the quantity and suitability of the physical river habitat for the target biota. Thus, environmental flow recommendations are based on the integration of hydrological, hydraulic and biological response data. Flow-related changes in physical microhabitat are modelled in various hydraulic programmes, typically using data on depth, velocity, substratum composition and cover, and more recently, complex hydraulic indices (e.g. benthic shear stress), collected at multiple cross-sections within each representative river reach. Simulated information on available habitat is linked with seasonal information on the range of habitat conditions used by target fish or invertebrate species (or life-history stages, assemblages and/ or activities), commonly using habitat suitability index curves. The resultant outputs, in the form of habitat-discharge curves for specific biota, or extended as habitat time and exceedence series, are used to derive optimum environmental flows. The habitat simulation modelling package PHABSIM (Bovee, 1982; Milhous, 1998; Stalnaker et al., 1994), housed within the Instream Flow Incremental Methodology (IFIM), is the pre-eminent model of this type. The relative strengths and limitations of such methodologies are described in King and Tharme (1994), Tharme (1996), Arthington and Zalucki (1998), Pusey (1998), and are compared with the other approaches in Tharme (2003).

Jowett (1997) concluded that:

(1) *hydrological methods* are easy to apply and produce a single flow assessment. However, in most cases the relationships between flow and the state of the ecosystem are poorly established;
(2) the ecological aim of the *hydraulic rating methods* is to retain the wetted perimeter and thus the productive area of a stream, however the levels of protection are unlikely to be closely related to the state of the ecosystem; and
(3) *habitat simulation methods* provide the most flexible approach to flow assessment but can be difficult to apply. Moreover, the outcome depends critically on how the method is applied and what species are considered. *Habitat simulation methods* are most suited to situations where there are clear management goals and defined levels of protection.

Jowett (1997) did not appreciate a fourth category of more recent environmental flow methods, termed *Holistic Methods* (Arthington, 1998; Tharme, 2003), which are described below.

2.5.1 Holistic Methods

Over the past decade, river ecologists have realised a broader approach to the definition of environmental flows which encompasses the whole river ecosystem rather than just a few target species (Arthington and Pusey, 1993; King and Tharme, 1994; Sparks, 1992, 1995; Richter et al., 1996; Poff et al., 1997). From the conceptual foundations of a proposed holistic approach (Arthington et al., 1992), a wide diversity of *holistic methods* has been developed and applied in Australia and South Africa, and more recently in the United Kingdom. This type of approach reasons that if certain features of the natural hydrological regime can be identified and adequately incorporated into a modified flow regime, then all other things being equal, the extant biota and functional integrity of the ecosystem should be maintained (Arthington et al., 1992; King and Tharme, 1994). Likewise, Sparks (1992, 1995) suggested that, rather than optimising water regimes for one or a few species, a better approach is to try to approximate the natural flow regime that maintains the "entire panoply of species".

Holistic methods are underpinned by the concept of the "natural flows paradigm" (Poff et al., 1997), and basic principles guiding river corridor restoration (Ward et al., 2001). They share a common objective – to maintain or restore the flow-related biophysical components and ecological processes of instream systems, floodplains and downstream receiving waters (e.g. terminal lakes and wetlands, estuaries and near-shore marine ecosystems).

Components that are commonly considered in holistic assessments include geomorphology, hydraulic habitat, water quality, riparian and aquatic vegetation, macroinvertebrates, fish, and other vertebrates that have some dependency on the river/riparian ecosystem (i.e. amphibians, reptiles, birds and mammals). Each of these components can be evaluated using a range of field and desktop techniques (see Tharme, 1996; Arthington and Zalucki, 1998 for reviews), and their flow requirements can then be incorporated into environmental flow assessment recommendations, as discussed in more detail below.

Holistic methods currently represent around 8 percent of the global total of EFA methods. At least 16 extant methods have been developed over the last ten years (Tharme, 2003). Recently, such methods have begun to attract growing international interest in both developed and developing regions of the world, with strong expressions of interest in more than twelve countries in Europe, Latin America, Asia and Africa (Tharme, 2003). These holistic approaches have been described (see Arthington et al., 1998b) as either *bottom-up* methods, designed to construct a modified flow regime by adding flow components to a baseline of zero flows, or *top-down* methods, addressing the question – how much can we modify a river's flow regime before the aquatic ecosystem begins to noticeably change or becomes seriously degraded?

The South African Building Block Methodology (or BBM) (King and Tharme, 1994; King and Louw, 1998; King et al., 2002) was the first structured approach of this type. It began as a *bottom-up* method, and more recently incorporating the Flow Stress-Response Method (O'Keeffe and Hughes, 2002). In this modified form, the BBM is legally required for intermediate and comprehensive determinations of the South African Ecological Reserve (DWAF, 1999). Other essentially *bottom-up* methodologies include *expert* and *scientific panel* methods developed and applied in Australia (reviewed in Cottingham et al., 2002).

Examples of *top-down* methods are the Benchmarking Methodology (Brizga et al., 2002) used routinely in Queensland, Australia, at the planning stage of new developments to assess the environmental impacts likely to result from future water resource developments, and DRIFT – Downstream Response to Imposed Flow Transformations (King et al., 2003), a scenario-based approach which also predicts the probable ecological impacts of various scenarios of flow regime change. The Flow Restoration Methodology (Arthington et al., 2000) is a *bottom-up* approach with the objective of shifting a regulated flow regime more towards its natural state, combined with a simple *top-down* appraisal of the probable ecological consequences of not restoring certain features of the pre-regulation flow regime. The Flow Events Method (Stewardson, 2001; Stewardson and Cottingham, 2002) is a rather similar *top-down* approach, usually linked to a *scientific panel* method. Additional *holistic methods* developed and applied elsewhere include the River Babingley Method (Petts et al., 1999) developed in England, and the Adapted BBM-DRIFT Methodology developed in Zimbabwe (Steward et al., 2002).

In applications of *holistic methods* to date, the focus has almost entirely been on river systems, with most effort focussed on the main river channel and its tributaries. It is only recently that specialist methods have been proposed to address the freshwater flow requirements of downstream receiving waterbodies such as floodplains and terminal lakes in large arid-zone and tropical rivers, and estuaries (Loneragan and Bunn, 1999). Further, methodologies to integrate the dynamic interactions of surface and groundwater systems into existing holistic methodologies are at a fairly immature stage of development, with none routinely applied as part of holistic assessments (King et al., 1999).

2.5.2 Basic Principles of Flow–Ecology Relationships

As the flow regime is regarded by many aquatic ecologists to be the key driver of river and floodplain wetland ecosystems, Bunn and Arthington (2002) focus their literature review on four key principles to highlight the important mechanisms that link hydrology and aquatic biodiversity, and to illustrate the consequent impacts of altered flow regimes. The four key principles are:

- **Principle 1:** Flow is a major determinant of physical habitats in streams, which in turn is a major determinant of biotic composition.
- **Principle 2:** Aquatic species have evolved life history strategies primarily in direct response to the natural flow regimes.
- **Principle 3:** Maintenance of natural patterns on longitudinal and lateral connectivity is essential to the viability of populations of many riverine species.
- **Principle 4:** The invasion and success of exotic and introduced species in rivers is facilitated by the alteration of flow regimes.

Bunn and Arthington (2002) conclude that, currently, evidence about how rivers function in relation to flow regime, and the flows that aquatic organisms need, exists largely as a series of untested hypotheses. To overcome these problems, aquatic science needs to move quickly into a manipulative or experimental phase, preferably with the aims of restoration and measuring ecosystem response.

Poff et al. (2003) describe conflicts between perceived needs of the ecosystem and of humans for freshwater and argue that currently river scientists are faced with the challenge of clearly defining the ecosystem needs. This is required to guide policy formulation and management actions in order to satisfy competing demands. They argue that numerous conflicts around the world provide dramatic evidence that water managers and other stakeholders are now demanding more than just a strong conceptual understanding to guide the management of individual rivers. They are asking – how much flow restoration is necessary to ensure ecological sustainability, and how does flow quantity, seasonal timing, and water quality need to be managed to achieve the desired ecological outcomes?

Based on the growing recognition that more effective approaches are needed, Poff et al. (2003) propose four steps to strengthen the role of science and society in managing rivers to meet human and ecosystem needs. These steps are:

- **Step 1:** Implement more large-scale river ecosystem experiments on existing and planned water management projects through controlled river flow manipulations.
- **Step 2:** Engage the problem through a collaborative process involving scientists, water resources managers and other stakeholders.
- **Step 3:** Integrate case specific studies into broader scientific understanding and generalisation.
- **Step 4:** Forge new and innovative funding partnerships to engage scientists, government agencies, private sector and NGOs.

2.5.3 Environmental Flow Assessment in Australia

To address the environmental flow assessment issue in Australia, a project entitled "Comparative Evaluation of Environmental Flow Assessment Techniques" was funded by Environment Australia, Land and Water Resources Research and Development Corporation, and the National Landcare Program. The objectives of the project were as follows:

- Undertake a review of the currently used and available techniques for assessing flow requirements, so that water managers can determine which techniques are most suitable for their river system, in terms of environmental values, limitations, advantages and cost-effectiveness.
- Propose *a best practice framework* for the application of techniques to environmental flow assessment.
- Provide research and development priorities for the refinement, development and integration of the techniques to facilitate their use in water allocation and water reform.

The outcomes of the project are described in four publications, Arthington (1998), Arthington et al. (1998a,b) and Arthington and Zalucki (1998). Provision of the environmental flow must be assessed and provided in many different contexts (Arthington et al., 1998b, pp. 3, 4). The two fundamental contexts are regulated and unregulated river systems. Moreover, frameworks for assessing environmental flows must have a number of properties to address various circumstances at the most relevant spatial and temporal scales. *Holistic methodologies* reviewed by Arthington (1998) are a significant improvement of traditional single-issue methods because they aim to consider the needs of the entire riverine ecosystem. However, no existing method is entirely suited to all circumstances. Most methods do not give adequate consideration to management issues that are not related to river flows. Each method has its strengths and original elements, and these need to be evaluated and incorporated into *a best practice framework* for river flow management.

Best Practice Framework

Arthington *et al.* (1998b, p. 6) describes a single overarching *best practice framework* for environmental flow assessment within a three-tiered hierarchy of environmental flow assessment. These levels are:

- *Level 1:* Basin-wide reconnaissance of development options, opportunities for restoration of regulated systems, and preliminary assessment of environmental flows. This assessment level may take up to one year.
- *Level 2:* Catchment or sub-catchment scale assessment of environmental flows for feasible development options such as new dam and increased allocations, and/or restoration of regulated systems. This level of assessment requires a minimum of two years.
- *Level 3:* Detailed assessment of special issues at all spatial scales including fish passage requirements, experimental dam release to stimulate fish migration and spawning, and channel management studies. This level of assessment may require a variable time scale of two to more than 5 years and could be undertaken concurrently with Level 2.

This framework adds a new dimension by including factors other than flow which may influence river condition or the effectiveness of flow management, and also provides a process for addressing human use constraints and their impacts on the provision of environmental flows (Arthington *et al.*, 1998b, p. 22). More recently, Cottingham *et al.* (2002) called for best practice guidelines for the conduct of the *scientific panel* method of environmental flow assessment used in Australia, and made several recommendations supporting those of Arthington *et al.* (1998b). These were:

- Clear processes for selecting panel members and protocols to guide the conduct of panels and the interactions between members.
- Guidelines for developing a "vision statement" and explicit ecological objectives, so that any ecosystem response to environmental flow provisions can be measured against the desired outcomes in an adaptive management framework.
- More explicit guidelines regarding the selection of field sites and the collection of new field data.
- Procedures for recording the strengths and limitations of evidence used to make environmental flow recommendations.
- Consideration of the social and economic implications of environmental flow recommendations.
- A standard process for presentation and documentation of findings.
- An opportunity to make recommendations on the additional information required to support or improve decisions relating to water management and, particularly, to strengthen the scientific basis of environmental flow assessments.

Recently, Arthington and Pusey (2003) provided a review of flow restoration and protection in Australian rivers. They concluded that water reform in Australia, and the retrieval of water for rivers and wetlands, are likely to take a long time to have some impact on the health of the river systems. During this period rivers and wetlands will continue to degrade. To overcome this problem they suggest establishment of *Water Banks* with the power and legal right to buy and sell water on behalf of the environment. Potential mechanisms to raise funds could include the establishment of a system giving citizens the option to make voluntary contributions on each water bill, and raising large national and international donations.[6] It should be noted that the concept of *Water Banks* has several precedents internationally (e.g. in California). In Australia this could operate independently or as part of the National Commission suggested by the Wentworth Group of Concerned Scientists (2002, pp. 18–21) to raise funds and support environmental management reforms.

2.6 SOCIAL AND CULTURAL ISSUES

The objective of large-scale land and water resources development is to modify the natural environment in order to enhance the economic and social benefits. However, achieving these objectives may also have some unforeseen

---

[6] An alternative tried in the Australian Capital Territory and South Australia is the imposition of a volumetric charge on all water users. This charge, theoretically covers the environmental costs of water abstractions.

social impacts. As part of any environmental assessment, a critical examination of the long-term social impacts is necessary and should include strategies to minimise or avoid the social costs of the project. Major issues include threats to indigenous peoples, loss of cultural properties, involuntary resettlement, new land settlement, and induced development. These issues will be discussed briefly, based on Chapter 3 of the World Bank Environmental Assessment Sourcebook 1999.[7] Further detailed information on social and cultural issues in development projects is available in World Bank (1991a, pp. 107–136).

2.6.1 Indigenous Peoples

Special action is required when developments affect local communities composed partly or entirely of indigenous, tribal, or ethnic minority groups. Such groups are vulnerable during dislocations and times of rapid socio-economic changes. Any significant impacts that may affect these social groups need to be specifically addressed. The needs of indigenous people are best taken into account during the formulation of development plans. The assessment of the project impacts on vulnerable groups requires information such as:

- **Formal legal and customary use-right.** Determine actual workings of constitutional, legislative, administrative, contractual or customary rights to use natural resources.
- **Resource use pattern.** Assess changes in patterns of access to, or use of, land, water, forest or other natural resources affected by the project plan.
- **Use of area by non-residents.** Analyse data on use of seasonal resources by graziers, fishermen, collectors of forest products, logging companies, and others.
- **Community participation.** Determine the extent to which indigenous groups feel that the proposed development is environmentally sound and culturally appropriate, which environmental constraints are to be addressed in project design and implementation, and which environmental opportunities are to be enhanced.
- **Identification, demarcation and registry of area.** Evaluate effectiveness of local mechanisms to resolve territorial disputes, and establishment of boundaries and buffer zones.
- **Inventory of flora and fauna.** Survey and analyse fauna and flora habitats, particularly endangered species, and those plants and animals used by local communities.
- **Social infrastructure.** Evaluate impacts on housing, schools, medical facilities, communications, transport networks, market, water supply, and waste disposal systems.
- **Public health conditions.** Evaluate health risks and diseases in the area, environmental pollution, health situation and hygienic conditions, traditional medicine and practices.
- **Institutional assessment.** Determine capacity of local organisations and indigenous groups to participate in decision-making, implementation, operation and evaluation.

2.6.2 Cultural Properties

Cultural property refers to sites, structures and remains that are of archaeological, historical, religious, cultural or aesthetic value. Water resource development projects, including inter-basin water transfer or their components, may have impacts on these cultural properties. Government agencies, museums, relevant university departments (archaeology, art, history, architecture and others) and interested NGOs can provide information and expertise about the cultural resources of the project area. Collected cultural information can be used to review project alternatives and devise mitigation measures.

2.6.3 Involuntary Resettlement

Some inter-basin water transfer projects may require involuntary resettlement of local people from the areas to be inundated by the reservoirs, or for development of irrigation and hydro-electric power generation facilities. Goldsmith and Hildyard (1984, Volume 3, pp. 77–95) review resettlement in a number of projects in various countries, while descriptions of resettlement projects in Turkey[8] are available in Altinbilek et al. (1999). Nakayama et al. (1999) examine the performance of the involuntary resettlement scheme for the Cirata Dam in Java, Indonesia. The largest resettlement plan ever attempted is the construction of the Three Gorges Dam in China, where the number of displaced people is estimated at about 1 200 000 (Fearnside, 1993; Heggelund, 2004).

---

[7] http://lnweb18.worldbank.org/ESSD/essdext.nsf/47DocByUnid/B8632ADEB9069BEC85256B9C00662D5E/$FILE/Chapter3SocialAndCulturalIssuesInEA.pdf (visited in June 2005).

[8] These consist of resettlements due to construction of: (1) Keban Dam, completed in 1973, affecting 126 villages, displacing 20 000 people; (2) Karakaya Dam, completed in 1987, affecting 105 villages with 249 families; (3) Atatürk Dam, completed in 1992, affecting 719 families; (4) Tahtali Dam, completed in 1996, affecting 7800 people; and (5) Çat Dam, completed in 1997, affecting 400 people.

2.6.4 New Land Settlement

Some inter-basin water transfer schemes may involve settlement of new lands,[9] where the proposed tenancy arrangements in relation to their effects on the environment are important. The adopted land tenure policy must provide settlers with a sufficient degree of security to provide the motivation to conserve their land as a valuable asset and to invest their own resources in its improvement. Either the title itself, or legislation, can be used to discourage or prohibit the subdivision of farm holding below the point that they can remain viable.

2.6.5 Induced Development

In some cases indirect social impacts have been overlooked during project design. A common problem is the failure to plan for the influx of voluntary migrants who take advantage of the new economic opportunities created. The influx of large numbers of people who seek employment or who come to provide services puts pressure on housing, food, school, water, electricity, health and sanitary facilities, and other essential services.

2.7 ECONOMIC APPRAISAL

A variety of techniques are available for analysing the costs and benefits associated with a particular project. These are cost–benefit analysis, incidence analysis, input–output analysis, and multiple objective programming (NSW Treasury, 1997). Cost–benefit analysis, which is the most comprehensive economic appraisal technique, quantifies all identifiable major costs and benefits (Ray, 1984; Layard and Glaister, 1994). Its aim is to show:

(1) whether the benefits of a project exceed its costs;
(2) among a range of options, which one has the highest net benefit; and
(3) which option is the most cost effective if project benefits are equivalent.

Economic appraisal assists decision-making among projects competing for limited government funds. Various criteria such as cost–benefit ratio, net present value, and the internal rate of return are used to assist the decision-making process.

Inter-basin water transfer proposals require detailed cost–benefit analysis to ensure that the proposals are economically feasible. The capital cost for construction of dams, tunnels, aqueducts, pipelines, pumping stations, access roads and hydro-power plants, as well as the annual operating cost of the project, including loan repayments, operation and maintenance, energy, and administration require assessment. Consideration of the discount rate, period of construction, and period of repayment of the investment are important in any economic appraisal.

The cost–benefit analyses of the project now require an economic assessment of the potential environmental impacts. The major problem in valuing environmental impacts is that they are generally not traded and therefore do not have a market value. However, numerous methodologies have been developed for the economic appraisal of the environmental impacts. These include multi-criteria analysis, decision analysis, and contingent valuation method, choice modelling and hedonic methods. Grafton *et al.* (2004, Chapter Seven: Economics of Non-Renewable Resources) provide an introduction to environmental valuation and conclude that contingent valuation is the most commonly used technique in environmental economics.

Economic appraisals also require assessment of risk and uncertainties (Dixon *et al.*, 1994, pp. 107–108; Arrow and Lind, 1994). Risk refers to situations with known probabilities. This means that the number and size of each possible outcome is known and the chance of each outcome occurring can be objectively determined. On the other hand, uncertainty refers to situations with unknown probabilities. That is, the number and size of each outcome may or may not be known, but the chance of any single outcome occurring cannot be objectively determined. Methods for assessment of risk and uncertainty include sensitivity analysis and scenario planning.

Although there are numerous publications about various aspects of water resources economics, such as water allocation (domestic and industrial water supply, irrigation, flood control, hydro-electric power generation, and navigation), water markets, and water quality (Young and Haveman, 1985; Grafton *et al.*, 2004, Chapter 6: Water Economics), there are no recent publications on the economic appraisal of inter-basin water transfer projects. Howe and Easter (1971) is still the most comprehensive publication available about this issue. They analysed the economics of proposals for inter-basin water diversion in the western United States. At that time issues related to the potential environmental impacts were not considered. Davidson (1984) provided a preliminary cost–benefit analysis of the inland diversion of the coastal rivers of New South Wales (see Chapter 5, pp. 116–117). Scott (1985)

---

[9] Such as the resettlement of returned soldiers in Australia after World War II.

described the economics of the water export policy from Canada to the United States using cost–benefit analysis. His analysis includes some very approximate estimates of the environmental costs.

2.8 WATER RIGHTS

A water right is the right to use water from a stream, lake, or groundwater system. The intense competition for water between agriculture, industry, urban water authorities, the environment and indigenous people has made the issue of water rights very complex, particularly in inter-basin water transfers. Legislation regarding water rights and water allocation vary from country to country and even between jurisdictions within the same country. Bruns and Meinzen-Dick (2000) provide information on water rights in various countries such as Bangladesh, India, Indonesia, Nepal, Pakistan, Spain and Sri Lanka. The problem of water rights becomes particularly complex if the water is to be transferred from one jurisdiction to another. Scott and Coustalin (1995) provide a comprehensive overview of the evolution of water rights, commencing with the Roman era and through:

- Medieval period of water law in England (1066–1600).
- Early industrial revolution in England (1600–1850).
- The modern doctrine of riparian rights in England (1851–1900).
- Industrial period in the United States (1827–1900).
- The prior appropriation phase: introduction of individual and use-based water rights in the United States (1850–1900).
- Modern period in the United States and England (1900–1995).

In England and the United States, there have been two historically common sets of rules for allocation of water. These are *riparian rights* and the *doctrine of prior appropriation*. A riparian right confers the right to the owner of land adjacent to a waterbody the right to use that water. This right is tied to the owner of land and cannot be transferred to a land owner not adjacent to a waterbody. The inadequacies of the riparian rights became apparent as population growth and economic development increased the demand for water in areas that were not immediately adjacent to waterbodies. As a result, a *doctrine of prior appropriation* emerged (particularly in the American west). Under this allocation rule, the priority or ranking of a claim to water use was based on the time at which it was first made, relative to other claims. An important difference between this rule and riparian rights was that the right to water use was no longer tied to the ownership of riparian land. In many jurisdictions, private rights (riparian and prior appropriation) to water use were at one time or another constrained by governments asserting their ownership over water. Governments have retained ownership of water and granted users the right to use, but not own the water.

Any efficient mechanism of water allocation must ensure that the following conditions are satisfied. First, it must base allocation on the relative value of water for different uses. Second, it must provide some security of tenure. This means that rights must be clearly defined and enforced. Third, it must allow for the transfer of water from relatively low-value to high-value applications. Finally, it must protect instream water needs, and it must ensure that water use and the trading of water rights do not generate negative environmental impacts (Grafton et al., 2004, p. 172). Not surprisingly, no mechanism for water allocation exists that meets all of these requirements. Markets for water rights are emerging in a number of countries, such as the United States and Australia. It is envisaged that these will help move water from lower value to higher value uses. However, the hope that the water market will solve all water problems appears misleading as is becoming apparent in California (see section 11.15.3).

An additional set of challenges arises when the efficient allocation of transboundary water is considered. This is an area where international and national laws are still evolving. Utton (1996) provides a review of international water laws based on the established doctrine of *equitable utilisation* developed out of water quantity allocations and the rule of *no significant harm*, which has its origin in environmental protection.

2.8.1 Water Rights and Water Legislations in Australia

In Australia, indigenous culture and tradition is closely linked to the biophysical environment including its water resources (Magowan, 2002). Australian indigenous people consider water to be sacred (Rose, 2005). They have developed an intimate understanding of water and its capacity to support life, and have adapted themselves to its unpredictability. Throughout the arid regions of the country they have a finely detailed knowledge of natural springs, of sites where one can dig to find water (soaks), of trees that hold water, and of roots that can be dug up for water. Water in Aboriginal Australia exists within a system of rights and responsibilities that is usually referred to as ownership. Groups of people belong to and own their country, including their water. Individual rights and responsibilities arise from their bodies of knowledge. The general rule articulated in simple terms is *always ask*. This rule identifies the right of the owners of the country to make a managerial decision and to say yes or no (Rose, 2005).

Since the arrival of European settlers with the First Fleet in 1788, access to and control of water resources (especially in the drier areas of the country) has been at the centre of many conflicts between indigenous people and the white settlers (Smith, 1998, pp. 139, 140). Many battles were fought over the right to control and manage water. There were many massacres in rivers and creek beds, as these were central camping and meeting places for aboriginal people (ATSIC, 2002a, p. 7). While in some countries such as Canada, United States and New Zealand, indigenous rights to water have been recognised for many years, in Australia this was acknowledged by the Council of the Australian Governments in June 2004, through the National Water Initiative.[10] More information on the historical, legal, cultural and other aspects of aboriginal rights to water is available in ATSIC (2000b).

After British settlement of Australia, the common law of England constituted the foundation for the management of Australia's water resources. Owners of the land had limited riparian rights to use water for all ordinary and domestic purposes. In contrast, landowners had unlimited right to use groundwater and surface run-off that flowed over their land. Common law rules were found to be inadequate to secure water supplies for mining, pastoral pursuits and town use. Therefore, common law was modified to suit Australia's conditions (Fisher, 2000, p. 3).

The Victorian Royal Commission on water supply in 1884, headed by Alfred Deakin, laid the foundation of present water legislation and the State control of water resources (Tan, 2001). In 1886, Victorian legislation was passed vesting the ownership of water resources in the Crown. This legislation tied the granting of water rights to land ownership, thus preventing transfer of water rights independently of the sale of land. This was a deliberate strategy to prevent water rights being controlled by a few wealthy individuals. In New South Wales, to overcome complications of the common law and to establish State ownership of water resources, the *Water Rights Act* of 1896 was passed by the legislature. The model of state control was followed by the other states and territories: Queensland (*Rights in Water and Water Conservation and Utilization Act* 1910); Western Australia (*Rights in Water and Irrigation Act* 1914); South Australia (*Control of Waters Act* 1919); Tasmania (*Water Act* 1957); Northern Territory (*Control of Water Ordinance* 1938); and the Australian Capital Territory (*Water Resources Act* 1998).

The wide variety of water entitlement types and water management practices in the States and Territories of Australia reflects the legislative, geographic, environmental, economic and political diversity of the country (Claydon, 1995, p. 3). In general, the ownership of all groundwater and surface water in watercourses, lakes and springs in Australia are vested in the Crown. State or Territory governments, on behalf of the Crown, generally confer entitlements to owners or occupiers of land to use water through licensing, issuing of permits and/or entering into agreements. Claydon (1995, pp. 4, 5) argued that although the existing system of water entitlement has enabled the development and use of water resources over many years, serious limitations and problems have emerged as competition for scarce water resources increases. These problems include:

- Lack of incentives to move water to higher value uses[11] and to increase investments.
- Lack of signals as to the true worth of water, leading to inappropriate water pricing.
- Conflict between resource management and commercial activities.
- Need for self-adjusting reform in the water industry.
- Lack of security in tenure and specification of rights over time, leading to missed investment opportunities.
- Lack of clarity regarding rights or entitlements.

In 1994, the Council of the Australian Governments (COAG, 1994) undertook major reform of water legislation with respect to issues such as water pricing, water allocation and entitlements, allocation of water for the environment, and institutional reforms (see section 3.11). While the ownership and management of water resources has been conferred to the States and Territories, the Commonwealth Government has assumed an important role through policy formulation and provision of financial assistance. Moreover, because of jurisdictional differences, several cooperative schemes for the management of water resources came into existence. This pattern originated from the River Murray Waters Agreement between New South Wales, Victoria and South Australia in 1915, which paved the way for the establishment of the River Murray Commission, and later evolved into the Murray–Darling Basin Commission and the Murray–Darling Basin Agreement (see section 3.11).

There was a rapid acceleration in the development of water resources for irrigation, resettlement of returned

---

[10] http://www.coag.gov.au/meetings/250604/iga_national_water_initiative.pdf (visited in June 2005).

[11] This refers not only to economic returns from consumptive water use, but also to the value to society from environmental and other non-consumptive uses (Claydon, 1995, p. xi).

soldiers, as well as urban and industrial water supplies after the World War II (see sections 3.8.1 and 3.9). The end of the 1980s marked the end of the major dam development phase, and the beginning of the economic era in water resources policy, followed closely by concerns over the impacts of water resources development on the environment and the sustainability of water resources schemes.

More information regarding the evolution of water law and water rights in the various jurisdictions of Australia, from the early years of the European settlement up to 30 June 1999, is available in Fisher (2000). Tan (2001) provided a critical analysis of water law and water rights reform in Australia over the period of 1989–1999. He argued that ecosystems have been allocated fairly vague water rights and suggested that more amendments will be required in Australian water legislation to protect environmental requirements and ecosystems. These issues have been addressed in the National Water Initiative introduced in 2004 (see section 3.13).

2.9 CONFLICTS AND THEIR RESOLUTION

Conflicts are widespread in all aspects of life. They can be resolved through lengthy and expensive court hearings or by Alternative Dispute Resolution (ADR) (Mackie, 1991; Kovach, 1994; Kheel, 1999). ADR procedures include negotiation, mediation, arbitration, case evaluation techniques and private judging (Kovach, 1994, p. 6). These processes have been developed to assist those involved in a dispute to arrive expeditiously at a mutually satisfactory resolution of a matter. The main goals of ADR are to (Mackie, 1991, p. 3):

(1) relieve court congestion as well as undue costs and delays;
(2) enhance community involvement in the dispute resolution process;
(3) facilitate access to justice; and
(4) provide more effective dispute resolution.

Water has often been the source of conflicts and disputes at various levels, from disputes between neighbouring farmers and landowners, disputes between States within a Federation, to disputes between independent nations. Wolf et al. (1999) provide a register of 261 international river basins, covering 45 percent of the land surface of the earth (excluding Antarctica). Overall, 145 nations have part of their territories within these international river basins. Disagreement over the sharing of water within these basins has been the cause of many international disputes.

Because of potential international conflicts, many nations have negotiated agreements about the use of water. Over 3600 separate treaties have been identified on international water bodies for the period 800 AD and 1985 (Biswas, 1999). The majority of these treaties deal with the navigation and passage of ships, which was the primary form of transportation during earlier times. However, after World War II, many treaties were negotiated for other uses like flood control, hydro-power generation, water quality management and water allocation.

A chronological list of more than 100 water conflicts, from 3000 BC up to the sabotage of water pipelines in Baghdad during 2003, is available at the URL provided in the footnote.[12] The categories of conflicts include:

- **Control of water resources:** where water supplies or access to water is at the root of tensions.
- **Political tool:** where water resources, or water systems are used for a political goal.
- **Terrorism:** where water resources, or water systems, are either targets of violence or coercion by individuals or groups.
- **Military tool:** where water resources, or water systems themselves, are used by a nation or nations as a weapon during a military action.
- **Military target:** where water resource systems are targets of military actions.
- **Development disputes:** where water resources or water systems are a major source of contention and dispute in the context of economic and social development.

Within the framework of the Transboundary Fresh Water Disputes Project,[13] Beach et al. (2000) provide a comprehensive review of the literature on conflicts and conflict resolution issues stemming from water quantity and quality problems around the world. This review examined issues such as:

(1) institutions and law;
(2) international water law and negotiation theory dealing with diagnosis of conflicts, and conflict resolution;
(3) economic theories in water allocation;
(4) water treaties;
(5) environmental disputes;

---

[12] http://www.worldwater.org/conflict.html (visited in June 2005).
[13] The objectives of the project are: (1) to undertake a qualitative and quantitative analysis of transboundary water conflict; and (2) to develop procedural and strategic templates for early intervention in order to help contain and manage conflicts.

(6) case studies of transboundary dispute resolution related to watersheds (Danube, Euphrates, Jordan, Ganges, Indus, Mekong, Nile, Plata, and Salween in Myanmar), aquifer systems (US–Mexico and West Bank aquifers), and lakes (Aral Sea and Great Lakes);

(7) summaries of more than 140 international water treaties, from the 1874 agreement regarding construction of a weir in a sub-basin of the Indus Basin, to the 1996 treaty between India and Bangladesh on sharing of the Ganga (Ganges) water; and

(8) about 280 annotated references.

The focus of this publication is on the political and social aspects and the skills for dispute resolution, rather than the hydrologic, engineering and organisational aspects of the problem.

In a more recent publication, the International Bureau of the Permanent Court of Arbitration (2003) provides papers from the Sixth Permanent Court of Arbitration International Law Seminar, which was held in the Netherlands in November 2002. It consists of approximately 15 papers presented by the world's most learned scholars and practitioners in the area of international law, environment and water dispute resolution. Some of the concluding remarks of the seminar are:

- In recent decades there has been a remarkable increase in the quantity of international legal rules relating to international watercourses.
- The number of disputes regarding watercourses, natural resources and the environment are also on the rise.
- The number of institutions and projects dealing with water and environmental disputes has also increased. These include the International Court of Justice, the European Court of Justice, the European Commission on Human Rights, International Tribunal for the Law of the Sea, the International Law Commission, the Permanent Court of Arbitration, and UNESCO's PCCP[14] project (from Potential Conflict to Co-operation Potential) which addresses the challenge of sharing water resources from a government perspective.
- Each dispute has its own facts and circumstances. Therefore, there is no general template that can be applied to different watercourse disputes.
- There is a broad recognition of the importance of a multidisciplinary approach to the settlement of watercourse disputes. Lawyers must be prepared to engage fully with engineers, scientists, economists, interest groups and others.
- Although judicial settlement or arbitration is a last resort, the threat of recourse to a court or arbitration tribunal may often be sufficient to encourage two parties to reach agreement. This is because parties understand that once they have gone into third-party adjudication, they have lost control of the process, and hence the outcome.

This publication also contains the text of the 1997 *United Nations Convention on the Law of the Non-Navigational Uses of International Watercourses*. It consists of 37 articles dealing with issues such as:

(1) the scope of the convention;
(2) equitable and reasonable utilisation of water resources;
(3) exchange of data and information;
(4) notification concerning planned measures;
(5) consultations and negotiations concerning planned measures;
(6) protection and preservation of ecosystems;
(7) management issues; and
(8) settlement of disputes.

2.9.1 Examples of Agreements and Treaties

Water resources development in multi-jurisdictional river basins creates tensions and disputes between jurisdictions. Some examples of agreements and treaties to ease tensions and settle disputes are:

- **Snowy Mountains Hydro-electric Agreement 1957.** The proposal for construction of the Snowy Scheme and inter-basin transfer of Snowy River waters to the Murray and Murrumbidgee rivers was a source of dispute between the Australian Federal Government and the states of New South Wales and Victoria. Federal Parliaments passed the *Snowy Mountains Hydro-electric Power Act* in 1949 without any agreement between these three governments. While construction of the Scheme was in progress, the *Snowy Mountains Hydro-electric Agreement 1957* was finally signed by the three parties in September 1957 and became formally effective on the 2nd January 1959 (see section 4.5).
- **Murray–Darling Basin Agreement.** Water in the Murray–Darling Basin in south-east Australia is shared between five states and territory each with sovereignty over water in their states and territory. Their agreement is described in section 3.11.

---

[14] UNESCO's International Hydrological Programme, within the framework of the United Nations World Water Assessment Programme, implements PCCP.

- **Colorado River Compact.** The Colorado River drains seven states in the United States before entering Mexico. The *Colorado River Compact* between these states and the 1944 treaty between the United States and Mexico are described in section 11.16.
- **United-States–Canada Boundary Water Treaty.** The 1909 *Boundary Waters Treaty* between the United States and Canada was introduced to prevent and resolve water disputes between two countries. It established a number of joint institutions to protect water quality and quantity of their shared water resources. These are described in section 12.11.
- **Indus Waters Treaty.** Partition of the Indian subcontinent on 15 August 1947 created independent Pakistan. On 1 April 1948, India stopped Indus water from flowing to Pakistan. Pakistan protested and India finally conceded to an interim agreement, which was not a permanent solution. Pakistan approached the World Bank in 1952 to help find a long-term solution. The *Indus Water Treaty* between India and Pakistan was finally signed on 19 September 1960. The treaty divided the use of rivers and canals between the two countries. Pakistan obtained exclusive rights for the three western rivers (Indus, Jhelum and Chenab) and India retained rights to the three eastern rivers (Ravi, Beas and Sutlej). The treaty included: (1) a 10-year transitional period from 1960 to 1970, during which water would continue to be supplied to Pakistan according to a detailed schedule; and (2) establishment of the Permanent Indus Commission. It is of interest that this Treaty has been robust to survive intact through several India–Pakistan wars. Further information about this treaty is available in Gulhati (1973), Yunus Khan (1990), Biswas (1992), and URLs provided in the footnote.[15]
- **Nile Waters Treaty.** The Nile River is shared by nine countries (Egypt, Sudan, Ethiopia, Uganda, Kenya, Tanzania, Rwanda, Burundi and Zaire). Its catchment area covers $2\,900\,000\,km^2$, nearly 10 percent of the land area of Africa. Construction of the Aswan High Dam commenced in 1960 and completed in 1968. The *Nile Waters Treaty* was signed by Egypt and Sudan prior to construction on the 8th November 1959. In this Treaty, the average annual flow of the Nile ($84 \times 10^9\,m^3$) is divided into three parts: $55.5 \times 10^9\,m^3$ for use by Egypt, $18.5 \times 10^9\,m^3$ for use by Sudan and the remaining $10 \times 10^9\,m^3$ is accounted for by evaporation and seepage losses from Lake Nasser created by the Aswan High Dam.[16] Ethiopia, which had not been a major player in the 1959 Treaty, has an estimated 75 percent share of the Nile average annual flow. It has been recently suggested that Ethiopia may eventually claim $40 \times 10^9\,m^3$ per year of the Nile waters for its irrigation needs. No other riparian state has exercised a legal claim to the Nile waters allocated by the 1959 Treaty.
- **Mekong River Commission.** The Mekong River rises at a height of 5100 m in China and then flows 4880 km through Myanmar, Laos, Thailand, Cambodia and Vietnam into the South China Sea. Its total drainage area is approximately $795\,000\,km^2$ (Osborne, 2004). The Mekong River Committee was established on 17 September 1957. It was composed of representatives from the four lower riparian states (Vietnam, Cambodia, Thailand and Laos). The Committee was authorised to promote, coordinate, supervise, plan and investigate water resources development projects in the Lower Mekong Basin. The early years were very productive and the Committee also helped overcome political suspicion through increased integration. By the 1970s the early momentum of the Mekong Committee began to subside due to instability and conflicts in the region (Beach *et al.*, 2000, pp. 107–110). On 5 April 1995 the four lower riparian countries signed a new agreement and the Mekong River Committee became the Mekong River Commission. Since the 1995 Agreement, the Mekong River Commission has launched a process to ensure reasonable and equitable use of the Mekong River System through a participatory process. National Mekong Committees in each country develop rules and procedures for water use. The Commission is also involved in fisheries management, navigation, irrigated agriculture, watershed management, environmental monitoring, flood management and exploring hydro-power options (Mekong River Commission, 2004). The two upper states (China and Myanmar) are dialogue partners to the Commission.

These examples show that some of these agreements are remarkably robust, able to operate under periods when signatories are in conflict. The factors which have led to the success of these treaties are worthy of study.

---

[15] <http://www.waterinfo.net.pk/pdf/iwt.pdf>, <http://www.transboundarywaters.orst.edu/projects/casestudies/indus.html> and <http://wrmin.nic.in/international/industreaty.htm> (all three visited in June 2005).

[16] http://www.transboundarywaters.orst.edu/projects/casestudies/nile_agreement.html (visited in June 2005).

2.9.2 Conflict Resolution in Inter-Basin Water Transfer

Inter-basin water transfer creates numerous conflicts among a wide range of stakeholders regarding issues such as water rights, water allocation, eventual changes in water quality, land use, protection of the environment, loss of flora and fauna, change of lifestyle, loss of income, resettlement, compensation and others. The extent of these problems depends on the extent of water transfer between basins.

In order to resolve controversies about inter-basin water transfer, three basic solution methods are available (Yevjevich, 2001):

(1) a market approach with negotiation;
(2) arbitration; and
(3) legal avenues.

Attempts to resolve conflicts are best in this sequence of methods. The market mechanism attempts to resolve the conflicts by allowing prices to balance supply and demand. In this case, representatives of institutions and individuals in the basin of origin outline the losses and damages to the basin and propose how to be compensated for both the immediate and the future losses. Then, direct negotiations can result in an agreement. If not, arbitration is a logical next step. Arbitration is the most practical method, which often leads to the final decision, provided that all parties in the conflict accept the arbitration procedure as final. In smaller issues related to inter-basin water transfers, arbitration may be the best method of finding a quick solution to avoid project delays and keep the cost of conflict to a minimum. If arbitration does not succeed, legal avenues may be the last option.

When inter-basin water transfers are imposed on the people and institutions of the basin from which water is diverted, one of the methods for resolving any conflicts is to compensate these people and institutions for their present and future losses. One of the approaches for raising funds required for compensation is to create the *Basin of Origin Equity Fund*. This fund can either receive income from all water users in a state or from those in the receiving basin who are benefiting from the extra water (Yevjevich, 2001).

A case study of the conflicts surrounding the proposed inter-basin transfer of water to provide extra drinking water for Sofia in Bulgaria is described by Clark and Wang (2003). In the 1970s, government officials advocated the construction of an inter-basin water transfer project consisting of dams, tunnels and diversion structures to overcome water shortages in Sofia. This controversial project sparked a high degree of polarisation between government officials and water scientists, as well as environmentalists, who viewed inter-basin transfer as being unnecessary, prohibitively expensive and detrimental to both the local population and the environment. All participants in the study, which used a qualitative approach employing focus groups and interviews of the stakeholders, agreed that Sofia's water supply network was highly inefficient and required modernisation. It was finally agreed that modernisation of the network would conserve enough water to supply Sofia without undertaking an expensive and environmentally damaging inter-basin water transfer project.

2.10 INTEGRATED ASSESSMENT AND MODELLING

In the previous sections, various physical, social, economic and environmental issues in inter-basin water transfer have been described. It is important to address all these issues when assessing a new project, using an integrative approach and allowing trade-offs and compromises to be agreed by all stakeholders. The following description of integrated assessment and modelling, including its aims, limitations and applications, is summarised from Jakeman and Letcher (2003).

Meeting the challenges of sustainability and catchment management requires an integrated approach that can assess different options for resource use, combined with several dimensions of ecological, social and economic values. This approach is the result of the increasing dissatisfaction of decision makers and the community with the outcomes of narrowly focused environmental management decisions, which have failed to deal with many interconnections and complexities between the physical and human environments. However, it is now possible to assess the effects of resource use and management in an integrated way that provides better guidance for decision-making. The increasing availability of spatial databases and rapidly improving computer software are facilitating such assessments. More importantly, the science of integrated assessment (IA) is maturing to the point where knowledge and practice of this discipline should now accelerate. The key features of IA include:

- A problem-focussed activity using an iterative, adaptive approach that links research to policy.
- An interactive, transparent framework that enhances communication.
- A process enriched by stakeholder involvement and dedicated to adoption.

- Provides the connections between the natural and human environment, recognising spatial dependencies, feedbacks and impediments.
- Recognition of essential missing knowledge for future inclusion.
- Characterisation and reduction of uncertainty in predictions.

In natural resources assessment, integration has several dimensions. These dimensions include the consideration of multiple issues and stakeholders, the key disciplines within and between the natural and human sciences, and multiple spatial and temporal scales of system representation and behaviour. The development and use of models are major activities of IA, as is their incorporation in environmental information systems and computer-based decision support systems. This is referred to as Integrated Assessment Modelling (IAM). The types of models include:

- Data models that are representations of measurements and experiments.
- Qualitative conceptual models as verbal or visual descriptions of systems and processes.
- Quantitative numerical models.
- Mathematical methods used to analyse the numerical models and to interpret the results.
- Decision-making models that transform the values and knowledge into action.

Documenting models makes their nature and assumptions more explicit and allows integration with other models. It is fundamentally important that both the assumptions and the procedures in such models are transparent and available so that stakeholders can assess reliability of model predictions. When incorporated into computer software, models allow scenarios to be tested more efficiently, and in particular allow the calculation and assessment of the ensuing trade-offs among environmental, economic and social outcomes.

Once models are constructed and parameterised, the question arises as to how to produce the information needed to assist the making of decisions. This could be achieved by implementing optimisation of the system with respect to a specified or preferred set of outcomes. In this situation the controllable variables in the system are allowed to vary within a practical range to best meet those outcomes. The main problem with this approach is that it tends to obscure the complexities and dependencies in the system being modelled, and hence overly simplifies the process.

An alternative is to use simulation as a way of more fully exploring the effects of controllable and uncontrollable scenario variables in terms of indicators of system response.

This approach allows one to develop a better understanding of interdependencies and may lead to the identification of outcomes, which are considered better trade-offs than pre-specified optima. Initially it is a more cumbersome approach, but the range of input variations explored can be reduced as understanding of the modelled system behaviour accrues.

Given the complexities and uncertainties of integrated modelling, its broad objective should be to increase understanding of the directions and magnitudes of changes under different options. Typically it should not be about accepting or treating system outputs as accurate predictions. It should be aimed at allowing differentiation between outcomes, at least with a qualitative confidence. For example, a particular set of outcomes or indicator values are overall better than another set, with good confidence, reasonable confidence, or little confidence. This accordingly facilitates a decision as to the worth of adopting a certain policy.

Credibility requires that model components are identifiable, plausible and explain system output behaviour satisfactorily. However, integrated models possess many uncertainties. These include measurement and sampling errors in data, model structure assumptions, model parameter values and assumed constants. These uncertainties must be characterised and preferably reduced as far as possible.

Jakeman and Letcher (2003) also provide three examples of IA application in land and water resources development: one in the Mae Chaem catchment, Northern Thailand; and two in the Yass and Namoi catchments, both in New South Wales, Australia.

In the case of the Mae Chaem catchment, agricultural expansion has produced competition for water at various scales and has resulted in erosion problems, downstream water quality deterioration, groundwater depletion, biodiversity loss, and shifts in the distribution of economic and social well-being and equity. Resolving this problem required integration of various disciplinary contributions including agronomy, climatology, economics, hydrology and soil science. An integrated modelling toolkit was developed and imbedded within a decision support system (DSS). The model components included a crop model, a rainfall-run-off model, a sheet erosion model and an economic model.

The aim of the Yass catchment study was to examine the effects of water resources policy and substantial changes in land use. Policies applied were those for volumetric conversions of licences on unregulated rivers, farm dam capture limits and expansion of viticulture on land previously used for grazing and rural residential subdivision.

The aim of the Namoi catchment study was to develop a tool for investigating the catchment scale trade-offs involved with various options for water allocation. The development of this tool was undertaken in response to needs expressed by stakeholder groups in the catchment, and was focussed by stakeholder input at various stages of model development.

Jakeman and Letcher (2003) conclude that the three examples presented illustrate the potential value of Integrated Assessment Modelling for quantifying the biophysical and socio-economic impacts that resulted from management interventions and uncontrollable factors. Finally, Jakeman et al. (2003) discuss difficulties and limitations of IA studies and describe the concept of Integrated Scenario Modelling (ISM).

2.11 CONCLUSIONS

This chapter has described major issues which should be considered in the development of water resources, such as dam building, hydro-electric schemes and the transfer of water from one basin (or catchment) to another. This includes biophysical aspects such as topography, geology, and hydrology, as well as engineering and geotechnical investigations. It also includes a detailed social, environmental and economic assessment of both the impacts and benefits that can be expected from the project.

In the past, water resources development projects only required technical investigations, project design and a limited economic feasibility study. No environmental and social impact assessments were required. As a result many of these projects have damaged the environment, caused unnecessary displacement of local people, and had shortfalls in achieving their physical, economic and social targets. In particular, water resources developments were heavily subsidised and the cheap water was wasted, causing land degradation and other environmental damages. In future, water resources development projects, including inter-basin water transfer projects, require an integrated approach regarding assessment of the physical, economic, social and environmental impacts, full cost recovery and implementation of the World Commission on Dams recommendations described in this chapter. It is important to note that the investigation of all alternative options, such as increased water use efficiency, water trading, and water pricing should be considered before any new dam building or inter-basin water transfers are initiated. In this respect, the example of water supply to the city of Sofia in Bulgaria described in Section 2.9.2 is a good example.

References

Altinbilek, H. D., Bayram, M. and Hazar, T. (1999). The new approach to development project-induced resettlement in Turkey. *International Journal of Water Resources Development* **15**(3): 291–300.

Anderson, M. P. and Woessner, W. W. (1992). *Applied Groundwater Modelling: Simulation of Flow and Advective Transport.* San Diego, California: Academic Press.

Annandale, G. W. (1987). *Reservoir Sedimentation.* Amsterdam: Elsevier.

Arrow, K. J. and Lind, R. C. (1994). Risk and uncertainty: uncertainty and the evaluation of public investment decision. In Layard, R. and Glaister, S. eds. *Cost-Benefit Analysis.* Second Edition. Cambridge: Cambridge University Press, pp. 160–178.

Arthington, A. H., King, J. M., O'Keeffe, J. H., Bunn, S. E., Day, J. A., Pusey, B. J., Bluhdorn, D. R. and Tharme, R. E. (1992). Development of a holistic approach for assessing environmental flow requirements of riverine ecosystems. In Pigram, J. J. and Hooper, B. P. eds. *Proceedings of an International Seminar and Workshop on Water Allocation for the Environment.* Centre for Water Policy Research, University of New England, Armidale, pp. 69–76.

Arthington, A. H. and Pusey, B. J. (1993). In-stream flow management in Australia: methods, deficiencies and future directions. *Australian Biologist* **6**: 52–60.

Arthington, A. H. (1998). *Comparative Evaluation of Environmental Flow Assessment Techniques: Review of Holistic Methodologies.* Occasional Paper No. 26/98. Canberra: Land and Water Resources Research and Development Corporation.

Arthington, A. H. and Zalucki, J. M. eds. (1998). *Comparative Evaluation of Environmental Flow Assessment Techniques: Review of Methods.* Occasional Paper No. 27/98. Canberra: Land and Water Resources Research and Development Corporation.

Arthington, A. H., Pausy, B. J., Brizga, S. O., McCosker, R. O., Bunn, S. E. and Growns, I. O. (1998a). *Comparative Evaluation of Environmental Flow Assessment Techniques: R and D Requirements.* Occasional Paper No. 24/98. Canberra: Land and Water Resources Research and Development Corporation.

Arthington, A. H., Brizga, S. O. and Kennard, M. J. (1998b). *Comparative Evaluation of Environmental Flow Assessment Techniques: Best Practice Framework.* Occasional Paper No. 25/98. Canberra: Land and Water Resources Research and Development Corporation.

Arthington, A. H. and Pusey, B. J. (2003). Flow restoration and protection in Australian Rivers. *River Research and Applications* **19**: 377–395.

Arthington, A. H., Brizga, S. O., Choy, S. C., Kennard, M. J., Mackay, S. J., McCosker, R. O., Ruffini, J. L. and Zalucki, J. M. (2000). *Environmental Flow Requirements of the Brisbane River Downstream From Wivenhoe Dam.* Brisbane: Centre for Catchment and In-stream Research, Griffith University.

ATSIC (2002a). *Onshore Water Rights Discussion Booklet.* Broome, Western Australia: Lingiari Foundation.

ATSIC (2002b). *Background Briefing Papers.* Broome, Western Australia: Lingiari Foundation.

Bates, G. (2002). *Environmental Law in Australia.* Chatswood, NSW: LexisNexis Butterworth.

Beach, H. L., Hammer, J., Hewitt, J. J., Kaufman, E., Kurki, A., Oppenheimer, J. A. and Wolf, A. T. (2000). *Transboundary Freshwater Dispute Resolution: Theory, Practice, and Annotated References.* Tokyo: United Nations University Press.

Best, E. (1984). Dams, engineering geology. In Finkl, Jnr, C. W. ed. *The Encyclopedia of Applied Geology.* New York: Van Nostrand Reinhold Company, pp. 111–123.

Biswas, A. K. (1992). Indus water treaty: the negotiating process. *Water International* **17**(4): 201–209.

Biswas, A. K. (1999). Management of international waters: opportunities and constraints. *International Journal of Water Resources Development* **15**(4): 429–441.

Blackwelder, B. and Carlson, P. (1986). *Disasters in International Water Development.* Washington D.C.: Environmental Policy Institute.

Bovee, K. D. (1982). *A Guide to Stream Habitat Analysis Using the Instream Flow Incremental Methodology*. Instream Flow Information Paper 12, FWS/OBS-82/26. Fort Collins: U.S. Department of Fisheries and Wildlife Service.

Brizga, S. O. (1998). Methods addressing flow requirements for geomorphological purposes. In Arthington, A. H. and Zallucki, J. M. eds. *Comparative Evaluation of Environmental Flow Assessment Techniques: Review of Methods*. Occasional Paper No. 27/98. Canberra: Land and Water Resources Research and Development Corporation, pp. 8–46.

Brizga, S. and Finlayson, B. (2000). *River Management: The Australian Experience*. Chichester: John Wiley and Sons.

Brizga, S. O., Arthington, A. H., Choy, S., Craigie, N. M., Kennard, M. J., Mackay, S. J., Pusey, B. J. and Werren, G. L. (2002). Benchmarking, a 'top-down' methodology for assessing environmental flows in Australian rivers. *Proceedings of the International Conference on Environmental Flows for Rivers*. Cape Town, South Africa: University of Cape Town (Published on CD).

Bruns, B. R. and Meinzen-Dick, R. S. eds. (2000). *Negotiating Water Rights*. London: ITDG Publishing.

Bunn, S. T and Arthington, A. H. (2002). Basic principles and ecological consequences of altered flow regimes for aquatic biodiversity. *Environmental Management* 30(4): 492–507.

Christen, E. W. and Hornbuckle, J. W. eds. (2001). *Subsurface Drainage Design and Management Practices in Irrigated Areas of Australia*. Canberra: Land & Water Australia.

Clark, W. A. and Wang, G. A. (2003). Conflicting attitudes toward inter-basin water transfers in Bulgaria. *Water International* 28(1): 79–89.

Claydon, G. ed. (1995). *Water Allocations and Entitlements: Towards a National Framework for Property Rights in Water, Review of Policy Issues and Options*. Canberra: Department of Primary Industries and Energy.

COAG (1994). *Communiqué from the COAG Meeting, Hobart, 25 February 1994*. Canberra: Department of the Prime Minister and Cabinet.

Committee on Western Water Management (1992). *Water Transfer in the West: Efficiencies, Equity and Environment*. National Research Council. The National Academies Press.

Cottingham, P., Thoms, M. C. and Quinn, G. P. (2002). Scientific panels and their use in environmental flow assessment in Australia. *Australian Journal of Water Resources* 5: 103–111.

Croke, B. F. W. and Jakeman, A. J. (2001). Predictions in catchment hydrology: an Australian Perspective. *Australian Journal of Marine and Freshwater Research* 52: 65–79.

Davidson, B. R. (1984). A preliminary benefit cost analysis of the inland diversion of the coastal rivers of New South Wales. *Review of Marketing and Agricultural Economics* 52(1): 23–47.

Dixon, J. A., Talbot, L. M. and Le Moigne, Guy, J.-M. (1989). *Dams and the Environment: Considerations in World Bank Projects*. Technical Paper No. 110. Washington D. C.: The World Bank.

Dixon, J. A. Scura, L. F., Carpenter, R. A. and Sherman, P. B. (1994). *Economic Analysis of Environmental Impacts*. Second Edition. London: Earthscan.

DWAF (1999). *Resource Directed Measures for Protection of Water Resources. Volume 2: Integrated Manual (Version 1.0)*. Pretoria, South Africa: Department of Water Affairs and Forestry, Institute for Water Quality Studies.

Fearnside, P. M. (1993). Resettlement plans for China's Three Gorges Dam. In Barber, M. and Ryder, G. eds. *Damming the Three Gorges: What Dam Builders Don't Want You to Know*. Second edition. London: Earthscan, pp. 34–58.

Fetter, C. W. (1993). *Contaminant Hydrogeology*. New York: Macmillan Publishing Company.

Fetter, C. W. (2001). *Applied Hydrogeology*. Fourth Edition. Upper Saddle River, N.J.: Prentice Hall.

Fisher, D. E. (2000). *Water Law*. Sydney: LBC Information Services.

Ghadiri, H. and Rose, C. W. (1992). *Modeling Chemical Transport in Soils: Natural and Applied Contaminants*. Boca Raton, Florida: CRC Press.

Ghassemi, F., Jakeman, A. J. and Nix, H. A. (1995). *Salinisation of Land and Water Resources: Human Causes, Extent, Management and Case Studies*. Sydney: University of New South Wales Press.

Gippel, C. J. and Stewardson, M. J. (1998). Use of wetted perimeter in defining minimum environmental flows. *Regulated Rivers: Research and Management* 14: 53–67.

Gleick, P. H. ed. (1993). *Water in Crisis: A Guide to the World's Freshwater Resources*. New York: Oxford University Press.

Goldsmith, E. and Hildyard, N. (1984). *The Social and Environmental Effects of Large Dams*. Volume 1: Overview; Volume 2 (1986): Case studies; and Volume 3 (1992): A Review of the Literature. Worthyvale Manore, Camelford, U.K.: Wadebridge Ecological Centre.

Grafton, R. Q., Adamowicz, W. A., Dupont, D., Nelson, H., Hill, R. J. and Renzetti, S. (2004). *The Economics of the Environment and Natural Resources*. Malden, M.A.: Blackwell Publishing.

Gulhati, N. D. (1973). *Indus Waters Treaty: An Exercise in International Mediation*. Bombay: Allied Publishers.

Hamblin, A. (2001). *Land, Australia State of the Environment Report 2001*. Collingwood, Victoria: CSIRO Publishing.

Harbaugh, A. W., Banta, E. R., Hill, M. C. and McDonald, M. G. (2000). MODFLOW-2000, the U.S. Geological Survey modular ground-water model – user guide to modularization concepts and the ground-water flow process. U.S. Geological Survey. Open File Report 00–92.

Harrop, D. O. and Nixon, J. A. (1999). *Environmental Assessment in Practice*. London: Routledge.

Heggelund, G. (2004). *Environment and Resettlement Politics in China: The Three Gorges Project*. Hants, England: Ashgate Publishing Limited.

Hill, M. C., Banta, E. R., Harbough, A. W. and Anderman, E. R. (2000). MODFLOW-2000, the U.S. Geological Survey modular ground-water model – user guide to observation, sensitivity, and parameter estimation processes and three post-processing programs. U.S. Geological Survey. Open File Report 00–184.

Holzbecher, E. (2002). *Groundwater Modeling: Computer Simulation of Groundwater Flow and Pollution*. Fermont, California: FiatLux Publications (Electronic Book Series). Published on CD.

Howe, C. W. and Easter, K. W. (1971). *Interbasin Transfer of Water: Economic Issues and Impacts*. Baltimore, Maryland: The John Hopkins Press.

Hudson, N. (1995). *Soil Conservation*. Third edition. London: Bastford.

International Bureau of the Permanent Court of Arbitration, ed. (2003). *Resolution of International Water Disputes*. The Hague: Kluwer Law International.

Jakeman, A. J. and Letcher, R. A. (2003). Integrated assessment and modelling: features, principles and examples for catchment management. *Environmental Modelling and Software* 18: 491–501.

Jakeman, A. J., Letcher, R. A. and Cuddy, S. M. (2003). Linking science and social science for sustainable catchment management: a modelling perspective. *Development Bulletin* 63: 24–27.

Jansen, R. B. ed. (1988). *Advance Dam Engineering for Design, Construction, and Rehabilitation*. New York: Van Nostrand Reinhold.

Jonas, D. and Byrne, P. (2003). *Airborne Laser Scanning: Beyond its Formative Years*.[17]

Jowett, I. G. (1997). Instream flow methods: a comparison of approaches. *Regulated Rivers: Research and Management* 13: 115–127.

Kheel, T. W. (1999). *The Keys to Conflict Resolution*. New York: Four Walls Eight Windows.

King, J. M. and Louw, D. (1998). Instream flow assessments for regulated rivers in South Africa using the Building Block Methodology. *Aquatic Ecosystem Health and Restoration* 1: 109–124.

King, J. M. and Tharme, R. E. (1994). *Assessment of the Instream Flow Incremental Methodology and Initial Development of Alternative Instream Flow Methodologies for South Africa*. WRC Report No. 295/1/94. Pretoria, South Africa: Water Research Commission.

---

[17] This paper and a number of others are available at: http://www.aamhatch.com.au/ linked to "Technical Papers" (visited in June 2005).

King, J. M., Tharme, R. E. and Brown, C. A. (1999). *Definition and Implementation of Instream Flows*. Thematic Report for the World Commission on Dams. Cape Town, South Africa: Southern Waters Ecological Research and Consulting.

King, J. M., Tharme, R. E. and De Villiers, M. S. eds. (2002). *Environmental Flow Assessments for Rivers: Manual for the Building Block Methodology*. Water Research Commission Technology Transfer Report No. TT131/00. Pretoria, South Africa: Water Research Commission.

King, J. M., Brown, C. A. and Sabet, H. (2003). A scenario-based holistic approach to environmental flow assessments for rivers. *River Research and Applications* **19**: 619–640.

Kipp, K. L. (1987).[18] HST3D: a computer code for simulation of heat and solute transport in three dimensional ground-water flow system. U.S. Geological Survey, Water Resources Investigation. Report 86–4095.

Kondolf, G. M. (1997). Hungry water: effects of dams and gravel mining on river channels. *Environmental Management* **21**(4): 533–551.

Kovach, K. K. (1994). *Mediation: Principles and Practices*. St. Paul, Minnesota: West Publishing Co.

Layard, R. and Glaister, S. eds. (1994). *Cost-Benefit Analysis*. Second Edition. Cambridge: Cambridge University Press.

Letcher, R. A. and Jakeman, A. J. (2002). Catchment hydrology. In El-Shaarawi, A. H. and Piegorsch, W. W. eds. *Encyclopedia of Environmetrics*. Chichester: John Wiley and Sons, pp. 281–290.

Loneragan, N. R. and Bunn, S. E. (1999). River flows and estuarine ecosystems: implications for coastal fisheries from a review and a case study of the Logan River, southeast Queensland. *Australian Journal of Ecology* **24**(4): 431–40.

Mackie, K. J. (1991). *A Handbook of Dispute Resolution: ADR in Action*. London: Routledge and Sweet & Maxwell.

Magowan, F. (2002). Negotiating indigenous water knowledge in a global water crisis. *Cultural Survival Quarterly* **26**(2): 18–20.

Mahmood, K. (1987). *Reservoir Sedimentation: Impact, Extent and Mitigation*. World Bank Technical Paper No. 71. Washington D.C.: World Bank.

Marshall, T. J., Holmes, J. W. and Rose, C. W. (1996). *Soil Physics*. Third Edition. Cambridge: Cambridge University Press.

McDonald, M. G. and Harbaugh, A. W. (1984). *A Modular Three-dimensional Finite Difference Ground-water Flow Model*. Open File Report 83–875. U.S. Geological Survey.

McDonald, M. G. and Harbaugh, A. W. (1988). *A Modular Three-dimensional Finite Difference Ground-water Flow Model*. U.S. Geological Survey. Techniques of Water Resources Investigations, Book 6, Chapter A1.

McLaughlin, D., Kinzelbach, W. and Ghassemi, F. (1993). Modelling subsurface flow and transport. In Jakeman, A. J., Beck, M. B. and McAleer, M. J. eds. *Modelling Change in Environmental Systems*. Chichester: John Wiley and Sons, pp. 133–161.

Mekong River Commission (2004). *Annual Report 2004: Mekong River Commission*. Vientiane, Laos: MRC.

Milhous, R. T. (1998). A review of the physical habitat simulation system. In Blažková, Ŝ., Stalnaker, C. and Novický, O. eds. *Hydroecological Modelling: Research, Practice, Legislation and Decision-making*. Report by U.S. Geological Survey, Biological Research Division and Water Research Institute. Fort Collins: U.S. Geological Survey, Biological Research Division, pp. 7–8.

Milson, J. (1989). *Field Geophysics*. Milton Keynes, England: Open University Press.

Morgan, R. P. C. (1986). *Soil Erosion and Conservation*. Essex, England: Longman Scientific and Technical.

Nakayama, M., Gunawan, B., Yoshida, T. and Asaeda, T. (1999). Resettlement issues of Cirata Dam Project: A post-project review. *International Journal of Water Resources Development* **15**(4): 443–458.

Nancarrow, B. E. ed. (1994). *International and National Trends in Public Involvement: Where to Next for the Australian Water Industry?* Proceedings of a Review sponsored by Australian Research Centre for Water in Society and International Association of Public Participation Practitioners, Sydney, 11 April 1994. Canberra: CSIRO Division of Water Resources.

Nielsen, D. M. ed. (1991). *Practical Handbook of Ground-water Monitoring*. Chelsea, Michigan: Lewis Publishers.

NSW Treasury (1997).[19] *Guidelines for Economic Appraisal*. Treasury Policy and Guideline Paper TPP97–2. Sydney: NSW Treasury, Office of Financial Management.

O'Keeffe, J. H. and Hughes, D. A. (2002). The Flow Stress-Response method for analysing flow modifications: applications and developments. In *Proceedings of International Conference on Environmental Flows for Rivers*. Cape Town: University of Cape Town (published on CD).

Olive, L. J., Loughran, R. J. and Kesby, J. A. (1994). *Variability in Stream Erosion and Sediment Transport*. Wallingford: International Association of Hydrological Sciences. Publication no. 224.

Osborne, M. (2004). *River at Risk: The Mekong and the Water Politics on China and Southeast Asia*. Sydney: Lowy Institute for International Policy.

Parasnis, D. S. (1986). *Principles of Applied Geophysics*. Fourth Edition. New York: Chapman and Hall.

Pearce, F. (1992). *The Dammed: Rivers, Dams, and the Coming World Water Crisis*. London: The Bodley Head.

Petts, G. E. (1984). *Impounded Rivers: Perspectives for Ecological Management*. Chichester: John Wiley & Sons.

Petts, G. E., Bickerton, M. A., Crawford, C., Lerner, D. N. and Evans, D. (1999). Flow management to sustain groundwater-dominated stream ecosystems. *Hydrological Processes* **13**: 497–513.

Poff, N. L., Allan, J. D., Bain, M. B., Karr, J. R., Prestegaard, K. L., Richter, B. D., Sparks, R. E. and Stromberg, J. C. (1997). The natural flow regime. *BioScience* **47**: 769–784.

Poff, N. L., Allan, J. D., Palmer, M. A., Hart, D. D., Richter, B. D., Arthington, A. H., Rogers, K. H., Meyer, J. L. and Stanford, J. A. (2003). River flows and water wars: emerging science for environmental decision making. *Frontiers in Ecological and the Environment* **1**(6): 298–306.

Pusey, B. J. (1998). Methods addressing the flow requirements of fish. In Arthington, A. H. and Zalucki, J. M. eds. *Comparative Evaluation of Environmental Flow Assessment Techniques: Review of Methods*. Occasional Paper 27/98. Canberra: Land and Water Resources Research and Development Corporation, pp. 64–103.

Ray, A. (1984). *Cost-Benefit Analysis: Issues and Methodologies*. Washington D.C.: The World Bank.

Richter, B. D., Baumgartner, J. V., Powell, J. and Braun, D. P. (1996). A method for assessing hydrologic alteration within ecosystems. *Conservation Biology* **10**: 1–12.

Riquier, J. (1982). A world assessment of soil degradation. *Nature and Resources* **18**(2): 18–21.

Rose, C. W. (1993). Soil erosion by water. In McTainsh, G. H. and Boughton, W. C. eds. *Land Degradation Processes in Australia*. Melbourne: Longman Cheshire, pp. 149–187.

Rose, D. B. (2005). Indigenous water philosophy in an uncertain land. In Botterill, L. C. and Wilhite, D. A. eds. *From Diaster Response to Risk Management: Australia's National Drought Policy*. Dordrecht, The Netherlands: Springer, pp. 37–50.

Rural Water Commission of Victoria (1988). *Irrigation and Drainage Practice*. Armadale, Victoria: Rural Water Commission of Victoria.

Schwartz, F. W. and Zhang, H. (2003). *Fundamentals of Ground Water*. New York: John Wiley and Sons.

Schwass, R. D. (1985). *Public Information and Public Participation in Water Resources Policy*. Toronto: Inquiry on Federal Water Policy.

Scott, A. (1985). *The Economics of Water Export Policy*. Ottawa, Canada: Department of Economics, The University of British Columbia.

Scott, A. and Coustalin, G. (1995). The evolution of water rights. *Natural Resources Journal* **35**(4): 821–979.

---

[18] Information regarding the latest version of the HST3D is available at the URL: http://wwwbrr.cr.usgs.gov/projects/GW_Solute/hst/index.shtml (visited in June 2005).

[19] This document is also available at: http://www.treasury.nsw.gov.au/pubs/tpp97_2/ea-index.htm (visited in June 2005).

Smith, D. I. (1998). *Water in Australia*. Melbourne: Oxford University Press.
Sombatpanit, S., Zöbisch, M. A., Sanders, D. W. and Cook, M. G. (1997). *Soil Conservation Extension From Concepts to Adoption*. Enfield, New Hampshire: Science Publishers.
Sparks, R. E. (1992). Risks of altering the hydrologic regime of large rivers. In Cairns, J., Niederlehner, B. R. and Orvos, D. R. eds. *Predicting Ecosystem Risk*, v. xx. Princeton, N.J.: Princeton Scientific Publication, pp. 119–152.
Sparks, R. E. (1995). Need for ecosystem management of large rivers and floodplains. *BioScience* **45**: 168–182.
Spitz, K. and Moreno, J. (1996). *A Practical Guide to Groundwater and Solute Transport Modeling*. New York: John Wiley and Sons.
Stalnaker, C. B., Lamb, B. L., Henriksen, J., Bovee, K. D. and Bartholow, J. (1994). *The Instream Flow Incremental Methodology : A Primer for IFIM*. National Ecology Research Centre, Internal publication. Fort Collins: National Biological Service.
Steward, H. J., Madamombe, E. K. and Topping, C. C. (2002). *Adapting Environmental Flow Methodologies for Zimbabwe*. In Unpublished Proceedings of the International Conference on Environmental Flows for Rivers, 3–8 March 2002, Cape Town, University of Cape Town.
Stewardson, M. (2001). The flow events method for developing environmental flow regimes. In Rutherfurd, I., et al. eds. *The Value of Healthy Rivers*. Proceedings of the 3rd Australian Stream Management Conference, 27–29 August 2001, Brisbane, Queensland. pp. 577–582.
Stewardson, M. J. and Cottingham, P. (2002). A demonstration of the flow events method: environmental flow requirements of the Broken River. *Australian Journal of Water Resources* **5**: 33–47.
Tan, P. L. (2001). *Dividing the Waters: A Critical Analysis of Law Reform in Water Allocation and Management in Australia from 1989–1990*. PhD Thesis. Canberra: The Australian National University.
Tharme R. E. (1996). *Review of International Methodologies for the Quantification of the Instream Flow Requirements of Rivers*. Water law review final report for policy development, for the Department of Water Affairs and Forestry. Cape Town, South Africa: Freshwater Research Unit, University of Cape Town.
Tharme, R. E. and King, J. M. (1998). *Development of the Building Block Methodology for Instream Flow Assessments, and Supporting Research on the Effects of Different Magnitude Flows on Riverine Ecosystems*. Report No. 576/1/98. Pretoria, South Africa: Water Research Commission.
Tharme, R. E. (2003). A global perspective on environmental flow assessment: emerging trends in the development and application of environmental flow methodologies for rivers. *River Research and Applications* **19**: 379–442.
Tinkler, K. J. and Wohl, E. E. eds. (1998). *Rivers Over Rock: Fluvial Processes in Bedrock Channels*. Geophysical Monograph 107. Washington D.C.: American Geophysical Union.
UNESCO (1990). *The Impact of Large Water Projects on the Environment*. Paris: UNESCO.
United States Bureau of Reclamation (1987). *Design of Small Dams*. Third Edition. Washington D.C.: United States Government Printing Office.
Utton, A. E. (1996). Which rule should prevail in international water disputes: that of reasonableness or that of no harm? *Natural Resources Journal* **36**(3): 635–41.
Voss, C. I. (1984).[20] *SUTRA, Saturated-Unsaturated TRAnsport: A Finite Element Simulation Model for Saturated-Unsaturated Fluid Density Dependent Groundwater Flow with Energy Transport or Chemically Reactive Single-Species Solute Transport. Report 84–4369*. Reston, Virginia: U.S. Geological Survey, Water Resources Investigation.
Voss, C. I, Boldt, D. and Shapiro, A. M. (1997). *SUTRA-GUI: A Graphical-User Interface for U.S. Geological Survey's SUTRA Code Using Argus ONE for Simulation of Variable-Density Saturated-Unsaturated Ground-Water Flow with Solute or Energy Transport. Report 87–421*. Reston, Virginia: U.S. Geological Survey.
Walhstrom, E. E. (1974). *Developments in Geotechnical Engineering 6: Dams, Dam Foundations, and Reservoir Sites*. Amsterdam: Elsevier.
Ward, J. V., Tockner, U., Uehlinger, U. and Malard, F. (2001). Understanding natural patterns and processes in river corridors as the basis for effective river restoration. *Regulated Rivers Research and Management* **117**: 311–323.
Waterloo Hydrogeologic (2000). *Visual MODFLOW User's Manual*. Waterloo, Ontario, Canada. Waterloo Hydrogeologic.
Wentworth Group of Concerned Scientists (2002). *Blueprint for a Living Continent: A Way Forward from the Wentworth Group of Concerned Scientists*. Sydney: World Wide Fund for Nature, Australia.
Wheater, H. S., Jakeman, A. J. and Beven, K. J. (1993). Processes and directions in rainfall-runoff modelling. In Jakeman, A. J., Beck, M. B. and McAleer, M. J. eds. *Modelling Change in Environmental Systems*. Chichester: John Wiley and Sons. pp. 101–132.
Withers, B. and Vipond, S. (1974). *Irrigation Design and Practice*. London: Batsford.
Wolf, A. T., Natharius, J. A., Danielson, J. J., Ward, B. S. and Pender, J. K. (1999). International river basins of the world. *International Journal of Water Resources Development* **15**(4): 387–427.
World Bank (1991a). *Environmental Assessment Sourcebook. Volume I: Policies, Procedure, and Cross-Sectoral Issues*. Technical Paper No. 139. Washington D.C.: The World Bank.
World Bank (1991b). *Environmental Assessment Sourcebook. Volume II: Sectoral Guidelines*. Technical Paper No. 140. Washington D.C.: The World Bank.
World Bank (1991c). *Environmental Assessment Sourcebook. Volume III: Guidelines for Environmental Assessment of Energy and Industry Projects*. Technical Paper No. 154. Washington D.C.: The World Bank.
World Commission on Dams (2000). *Dams and Development: A New Framework for Decision-Making*. London: Earthscan Publications.
Yevjevich, V. (2001). Water diversions and interbasin transfers. *Water International* **26**(3): 342–348.
Young, R. A. and Haveman, R. H. (1985). Economics of water resources: a survey. In Kneese, A. V. and Sweeney, J. L. *Handbook of Natural Resources and Energy Economics*, Volume **II**. Amsterdam: North-Holland, pp. 465–529.
Yunus Khan, M. (1990). Boundary water conflict between India and Pakistan. *Water International* **15**(4): 195–199.
Zheng, C. and Bennett, G. D. (1995a). *Applied Contaminant Transport Modeling: Theory and Practice*. New York: Van Nostrand Reinhold.
Zheng, C. and Bennett, G. D. (1995b). *MT3D: A Modular Three Dimensional Transport Model for Simulation of Advection, Dispersion and Chemical Reactions of Contaminants in Ground-water Systems*. Bethesda, Maryland: S.S. Papadopulos and Associate Inc.

---

[20] Information regarding the latest version of SUTRA is available at: http://water.usgs.gov/software/sutra.html (visited in June 2005).

Part II
Inter-basin Water Transfer in Australia

3 Land and Water Resources of Australia

3.1 GEOGRAPHY

The Australian landmass covers 7 682 300 km^2, with maximum dimensions of 3680 km from north to south and 4000 km from east to west. The area of Australia is as great as that of the United States of America (excluding Alaska), about 50 percent greater than Europe (excluding the former Soviet Union) and 32 times greater than the United Kingdom. Australia is the lowest and flattest of the continents. The average surface altitude is only 300 m above mean sea level. Approximately 87 percent of the landmass has an altitude of less than 500 m and 99.5 percent is less than 1000 m. The present topography is due to a long landscape history that started in the Permian period (about 290 million years ago) when much of Australia was glaciated by a huge icecap. After the ice melted, parts of the continent subsided and were covered by sediments to form numerous sedimentary basins, including the Great Artesian Basin. About 55 million years ago, Australia started to drift northward, moving from a position adjacent to Antarctica.

Australia is divided into three physiographic units (Figure 3.1). These are (Australian Geographic Society, 1988):

- **The Eastern Highlands Belt** extends along the east coast of Australia and is a series of ranges of varying height and includes the main divide of the Great Dividing Range. The divide passes through Australia's highest point, Mt Kosciuszko, with an elevation of 2228 m. The western slopes of the Great Dividing Range are generally gradual, while the eastern descent to coastal lowlands is much steeper.
- **The Central Eastern Lowlands** stretches across the continent from the Gulf of Carpentaria to the coast of South Australia. Most of the lowlands are below 150 m and at Lake Eyre descend to about 15 m below mean sea level.
- **The Great Western Plateau** includes most of Western Australia and the Northern Territory. It consists of Precambrian rocks (some over 3000 million years old) with an average elevation of about 300 m. The surface is flat and uniform over much of Western Australia.

Australia is a federation of six states (Figure 3.1): New South Wales (NSW), Victoria (VIC), Queensland (QLD), South Australia (SA), Western Australia (WA) and Tasmania (TAS); two mainland territories, the Northern Territory (NT) and the Australian Capital Territory (ACT); and several small island territories. Apart from off-shore marine environment, resource ownership is vested in the states and territories.

3.2 POPULATION

At 30th June 2003, Australia supported a growing population of just under 19.9 million, an increase of 1.2 percent[1] over the previous year. This was 4.7 million greater than in 1982 and about 16.1 million more than the 1901 population of 3.8 million (Table 3.1). The second half of the twentieth century has seen higher rates of growth than the first due to a natural increase as well as increased net overseas migration. Most of the Australian population is concentrated in two widely separated coastal or near coastal regions. By far the largest of these, in terms of area and population, lies in the south-east and east. The smaller of the two regions is the south-west of the continent. In both regions the population is concentrated in urban centres close to the coast, particularly states' capital cities such as Sydney, Melbourne, Perth and

---

[1] This was equal to the overall world growth rate of 1.2 percent (Australian Bureau of Statistics, 2005, Table 5.2). Growth rates for a number of countries were: Germany (0.1%); Japan (0.1%); the United Kingdom (0.3%); New Zealand (1.1%); India (1.5%); and Singapore (1.9%).

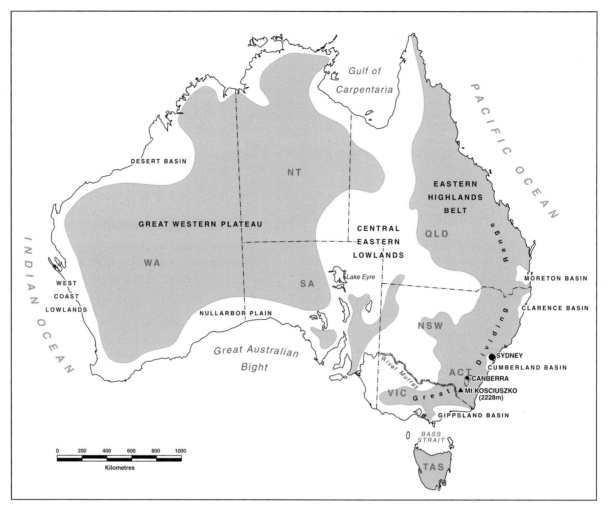

Figure 3.1 Physiographic division of Australia (Australian Geographic Society, 1988, Volume 6, p. 2295).

Table 3.1. *Population of Australia's states and territories from 1901 to 2003 (in 000s)*

Year	NSW	VIC.	QLD.	SA	WA	TAS.	NT	ACT	Total
1901	1361.7	1203.0	502.3	356.1	188.6	171.7	4.8	–	3788.1
1922	2154.4	1570.4	779.3	503.7	341.5	213.9	3.7	2.9	5569.9
1942	2828.7	1962.6	1039.8	608.3	477.0	240.9	9.1	14.4	7180.7
1962	3986.9	2983.1	1551.0	987.5	766.0	355.7	46.0	66.2	10 742.3
1982	5303.6	3992.9	2424.6	1331.1	1338.9	429.8	130.3	233.0	15 148.2
2000	6486.2	4741.3	3561.5	1505.0	1874.5	471.4	195.6	315.2	19 153.4
2002	6640.4	4872.5	3707.2	1520.2	1927.3	472.7	198.0	321.8	19 662.8
2003	6686.8	4917.3	3796.2	1527.1	1952.2	477.1	198.3	322.8	19 880.6

Source: Australian Bureau of Statistics (2004, 2005, Table 5.4).

Brisbane. Half the area of the continent contains only three percent of the population and the most densely populated one percent of the continent contains 84 percent of the population (Australian Bureau of Statistics, 2002, p. 79).

Using assumptions on future levels of birth, death and migration, the Australian Bureau of Statistics has projected the Australian population to the year 2050. Three main series of projections were produced (Australian Bureau of

Statistics, 2004, pp. 85–87). *Series A* assumes an annual net overseas migration of 125 000 per year from 2005 through to 2050 and a total fertility rate[2] of 1.8 babies per woman from 2010, remaining constant through to 2050. *Series B* assumes an annual net overseas migration of 100 000 per year from 2005 through to 2050, and a total fertility rate of 1.6 babies per woman from 2010, remaining constant through to 2050. *Series C* assumes an annual net overseas migration of 70 000 from 2005 through to 2050 and a total fertility rate of 1.4 babies per woman by 2010, remaining constant through to 2050. According to these projections, by 2050 the Australian population would be 31.4 million (*Series A*), 26.4 million (*Series B*) and 23 million (*Series C*). Under *Series B* (medium series) the population will increase in all states and territories except Tasmania and South Australia. The largest increases would be for the Northern Territory (92 percent), followed by Queensland (74 percent) and Western Australia (34 percent).

While population of numerous coastal regions of Australia over the period of 2001 to 2011 will continue to increase,[3] in large inland regions of NSW, QLD, SA and WA population will decline (Australian State of the Environment Committee, 2001, p. 38). Although no reason for such an important population decline has been provided, it seems that this is mainly due to harsh environmental conditions in these areas and much better living opportunities close to major capital cities which are located in coastal areas of the country. Using non-market environmental valuation, Bennett *et al.* (2004) estimated the value of the benefits associated with the maintenance of rural populations. They found that a significant proportion of Australian households are willing to pay to maintain rural population levels.

Although the Australian Bureau of Statistics has estimated the Australian population for the year 2050, there is no consensus among the experts and policy makers regarding the sustainable population for the country (Dovers *et al.* 1992; Norton *et al.* 1994; and Dovers, 1997). While some researchers argue that an increased population will contribute to economic development of the country and will improve defence capability of Australia, opponents are concerned about the adverse impacts of the population increase on further environmental degradation (Day and Rowland, 1988; Cocks, 1992, 1996).

Gifford *et al.* (1975) examined the implications for population size on the potential production of food and water availability in Australia. They concluded that with food export eliminated, the country could cater for 60 million people without excessive risk of agricultural instability, assuming climate does not deteriorate.

Less than a third of Australia (237 Mha) has climatic conditions that can support crop and pasture (Nix, 1988). Terrain constraints reduce this to 132 Mha and soil constraints to 77 Mha or 10 percent of total area of the Australia. Of this, 55 Mha are already under cultivation or are occupied by developments such as cities and roads. Therefore, only 22 Mha are potentially available for further development. If all potentially arable land and regulated water supplies were developed, and food export phased out, it is possible that population could reach 50 million. However, without major effort to prevent further land degradation it is unlikely that such a population could be sustainable (Nix, 1988).

Newman *et al.* (1994) estimated the carrying capacity of Australia with respect to eight constraints. Their estimates (in millions) for 2040 with respect to each constraint are: water (150); forests and timber products (23); cotton (100); wool (346); food (45); marine resources (23); coal for electricity generation (355); and petroleum (1). Overall, they suggested that the country's population should be limited to 23 million by 2040.

In 1994, The House of Representatives Standing Committee for Long Term Strategies conducted an inquiry into Australia's population "*carrying capacity*" defined as: *that combination of population, location and demographic characteristics which best serve Australia's national interest, and which allow individuals in the society to live long, self-fulfilling lives.* The Terms of Reference for the inquiry included the following: (1) the population which can be supported in Australia within and then beyond the next 50 years, taking account of technology options, possible patterns of resource use and quality of life considerations; and (2) the range of community views on population size and its political, social, economic and environmental significance. The Committee received 271 written submissions from all over Australia, examined 23 witnesses who were representative of a wide range of expertise and community opinions, and held five public hearings in Canberra, Melbourne and Sydney. In its conclusions, the committee rejected options for the high range (50 to more than 100 million) and low range (less than 9 million) as illusory, unattainable

---

[2] Fertility rate was 3.93 babies per woman in 1901. It declined to 2.17 in 1933 because of the Great Depression. It then increased to 3.55 in 1961. After that it fell rapidly to 2.9 in 1966 and 2.06 by 1976. Since then, it has been declining slowly and was 1.80 in 1996 and 1.73 in 2001 (Australian Bureau of Statistics, 2004, pp. 97–100).

[3] This increase is particularly significant in coastal areas around Sydney, Brisbane, Melbourne and Perth.

and undesirable. It put forward the following options for community discussion and political decisions:

- Relatively high population growth (30 to 50 million).
- Moderate population growth (23 to 30 million).
- Population stabilisation (17 to 23 million).
- Moderate to major population reduction (5 to 17 million).

Also, the Committee made 15 recommendations including the following:

Recommendation 2: The Australian Government should adopt a population policy which explicitly set out options for long-term population change, in preference to the existing situation where a *de facto* population policy emerges as a consequence of year by year decisions on immigration intake taken in an *ad hoc* fashion; such decisions being largely determined by the state of the economy in the particular year and with little consideration of the long-term effects. Population policy is central to establishing national goals and must involve the Prime Minister directly.

The most recent study of the population options for 2050 and 2100, which considered technology, resources and environmental issues, was undertaken by CSIRO Sustainable Ecosystems and sponsored by the Department of Immigration and Multicultural and Indigenous Affairs (Foran and Poldy, 2002). The purpose of the study was to increase the range and depth of insights into the effect of future population size on infrastructure and environmental issues within Australia's economy. The study brought together many issues in demography, physical science, ecology and economics, and considered three population scenarios determined by the net immigration rates. All three scenarios are considered as physically feasible. However, they carry with them a number of rewards and risks. The authors used the *Whatif* analytical platform for simulation of various options. The three scenarios are:

- **The low population scenario** (zero immigration intake per year); under this scenario the Australian population will be 20 million by 2050 and 17 million by 2100. The rewards of this scenario include smaller increase in energy usage and subsequent emissions, potentially more robust trade balance and opportunities to refurbish national infrastructure. The risks include the possibility that mature aged workers may have to work longer and harder, a potential loss in economic confidence, and the decline in rural areas and some regional cities.
- **The medium population scenario** (an immigration intake of 70 000 per year); this case will give a population of 25 million by 2050 and 32 million by 2100. This scenario results in continuing growth in a number of major cities and regions, a steady progress to a balanced population size and structure, and options to enhance the transition from the old to the new economy. The risks include the potential to stay with moderate but inadequate measures of environmental management and retention of old industries.
- **The high population scenario** (immigration intake of 0.7 percent of the population size in each year); under this scenario, Australia's population will increase to 32 million by 2050 and 50 million by 2100. The rewards of this scenario lie with a growing economy, strong export industries, and formation of potential world-sized cities. The risks lie with a potential de-coupling of the large urban agglomerations from the base of ecosystem services that support their lifestyle and function. This option requires a decision on whether to grow the current cities on their margins, or form new cities. If the later option were chosen, by 2100 the equivalent of 90 cities the size of Canberra, with a population of 310 000, would have to be located and established.

In spite of numerous studies and inquiries over the past decades, Federal Government has not adopted any declared population policy, because this is politically a highly sensitive and divisive issue.

3.3 CLIMATE

Australia features a wide range of climatic zones from tropical to sub-arctic. A large proportion of the central and western parts of the country have arid and semi-arid climatic conditions. The northern part has a tropical climate. The south-east and the south-west portions enjoy a temperate climate due to the moderating influence of the oceans that surround these parts of the continent.

The climate of eastern and northern Australia is influenced by the Southern Oscillation, which is a seesawing of atmospheric pressure between northern Australia and the Central Pacific Ocean (Australian Bureau of Statistics, 2001, pp. 7 and 8). Since the mid 1970s the Southern Oscillation phenomenon has received considerable attention worldwide (Allan *et al.*, 1996a). This interest has spread among scientists, primary producers, policy analysts, management specialists and the wider community. Considerable climatic, marine and terrestrial impacts are connected to it, particularly in the Indo-Pacific region.

The Southern Oscillation Index (SOI) defines the strength of the Southern Oscillation, which is a normalised measure of the difference in sea level atmospheric pressure between Tahiti in the Central Pacific Ocean and Darwin in northern Australia. Following the work of Troup (1965), the SOI has frequently been referred to as the Troup's SOI Index (Allen et al., 1996b). At one extreme, when the pressure is abnormally high at Darwin and abnormally low at Tahiti, SOI is negative (El Niño event). In this case, severe and widespread drought is likely to occur over eastern and northern Australia. The opposite extreme (La Niña event) is when the pressure at Darwin is abnormally low and that of Tahiti is abnormally high (positive SOI). In this case, rainfall is generally above average over eastern and northern Australia, and may cause major floods.[4]

3.3.1 Temperature

Australia is not only a dry country, but is also subject to extreme high and low temperatures. The average annual air temperature ranges from 28°C along the Kimberley coast in the extreme north of Western Australia, to 4°C in the alpine areas of south-eastern Australia (Australian Bureau of Statistics, 2002, pp. 13–15). In January, the average maximum temperature exceeds 35°C over a vast area of the interior and exceeds 40°C over appreciable areas of the north-west. In the same month, the average minimum ranges from 27°C on the north-west coast to 5°C in the alpine areas of the south-east. July is the month with the lowest average temperature in all parts of the Australian continent. The average minimum in July falls below 5°C in areas south of the tropics (away from the coast). Alpine areas record the lowest temperatures, with the July average being a minimum of −5°C. The extreme maximum and minimum temperatures recorded are 50.7°C (at Oodnadatta, South Australia on 2 January 1960) and −23.0°C (at Charlotte Pass, New South Wales, on 18 June 1994).

The analysis of the average annual temperatures over Australia for the period of 1910–99 indicates an overall warming trend during the second half of the twentieth century. However, some parts of New South Wales and Queensland experienced a slight cooling trend (−0.5 to −1°C), while the temperature rise in a large part of Western Australia exceeded 1 to 1.5°C (Australian Bureau of Statistics, 2001, p. 40).

3.3.2 Rainfall

Australia is a dry continent and a significant part of its atmospheric precipitation is rainfall. Generally, snow covers small areas of the Great Dividing Range above 1400 to 1500 m from late autumn to early spring. Similarly, in Tasmania, the elevated mountains are frequently covered with snow above 1000 m over the same seasons.

About 80 percent of Australia receives an average annual rainfall lower than 600 mm and 50 percent lower than 300 mm, with the overall average being 450 mm (Australian Bureau of Statistics, 2002). Less than 4 percent of the continent has a rainfall higher than 1200 mm per year (Figure 3.2). The area of lowest rainfall is in the vicinity of Lake Eyre in South Australia, where the average annual rainfall is about 100 mm. The region with the highest rainfall is the east coast of Queensland between Cairns and Cardwell, where Happy Valley has an average annual rainfall of 4436 mm for the 43 years from 1956 to 2000 inclusive, and Babinda an average of 4092 mm over 84 years from 1911 to 2000 inclusive. The mountainous region of western Tasmania also has a high annual rainfall with Lake Margaret having an average of 3565 mm for 76 years to 1987 inclusive. In the mountainous area of north-east Victoria and some parts of the eastern coastal slopes there are small pockets with average annual precipitation of greater than 2500 mm. In the Snowy Mountains area of New South Wales, the highest average annual precipitation exceeds 2000 mm. The rainfall pattern of Australia is strongly seasonal in character with a winter rainfall region in the south and a summer monsoonal rainfall region in the north.

An analysis of the average annual rainfall over Australia from 1900 to 1999 indicated a very weak increase in average annual rainfall during the twentieth century, although individual areas have experienced much stronger trends, both positive and negative. Positive trends have been observed in the Northern Territory, particularly in the Darwin region (with an increase of more than 300 mm) and the coastal areas of New South Wales (with increases of 200 to 300 mm). Negative trends have been identified in the coastal areas of Queensland (with decreases of more than 300 mm), the south-west corner of Western Australia (with decreases of 100 to 200 mm) and the coastal areas of South Australia, as well as a large part of Tasmania where the rainfall decreased up to 100 mm (Australian Bureau of Statistics, 2001, p. 37).

3.3.3 Evaporation

The average annual Class A pan evaporation ranges from more than 4000 mm over central Western Australia to less than 1000 mm in alpine areas of the south-east and in much

---

[4] For the annual variations of the SOI and its correlation with flood and drought during the twentieth century, visit http://www.bom.gov.au/lam/climate/levelthree/c20thc/flood.htm (visited in June 2005).

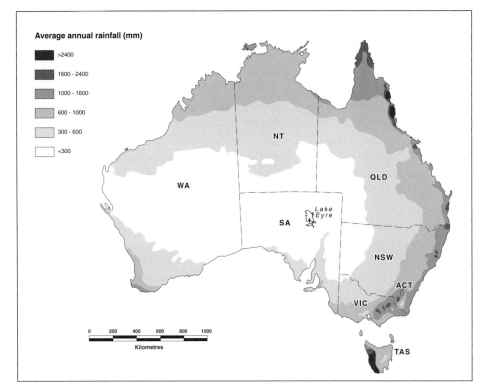

Figure 3.2 Average annual rainfall in Australia (Bureau of Meteorology website).[5]

of Tasmania (Australian Bureau of Statistics, 1988, pp. 228–30). In areas south of the tropics, average monthly evaporation follows seasonal changes in solar radiation, giving highest evaporation in December and January, and lowest in June and July. The average annual pan evaporation is below 1000 mm in the western part of Tasmania and the alpine region of south-east Australia. It gradually increases towards the north-west. Its value exceeds 2500 mm in a large part on the country and reaches values higher than 3200 mm in the northern parts of the Western Australia.[6]

Figure 3.3 shows the rainfall deficit map of Australia. It indicates that almost over the entire continent pan evaporation is higher than the average annual rainfall. The exceptions are the small areas of Tasmania, Australian Alps, some coastal areas of New South Wales, and Queensland as well as the south-western corner of Western Australia. The deficit is the highest and exceeds 3000 mm yr^{-1} in three large areas of Western Australia, central Australia and the Northern Territory.

Dunlop *et al.* (2001a, pp. 13 and 14) estimated the annual Australian transpiration (water used by plants) at the time of European settlement in 1788, to about 3.44×10^{12} m^3. This is about 10 percent higher than their estimate for 2000, which is approximately 3.10×10^{12} m^3. The reduction is due to substantial clearing of open forests and woodlands for cropping and grazing since settlement.

3.4 CLIMATE CHANGE

It is most likely that the increasing concentrations of carbon dioxide and other gases in the atmosphere are causing air temperatures to rise and the world's climate to change (see section 1.9.2). In Australia, climate changes have been estimated for the years 2030 and 2070, using climate simulation models. Following is a brief description of the simulation results summarised from Whetton and Hennessy (2001) and the CSIRO website.[7]

By 2030, annual average temperatures are expected to be 0.4 to 2°C higher over most of Australia, with slightly less warming in some coastal areas and Tasmania, and the potential for greater warming in the north-west (Figure 3.4). By 2070, annual average temperatures are expected to increase by 1.0 to 6.0°C over most of Australia with spatial variation similar to those for 2030. The range of warming is greatest in spring and less in winter. In the north-west, the greatest potential warming occurs in summer. Increases in

---

[5] Rainfall and temperature at: http://www.bom.gov.au/climate/averages/index.shtml (visited in June 2005).
[6] Evaporation at http://www.bom.gov.au/climate/averages/index.shtml (visited in June 2005).
[7] http://www.dar.csiro.au/publications/projections2001.pdf (visited in June 2005).

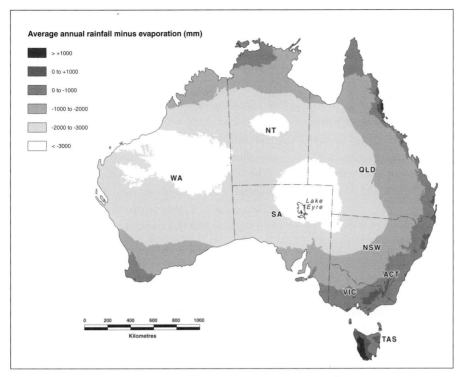

Figure 3.3 Rainfall deficit (rainfall minus pan evaporation) map of Australia.[8]

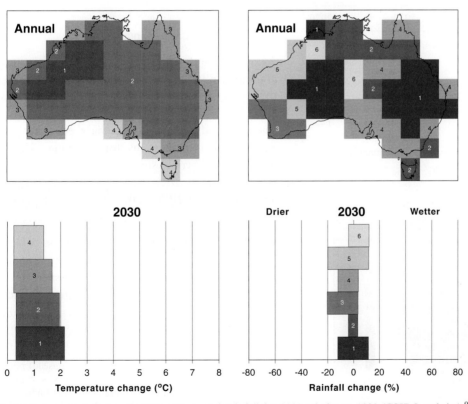

Figure 3.4 Average annual changes in temperature and rainfall for 2030 relative to 1990 (CSIRO website).[9]

[8] Produced by Janet Stein, CRES, ANU with the ANUCLIM package http://cres.anu.edu.au/outputs/anuclim.php (visited in June 2005).
[9] http://www.dar.csiro.au/publications/projections2001.pdf (visited in June 2005).

average annual temperature can lead to large changes in the frequency of extremely hot and cold days. As shown in Table 3.2, the average number of summer days over 35°C in Melbourne would increase from 8 at present to 9–12 by 2030 and 10–20 by 2070. In Perth, such hot days would rise from 15 at present to 16–22 by 2030 and 18–39 by 2070. On the other hand, locations that currently experience frequent frosty days are expected to have less cold mornings or even become frost-free by late in the century. For example, Canberra has on average 44 days of below 0°C minimum temperature. It is expected that this will reduce to 31–42 by 2030 and 6–38 by 2070.

Projected annual averages for rainfall (Figure 3.4) tend towards a decrease in the south-west, from −20% to +5% by 2030 and from −60% to +10% by 2070. In parts of the south-east and in Queensland the ranges are −10% to +5% by 2030 and −35% to +10% by 2070. In some other areas, including much of eastern Australia, projected ranges are −10% to +10% by 2030 and −35% to +35% by 2070. The ranges for the tropical north are −5% to +5% by 2030 and −10% to +10% by 2070.

Most models predict an increase in extreme daily rainfall leading to more frequent heavy rainfall events and flooding. This occurs not only where average rainfall increases but also where average rainfall decreases slightly. Reductions in extreme daily rainfall are predicted to occur only where average rainfall declines significantly.

The climate change simulations predict increases in potential evaporation across Australia, primarily as a consequence of higher temperatures.[11] The increases occur in all seasons and when annually averaged range from 0 to 8 percent per degree of global warming over most of Australia, and up to 12 percent over the eastern highlands and Tasmania. When the simulated increases in potential evaporation are considered in combination with simulated rainfall change, the overall pattern shows decreases in the moisture balance on a nationwide basis. Average decreases in the annual atmospheric moisture balance range from about 40 to 120 mm per °C of global warming (Figure 3.5). These decreases in moisture balance would mean greater moisture stress for Australia.

Table 3.2. *The average number of summer days over 35°C at capital cities for present conditions and those predicted for 2030 and 2070*

Capital city	Present	2030	2070
Hobart	1	1–2	1–4
Sydney	2	2–4	3–11
Brisbane	3	3–6	4–35
Canberra	4	6–10	7–30
Melbourne	8	9–12	10–20
Adelaide	10	11–16	13–28
Perth	15	16–22	18–39

Source: Whetton and Hennessy (2001, Table 2).

3.4.1 Climate Change Impacts

Agriculture, water resources, forests, and the energy sector are sensitive to climate changes. All natural ecosystems of the country are also vulnerable. Those particularly at risk are coral reefs, alpine ecosystems, mangroves and wetlands. The predicted climate changes have economic, social and environmental implications for Australia. The following paragraphs describe some of the potential impacts of these changes (CSIRO website).[12] Further information is available in the IPCC (2001, Chapter 12: Australia and New Zealand).

- **Water resources:** Increased water stress is likely due to higher temperatures and evaporation. Although increases in streamflow are possible in northern Australia if summer rainfall increases, decreases in streamflow seem likely for southern Australia due to

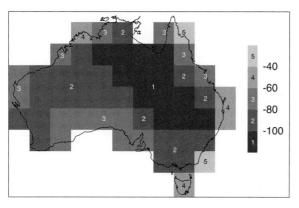

Figure 3.5 Average annual changes (mm) in atmospheric moisture balance per 1°C global average warming (CSIRO website).[10]

---

[10] http://www.dar.csiro.au/publications/projections2001.pdf (visited in June 2005).

[11] These simulated results are in contradiction with findings of Roderick and Farquhar (2004), which show that pan evaporation measurements in Australia have been declining since early 1970s.

[12] http://www.marine.csiro.au/iawg/impacts2001.pdf (visited in June 2005).

reductions in rainfall. Estimated changes in streamflow in the east-central Murray–Darling Basin ranges from 0 to −20 percent in 2030, and +5 to −45 percent in 2070. This would result in water shortages, particularly in winter rain-fed systems that are already under stress and would sharpen competition between different water users. In south-western Australia, a reduction in rainfall would adversely affect water supplies for both agriculture and urban communities. More frequent high-intensity rain in some other areas may have some benefits, contributing to groundwater recharge and filling dams, but would also increase the risk of flooding, landslides and erosion.

- **Agriculture:** In dryland farming and grazing lands, higher carbon dioxide concentrations may increase plant productivity. However, if warmer conditions are accompanied by rainfall decreases in agricultural regions, particularly in winter and spring, the benefits of higher carbon dioxide levels will be limited.

Impacts on wheat and fruit production, high rainfall pastures, and rangelands are:

(1) For warming of 1–4°C with no changes in rainfall, wheat yields would increase. However, if rainfall declines by 20 percent, yield would increase for a 1°C warming but would decline for greater warming. The positive response of wheat to higher carbon dioxide levels may come at the price of lower protein contents (9 to 15 percent reduction for a carbon dioxide level of 700 ppm).

(2) Temperate fruits need winter chilling to ensure normal bud-burst and fruit set. Warmer winters will reduce the accumulated chilling, leading to lower yields and reduced fruit quality. Stone fruit and apples in southern Australia are particularly vulnerable.

(3) Temperate pastures in high-rainfall regions are largely found in NSW and Victoria. The positive impact of elevated carbon dioxide levels and the negative impact of warming are likely to cancel each other out in this pasture zone. However, likely decreases in winter and spring rainfall in southern Australia would greatly reduce plant production, significantly constraining animal production. Rising temperatures are also likely to lower milk yield from cows.

(4) Nearly three quarters of Australia is rangeland where cattle and sheep grazing are the main land use. If rainfall decreases in southern Australia by more than 10 percent in winter and spring, then forage and animal production will be reduced, despite the benefits of an increased carbon dioxide level.

- **Dryland salinity:** Reduced rainfall in the south-west and the Murray–Darling Basin would reduce recharge to groundwater, slowing the expansion of dryland salinity problems.
- **Forestry:** A doubling of carbon dioxide with a warming of 3°C and no rainfall change, would encourage tree growth across much of southern Australia. However, a reduction in winter and spring rainfall in southern Australia, and increased fire frequency, would offset some of these benefits. The benefits will also be affected by changes in pest activity.
- **Natural ecosystems:** (1) The wetlands of the country are already under threat. Climate change and sea level rise would add to their vulnerability. The vast freshwater floodplains of northern Australia will be subject to significant saltwater intrusion because of sea level rise. (2) Less snow and a shorter snow season will affect alpine ecosystems. (3) Coral reefs would be impacted by bleaching due to warmer water temperatures and damage from tropical cyclones.

3.5 DROUGHT

Drought is a major challenge in Australia, its occurrence often coincides with El Niño events in eastern Australia. However, drought does not affect the whole continent simultaneously. Fraser (1984, pp. 566–571) provides details of about 40 droughts since the foundation of the colony in 1788 until 1982. The first reported drought in New South Wales was in 1789–91. Since the 1860s there have been ten major droughts (Table 3.3). Apart from the recent 2002–2006 drought, the drought periods of 1895–1903 (the so-called Federation Drought), 1963–68, 1982–83 and 1991–95 were the most devastating in terms of their extent and effects on primary production. Reynolds et al. (1983) provides 41 small-scale maps showing the extent of the drought affected areas from 1888 to 1982, while the major Australian drought years and the spatial variations in the intensity of El Niño induced droughts for 1954, 1982–83 and 1991–95 are available at the website of the Bureau of Meteorology.[13]

During the Federation Drought (1895–1903), a significant part of Australia and in particular, the coastal area of Queensland, inland areas of New South Wales, South Australia and central Australia were affected. This was probably Australia's worst drought in terms of severity and

---

[13] http://www.bom.gov.au/climate/drought/livedrought.shtml (visited in June 2005).

Table 3.3. *Effects of major droughts from 1864 to 1995*

Year	Effect	Year	Effect
1864–66	All states affected except Tasmania.	1939–45	Loss of nearly 30 million sheep between 1942 and 1945.
1880–86	Southern and eastern states affected.	1963–68	Widespread drought. Also longest drought in arid central Australia: 1958–67. The last two years saw a 40 percent drop in wheat harvest, a loss of 20 million sheep, and a decrease in farm income of $300–$500 million.
1895–1903	Sheep numbers halved and more than 40 percent loss of cattle. Most devastating drought in terms of stock losses.	1972–73	Mainly in eastern Australia.
1911–16	Loss of 19 million sheep and 2 million cattle.	1982–83	Total loss estimated in excess of $3000 million. Most intense drought in terms of vast areas affected.
1918–20	Only parts of Western Australia were free from drought.	1991–95	Average production by rural industries fell about 10 percent, resulting in a possible $5000 million cost to the Australian economy, $590 million drought relief provided by the Federal Government between September 1992 and December 1995.

Source: Bureau of Meteorology.[14]

extent. Sheep numbers, which had reached more than 100 million, were reduced by approximately half and cattle numbers by more than 40 percent. Average wheat harvest dropped to only 210 kg ha^{-1} in 1902 (Australian Bureau of Statistics, 1988, p. 622).

The 1982–83 drought was most severe in the south-eastern part of the continent. Virtually all of Victoria and southern New South Wales had registered record low rainfall for the eleven months from April 1982 to February 1983. Also, much of the settled areas of South Australia recorded their lowest ever rainfall for the ten months from May 1982 to February 1983 inclusive. Total drought losses were estimated by the Australian Government to exceed $3000 million (Australian Bureau of Statistics, 1988, p. 622).

The 1991–95 drought resulted in a possible $5 billion cost to Australia's economy and allocation of $590 million drought relief by the Federal Government (Table 3.3). The most recent and severe drought of the past 100 years occurred in 2002–2003,[15] which affected large parts of New South Wales, Queensland, Victoria and Western Australia. It reduced the economic growth by about $7 billion and the Federal Government committed more than $1 billion in assistance to farm families (Botterill and Fisher, 2003).

Drought has a substantial adverse effect on farm enterprises in terms of production, profitability and viability, resulting in great hardship for individuals and communities. To alleviate these hardships, both the federal and state governments provide drought assistance, including: Drought Relief Payment; Drought Investment Allowance; and funding for Research and Development. The Federal Government expenditure related to drought for the period of 1978 to 1996 was $1.0 billion, with a 1996 net present value of $1.6 billion, while the states' expenditure from 1991–92 to January 1996 was $190 million (Munro and Lembit, 1997, Tables 2 and 3).

Further information regarding drought and drought policy in Australia is available in numerous publications, including: Foley (1957); Reynolds *et al.* (1983); Hamer (1985); Camm *et al.* (1987, pp. 20–22), Wilhite (1993), Daly (1994), O'Meagher *et al.* (1998); Smith (1998); White *et al.* (1998), O'Meagher *et al.* (2000), Botterill (2003), and Botterill and Wilhite (2005).

3.6 FLOOD

The most common and significant threats to the social and economic well-being of flood-prone communities arise from heavy rainfall and storm surge flooding (Emergency

---

[14] http://www.bom.gov.au/lam/climate/levelthree/cpeople/drought4.htm (visited in June 2005).
[15] This drought continued in 2004, 2005 and 2006.

Management Australia, 1999, p. 2). The early European settlers had not been long in New South Wales before they became aware of the flooding hazard. In 1789, explorers of the Hawkesbury River saw flood debris 7 to 12 m above the existing water level and in 1806, flooding of the same river almost reduced the colony to starvation (Australian Geographic Society, 1988, p. 1286).

Widespread flooding may occur throughout Australia. However, it has a higher incidence in the north and in the eastern coastal areas. It is most damaging economically along the streams flowing eastward from uplands to the seaboard of Queensland and New South Wales. These floods are destructive in the more densely populated coastal river valleys of New South Wales such as Tweed, Richmond, Clarence, Macleay, Hunter and Nepean-Hawkesbury (see Figure 5.1).

Fraser (1984, pp. 572–578) describes more than 70 floods, storms and cyclones for the period of March 1799 to March 1983. In the twentieth century, major floods occurred in 1916, 1917, 1950, 1954–56 and 1973–75. More recent floods occurred in 1988–89 and 1999. Flood years of the twentieth century coincide with the strongly positive SOI during La Niña events.[16] The year-long flood of July 1973 to June 1974 was extensive in Queensland, New South Wales, Victoria and Tasmania. During the 1974 flood, almost one third of New South Wales and Queensland were inundated, with Brisbane being the hardest hit city. In Brisbane, with a population of about 911 000, flooding occurred during the Australian Day holiday weekend (25–29 January 1974). At least 6700 homes were partially or totally flooded in the Brisbane metropolitan area and floodwaters entered the gardens of about 6000 other houses. In the nearby city of Ipswich, with a population of 65 000, about 1800 residential or commercial premises were partially or totally inundated. In the Brisbane–Ipswich area, 14 people lost their lives and the total damage was estimated to be about $200 million (Bureau of Meteorology, 1974).

The Fitzroy and Burdekin river basins of Queensland flood during the summer-wet seasons. Much of the run-off due to heavy rain in central and western Queensland flows southward through the normally dry channels of the network of rivers draining the interior lowlands into Lake Eyre. This widespread rain may cause floods over an extensive area, which generally seep away or evaporate. Occasionally, however, floodwaters reach the lake in significant volume, and life returns to the whole area with the appearance of numerous fish, wildlife and native vegetation. The Condamine River and other northern tributaries of the Darling River also carry large volumes of floodwater to the south, and extensive flooding occurs along their courses in sparsely populated areas. Floods also occur at irregular intervals in the Murray–Murrumbidgee system of New South Wales and Victoria, the coastal streams of southern Victoria and the north coast streams of Tasmania.

Large numbers of people live in the flood-prone areas in eastern Australia, and the value of agricultural, industrial, commercial, residential and public assets, which are at risk from floodwaters, is very large. The average annual cost of floods in Australia has been estimated to be at least $350 million, and a minimum of 200 000 homes and offices are prone to 1 in 100 year flood events (Standing Committee on Agricultural and Resource Management, 2000, p. xi). The two states of New South Wales and Queensland account for about 80 percent of the annual cost in approximately equal proportions (Australian Water Resources Council, 1992, p. xv). Nearly 46 percent of the average annual flood bill is caused by mainstream flood damage to urban properties. Stormwater damage to urban properties accounts for a further 10 percent of the bill, with the remainder incurred by farming enterprises (26 percent) and public infrastructure in rural areas (17 percent). Further information regarding floods and its mitigation is available in Australian Geographic Society (1988, pp. 1286–1289), Smith (1988, pp. 213–243), and Smith (2002).

3.7 SOIL RESOURCES

In recent geological times, there has been a general absence across the Australian continent of major processes that renew soils such as mountain building, volcanic activities and glaciation. The land surface across many parts of Australia is therefore ancient and its soils are strongly weathered and infertile. The first comprehensive soil map of Australia was published by Northcote *et al.* (1960–68). This set of 10 sheets at the scale of 1:2 000 000 contained an explanatory publication for each sheet. In 1991, these sheets were digitised by the National Resource Information Centre (NRIC), now the Science Secretariat, within the Bureau of Rural Sciences (BRS), and are now the Digital Atlas of Australian Soils. The current version of the Digital Atlas was released on the Internet on 2nd November 1998. Figure 3.6 shows a simplified soil map of Australia.

Australian soils in general have many distinctive features. Surface layers have low contents of organic matter and are often poorly structured. Subsurface layers with a sharp increase in clay content are widespread (Kurosol, Chromosol and Sodosol soil orders) and can restrict drainage and root

---

[16] http://www.bom.gov.au/lam/climate/levelthree/c20thc/flood.htm (visited in June 2005).

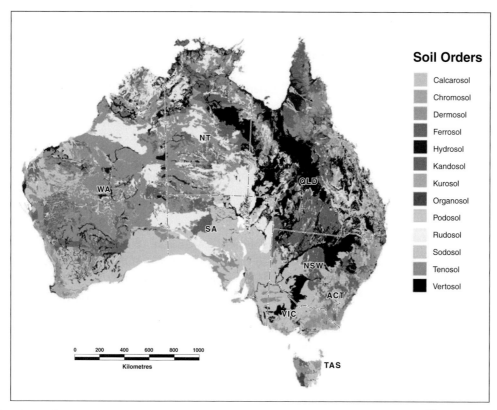

Figure 3.6 Generalised soil map of Australia (website of the Australian Natural Resources Atlas).[17]

growth. In these soils, bleached layers with very low nutrient levels are also common. Soils affected by salinity, sodicity and acidity, either now or in earlier geological times, cover large portions of the arable lands of the continent and they have various nutrient and physical limitations. Australia has also very large areas of cracking clay soils (Vertosols). These are relatively fertile but have physical limitations. Soils formed in aeolian sands (Rudosols and Tenosols) fringe the southern cropping lands, but are more extensive in the arid zone. Finally, the remaining ancient land surfaces, particularly in northern Australia, have very deep and strongly weathered soils (Kandosols) with very low level of nutrients (McKenzie et al., 2001).

The Australian Soil Resources Information System (National Land & Water Resources Audit 2001c, pp. A35–A73) provides information on various soil characteristics such as depth, phosphorus and nitrogen contents, organic carbon content, bulk density, available water capacity, hydraulic conductivity, pH, and soil erodibility, and for each soil character, a distribution map across Australia is provided.

Dunlop et al. (1999) provide a continent-wide assessment of the quality of Australia's soils for broadacre, rainfed agriculture. Their assessment is based on the Digital Atlas of Australian Soils and a database of 7000 soil profiles held by CSIRO Land and Water. The quality scores are derived from a number of interpreted soil variables that are important for successful plant growth. The variables are:

(1) hydraulic conductivity of the A and B soil horizons;
(2) bulk density of the A and B horizons;
(3) depth of the solum (A and B horizons together);
(4) gross nutrient levels in the soil;
(5) proportion of gravel in the soil; and
(6) soil texture.

Also, information on climate and terrain are included to give a more complete picture of rainfed cropping potential. It should be noted that although this information is extensive, there are a number of limitations to the spatial coverage of the soil datasets, hence interpretation should be regarded as indicative pending more detailed soil mapping. Figure 3.7 shows the quality of dominant soils of Australia.[18]

---

[17] http://audit.ea.gov.au/ANRA/land/land_frame.cfm?region_type=AUS&region_code=AUS&info=soil_ overview (visited in June 2005).

[18] In the Northcote soil classification scheme, the description of each mapping unit includes a dominant and several sub-dominant soil types. Soil quality maps were constructed based on the interpreted properties of the dominant soil and up to five sub-dominant soils in each mapping unit.

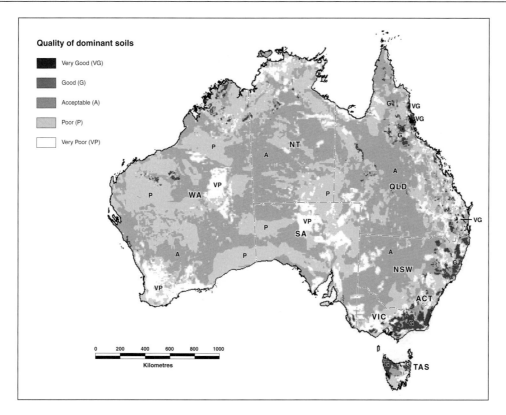

Figure 3.7 Quality of dominant soils of Australia (Dunlop et al., 1999).

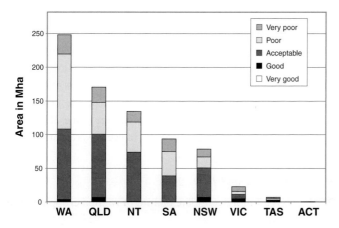

Figure 3.8 Area of land in each soil quality class for various states and territories (Dunlop et al., 1999).[19]

Only 0.6 Mha are very good and as little as 23.5 Mha are of good quality. The majority of soils are acceptable (361.3 Mha) or poor (262.3 Mha), while 106 Mha are very poor. The main cropping regions of Australia, in the east and south of the country, contain most of the good and very good quality soils, but also includes large regions of acceptable and very poor quality soils. The semi-arid and tropical rangelands, containing about two-thirds of Australia's agricultural land, are mainly poor or acceptable soils. The majority of the good and very good soils are in higher rainfall regions in the east, however much of the better rated soils are in hilly or mountainous terrain where slopes preclude agriculture.

Figure 3.8 shows the extent of quality soils in various states and territories. The extent of very good soils is limited to 0.6 Mha, which is mainly in Queensland. However, because of scale restrictions it is not shown on Figure 3.8. Good quality soils are mainly in New South Wales (6.7 Mha),

[19] Very good soils can't be seen because of the figure's scale.

Queensland (6 Mha), Victoria (4.5 Mha), and Western Australia (3.4 Mha). The quality of most soils in the various states and territories ranges from acceptable to very poor (Figure 3.8). In the case of Australian Capital Territory, 0.07 Mha are of good quality, 0.07 Mha are acceptable, 0.09 Mha are poor and 0.01 Mha are very poor.

3.8 AGRICULTURAL LAND USE

Agriculture is the most extensive form of land use in Australia in spite of its harsh environment. The estimated total area of agricultural activities was 456 Mha at 30 June 2000, representing about 59 percent of the total land area of the country (Australian Bureau of Statistics, 2002, p. 452). This consisted of 24 Mha for crops, 24 Mha for sown pasture and grasses, and the remaining 408 Mha was arid or rugged land held under grazing licences or fallow land. Distribution of the agricultural land in Australian states and the Northern Territory is shown in Figure 3.9. The rest of the Australian land area is unoccupied land, mainly desert in western and central Australia, Aboriginal land reserves, forests, mining leases, national parks and urban areas.

Livestock grazing uses the largest area of land use in Australian agriculture. Grazing has led to the replacement of large areas of native vegetation by introduced grasses in higher rainfall and irrigated areas.

3.8.1 Irrigated Land

Irrigated land in 1999–2000 occupied 2.3 MhaL, which has increased by 0.26 Mha from 1996–97 (Table 3.4). Major increases occurred for cotton production (102 000 ha), and grapes (43 400 ha), while the area of land used for rice decreased by 25 200 ha. The largest area of irrigated land is in New South Wales (925 000 ha), followed by Victoria and Queensland (Figure 3.10).

Although the reduction in area of land used for rice (a high water use and low return crop) is a positive development, the increased area of land used for pasture and grains is problematic. Water reforms in Australia were meant to move irrigation water use from low to high value crops. The huge increase of 62 percent in irrigated grapes area could be viewed as a success of this policy. However, many small and medium sized grape growers are in financial trouble. While a decade ago there were 540 grape producers, currently 1400 producers are competing for the same market. Growers are finding that their products were either surplus to customer's needs or attracted only a portion of the price paid several years ago. During the 2002 vintage some 250 000 tonnes of grapes or 15 percent of the harvest were not picked or were

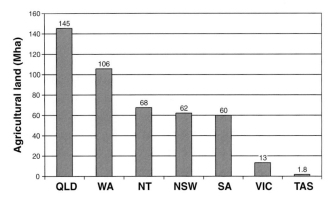

Figure 3.9 Extent of the agricultural land in various states and the Northern Territory in 1999–2000 (Australian Bureau of Statistics, 2002, p. 452).

Table 3.4. *The extent and changes in the irrigated land between 1996–97 and 1999–2000*

Crop	1996–97[a] (ha)	1999–2000[b] (ha)	Changes (ha)	(%)
Pasture, grains and others	1 174 700	1 260 900	+86 200	+7.3
Cotton	315 000	417 000	+102 000	+32.4
Sugarcane	173 200	201 900	+28 700	+16.6
Rice	152 400	127 200	−25 200	−16.5
Grapes	70 200	113 600	+43 400	+61.8
Vegetables	88 800	99 200	+10 400	+11.7
Fruit	82 300	96 600	+14 300	+17.4
Total	2 056 600	2 316 400	+259 800	+12.6

Source: [a] National Land & Water Resources Audit (2001a, Table 26); [b] Australian Bureau of Statistics (2002, Table 16.3 with minor corrections).

sold below the production cost (*Weekend Australian*, 6–7 July 2002, p. 29).

3.8.2 Share of Agriculture in Australian Economy

Agriculture supplies most of Australia's domestic requirements for meat, fresh fruit, vegetables, sugar, rice, wine and dairy products. It is also one of the main export sectors of the Australian economy. In 1999–2000, the gross value of farm products was $30.2 billion. In terms of employment, the agricultural industry and related services employed 409 200 people in 2000 (Australian Bureau of Statistics, 2002, pp. 453–455). In spite of these statistics, because of the rapid growth of the other sectors of the Australian economy, the contribution of agricultural products to export and the GDP have continuously declined over the past five decades (Figure 3.11). In line with this, the percentage of the workforce in the agricultural sector has also declined. The social landscape of Australian agriculture has changed with a decline in number of farms and farmers, aging of the rural population, and increased dependence of farmers on off-farm incomes (National Land & Water Resources Audit, 2002a, pp. 74–81).

3.9 WATER RESOURCES

3.9.1 Surface Water

Australia is the driest inhabited continent in terms of the ratio of run-off to rainfall. The average rainfall over the continent is 450 mm but this results in only 52 mm of run-off. Only about 11 percent of the rainfall in Australia emerges as run-off, and with the exception of a minor fraction of less than 1 percent which recharges the aquifers, almost all the remaining 89 percent returns to the atmosphere by evapo-transpiration (Brown, 1983). In North America, Asia, South America and Europe, run-off values are 45, 44, 43, and 36 percent of the rainfall respectively (Gleick, 1993, Table 2.3). Australia has a contribution of about 1 percent in the world's river run-off, while the major contributors are: Asia, 31 percent; South America, 25 percent; North America, 17 percent; Africa 10 percent; and Europe, 7 percent (Gleick, 1993, Table 2.4).

Australia is divided into 12 drainage divisions (Figure 3.12 and Table 3.5). The longest river system is the Murray–Darling, which drains parts of Queensland, the major part of New South Wales, a large part of Victoria and part of South Australia. The River Murray is about 2520 km in length and the Darling and Upper Darling together are also just over 2520 km long. The total continental run-off of $387 \times 10^9 \, m^3$, which is slightly higher than $19\,300 \, m^3$ per head of population, is unevenly distributed across Australia. About 88 percent of the run-off is contained in the five drainage divisions (I, II, III, VIII and IX) along the north and east coasts of the mainland and Tasmania. Approximately 25 percent of surface run-off can be diverted on a sustained basis into conventional water supply systems,

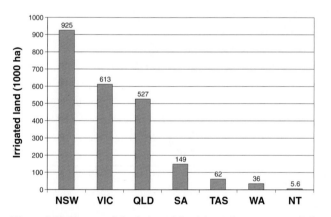

Figure 3.10 Extent of the irrigated land in various states and the Northern Territory in 1999–2000 (Australian Bureau of Statistics, 2002, Table 16.3).

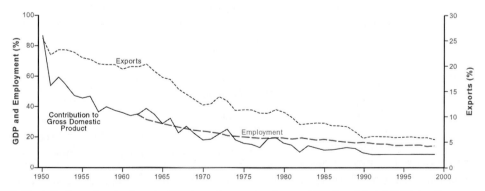

Figure 3.11 Contribution of agriculture to GDP, employment and export (National Land & Water Resources Audit, 2002a, p. 32).

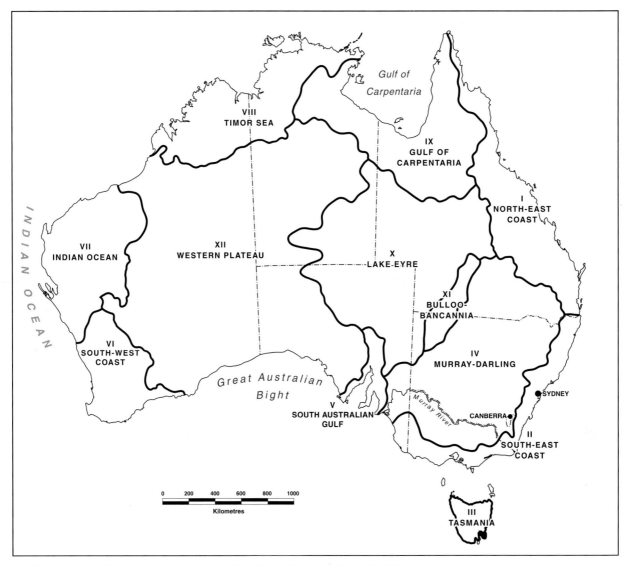

Figure 3.12 Drainage divisions of Australia (Australian Water Resources Council, 1987).

using existing storages and potential dam sites (Australian Water Resources Council, 1987). Currently, less than 5 percent of these resources have been developed. Some 66 percent or $12 \times 10^9 \, m^3$ of the developed resources are in the Murray–Darling Basin (Drainage division IV). Approximately 98 percent of the divertible surface water resources of the country have a total dissolved solids content of less than $1500 \, mg \, L^{-1}$, making it suitable for both human consumption and agricultural use.

In 2002, Australia had 499 large dams with heights of more than 10 m and reservoir capacities greater than $10^6 \, m^3$. These dams had a combined capacity of approximately $93 \times 10^9 \, m^3$, and were developed mainly for urban, irrigation, industrial and hydro-electric power users. Tasmania has the largest storage capacity of $32 \times 10^9 \, m^3$, followed by New South Wales,[20] Victoria, Western Australia, and Queensland (Figure 3.13). The remaining capacity was in the Northern Territory, and South Australia. Australia's several million farm-dams account for an estimated 9 percent of the total water stored (National Land & Water Resources Audit, 2001a, p. 27).

3.9.2 Groundwater

Australia has been divided into 61 groundwater provinces (Figure 3.14). Each province has a broad uniformity of hydrogeological conditions and water-bearing

---

[20] The NSW's dam statistics includes three dams in the Australian Capital Territory with a total capacity of $89.7 \times 10^6 \, m^3$.

Table 3.5. *Surface water resources of Australia (in $10^9 m^3$)*

Drainage division		Area (km²)	Mean annual run-off	Percent mean annual run-off	Mean annual outflow	Developed volume
North-east Coast	I	451 000	73.44	19.0	69.58	3.182
South-east Coast	II	274 000	42.39	10.9	40.37	1.825
Tasmania	III	68 200	45.58	11.8	45.34	0.451
Murray–Darling	IV	1 060 000	23.85	6.2	5.75	12.051
South Australian Gulf	V	82 300	0.95	0.2	0.79	0.144
South-west Coast	VI	315 000	6.79	1.8	5.93	0.373
Indian Ocean	VII	519 000	4.61	1.2	3.48	0.012
Timor Sea	VIII	547 000	83.32	21.5	81.46	0.048
Gulf of Carpentaria	IX	641 000	95.61	24.7	96.07	0.052
Lake Eyre	X	1 170 000	8.63	2.2	NA	0.007
Bulloo–Bancannia	XI	101 000	0.55	0.1	–	<0.001
Western Plateau	XII	2 450 000	1.49	0.4	NA	0.001
Total		7 680 000	387.21	100		18.147

Source: National Land & Water Resources Audit (2001a, Table 3).

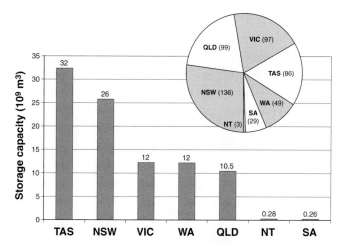

Figure 3.13 Number of dams higher than 10 m and their storage capacities in Australian states and the Northern Territory (Register of Large Dams in Australia).[21]

characteristics. They are identified as either predominantly sedimentary or fractured rocks (Australian Water Resources Council, 1987).

Australia has an estimated $25.8 \times 10^9 m^3$ of groundwater with a total dissolved solids (TDS) content of less than $5000 mg L^{-1}$ that can be extracted on a sustainable basis (Figure 3.15). Northern Territory has $5.8 \times 10^9 m^3$ of fresh groundwater with a TDS of less than $500 mg L^{-1}$. It is followed by Tasmania, Queensland, New South Wales, Western Australia, Victoria and the Australian Capital Territory (Figure 3.16). Australia has one of the world's largest aquifer systems. The Great Artesian Basin (Figure 3.14 Province 55 S) has an area of 1.7 million km² or 22 percent of Australia and is about 3000 m thick. Each year, the Basin supplies $586 \times 10^6 m^3$ of water for a variety of uses such as: pastoral, 93.5 percent; oil and gas, 4.3 percent; and mining industries 2.2 percent (Habermehl, 2001).

3.9.3 Water Use

In 1996–97, the annual water use in Australia was about $24 \times 10^9 m^3$ (Table 3.6), of which 79 percent was supplied by surface water (73 percent by rivers and 6 percent by harvesting the overland flows) and 21 percent by groundwater (National Land & Water Resources Audit, 2001a). Although the contribution of groundwater to total water use seems relatively small, many areas are totally or at least heavily dependent on groundwater as a source of water supply. These areas occur mainly in the arid or semi-arid zones, but also in temperate and tropical zones such as Perth metropolitan area and the coastal plains of Queensland.

Water use has steadily increased over the past few decades. Table 3.7 shows the changes in mean annual surface water use between 1983–84 and 1996–97. During this 13-year period surface water use increased by 59 percent. The greatest percentage increases were in Tasmania (173 percent) and Queensland (145 percent). By volume, New South Wales had the largest increase in water use of about $3 \times 10^9 m^3$.

Table 3.8 shows that irrigation is the major user of surface water resources (80.3 percent), followed by urban/industrial

[21] http://www.ancold.org.au/dam_register.html (visited in June 2005).

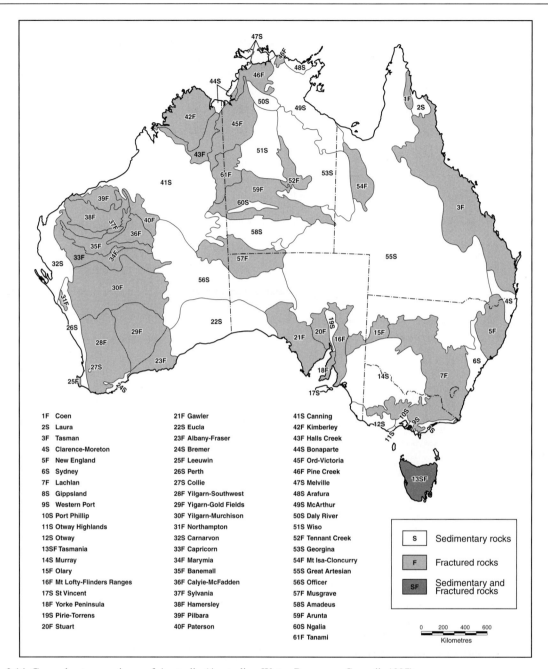

Figure 3.14 Groundwater provinces of Australia (Australian Water Resources Council, 1987).

(17.3 percent) and others (2.4 percent). The three states using the major part of the irrigation water are: New South Wales, Victoria and Queensland respectively. Irrigated pasture, livestock, and grain (excluding rice) use about $8.8 \times 10^9 \, m^3$ of water or 57 percent of total water use for irrigation (National Land & Water Audit, 2001a, Table 26).

In terms of irrigation water use per hectare, rice was the highest user, followed by grapes, fruit, pasture, vegetables, sugarcane, and cotton (Figure 3.17).

The gross value from irrigated agriculture was $7.25 billion for 1996–97. The highest financial return per hectare came from vegetables ($12 604 per ha), followed by fruit, grapes, cotton, sugarcane, pasture and grains, and rice (Figure 3.18). The financial return on water use was the highest for vegetables ($1762 per $10^3 \, m^3$), followed by fruits, grapes, cotton, sugarcane, pasture and grain, and rice (Figure 3.19).

Groundwater use has also increased dramatically between 1983–84 and 1996–97 by 88 percent (Table 3.9).

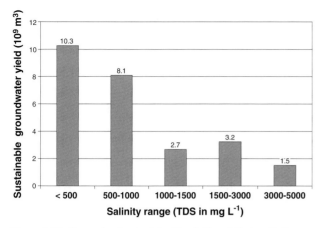

Figure 3.15 Groundwater sustainable yield and their salinity ranges (National Land & Water Resources Audit, 2001a, Table 12).

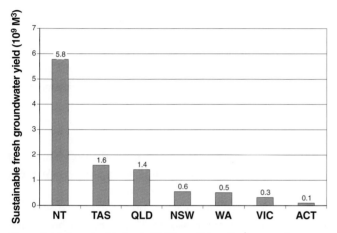

Figure 3.16 Sustainable fresh (TDS < 500 mg L^{-1}) groundwater yield of the Australian states and territories (National Land & Water Resources Audit, 2001a, Table 12).

Table 3.6. *Surface water and groundwater use in Australia for 1996–97 (in 10^6 m^3)*

States and territories	Surface water	Ground water	Ratio of surface water to groundwater
New South Wales	9000	1008	9.0
Victoria	5166	622	8.3
Queensland	2969	1622	1.8
Western Australia	658	1138	0.6
South Australia	746	419	1.8
Tasmania	451	20	22.6
Northern Territory	51	128	0.4
Australian Capital Territory	68	5	13.6
Sub-total	19 109	4962	3.9
Total		24 071	–

Source: National Land & Water Resources Audit (2001a, Table 13).

Table 3.7. *Change in average annual surface water use between 1983–84 and 1996–97 (in 10^6 m^3)*

States and territories	Surface water use in: 1983–84	Surface water use in: 1996–97	Increase in surface water use (%)
New South Wales	5932	9000	52
Victoria	3714	5166	39
Queensland	1209	2969	145
Western Australia	461	658	43
South Australia	498	746	50
Tasmania	165	451	173
Northern Territory	29	51	76
Australian Capital Territory	NA	68	–
Total	12 008	19 109	59

Source: National Land & Water Resources Audit (2001a, Table 19).

Table 3.8. *Annual surface water use by major user categories in 1996–97 (in 10^6 m^3)*

States and territories	Irrigation	Urban/Industrial	Others	Total
New South Wales	8000	900	100	9 000
Victoria	4021	860	285	5 166
Queensland	2162	787	20	2 969
Western Australia	430	206	22	658
South Australia	465	269	12	746
Tasmania	266	179	5	450
Northern Territory	6	39	6	51
Australian Capital Territory	4	63	1	68
Total	15 354	3 303	451	19 108

Source: National Land & Water Audit (2001a, Table 21).

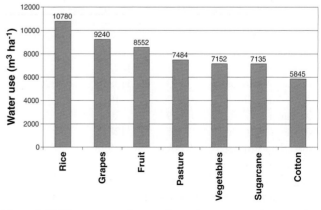

Figure 3.17 Water use per hectare for various crops and pasture in 1996–97 (National Land & Water Resources Audit, 2001a, Table 26).

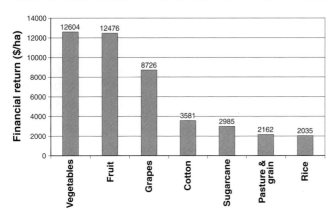

Figure 3.18 Financial return in terms of $ per hectare for various crops and pasture (National Land & Water Resources Audit, 2001a, Table 26).

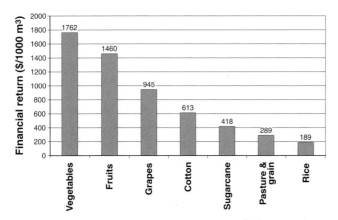

Figure 3.19 Financial return in terms of $ per $10^3 \, m^3$ of water used for various crops and pasture (National Land & Water Resources Audit, 2001a, Table 26).

Table 3.9. *Change in mean annual groundwater use between 1983–84 and 1996–97 (in $10^6 \, m^3$)*

States and territories	Groundwater use in: 1983–84	1996–97	Percent change in groundwater use
New South Wales	318	1008	217
Victoria	206	622	202
Queensland	1121	1622	45
Western Australia	373	1138	205
South Australia	542	419	−22
Tasmania	9	20	122
Northern Territory	65	128	97
Australian Capital Territory	NA	5	—
Total	2634	4962	88

Source: National Land & Water Resources Audit (2001a, Table 22).

In New South Wales, Victoria and Western Australia it increased by more than 200 percent over this 13-year period. In volumetric terms, groundwater use in Western Australia increased by about $765 \times 10^6 \, m^3$ between 1983–84 and 1996–97, followed by New South Wales ($690 \times 10^6 \, m^3$) and Victoria ($416 \times 10^6 \, m^3$). Part of this increase in New South Wales and Victoria was due to the 1995 cap on surface water diversion in the Murray–Darling Basin (see Section 3.11.1).

In 1996–97, out of the $4962 \times 10^6 \, m^3$ of total groundwater extracted, $2582 \times 10^6 \, m^3$ was used for irrigation, followed by $1451 \times 10^6 \, m^3$ for urban/industrial and $919 \times 10^6 \, m^3$ for other rural uses (National Land & Water Resources Audit, 2001a, Table 23).

3.9.4 Price of Water

In Australia, the price of water for irrigation is low, because in the past low priced water was seen as an investment for promoting inland settlement and regional development. In 2000, it was less than $0.05 per m^3 in most large irrigation districts of New South Wales and Victoria (Osborne and Dunn, 2004, Figure 121). In small irrigation areas such as south-east Tasmania ($0.2 per m^3), Pioneer Valley in Queensland ($0.7 per m^3) and in Carnarvon in Western Australia ($0.8 per m^3) prices were higher.

At 30 June 2001, the average price for treated water in major towns and cities was $0.73 per m^3 (Osborne and Dunn, 2004, Figure 120). This varied from $0.32 per m^3 in Shepparton to $0.93 per m^3 in Sydney. In Western Europe, price of water is generally much higher. For example it is $2.96 per m^3 in Germany, $2.86 per m^3 in Denmark and $2.05 per m^3 in the United Kingdom (Osborne and Dunn, 2004, p. 148).

Some jurisdictions in Australia charge a water abstraction levy to cover the environmental cost of water abstraction. For example in 2005, in the Australian Capital Territory this is currently $0.25 per m^3, and the water price for annual usage of 100–300 m^3 was $1.14 per m^3, while for excess of 300 m^3 it was $1.53 per m^3.

3.10 ENVIRONMENTAL DEGRADATION

In Australia, ownership and management of land and water resources is vested in the states and territories. Approximately, for 200 years since European settlement, development of land and water resources of the country took place without any major national coordination or consideration of environmental impacts. Large tracts of land were cleared and used for agricultural production. These developments caused: large-scale environmental degradation such as irrigated and dryland salinity, erosion, and soil acidification; deterioration of water quality through increased salinity, turbidity, and nutrient loads; and loss of biodiversity. Most of this damage was caused by a lack of understanding of the uniquely different and fragile environment of the Australian continent. It is only in the past three decades that coordinated attempts have been made to understand and address the consequences of these poorly planned developments. In the following sections, some of these issues will be discussed. Readers are referred to the *Australia State of the Environment 2001* (Australian State of the Environment Committee, 2001) for further information.

3.10.1 Land Clearing

The extent of land clearing in Australia and its subsequent impacts on environmental degradation is one of the most striking features of the country's development. Before European settlement, forest and woodlands covered 33 percent of the country or about 250 Mha. It has been estimated that 50 percent of the original tall and medium forests and 35 percent of the woodlands have been cleared or severely modified (Prinsley, 1991). In the Murray–Darling Basin alone, perhaps as many as 20 billion trees have been removed (Nicoll, 1994). Another estimate indicates that in 1788, Australia had about 60 billion trees taller than 10 m. Since then, 20 billion trees have been axed, sawn, ringbarked, poisoned, burnt and bulldozed (Beale and Fray, 1990, p. 33). This means a loss of about 1000 trees for each man, woman and child alive in Australia today. Land clearing is still taking place, although its annual rate has decreased. The average annual rate of land clearing decreased from the 1970s to the early 1990s, but increased in the late 1990s, mainly due to an increased rate of clearing in Queensland (Table 3.10).

To reduce the extent of land clearing and protect the native vegetation on private land, Australian states have progressively introduced regulations. Bartel (2003a, b) compares the administration of land clearing legislation in different parts of Australia, and in particular for NSW.

3.10.2 Salinity

Widespread dryland salinisation is one of the consequences of land clearing. Replacement of deep-rooted native vegetation with grass species has increased recharge to the aquifer systems and raised watertables close to the land surface. Once the watertable reaches the critical depth of 2 to 3 m from the soil surface, appreciable upward movement of groundwater takes place via capillary rise. Evaporation of the water accumulates salt at the land surface. In the irrigated areas, watertables rise due to excessive recharge caused by application of excess irrigation water and lack of drainage facilities (see Ghassemi *et al.*, 1995, for description of salinisation process and global extent of land salinisation).

The National Land & Water Resources Audit (2001b) provides the most recent estimate of the extent of dryland salinisation in Australia, and its expansion by the year 2050 if the current trend continues. These estimates are:

- Approximately 5.7 Mha are at risk or already affected by dryland salinity. By the year 2050 this area will increase three-fold to 17 Mha.
- Some 20 000 km of major roads and 1600 km of railways are at high risk. These could increase to 52 000 km and 3600 km respectively by the year 2050.

Table 3.10. *Annual land clearing rates (ha per year) from 1971 to 1999*

State	1971–1980	1981–1990	1991–1995	1996–1999
Queensland	886 257	350 791	289 000	382 500
New South Wales	428 151	52 232	19 120	30 000
Western Australia	92 464	92 562	21 150	3 145
Victoria	21 200	10 766	2 450	2 450
Northern Territory	21 090	12 843	3 320	3 320
Tasmania	11 817	2 413	940	940
South Australia	4 171	28 797	1 370	2 088
Australian Capital Territory	–	163	–	–
Total	1 465 150	550 567	337 350	424 443

Source: Australian Greenhouse Office (2001, Table 7).

- Up to 20 000 km of streams could be significantly salt affected.
- About 630 000 ha of the remnant native vegetation and associated ecosystems are currently at risk. These areas are projected to increase to 2 Mha over the next 50 years.
- Over 200 towns could suffer damage to infrastructure and community assets by 2050.

Stream salinisation is one of the largest effects of land salinisation. Jolly *et al.* (1997) and Walker *et al.* (1998) provide historical trends in stream salinity and catchment salt balances for the Murray–Darling Basin, while the Murray–Darling Basin Ministerial Council (1999a) provides a 100-year perspective for river and land salinisation in the Basin. For Western Australia, a review of land and stream salinisation is available in Ghassemi *et al.* (1995, pp. 154–176).

3.10.3 Acid Sulphate Soils

Million hectares of Holocene-age (10 000 years BC) coastal floodplain lowlands, swamps and wetlands throughout the world have developed on sediments that contain iron sulphide minerals. These sediments, predominantly deposited during the last major sea-level rise, are known as acid sulphate soils (Dent, 1986). Australia has an estimated 5 Mha of coastal acid sulphate soils under a variety of land-uses. In Australia, recognition of their importance has lagged behind the rest of the world. The first Australian maps, showing their widespread distribution, were only published in 1995.

Sulphide minerals pose no problems provided the sediments remain beneath the watertable. When the watertable falls below the sulphide layer, either naturally or by drainage works, sulphides oxidise producing sulphuric acid. Acid produced in the soil-water attacks soil minerals, dissolving aluminium, iron, silica, manganese and other species. Rain causes the export of acidic groundwater into streams. Drainage and flood mitigation schemes have promoted the partial acidification of many coastal streams. Thousands of tonnes of sulphuric acid are discharged annually into coastal streams (White *et al.*, 1997). The ecological consequences of soil acidification and export can be severe. Introduced crops and pasture, which are not acid-tolerant die or are stunted, leaving exposed, bare acid scalds, which are strongly acidified, right to the surface.

Acidic waters flow into streams causing fish kills, fish diseases and dramatic changes in aquatic communities. Estuaries can be rendered sterile for months at a time. Aquaculture and estuarine fishing is adversely affected and aquaculture (particularly oysters) can be severely impacted. Once the process has started, it is difficult to treat because of the large costs and complex social, institutional, legal and political factors. However, significant progress has been made in NSW under the inter-governmental and industry based Acid Sulphate Soil Management Advisory Committee. Australia is one of the few countries in the world with an Acid Sulphate Soil National Strategy.

3.10.4 Other Forms of Land Degradation

Apart from salinisation, land clearing and acid sulphate soils, the other forms of land degradation include soil erosion by water and wind, increasing soil acidification in agricultural areas, increase in soil sodicity, loss of organic matter, loss of soil nutrients, loss of soil structure and accumulation of heavy metals and pesticides (Conacher and Conacher, 1995; Hamblin, 2001).

3.10.5 Loss of Biodiversity

Life in Australia evolved in relative isolation from other landmasses for at least 50 million years. As a result, Australia possesses a rich diversity of unique and unusual plants, animals and microorganisms. About 85 percent of flowering plants, 83 percent of mammals, 45 percent of birds, 89 percent of reptiles, 93 percent of frogs and 90 percent of freshwater fish are endemic to Australia (Williams *et al.*, 2001, Table 46).

A number of problems threaten this diversity. These include land clearing, salinity, pollution, nutrient loading, sedimentation of waterways and coastal areas, degradation of wetlands, climate change, diseases, and introduced plants and animals. In particular, land clearing has destroyed habitat for many species. For example, 1000 to 2000 birds permanently lose their habitat for every 100 ha of woodland cleared (Australian State of the Environment Committee, 2001, p. 74).

The River Disturbance Index, derived for over three million kilometres of streams in Australia, indicates that very few large rivers remain undisturbed (Stein *et al.*, 2001; website of Australia's Natural Lands and Rivers).[22]

Construction of some 4000 weirs in the Murray–Darling Basin, and as many as 20 000 across the south-east of the continent, transformed many rivers into a series of stepped lakes. The effects of weirs are numerous, and include changes to temporal flow patterns, changes to water temperature,

---

[22] http://www.heritage.gov.au/anlr/code/rivmaps.html (visited in June 2005).

drowning of riparian and floodplain areas, reduced flow variability, sediment capture, obstruction of passage by fish and other aquatic biota, and benefiting carp and blue-green algae (Blanch, 2001, p. 5).

Numerous species have become endangered or vulnerable (Table 3.11). Moreover, 4 species of frogs, 23 birds, 27 mammals and 63 flowering plants are now presumed extinct.

The National Land & Water Resources Audit (2002b) provides an assessment of terrestrial biodiversity in Australia. The assessment:

(1) details the condition and trend of wetlands, riparian areas, threatened species and ecosystems across Australia;
(2) provides a listing of Australia's threatened ecosystems;
(3) identifies the processes that threaten biodiversity and their relative frequency in each of the bioregions;
(4) details special values, patterns and trends for Australia's birds and mammals, eucalypts and acacias; and
(5) presents conservation opportunities at both Australia-wide and regional scales.

The publication identifies the need to significantly enhance biodiversity conservation through a number of measures including:

(1) investing in protective management to minimise threatening processes such as clearing of vegetation, overgrazing, the spread of weeds and feral animals, and inappropriate fire regimes;
(2) engaging the broader community in the recovery of threatened species and ecosystems; and
(3) identifying specific biodiversity conservation objectives as part of integrated natural resources management.

3.10.6 Water Quality Deterioration

Australia's limited surface water and groundwater resources means that it is important to protect the quality of these vital resources for various human uses (human consumption, agricultural, industrial and others) and for maintaining the health of the aquatic ecosystems and endemic flora and fauna. In the past four decades water quality has been steadily deteriorating due to increased salinity, industrial and agricultural contaminants, acidification, nutrients and turbidity. Following is a brief description of some of these issues, mainly based on Ball *et al.* (2001, pp. 44–86).

Salinity. Stream salinity is a major issue in numerous catchments of south-east and south-west Australia. The increase in stream salinity of the Murray–Darling Basin has been recognised as a problem for over 40 years. Along the river, salinity increases from less than $40\,\mu S\,cm^{-1}$ in the headwaters to about $600\,\mu S\,cm^{-1}$ at Morgan in South Australia. It is predicted that without implementation of adequate management strategies, salinity of the River Murray at Morgan will rise from its current level to $790\,\mu S\,cm^{-1}$ in 2050 and $900\,\mu S\,cm^{-1}$ by 2100. This increase will cause serious problems for irrigators and other water users along the river and for the water supply of Adelaide (see Section A in Chapter 10). Salinity of many inland waters is predicted to increase over the next 100 years to levels that are likely to affect aquatic ecosystems and reduce the quality of water for drinking and irrigation. River systems under the greatest threat from increasing salinity include the Warrego, Condamine–Balonne, Border, Lachlan, Bogan, Macquarie, Castlereagh, Namoi, Murray, Avon and Loddon rivers in the Murray–Darling Basin and most rivers in south-west Western Australia.

Table 3.11. *Estimated number of species in Australia and the number of endangered, vulnerable and presumed extinct in 1993 and 2001*

Major group	Estimated number (2001)	Endangered 1993	Endangered 2001	Vulnerable 1993	Vulnerable 2001	Presumed extinct 1993	Presumed extinct 2001
Fish	5 250	7	13	6	17	–	–
Amphibians (frogs)	176	7	15	2	12	–	4
Invertebrates	192 700	–	–	–	4	–	–
Reptiles (snakes, lizards)	63	6	11	15	38	–	–
Birds	825	26	33	25	61	20	23
Mammals	363	28	29	19	61	21	27
Flowering plants, cycads, conifers, ferns and fern allies	20 000 to 25 000	226	517	661	654	74	63

Source: Williams *et al.* (2001, Tables 46 and 50).

Contaminants. Contaminants include pesticides, pathogens (bacteria, viruses and others), heavy metals, endocrine disruptors and organic chemicals. Possibly the most widespread contaminants are pesticides, which are used extensively in agriculture, especially for the cultivation of cotton, rice, sugarcane and horticultural crops. In urban areas, pesticides are used in the garden and for insect control. It is likely that there are other chemicals affecting aquatic flora and fauna, especially in river systems draining urban and irrigated areas where chemical use is high. However, there is little information on the occurrence and ecological affects of these chemicals in the environment.

Contamination from mining. There are over 300 derelict mines in New South Wales, 100 in Queensland and an unknown number in other states and territories requiring rehabilitation. Abandoned mines often discharge acidic waters due to exposure of sulphides in the rocks to the air. This acidic water can eventually find its way into rivers and groundwater systems. Tailings from processing of ores can also be improperly stored and can spill into rivers or leach into groundwater. Tailings can contain heavy metals or other toxic substances such as arsenic or cyanide used in the processing of ore.

Acidification. Acidification of water resources may occur due to exposure of acid sulphate soils, acidic discharges from mine sites or acidification of soils in agricultural areas. The periodic acidification of many coastal streams has been attributed to acid sulphate soils. For example, Heath (2004) estimated that the annual sulphuric acid load for the Tweed River (in NSW) where the river meets the ocean is approximately 2800 tonnes per year. Although acidification due to other processes is currently a localised problem, there is some evidence that this could become a significant problem in future. As well as having direct impacts on aquatic flora and fauna, acidification can increase the leaching of pollutants and nutrients from contaminated sediments.

Turbidity. Turbidity is an indicator of the amount of soil particulate and organic matter suspended in water. It is the result of soil erosion by water which causes sedimentation in inland waters and has significant impacts such as siltation of river channels, infilling of reservoirs and wetlands, smothering of aquatic flora and fauna, reduced light penetration, and the development of algal blooms. Many of Australia's inland rivers and wetlands have high turbidity, whereas the coastal systems generally have a low turbidity. Turbidity in rivers, streams and dams regularly exceed Australian water quality guidelines in many catchments of the Murray–Darling Basin. Other river systems where turbidity is a major issue are the coastal catchments of Melbourne and those immediately west to the Victoria–South Australia border, Sydney and surrounding catchments, central and mid-north Queensland coastal catchments, and the Brisbane region.

Eutrophication and algal blooms. Eutrophication occurs when the major nutrients such as nitrogen and phosphorus accumulate in water or sediments. Nutrients can originate from diffuse sources (65–95 percent), or point sources such as effluent outfalls. Given the right conditions, elevated concentrations of nutrients stimulate the growth of aquatic flora to nuisance levels. Examples include microscopic algae in the water column, which may result in algal blooms. Other factors such as water temperature, flow, turbidity and the condition of riparian vegetation are also important in the development of algal blooms. Blue-green algal blooms are of most concern in inland waters as certain species produce toxins that may cause skin irritations, gastrointestinal disorders, influenza-like symptoms, and in extreme cases, permanent organ damage and death. There have been widespread outbreaks of algal blooms in Australian freshwater and estuarine systems. For example, a massive bloom of toxic blue-green algae (*cyanobacteria*) occurred along 1 000 km of the Darling River in the late spring of 1991.

Groundwater contamination. The most significant diffuse contaminant of groundwater in Australia is nitrates. It is widespread in the Northern Territory, Victoria and Western Australia. In many areas, the concentration is greater than the recently revised Australian Drinking Water Quality Guidelines level of $50\,\text{mg}\,\text{L}^{-1}$ (as nitrate), resulting in groundwater that is unfit for drinking. In some of the more contaminated areas, the concentration is in excess of $100\,\text{mg}\,\text{L}^{-1}$. In sparsely settled areas, high nitrate concentrations have been attributed to leaching from termite mounds. Leaching of weed control chemicals into groundwater is also becoming a problem. The most significant point source contaminants include underground hydrocarbons storage tanks, septic tanks, landfills, intensive rural industries (nitrate), cattle and sheep dips (pesticides), manufacturing spills, and mining-related activities (heavy metals, acid, hydrocarbons).

3.11 MANAGEMENT REFORMS AND PROGRAMMES

Since the beginning of the 1990s a number of strategic plans and agreements have been developed to better manage natural resources of the country and in particular its water resources. These include:

- **The National Strategy for Ecologically Sustainable Development** (Council of the Australian Governments,

1992). The goal of the strategy is to achieve *development that improves the total quality of life, both now and in the future, in a way that maintains the ecological processes on which life depends.* The strategy covers eight sectoral issues: (1) agriculture; (2) fisheries; (3) forestry; (4) manufacturing; (5) mining; (6) urban planning; (7) tourism; and (8) energy. Although water is not one of the eight issues, the strategy considers *Water Resources Management* as one of the 22 inter-sectoral issues.

- **The Murray–Darling Basin Agreement 1992.** Prior to Australian Federation in 1901, the River Murray was the source of considerable conflicts between the colonies of New South Wales, Victoria and South Australia. After many years of dispute and negotiations, the Federal Government and state governments of New South Wales, Victoria and South Australia signed the River Murray Waters Agreement (RMWA) in 1915. To implement the agreement the River Murray Commission (RMC) was established and took up its responsibilities in 1917. In 1976, an interim agreement was reached that permitted the RMC to consider water quality as well as water quantity. The formal amendments to the RMWA were agreed in 1981 and the legislation passed through the federal and state parliaments, and came into effect on 1 January 1988. The RMWA and the 1988 agreement have been replaced by a totally revised Murray–Darling Basin Agreement, which came into operation on 1 July 1992. The purpose of the agreement is to *promote and co-ordinate the equitable, efficient and sustainable use of the water, land and other environmental resources of the Murray–Darling Basin* (Murray–Darling Basin Ministerial Council, 1992). To achieve its objectives, the Agreement established the Murray–Darling Basin Ministerial Council, the Murray–Darling Basin Commission, and the Community Advisory Committee.
- **Strategic Water Resources Policy** (Council of the Australian Governments, 1994). This policy covers a number of issues including:
  - Water pricing based on full-cost recovery and the removal of cross-subsidies.
  - Water allocation or entitlements backed by separation of water property rights from land title.
  - Allocation of water for the environment.
  - Trading in water allocations or entitlements to maximise its contributions to national income and welfare, within the social, physical and ecological constraints of catchments.
  - Institutional reforms to ensure an integrated approach to natural resources management.
  - Public consultation, participation and education.
- **Great Artesian Basin Strategic Management Plan, 2000.** Since 1878 when the first artesian bore was drilled near Bourke, groundwater has contributed to pastoral, mining and community development of the Basin. More than 100 000 people live in the Basin, of which 24 000 work in agriculture, 2500 work in the minerals and petroleum industries and an additional 27 000 work in service industries. Since 1878, more than 5000 bores have been drilled in the Basin and free flowing groundwater has significantly reduced the pressure heads of the aquifer. The Great Artesian Basin Consultative Council was established in 1997 under the auspice of Ministerial Council consisting of Federal ministers for Agriculture and Environment, and state ministers from Queensland, New South Wales, South Australia and Northern Territory. A Strategic Management Plan was prepared in September 2000, which highlights the relationships between the technological, social, environmental, physical and financial issues in the Basin. The objective is to *reduce groundwater wastage, control the Basin's discharge, and restore the aquifer's pressure.*[23]
- **Lake Eyre Basin Agreement, 2000.** The Lake Eyre Basin covers 1 140 000 km^2 or 15 percent of the Australian continent. It is the world's largest internal drainage system and includes catchments of all rivers draining into Lake Eyre. The Basin is located in the arid part of the country and major dams, weirs, and diversions have not substantially altered its rivers. During dry periods, wetlands of the Basin are vital refuges for wildlife. The Coongie Lakes, with an area of 19 800 km^2, support 73 species of waterbirds and 13 wetland dependent species. Oil and gas reserves lie beneath a large segment of the Basin. The Basin has a population of about 60 000. They live on grazing properties, at mine sites and in small, scattered settlements and towns. The South Australian Government took the lead in developing an agreement with Queensland and the Federal Government. Widespread concern about the impacts of a proposal to grow irrigated cotton at Windorah on Coopers Creek in Queensland provided the trigger for developing an agreement. The Lake Eyre Basin Agreement was negotiated over a period of 4 years and was signed by the three governments on 21 October 2000. The objective of the Agreement is to jointly address issues

---

[23] Further information is available at: http://www.gab.org.au (visited in June 2005).

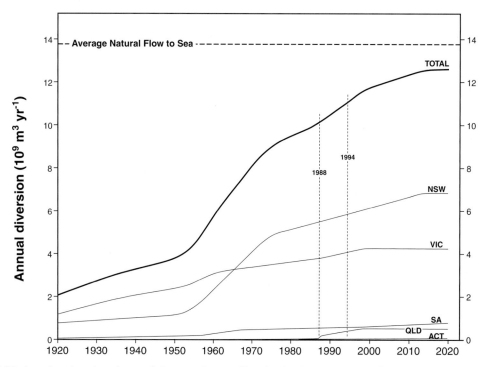

Figure 3.20 Actual and projected growth in annual water diversion in the Murray–Darling Basin (Murray–Darling Basin Ministerial Council, 1995, Figure 1).

about management of water and related natural resources of the Basin. This will be overseen by the Lake Eyre Basin Ministerial Forum which consists of one minister from Queensland, one from South Australia and one from the Federal Government. The first meeting of the Forum was held in Longreach, Queensland on 26 May 2001 and resulted in: approval of an annual budget of $500 000; appointment of the Basin's Coordinating Group; establishment of a Scientific Advisory Panel; establishment of a secretariat; and a process and timetable for the development of policies and associated strategies regarding the management of water and natural resources of the Basin.[24]

In addition to the various national strategies and agreements that aim to prevent further environmental degradation, states and federal governments have developed numerous other programs. Federal Government has four major programs and numerous Action Plans funded through *Envirofund* and managed by the *Natural Heritage Trust*. Major programs are *Landcare Program*, *Bushcare Program*, *Rivercare Program* and *Coastcare Program*. In October 2000, the Prime Minister announced a *National Action Plan for Salinity and Water Quality in Australia*. This plan was subsequently endorsed by states and territory governments with funding of $1.4 billion over seven years, comprising 50–50 federal and states contributions. Further information regarding the environmental management programs are available at various websites including the Natural Heritage Trust,[25] Murray–Darling Basin Commission,[26] and numerous state agencies dealing with environmental issues and management of land and water resources.

3.11.1 Cap on Surface Water Development in the Murray–Darling Basin (1995)

The last 100 years has seen large increases in the extraction of water from the Murray–Darling river system. Diversions commenced in the 1870s, and the rate of growth increased sharply in the 1950s and 1960s (Figure 3.20). Water extractions tripled in the 50 years to 1994. The state governments, Federal Government and the community all supported these developments that brought prosperity to about 1.9 million people living in the Basin. Moreover, more than one million people in Adelaide are heavily dependent on the River Murray for their drinking water supply.

The Basin enriches Australia by an estimated $23 billion a year. The annual value of its agricultural products exceeds $10 billion, which is about 30 percent of

---

[24] For further information visit http://www.dwr.sa.gov.au/files/lebagreement2.pdf (visited in June 2005).
[25] http://www.nht.gov.au (visited in June 2005).
[26] http://www.mdbc.gov.au (visited in June 2005).

Australia's gross value of agricultural production. Of this, $5 billion is derived from irrigation. Tourism and leisure have an annual earning of $6.5 billion, mining $3 billion, electricity $0.3 billion, and commercial fishing and other industries $2.5 billion. The food processing industry, which created large numbers of jobs in rural and urban areas of the Basin, depends heavily on irrigation for a steady supply of quality products. The wider value of the Basin to the national economy is around $75 billion a year and supports 1.5 million jobs (Murray−Darling Basin Ministerial Council, 2002b, p. 8). However, some of these developments have caused significant changes to the river flow regimes that have impacted on the river health. The major problems are:

(1) reduction in the areas of healthy wetlands and red gum forests;
(2) reduction in the number of native fish;
(3) increase in frequency of low flow periods;
(4) increase in the frequency of algal blooms; and
(5) increase in salinity levels.

In June 1993, the Murray−Darling Basin Ministerial Council initiated an audit of water use in the Murray−Darling Basin (Murray−Darling Basin Ministerial Council, 1995). In response to the issues raised by the audit, the Ministerial Council, at its June 1995 meeting, decided to introduce an interim Cap on diversion of the Basin's surface water resources. In December 1996, this was declared as a permanent Cap effective from 1 July 1997. The Ministerial Council agreed that the Cap be defined as: *"The volume of water that would have been diverted under 1993−94 levels of development. In unregulated rivers this Cap may be expressed as an end-of-valley flow regime"*. This decision was seen as an essential first step in establishing management systems to achieve healthy rivers and sustainable consumptive use. In other words, the Council determined that a balance needed to be struck between the significant economic and social benefits that have been obtained from the development of the Basin's water resources on one hand, and environmental uses in the river on the other.

The Cap is in accordance with the COAG water reform strategy. Its benefits include:

- A greater emphasis on achieving water use efficiency as a means to obtain water for further development.
- Security of water supplies to the existing water users.
- Reduction in accessions to the watertable with fewer consequent problems from waterlogging and soil salinisation.
- Better framework for trading in water entitlements both within states and between individuals in different states.
- Less deterioration in water quality.
- Less deterioration in the health of natural ecosystems.

The Cap does not attempt to reduce the Basin's water diversions, instead it aims to prevent them from being increased. The Ministerial Council decided that preventing any increase in diversion from the Basin was essential to arrest future decline in both river health and the security of supply to the existing water users. In 1996, an Independent Audit Group was set up by the Ministerial Council to review progress towards implementation of an operational Cap and to consider ways to resolve inconsistencies and equity in water use (Murray−Darling Basin Ministerial Council, 1996a,b). Subsequent reviews of Cap Implementations for the water years 1996−97 to 2003−04 were published by the Murray−Darling Basin Ministerial Council (1997, 1998, 1999b, 2001, 2002a, 2003, 2004, 2005), as were Water Audit Monitoring Reports for 1994−95 and 1996−97 to 2002−03 (Murray−Darling Basin Commission, 1997, 1998, 1999, 2000, 2001, 2002, 2003, 2004).

As part of the decision by the Ministerial Council to introduce a permanent Basin-wide Cap on diversion of surface water resources, a major review of the operation of the Cap was implemented in 2000, which marked 5 years of Cap implementation (Murray−Darling Basin Ministerial Council, 2000a,b,c). The review was conducted in four components to cover all aspects of the Cap's operation. The four components were:

- Ecological sustainability of rivers.
- Economic and social impacts.
- Equity.
- Implementation and compliance.

A total of 23 recommendations were made on these components (Murray−Darling Basin Ministerial Council, 2000c). Examples are:

- **Recommendation 1.** The Cap should be maintained.
- **Recommendation 4.** A high priority be given to further improving the knowledge base available to natural resource managers, especially our understanding of the ecology of the Basin.
- **Recommendation 5.** As better information informs our management of the Basin's resources, the level at which the Cap is set should continue to be refined to reflect our increased understanding.
- **Recommendation 6.** A Sustainable Rivers Audit, casting the Cap as an input to Basin health rather than an outcome in itself, be developed and implemented.

- **Recommendation 7.** Partner governments consider the opportunity provided by the 2001 Census to supplement the socio-demographic data it will provide with further information to improve our understanding of the social impacts of the Cap.
- **Recommendation 8.** Partner governments strive to better inform and engage stakeholders on major policy initiatives such as the Cap through improved communication.
- **Recommendation 13.** An integrated catchment management approach is required to address the equity issues associated with the Cap.
- **Recommendation 14.** Diversions from floodplain and overland flows be included in Cap accounting arrangements as a matter of priority.
- **Recommendation 15.** Farm dam water use should be included in Cap accounting arrangements as soon as practicable and all future administrative arrangements should support this outcome.
- **Recommendation 20.** Groundwater be managed on an integrated basis with surface water within the spirit of the Cap.

The most recent Review of the Cap Implementation has identified progress in each of the states and the Australian Capital Territory in establishment and operation of the Cap (Murray–Darling Basin Ministerial Council, 2005). However, a number of strategic issues that need to be addressed have been identified. These are:

(1) establishment of Cap targets in Queensland, New South Wales and the Australian Capital Territory;
(2) measurement of diversions; and
(3) accreditation of models for Cap measurement.

Imposition of the Cap on further development of surface water resources in the Basin beyond its 1993–94 level would improve both the health of riverine ecosystems and the management of scarce water resources in the next two to three decades. However, after this adjustment period, shortage of water in the Basin for further development will become a major issue for managers of land and water resources.

3.11.2 Environmental Flows for the River Murray

The Murray–Darling Basin Ministerial Council met in Corowa on 12 April 2002, on the one-hundredth anniversary of the first interstate meeting on cooperative management of the River Murray, also held in the same town. In order to improve understanding of the costs and benefits of recovering water for the environment, the Council directed its executive arm (the Murray–Darling Basin Commission) to identify and address key issues such as equity, property rights and water trading, through the development of a business case for the recovery of 350, 750 and $1500 \times 10^6 \, \text{m}^3$ of water per year for the River Murray. These volumes of water were intended to be used as reference points for further analysis and community consultation.

As a first stage of this process a discussion paper entitled *The Living Murray* was released (Murray–Darling Basin Ministerial Council, 2002b). This was in response to substantial evidence that the River Murray system is degraded and that this degradation threatens the Basin's agricultural industries, communities, natural and cultural values, and national prosperity.

In August 2003, the Council of the Australian Governments (COAG) committed $500 million to address the over-allocation of water to consumptive use in the Murray–Darling Basin. This financial commitment could, among other things, be used to purchase water for future use by the environment (Australian Bureau of Agricultural and Resource Economics, 2003). On 14 November 2003, the Murray–Darling Basin Ministerial Council took a historic First Step decision to address the declining health of the River Murray System. The decision consisted of returning on average $500 \times 10^6 \, \text{m}^3$ of water per year over the next 5 years to the river system at six key areas that have been designated as ecologically significant assets. These are:

(1) Barmah-Millewa Forest;
(2) Gunbower and Koondrook-Perricoota Forests;
(3) Hattah Lakes;
(4) Chowilla Floodplain (including Lindsay–Wallpolla);
(5) Murray Mouth, Coorong and Lower Lakes; and
(6) River Murray Channel (Figure 3.21).

The interim ecological objectives and outcomes for each one of the six significant assets are listed in Table 3.12.

Water for this First Step will come from a range of options, with a priority for on-farm initiatives, efficiency gains, infrastructure improvements and rationalisation, and some purchase of water from willing sellers, rather than by way of compulsory acquisition. Implementation of this First Step would cost $500 million, supplied from financial year 2004–05, based on the COAG commitment of August 2003.

Although environmentalists have welcomed the above-mentioned decision, they requested that the state and federal governments commit more funds for a further annual allocation of $1000 \times 10^6 \, \text{m}^3$ for the river in order to increase the level of allocation to $1500 \times 10^6 \, \text{m}^3$ recommended by the Wentworth Group of Concerned Scientists (2002).

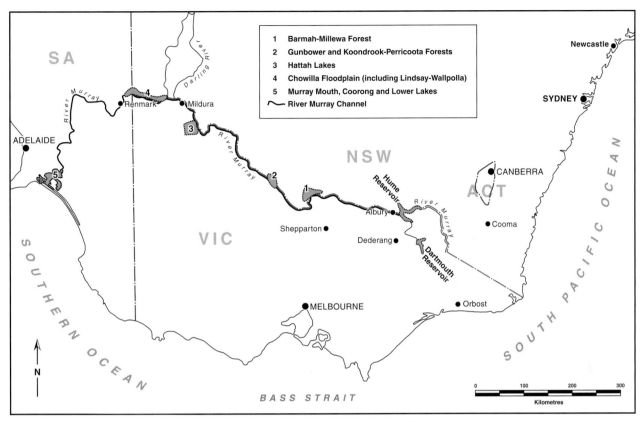

Figure 3.21 Location of the six significant ecological assets for the Living Murray.[27]

3.12 ESTIMATES OF FUTURE WATER REQUIREMENTS

Over the period of 1995–96, two reports attempting to address the long-term water requirement of the country were published. These are:

(1) *Curbing our Thirst: Possible Future for Australia's Urban Water System in the 21st Century*, published by the Australian Science and Technology Council (1995); and

(2) *Managing Australia's Inland Waters: Role for Science and Technology* (Department of Industry, Science and Tourism, 1996).

Smith (1998) provides a review of these reports. More recent estimates of the future water requirements are provided in the following sections.

3.12.1 The Australian Academy of Technological Sciences and Engineering Estimates

In 1999, the Australian Academy of Technological Sciences and Engineering used the MONASH general equilibrium model (Adams *et al.*, 1994) to simulate economic development to the year 2020–21 and to estimate its water requirements. Water estimates are based on the estimated water supply of $20 \times 10^9\,\text{m}^3$ for 1995–96 and the following scenarios.

Trend Scenario assumes that the rate of growth, structural characteristics and regional distribution of the economy continue to change in a similar fashion to trends in the last 10 years before the study. Under this scenario the annual water use could reach $33 \times 10^9\,\text{m}^3$, representing an increase of 65 percent. Compared with water availability data, it is clear that such growth is unsustainable in many regions of Australia. In particular, the shortfall is most serious in the Murray–Darling Basin where water resources are already fully committed.

Non-adaptive Scenario assumes that given the resource constraints in the *Trend Scenario*, water remains in its current use without major improvement in water use efficiency and progressive water allocation. Environmental problems will worsen, and new developments can only occur

---

[27] http://www.thelivingmurray.mdbc.gov.au/implementing/six_significant_ecological_assets (visited in June 2005).

Table 3.12. *The interim ecological objectives and expected outcomes of the increased flow at the six significant ecological assets*

Asset	Interim ecological objectives and expected outcomes
Barmah-Millewa Forest	Enhance forest, fish and wildlife values. – Achieve successful breeding of thousands of colonial waterbirds in at least three years in ten. – Maintain healthy vegetation in at least 55 percent of the forest area.
Gunbower and Koondrook–Perricoota Forests	Maintain and restore a mosaic of healthy floodplain communities. – Reinstate at least 80% of permanent and semi-permanent wetlands in healthy condition. – Maintain at least 30 percent of river red gum forest in healthy condition. – Successful breeding of thousands of colonial waterbirds in at least three years in ten. – Healthy populations of resident native fish in wetlands.
Hattah Lakes	Restore healthy examples of all original wetland and floodplain communities. – Restore the aquatic vegetation zone in and around at least 50 percent of the lakes to increase fish and bird breeding and survival. – Increase successful breeding events of threatened colonial waterbirds to at least two in ten years. – Increase the population size of and breeding events of the endangered Murray Hardyhead, Australian smelt, Gudgeons and other wetland fish.
Chowilla Floodplain (including Lindsay-Wallpolla)	Maintain high biodiversity values of the Chowilla Floodplain. – Maintain high value wetlands. – Maintain current area of river red gum. – Maintain at least 20 percent of the original area of black box vegetation.
Murray Mouth, Coorong and Lower Lakes	A healthier lower lakes and Coorong estuarine environment. – Keep the Murray mouth open. – Provide more frequent conditions for estuarine fish spawning. – Enhance migratory waterbird habitat in the Lower Lakes.
River Murray Channel	To increase the frequency of higher flows in spring that are ecologically significant. – To overcome barriers to migration of native fish species between the sea and Hume Dam. – Expand ranges of many species of migratory fishes. – Maintain current levels of channel stability.

Source: Website of the Living Murray.[28]

[28] http://www.thelivingmurray.mdbc.gov.au/implementing/six_significant_ecological_assets (visited in June 2005).

Table 3.13. *Estimates of water use for various scenarios (in $10^9 m^3$)*

	Base year 1995–96	Trend scenario 2020–21	Non-adaptive scenario 2020–21	Adaptive management scenario 2020–21
Murray–Darling Basin	11.5	18.0	10.8	11.4
Rest of Australia	8.5	15.0	15.1	16.0
Total	20.0	33.0	25.9	27.4

Source: Australian Academy of Technological Sciences and Engineering (1999, p. IX).

in regions with undeveloped resources. Under this scenario the annual water use could reach $25.9 \times 10^9 m^3$.

Adaptive Management Scenario assumes that through a number of initiatives, water is transferred to its most profitable uses. These initiatives include the extension of water trading arrangements, progressive changes in water pricing, and much improved water use efficiency within distribution systems and among end-users. On these bases, national water use in 2020–21 is estimated at $27.4 \times 10^9 m^3$ with no increase in water use in the Murray–Darling Basin. Regions with the fastest growth in water use would be the northern coasts of Queensland, the Kimberley region in Western Australia and South Australia. Table 3.13 compares results of the three scenarios in terms of water use.

The report by the Australian Academy of Technological Sciences and Engineering (1999) demonstrates ways in which the economy can grow, despite water resources limitations. It also shows that some water can be re-allocated to the environment without great economic penalty. This is particularly true if the reallocated water comes from efficiency gains and reduction in low value-adding water uses.[29] However, the publication has not considered the social costs associated with the investigated scenarios, in particular the *Adaptive Management*.

3.12.2 The 2001 CSIRO Sustainable Ecosystems Estimates

In 2001, the Resource Futures Program in CSIRO Sustainable Ecosystems, as part of the project *Decision Points for Land and Water Futures*, published the following reports in a series on Australian water futures, which intend to explore different scenarios for the development of Australia's water resources over the next 50 years.

Report I, Water Use in Australia (Dunlop *et al.* 2001a) consists of two sections. The first section contains some discussion of general and broad-scale water issues such as water availability, water quality and quantity, water rights and equity, climate change, and the national water policy issues. The second section contains an overview of a number of issues for each state and territory. These issues are water institutions, water practice, water reform, water hot-spots, water futures, and urban water issues.

Report II, Australian Water Use Statistics (Dunlop, 2001) contains statistics regarding irrigated area, water use (agricultural, domestic and others), and available water resources (surface water and groundwater) for Australia, states and territories, and 58 statistical divisions in Australia.

Report III, Water Futures Workshop: Issues and Drivers (Dunlop and Foran, 2001) describes the outcomes of an expert workshop on water use and resource futures which was held on 20 February 2001 at the CSIRO Sustainable Ecosystems in Canberra. The aim of the workshop was to identify and discuss issues that are likely to affect the availability and use of water at a regional level around Australia in the twenty-first century.

Report IV, Scenarios of Future Water Use (Dunlop *et al.* 2001b) divides Australia into three northern, central and southern broad regions and describes a "starting" scenario and three alternative water use scenarios. These are based on the information from Report I, knowledge and opinions collected at the workshop (Report III), and considering the fact that currently a significant volume of water is available in the Murray–Darling Basin for economic growth and the environment, if the federal and state governments and the community are willing to redirect these waters away from pasture and rice production. The scenarios have a national focus and 50-year time frame, and have been developed to

---

[29] The fallacy of this argument is that private farmers who have made the efficiency gain will be willing to transfer water to environmental use. Studies have shown that farmers are in general not willing to trade their excess water rights. While some moves to certain high value water use crops have been achieved since 1996 (see Table 3.4), there has been none the less a large increase in irrigated pasture due to dairy deregulations.

Table 3.14. *The historic and estimated water use for the four scenarios in 2050 (in $10^9 m^3$)*

History	2050			
1996	Starting scenario	Commodity J-curve scenario	Urban extravagance scenario	Climate change scenario
22.2[a]	27.4	28.8	27.0	26.6

[a] Estimated by The Australian Bureau of Statistics.
Source: Dunlop et al. (2001b).

provide context for long-term policy development and to provoke discussion about future water use issues.

The following paragraphs summarise the scenarios developed by Dunlop et al. (2001b) and their outcomes.

The Starting Scenario incorporates the effects of the "given" drivers. Its primary role is the starting point (or reference) for use in the modelling of the three alternative scenarios. The scenario sees considerable expansion of irrigated sugarcane and cotton in northern Australia. In southern Australia there is no increase in total water use. However, increasing water use efficiencies provide enough water for moderate growth in most higher value commodities. In south-eastern Australia there is a slight reduction in water use, due to increased irrigation water use efficiency. Under this scenario, the total water use will increase from $22.2 \times 10^9 m^3$ in 1996 to $27.4 \times 10^9 m^3$ in 2050 (Table 3.14).

The Commodity J-curve Scenario assumes expanded export markets for irrigated products. Under this scenario, there would be considerable expansion in irrigated pasture, fodder and grains, while there would be some contraction of sugarcane and cotton as a result of the expansion in other crops. In south-western Western Australia, there would be considerable expansion of wine grapes and vegetables. In south-eastern Australia, there would be marked expansion in vines, citrus, and other fruits and vegetables. Southern Queensland and northern New South Wales would see some reductions in cotton and sugarcane and some expansion of fruits, vegetables and beef cattle feedlots. For this scenario, the total water use would be $28.8 \times 10^9 m^3$ in 2050 and in the central and southern regions $3.3 \times 10^9 m^3$ would be returned to the environment as environmental flows.

The Urban Extravagance Scenario assumes that population would grow to 32 million by 2050 with substantial growth in the size of most Australian cities (9.9 million for Sydney, 5.8 million for Melbourne, 3.1 million for Brisbane, 2.8 million for Perth and 2 million for Adelaide). Under this scenario, the total water use in 2050 would be $27 \times 10^9 m^3$. For this scenario increase in water consumption in big cities would be met by purchasing water from lower value agricultural uses.

The Climate Change Scenarios considers two impacts of climate change both due to increases in temperature and decreases in average rainfall. These are:

(1) an increase of one third in the demand for water by native vegetation, crops, and humans; and
(2) a decrease by one third in run-off and catchment water yields.

Extra demands for urban and irrigation water are met by significant reduction in the area irrigated. The total water use for this scenario would be $26.6 \times 10^9 m^3$ by 2050. The environmental flows will be just maintained in the more stressed systems and will be reduced in others. Table 3.14 compares results of various scenarios with the 1996 water use.

3.12.3 The 2002 CSIRO-DIMIA Estimates

As part of a study regarding the options for Australia's population to 2050, Foran and Poldy (2002) analysed water requirements of the country under various population scenarios.[30] They estimated that without considering the effect of global climate change on hydrological regimes and water use, by 2050 water requirements could be about $40 \times 10^9 m^3$ per annum for an estimated population of 25 million. The main users of water would be New South Wales, Victoria and Queensland. Agriculture would require about 80 percent of total water use ($32 \times 10^9 m^3$). This is partly due to the development of more than one million hectares of irrigated agriculture in northern Australia. The water requirements for manufacturing industry, mining and domestic use would represent about 20 percent of the total. The direct population effect for total urban requirements is 5, 6 and $7.5 \times 10^9 m^3$ per year for the low (20 million), medium (25 million), and high (32 million) population scenarios respectively.

It is possible that improvements in the efficiency of water use could allow the higher population levels to function comfortably and equitably for the same water requirements as the low population scenario (Foran and Poldy, 2002). Apart from the increased efficiency of water use in agriculture, which is the largest water user in the country, the options available for improving the efficiency of water use for urban areas could be technological such as better designed water appliances, low flow shower heads, and drip

---

[30] For the description of these scenarios, see section 3.2.

irrigation, while some others are behavioural such as watering the garden at night, shorter showers, and dryland gardens.

3.12.4 The 2002 CSIRO Sustainable Ecosystems Estimates

Within the framework of the *Decision Points for Land and Water Futures* project of the CSIRO Sustainable Ecosystems, Dunlop *et al.* (2002) developed three scenarios for the future se of land and water in Australia.[31] The scenarios have been designed to ensure that Australia's natural resources are managed for the greatest possible long-term social, environmental and economic benefits for all Australians. Each scenario has different emphases, different challenges, and threats. These scenarios are briefly described below.

Dryland Agriculture Scenario recognises that currently rainfall is not adequately used for dryland agriculture and that more natural flow regimes are required to restore river and wetland health. This scenario is characterised by increases in production from dryland agriculture, decreases in the extraction of water for irrigation, a slowing in the rate of land degradation, and significant increases in river health.

The scenario sees retirement of 18 Mha of low productive croplands and sown pasture, and 9 Mha expansion of cropping and intensive pasture production across northern Australia. Better use of soil-water and increased inputs were predicted to lead to substantial increases in yields and crop production. A reduction of 40 percent in the area irrigated and moderate increases in water use efficiency see increases in environmental flow of more than $6.6 \times 10^9 \, m^3$ compared to 1996. Increased stresses on land resources due to intensification of dryland production may be offset by the use of soil conditioners, improved cropping systems and transferring of some agricultural land to perennial vegetation. Land degradation is not eliminated but its rate of increase is slowed. Increased environmental flows and protection of riparian zones halts the decline in river health in southern Australia. Salt loads in river systems continue to increase, but increased dilution flows result in salt concentrations being contained in the medium term.

Irrigation is a scenario that sees substantial value in continuing to shift the emphasis of agricultural production from low-value, highly variable dryland production to high-value irrigation. This scenario is characterised by:

(1) significant reorganisation of irrigated agriculture, bringing increased production of higher value crops;
(2) expansion of irrigation in northern Australia;
(3) a steady replacement of dryland crops and pastures with forestry and conservation planting;
(4) continuing stress on southern river systems despite some increase in environmental flows; and
(5) significant reduction in most land degradation.

About 29 Mha of dryland crops and sown pastures are retired with 15 Mha used for forestry or converted to native vegetation. The area under irrigation increases by about 2 Mha with just over half of this increase spread across northern Australia. In southern Australia increases in water use efficiency from restructuring of irrigation systems and renewal of aging infrastructure provide sufficient saving to supply the increases in irrigation and about $0.8 \times 10^9 \, m^3$ of additional environmental flows. With revegetation and contraction of farming to the more resilient areas, the area of land affected by degradation eventually declines. Similarly, the protection of remnants and establishment of corridors and new areas of native vegetation gradually improves the conservation status of many native plants, animals and communities. However, ecosystems and species that depend on groundwater, rivers or wetlands remain under pressure in southern Australia.

Post-agriculture is a scenario that recognises and capitalises on the diverse values people see in our landscapes, and seeks to take regional Australia beyond its dependence on European agriculture. This scenario is characterised by reinvented agricultural systems, significant decreases in areas of dryland and irrigated farming, and expansion of new service and manufacturing industries.

Under this scenario, the area of rainfed crops and sown pastures falls by 48 percent with 19 Mha revegetated for forestry and conservation, the area of irrigated land falls by 60 percent providing $8.7 \times 10^9 \, m^3$ for additional environmental flows. Farming systems on the remaining cropland are gradually redesigned to better suit Australia's poor soils and variable climate, leading to substantial decreases in land degradation and continuing increases in yields. Despite substantial reductions in the area of most crops, production continues to exceed domestic demand by a factor of two or more for all commodities (cereals, legumes, oil crops, cotton, fruit, etc.) in the scenario. Success depends on improving the net value of agriculture, forestry and native ecosystems through value adding and marketing of new products

---

[31] Dunlop *et al.* (2002) emphasise that scenarios are not predictions, nor options from which a preferred choice should be made. Rather, they are examples distilled from many possible futures. They serve to challenge preconceptions about the future, highlight opportunities and challenges, and provide a platform for testing specific strategies or policies.

Table 3.15. *The historic and modelled water use for three scenarios in 2050 (in $10^9 m^3$)*

Water use	History		2050		
	1996[a]	2001[b]	Dryland agriculture	Irrigation	Post-agriculture
Irrigation	17.9	17.3	10.3	25.6	8.1
Domestic, rural, and industrial	6.2	6.6	7.2	7.2	7.2
Total	24.1	23.9	17.5	32.8	15.4

Note: [a] Estimated by The National Land & Water Resources Audit (2001a); [b] Modelled.
Source: Dunlop et al. (2002, Table 6.1).

(forestry, pharmaceutical, etc.) and services (tourism, ecosystem services, etc.), and attracting non-rural businesses to regional Australia.

In the dryland and post-agriculture scenarios, water use drops significantly to $17.5 \times 10^9 m^3$ and $15.4 \times 10^9 m^3$ respectively (Table 3.15). In the irrigation scenario, water use continues to increase due to considerable expansion of irrigation in northern Australia.

3.13 NATIONAL WATER INITIATIVE

The Council of Australian Governments (COAG) held its fourteenth meeting on 25 June 2004 in Canberra. COAG reached a National Water Initiative (NWI) Agreement and establishment of the National Water Commission (NWC).[32] The signatories[33] agreed to implement the NWI in recognition of the continuing national imperatives to increase the productivity and efficiency of Australia's water use, the need to service rural and urban communities, and to ensure the health of river and groundwater systems by establishing clear pathways to return all systems to environmentally sustainable levels of extraction. The objective of the NWI is to provide greater certainty for investment and the environment, and to underpin the capacity of Australia's water management regimes to deal with change responsively and fairly.

Key elements of the NWI Agreement and their expected outcomes include:

- **Water access entitlements and planning framework to:** enhance the security of water access entitlements; provide a statutory basis for environmental and other public benefit outcomes in surface and groundwater systems to protect water sources and their dependent ecosystems; provide for adaptive management of surface and groundwater systems in order to meet productive, environmental and other public benefit outcomes; and recognise indigenous needs in relation to water access and management.

- **Water markets and trading to:** facilitate the operation of efficient water markets and the opportunities for trading, within and between states and territories; minimise costs on water trades; recognise and protect the needs of the environment; and provide appropriate protection to third-party interests.

- **Best practice water pricing and institutional arrangements to:** promote economically efficient and sustainable use of water resources, water infrastructure assets, and government resources devoted to the management of water; ensure sufficient revenue streams to allow efficient delivery of required services; facilitate the efficient functioning of water market; and give effect to the principles of user-pay and achieve pricing transparency.

- **Integrated management of water for environment and other public benefits to:** identify the desired environmental and other public benefit outcomes with as much specificity as possible; establish and equip accountable environmental water managers with the necessary authority and resources to provide sufficient water at the right times and places; and optimise the cost effectiveness of measures to provide water for these outcomes.

- **Water resources accounting to:** ensure that adequate measurement, monitoring and reporting systems are in place in all jurisdictions to support public and investor confidence in the amount of water being traded,

---

[32] Full text of the National Water Initiative Agreement, and establishment of the National Water Commission is available at http://www.coag.gov.au/meetings/250604/iga_national_water_initiative.pdf (visited in June 2005).

[33] Prime Minister, and Premiers of NSW, VIC, QLD and SA, plus Chief Ministers of the ACT and NT signed the agreement. However, Premier of WA refused to sign insisting that the Agreement did nothing to address problems of his State. Premier of Tasmania was not able to sign, but declared that he will continue to seek productive discussions with the Federal Government. Tasmania signed the Agreement in 2005, but the position of WA remained unchanged.

extracted for consumptive use, and recovered and managed for environmental and other public benefit outcomes.
- **Urban water reform to:** provide healthy, safe and reliable water supplies; increase water use efficiency in domestic and commercial settings; encourage the reuse and recycling of wastewater; facilitate water trading between and within the urban and rural sectors; encourage innovation in water supply sourcing, treatment, storage and discharge; and achieve improved pricing for metropolitan water.
- **Community partnerships and adjustment to:** engage water users and other stakeholders in achieving the objectives of the Agreement by improving certainty and building confidence in reform processes; transparency in decision-making; and ensuring that sound information is available to all sectors at key decision points.

COAG also agreed to establish a National Water Commission (NWC). The NWC will report to COAG. It will assess progress in implementing the NWI Agreement and advise on actions required to better realise the objectives of the Agreement. The Commission will be funded by the Federal Government, and will be made up of seven members with relevant expertise, of whom four (including the Chair) will be appointed by the Federal Government and three by state/territory governments.

Commencing in 2006–07, the NWC will undertake:

- Biennial assessments of progress with the NWI Agreements and state and territory implementation plans, and advice on actions required to better realise the objectives and outcomes of the Agreement.
- A third biennial assessment in 2010–11 in the form of a comprehensive review of the Agreement against the indicators (to be developed by the Natural Resource Management Ministerial Council and NWC) and an assessment of the extent to which actions undertaken in the NWI Agreement contribute to the national interest and their impacts on regional, rural and urban communities.
- Biennial assessments of the performance of the water industry against national benchmarks, in areas such as irrigation efficiency, water management costs and water pricing.

Drawing on the NWC assessment in 2010–11, COAG will review the objectives and operations of the NWC in 2011.

The NWI Agreement is a significant innovation and a major step forward in Australian water policy. It intends to implement a number of policies for satisfying demand, better management of developed water resources, and protection of the environment. However, it seems that it may not easily achieve its objectives because of a number of problems including the following:

- The NWI has been developed without a plan for the optimal population for the country, an estimate of the long-term future water requirements considering potential impacts of global warming, and priority areas for development of land and water resources of the country (see the following section).
- It is unlikely that if irrigators, who are the largest water users in the country, invest in increasing their water use efficiency, they would be prepared to return the saved water to the environment or trade it with other water users. It is more likely that they may use it for expansion of their irrigation activities.
- No clear arrangement is provided for funding implementation projects and conflict resolution processes.
- Water trading is a key element of the NWI. However, this may have adverse social and economic impacts on the communities which trade their water with other water users. No mechanism has been considered to prevent such adverse impacts.[34]

Connell *et al.* (2005) and Connell (2005) provide critical analysis of the NWI. They describe the NWI within its broader context and discuss the particularities involved in its implementation. They conclude that NWI is an ambitious document that combines a number of antagonistic stands. These will generate considerable tension and conflicts that are unlikely to be resolved by the very weak processes for coordination and compliance within them. Moreover, NWI has significant implications for the working of the Australian federal system in relation to environmental matters.

3.14 POTENTIAL ROLE OF INTER-BASIN WATER TRANSFER

Based on various estimates for future water requirements of Australia described in this chapter, the following observations can be made:

- The Adaptive Management Scenario developed by the Australian Academy of Technological Science and

---

[34] In California, 22 of the 58 State's Counties have adopted water export restrictions because of their social and economic adverse impacts (see section 11.15.3).

Engineering (1999) requires $27.4 \times 10^9 \, m^3$ of water in 2020. Extrapolating this estimate indicates that water requirements of this scenario for 2050 would be $36.3 \times 10^9 \, m^3$.
- The four scenarios developed by Dunlop et al. (2001b) require 27.4, 28.8, 27 and $26.6 \times 10^9 \, m^3$ in 2050 respectively.
- The CSIRO-DIMIA study (Foran and Poldy, 2002) estimated that the water requirements of a population of 25 million in 2050 would be about $40 \times 10^9 \, m^3$.
- The three scenarios developed by Dunlop et al. (2002) require $17.5 \times 10^9 \, m^3$, $32.8 \times 10^9 \, m^3$, and $15.4 \times 10^9 \, m^3$ in 2050 respectively.

The majority of the above estimates, which are based on fairly simplistic assumptions, are higher than the 1996–97 level of $24 \times 10^9 \, m^3$ water use, consisting of $19 \times 10^9 \, m^3$ from surface water and $5 \times 10^9 \, m^3$ of groundwater resources (see Table 3.6). Currently, surface water resources of almost all inland catchments (particularly catchments of the Murray–Darling Basin) are fully or even over allocated to the extent that there is insufficient water to satisfy environmental requirements. Important policies developed within the framework of the *National Water Initiative*, such as increasing efficiencies of agricultural and urban water use, water use restriction in urban areas, water trading, and reuse of wastewater, will change the situation in favour of better management of currently developed water resources and to protect the aquatic environment. However, these measures will be able to alleviate our water problems for the next two to three decades only. Longer-term measures require an accurate estimate of long-term future water requirements of the country. Providing such an estimate requires answers to a number of fundamental questions:

- What is the country's optimum population?[35]
- When is the country expected to reach its optimum population?
- Does the country prefer continuation of current population migration from inland to the coastal areas,[36] or would it be wise to assist development of the inland population centres?
- What is the target in terms of developing land and water resources of the country for exporting food for the world's growing population?[37]
- What will be the long-term impact of climate change (changes in temperature and rainfall) on the country's water resources, water requirements and development of land and water resources?
- Where are the priority areas for further water and land resources developments?
- How can the shortfall in projected water requirements be met?
- How much is the country prepared to invest in alternative sources of water supply such as increasing water use efficiency in all sectors of the economy (particularly in agricultural industry), reuse of treated waste water, water trading, and desalination of brackish and saline water?
- What would be the respective shares of surface and groundwater in future water requirements of the country?
- What would be the economic and social impacts of water trading on communities selling their water entitlements to other water users?
- What would be the reasonable compromise in terms of sharing water resources for economic development and environmental requirements?
- Can the state and federal governments deliver a 10 to 20 year adjustment process in regions that would be most affected by future changes in land and water use?

There are no easy or straightforward answers to the above questions. Therefore, it would be important that the federal and state governments and the community (irrigators, farmers, urban water users, industry, economists, social scientists, environmentalists, and others) jointly prepare a long-term *Visionary Master Plan* for the development and management of land and water resources of the country. This plan should be prepared within the long-term economic and social development objective of the country and protection of its environment. If the current developed water resources and options such as increasing water use efficiency, reuse of treated wastewaters, water trading, and desalination, prove not to be sufficient for future requirements of the country, inland diversion of a reasonable fraction of surface water resources from suitable catchments located on the eastern side of the Great Dividing Ranges, as well as the northern part of the country, have the potential to satisfy the shortfall. However, this would depend on the long-term availability of excess water in coastal catchments due to impacts of climate changes and current trend of population migration from inland towards the costal areas and an assessment of the environmental impacts. If such diversions are to be implemented, the community should

---

[35] See section 3.2.
[36] See section 3.2.
[37] See sections 1.1 and 1.6–1.9 in Chapter 1.

accept their economic, social and environmental consequences.

Priority areas need to be identified and detailed investigations need to be undertaken to provide suitable answers to the above questions. By its nature, this would be a long planning process that may take a decade or more. Therefore, it is necessary to commence the process as soon as possible. Perhaps the *National Water Commission* could undertake this activity within the framework of the *National Water Initiative*.

3.15 CONCLUSIONS

Australia is a vast and dry continent with a limited population of 20 million. It has an average annual run-off of $387 \times 10^9 \, m^3$, which is about $19\,300 \, m^3$ per head of population. However, this run-off is unevenly distributed and approximately 88 percent is contained in five relatively narrow coastal drainage divisions in the north, east, and south-east of the country. This chapter has provided background information on the land and water resources of Australia. It has reviewed long-term predictions for the population of Australia, the extent of land and water resources development, the natural cycles of flood and drought, global warming and its predicted impacts, and long-term future water requirements. Development of land and water resources of the country over the past 200 years has resulted in some serious problems such as large-scale salinisation of land and water resources, water quality deterioration, and environmental degradation.

Over the past two decades numerous measures have been undertaken to better manage land and water resources of the country and protect its fragile environment. In this respect a number of significant intergovernmental agreements have been reached and policies have been developed. These include:

- The National Strategy for Ecologically Sustainable Development, 1992.
- The Murray–Darling Basin Agreement, 1992.
- Strategic Water Resources Policy, 1994.
- Cap on Surface Water Development in the Murray–Darling Basin, 1995.
- Great Artesian Basin Strategic Management Plan, 2000.
- Lake Eyre Basin Agreement, 2000.
- Environmental Flow for the River Murray, 2003.

The most important and the most recent policy is the 2004 National Water Initiative. It is hoped that implementation of this policy will further improve water resources management of the country. However, this Initiative has been developed without a *Visionary Master Plan* for long-term management and development of land and water resources of Australia, which takes into account optimum population size and the nation's economic development objectives. Overall, this initiative is a major step forward if the federal and state governments coordinate their activities and adequately fund projects in the areas of water saving measures, water pricing and protection of the environment.

References

Adams, P. D., Dixon, P. B. and McDonald, D. (1994). MONASH forecasts of output and employment for Australian industries: 1992–93 to 2000–01. *Australian Bulletin of Labour* **20**(2): 85–109.

Allan, R., Lindesay, J. and Parker, D. (1996a). *El Niño Southern Oscillation and Climate Variability*. Collingwood, Victoria: CSIRO Publishing.

Allan, R. J., Beard, G. S., Close, A., Herczeg, A. L., Jones, P. D. and Simpson, H. J. (1996b). *Mean Sea Level Pressure Indices of the El Nino Southern Oscillation: Relevance to Stream Discharge in South-Eastern Australia*. Report No. 96/1. Canberra: CSIRO Division of Water Resources.

Australian Academy of Technological Sciences and Engineering (1999). *Water and the Australian Economy*. Parkville, Victoria: AATSE.

Australian Bureau of Agricultural and Resource Economics (2003). *Government Purchase of Water for Environmental Outcomes*. Canberra. ABARE.

Australian Bureau of Statistics (1988). *Year Book Australia 1988*. Canberra: ABS.

Australian Bureau of Statistics (2001). *Year Book Australia 2001*. Canberra: ABS.

Australian Bureau of Statistics (2002). *Year Book Australia 2002*. Canberra: ABS.

Australian Bureau of Statistics (2004). *Year Book Australia 2004*. Canberra: ABS.

Australian Bureau of Statistics (2005). *Year Book Australia 2005*. Canberra: ABS.

Australian Geographic Society (1988). *The Australian Encyclopaedia*. Fifth Edition. Terry Hills, NSW: Australian Geographic Pty. Ltd.

Australian Greenhouse Office (2001). *National Greenhouse Gas Inventory*. Canberra: AGO.

Australian Science and Technology Council (1995). *Curbing our Thirst: Possible Futures for Australia's Urban Water System in the 21st Century*. Canberra: Australian Government Publishing Service.

Australian State of the Environment Committee (2001). *Australia State of the Environment 2001*. Collingwood, Victoria: CSIRO Publishing.

Australian Water Resources Council (1987). *1985 Review of Australia's Water Resources and Water Use*. Canberra: Australian Government Publishing Service. v.1: Water Resources Data Set; v.2: Water Use Data Set.

Australian Water Resources Council (1992). *Floodplain Management in Australia*. Water Management Series No. 21. Canberra: Department of Primary Industries and Energy.

Ball, J., Donnelley, L., Erlanger, P., Evans, R., Kollmorgen, A., Neal, B. and Shirley, M. (2001). *Inland Waters, Australia State of the Environment Report 2001 (Theme Report)*. Collingwood, Victoria: CSIRO Publishing.

Bartel, R. L. (2003a). Bypassing the local: a comparison of the administration of land clearance legislation in Australia. *Local Government Law Journal* **8**(4): 184–213.

Bartel, R. L. (2003b). Compliance and complicity: an assessment of the success of land clearance legislation in New South Wales. *Environmental and Planning Law Journal* **20**(2): 116–141.

Beale, B. and Fray, P. (1990). *The Vanishing Continent: Australia's Degraded Environment*. Sydney: Hodder & Stoughton.

Bennett, J., van Bueren, M. and Whitten, S. (2004). Estimating society's willingness to pay to maintain viable rural communities. *The Australian Journal of Agricultural and Resource Economics* **48**(3): 487–512.

Blanch, S. ed. (2001). *The Way Forward on Weirs Conference Proceedings*. Presented on 18th–19th August 2000, Royal North Shore Hospital, St Leonard, NSW, Sydney: Inland Rivers Network.

Botterill, L. C. (2003). Uncertain climate: the recent history of drought policy in Australia. *Australian Journal of Politics and History* **49** (1): 61–74.

Botterill, L. C. and Fisher, M. eds. (2003). *Beyond Drought: People, Policy and Perspectives*. Collingwood, Victoria.: CSIRO Publishing.

Botterill, L. C. and Wilhite, D. A. eds. (2005). *From Disaster Response to Risk Management: Australia's National Drought Policy*. Dordrecht, The Netherlands: Springer.

Brown, J. A. H. (1983). *Australia's Surface Water Resources: Water 2000 Consultants Report No. 1*. Canberra: Australian Government Publishing Service.

Bureau of Meteorology (1974). *Brisbane Floods: January 1974*. Canberra: Australian Government Publishing Service.

Camm, J. C. R., McQuilton, J., Plumb, T. W. and Yorke, S. G. (1987). *Australians: A Historical Atlas*. Sydney: Fairfax, Syme & Weldon Associates.

Cocks, D. (1992). *Use with Care: Managing Australia's Natural Resources in the Twenty First Century*. Sydney: University of New South Wales Press.

Cocks, D. (1996). *People Policy: Australia's Population Choices*. Sydney: University of New South Wales Press.

Conacher, A. and Conacher, J. (1995). *Rural Land Degradation in Australia*. Melbourne: Oxford University Press.

Connell, D. (2005). The Chariot Wheels of the Commonwealth: The Past, Present and Future of Inter-Jurisdictional Water Management in the Murray–Darling Basin. Canberra: Centre for Resource and Environmental Studies, The Australian National University. PhD dissertation.

Connell, D., Dovers, S. and Grafton, R. Q. (2005). A critical analysis of the National Water Initiative. *Australian Journal of Natural Resources Law and Policy* **10**(1): 81–107.

Council of the Australian Governments (1992). *National Strategy for Ecologically Sustainable Development*. Canberra: Australian Government Publishing Service.

Council of the Australian Governments (1994). *Communiqué from the COAG Meeting, Hobart, 25 February 1994*. Canberra: Department of the Prime Minister and Cabinet.

Daly, D. (1994). *Wet as a Shag, Dry as a Bone: Drought in a Variable Climate*. Information Series Q193028. Brisbane: Queensland Department of Primary Industry.

Day, L. H. and Rowland, D. T. eds. (1988). *How Many More Australians? The Resource and Environmental Conflicts*. Melbourne: Longman Cheshire.

Dent, D. L. (1986). *Acid Sulphate Soils: A Baseline for Research and Development*. Wageningen, Netherlands: International Institute for Land Reclamation and Improvement (ILRI). Pub. No. 39.

Department of Industry, Science and Tourism (1996). *Managing Australia's Inland Waters: Roles for Science and Technology*. Canberra: DIST.

Dovers, S., Norton, T., Hughes, I. and Day, L. (1992). *Population Growth and Australian Regional Environments*. Bureau of Immigration Research's publication. Canberra: Australian Government Publishing Service.

Dovers, S. (1997). Dimensions of the Australian population-environment debate. *Development Bulletin*, **41**: 50–53.

Dunlop, M., McKenzie, N., Jacquier, D., Ashton, L. and Foran, B. (1999). *Australia's Stocks of Quality Soils*. Working Paper Series 99/06. Canberra: CSIRO Sustainable Ecosystems.

Dunlop, M. (2001). Australian Water Use Statistics: Report II of IV in a Series on Australian Water Futures. Canberra: CSIRO Sustainable Ecosystems.

Dunlop, M. and Foran, B. (2001). *Water Futures Workshop-Issues and Drivers: Report III of IV in a Series on Australian Water Futures*. Canberra: CSIRO Sustainable Ecosystems.

Dunlop, M., Hall, N., Watson, B., Gordon, L. and Foran, B. (2001a). *Water Use in Australia: Report I of IV in a Series on Australian Water Futures*. Canberra: CSIRO Sustainable Ecosystems.

Dunlop, M., Foran, B. and Poldy, F. (2001b). *Scenarios of Future Water Use: Report IV of IV in a Series on Australian Water Futures*. Canberra: CSIRO Sustainable Ecosystems.

Dunlop, M., Turner, G., Foran, B. and Poldy, F. (2002). *Decision Points for Land and Water Futures*. Canberra: CSIRO Sustainable Ecosystems.

Emergency Management Australia (1999). *Managing the Floodplain*. Canberra: Emergency Management Australia.

Foley, J. C. (1957). *Drought in Australia: Review of Records from Earliest Years of Settlement to 1955*. Bulletin No. 43. Melbourne: Bureau of Meteorology.

Foran, B. and Poldy, F. (2002). *Future Dilemmas: Options to 2050 for Australia's Population, Technology, Resources and Environment*. Canberra: CSIRO Sustainable Ecosystems.

Fraser, B. ed. (1984). *The Macquarie Book of Events*. Macquarie Library.

Ghassemi, F., Jakeman, A. J. and Nix, H. A. (1995). *Salinisation of Land and Water Resources: Human Causes, Extent, Management and Case Studies*. Sydney: University of New South Wales Press.

Gifford, R. M., Kalma, J. D., Aston, A. R. and Millington, R. J. (1975). Biophysical constraints in Australian food production: implications for population policy. *Search* **6**(6): 212–223.

Gleick, P. H. ed. (1993). *Water in Crisis: A Guide to the World's Fresh Water Resources*. New York: Oxford University Press.

Habermehl, M. A. (2001). Hydrogeology and environmental geology of the Great Artesian Basin, Australia. In Gostin, V. A. ed. *Gondwana to Greenhouse: Australian Environmental Geoscience*. Special Publication No. 21. Geological Society of Australia Inc, pp. 127–143.

Hamblin, A. (2001). *Land, Australia State of the Environment Report 2001 (Theme Report)*. Collingwood, Victoria: CSIRO Publishing.

Hamer, W. I. (1985). *A Review of the 1982/1983 Drought*. Melbourne: Department of Agriculture.

Heath, L. C. (2004). An Integrated Approach to the Remediation and Management of Coastal Acid Sulfate Soils. PhD Thesis. Canberra: Centre for Resource and Environmental Studies, The Australian National University.

House of Representatives Standing Committee on Long Term Strategies (1994). *Australias Population Carrying Capacity: One Nation – Two Ecologies*. Canberra: Australian Government Publishing Service.

IPCC (2001). *Climate Change 2001: Impacts, Adaptation and Vulnerability*. Cambridge: Cambridge University Press.

Jolly, I. D., Dowling, T. I., Zhang, L., Williamson, D. R. and Walker, G. R. (1997). *Water and Salt Balances of the Catchments of the Murray–Darling Basin*. Canberra: CSIRO Land and Water.

McKenzie, N., Isbell, R., Brown, K. and Jacquier, D. (2001). Major soils used for agriculture in Australia. In National Land and Water Resources Audit. *Australian Agriculture Assessment 2001*. Canberra: Natural Heritage Trust. v.2, pp. A75–A120.

Munro, R. K. and Lembit, M. J. (1997). Managing climate variability in the national interest: needs and objectives. In Munro, R. K. and Leslie, L. M. eds. *Climate Prediction for Agricultural and Resource Management*. Canberra: Bureau of Resource Sciences.

Murray–Darling Basin Commission (1997). *Water Audit Monitoring Report 1994/95: A Preliminary Report of the Murray–Darling Basin Commission on Diversions*. Canberra: MDBC.

Murray–Darling Basin Commission (1998). *Water Audit Monitoring Report 1996/97: Report of the Murray–Darling Basin Commission on the Final Year of the Interim Cap in the Murray–Darling Basin*. Canberra: MDBC.

Murray–Darling Basin Commission (1999). *Water Audit Monitoring Report 1997/98: Report of the Murray–Darling Basin Commission on the Cap on Diversions*. Canberra: MDBC.

Murray–Darling Basin Commission (2000). *Water Audit Monitoring Report 1998/99: Report of the Murray–Darling Basin Commission on the Cap on Diversions*. Canberra: MDBC.

Murray–Darling Basin Commission (2001). *Water Audit Monitoring Report 1999/00: Report of the Murray–Darling Basin Commission on the Cap on Diversions*. Canberra: MDBC.

Murray–Darling Basin Commission (2002). *Water Audit Monitoring Report 2000/01: Report of the Murray–Darling Basin Commission on the Cap on Diversions*. Canberra: MDBC.

Murray–Darling Basin Commission (2003). *Water Audit Monitoring Report 2001/02: Report of the Murray–Darling Basin Commission on the Cap on Diversions*. Canberra: MDBC.

Murray–Darling Basin Commission (2004). *Water Audit Monitoring Report 2002/03: Report of the Murray–Darling Basin Commission on the Cap on Diversions*. Canberra: MDBC.

Murray–Darling Basin Ministerial Council (1992). *Murray–Darling Basin Agreement*. Canberra: MDBMC.

Murray–Darling Basin Ministerial Council (1995). *An Audit of Water Use in the Murray–Darling Basin*. Canberra: MDBMC.

Murray–Darling Basin Ministerial Council (1996a). *Setting the Cap: Report of the Independent Audit Group*. Canberra: MDBMC.

Murray–Darling Basin Ministerial Council (1996b). *Setting the Cap: Report of the Independent Audit Group*. Executive Summary. Canberra: MDBMC.

Murray–Darling Basin Ministerial Council (1997). *Review of Cap Implementation 1996/97: Report of the Independent Audit Group*. Canberra: MDBMC.

Murray–Darling Basin Ministerial Council (1998). *Review of Cap Implementation 1997/98: Report of the Independent Audit Group Including Responses by the Four State Governments*. Canberra: MDBMC.

Murray–Darling Basin Ministerial Council (1999a). *The Salinity Audit of the Murray–Darling Basin: A 100-Year Perspective*. Canberra: MDBMC.

Murray–Darling Basin Ministerial Council (1999b). *Review of Cap Implementation 1998/99: Report of the Independent Audit Group Including Responses by the Five State and Territory Governments*. Canberra: MDBMC.

Murray–Darling Basin Ministerial Council (2000a). *Review of the Operation of the Cap: Draft Overview Report of the Cap Project Board, Including the four Companion Papers*. Canberra: MDBMC.

Murray–Darling Basin Ministerial Council (2000b). *Review of the Operation of the Cap: Draft Overview Report on the Cap Project Board*. Canberra: MDBMC.

Murray–Darling Basin Ministerial Council (2000c). *Review of the Operation of the Cap: Overview Report of the Murray–Darling Basin Commission*. Canberra: MDBMC.

Murray–Darling Basin Ministerial Council (2001). *Review of Cap Implementation 1999/00: Report of the Independent Audit Group Including Responses by Five State and Territory Governments*. Canberra: MDBMC.

Murray–Darling Basin Ministerial Council (2002a). *Review of Cap Implementation 2000/01: Report of the Independent Audit Group Including Responses by Five States and Territory Governments*. Canberra: MDBMC.

Murray–Darling Basin Ministerial Council (2002b). *The Living Murray: A Discussion Paper on Restoring the Health of the River Murray*. Canberra: MDBMC.

Murray–Darling Basin Ministerial Council (2003). *Review of Cap Implementation 2001/02: Report of the Independent Audit Group Including Special Audits of the Lachlan and Gwydir Valleys and Responses by the Five States and Territory Governments*. Canberra: MDBMC.

Murray–Darling Basin Ministerial Council (2004). *Review of Cap Implementation 2002/03: Report of the Independent Audit Group Including Special Audits of the Lachlan and Macquarie Valleys and Responses by the Five States and Territory Governments*. Canberra: MDBMC.

Murray–Darling Basin Ministerial Council (2005). *Review of Cap Implementation 2003/04: Report of the Independent Audit Group*. Canberra: MDBMC.

National Land & Water Resources Audit (2001a). *Australian Water Resources Assessment 2000*. Canberra: Natural Heritage Trust.

National Land & Water Resources Audit (2001b). *Australian Dryland Salinity Assessment 2000*. Canberra: Natural Heritage Trust.

National Land & Water Resources Audit (2001c). *Australian Agriculture Assessment 2001*, Canberra: Natural Heritage Trust. v.2., pp. 219–318 and A35–A120.

National Land & Water Resources Audit (2002a). *Australians and Natural Resources Management 2002*. Canberra: Natural Heritage Trust.

National Land & Water Resources Audit (2002b). *Australians Terrestrial Biodiversity Assessment 2002*. Canberra: Natural Heritage Trust.

Newman, P., Bathgate, C., Bell, K., Dawson, M., Hudson, K., Lehman, J., Mason, S., Myhill, P., McEnaney, L., McCreddin, C., Pidgeon, C., Pratt, C., Ringrose, C., Ringvall, K. and Sargent, M. (1994). *Australia's Population Carrying Capacity: An Analysis of Eight Natural Resources*. Murdoch, Western Australia: Institute for Sustainability and Technology Policy, Murdoch University.

Nicoll, C. (1994). *A Directory and Bibliography of Tree Water Use Research*. Canberra: CSIRO Division of Water Resources.

Nix, H. (1988). Australia's renewable resources. In Day, L. H. and Rowland, D. T. eds. *How Many More Australians? The Resource and Environmental Conflicts*. Melbourne: Longman Cheshire. pp. 64–76.

Northcote, K. H., Beckmann, G. G., Bettenay, E., Churchward, H. M., van Dijk, D. C., Dimmock, G. M., Hubble, G. D., Isbell, R. F., McArthur, W. M., Murtha, G. G., Nicolls, K. D., Paton, T. R., Thompson, C. H., Webb, A. A. and Wright, M. J. (1960-68). *Atlas of Australian Soils, Sheets 1 to 10 with Explanatory Data*. Melbourne: CSIRO and Melbourne University Press.

Norton, T., Dovers, S., Nix, H. and Elias, D. (1994). *An Overview of Research on the Links Between Human Population and the Environment*. Canberra: Australian Government Publishing Service.

O'Meagher, B., du Pisani, L. G. and White, D. H. (1998). Evolution of drought policy and related science in Australia and South Africa. *Agricultural Systems*. **57**: 231–258.

O'Meagher, B., Stanford Smith, M. and White, D. H. (2000). Approaches to integrated Drought risk management: Australia's national drought police. In Wilhite D. A. ed. *Hazards and Disasters: A Series of Definitive Major Works*. London: Routledge Publishers, v.2., pp. 115–128.

Osborne, M. and Dunn, C. (2004).[38] *Talking Water: An Australian Guidebook for the 21st Century*. Sydney: Farmhand Foundation.

Prinsley, R. T. (1991). *Australian Agroforestry: Setting the Scene for Future Research*. Canberra: Rural Industries Research and Development Corporation.

Reynolds, R. G., Watson, W. D. and Collins, D. J. (1983). *Water Resources Aspects of Drought in Australia*. Water 2000: Consultants Report No. 13. Department of Resource and Energy. Canberra: Australian Government Publishing Service.

Roderick, M. L. and Farquhar, G. D. (2004). Changes in Australian pan evaporation from 1970 to 2002. *International Journal of Climatology*. **24**: 1077–1090.

Smith, D. I. (1998). *Water in Australia: Resources and Management*. Melbourne: Oxford University Press.

Smith, D. I. (2002). The what, where, and how much of flooding in Australia. In Smith, D. I. and Handmer, J. eds. *Residential Flood Insurance: The Implications for Floodplain Management Policy*. Canberra: Water Research Foundation of Australia. pp. 73–98.

Standing Committee on Agriculture and Resource Management (2000). *Floodplain Management in Australia: Best Practice Principles and Guidelines*. SCARM Report 73. Collingwood, Victoria: CSIRO Publishing.

Stein, J. L., Stein, J. A. and Nix, H. A. (2001). Wild rivers in Australia. *International Journal of Wilderness*. **7**(1): 20–24.

Troup, A. J. (1965). The southern oscillation. *Quarterly Journal of the Royal Meteorological Society*. **91**: 490–506.

Walker, G. R., Jolly, I. D., Williamson, D. R., Gilfedder, M., Morton, R. and 13 others (1998). *Historical Stream Salinity Trends and Catchment Salt Balances in the Murray–Darling Basin: Final Report*. Canberra: MDBC.

Wentworth Group of Concerned Scientists (2002). *Blueprint for a Living Continent: A Way Forward from the Wentworth Group of Concerned Scientists*. Sydney: World Wide Fund for Nature, Australia.

---

[38] Available at the URL http://www.farmhand.org.au/press.html (visited in June 2005).

Whetton, P. H. and Hennessy, K. H. (2001). Climate change projections for the Australian region. In *Proceedings of the MODSIM 2001 International Congress*, Canberra, 10–13 December 2001. Canberra: Modelling and Simulation Society of Australia and New Zealand, pp. 647–654.

White, D. H., Howden, S. M., Walcott, J. J. and Cannon, R. M. (1998). A framework for estimating the extent and severity of drought, based on grazing system in south-eastern Australia. *Agricultural Systems*, **57**: 259–270.

White, I., Melville, M. D., Wilson, B. P. and Sammut, J. (1997). Reducing acid discharge from coastal wetlands in eastern Australia. *Wetland Ecology and Management*, **5**:55–72.

Wilhite, D. A. ed. (1993). *Drought Assessment, Management and Planning: Theory and Case Studies*. Boston: Kluwer Academic Publishers.

Williams, J., Read, C., Norton, A., Dovers, S., Burgman, M., Proctor, W. and Anderson, H. (2001). *Biodiversity, Australia State of the Environment Report 2001 (Theme Report)*. Collingwood, Victoria: CSIRO Publishing.

4 The Snowy Mountains Hydro-electric Scheme

4.1 LOCATION

The Snowy Mountains area is located in the south-east corner of the State of New South Wales (Figure 4.1). It is the most elevated area of the continent and includes Australia's highest peak Mt Kosciuszko with an elevation of 2228 m. The average annual precipitation for most of the area ranges between 1200 mm to 1600 mm with small areas exceeding 2000 mm (Figure 3.2). Above an elevation of 1400 m, precipitation in winter often falls as snow and extensive areas are covered with a thick mantel. The major rivers of the area are:

- **The Snowy River**,[1] which originates at an elevation of 1770 m, drains the south-eastern part of the area along its 300 km length, and discharges to Bass Strait at Marlo (south-east of Orbost) in Victoria. One of its major tributaries is the Eucumbene River (see Figure 4.3).
- **The Murrumbidgee River**, which drains the north-eastern corner of the area. Its major tributary is the Tumut River.
- **The River Murray**, which has its source in the western part of the region.

The Snowy Mountains Hydro-electric Scheme, which diverted flows from the Snowy and Murrumbidgee rivers is located in the Snowy Mountains area.

4.2 HYDROLOGY

The Snowy River has an average annual natural flow of $1177 \times 10^6 \, m^3$ at Jindabyne (see Figure 4.3) and $2150 \times 10^6 \, m^3$ at its mouth. The Snowy catchment above Jindabyne represents approximately 15 percent of the total area of the catchment, but because of the high precipitation, it contributes approximately 50 percent of the total annual run-off of the catchment. Construction of the Snowy Mountains Hydro-electric Scheme during the 1950s to 1970s (Appendix B) has diverted 99 percent of flows in the Snowy River upstream of Jindabyne Dam (Snowy Water Inquiry, 1998d, p. 4). Under natural conditions, the Snowy River at Dalgety (Figures 4.2 and 4.3) had a strongly seasonal flow, due largely to the spring (September to November) snowmelt. Flood frequency was influenced by winter rainfall and spring snowmelt. Under unregulated conditions, multiple flood events were quite common, especially in spring. At Dalgety, floods with flows between 15 and $20 \times 10^6 \, m^3$ per day occurred 1.7 times per year prior to the Scheme. Now the occurrence is on average once every 10 years (Snowy Water Inquiry, 1998c, Appendix 2, p. 18).

In the lower section of the Snowy River, a smaller portion of the catchment is affected by diversion. Snowmelt in the upper catchment formerly contributed large flows in the months of August and September. Currently, the mean annual flow volumes for the River at Jarrahmond (Figure 4.3) are about 53 percent of the pre-Scheme flows (Figure 4.4). Floods with peak discharges between about $20 \times 10^6 \, m^3 \, d^{-1}$ and $110 \times 10^6 \, m^3 \, d^{-1}$ have been reduced in frequency (Snowy Water Inquiry, 1998c, Appendix 2, pp. 25–26).

The low flow regime in the Snowy River has allowed saline water from the estuary to progress some 7 to 10 km upstream from its pre-Scheme location (Snowy Water Inquiry, 1998c, Appendix 2, pp. 25–26). Further information regarding pre- and post-Scheme flow regimes of the river systems in the Snowy Mountains area are available from the Snowy Water Inquiry (1998c, d).

4.3 DECLINE IN PRECIPITATION

One of the important aspects of the hydrology of the Snowy Scheme area is the decline in precipitation that has

---

[1] As a result of Andrew Barton "Banjo" Paterson's poem *The Man from Snowy River*, the Snowy River has become part of the Australian folklore.

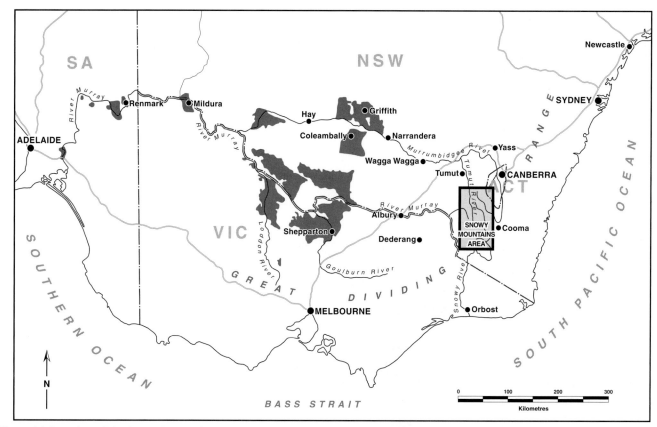

Figure 4.1 Location of the Snowy Mountains Hydro-electric Scheme (Snowy Mountains Hydro-electric Authority, 1993, p. 8).

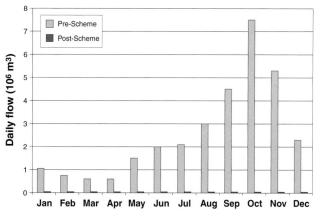

Figure 4.2 Mean daily flow of the Snowy River at Dalgety, for January to December (Snowy Water Inquiry, 1998c, Appendix 2, Figure 10).

occurred approximately over the last 140 years. Kraus (1954) studied the secular changes in the rainfall regime of southeastern Australia covering the southern part of New South Wales and Victoria. He demonstrated that the average annual precipitation declined over the period of 1869–1943. The study of rainfall records suggests that for the 70 years prior to 1945 the climate of the Snowy Mountains area became steadily drier (Pender et al., 1956). A recent investigation (Environment Australia, 2000, pp. 25 and 26) indicates that from 1901 to 1991, the amount of water entering the Scheme's catchment has declined by approximately $100 \times 10^6 \, m^3$ or $1.1 \times 10^6 \, m^3 \, yr^{-1}$. In the long-term, continuation of this trend and the prospect of global warming should reduce snow-cover in the area and affect Alpine ecosystems, hydrology and irrigation water supply (Whetton et al., 1996 and Whetton, 1998).

4.4 HISTORICAL BACKGROUND

The Snowy Mountains Hydro-electric Scheme is the result of visionary planners and courageous politicians, who proposed, designed, legislated and implemented one of the world's major engineering projects. These include the Labor Prime Minister *Ben Chifley*,[2] his Minister for Public Works and Housing *Nelson Lemmon* and *William Hudson* who was the first Commissioner of the Snowy Mountains Hydro-electric Authority. The Scheme is one of the engineering wonders of the world and the American Society

[2] See his biography in Appendix A.

THE SNOWY MOUNTAINS HYDRO-ELECTRIC SCHEME

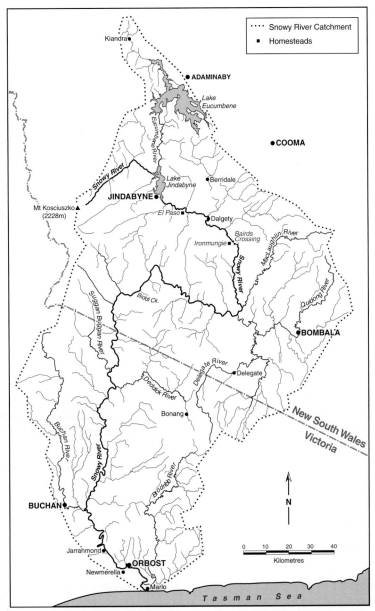

Figure 4.3 Snowy River catchment (Seddon, 1994, Figure 1.1).

of Civil Engineers on two occasions (1967 and 1997) ranked it as one of the greatest engineering achievements of the twentieth century (Collis, 1999, p. 8).

The history of water diversion from the Snowy River and development of the Scheme can be found in Hudson (1956); *Struggle for the Snowy* (Wigmore, 1968); *Snowy: The Making of Modern Australia* (Collis, 1999); and *Schemes of Nation: A Planning History of the Snowy Mountains Scheme* (Byrne, 2000). The following short history of the Scheme is mainly based on Wigmore (1968).

The potential for inland diversion of the Snowy River waters was first recognised by the Polish geologist Sir Paul Edmund Strzelecki, who explored the Australian Alps and named Mt Kosciuszko in 1840 (Shaw, 1984, p. 612). His idea was ignored until the severe drought of 1882–83 which killed 24 million sheep and destroyed the country's wheat crop. In 1884 a New South Wales Royal Commission, charged with the task of investigating the State's water resources, reported that for both the Murray and Murrumbidgee rivers, water could be stored in the mountains near their sources and then released down the rivers when the supply was low. The Commission also reported a much more visionary concept of turning the Snowy waters inland (Raymond, 1999, pp. 8 and 9). However, when the drought abated and better seasons returned, the proposal was shelved.

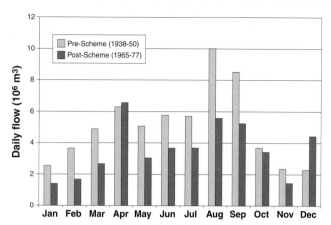

Figure 4.4 Mean daily flow of the Snowy River at Jarrahmond for January to December (Snowy Water Inquiry, 1998c, Appendix 2, Figure 12).

Between 1915 and 1918 the NSW Department of Public Works investigated the feasibility of using the Snowy River waters to generate electricity for Sydney and the south-east corner of the state. This proposal involved the construction of a dam at Jindabyne and an aqueduct leading to a power station with a capacity of 150 MW. There was no consideration of irrigation and the plan did not progress further. However, the scheme did raise the possibility of hydro-electric power generation in the Snowy area.

Various proposals to use the waters of the Snowy River were advanced in the 1920s, each one becoming more ambitious. The emphasis was then on irrigation but nothing eventuated. In 1937 the hydro-electric scheme was reviewed. Overseas consultants retained by the Department of Public Works recommended a more extensive scheme in which the aqueduct was replaced by a tunnel and the capacity increased to 250 MW.

In 1942, the NSW Government appointed a Committee under the chairmanship of the Chief Engineer of the Public Works Department to investigate the possible use of the Snowy River waters. In 1944, the Committee recommended diversion of the Snowy waters inland to the Murrumbidgee for irrigation.

The New South Wales Government adopted the recommendation, but its implementation would have prevented the Victorian Government from going ahead with a proposed hydro-electric scheme on the Lower Snowy River and diversion of Snowy waters to the River Murray. The Commonwealth Government was also concerned over its right to use the Snowy River for production of hydro-electricity for the Australian Capital Territory. A conference of Ministers of the Commonwealth and the two states was called in 1946 to consider the two proposals. The conference recognised that the Victorian Government's proposal had not been investigated in any detail. Moreover, both states insisted on their sovereignty over management of their water resources. The conference finally agreed that the Commonwealth should undertake a preliminary investigation into the practicality and implications of diversion of the Snowy waters. The Department of Works and Housing carried out the engineering investigations and the Department of Post-War Reconstruction examined the agricultural aspects. In 1947, both Departments jointly reported their findings to the Conference of the Commonwealth and State Premiers. This was the first proposal for both hydro-electric power generation and irrigation. The report stated that:

(1) based on the information available, there was no reason to doubt the practicality of the proposal;
(2) diversion of Snowy waters to the River Murray would generate more hydro-electric power than diversion to the Murrumbidgee River; and
(3) from the agricultural point of view diversion to either the Murray or the Murrumbidgee was possible, however better outcomes would flow from diversion to the Murrumbidgee.

The Premiers' Conference was undecided because the report had not considered the practicality of diversions to both the Murray and Murrumbidgee rivers. Instead, they asked that further investigations be carried out. As a result, a technical committee known as the *Commonwealth and States Snowy River Committee* was established. The Committee extended its investigations to the whole Snowy Mountains area and submitted reports in November 1948 and June 1949.

The Committee recommended to the Meeting of the Commonwealth and State Ministers in February 1949 that neither of the two original schemes should be adopted. It recommended that a totally new and much more comprehensive scheme, involving the use of Snowy waters for diversion to both the Murray and Murrumbidgee rivers, be considered. The scheme would divert 2.2×10^9 m^3 of water to both rivers and would generate 1720 MW of electricity. This capacity was nearly equal to the total capacity of all power plants in Australia. The equivalent production from thermal power stations would have required 4 Mt per year of black coal, approximately one-third of the country's total production. The generated electricity would be delivered to the capital cities of Sydney, Melbourne and Canberra and also to defence industries, at about half of the cost of producing electricity by conventional power stations using coal or oil. The Commonwealth Government intended to use the new supplies of water and electricity for decentralisation and development of large inland cities with populations of about one million in the Murrumbidgee and Murray valleys.

These plans were the largest and most significant development scheme ever envisaged in Australia. In May 1949, in a Broadcast Report to the Nation, Prime Minister Ben Chifley declared (Wigmore, 1968, p. 147):

> The Snowy Mountains plan is the greatest single project in our history. It is a plan for the whole nation, belonging to no one state nor to any group or section. It is a two-sided plan, because it provides not only for the provision of vast supplies of new power but also for an immense decentralisation of industry and population. This is a plan for the nation and it needs the nation to back it. I trust that you will all keep yourselves informed of its progress. I recommend that you listen to the discussions on it when they take place in the Parliament.

The legislation for the establishment of an Authority to implement the proposal was debated in the Commonwealth Parliament. Nelson Lemmon, the Commonwealth Minister for Public Works and Housing, in his concluding remarks on the Second Reading of the Snowy Mountains Hydro-electric Bill 1949 at the Commonwealth Parliament, said (Commonwealth of Australia, 1949, p. 250):

> I believe that this scheme will be one of the greatest single factors that will make it possible for this country to carry a greatly increased population. It will also be one of the greatest factors to bring about a permanent and effective policy of decentralisation. Because of the great quantity of water that will be made available for irrigation and the cheap power that will be generated, there is no reason why we should not see, in our time, inland cities in the Murray Valley and Murrumbidgee Valley that should carry a population of 1 000 000 people. The Government faces this task of construction in the knowledge that it will encounter many problems. No doubt, at times, we shall meet with some disappointments, and the work will have many critics. However, those critics, in the main, will be people who have little faith. This Government has faith in its engineers, its people and the future of Australia.

However, there was strong opposition to the proposed legislation. The NSW Government felt that the Commonwealth's action in legislating to divert water from the Snowy to the River Murray would seriously interfere with its right regarding its water resources, and that the legal basis of the Commonwealth action was not beyond the possibility of a challenge in the High Court. The Federal Opposition was also concerned about the legislation. During the parliamentary debate, the leader of the Opposition, Robert Menzies, referred to the importance of the Scheme, but he said (Wigmore, 1968, p. 149):

> It is a great pity that the Government should think this to be a proper occasion for brushing the states on one side, for assuming a power which it does not possess, and for setting out on an undertaking as a result of legislation which is tainted with serious illegality.

4.5 SNOWY MOUNTAINS ACT

The Commonwealth Government passed the legislation by using its defence power as defined under Section 51 of the Australian Constitution (Attorney-General's Department and Australian Government Solicitor, 1999, p. 32), and the *Snowy Mountains Hydro-electric Power Act* of 1949 became effective from 7 July 1949.[3] The argument for using the defence power was that most power stations were located close to the coast and were vulnerable to enemy attack, while the Snowy Scheme's planned power stations were far away from the coast in a mountainous region, and even some of them were to be located underground to increase their safety against any attack. Moreover, the electrical output was expected to be used for defence industries as well as civilian requirements. Under the *Act*, the Snowy Mountains Hydro-electric Authority was established and empowered to construct, maintain, operate, protect, manage and control works for the collection, diversion and storage of water, and for the generation of electricity in the Snowy Mountains area. The Authority was required to supply electricity to the Commonwealth for defence purposes and for consumption in the Australian Capital Territory and other capital cities. A Commissioner and two Associate Commissioners to be appointed by the Governor-General, constituted the Authority.

William Hudson,[4] formerly Engineer-in-Chief of the Sydney Water Board, was appointed as the first Commissioner of the Authority and took up his duties in August 1949. His Associate Commissioners were T. A. Lang, a civil engineer who at the time was the Commissioner of Irrigation and Water Supply in Queensland, and E. L. Merigan, Chief Electrical Engineer of the State Electricity Commission in Victoria.

On 17 October 1949, about three months after the *Act* came into force, construction was launched by the Governor-General Sir William McKell at the proposed dam site in the

---

[3] Text of the amended legislation is available at: http://scalePlus.law.gov.au (visited in July 2005).
[4] See his biography in Appendix A.

bed of Eucumbene River. The ceremony took place in the presence of the Prime Minister Ben Chifley and other dignitaries. However, the opposition boycotted the launch ceremony for political reasons. Unfortunately for Ben Chifley, the sense of fulfilment was short. His government was defeated two months later by Robert Menzies,[5] and he died not long after (Raymond, 1999, p. 15).

The change of the Commonwealth Government from a Labor to a Liberal-Country Party coalition in December 1949, intensified anxiety about the future of the Scheme, then less than 6 months old. An appeal to the High Court could have resulted in a judgement declaring the Scheme unconstitutional. Alternatively, the Commonwealth Government could starve it of the very substantial funds it needed and let it die from financial problems. However, because of the merit and significance of the Scheme, its appeal to the public, and the extent of the financial commitment, it was hardly likely that the new government would abandon it (Wigmore, 1968, pp. 164–165).

In 1951, there was a mild economic depression, and criticism of Commonwealth Government spending led to calls in the press for the Scheme to be shut down. The newly appointed Minister for National Development, Senator William Spooner, visited the Scheme to assess its progress and was impressed by Hudson's planning, and the Government decided to continue support for the project (Raymond, 1999, p. 31).

The first major component of the Scheme was the Guthega Project, consisting of a dam on the upper Snowy River, a tunnel and a power station. Construction began in 1951 and on 23 April 1955 the Prime Minister, Robert Menzies, opened the power station (Raymond, 1999, pp. 31 and 32).

At a function marking an advanced stage of construction at the Tumut Pond Dam in 1958, the Prime Minister, Robert Menzies, who by then had revised his opinion of the project, spoke of the triumph of the Scheme and added (Collis, 1999, p. 45):

> In a period in which we in Australia are still, I think, handicapped by parochialism, by a slight distrust of big ideas and of big people or of big enterprises.... This Scheme is teaching us and everybody in Australia to think in a big way, to be thankful for big things, to be proud of big enterprises and ... to be thankful for big men.

Following years of dispute between the Commonwealth and State Governments regarding the legality of the 1949 *Act*, the *Snowy Mountains Hydro-electric Agreement 1957*, between the States of New South Wales and Victoria and the Commonwealth, was signed in September 1957 and became formally effective on 2 January 1959 (Shellshear, 1962, p. 2).

4.6 COST OF THE SCHEME

Total expenditure of the Scheme was initially estimated at £225 million (Commonwealth Bureau of Census and Statistics, 1951, p. 1158). This was at a time when the country had a population of about 8 million, the Commonwealth Consolidated annual revenue was £554.4 million and the gross value of agricultural production was approximately £265 million (Commonwealth Bureau of Census and Statistics, 1951, pp. 758 and 938). In 1953, following preparation of detailed plans of the Scheme, the cost estimate increased to £400 million (Unger, 1989, p. 224). Finance for construction of the Scheme was provided by loans from the Commonwealth Government with the interest rate of a long-term bond. Initially this was about 3.1 percent but gradually increased over the years. Loans were to be paid back over a period of 70 years from annual incomes generated by the sale of electricity to New South Wales, Victoria and the Commonwealth. However, it was agreed to deliver water to the States free of charge for their own use (Raymond, 1999, p. 14). The Scheme was finally completed at a cost of $820 million (approximately $9 billion in 2002 prices),[6] only slightly over the 1953 estimate of £400 million (Unger, 1989, p. 224).

4.7 TECHNICAL FEATURES OF THE SCHEME

The Snowy Mountains Hydro-electric Scheme consists of two major developments:

(1) the Snowy–Tumut Diversion which diverts waters of the upper Murrumbidgee, Eucumbene, upper Tooma, and upper Tumut rivers, to the Murrumbidgee River; and

(2) the Snowy–Murray Diversion which mainly consists of the western diversion of the Snowy River waters at Jindabyne to the catchment of the upper River Murray.

The main engineering features of the Snowy Mountains Scheme are:

- 16 dams with a total capacity of $8.47 \times 10^9 \, m^3$ (Table 4.1). The most elevated reservoir is Guthega Pondage with a full supply level (FSL) of 1582 m, while the lowest one is the Blowering Dam at the FSL of 380 m (Figure 4.5).

---

[5] See his biography in Appendix A.
[6] In February 1966 the Australian currency system was converted from the British system of pounds to Australian dollars which were worth half a pound.

Table 4.1. *Dams of the Snowy Mountains Scheme in order of capacity*

Name	Type	Height (m)	Crest length (m)	Capacity (10^6 m^3)	Year of completion
Eucumbene	Earthfill	116.1	579.1	4798.40	1958
Blowering	Rockfill	112.2	807.7	1632.40	1968
Talbingo	Rockfill	161.5	701.0	920.60	1970
Jindabyne	Rockfill	71.6	335.3	689.90	1967
Tantangara	Concrete gravity	45.1	216.4	254.10	1960
Tumut Pond	Concrete arch	86.3	217.9	52.80	1959
Jounama	Rockfill	43.9	518.2	43.50	1968
Tooma	Earthfill	67.1	304.8	28.10	1961
Khancoban	Earthfill	18.3	1066.8	21.50	1966
Geehi	Rockfill	91.4	265.2	21.10	1966
Island Bend	Concrete gravity	48.8	146.3	3.02	1965
Tumut-2	Concrete gravity	46.3	118.9	2.70	1961
Murray-2	Concrete arch	42.7	131.1	2.30	1968
Guthega	Concrete gravity	33.5	139.0	1.55	1955
Happy Jacks	Concrete gravity	29.0	76.2	0.27	1959
Deep Creek	Concrete gravity	21.3	54.9	0.005	1961
Total				8472.245	

Source: Snowy Mountains Hydro-electric Authority (1993, p. 172).

- 135 km of tunnels (Table 4.2).
- Two pumping stations at Tumut 3 and Jindabyne with respective capacities of 297.3 m^3 s^{-1} and 25.5 m^3 s^{-1} (Table 4.3).
- Seven power stations with a total generating capacity of 3756 MW (Table 4.4).
- Twenty aqueducts with a total length of 80 km.

The Scheme's network of dams and reservoirs, with Lake Eucumbene as the largest storage facility (Figures 4.5 and 4.6), collects and stores a high percentage of the precipitation that falls in the Snowy Mountains Scheme area.

The Snowy Scheme power stations are designed to provide a large generating capacity with a rapid response when needed at peak hours or at emergency cases. The Scheme can start its generators in less than 90 seconds, and provide power to the transmission grid. This is not possible for the thermal power stations, which require much longer times to start generating electricity.

Further technical and engineering features of the Scheme, including maps, cross-sections, diagrams and photos of all dams, power stations, tunnels, and aqueducts are available in the Snowy Mountains Hydro-electric Authority (1993). Also, McLean (1999) provides a comprehensive list of the technical writings about the Scheme. Appendix B shows a timetable of the Scheme's construction.

4.8 WATER RELEASES

Since its completion, the Scheme has released an average of 2.41×10^9 m^3 yr^{-1} with approximately equal volumes to both the Murray and Tumut developments (Table 4.5). Water sourced from the Snowy catchment increases the average catchment yield in the Upper Murray from 0.62×10^9 m^3 yr^{-1} to 1.2×10^9 m^3 yr^{-1} and the Murrumbidgee from 0.65×10^9 m^3 yr^{-1} to 1.21×10^9 m^3 yr^{-1}. The additional water for the Murrumbidgee River is released from Blowering Dam via the Tumut River. Flows in the upper Murrumbidgee are on the other hand reduced on average by 0.29×10^9 m^3 yr^{-1} as a result of diversion of water from Tantangara Dam to Lake Eucumbene (Environment Australia, 2000, pp. 11 and 12). A small level of planned riparian releases and spills occur in the majority of rivers and streams within the Scheme area (Table 4.6).

4.9 ELECTRICITY PRODUCTION

In 1949, the planned power generation of the Scheme (1720 MW) was equal to the total capacity of Australia. However, the share of hydro-electricity in total electricity generation of the country dropped gradually to 7 percent in 1974–75 and to 3 percent in 2000–01 (Australian Bureau of

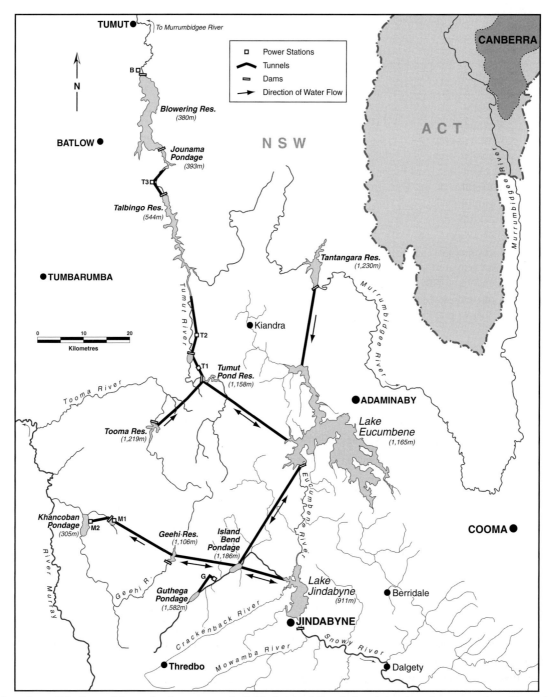

Figure 4.5 Locations of the Snowy Mountains dams, reservoirs (with their Full Supply Levels in m), and power stations (Shaw, 1984, p. 697).

Statistics, 2004, p. 509). The Snowy Scheme is still the largest producer of hydro-electricity in Australia contributing about 50 percent.[7]

Formerly, state-owned authorities handled the generation, transmission and distribution of electricity to the consumers at a price fixed by the state governments. Over the last 20 years, these state monopolies have been broken down into power producers, transmission companies and retailers (suppliers to the consumers). When demand is steady and supply is in excess of demand, prices are low. However, when demand exceeds supplies, due to weather conditions or

---

[7] Hydro-electric power producers in Australia are: Snowy Mountains (3756 MW); Tasmania (2265 MW); Queensland (632 MW); Victoria (467 MW); New South Wales (423 MW); and Western Australia (32 MW). **Source:** http://www.actewagl.com.au/education/electricity/generation/hydroelectric.cfm (visited in July 2005).

Table 4.2. *Tunnels of the Snowy Mountains Scheme*

Tunnel	Length (km)	Excavated section (m × m)	Lined section (m × m)	Percentage lined	Year of completion
Eucumbene – Snowy	23.5	6.30 × 6.35	6.10 × 6.10	19.7	1965
Eucumbene – Tumut	22.2	6.91	6.40	28.3	1959
Murrumbidgee – Eucumbene	16.6	3.35 × 3.35	3.10 × 3.10	17.7	1961
Snowy – Geehi	14.5	6.30 × 6.30	6.10 × 6.10	13.3	1966
Tooma – Tumut	14.3	3.79 × 3.71	3.43	20.0	1961
Murray 1 Pressure	11.8	–	6.93 × 6.93	100	1966
Tumut 2 Pressure and Tailwater	11.3	–	6.40	100	1961
Jindabyne – Island Bend	9.8	3.96 × 3.96	3.76	10.6	1968
Guthega	4.7	5.87 × 5.74	5.26 × 5.05	11.6	1955
Murray 2 Pressure	2.4	–	7.47 × 7.47	100	1969
Tumut 1 Pressure	2.4	–	6.40	100	1959
Tumut 1 Tailwater	1.3	8.53 × 7.77	7.93 × 7.49	54.5	1959
Total	134.8	–	–	–	–

Source: Snowy Mountains Hydro-electric Authority (1993, p. 172).

Table 4.3. *Pumping stations of the Snowy Mountains Scheme*

Pumping station	Number of units	Capacity ($m^3 s^{-1}$)	Total pumping head (m)	Year of completion
Tumut 3	3	297.3	155.1	1973
Jindabyne	2	25.5	231.6	1969

Source: Snowy Mountains Hydro-electric Authority (1993, p. 172).

Table 4.4. *Power stations of the Snowy Mountains Scheme*

Power station	Installed capacity (MW)	Number of units	Head (m)	Year of completion
Tumut 3	1500	6	150.9	1973
Murray 1	950	10	460.2	1967
Murray 2	550	4	264.3	1969
Tumut 1	330	4	292.6	1959
Tumut 2	286	4	262.1	1962
Blowering	80	1	86.6	1969
Guthega	60	2	246.9	1955
Total	3756	31	–	–

Source: Snowy Mountains Hydro-electric Authority (1993, p. 172).

mechanical breakdowns, prices can rise considerably. The Snowy's quick-start capacity can generate power within minutes and gives it the ability to access higher prices (Raymond, 1999, p. 113).

4.10 WORKFORCE

For the first decade, engineers from the US Bureau of Reclamation designed major components of the Scheme and monitored their construction (Raymond, 1999, p. 20). The Bureau had unequalled experience in water resources development projects, such as the giant Hoover Dam on the Colorado River, and the Central Valley Project (see Section 11.14.3). Each year a group of young Snowy Scheme engineers went to the United States for job training with the Bureau. By 1958 the Authority was confident enough to take over the design and supervision of all remaining dams, tunnels, power stations and transmission lines (Raymond, 1999, p. 20).

Construction of the Scheme, which started on 17 October 1949, was completed in 1974. The number of people employed by the Authority and its contractors in the Snowy Mountains area reached a peak of 7300 in 1959 and gradually decreased as the remaining projects were completed (Snowy Mountains Hydro-electric Authority, 1993, p. 7). During the 25-year construction period, 100 000 men and women from 33 countries worked on the Scheme (Collis, 1999, pp. 38–39). Two thirds of them were migrant workers. Despite all precautions, 121 men were killed during construction with at least half of these deaths occurring in the tunnels. During the construction period, seven regional townships, 100 camps and 1600 km of roads and tracks were constructed. Moreover, the project generated a major influx of skilled migrants into the Australian workforce. Numerous contractors from various countries, including Australia, Belgium, France, Italy,

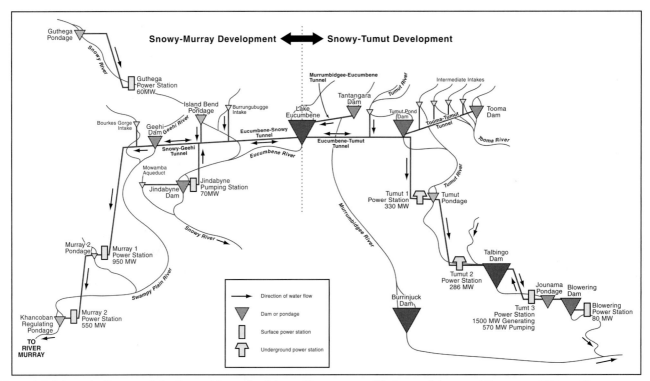

Figure 4.6 Schematic diagram of the Snowy Mountains Scheme (Department of Industry, Science and Resources 2000, p. 38, Figure 4.1).

Table 4.5. *Catchment contributions to annual flows*

Development	Flow contribution (10^9 m^3 yr^{-1})
1. Snowy–Murray development	
Geehi	0.35
Jindabyne	0.36
Island Bend	0.54
Sub-total	*1.25*
Losses[a]	*0.05*
Net average release	*1.20*
2. Snowy–Tumut development	
Eucumbene	0.29
Tooma	0.30
Tumut Pond	0.36
Tantangara	0.30
Sub-total	*1.25*
Losses[a]	*0.04*
Net average release	*1.21*
Total net average releases	2.41[b]

Note: [a] Losses are mostly due to evaporation; [b] This includes 1.14×10^9 m^3 diversion from the Snowy River.
Source: Department of Industry, Science and Resources (2000, Table 4.1).

Japan, Switzerland, United States of America and the United Kingdom, contributed to the construction of the Scheme.

From 1960, the Authority used its pool of skilled people to engage in other projects in Australia and overseas. This led to the formation of the Snowy Mountains Engineering Corporation (SMEC) in 1970 to undertake external projects. SMEC was a highly professional public owned consulting organisation which worked in more than 60 countries on the investigation, design and construction of engineering projects. SMEC was privatised in 1993 and is wholly owned by its staff (Raymond, 1999, p. 17).

The fascinating human story of the Snowy Scheme is described in numerous publications including *Voices from the Snowy* (Unger, 1989),[8] *The Snowy: The People Behind the Power* (McHugh, 1989), *Mud, Sweat and Snow* (Gough, 1994), and *The Snowy: Cradle of a New Australia* (Kobal, 1999).

---

[8] Margaret Unger was the daughter of William Hudson. She worked as a junior member of the Authority's public relations team from 1952 to 1956.

Table 4.6. *Existing riparian releases and spills in the Snowy Mountains Scheme area*

Catchment	Annual flows ($10^6 \, m^3$)	Percentage of the average natural low
Snowy River		
Jindabyne dam	10	1
Mowamba river	22	2
Sub-total	32	
Montane Rivers		
Upper Murrumbidgee river	6	2
Geehi river	17	9
Tooma river	10	5
Goodradigbee river	<1	2
Sub-total	33	
Upper Snowy River		
Guthega	32	14
Island Bend	37	9
Sub-total	69	23
Total	134	

Source: Department of Industry, Science and Resources (2000, Table 4.2).

4.11 ENVIRONMENTAL IMPACTS OF THE SCHEME

Prior to the commencement of the Scheme, activities such as grazing, burning of the native vegetation and introduction of animals (rabbits, goats, pigs and horses), fish and weeds (including willow and blackberry) had degraded the catchments of the Snowy and other rivers in the area (Snowy Water Inquiry, 1998d, Appendix A2, pp. 1 and 2). Their impacts on the rivers and streams included:

- Marked increase in the amount of sediment and nutrients transported.
- Damaged or changed riparian vegetation.
- Reduced populations of native fish and probably invertebrates by the introduction of fish such as brown and rainbow trout.
- Reduced aquatic habitat by changing riparian and aquatic vegetation.

From its establishment, the Authority attempted to implement a number of measures to minimise the environmental impact of the Scheme. In 1955, the Authority's budget for soil conservation was £200 000, the equivalent of $7.8 million in 1999 values (Snowy Mountains Hydro-Electric Authority, 1999, p. 2). Since 1967, the Authority has been involved in the establishment and management of the Kosciuszko National Park, as the majority of the Scheme's infrastructure lies within the Park.

The biggest environmental impact of the Snowy Scheme has been the diversion of flow from the Snowy River, which has changed its natural flow pattern and riverine environment (Snowy Water Inquiry, 1998d, Appendix A2, pp. 4 and 5). Some of the issues are:

- Reduced sediment carrying capacity leading to sediment deposition in channels, pools along the river, and adjacent to the coastal floodplain.
- Isolation of the river channel from riparian vegetation in some reaches, and invasion of the channel by vegetation in other reaches.
- Water quality changes (in particular in the pooled sections of the rivers), such as higher summer water temperatures and lower winter temperatures because of reduced flows, lower dissolved oxygen, higher nutrient concentrations, and higher algal productivity.
- Changes in the macro-invertebrate communities in many reaches from those favouring cold, fast-flowing water to still water varieties.
- Changes to the populations of frogs and native fish, with some increasing and others decreasing in different areas.
- Reduced overbank flooding on the floodplain from upstream of Orbost to Marlo (Figure 4.3).

In irrigated areas, diverted waters of the Snowy River have made major contributions to the economic and social welfare of numerous communities along the Murrumbidgee and River Murray valleys. However, this was not without serious side effects. In these areas, inefficient use of water has lead to widespread watertable rise and salinisation. On the positive side, unlike power stations which burn fossil fuels, the Snowy Scheme has the advantage of not producing greenhouse gases. The Snowy's annual output of electricity saves the release of 5 000 000 tonnes of carbon dioxide, 15 000 tonnes of sulphur and 10 500 tonnes of nitrous oxides (Raymond, 1999, p. 112).

4.12 CORPORATISATION OF THE SCHEME

In 1997, the Snowy Mountains Hydro-electric Scheme was corporatised under the *NSW Snowy Hydro Corporation Act 1997*. The aim of corporatisation was to establish a commercially viable electricity generation business, *Snowy Hydro Ltd (SHL)*, which can compete in the National Electricity Market. Its owners are the New South Wales

Government (58 percent), Victorian Government (29 percent) and the Commonwealth Government (13 percent). These percentages reflect the previously agreed entitlements of the three governments to the electricity supplies of the Scheme (Snowy Water Inquiry, 1998a).

As part of the corporatisation process, in 1998 the Snowy Water Inquiry examined the environmental issues arising from the pattern of water flows caused by the operation of the Scheme (see the following section). Among other things, this was followed by the release of the Draft Environmental Impact Statement (Department of Industry, Science and Resources, 2000) and the Environment Assessment Report (Environment Australia, 2000), which assessed the environmental impacts of the proposed corporatisation. *Snowy Hydro Ltd* became official on 28 June 2002. Its general operations are subject to New South Wales' laws and regulations. It is licensed for 75 years to collect and use water for the purposes of power generation.

4.13 THE SNOWY WATER INQUIRY

The Snowy Water Inquiry was initiated by the New South Wales, Victorian and Commonwealth Governments under the *NSW Snowy Hydro Corporation Act 1997*, to address concerns about the impacts of water diversions on the health of the Snowy River. The Inquiry's full Terms of Reference are available in the Snowy Water Inquiry (1998b, c). These can be summarised as:

- Examine environmental issues arising from the pattern of water flows caused by the operation of the Snowy Mountains Scheme in rivers and streams in the Snowy water catchment.
- Develop and submit to the governments a range of comprehensive, fully costed options to address those environmental issues.
- Consider the environmental, economic, agricultural, social, heritage and other impacts of the options.
- Identify costs and benefits of each option, so far as practicable, in terms of currently available information.
- Consider the full range of options available to address the environmental issues including: (1) environmental flows in all rivers and streams; (2) altered catchment practices; and (3) river remediation works.
- Identify the environmental management objectives of each option.
- In developing the options, the Inquiry must have regard to: (1) the environmental and water management legislation of the Commonwealth, New South Wales and Victoria; (2) the policies of the three governments in relation to streams and rivers affected by the operation of the Scheme; and (3) the Murray–Darling Basin Agreement and Ministerial Council policies and various Acts.

The Inquiry involved extensive community consultation including receipt of 473 submissions and numerous public hearings, and expert analysis. The final report of the Inquiry was submitted to the respective governments on the 23rd October 1998 (Snowy Water Inquiry, 1998c). It describes 23 Options (Options 1 to 23) and seven Composite Options (Options A, B, C, D, E, F and G). Table 4.7 shows a summary of the composite options. The seven Composite Options are:

- **Option A: Sustained current environmental condition of all rivers in the inquiry area.** The aim of this option is to sustain the current modified condition of the rivers. Its purpose is to demonstrate the nature and range of activities necessary to prevent further deterioration of the environmental conditions due to the impacts of the Scheme and the modifications of the riverine environment.
- **Option B: Improve environmental conditions in rivers of the Scheme.** This option builds on the current sustainable conditions for all rivers in Option A. Additional flows, catchment and riverine works and measures are provided for the Snowy River below Jindabyne and the upper Murrumbidgee River.
- **Option C: Further improve environmental conditions in rivers downstream of the Scheme.** This option builds on Options A and B. Additional flows are provided to the Snowy River below Jindabyne and the Upper Murrumbidgee River.
- **Option D: Improve conservation values including the environment of high mountain rivers.** This option builds on Option B. Additional flows, catchment and riverine works and measures are provided to the upper Murrumbidgee River and the Snowy River.
- **Option E: Further improve conservation values, extending the area of high mountain rivers.** This option builds on Option D and the upper Snowy River. Additional flows and catchment and riverine works and measures are provided to the Upper Murrumbidgee and the Tooma River.
- **Option F: Restore pools in the Snowy floodplain reaches.** This Option builds on Option E. Additional channel maintenance flows and physical in-stream works in the Orbost Reach are provided to the Snowy River.

Table 4.7. *Composite options for the Snowy*

Composite option	A	B	C	D	E	F	G
1. Flows ($10^6 \text{ m}^3 \text{ yr}^{-1}$)							
Within Scheme	15	30	44	96	152	152	169
Out of Scheme	38	140	198	140	140	256	140
Provision for max flow	8	20	40	20	140	51	140
2. Costs – Water ($m)							
Catchment works	23	23	24	24	24	32	23
Water efficiency investment	9	42	42	42	42	42	42
3. Cost – Electricity ($m)							
Greenhouse abatement	6	23	35	41	53	70	49
Fuel substitution	6	36	54	54	67	98	64
Outlet works	1	22	32	27	27	38	27
4. Cost – Economic ($m)							
Salinity	0	3	3	3	3	3	3
Agriculture	0	3	27	3	10	60	6
Total Costs of 2, 3 and 4 ($m)	45	152	217	194	226	343	214
5. Benefit – Use Values ($m)							
Canoeing and rafting	0	3	3	3	3	4	3
Fishing	0	23	30	45	46	57	23
Total Benefits ($m)	0	26	33	48	49	61	26
Net Threshold Cost ($m)	45	126	184	146	177	282	188
Net additional cost – max flow ($m)	5	17	38	22	22	45	21
Environment/Social Values ($m)							
Impact on Snowy Hydro ($m)	28	83	119	108	130	188	121
Additional impact of max flow ($m)	0	1	3	1	1	4	1

Source: Snowy Water Inquiry (1998c, Table 1).

- **Option G: Maintain and incrementally improve the environmental condition of all rivers in the Inquiry area.** This option builds on the current sustainable conditions for all rivers in Option A. Additional flows, and where appropriate catchment and riverine works and measures, are provided to the Snowy River below Jindabyne, the Snowy River above Jindabyne, the Upper Murrumbidgee River, the Geehi and Upper Swampy Plain rivers, additional flow in the Eucumbene, Tooma and Tumut rivers and current flows in the lower Swampy Plain and Murray rivers.

The development of the Composite Options was intended to provide Governments with a reduced number of options to assist their choice of the most suitable one. However, the Commissioner of the Inquiry (Robert Webster) nominated the Composite Option D as his preferred option for implementation by the Governments. This option recommends returning $140 \times 10^6 \text{ m}^3$ per annum of irrigation water lost by leakage from the Murray and Murrumbidgee Rivers to the Snowy River at the Jindabyne Gorge. This volume of water is about 15 percent of the average natural flow of the Snowy River.

The Composite Option D in Table 4.7 would cost $194 million including $72 million in water related costs ($24 million for Alpine catchment and river works; $42 million in works to make the necessary water savings; and $6 million in costs to agriculture and increased salinity). The other $122 million are costs associated with electricity-generation ($27 million for works on the Snowy's dams; $54 million to substitute fossil fuels for electricity generation; and $41 million in greenhouse abatement costs). This option has been claimed to be a balanced option for the requirements of:

(1) the Snowy River;
(2) the Snowy Scheme; and
(3) the Murray and Murrumbidgee irrigation areas.

4.14 THE ENVIRONMENTAL FLOW AGREEMENT

Submission of the Final Report of the Snowy Water Inquiry and its recommendation of a 15 percent flow return to the Snowy River was not satisfactory to either the irrigation or environmental lobby groups. Irrigators claimed that a 15 percent return to the Snowy River posed a threat to River Murray water users, while the environmental groups claimed a 28 percent return of flow to the Snowy River was required to restore the River's ecology.

Finally, the three governments (New South Wales, Victoria and the Commonwealth) agreed to increase the environmental flows of the Snowy and the Murray rivers as well as the other rivers in the Snowy Scheme area. This would improve the currently degraded riverine environments and the habitat for a diverse range of species through a combination of: improving the temperature regime of river water; achieving channel maintenance and flushing flows within rivers; restoring connectivity within rivers for migratory species and for dispersion; improving triggers for fish spawning; and improving the aesthetics of the riverine environments. Moreover, the agreement will safeguard interests of irrigators and will maintain the quantity and quality of South Australia's water supply (Environment Australia, 2000, p. 39). Water to be released to the environment is expected to come from decommissioning of various aqueducts in the Snowy Mountains area and water saving projects (Environment Australia, 2000, pp. 20–22 and 35–36).

The agreement is to increase the environmental flows of the Snowy River by up to 28 percent of its Average Natural Flow (ANF) below the Jindabyne Dam at the confluence of the Mowamba and Snowy rivers (Figure 4.5) at a cost of $300 million, shared equally by the two governments of NSW and Victoria. The Premier of NSW and his Victorian counterpart opened the first stage of the agreement on 28 August 2002 by decommissioning the Mowamba River Aqueduct. This release has increased the environmental flow of the Snowy River by $38 \times 10^6 \, m^3 \, yr^{-1}$. This volume plus the Average Base Passing Flow (ABPF) of $35 \times 10^6 \, m^3 \, yr^{-1}$ (consisting of $9 \times 10^6 \, m^3 \, yr^{-1}$ release from Jindabyne Dam plus $26 \times 10^6 \, m^3 \, yr^{-1}$ of the non-regulated flow past the Works on the Mowamba River and Cobbon Creek) at the same reference point has increased the environmental flow of the river to about 6 percent of its ANF ($1177 \times 10^6 \, m^3 \, yr^{-1}$). Other targets are:

- **The 7-year target** is to increase the environmental flow by $142 \times 10^6 \, m^3$ per year. This volume plus the ABPF of $35 \times 10^6 \, m^3 \, yr^{-1}$ would increase the environmental flow to a level of $177 \times 10^6 \, m^3 \, yr^{-1}$ or 15 percent of the ANF of the Snowy River.
- **The 10-year target** is an increase of $212 \times 10^6 \, m^3 \, yr^{-1}$, plus the ABPF volume of $35 \times 10^6 \, m^3 \, yr^{-1}$, increasing the environmental flow in the Snowy River to $247 \times 10^6 \, m^3 \, yr^{-1}$ or 21 percent of its ANF.
- **The long-term (>10 years) target** is to increase the environmental flow by $294 \times 10^6 \, m^3$ per annum. By including the ABPF, this will constitute a total volume of $329 \times 10^6 \, m^3 \, yr^{-1}$ or 28 percent of the Snowy River's ANF. However, achievement of this target would depend on the success of water saving projects.

Within the framework of the agreement, the environmental flows of the upper Murrumbidgee River, and key Alpine rivers of the Kosciuszko National Park will also be enhanced. It was also agreed to increase the environmental flow for the River Murray by $70 \times 10^6 \, m^3$ per year at a cost of $75 million, funded by the Federal Government over 10 years from 2002.

4.15 PRECIPITATION ENHANCEMENT PROJECT

Atmospheric precipitation over the Snowy Scheme area has been declining for about 140 years (see section 4.3). In 2004, the Snowy Mountains region had been in a severe drought for seven years, which was considered the worst drought in the area for 20 years. As a result, in July 2004 water storages in major reservoirs of the Scheme were approximately 50 percent for Lakes Eucumbene and Jindabyne, and only 8 percent for Tantangara Reservoir.[9] Projections indicated that under continuing dry weather patterns, water storages in major reservoirs of the Scheme would decline even further. However, precipitation increased in the second half of 2004. Subsequently, by the end of the year water storage increased to about 65 percent in Lake Eucumbene, 63 percent in Lake Jindabyne, and 10 percent in Tantangara Reservoir.

The *Snowy Hydro Ltd* is conducting a six-year research project of winter cloud seeding to assess the feasibility of increasing snow precipitation in the Snowy Mountains at a cost of $20 million. The New South Wales Parliament approved the *Snowy Mountains Cloud Seeding Trial*

---

[9] For plots of storage levels in Lake Eucumbene, Lake Jindabyne and Tantangara Reservoir since 1997 visit: http://www.snowyhydro.com.au/lakeLevels.asp?pageID=47&parentID=6 (visited in July 2005).

Bill 2004 and the cloud seeding research commenced in the winter of 2004. The target area covers 1000 km^2 of the Kosciuszko National Park above 1400 m elevation. The expected average annual increase in snowfall over the target area is approximately 10 percent. This increase in snowfall is equal to an approximate 70×10^6 m^3 increase in annual water yield in the area. This would partially offset the impacts of drought conditions on the irrigators and would assist hydro-power production. Also, increased snowfall in the region would be beneficial to flora and fauna of the Alpine regions and the tourism industry during the ski seasons.[10]

4.16 CONCLUSIONS

The Snowy Mountains Hydro-electric Scheme launched in October 1949 is the largest engineering project ever undertaken in Australia. Australia had a population of 8 million at the time, and the initial estimated cost of the project was approximately 40 percent of the Commonwealth annual consolidated revenue. The project took 25 years to complete, employed a workforce of 100 000 from 33 countries at a total cost of $9 billion in current values. The Federal Government intended to use the water from the Snowy River to boost economic activities in inland areas of New South Wales and Victoria, facilitate settlement of Returned Soldiers, and create a renewable energy source equivalent to the total existing electricity generating capacity of the country.

The Scheme was designed at a time when only engineering issues were considered. It has made a significant contribution to the economic and social development of the country. The project was highly subsidised and farmers received water at a very low cost. Despite its benefits, there have also been negative impacts. These include a contribution to the development of irrigated land salinity across large tracts of the Murray and Murrumbidgee valleys. The legendary Snowy River has become degraded as a result of diversion of 99 percent of its headwaters. Under pressure from the community and environmentalists, the New South Wales, Victoria and Federal governments have agreed recently to alleviate problems and to increase the flow rate of the Snowy River at Dalgety to 15 percent and ultimately to 28 percent of its pre-Scheme rates.

Despite the enormous benefits and knowledge gained from the Scheme, under the present state of environmental knowledge and economic rationale, it is unlikely that a similar project (or even a smaller one) will be approved in the foreseeable future, unless water scarcity becomes serious.

References

Attorney-General's Department and Australian Government Solicitor (1999). *The Constitution: As in Force on 1 July 1999*. Canberra: Office of Legislative Drafting, Attorney-General's Department.

Australian Bureau of Statistics (2004). *Year Book Australia 2004*. Canberra: ABS.

Byrne, G. (2000). *Schemes of Nation: A Planning History of the Snowy Mountains Scheme*. Sydney: Department of Art History and Theory, University of Sydney. PhD dissertation.

Collis, B. (1999). *Snowy: The Making of Modern Australia*. Canberra: Tabletop Press.

Commonwealth Bureau of Census and Statistics (1951). *Official Year Book of the Commonwealth of Australia*. Canberra: Commonwealth Government Printer.

Commonwealth of Australia (1949). *Parliamentary Debates, Sessions 1948–49: Second Session of the Eighteenth Parliament*, Third Period, from 18th May 1949 to 7th July 1949. Canberra: Commonwealth Government Printer.

Department of Industry, Science and Resources (2000). *Corporatisation of the Snowy Mountains Hydro-electric Authority: Draft Environmental Impact Statement*. Canberra: Department of Industry, Science and Resources.

Environment Australia (2000). *Environmental Assessment Report: Proposed Corporatisation of the Snowy Mountains Hydro-electric Authority*. Canberra: Department of Environment and Heritage.

Gough, N. (1994). *Mud, Sweat and Snow: Memories of Snowy Workers 1949–1959*. Tallangatta, Victoria: Noel Gough.

Hudson, W. (1956). The Snowy Mountains Scheme. In *Year Book of the Commonwealth of Australia*. Canberra: Commonwealth Government Printer, pp. 1103–1130.

Kobal, I. (1999). *The Snowy: Cradle of a New Australia*. Rydalmere, NSW: Ivan Kobal.

Kraus, E. B. (1954). Secular changes in the rainfall regime of S.E. Australia. *The Quarterly Journal of the Royal Meteorological Society* **80**: 591–601.

McHugh, S. (1989). *The Snowy: The People Behind the Power*. Port Melbourne: William Heinemann.

McLean, J. N. ed. (1999). *Technical Writings about the Snowy Mountains Scheme: A Bibliography of Works Published or Presented*: 1941–1998. Cooma, NSW: Snowy Mountains Hydro-electric Authority.

Pender, E. B., Walsh, D. T. and Anderson, D. (1956). Hydrology in the Snowy Mountains area. *Journal of the Institutions of Engineers Australia* **28**(3): 51–66.

Raymond, R. (1999). *A Vision for Australia: The Snowy Mountains Scheme 1949–1999*. Edgecliff, NSW: Focus Publishing.

Ryan, B. F. and Sadler, B. S. (1995).[11] *Guidelines for the Utilisation of Cloud Seeding as a Tool for Water Management in Australia*. Canberra: Agricultural and Resource Management Council of Australia and New Zealand.

Seddon, G. (1994). *Searching for the Snowy: An Environmental History*. St Leonard, NSW: Allen & Unwin.

Shaw, J. ed. (1984). *Collins Australian Encyclopedia*. Sydney: William Collins.

Shellshear, W. M. (1962). *The Snowy Scheme*. Sydney: Horwitz Publications.

---

[10] A. Further information about the Snowy Precipitation Enhancement Research Project is available at: http://www.snowyhydro.com.au/levelThree.asp?pageID=85&parentID=254&grandParentID=3 and http://www.snowyhydro.com.au/files/SPET.pdf (both visited in July 2005).

B. For information regarding California's experience in precipitation enhancement see the relevant section 11.15.8.

C. Information regarding cloud seeding in Australia is available in Ryan and Sadler (1995) and the URL http://www.hydro.com.au/home/Energy/Cloud+Seeding/ (visited in July 2005).

[11] http://www.dar.csiro.au/publications/cloud.htm (visited in July 2005).

Snowy Mountains Hydro-electric Authority (1993). *Engineering Features of the Snowy Mountains Scheme*, Third edition. Comma, NSW: SMHEA.

Snowy Mountains Hydro-electric Authority (1999). *Meeting the Environmental Challenge, 1999*. SMHEA.

Snowy Water Inquiry (1998a). *Snowy Water Inquiry: Issues Paper*. Sydney: Snowy Water Inquiry.

Snowy Water Inquiry (1998b). *A Guide to the Snowy Water Inquiry*. Sydney: Snowy Water Inquiry.

Snowy Water Inquiry (1998c). *Snowy Water Inquiry: Final Report*. Sydney: Snowy Water Inquiry.

Snowy Water Inquiry (1998d). *Appendix of Resource Materials: Part 1*. Sydney: Snowy Water Inquiry.

Unger, M. (1989). *Voices from the Snowy*. Sydney: New South Wales University Press.

Whetton, P.H., Haylock, M.R. and Galloway, R. (1996). Climate change and snow-cover duration in the Australian Alps. *Climate Change* **32**: 447–479.

Whetton, P. (1998). Climate change impacts on the spatial extent of snow-cover in the Australian Alps. In Green, K. ed. *Snow: A Natural History – An Uncertain Future*. Canberra: Australian Alps Liaison Committee, pp. 195–206.

Wigmore, L. (1968). *Struggle for the Snowy: The Background of the Snowy Mountains Scheme*. Melbourne: Oxford University Press.

5 Inter-Basin Water Transfer from Coastal Basins of New South Wales

5.1 INTRODUCTION

Some of the coastal New South Wales rivers appear to have the potential for diverting part of their flow into the inland basins. However, the New South Wales Department of Infrastructure, Planning and Natural Resources (DIPNR) has no plans for inter-basin water transfer for the near future, although the Department's current water sharing policy under the *Water Management Act 2000* (DLWC, 2001) does not exclude inter-basin water transfer.[1]

Out of 22 coastal river basins in NSW, only nine have their western boundaries on the Great Dividing Range and are suitably located for inland water diversion (Figure 5.1). Five of these basins: the Hunter (II-10), Hawkesbury (II-12), Shoalhaven (II-15), Bega (II-19), and Snowy[2] (II-22) do not have excess water for diversion (Water Resources Commission of NSW, 1981, Table 3.1).

In the absence of recent published data for the NSW river basins (National Land & Water Resources Audit, 2001, Appendix 1), Table 5.1 provides characteristics of the four remaining basins in 1983–84. It indicates that water use in these basins (from both surface and groundwater resources) was less than one percent of their respective average annual run-off. Assuming doubling water consumption since 1984 and further doubling in the following two decades, and ignoring the environmental flow requirements, it seems that a considerable volume of water may be uncommitted in these basins for possible inland diversion. However, the issue is far more complex because of the high cost, environmental concerns and opposition of the community to large-scale inland water transfer projects.

5.2 ENVIRONMENTAL PROBLEMS OF THE NORTH COAST RIVER BASINS

Agriculture is the single largest contributor to the North Coast economy. The farm gate value of agricultural production is in excess of $650 million per annum. This rises to about $1 billion when processing is included, particularly in the meat, milk, sugar and macadamia industries. However, demographic studies indicate that the population of the North Coast is set to rise 30 percent by 2030, especially along the coastal fringe. Tourism is steadily overtaking agriculture as the largest industry in the region (Healthy Rivers Commission, 2003a, p. 43).

In the South East Coast Division II (Figure 5.1), the Clarence River Basin is the largest in terms of basin size and river flows. About 20 percent of the basin is national park, 30 percent is managed by State Forests, around 49 percent is agricultural land, and less than one percent is urbanised. It is unique because it lies in a transition zone between temperate and tropical flora, making it one of the most biodiverse regions in Australia. Approximately 60 percent of the Clarence River Basin satisfies the National Parks and Wildlife Service's criteria for high conservation value (Healthy Rivers Commission, 1999). Although many parts of the Clarence River Basin are in relatively good condition, some parts of the Basin are affected by significant land, water and environmental degradation. Environmental problems of the Basin are caused by various activities including the following (Healthy Rivers Commission, 1999):

- **Water extraction.** Increase in the volume and change in the pattern of extractions for town water supply, irrigation and hydro-power generation.
- **Agricultural activities.** Agricultural practices associated with grazing, horticulture and irrigation are seen as contributing to erosion, sedimentation, loss of habitat and nutrient inputs.
- **Wastewater.** Sewage effluent from sewage treatment plants, thirty unsewered towns, rural residential developments, boats (particularly during summer), and storm run-off from urban areas.

---
[1] http://www.dipnr.nsw.gov.au/water/sharing/ (visited in July 2005).
[2] See Chapter 4 for the Snowy Scheme.

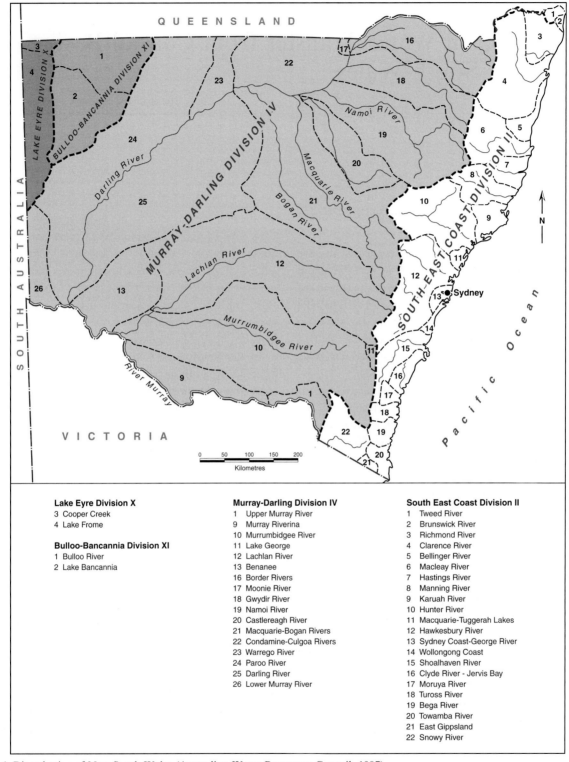

Figure 5.1 River basins of New South Wales (Australian Water Resources Council, 1987).

Table 5.1. *Characteristics of four coastal river basins of New South Wales in 1983–84*

River basin	Basin No.	Area (km^2)	Mean annual run-off (10^6 m^3)	Population[a]	Area irrigated (ha)	Annual water use[b] (10^6 m^3) SW[c]	GW[c]	Total
Clarence	II-4	22 700	5000	60 000	725	26.3	0.7	27.0
Macleay	II-6	11 500	2000	50 000	700	9.0	10.0	19.0
Manning	II-8	8420	2500	35 000	2050	11.3	0.4	11.7
Tuross	II-18	2180	610	8000	4	1.7	–	1.7

Note: [a] Not representative of summer population; [b] From 1 July 1983 to 30 June 1984; [c] SW (surface water) and GW (groundwater).
Source: Australian Water Resources Council (1987).

- **Water quality.** High bacterial and viral contamination in the Clarence River at Grafton, pollutants from diffuse sources, algal problems in parts of the Clarence and Nymboida rivers, and high levels of turbidity in some areas.
- **Mining activities.** Erosion and sedimentation problems in parts of the Clarence River and its tributaries due to past mining activities, and opposition to establishment of a new gold mine and its potential impact on the human health and the environment if an accident occurs at the mine resulting in the release of cyanide into the river system.
- **Drainage and dredging.** A range of activities such as drainage of acid sulphate soils, accumulation of sediment and dredging, affects the health of wetlands and the estuarine ecosystems.
- **Fishing.** Commercial and recreational fishing in the estuary impacts on fish and prawn stocks, seagrass beds and benthic species.
- **Vegetation.** Basin has degraded riverbanks, lack of native vegetation or major infestations of weeds.
- **Soil erosion.** Clearing of vegetation for grazing and cropping is the probable cause of gully erosion and stream/riverbank erosion.
- **Aquaculture.** Although the aquaculture industry is small, there have already been a number of incidents, which have included the discharge of polluted water to the Clarence estuary.
- **Rural residential development.** Rural residential developments have impacted on river health, particularly with regard to sewage disposal, and water supply.

The Healthy Rivers Commission (1999, Section 8.4.4) argued that any proposal to divert substantial quantities of water from the Clarence River Basin would present significant risk to the health of riverine ecosystems, and those activities and values dependent on them. Moreover, the community is opposed to such diversions.

Other North Coast basins of NSW such as Tweed, Brunswick, Richmond, Macleay, and Manning are faced with a number of environmental problems caused by the increasing demand for water in various sectors, agricultural activities, wastewater disposal, urban and rural residential development, aquaculture, fishing, dredging, and drainage of acid sulphate soils.

The current river health of the North Coast region can be summarised as (Healthy Rivers Commission, 2003a, pp. 36 and 37):

- Most North Coast river basins are in a better than average condition compared to other NSW coastal river basins. However, half of the 159 North Coast sub-basins are under high environmental stress, while one in six has been identified as having high conservation values.
- Most basins have more than 50 percent tree cover. Riverside vegetation is generally in moderate condition with about 70 percent of stream and estuary length vegetated.
- One third of sub-basins are considered to be highly stressed because of water extraction.
- One third of tidal streams and one fifth of non-tidal streams have high levels of bank erosion. Bed instability is generally less severe, with one in six streams significantly affected.
- Water quality is variable when measured against guidelines for recreation, edible seafood, potable water and protection of aquatic ecosystems. Poorly flushed tidal streams are worst affected, with only 15 percent meeting guidelines more than half of the time, and many of those are affected by high algal levels. Several estuaries

are affected by acid run-off or water with low dissolved oxygen, high dissolved aluminium and iron concentrations, leading to major fish kills in wet periods. About half of the non-tidal streams meet guidelines more than half the time. The main identified problems are related to excessive turbidity, bacteria and nutrients.

- Macroinvertebrate communities are in relatively poor condition in extensively cleared basins such as the Tweed and Richmond, and in relatively good condition in basins with good forest cover, such as Bellinger and Hastings.
- The state of fish populations in many of the North Coast rivers is relatively poor and possibly declining. Most native species are now restricted in location, and several are endangered. Reduced areas of wetlands have caused a loss of organic carbon to estuaries with adverse impacts on the food chain and fish population.
- The North Coast region contains 12 wetlands that are listed as important (Environment Australia, 2001). Coastal wetlands are present in 30 percent of the sub-basins, but to date no wetland has been given Ramsar status. Since European settlement, a large majority of coastal wetlands have been drained and reclaimed for agricultural production and other developments, often with adverse impacts on downstream water quality.
- Seagrass and salt-marsh habitats have generally declined, particularly near major population centres. Mangrove habitat has declined where tidal exchange has been impeded by foreshore structures such as floodgates and seawalls, but may have expanded in areas along the coast with increased sedimentation.

5.2.1 Coastal Lakes

Many lakes along the NSW coast generate ecological, social and economic benefits that are enjoyed by tourists, as well as the local communities that live, work or play near them (Healthy Rivers Commission, 2002). One of the features common to coastal lakes is that there is increasingly pressure on them to the extent that many of them are highly degraded. It is not only the environmental values of coastal lakes that are being threatened. The various activities such as tourism, fishing and oyster growing, are also being placed at risk. The degradation trend cannot continue, and communities expect strong and decisive action by all levels of government to protect remaining values, contain further damage and undertake targeted repair with a sensible ordering of priorities.

5.2.2 Oyster Industry

The NSW oyster industry is 130 years old and is the oldest aquaculture industry in Australia. It is by far the most valuable aquaculture industry in NSW, producing 80 percent of the total value of aquaculture product. The long-term average present value of production from oysters across the state is $8000 per ha. Some estuaries have returns as high as $35 000 per ha (White, 2001). The oyster industry is important to NSW from social, economic and environmental perspectives. Oyster farming occurs in 30 estuaries. The general conditions required for growing healthy oysters are well oxygenated, clear, brackish to saline waters, with pH in the range of 6.75 to 8.75, suitable tidal exchange, adequate phytoplankton supply and control of upstream sources of run-off and pollution. Oysters are critical indicators of river health because they are filter feeders, extracting phytoplankton, bacteria, suspended solids and inorganic particles from the surrounding water as their food source. Therefore, it is important to maintain clean and unpolluted water in the estuaries to sustain production of healthy oysters acceptable to the domestic and overseas market. This requires adequate management of land and water resources upstream of oyster cultivation areas (Healthy Rivers Commission, 2003b). The outbreak of hepatitis A in early 1997, which resulted from 450 people eating contaminated oysters grown in Wallis Lake (between Newcastle and Port Macquarie) highlights the susceptibility of farmed oysters to river health problems.

5.3 PROPOSED DIVERSION SCHEMES

The following sections summarise diversion schemes developed up to 1982 for the Clarence, Macleay, Manning and Tuross river basins. Brief descriptions of three schemes developed for the Clarence River in 1984 (The Newton Boyd Scheme), 1985 (The Water Research Foundation of Australia's Scheme), and 2002 (The White Scheme) are also provided.

5.3.1 Pre-1982 Proposed Diversion Schemes

The former New South Wales Water Resources Commission and its predecessor, the Water Conservation and Irrigation Commission had, from time to time, considered preliminary proposals for the inland diversion of water from coastal river basins. These occurred both during routine reviews and updates of information, and also in response to representations from interested members of the community. Drought has always been a stimulus to public enquiries and reviews of new water schemes. For example, a general review of inland diversion schemes was carried out in 1967. Some individual basins such as the Shoalhaven and Clarence were examined again in 1971 and 1979. In the early 1980s, the then Water Resources Commission of New South Wales engaged Rankine & Hill Pty Ltd Consulting Engineers to undertake a preliminary review and update of previously identified inland diversion schemes (Water Resources Commission of NSW, 1981). The same consultants simultaneously investigated the possibilities for inland diversion of surplus water from the Clarence River Basin (Clarence Valley Inter-Departmental Committee on Water Resources, 1982).

The following description of the schemes proposed for the Clarence, Macleay, Manning and Tuross river basins are based on the Water Resources Commission of NSW (1981), and the Clarence Valley Inter-Departmental Committee on Water Resources (1982), both prepared by Rankine & Hill Pty Ltd Consulting Engineers. They contain a number of assumptions regarding the cost estimations, which are described below.

Assumptions for Capital Cost Estimations

The Consultant's review of possible inland diversion schemes was based on simple procedures and was concerned mainly with identifying the more favourable schemes. The estimated capital costs[3] consisted of construction costs of dams, tunnels, pipelines and pumping stations. The annual costs included loan repayments, operation and maintenance works and electricity consumption. An interest rate of 10.4 percent per annum over a 40 year period was used for calculation of loan repayments. However, the adopted approach, which relied on readily available information without any additional fieldwork, neglected several aspects of inland water diversion and their associated costs. These include:

- The bed and bank erosion caused by diverted waters in the headwaters of the inland stream.
- The need to regulate diverted flows in conjunction with flows in the inland streams in order to maximise overall water productivity.
- Infrastructure costs for irrigation development (canals, drains, pumping stations, etc.).
- Storage facilities on the inland streams might be absent or have insufficient capacity to store and regulate extra flows.
- The increased flow in the inland streams would be associated with increased evaporation and streambed losses, reducing the net quantity of diverted water for use.

- Detailed assessment of potential benefits to inland regions from the additional water for irrigation, hydro-electric power generation, increased town and industrial water supplies, as well as flood mitigation in the coastal streams.
- Comprehensive studies of the potential impacts of water diversion on the social, physical, biological and environmental aspects of both the coastal and inland basins.

The estimated costs for the reviewed schemes given here are based on the 1981 prices. These have been multiplied by a factor of 2.5, in order to convert them to 2002 values.

Clarence River Basin

The Clarence Valley Inter-Departmental Committee on Water Resources was constituted in 1968 following the approval of the State Premier. The Committee consisted of representatives of the Water Resources Commission, the Electricity Commission, and the Department of Public Works. Its objective was to inquire into the control, development and use of the Clarence River Basin's water resources. The Committee had several specific terms of reference, one of which was to report on the possible diversion of Clarence water inland. In 1975 the Committee reported on the feasibility of a large dam on the Mann River for flood mitigation, irrigation and hydro-power generation. The report concluded that the proposal was not economic in the immediate future but it should be reassessed within 10 years. In April 1980, the then Minister for Water Resources announced that the Committee would review and update the dam proposal and the possibility of inland diversions from the Clarence Basin. In 1982, the Rankine & Hill Pty Ltd Consulting Engineers presented the results of a preliminary investigation undertaken on behalf of the Committee for diverting water inland from the Clarence River Basin (Clarence Valley Inter-Departmental Committee on Water Resources, 1982). The report describes fourteen basic schemes (CLA-1 to CLA-14) together with a number of supplementary schemes (CLA-SUP-A, B and C). These schemes, which would divert water by gravity, pumping or a combination of both, are listed below. Their locations are shown in Figure 5.2 and details are in Appendix C1.

- The Clarence River to Condamine River Scheme (CLA-1)

---

[3] The environmental and social costs have not been considered.

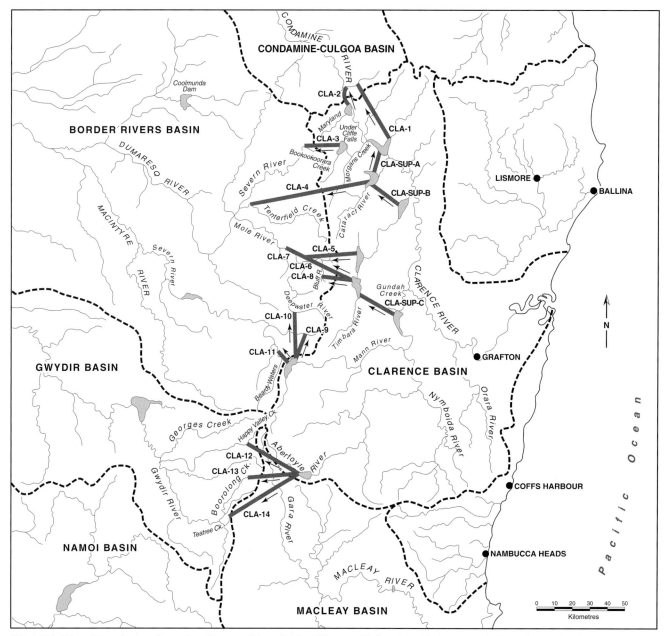

Figure 5.2 Diversion schemes from the Clarence River Basin (Clarence Valley Inter-Departmental Committee on Water Resources, 1982).

- The Maryland River to Condamine River Scheme (CLA-2)
- The Bookookoorara Creek to Severn River Scheme (CLA-3)
- The Cataract River to Tenterfield Creek Scheme (CLA-4)
- The Timbarra River to Bluff River Scheme (CLA-5)
- The Timbarra River to Bluff River Scheme (CAL-6)
- The Timbarra River to Mole River Scheme (CAL-7)
- The Timbarra River to Bluff River Scheme (CLA-8)
- The Mann River to Severn River Scheme (CLA-9)
- The Mann River to Deepwater River Scheme (CLA-10)
- The Mann River to Beardy Waters Scheme (CLA-11)
- The Aberfoyle River to Happy Valley Creek Scheme (CLA-12)
- The Aberfoyle River to Boorolong Creek Scheme (CLA-13)
- The Aberfoyle River to Teatree Creek Scheme (CLA-14)
- The Clarence River to Morgans Creek Scheme (CLA-SUP-A)

Table 5.2. *Summary of more favourable diversion schemes from the Clarence River Basin, in order of annual cost of water per $10^3 m^3$*

Scheme reference number	Inland basin	Diversion method	Annual diversion volume ($10^6 m^3$)	Capital cost[a] ($M)	Annual cost[a] ($M)	Cost[a] of water Capital cost ($ per $10^3 m^3$)	Annual cost ($ per $10^3 m^3$)
CLA-7C[b]	Border rivers	Pumping/gravity	755	1640	248	2170	330
CLA-6C[b]	Border rivers	Pumping/gravity	755	1548	261	2050	345
CLA-1	Condamine–Culgoa	Pumping	67	228	32	3400	480
CLA-6	Border rivers	Pumping/gravity	89	463	55	5200	620
CLA-7	Border rivers	Pumping/gravity	89	500	57	5620	640
CLA-4AB	Border rivers	Pumping/gravity	242	1478	190	6100	785
CLA-4A	Border rivers	Pumping/gravity	113	783	95	6930	840
CLA-9	Border rivers	Gravity	21	163	18	7760	860
CLA-11	Border rivers	Gravity	13	108	12	8310	925

Note: [a] The 1981 estimates of costs have been multiplied by a factor of 2.5 in order to represent the 2002 prices; [b] For details see Appendix C1.
Source: Clarence Valley Inter-Departmental Committee on Water Resources (1982).

- The Timbarra River to Cataract River Scheme (CLA-SUP-B)
- The Mann River to Timbarra River Scheme (CLA-SUP-C)

More Favourable Schemes

Table 5.2 summarises details of the more cost effective diversion schemes from the Clarence River Basin in order of the annual cost of water per $10^3 m^3$. Although a number of schemes were physically practical, the costs were too high to justify their construction. However, in case of CLA-7C, $755 \times 10^6 m^3 yr^{-1}$ could be diverted from the Clarence River Basin to the Border Rivers Basin at a capital cost of $1640 million and an annual cost of $330 per $10^3 m^3$. Another favourable scheme is the CLA-6C, which would divert $755 \times 10^6 m^3 yr^{-1}$ (the same as CLA-7C) at slightly lower capital cost of $1548 million but at a higher annual cost of $345 per $10^3 m^3$.

Macleay River Basin

In the Macleay River Basin, the average annual discharge from all streams is $2 \times 10^9 m^3$. A considerable volume of water is potentially available for inland diversion (Table 5.1). Water can be diverted by gravity and/or pumping to the Gwydir and Namoi River basins. These schemes are listed below. Their locations are shown in Figure 5.3, and their technical and cost estimate details are presented in Appendix C2. In general, these schemes are much smaller than those proposed for the Clarence River Basin.

- The Gara River to Teatree Creek Scheme (MAC-1)
- The Gara River to Rocky River Scheme (MAC-2)
- The Gara River to Roumalla Creek Scheme (MAC-3)
- The Gara River to Bough Gully Scheme (MAC-4)
- The Apsley River to MacDonald River Scheme (MAC-5)
- The Chandler River to Gara River Scheme (MAC-SUP-A)
- The Styx River to Chandler River Scheme (MAC-SUP-B)
- The Styx River to Chandler River Scheme (MAC-SUP-C)
- The Roumalla Creek to Hell Hole Creek (MAC-SUP-D)
- The Aberfoyle River to Gara River Scheme (MAC-SUP-E)

Manning River Basin

The average annual discharge of all streams in the Manning River Basin is estimated to be about $2.5 \times 10^9 m^3$. Apparently, the Basin has a significant potential for further water use and inland water diversion (Table 5.1).

Investigations were carried out in 1971 for the diversion of water from the Barnard River to the Peel River

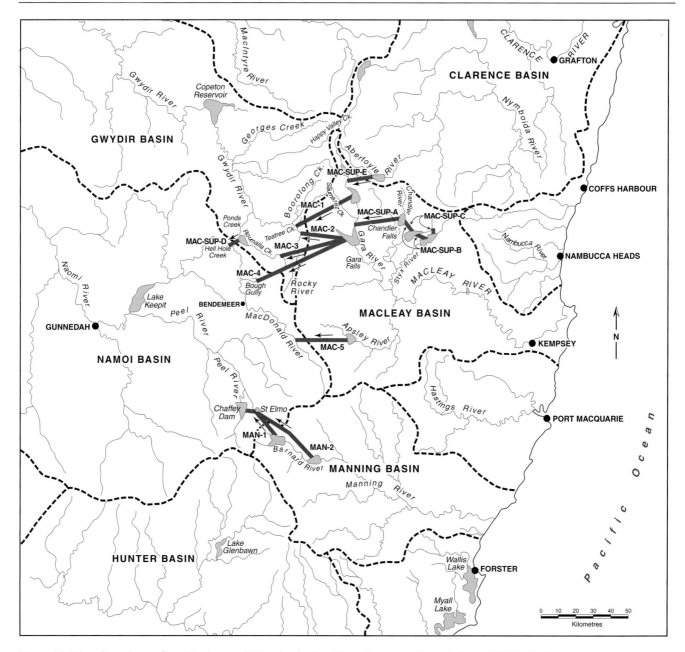

Figure 5.3 Diversion schemes from Macleay and Manning basins (Water Resources Commission of NSW, 1981).

(Water Resources Commission of NSW, 1981). Two schemes were considered: a gravity scheme (MAN-1), and a pumped/gravity scheme (MAN-2). Their locations are shown in Figure 5.3 and details are presented in Table 5.3.

Tuross River Basin

The average annual discharge of all streams in the Tuross River Basin is estimated at about $0.61 \times 10^9 \, m^3$. However, only a small percentage of this water was being used in the mid 1980s for urban, industrial and irrigation water supply (Table 5.1). The two small schemes (TUR-1 and TUR-2) were considered to divert water into inland basins. Their locations are shown in Figure 5.4 and their details are presented in Table 5.3.

More Favourable Diversion Schemes from the Macleay, Manning, and Tuross Basins

Table 5.4 summarises details of the more favourable diversion schemes from the Macleay, Manning, and Tuross River basins.

INTER-BASIN WATER TRANSFER FROM COASTAL BASINS OF NEW SOUTH WALES 115

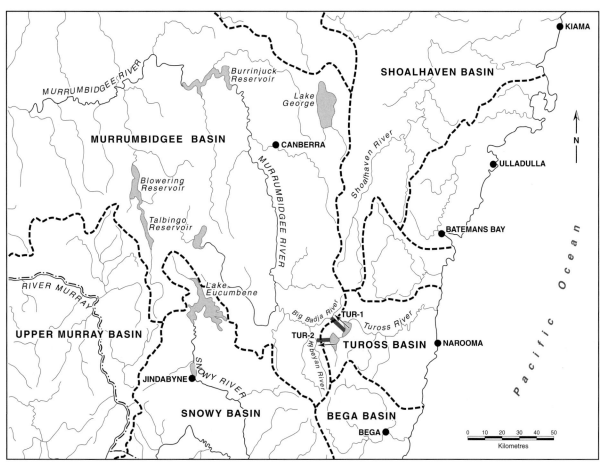

Figure 5.4 Diversion schemes from Tuross River Basin (Water Resources Commission of NSW, 1981).

Table 5.3. *Details of diversion schemes from the Manning and Tuross River Basins*

Diversion details		Manning River			Tuross River		
		MAN-1	MAN-2		TUR-1		TUR-2
Inland basin		Namoi	Namoi		Murrumbidgee		Murrumbidgee
Diversion method	Pumping	—	×		×		—
	Gravity	×	×		×		×
Diversion volume ($10^6\,\mathrm{m}^3\,\mathrm{yr}^{-1}$)		41	24		10.5		14
Tunnel/pipeline details	Length (km)	30 (T)	0.8 (P)	51 (T)	1.5 (P)	11 (T)	8.3 (T)
	Diameter (m)	2.7	0.6	2.7	0.8	2.7	2.7
	Capacity ($\mathrm{m}^3\,\mathrm{s}^{-1}$)	3.8	1.5	2.3	0.7	4.7	3.7
Pumping capacity ($\mathrm{m}^3\,\mathrm{s}^{-1}$)		—	1.5		0.7		—
Total capital cost[a] ($M)		290	285		115		68
Total annual cost[a] ($M)		32	33		15		8
Cost[a] of water	Capital cost ($ per $10^3\,\mathrm{m}^3$)	7075	11 875		10 950		4860
	Annual cost ($ per $10^3\,\mathrm{m}^3$)	780	1375		1430		570

Note: [a] The 1981 estimates of costs have been multiplied by a factor of 2.5 in order to represent the 2002 prices.
Source: Water Resources Commission of NSW (1981, Tables 3.4 and 3.5).

Table 5.4. *Summary of more favourable diversion schemes from Macleay, Manning and Tuross River Basins*

Scheme reference number	Coastal basin	Inland basin	Diversion method	Annual diversion volume (10^6 m^3)	Capital cost[a] ($M)	Annual cost[a] ($M)	Cost[a] of water Capital cost ($/$10^3$ m^3)	Annual cost ($/$10^3$ m^3)
MAC-2	Macleay	Gwydir	Gravity	33	275	30	8330	910
MAC-2A	Macleay	Gwydir	Gravity	56	453	49	8100	875
MAC-2AE	Macleay/Clarence	Gwydir	Gravity	73	598	65	8190	890
MAC-5	Macleay	Namoi	Gravity	22	150	17	6820	770
MAN-1	Manning	Namoi	Gravity	41	290	32	7050	780
TUR-2	Tuross	Murrumbidgee	Gravity	14	68	8	4860	570

Note: [a] The 1981 estimates of costs have been multiplied by a factor of 2.5 in order to represent the 2002 prices.
Source: Water Resources Commission of NSW (1981).

For the Macleay River Basin, Scheme MAC-2 would divert 33×10^6 m^3 yr^{-1} of water at an annual cost of $910 per 10^3 m^3. With the addition of Supplementary Scheme A, Scheme MAC-2A would divert 56×10^6 m^3 yr^{-1} at an annual cost of $875 per 10^3 m^3. Another favourable scheme is MAC-2AE, which would divert 73×10^6 m^3 yr^{-1} at an annual cost of $890 per 10^3 m^3. Finally, Scheme MAC-5 is capable of diverting 22×10^6 m^3 yr^{-1} at an annual cost of $770 per 10^3 m^3.

In the case of the Manning River Basin, only Scheme MAN-1 is more favourable. This Scheme would divert 41×10^6 m^3 yr^{-1} of water at an annual cost of $780 per 10^3 m^3. Scheme TUR-2 is the more favourable diversion scheme from the Tuross River Basin. This Scheme would divert 14×10^6 m^3 yr^{-1} of water at an annual cost of $570 per 10^3 m^3.

Benefit–Cost Analysis of the Diversion Schemes

The diverted waters were expected to be used for irrigation developments in inland basins. Davidson (1984) described the critical factors, which should be taken into consideration for benefit–cost analysis of the irrigation projects. These included:

(1) the capital invested by the government and farmers;
(2) water transmission losses;
(3) the rate of return from utilisation of the additional irrigation water;
(4) the secondary benefits from the project;
(5) the discount rate; and
(6) the life of project.

Davidson (1984) analysed the benefit–costs of the more favourable diversion schemes (Table 5.5), considering a number of assumptions including the following:

- Capital works will be constructed over periods of either 4 or 6 years.
- Eighty percent of the diverted water will reach irrigation farm boundaries.
- Irrigated cropping would replace dryland wheat growing on all areas.
- In all areas, land irrigated with diverted water will have to be levelled and farm channels constructed at a cost of $1000 per hectare ($2500 in 2002 prices).
- On all rivers, the cost of distributing additional water would be equal to the price paid by farmers in 1982.
- On the northern rivers (Condamine, MacIntyre, Gwydir and Namoi) water would be used by farmers already irrigating, and no additional labour, pumping equipment or farm machinery would be required.
- As future prices and yields were unknown, benefits were calculated assuming 1982 prices and yields, and also with prices and yields of 10, 20 and 50 percent higher than those obtained in 1982.
- To examine the sensitivity of benefit–costs to interest rates, discount rates of 3, 5 and 7 percent were used for all favourable schemes, assuming a lifetime of 50 years.
- As the rate of uptake of diverted water and the purpose for which it will be used were unknown, the benefits and costs were calculated on three distinct bases. These were:
  - *Immediate Uptake Optimum Use (IO)*, in which all water will be used in the year it becomes available and all of it will be used for the activity giving the highest gross margin.

Table 5.5. *Benefit–cost ratios of the most favourable diversion schemes, considering a 4-year construction period*

Favourable schemes[a]	Water uptake and usage[b]	Discount rate								
		3 percent			5 percent			7 percent		
		Yield or price ratio to 1982 yields or prices								
		1	1.2	1.5	1	1.2	1.5	1	1.2	1.5
CLA-1	IO	0.64	1.32	2.34	0.44	0.91	1.61	0.32	0.67	1.18
	IN	0.41	0.96	1.78	0.28	0.66	1.22	0.21	0.48	0.90
	NN	0.37	0.86	1.59	0.24	0.56	1.05	0.17	0.40	0.73
CLA-7C	IO	0.93	1.74	2.96	0.68	1.27	2.16	0.51	0.97	1.65
	IN	0.69	1.38	2.41	0.50	0.99	1.74	0.38	0.75	1.31
	NN	0.62	1.23	2.15	0.43	0.86	1.50	0.31	0.62	1.08
MAC-2A	IO	0.57	0.87	1.32	0.39	0.60	0.91	0.29	0.44	0.67
	IN	0.47	0.71	1.07	0.32	0.49	0.74	0.24	0.36	0.54
	NN	0.41	0.62	0.94	0.27	0.41	0.62	0.19	0.29	0.43
MAC-2AE	IO	0.55	0.85	1.30	0.38	0.59	0.89	0.28	0.43	0.65
	IN	0.47	0.72	1.10	0.32	0.49	0.75	0.24	0.36	0.55
	NN	0.40	0.61	0.93	0.26	0.40	0.61	0.18	0.28	0.43
MAC-5	IO	0.66	1.02	1.55	0.46	0.70	1.07	0.34	0.51	0.78
	IN	0.55	0.83	1.26	0.38	0.57	0.87	0.28	0.42	0.64
	NN	0.48	0.73	1.11	0.32	0.48	0.73	0.22	0.34	0.51
MAN-1	IO	0.64	0.89	1.49	0.44	0.67	1.03	0.32	0.49	0.75
	IN	0.52	0.80	1.21	0.36	0.55	0.83	0.26	0.40	0.61
	NN	0.46	0.70	1.07	0.30	0.46	0.70	0.21	0.32	0.49
TUR-2	IO	0.52	0.76	1.12	0.36	0.52	0.77	0.26	0.38	0.57
	IN	0.28	0.38	0.52	0.19	0.26	0.36	0.14	0.19	0.26
	NN	0.25	0.33	0.46	0.16	0.22	0.30	0.12	0.15	0.21

Note: [a] See Tables 5.2 and 5.4 for details. [b] *(IO): Immediate uptake optimum use, (IN) Immediate uptake normal use, and (NN): Normal uptake normal use.*
Source: Davidson (1984, Tables 9 and 12).

- *Immediate Uptake Normal Use (IN)*. As in *(IO)* except that it was assumed that the additional water on the northern rivers would be used to irrigate a combination of crops in the same proportion as those irrigated on the Namoi in 1982. On the Murray and Murrumbidgee rivers it was assumed that diverted waters would be used for irrigating crop species and pastures in the same proportion as in 1982.
- *Normal Uptake Normal Use (NN)*. As in *(IN)* except that it was assumed that diverted water would be taken up by farmers at increasing rates of 15, 20, 30, 40, 60 and 100 percent from year one to year six.

The preliminary analyses of Davidson (1984) are shown in Table 5.5. Using a discount rate of 3 percent, Davidson (1984) indicates that at the 1982 prices and yields, none of the schemes had a benefit–cost ratio of greater than 0.93 (this was for the Scheme CLA-7C with a diversion capacity of $755 \times 10^6 \, \text{m}^3 \, \text{yr}^{-1}$), even if all of the diverted waters were used immediately by farmers for the most profitable activities. Using discount rates of 5 percent and 7 percent, the benefit–cost ratio declined to 0.68 and 0.51 respectively. Davidson (1984) concluded that none of the favourable schemes could be justified on economic grounds considering the 1982 prices and technologies.

5.3.2 Post-1982 Proposed Diversion Schemes

The Newton Boyd Scheme

Newton Boyd is located at about 65 km west of Grafton (Figure 5.5). David D. Coffey, a Consulting Engineer from Sydney, described two variants of a scheme for the inland diversion of $1100 \times 10^6 \, \text{m}^3$ of water per annum from tributaries of the Clarence River close to Newton Boyd

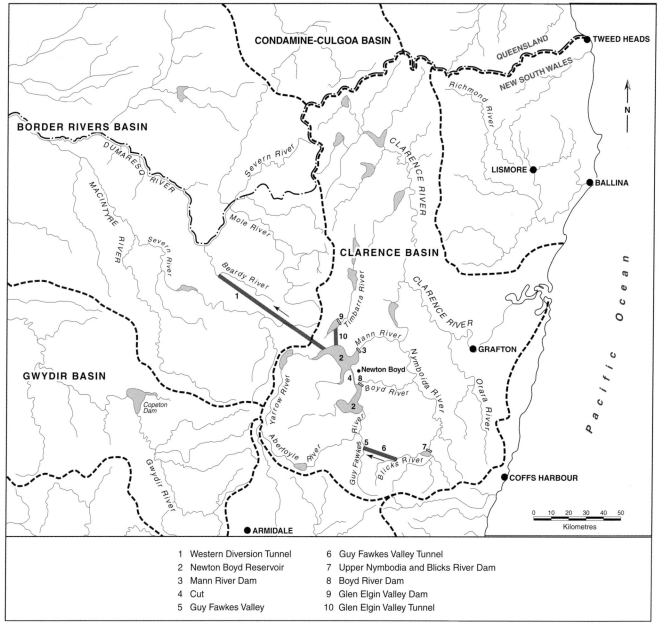

Figure 5.5 The Newton Boyd Scheme (Coffey, 1985).

(Coffey, 1985). This large Scheme required 15 to 25 years to be constructed and expected to divert water at lower prices than any other proposed schemes.

In the first variant, water from the Mann and Boyd rivers are impounded in two large reservoirs created by dams at points (3) and (8) in Figure 5.5 and joined by a canal or 'cut' at (4). A dam at (7) would collect water from the upper Nymboida and Blicks rivers. Then, a 28.8 km long tunnel (6) with a diameter of 3.6 m would divert water from this reservoir to the Guy Fawkes Valley at (5), which would conduct waters to the Boyd River Dam. Additionally, the upper Timbarra River would be dammed at (9). Then, a 9 km tunnel at (10) with a diameter of 2.7 m would divert water from this reservoir to the Newton Boyd Reservoir (2).

For the inland diversion, water in the Newton Boyd Reservoir would enter an 81 km tunnel (1) with a diameter of 6.9 m and would gravitate under the Great Dividing Range to the Beardy River of the Border Rivers Basin, upstream of its junction with the Dumaresq River. Hydro-electric generating stations would be installed at (5) and (10) with the respective capacities of 320 MW and 130 MW.

The total catchment area of the Scheme is 5350 km^2, which is about 23.6 percent of the total Clarence River Basin area (see Table 5.1). The annual quantity of water capable of being diverted is a quarter of the Clarence River flow at its mouth and it was claimed by Coffey that this diversion would not have significant impact on the lower Clarence River Basin.[4] The average annual water flows into the Newton Boyd Reservoir are shown in Table 5.6. Other technical details are in Tables 5.7 and 5.8.

The 1984 total capital cost of the scheme and the cost of water per 1000 m^3 were $1395 million and $136 respectively (their 2002 equivalent values are $2650 million and $260). Comparing these figures with those in Table 5.2 indicates that this scheme, with a gravity diverting capacity of 1100×10^6 m^3 per annum, has a higher capital cost but a lower water price.

The layout of the second variant of the Scheme is to some extent similar to the first variant shown in Figure 5.5, but water from a much larger reservoir at Glen Elgin (9) discharges northward to the headwaters of the Mole River.[5] Moreover, the capacity of the Newton Boyd reservoir would be reduced to 3800×10^6 m^3. Other details of the variant are shown in Tables 5.9 and 5.10.

The Scheme requires construction of five tunnels with a total length of 43.3 km and diameters of 3.6 m to 11 m, and would have the capacity of generating 1500 MW of electricity. Also, a smaller 320 MW hydro-electric power station would provide additional low cost power to the electricity grid. The 1985 estimated cost of the second variant of the Scheme is about $1662 million or $2975 million in 2002 prices, higher than the cost of the first variant by about $325 million. This is partly because of the hydro-electric component of the Scheme.

Table 5.6. *The average annual diversion into the proposed Newton Boyd Reservoir*

No.	River	Diversion (10^6 m^3)
1	Boyd	430
2	Mann	240
3	Nymboida	500
4	Timbarra	50
Total		1220

Source: Coffey (1985).

Table 5.7. *Characteristics of the dams for the first variant of the Newton Boyd Scheme*

Dams and their numbers in Figure 5.5	Type	Height of crest above the streambed (m)	Embankment volume (10^6 m^3)
Mann River Dam (3)	Rockfill	179	21.4
Boyd River Dam (8)	Rockfill	184	22.2
Nymboida/Blicks Rivers Dam (7)	Rockfill	160	10.0
Glen Elgin Valley Dam (9)	Concrete weir	15	—

Source: Coffey (1985).

Table 5.8. *Storage capacities of the three reservoirs for the first variant of the Newton Boyd Scheme*

Reservoir	Operating level (m AHD)	Capacity between operating levels (10^6 m^3)
Newton Boyd	372–408	5800
Nymboida/Blicks	500–550	680
Glen Elgin	850–860	30
Total		6510

Source: Coffey (1985).

Table 5.9. *Characteristics of the dams for the second variant of the Newton Boyd Scheme*

Dams and their numbers in Figure 5.5	Type	Height of crest above the streambed (m)	Embankment volume (10^6 m^3)
Mann River Dam (3)	Rockfill	130	9.7
Boyd River Dam (8)	Rockfill	134	10.5
Nymboida/Blicks Rivers Dam (7)	Rockfill	160	10.0
Glen Elgin Valley Dam (9)	Rockfill	111	11.5

Source: Coffey (1985).

The Water Research Foundation of Australia's Scheme

An article entitled "A Scheme to dwarf the Snowy" in *The Sun–Herald* dated 14 April 1985 by the Hon. Jack Beal,

---

[4] This claim is not justified, because as described earlier, wetlands and lakes of the lower Clarence River as well as oyster farms are sensitive to flow reduction.

[5] Figure 5.5 does not show the larger reservoir and the diversion channel to the Mole River.

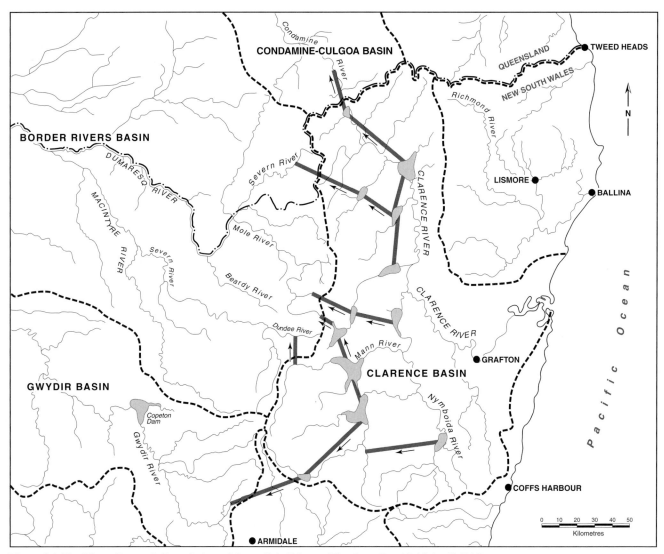

Figure 5.6 The Water Research Foundation of Australia's Scheme (*The Sun–Herald*, 14 April 1985).

Table 5.10. *Storage capacities of the three reservoirs for the second variant of the Newton Boyd Scheme*

Reservoir	Operating level (m AHD)	Capacity between operating levels (10^6 m^3)
Newton Boyd	320–362	3800
Nymboida/Blicks	500–550	680
Glen Elgin	925–948	2050
Total		6530

Source: Coffey (1985).

Chairman of the Water Research Foundation of Australia and former Minister for Water Resources in NSW, provided a brief description of a Scheme for the Clarence River Basin. The Scheme had the potential to divert 2×10^9 m^3 of water to the Murray–Darling Basin. The water would enter this Basin in the upper reaches of the Condamine River, Border Rivers and Gwydir River (Figure 5.6).

Some of the claimed features of the Scheme were:

- Water diverted westward would be 40 percent of the Clarence average annual discharge.
- Stored water could be returned to the Clarence system in drought periods.
- Large storages would make the Scheme substantially drought proof.
- 3000 MW of electricity could be provided to NSW and QLD at peak demand times.
- The 1985 estimated cost of the water component was $1.5 billion and the power components $1.8 billion, a total of $3.3 billion, which is about $5.9 billion in 2002 prices.

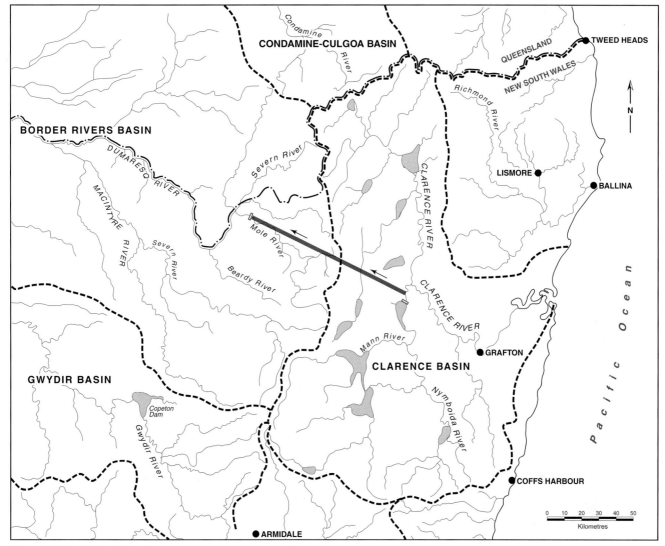

Figure 5.7 The White Scheme.

- A further $500 million ($895 million in 2002 prices) spent on inland regulating, monitoring and ancillary works could greatly increase the efficiency of water use in inland basins.
- The Scheme would be built over a period of 25 years, making the average annual expenditures affordable.

No cost–benefit analysis and assessment of environmental and social impacts of the Scheme on coastal communities were undertaken. Also, no estimate of the diverted water price per $1000 \, m^3$ was provided.

The White Scheme

G. B. White (Chairman of White Industries Australia Limited) at the Utility Congress (UTILICON), which was held in Melbourne 3–6 September 2002, presented the most recent proposal for the inland diversion of the Clarence River waters. The following description of the proposal is based on his paper.[6]

The proposal provides an opportunity for flood mitigation in the Clarence River Basin and would address over-allocation of water in the Murray–Darling Basin. It consisted of building a dam with a capacity of $900 \times 10^6 \, m^3$ on the Mann River, approximately 20 km upstream of the confluence of the Mann and Clarence rivers and 60 km north-west of Grafton (Figure 5.7).

In this Scheme, water would be pumped via a 70 km pressure tunnel to a $400 \times 10^6 \, m^3$ regulating weir on the

---

[6] http://home.iprimus.com.au/wial/watersaving/index.html (visited in July 2005).

Mole River, passing through a 28 MW hydro-power plant before flowing to the Dumaresq River. It was claimed that annually, this would deliver $950 \times 10^6\,\text{m}^3$ of water to the Border Rivers Basin. This volume of water would be sufficient to irrigate up to 130 000 ha of land in the Border Rivers region and downstream areas at a rate of $7300\,\text{m}^3$ per hectare.[7] It was estimated that this could produce $500 million of agricultural products per annum.

The cost of the Scheme was estimated at $1.52 billion, financed by the private sector against sales of water licences[8] and long-term operation and management rights. The project would be built in 4 years and during this period would create 1200 direct and 2000 indirect jobs.

It was claimed that the Scheme was planned on an integrated and sustainable basis. The volume of water, which would be released to the Mann River was claimed to be in excess of recommendations by the *NSW Healthy Rivers Commission* and would provide adequate flow for fish spawning and migration. Moreover, the Scheme had planned drainage facilities and salt interception infrastructure for the irrigated lands.

The proposal was rejected by the NSW Government for the following major reasons:

(1) the Clarence River already faces water quality and quantity problems particularly in the estuarine area, and diversion of an additional $950 \times 10^6\,\text{m}^3$ of water per annum would compound these problems;
(2) the proposal did not demonstrate that the Scheme is economically viable and environmentally sustainable; and
(3) the environmental effects of diverting such a large volume of water to the Mole River would be severe.

5.4 THE SCOPING STUDY

In 1989, a Scoping Study of the concept of inland diversion in north-east NSW was undertaken by Lyall & Macoun Consulting Engineers (1989). The study was commissioned by the then Department of Water Resources[9] in response to public interest in inland water diversion. The objective was not to identify any particular diversion scheme for implementation, but rather to identify the issues to be addressed in a more detailed investigation. Eight major hurdles to be overcome in an initial pre-feasibility study were identified:

(1) a narrowing of the range of feasible engineering options;
(2) firmer estimates of costs of implementing these schemes;
(3) firmer estimates of benefits, particularly in the agricultural sector;
(4) identification and assessment of the significance of environmental effects due to changes in flow regimes of coastal and inland rivers;
(5) determination of the potential for waterlogging and salinity problems that might emerge in irrigated lands;
(6) demonstration that inland diversion can be a form of sustainable development;
(7) identification of practical schemes for financing inland diversion; and
(8) identification of key issues and factors related to community acceptance of the scheme.

Five recommendations for the conduct of an initial pre-feasibility study were:

- A *Consultancy Team* composed of joint consultants should be engaged to carry out the study.
- A *Steering Committee* should be formed to arrange for the engagement of the consultancy team, and to provide a formal and comprehensive mechanism for maintaining contact between the client and consultant.
- A *Community Forum* should be constituted to facilitate a two-way flow of information between the consultants and the wider community.
- A *Study Team* should be formed within the Department of Water Resources to provide secretarial services to the *Steering Committee* and *Community Forum*, and technical support in the acquisition of data and liaison with other government authorities.
- A specialist consultant should be appointed in the role of Community Liaison Officer to assist the *Steering Committee*, *Community Forum* and *Study Team*.

5.5 CLEARANCE SCHEME AND WATER SUPPLY OF ADELAIDE

The River Murray is a major source of water supply for Adelaide. Water is pumped via a number of pipelines to Adelaide and neighbouring areas (see section A in Chapter 10). The Government of South Australia (1989) considered 21 different options to overcome domestic water shortages

---

[7] If the diverted water is fully allocated to irrigation of new lands, the proposal will not address the problem of water over-allocation in the Murray–Darling Basin.
[8] The price of water per $10^3\,\text{m}^3$ is not available in order to compare it with the water price of the other schemes described before.
[9] Currently, Department of Infrastructure, Planning and Natural Resources (DIPNR).

in Adelaide as a result of unforeseen circumstances such as inordinate increase in demand due to climate change or other reasons. These options included: groundwater extraction from the Great Artesian Basin and its pumping to Adelaide via an 800 km pipeline; seawater desalination; diversions from the Clarence River in NSW, the Ord River in Western Australia (see sections 9.5.4, 9.5.5 and Appendix F) and the Bradfield Scheme (see Chapter 6); effluent reuse; and harvesting Antarctic icebergs. For Clarence River waters, it was concluded that if such a scheme became a reality, an agreement between the South Australian and the NSW governments would be required allowing Clarence water to be pumped from the lower River Murray to metropolitan Adelaide and other areas. The maximum volume of water was expected to be $300 \times 10^6 \, m^3$ per annum with a total dissolved solids content of $380 \, mg \, L^{-1}$. The 1989 estimated capital cost was $2.2 billion and the cost of water was $1150 per $10^3 \, m^3$. The 2002 equivalent of these costs would be $3.2 billion and $1700 respectively. This option is no longer considered viable and is described here because of its historic value.

5.6 CONCLUSIONS

Numerous proposals for inland diversion from Clarence, Macleay, Manning and Tuross basins have been described in this chapter. Davidson (1984) analysed the benefit–costs of the more favourable diversion schemes proposed up to 1982, and concluded that they could not be justified on economic grounds.

Other schemes proposed after 1982 for inland water diversion from the Clarence River are:

- The Newton Boyd Scheme claimed to be able to divert $1100 \times 10^6 \, m^3$ of water per annum by gravity at a cost of $260 per $10^3 \, m^3$ in 2002 prices. This is the cheapest price compared to all other schemes developed so far for the Clarence River.
- The Water Research Foundation of Australia's Scheme planned to divert $2 \times 10^9 \, m^3$ of water per annum from the Clarence River to the Condamine, Border Rivers and the Gwydir basins. Also, it intended to generate 3000 MW of hydro-electric power.
- The White Scheme demonstrated the interest of the private sector in inland water diversion and claimed that it would divert water at an annual rate of $950 \times 10^6 \, m^3$.

None of these more recent proposals were supported by cost–benefit analysis and environmental and social impact assessment.

As described in this chapter, the population of the NSW North Coast basins is increasing and in economic terms tourism is taking over agriculture as the region's major activity. More water is required to satisfy residents and the tourism industry. Also, coastal basins, estuaries and lakes are facing serious environmental problems such as water quality deterioration, drainage of acid sulphate soils, sedimentation and dredging. Communities are opposed to the concept of large-scale inland water diversion, particularly from the Clarence River Basin. Their interests lie in fixing their environmental problems rather than aggravating them by inland diversion of their water resources.

There is little chance in the foreseeable future that any diversion proposal would be approved by the authorities and accepted by the community. Water users and water resources managers of inland basins will need to satisfy their long-term water requirements by increasing water use efficiency and other measures, rather than looking for extra sources of water from coastal rivers. If inland diversion of costal water is to be considered in the future, it would require detailed field investigations, feasibility studies and an integrated assessment of economic, environmental and social factors that consider the benefits and impacts on both the coastal and inland basins.

References

Australian Water Resources Council (1987). *1985 Review of Australia's Water Resources and Water Use*. Canberra: Australian Government Publishing Service. V. **1**: Water Resources Data Set; V. **2**: Water Use Data.

Clarence Valley Inter-Departmental Committee on Water Resources (1982). *Possibilities for Inland Diversion of Water from the Clarence Valley*. Rankine & Hill Pty Ltd Consulting Engineers.

Coffey, D. D. (1985). Diversion of Clarence River tributaries to the Murray–Darling Basin. In *Proceedings, Water Supply and Resources Conference, 28–29 November 1985, Sydney*. Sydney: New South Wales Water Supply and Resources Committee, pp. 124–133.

Davidson, B. R. (1984). A preliminary benefit cost analysis of the inland diversion of the coastal rivers of New South Wales. *Review of Marketing and Agricultural Economics* **52**(1): 23–47.

DLWC (2001). *Water Management Act 2000: What it Means for NSW*. Sydney: NSW Department of Land and Water Conservation.

Environment Australia (2001). *A Directory of Important Wetlands in Australia*. Third Edition. Canberra: Environment Australia.

Government of South Australia (1989). *South Australia Water Futures: 21 Options for the 21st Century*. Adelaide: Engineering and Water Supply Department.

Healthy Rivers Commission (1999). *Independent Inquiry into the Clarence River System*. Final Report. Sydney: HRC.

Healthy Rivers Commission (2002). *Coastal Lakes: Independent Inquiry into Coastal Lakes*. Final Report.

Healthy Rivers Commission (2003a). *North Coast Rivers: Independent Inquiry into the North Coast Rivers*. Final Report. Sydney: HRC.

Healthy Rivers Commission (2003b). *Oysters: Independent Review of the Relationship Between Healthy Oysters and Healthy Rivers*. Final Report. Sydney: HRC.

Lyall Macoun Consulting Engineers (1989). *Report on a Scoping Study for an Initial Pre-feasibility Study into the Concept of Inland Diversion in the North-east NSW*. Crows Nest, NSW: Lyall & Macoun Consulting Engineers.

National Land & Water Resources Audit (2001). *Australian Water Resources Assessment 2000*. Canberra: Natural Heritage Trust.

Water Resources Commission of NSW (1981). *Possibilities for Inland Diversion of N.S.W. Coastal Streams*. Rankine & Hill Pty Ltd Consulting Engineers.

White, I. (2001). Safeguarding environmental conditions for oyster cultivation in New South Wales. In *Oysters: Independent Review of the Relationship Between Healthy Oysters and Healthy Rivers* (2003). Final Report. Sydney: Healthy Rivers Commission. Appendix 2.

6 The Bradfield and Reid Schemes in Queensland

SECTION A: THE BRADFIELD SCHEME

6.1 INTRODUCTION

In March 1938, J.J.C. Bradfield[1] & Son Consulting Engineers from Sydney, submitted to the Queensland Government a report entitled *"Queensland: The Conservation and Utilization of Her Water Resources"*. The report described the rainfall pattern, the water resources of the State, and estimates of the damage caused by drought. It described also a proposal for the transfer of water from the coastal catchments of northern Queensland across the Great Dividing Range to inland rivers to increase the amount of water available for agriculture in central Queensland. The coastal rivers included in the proposal are located in a high annual rainfall area of Queensland. Their average annual flows at their mouths are listed in Table 6.1. Bradfield proposed to divert water from the Tully River to the Herbert River,[2] and then across to the Burdekin River. Water stored on the Upper Burdekin River would flow by gravity south-westerly to the upper reaches of the Flinders River, and then to the Thomson River, where it would flow to Cooper Creek (Figure 6.1).

The elevation measurements required for the development of the Scheme were taken with a barometer (Raxworthy, 1989, pp. 135–137). This led to inaccuracies in land elevations and some mistakes in the development of the proposal. The following sections describe some details of the Bradfield Scheme, based on the Bradfield (1938) report.

6.2 WATER AVAILABILITY

Because only a few, and in most cases no flow records for the Herbert, Burdekin and Flinders Rivers were available, Bradfield estimated the average annual run-off of these rivers based on the available rainfall data. According to Nimmo (1947), Bradfield used an empirical formula developed by Keller (1906) for flow estimates of ungauged rivers in Germany. This formula can be presented as:

$$R = 0.942P - 405$$

where:
R = Average annual run-off (in mm).
P = Average annual precipitation (in mm).

Table 6.2 shows the results of the Bradfield (1938) estimates converted to metric units.

If the total annual run-off presented in Table 6.2 could be stored and discharged at a constant rate over a year, it would generate a flow rate of about $204 \, m^3 \, s^{-1}$. From this flow, approximately $170 \, m^3 \, s^{-1}$ was considered appropriate for diversion. The remaining $34 \, m^3 \, s^{-1}$ was assumed to flow down the normal courses of the rivers. Assuming a 33 percent loss due to evaporation and seepage, some $114 \, m^3 \, s^{-1}$ would be available for irrigation and watering stock in the interior of Queensland.

6.3 OUTLINE OF THE BRADFIELD SCHEME

The diversion of the coastal rivers was planned to commence near the Tully Falls by impounding the Tully River (Figure 6.1). A short tunnel or open cut channel would divert the dammed waters to the Blunder Creek, a tributary of the Herbert River. The next impounding site would be on the Herbert River, where water would be diverted through a tunnel or open cut channel to the Upper Burdekin River. The site for the final impounding of water on the eastern side of the Great Dividing Range is at Hell's Gates, a gorge on the Burdekin River. The backing-up of water along the Clark River, a tributary of the Burdekin, meant that the upper

---

[1] See his biography in Appendix A.
[2] See Johnson and Murray (1997) for a description of the natural resources of the Herbert River Catchment.

Table 6.1. *Characteristics of the major rivers of the Bradfield Scheme*

River	Catchment area (km^2)	Mean annual run-off (10^6 m^3)	Divertible volume (10^6 m^3)	Developed volume (10^6 m^3)
Tully	1690	3680	476	236
Herbert	10 100	4990	1440	35
Burdekin	130 000	10 100	5350	1190
Flinders	109 000	3030	87	6

Source: Australian Water Resources Council (1987).

reach of this water storage was close to the Great Dividing Range allowing it to be diverted inland through a tunnel or open cut channel.

Once through the Divide, the waters would be discharged into the headwaters of the Flinders River or one of its tributaries and a suitable site would be dammed, to create a large artificial water reservoir, from which a constant flow of water would be diverted through a short open cut channel into Tower Hill Creek, a tributary of the Thomson River. By using existing streams and connecting these with tunnels and open-cut channels, diverted waters would be made available for irrigation as well as stock and domestic purposes over a large tract of land in the Thomson River Catchment around Longreach and eventually to Windorah on the Cooper Creek.

The diverted water was claimed to irrigate over one million hectares of agricultural land in Queensland as well as provide stock and domestic water supplies along the entire length of the rivers and diversion channels included in the scheme. There was also the potential for developing hydro-electric schemes which could then provide electricity for domestic and industrial purposes, and for the pumping of irrigation water into farms.

6.4 COSTS AND BENEFITS OF THE SCHEME

Bradfield (1938) estimated the cost of the Scheme to be £30 000 000 or approximately $1.7 billion in 2002 prices. He recommended that further investigations should be undertaken to accurately estimate the cost and feasibility. This included aerial surveys, accurate measurements of heights, river flow measurements, more detailed investigations of dam sites, as well as geological and engineering investigations. Bradfield suggested that the Scheme could be partially funded from the £35 000 000 (approximately $2 billion in 2002 prices) that had been provided free of interest by the British Government a few years earlier for Empire development in Australia.

Diverted waters would be available for increasing the supply of fodder and watering stock in times of drought, and the surplus could be used for irrigation. Under the Scheme, Bradfield estimated it would be possible to feed an additional 20 million sheep, which meant an increase of £10 000 000 ($570 million in 2002 prices) in annual income for the State.

It was claimed that the Scheme had the potential to generate over 370 MW and possibly up to 740 MW of hydro-electric power which could be used for electrification of the railways and other power requirements of the State.

6.5 THE 1947 REVIEW OF THE SCHEME

In February 1947, W. H. R. Nimmo, who was the Chief Engineer of the Stanley River[3] Works, critically reviewed the Bradfield Scheme. He was able to use additional information on topography and on river flows that was not available to Bradfield. Nimmo demonstrated that Bradfield had overestimated the quantities of water available from the Tully, Herbert and Burdekin rivers by about 250 percent. He also showed that water could not be diverted by gravity from the Burdekin to the Flinders catchments. These issues, and some others, are described in the following sections based on the Nimmo (1947) review report.

6.5.1 Water Availability for Diversion

The Nimmo (1947) estimates of the average annual run-off of the Tully, Herbert and Burdekin rivers are listed in Table 6.3.

Comparison of the data in Tables 6.2 and 6.3 shows that for the Tully River, the reduction in the area of the catchment is largely offset by the increase in the estimated rainfall. For the other two rivers, there is not a great

---

[3] A tributary of the Brisbane River. Its waters are captured by the Somerset Dam which has a capacity of 368 × 10^6 m^3.

THE BRADFIELD AND REID SCHEMES IN QUEENSLAND

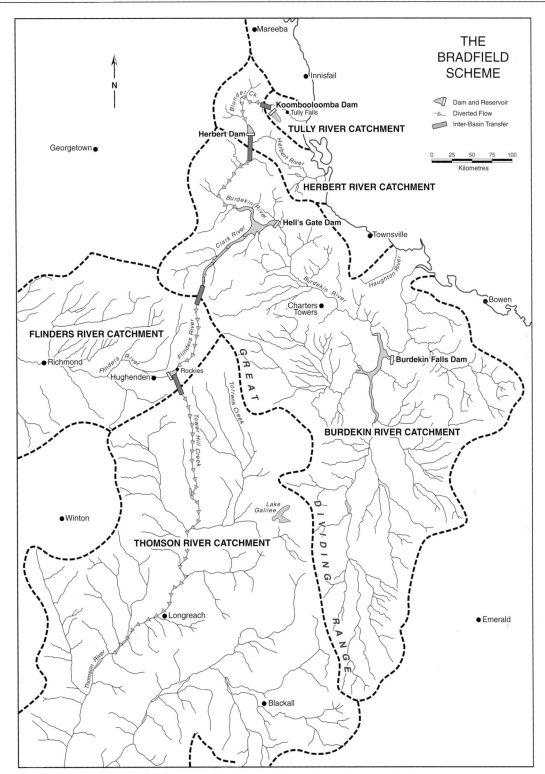

Figure 6.1 The Bradfield Scheme (Australian Bureau of Statistics, 1984, p. 109, and Nimmo, 1947).

difference in the total area of the catchments used by the Scheme, but the annual rainfall data are slightly different. The major difference is the estimated annual run-off. This is due to the fact that the formula used by Bradfield is applicable to the German catchment where the average annual temperature is about 9°C, while in the Queensland catchments average annual temperatures are in the order of 21°C. In addition, the Queensland catchments are all tropical

Table 6.2. *Bradfield's estimates of the annual run-off for the main rivers in his Scheme*

River	Catchment area (km^2)	Area of the catchment used by the Scheme (km^2)	Annual rainfall (mm)	Annual run-off (10^9 m^3)
Tully	1425	365	1905	0.50
Herbert	9065	5280	914	2.40
Burdekin	132 025	18 660	620	3.29
Flinders	108 780	2590	533	0.25
Total	251 295	26 895	–	6.44

Source: Bradfield (1938).

Table 6.3. *Nimmo's estimates of the annual run-off of the rivers in the Bradfield Scheme area*

River[a]	Area of the catchment used by the Scheme (km^2)	Annual rainfall (mm)	Annual run-off (10^9 m^3)
Tully	264	2340	0.4
Herbert	5268	880	0.9
Burdekin	18 363	735	1.2
Total	23 895	–	2.5

Note: [a] No estimate was provided for the Flinders River.
Source: Nimmo (1947).

or sub-tropical with pronounced wet and dry periods. Therefore, rainfall losses due to evaporation are much higher in Queensland and subsequently their annual run-offs are lower.

The average annual run-off of 2.5×10^9 m^3 is the average of all flows including the large floods, which cannot easily be retained by reservoirs and will be lost over the spillways (Nimmo, 1947). Of the remainder, much will be lost by evaporation from reservoirs and also as channel seepage. Therefore, not more than half of the average annual run-off would be available for diversion and not much more than half the quantity diverted, or approximately 625×10^6 m^3 per year could be available at the farm level. Nimmo (1947) also commented on the feasibility of building the proposed dams and conduits for the Bradfield Scheme. Some of his comments are summarised in the following sections.

6.5.2 The Tully Dam

A 37 m high dam could be built a short distance upstream of Tully Falls with a capacity of 245×10^6 m^3. This would produce a steady regulated flow of 8.5 m^3 s^{-1} throughout the year. However, Nimmo (1947) recommended that the final site should only be selected after further field investigations.

6.5.3 The Diversion Conduit to Blunder Creek

The topographic features of the divide between Tully and Blunder Creek catchments make it impossible to divert water to Blunder Creek in an open cut channel as proposed by Bradfield. A 12.8 km tunnel would be required for the inter-basin water diversion. This tunnel, with a regulated flow capacity of 8.5 m^3 s^{-1}, would have to be lined with concrete to prevent water losses.

6.5.4 The Herbert Dam

The Bradfield report proposed construction of a 30.5 m high dam immediately upstream of the Herbert Falls. As Bradfield provided no information about the volume of water to be impounded, Nimmo calculated that the capacity would be about 62×10^6 m^3. However, this volume would be too small to adequately regulate the flow of the Herbert River.

6.5.5 The Herbert–Burdekin Tunnel

This tunnel would be about 21 km long. However, the 14 m difference of level between the Herbert Falls and the diversion point on the Burdekin River, used by Bradfield, was wrong. This error was probably due to the elevation measurements being taken by altimeter. Nimmo recommended that these measurements be confirmed by surveying before determining the feasibility of this tunnel.

6.5.6 The Hell's Gates Dam

The Bradfield Scheme included raising the water level of the Burdekin River to 627 m by a 122 m high dam at the Hell's Gate with an assumed elevation of 505 m. Since the principal function of the dam was to raise the water level to permit water diversion to the west, only a very small proportion of the reservoir capacity could be used as a regulating storage. Moreover, the extensive surface of the reservoir would have to be maintained at its full level. This would generate very large evaporation loss, and during severe droughts this loss may exceed the inflows from the Burdekin upper catchment.

6.5.7 The Tunnel Under the Main Divide

A tunnel was planned for diverting water from the Burdekin to the Flinders River Catchment. The elevation of the Great Dividing Range, under which the tunnel would pass, approaches 900 m. The elevation of the tunnel's discharge point is about 427 m. Investigations at the Hell's Gates indicated that the maximum height of a dam at this location with an elevation of 305 m[4] could not exceed 91 m, raising the water level to 396 m. This means that the elevation of the tunnels intake point would be 31 m below its discharging point. Therefore, it would be impossible to divert water by gravity to the Flinders River from a reservoir at Hell's Gates through its associated tunnel.

6.5.8 Storage on the Flinders River

Examination of the Flinders River above the Rockies showed that the river runs through a narrow and rapidly rising gorge, which does not lend itself to the creation of large water storages. Since sufficient storage capacity could not be obtained to regulate any intermittent flow, any water diverted from the east of the Great Dividing Range must be regulated to an almost steady flow before diversion.

6.5.9 Diversion from Flinders to Thomson

From the elevation data measured by altimeter it appeared that water could be diverted from the Flinders River to the Thomson River via Tower Hill Creek.

6.5.10 Distribution of Water in West Queensland

The demand for irrigation water would not be uniform for all seasons of the year. The fluctuating demand could perhaps be met to a considerable extent by increasing the size of dams and tunnels east of the Great Dividing Range. However, the lack of a large storage on the western side of the Divide would increase the difficulty of regulation and distribution.

6.5.11 Cost of the Scheme

The 1947 cost of the modified scheme, to the point of delivery of water into the Flinders River, was estimated to be about £100 million ($4.2 billion in 2002 prices). For a fifty-year loan period, the annual charges would be about £6 million ($252 million in 2002 prices). If $620 \times 10^6 \, m^3$ per year, equivalent at a continuous flow $19.7 \, m^3 \, s^{-1}$, could be used for irrigation, the cost of water delivered at the Rockies would be £9.7 ($407 in 2002 prices) per $10^3 \, m^3$, to which the annual charge of a distribution system would have to be added.

6.6 THE EXPANDED BRADFIELD SCHEME

The Bradfield proposal to the Queensland Government was rejected because of its high cost. However, Bradfield continued to develop and promote his proposal in the press and in public lectures until his death in 1943. His expanded proposal was published in *Walkabout* in July 1941 (Bradfield, 1941a) and *Rydge's* in October 1941 (Bradfield, 1941b).

Bradfield (1941a) argued that rivers of the dry area of Australia could be dammed for the development of irrigation. He believed that suitable sites on the rivers flowing from the MacDonnell Ranges such as Glen Helen Gorge and Simpson's Gap, both to the west of Alice Springs, could be dammed. The impounded waters could be used to irrigate 1.3 Mha of land in the heart of the country. At a rate of $12\,200 \, m^3 \, ha^{-1}$, this would require $1.6 \times 10^9 \, m^3$ of water and represents a depth of water of 21 m over an area of 78 km². In some of the gorges water could be stored to a depth of 30.5 m. Annually, an extra water depth of 2.5 m should be considered to compensate the evaporation of a surface area of 78 km². This would represent an annual volume of $200 \times 10^6 \, m^3$. Irrigation of 1.3 Mha of permanent pasture would probably add some 3 million sheep to the area. Bradfield (1941a) also suggested further dams on the Fink, Georgina, and Diamantina rivers and the Cooper Creek for the development of the irrigation schemes (Figure 6.2). The content of Bradfield's (1941b) paper is very much similar to his previous publication (Bradfield, 1941a).

Table 6.4 provides the average annual run-off of the major rivers in his Expanded Scheme area. It clearly indicates that although the average annual run-off of the

---

[4] Bradfield's estimate was 505 m.

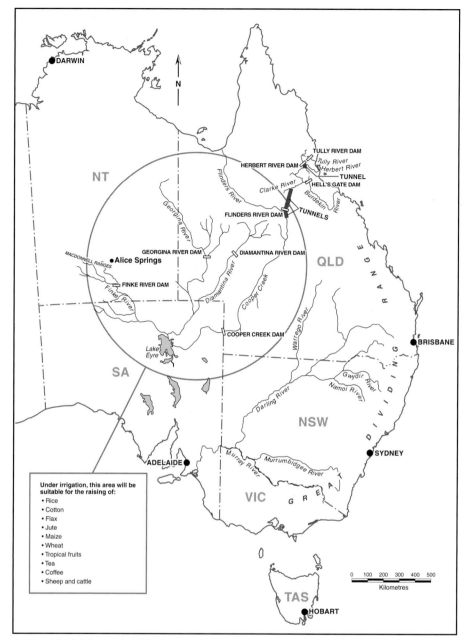

Figure 6.2 The Expanded Bradfield Scheme (Bradfield, 1941a).

Table 6.4. *Characteristics of the major rivers in the Expanded Bradfield Scheme*

Catchment	Area (km^2)	Mean annual run-off (10^6 m^3)	Divertible volume (10^6 m^3)
Georgina River	242 000	2000	48
Diamantina River	158 000	960	57
Cooper Creek	296 000	2330	37
Fink River	115 000	238	29

Source: Australian Water Resources Council (1987).

rivers are quite substantial, only a very small portion of their run-off could be diverted.[5] Another doubtful issue is the suitability of land in central Australia (see Figures 3.6 and 3.7) for cultivating rice, cotton, maize, wheat, tropical fruits and other crops, as suggested by Bradfield (1941a).

---

[5] The Department of Natural Resources and Mines developed Water Management Plan for the Cooper Creek in 1999, Burdekin in 2002, and Georgina-Diamantina in 2003 (see References). The main objectives of these plans are: sustainable management of water resources to meet future water requirements, protecting the natural ecosystems, and security of supply to water users.

6.6.1 Climatic Impacts of the Expanded Bradfield Scheme

Bradfield (1941a, b) claimed that the large bodies of water included in his Expanded Scheme, and development of associated irrigation, would change the climate of the project area by increasing humidity due to the annual 2500 mm of evaporation. This would result in an increased rainfall of about 100 mm per annum over an area of 1 300 000 km^2 of dry inland. In supporting this claim, Bradfield quoted papers by Quayle (1921, 1922) on rainfall improvements caused by human activities.

Quayle (1921) in his paper entitled *"Possibilities of modifying climate by human agency, with special application to south-eastern Australia"*, investigated the impact of various human activities on rainfall in Victoria. These included the substitution of crops and grass for the drought resistant forest cover in the Mallee, and development of irrigation. He investigated the possible effects of the Goulburn irrigation area on local rainfall. He analysed the average annual rainfall of a number of stations that were classified as "unaffected", "affected" and "check", for three successive decades (1885–94, 1895–1904 and 1905–1914). His results gave support to the assumption that stations located south-east of the main irrigation area benefited by an increase of at least 5 percent of the annual rainfall or by 25 mm. Quayle (1921) concluded that development of irrigation along the River Murray in Victoria and New South Wales had a noticeable effect in changing the climate of Northern Victoria, including the Mallee region. Quayle (1922) extended his analysis of rainfall data to the Lake Torrens and Lake Frome regions in South Australia, and again observed some improvement in rainfall. In the absence of any irrigation developments in these areas, he attributed the rainfall improvements to land use changes, which consisted of substitution of native Mallee vegetation with crops and grass.

Bradfield's reputation as an eminent Australian engineer meant that he attracted wide attention and support from the public, the National Council of Primary Producers and the press, not only because of the possibilities of the wide-scale water harvesting, irrigation and settlement involved, but also because of his assumption that the climate of the extensive dry area of central Australia could be modified. In order to investigate Bradfield's claim, the matter was referred to a Committee of meteorologists comprising of:

- **F. Loewe**, Lecturer in Meteorological Science at Melbourne University;
- **W/Cdr. H. M. Treloar**, Assistant Director (Technical), Australian Service;
- **J. C. Foley**, Supervising Meteorologist, Australian Climatology;
- **E. T. Quayle**, Former Senior Meteorologist at the Australian Service.

After close examination of available information from Australia, United States of America, Germany and some other countries,[6] the Committee concluded by a majority of 3 to 4 (Quayle refused to endorse the conclusions) that:

> In summarising, the majority of the Committee agree that the claim of Dr Bradfield that evaporation from a water surface of 20 000 square miles would cause a fall of rain of 4 inches over 500 000 square miles of the dry inland cannot be substantiated. We are of the opinion, further, that the results reached by Mr Quayle are not substantiated by present statistical methods and that in drawing his conclusions he has not taken into account other influences of far-reaching importance, which may possibly account for all features attributed to the effect of local evaporation.
>
> While the evidence which we have been able to collect does not entirely disprove the contention that some improvement in climatic conditions would follow the creation of a large inland lake, the extent of the improvement claimed is, we believe, very much over-estimated.
>
> The best that could be hoped for would be a slight amelioration of the climate, more particularly of temperature in the immediate neighbourhood of the storage area.

The information analysed, and the results of the Committee's investigations, are available in the Commonwealth Meteorological Bureau (1945).

As supporting evidence for the Committee's conclusions, it should be pointed out that the Royal Geographical Society of Australia (1955) had investigated the great flooding in 1949–50 of Lake Eyre. As a consequence of this flooding, Lake Eyre was filled in September 1950 and dried up in January 1953. The Royal Geographical Society (1955, p. 29) reported that the filling of the Lake, and subsequent evaporation of its large volume of water, had virtually no effect on the climate of the surrounding area.

6.7 THE 1982 REVIEW OF THE SCHEME

In late September 1982, Cameron McNamara Consultants commenced a review of the initial 1938 Bradfield Scheme for the Coordinator General's Department of the Queensland Government, and submitted their reports in December of

---

[6] Pomel (1872) was the first to say that flooding of the Sahara depression would not have any impact on the climate of the region (see Appendix E).

the same year (Cameron McNamara Consultants, 1982a, b). This was a desktop pre-feasibility study involving a two-day aerial reconnaissance of the regions involved. The consultants concentrated on the assessment of the following issues:

- Availability of suitable land in the receiving region.
- Availability of water for inter-basin transfer.
- The engineering aspects of inter-basin water transfer.
- Estimates of costs and benefits.

The following sections will expand these issues, based on the Consultants reports.

6.7.1 Land Resources

The net area of land suitable for irrigation, as determined from soil maps, is shown in Table 6.5. The four areas listed in this table all offered large areas apparently suitable for irrigation. However, the following problems could reduce the extent of suitable lands in these areas:

- Flooding may eliminate some of the lands on the Flinders and Thomson rivers.
- Lands on the Thomson River would involve significant extra costs for a delivery system and extra en-route water loss.
- Lands on Torrens Creek would also require added cost for delivery and there could be problems of induced salinity on the neighbouring down slope lands.
- Within the extensive area of the Downs, there is a mosaic of different soils, many of which are too shallow, too sloping or too saline for intensive irrigation. Thus, detailed surveys would be required to define areas suitable for irrigation.

6.7.2 Potential Crops

Within the limited scope of the study, the Consultants had no time to examine the problem of choosing crops that would ensure sustained profitable production on the large areas of land. They did however conclude that few of the crops currently grown profitably under irrigation are suited to the climatic conditions prevailing in these regions. An over-riding factor in the choice of crops would be the need to service the very high cost of irrigation water at the point of use. Therefore, the Consultants recommended that detailed agronomy and marketing studies of irrigated cropping for all potential areas be undertaken.

6.7.3 Industrial/Urban Water Demand

In 1982, the major industrial/urban water demand in the region covered by the Bradfield Scheme was for the coal mines of the Galilee Basin. Water requirements for these coal mines were estimated, assuming five mines operating simultaneously at an average production per mine of 4.5 million tonnes per year, plus associated urban water demand. Water supply for one thermal power station of 1400 MW capacity near Galilee Basin was also considered. This was in addition to a power station that was already planned, and would use water from the Burdekin Falls Dam.[7] Table 6.6 shows the summary of the estimated water requirements.

6.7.4 Available Water Resources

Cameron McNamara (1982b) estimated the available water resources of the Bradfield Scheme by using new data available since the Nimmo (1947) review of the Scheme. Because Nimmo (1947) did not provide an estimate of the average annual run-off for the Flinders River (Table 6.3), it is more appropriate to compare the estimates of the average annual run-off of the Tully, Herbert and Burdekin Rivers. The total average annual run-off of these three rivers is about $2.9 \times 10^9 \, m^3$ (Table 6.7), which is less than half of the $6.2 \times 10^9 \, m^3$ estimated by Bradfield (Table 6.2), but about 20 percent higher than Nimmo's estimate of $2.5 \times 10^6 \, m^3$ (Table 6.3).

Table 6.5. *Estimates of land suitable for irrigation within the Bradfield Scheme area*

No.	Area	Extent (ha)
1	Flinders River: Riparian zone to Richmond	60 000
2	Thomson River: Riparian zone to Longreach	125 000
3	The Downs: Richmond, Winton and Longreach	800 000
4	Torrens Creek: Northern part of the Torrens Creek Catchment	110 000
Total		1 095 000

Source: Cameron McNamara Consultants (1982a, Summary Table 1).

---

[7] Burdekin Falls Dam on the lower Burdekin River (Figure 6.1) with a capacity of $1.86 \times 10^9 \, m^3$ was completed in 1987. It is the largest water storage in Queensland.

The Cameron McNamara (1982b) estimate of the water available for diversion was $864 \times 10^6\,\text{m}^3$ per year (Table 6.7). Assuming that irrigated land would require $8000\,\text{m}^3\,\text{ha}^{-1}$ of water per year, and that distribution efficiency would be 67 percent in the irrigation network, the annual water requirement at the head of the irrigation scheme would be $12\,000\,\text{m}^3\,\text{ha}^{-1}$. This meant that the Scheme would be able to supply water to only 72 000 ha of irrigated land west of the Great Dividing Range.

6.7.5 Engineering Aspects of Inter-Basin Water Transfer

From their examination of various alternatives, Cameron McNamara Consultants (1982a, b) suggested a practical method of inter-basin transfer of water. This included a 13 km tunnel from Koombooloomba Dam on the Tully River to the Upper Herbert River (Figure 6.3). Another tunnel with a length of 18 km would transfer waters from the upper Herbert to the upper Burdekin River.

To transfer the water across the Great Dividing Range, a vertical lift of 334 m, plus the equivalent of an additional 75 m to overcome friction, was required (a total lift equivalent to 409 m). The lifting of water from the Clarke River (which contained water backed-up from the 76 m high Hell's Gates Dam on the Burdekin River) would be by three pumping stations and three sections of pipelines (with diameters of approximately 2.5 m) totalling 47 km, and interconnected by 69 km of canals (Figure 6.3). These would connect to a 34 km tunnel under the Mt Courtney area, for discharge into the upper Flinders River, which would convey waters to a possible dam at the Glendower site, upstream from Hughenden.

To implement the project, which would consist of about 40 construction units (dams, canals, pipelines, pumping stations, electricity supply facilities, and others), and develop the irrigation areas, Cameron McNamara Consultants (1982b) suggested a four-stage program over a period of 20 years. Their Stages 1 and 2 were based on the use of the Burdekin River waters only, which would supply water to 11 300 ha and 42 200 ha of irrigated lands respectively. Stage 3 involved addition of water from the diversion of the Herbert River and would increase the size of the irrigated lands to 58 500 ha. Stage 4 involved the addition of water from the Tully River and would supply water to a total of 72 000 ha of land. For each stage of the development, Cameron McNamara Consultants (1982b) provided details of the engineering works needed to be implemented as well as the water supply flow charts.

This tentative staging was based on the eventual gaining of water from the Upper Herbert and Tully rivers that was being used for, or proposed for hydro-power generation. If the allocation of these waters to hydro-power generation were to continue indefinitely, then the scope of the Scheme would be reduced to the completion of Stage 2, which would increase the Scheme's total cost per hectare of irrigated land.

Table 6.6. *Industrial/urban water demand in the receiving region of the Bradfield Scheme*

Type of water use	Annual demand ($10^6\,\text{m}^3$)
Coal mining	18
Thermal power station	30
Associated urban demand	12
Total	60

Source: Cameron McNamara (1982a, Table 2).

Table 6.7. *The 1982 estimate of the annual available water resources of the Bradfield Scheme*

River	Catchment area (km^2)	Average annual rainfall (mm)	Average annual run-off ($10^6\,\text{m}^3$)	Divertible water at source ($10^6\,\text{m}^3$)	Deductions[a] ($10^6\,\text{m}^3$)	Water available at each river basin ($10^6\,\text{m}^3$)	Net water available for diversion ($10^6\,\text{m}^3$)
Tully	163	2610	351.6	286	43	243	–
Herbert	5370	1034	1072.3	437	102	335	–
Burdekin	17 858	695	1520.0	740	160	580	–
Flinders	2150	630	89.0	30	264	−234	–
Total	25 541	–	3032.9	1493	569	924	864[b]

Note: [a] Evaporation, transmission losses and irrigation extraction; [b] This figure is derived from $924 \times 10^6\,\text{m}^3$ at Hughenden less the industrial/urban demand of $60 \times 10^6\,\text{m}^3$, shown in Table 6.6.
Source: Cameron McNamara (1982b, Table 5.5).

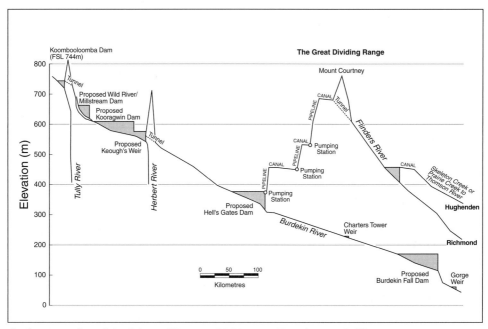

Figure 6.3 Longitudinal cross-section of the Scheme (Cameron McNamara Consultants, 1982b).

6.7.6 Estimates of Costs and Benefits

Construction of the scheme would cost about $1440 million (in 2002 prices) for Stage 1, which would irrigate some 11 300 ha, and $3450 million (in 2002 prices) for all four Stages, which would irrigate 72 000 ha (Table 6.8). This did not include the construction costs of the distribution systems that would deliver water to each of the irrigation farms. Moreover, the annual cost of electricity for pumping water would be about $42.5 million for Stage 1, and would increase to $142.5 million for all four Stages in 1982 prices.

According to Cameron McNamara Consultants (1982b, pp. 5–26), it was not within the scope of their study to assess in detail the benefits of the Scheme. However, they believed that construction of the project over a period of 20 years would create 20 000 man-years of direct employment. This figure did not include employment associated with the construction of water delivery systems to the farm gates. Also, it did not include any multiplier effects during construction, or continuing employment on irrigation farms after completion of the project.

6.8 BRADFIELD SCHEME AND WATER SUPPLY OF ADELAIDE

If the Bradfield Scheme ever proceeded, it might be feasible to pump the water from this Scheme some 800 km south, from a weir located on Cooper Creek near the Queensland and South Australian border, to metropolitan Adelaide

Table 6.8. *Estimated costs for the four Stages of construction*

Stage	Cumulative extent of the irrigated area (ha)	Order of cost in 1982 prices ($M)	Order of cost in 2002 prices[a] ($M)
1	11 300	576	1440
2	42 200	192	480
3	58 500	397	993
4	72 000	215	537
Total	72 000	1380	3450

Note: [a] The 1982 estimates of costs have been multiplied by a factor of 2.5 in order to represent 2002 prices.
Source: Cameron McNamara Consultants (1982b).

(Government of South Australia, 1989, p. 17).[8] In this case, the maximum volume of pumped water would be $200 \times 10^6 \, m^3$ per annum with a quality of 100 mg L^{-1} TDS. The 1989 estimated project cost would be $4.5 billion and the estimated cost of water would be $4300 per $10^3 \, m^3$. These costs would be $6.5 billion and $6200 respectively in 2002 prices.

This option was one of 21 suggestions for providing water to South Australia in an improbable, but not impossible water shortage. As with the case of water supply of Adelaide by diverted water from the Clarence River Basin (see section 5.5), description of this option here has only a historic interest.

---

[8] A short description of water supply of Adelaide via pipelines from the River Murray is provided in section A of Chapter 10.

SECTION B: THE REID SCHEME

6.9 INTRODUCTION

In 1946, L. B. S. Reid, an engineer from Queensland proposed a scheme for inland transfer of northern Queensland rivers. It was planned to use gravity diversion of the headwaters of the Mitchell and Gilbert rivers that normally discharge to the Gulf of Carpentaria (Figure 6.4). The diverted waters would be used for the development of irrigation in the Flinders River Basin as well as the Diamantina River Basin. Table 6.9 shows the major characteristics of these rivers. The Mitchell and Gilbert rivers are located in a high rainfall area of northern Queensland (see Figure 3.2). Solely based on hydrological considerations, they have the potential for diverting inland a significant portion of their waters. For example, the Mitchell River has a mean annual run-off of $12\,000 \times 10^6\,m^3$, while water use in its catchment in the year 2000 was only $55.23 \times 10^6\,m^3$. This is also the case for the Gilbert River, with a mean annual flow of $5580 \times 10^6\,m^3$ and water use of only $6.41 \times 10^6\,m^3$.

The soils of the Flinders and Diamantina basins, which would receive the water for irrigation, are mainly of the cracking clay soils (see Plumb, 1980 and soil maps of Australia, Figures 3.6 and 3.7), suitable for irrigation.

6.10 DESCRIPTION OF THE SCHEME

The Scheme has been described in Noakes (1946, pp. 96–106) and Noakes (1947, pp. 18–28). A brief description is provided below, based on these two almost identical references.

Table 6.9. *Characteristics of the major rivers of the Reid Scheme*

River	Catchment area[a] (km^2)	Mean annual run-off[a] ($10^6\,m^3$)	Year 2000 water use[b] ($10^6\,m^3$)
Mitchell River	71 800	12 000	55.23
Gilbert River	46 900	5580	6.41
Flinders River	109 000	3030	7.70
Diamantina River	158 000	960	0.02

Source: [a]Australian Water Resources Council (1987, pp. 56 and 57), and [b] National Land & Water Resources Audit (2001, pp. 100 and 101).

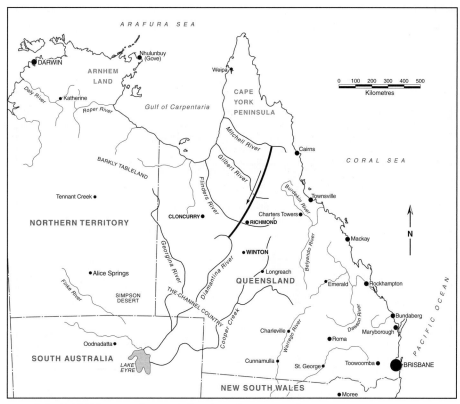

Figure 6.4 Location map of the Reid Scheme and general direction of water transfer.

The Walsh River is the Scheme's head of water supply (Figure 6.5) and the Scheme consists of:

- Ten diversion dams and reservoirs on the headwaters of the Mitchell and Gilbert Rivers (Walsh, Tate, Lynd, Einasleigh, Etheridge and Gilbert rivers).
- Some 274 km of canals, which would follow the natural contours of the land and would have concrete lining at appropriate locations.
- Four tunnels with a total length of 17 km and with dimensions of 22 m × 11 m. These tunnels would have reinforced concrete lining and would be fitted with safety gates.
- The main reservoir on the southern side of the Gregory Range would have a capacity of $7.5 \times 10^9 \, m^3$. The backed up waters of this reservoir would be diverted via a tunnel to a small reservoir at the headwaters of the

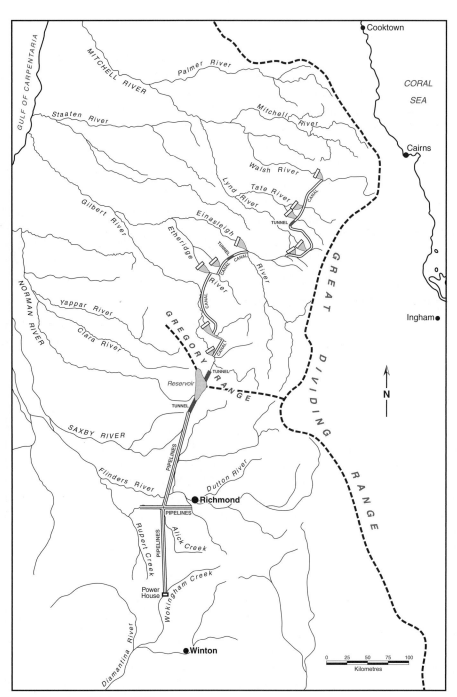

Figure 6.5 The Reid Scheme (based on Noakes, 1947).

Saxby River. This reservoir would feed a number of pipelines diverting water to the Richmond and Cloncurry (280 km west of Richmond) areas of the Flinders River Basin, as well as to Workingham Creek, which is a tributary of the Diamantina River.

The Scheme was expected to take 5 years to be built and create 50 000 direct and indirect jobs during the construction period.

6.11 COST OF THE SCHEME

No estimate was available for the capital and annual cost of the Scheme, or for the annual price of the diverted water. However, because the Scheme was proposed shortly after World War II, the author claims that the construction of the Scheme, as well as the Bradfield Scheme, would cost less than the cost of one year of war and would constitute a great asset for the nation.

6.12 EXPECTED BENEFITS OF THE SCHEME

While no estimates of the environmental impacts of this Scheme were considered, the following benefits were expected:

- Development of irrigation in the Flinders and part of the Diamantina River Basins, using sprinklers to irrigate fruit trees and lucerne. This was expected to provide farming opportunities to 5000 Returned Soldiers on farms of 40 to 80 ha.
- Freshwater fisheries could be developed in the impounded areas.
- Scenic resorts could be built on the highlands on each side of the lakes. This would be an added source of revenue as well as a great tourist attraction during the hot summer months.

6.13 CONCLUSIONS

Two schemes for the inland diversion of water from the Queensland coastal rivers that were proposed in 1938 and 1946 have been described here. Firstly, Bradfield (1938) overestimated the average annual volume of water available from the Tully, Herbert and Burdekin Rivers for diversion to the west of the Great Dividing Range, and secondly his assumption that the water could be transferred inland by gravity was incorrect. These errors were due to a lack of rainfall and topographic data at the time. Despite these errors, neither Nimmo nor Cameron McNamara Consultants were able to totally reject Bradfield's basic concept. Also, the general route considered by Bradfield for inter-basin transfer proved to be fundamentally correct.

The expanded Bradfield Scheme, which consisted of constructing dams on the Fink, Georgina, and Diamantina Rivers and Cooper Creek in central Australia, is unrealistic because:

(1) the divertible volume of water in these rivers is minimal compared to their average annual run-off;
(2) the suitability of soils in the area for growing crops is questionable; and
(3) implementation of the scheme would not increase the average annual rainfall of inland Australia.

Unlike the Bradfield Scheme, the Reid Scheme, is less well known or well developed. Some of the problems with this scheme are:

- No hydrological details about the expected volume of water from each of the rivers are provided.
- The sizes of the various diversion dams and reservoirs were not provided.
- No attempt was made to estimate the cost and benefits of the Scheme.

In spite of all the above-mentioned problems, the concept does have some merit, however no estimates of environmental impacts have been considered.

If any modified version of the Bradfield Scheme or the Reid Scheme should be considered, many further investigations would be required (see Chapter 2). These include assessment of their benefits, costs, engineering feasibility, and also their impacts on the environments of both the coastal and inland catchments affected by the schemes. Moreover, these schemes should be designed within the frameworks of water management plans developed for various river basins covering different parts of the Scheme.

References

Australian Bureau of Statistics (1984). *Queensland Year Book 1984*. Australian Bureau of Statistics, Queensland Office.

Australian Water Resources Council (1987). *1985 Review of Australia's Water Resources and Water Use*. Canberra: Australian Government Publishing Service. Volume 1: Water Resources Data Set; Volume 2: Water Use Data Set.

Bradfield, J.J.C. (1938). *Queensland: The Conservation and Utilization of Her Water Resources*. Sydney: Dr. J.J.C. Bradfield & Son Consulting Engineers.

Bradfield, J. J. C. (1941a). Rejuvenating Inland Australia. *Walkabout: Australia's Geographic Magazine.* July 1941, pp. 7–15.

Bradfield, J. J. C. (1941b). Watering Inland Australia. *Rydge's: The Business Management Monthly.* October 1941, pp. 586–589 and 606.

Cameron McNamara Consultants (1982a). *The Bradfield Concept Preliminary Study: Executive Summary of a Report Prepared for the Coordinator General.* Brisbane: Cameron McNamara Consultants.

Cameron McNamara Consultants (1982b). *The Bradfield Concept Preliminary Study: A Report Prepared for the Coordinator General.* Brisbane: Cameron McNamara Consultants.

Commonwealth Meteorological Bureau (1945). Bradfield Scheme for watering the inland: meteorological aspects. *Meteorological Bureau Bulletin,* No. 34.

Department of Natural Resources and Mines (1999). *Cooper Creek Draft Water Management Plan.* Revised Draft. Brisbane: Department of Natural Resources and Mines.

Department of Natural Resources and Mines (2002). *Burdekin Basin Proposal to Prepare a Draft Water Resource Plan.* Brisbane: Department of Natural Resources and Mines.

Department of Natural Resources and Mines (2003). *Overview Report and Draft Plan: Georgina and Diamantina Water Resources Planning.* Brisbane: Department of Natural Resources and Mines.

Geographical Society of Australia, South Australian Branch (1955). *Lake Eyre, South Australia: The Great Flooding of 1949–50.* The Report of the Lake Eyre Committee. Adelaide: Griffin Press.

Government of South Australia (1989). *South Australia Water Futures: 21 Options for the 21st Century.* Adelaide: Engineering and Water Supply Department.

Johnson, A. K. L. and Murray, A. E. (1997). *Herbert River Catchment Atlas.* Townsville: CSIRO Tropical Agriculture, Davies Laboratory.

Keller, H. (1906). Niederschlag Abfluss und Verdunstung in Mitteleuropa. *Jahrbuch für Gewässerkunde, Norddeutschland, Besondere Mitteilungen* (Rainfall, runoff and evaporation in Middle Europe). *Yearbook of Hydrography, North Germany).* Berlin **1**(4): 1–43.

Nimmo, W. H. R. (1947). *The Bradfield Scheme.* Brisbane: Stanley River Works Board.

Noakes, A. W. (1946). *The Life of a Policeman: A Comprehensive Work of Conditions in the Out-back of Queensland.* South Brisbane: Rallings & Rallings.

Noakes, A. W. (1947). *Water for the Inland: A Brief and Vivid Outline of Conditions in the Out-back of Queensland in which is Embodied the Reid and Dr. Bradfield Water Schemes.* South Brisbane: Rallings & Rallings.

Plumb, T. ed. (1980). *Atlas of Australian Resources.* Third Series, Volume **1**. Soils and Land Use. Canberra: Commonwealth Government Printer.

Pomel, A. (1872). Le Sahara, observations de géologie et de géographie physique et biologique, avec des aperçus sur l'Atlas et le Soudan, et discussion de l'hypothèse de la mer saharienne á l'époque préhistorique. *Bull. Soc. Climatologique de l'Algérie* **8**: 133–265.

Quayle, E. T. (1921). Possibilities of modifying climate by human agency, with special application to south-eastern Australia. *Proceedings of the Royal Society of Victoria.* Vol. **XXXIII** (New Series): 115–132.

Quayle, E. T. (1922). Local rain producing influences under human agency in Central Australia. *Proceedings of the Royal Society of Victoria.* Vol. **XXXIV** (New Series): 89–104.

Raxworthy, R. (1989). *The Unreasonable Man: The Life and Works of J. J. C. Bradfield.* Sydney: Hale & Iremongger Pty Limited.

Further Reading

Timbury F. R. V. (1944). *The Battle for the Inland: The Case for the Bradfield and Idriess Plans.* Sydney: Angus and Robertson.

7 Three Schemes for Flooding Lake Eyre

7.1 INTRODUCTION

Lake Eyre in central Australia was named after Edward John Eyre, the first European who sighted its shores in 1840. It is the largest ephemeral playa-lake in Australia with a total area of $9300\,km^2$, and is actually comprised of two lakes: Lake Eyre North and Lake Eyre South, which are connected by the Goyder Channel (Figure 7.1). The North Lake has a very gentle slope of about 3 cm per km from the north to the south, and has an elevation of $-15\,m$ AHD at its lowest point, while the minimum elevation of the South Lake is $-13\,m$. Before 1950, it was believed that the Lake was always dry and had not contained water since its first sighting by Europeans.

The Lake Eyre Basin has a total population of about 60 000 people of which 27 000 live in Alice Springs. Major economic activities of the Basin include grazing, mining, and oil and gas production. It contains substantial reserves of natural gas, copper, uranium, gold, silver and opal (ABARE, 1996).

This chapter describes three proposals to flood Lake Eyre by:

(1) seawater via a canal from Port Augusta in South Australia;
(2) diversion of Queensland coastal rivers towards Lake Eyre; and
(3) extracting groundwater from the Great Artesian Basin.

Before describing these proposals some of the important characteristics of the Lake Eyre Basin such as geology, climate, water resources, and flooding history are summarised.

7.2 CHARACTERISTICS OF THE LAKE EYRE BASIN

7.2.1 Geology

Drexel et al. (1993), and Drexel and Preiss (1995) describe the geological history of the Lake Eyre Basin, while Wells and Callen (1986) detail the Cainozoic sediments of the Lake. In the late Cainozoic, the Lake Eyre Basin was superimposed on the Tertiary Birdsville Basin and the even more extensive pre-existing Eromanga Basin. Its centre was located north of Lake Eyre in the Simpson Desert. During the Pleistocene, the centre of the Lake Eyre Basin moved to its present position in the southern part of Lake Eyre North. In the late Pleistocene and Holocene, aeolian forces were dominant, leading to the formation of large dune fields.

The stratigraphy of the Quaternary sequence in the Madigan Gulf at the southern end of Lake Eyre reveals a number of depositional environments (Magee et al., 1995). The wettest phase occurred during the last interglacial when an enlarged Lake Eyre was up to 25 m deep. There have been a number of subsequent dry periods, culminating in the deposition of a substantial salt crust around the time of the glacial maximum.

7.2.2 Lake Level Fluctuations and Environmental Changes

The level of Lake Eyre has fluctuated over the past 150 000 years (Magee et al., 2004). The Lake attained its maximum level of $+10\,m$ between 130 000 and 110 000 years ago, which coincided with a high sea level (Figure 7.2). Lower levels of $+5\,m$ and $-2\,m$ occurred between 100 000 to 80 000 years ago and between 65 000 to 60 000 years ago respectively. A major deflation[1] episode occurred around 50 000 years ago, which coincided with a relatively low sea level (Figure 7.2). This excavated the present Lake Eyre Basin and deposited gypsum and clay-rich aeolian sediments at a number of sites around the Lake (Magee and Miller, 1998). Between 30 000 and 12 000 years ago the Lake was at least as dry as it is today. About 10 000 years ago, a minor lacustral phase occurred until the modern ephemeral playa became established about 3000 to 4000 years ago.

---

[1] The removal of material from a beach or other land surface by wind action.

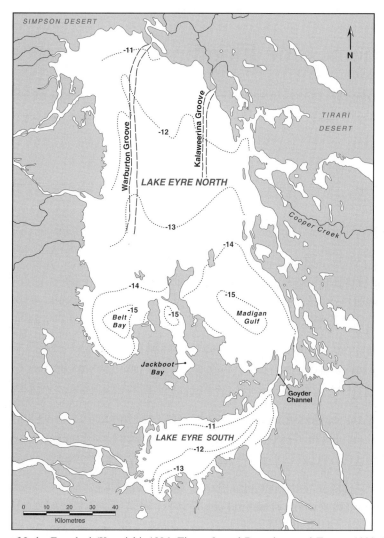

Figure 7.1 Elevation contours of Lake Eyre bed (Kotwicki, 1986, Figure 9, and Bonython and Fraser, 1989, Figure 5.1).

Croke et al. (1999) describe the influence of the Australian monsoon on the hydrology of the Lake between 90 000 and 130 000 years ago. This coincided with a time of high sea level and strong monsoonal activities in northern and western Australia. The large monsoon-fed river systems of the Cooper and Diamantina contributed most of the inflow to this mega lake. Also, western rivers had an enhanced fluvial activity between 100 000 and 110 000 years ago. Other examples of environmental changes in the Lake Eyre Basin are, vegetation change over the past 65 000 years in central Australia (Johnson et al., 1999), and the extinction of the large and medium sized land mammals in late Pleistocene (Miller et al., 1999).

7.2.3 Reconstruction of Late Quaternary Palaeohydrology of Lake Eyre

Comprehensive research in the Lake Eyre catchment over the last few decades has determined the ages and elevations of the Lake's shoreline deposits (DeVogel et al., 2004). Four prominent shorelines have been recognised around Lake Eyre, with the ages between 40 000 and 125 000 (Table 7.1). These high shorelines have been well preserved and mapped on the south side of the Lake, and to a lesser degree on the west side. DeVogel et al. (2004) used a Geographic Information System (GIS) to map the palaeo-lakes, and through a Digital Elevation Model, they were able to simulate the levels of historic and prehistoric floods. They found that at the peak (125 000 years ago), Lake Eyre covered more than 25 000 km^2. These Lake levels represent climatic conditions very different from the modern situation.

7.2.4 Climate

The Lake Eyre Drainage Basin (Figure 7.3) includes some of the most arid and sparsely populated lands of the continent. Rainfall in the north of the Basin is summer dominated,

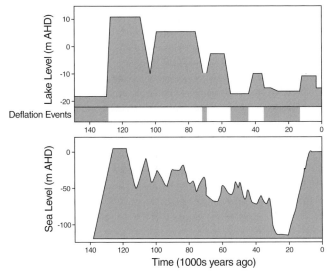

Figure 7.2 Fluctuation in the sea level and level of Lake Eyre during the last 150 000 years with respect to the Australian Height Datum (Magee *et al.*, 2004, Figure 3).

Table 7.1. *Some results from the GIS model for Lake Eyre*

Flood event	Flood level (m)	Area (km^2)	Volume (10^9 m^3)
40 000 years ago	−10.0	6740	20
65 000 years ago	−3.5	9920	74
80 000 years ago	+5.0	19 590	216
125 000 years ago	+10.0	25 260	332

Source: DeVogel *et al.* (2004, Table 1).

and is influenced by the El Niño Southern Oscillation. The mean annual rainfall of the area is low, while its evaporation rate is high. The annual rainfall deficit (rainfall minus evaporation) has maximum values of −3000 to −3500 mm over the Lake and its surrounding area (Figure 3.3). As an indication, Table 7.2, lists the climatic data for three stations around the Lake. It shows that:

(1) the mean daily maximum summer (January) temperatures are high, reaching about 38°C;
(2) for more than 30 days per year, the maximum daily temperature is above 40°C;
(3) the mean annual rainfall values are low (162.4 to 209.5 mm), and
(4) the annual evaporation rates are extremely high, exceeding 3 500 mm.

The evaporation rate from the Lake water surface was calculated in 1951 when the Lake was flooded (Lake Eyre Committee, 1955, pp. 50–56). The annual evaporation rate was about 1960 mm, with monthly values ranging from 30 mm in the winter month of June to 280 mm in the summer months of December and January.

7.2.5 Water Resources

The total area of the Lake Eyre Basin is 1 140 000 km^2 or about 15 percent of mainland Australia. The Basin receives 6.31×10^9 m^3 of run-off from the Georgina, Diamantina, Fink, Todd and Hay rivers and Cooper Creek (Figure 7.3 and Table 7.3). The National Land & Water Resources Audit (2001, Table 3) reported a value of 8.64×10^9 m^3, without providing any details of its estimation. In spite of this significant volume of run-off, the very high evaporation rates and the long time it takes for water to reach the Lake, means that it only fills with water during exceptionally wet years (see Section 7.2.6).

Groundwater occurs in sandstone aquifers of the Great Artesian Basin throughout much of the Lake Eyre Basin (Habermehl, 2001). Its waters are used for pastoral activities, domestic water supply, and the oil, gas and mining industries. In particular, it is the source of water for the Olympic Dam uranium and copper mine, and its associated town of Roxby Downs.

7.2.6 Flooding History of Lake Eyre

Apart from Edward John Eyre, prior to 1949 only two other European explorers had apparently seen water in Lake Eyre (Kotwicki, 1986, p. 50). The first was in 1896, when J. Ross reported water in what is now known as the Warburton Groove (Figure 7.1). The second was in 1922, when for the first time, G. H. Halligan observed the Lake from the air and ascertained that it was one-third covered with water. The first detailed description of flooding is from 1949–50 when heavy rainfall in central and western Queensland caused Lake Eyre and many other small lakes to fill with water (Lake Eyre Committee, 1955). The following description is based on this publication.

In February 1949, a tropical cyclone passed over the Queensland coast and moved inland producing flooding over a vast area. During the last few days of February another cyclone crossed the Queensland coast. Throughout March, rainfall continued and by the end of the month many weather stations had received the highest March rains on record. Over the next few months there was a temporary relief from rain. September 1949 saw some more heavy falls, whilst October was very wet throughout Queensland and rivers again rose rapidly, but November and December were relatively dry.

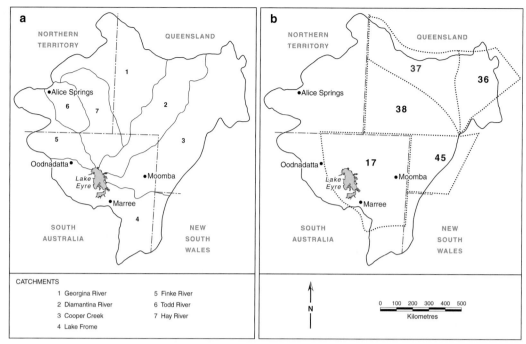

Figure 7.3 (a) Sub-basins of the Lake Eyre Drainage Division (Australian Water Resources Council, 1987), (b) Meteorological Districts (Lake Eyre Committee, 1955).

Table 7.2. *Climatic data for three stations around the Lake Eyre, for the period ended in December 2004*

Station name	Established in	Elev. (m)	Mean daily max. temp. in °C for:		Mean daily min. temp. in °C for:		Highest max. temp. (°C)	Lowest min. temp. (°C)	Mean number of days with:		Mean annual rainfall (mm)	Mean annual evapo. (mm)
			Jan.	Jul.	Jan.	Jul.			max. temp. >40°C	min. temp. <0°C		
Marree	1885	50.0	37.7	19.0	21.2	4.8	49.4	−2.8	32.9	2.7	162.4	N.A.
Moomba	1972	39.0	37.5	19.2	23.3	6.3	49.1	−1.4	31.0	0.2	209.5	3540
Oodnadatta	1939	116.5	37.6	19.5	22.7	5.8	50.7	−2.6	31.9	1.2	175.6	3760

Source: Bureau of Meteorology website.[2]

The year 1950 started with good rains in January and even better rains in February in Queensland. Heavy rainfall continued in March with three cyclones contributing to the flooding. April rainfall was above normal and good rains continued during the next three months. August and September were relatively dry, while October, November and December were wet and fairly general flooding was reported.

Lake Eyre reached its highest level at the end of September 1950, with water depth of 4 m measured in the Madigan Gulf. As a result of the flooding, thousands of seagulls and other waterbirds migrated there. Many of these birds fed on the abundant supply of fish while others consumed the diverse variety of insects and other macro-invertebrates that thrived after the waters arrived.

At the beginning of the year 1951 there was no inflow and the water level dropped due to high rates of evaporation during an exceptionally hot summer. The salinity rose and the larger species of fish died. The water level continued to drop and by 20 November 1951, the remaining waters were divided into two separate areas, one at Madigan Gulf

[2] http://www.bom.gov.au/climate/averages/tables/ca_sa_names.shtml (visited in July 2005).

Table 7.3. *Characteristics of the Lake Eyre sub-basins*

No.	Sub-basin	Area (km^2)	Mean annual run-off (10^6 m^3)
1	Georgina River	242 000	2000
2	Diamantina River	158 000	960
3	Cooper Creek	296 000	2330
4	Lake Frome	205 000	550
5	Fink River	115 000	238
6	Todd River	71 000	102
7	Hay River	83 000	127
Total		1 170 000	6307

Source: Australian Water Resources Council (1987, Table S15).

and the other in the Belt Bay. In January 1953 the lake was reported dry. The cycle of flooding and drying had taken about three and a half years. Minor flooding occurred again in 1953 and 1955. Table 7.4 shows the area of the Lake covered with water.

The estimated volume of water in the Lake at its highest level was about 21×10^9 m^3. Considering the evaporative losses, the quantity of water flowing into the Lake during 1949–50 was estimated at about 31×10^9 m^3 or approximately five times the average annual run-off of the Lake Eyre Drainage Division (Table 7.3). It was found that the 1950 flooding of the Lake had no impact on the climate of the surrounding area (Lake Eyre Committee, 1955, p. 29).

The Lake Eyre Committee (1955, pp. 11–26) provided an analysis of the 1949–50 rainfall records and compared them with the average rainfall records since 1880 for five meteorological districts; 17, 36, 37, 38 and 45 (Figure 7.3). Figure 7.4 shows the average annual rainfall of these districts for the period 1880–1950, and compares it with the wet years of 1949 and 1950. The 1950 rainfall was particularly significant for District 36 which is the closest district to the Queensland coastal area. The Lake Eyre Committee (1955, p. 26) concluded:

> With 1949 and 1950 as positive evidence, one can no longer say that Lake Eyre has never had water in it since white settlement began, and indeed it is quite likely that over the last one hundred years water has flooded into the lake on several occasions.

The next major flood of the Lake occurred during the period 1974–77 (Kotwicki, 1986, p. 54). During this flood, Lake Eyre North was at its highest recorded level of −9.09 m (a water depth of 5.91 m at Madigan Gulf) in May–June 1974. Goyder Channel started to flow on 19 March 1974 and flowed until October when the equilibrium level of −9.5 m between both lakes was attained. The water balance for the period of January 1974 to June 1976 indicated that the total intake was 56×10^9 m^3, which included 8×10^9 m^3 directly from rainfall on the Lake surface. During the same period, evaporation was estimated at 39.5×10^9 m^3. Therefore, the water storage in the Lake at June 1976 was 16.5×10^9 m^3.

The last major flood occurred in January 1984 (Kotwicki, 1986, p. 66). Widespread heavy rains ranging from 105 mm to 360 mm were recorded from 9 to 15 January in the far north of South Australia. Lake Eyre South filled independently and overflowed to Lake Eyre North. Analysis of the monthly LANDSAT photographs, taken between 20 December 1983 and 22 February 1984, showed that the Lake was completely dry on 20 December 1983, while on 21 January 1984 both Lakes were covered with water. By 22 February 1984, the whole system was in the process of drying up. By 4 March 1985 only a small volume of water remained in the southern part of the Lake Eyre North (see Kotwicki, 1986, pp. 69–73 for the LANDSAT photographs). Another major flood similar to the 1984 event occurred in March 1989, and minor floods occurred in 1991, 1992, 1997, and 2000.[3]

7.2.7 Modelling of the Flooding Events

Kotwicki (1986) divided the Lake Eyre Drainage Division into 22 sub-basins and reconstructed its flooding for the period 1885–1985, using a simple numerical run-off model and daily rainfall data. It was assumed that run-off from the Lake Eyre Drainage Basin is directly proportional to the precipitation. In 40 years of this period, drought conditions prevailed, so the remaining 60 years were simulated. The model was calibrated against the major flooding of 1949, 1950, 1955, 1963, 1967, 1974, 1977 and 1984. Simulation results indicated that there were four major periods of flooding of Lake Eyre North: 1885–94, 1916–21, 1949–50 and 1974–77. Periods without inflows ranged from 1 to 7 years. There were 24 such periods lasting on average 2 years. Table 7.5 indicates the major simulated fillings of Lake Eyre North and their return periods.

The simulation results and existing data suggest that Lake Eyre North more often contained some quantity of water rather than being dry and almost all its area is covered on average once every 8 years (Table 7.6).

Simulated results for Lake Eyre South were considered as inaccurate due to the paucity of data. Years of likely inflows included 1885, 1889, 1904, 1920, 1930, 1939 and 1941,

---

[3] http://lakeeyreyc.com/fldhist.html (visited in July 2005).

Table 7.4. *Area of water in Lake Eyre from September 1950 to October 1952*

No.	Date	Area (km^2)	No.	Date	Area (km^2)
1	End Sep. 1950	8000	4	20 Nov. 1951	2150
2	1 Aug. 1951	7000	5	7 Aug. 1952	1060
3	1 Nov. 1951	3500	6	10 Oct. 1952	290

Source: Lake Eyre Committee (1955).

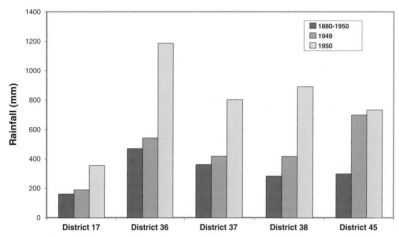

Figure 7.4 Selected average annual rainfalls of five meteorological districts covering the Lake Eyre Drainage Division (Lake Eyre Committee, 1955, Tables III and IV).

while known floods occurred in 1955, 1963, 1968, 1973, 1974, 1975, 1976 and 1984. It is estimated that the mean annual inflows may have been in the order of $0.15 \times 10^9 \, \text{m}^3$.

7.3 PORT AUGUSTA–LAKE EYRE CANAL SCHEME

The idea of flooding Lake Eyre by seawater was inspired by a French proposal in the 1870s to flood the Sahara depression in Africa by means of a canal from the Mediterranean Sea (see Appendix E). In the early 1880s, it was proposed to cut a canal from the sea at Port Augusta in South Australia to Lake Eyre (Figure 7.5), and flood its vast basin with seawater. The general assumption was that once the Lake was full of water, evaporation from its surface would be considerable. They assumed that the increased moisture in the atmosphere would increase the local rainfall near the Lake. However, as described in Chapter 6 (Section 6.6.1), this assumption was not correct.

This Scheme was discussed in 1883 in the Legislative Council of South Australia (Parliament of South Australia, 1883, pp. 261–264 and Parliament of South Australia, 1883–84). The Hon. Richard C. Baker moved that: *A flying survey and levels be taken by the Railway Department to show the practicability or otherwise, and the approximate cost of a canal to connect Lake Eyre with the sea.* In moving this motion, the Hon. R. C. Baker made it clear that he was not necessarily supporting the proposal. He only desired to ascertain the probable cost. He went on to describe the extent of the dry land in Australia and added that:

> If by any means the rainfall could be increased, the benefit to Australia would be incalculable because it contained magnificent soil just fitted for agriculture. All that was required was a sufficient rainfall, the absence of which was the cause of our present sparse population. If seawater could once be got into Lake Eyre the consequence would be that we would be able to have agricultural settlement extending over the whole of Australia.

The Commissioner of Public Works (Hon. J. G. Ramsay) replied to the Hon. R. C. Baker and said that *he quite agreed with the Hon. Member regarding the advantages to be derived from making an inland sea if it could be done at a reasonable cost.* Then he provided the following information:

- The level of Lake Eyre's shoreline surveyed by one of the railway surveyors was at 11.9 m below the sea level at Port Augusta.

Table 7.5. *Major simulated fillings of the Lake Eyre North and their return periods*

Year	Inflow volume ($10^9 m^3$)	Return period once every x (years)	Year	Inflow volume ($10^9 m^3$)	Return period once every x (years)
1974	39.3	150	1916	13.2	14
1950	27.2	55	1887	12.4	13
1890	20.4	26	1975	11.0	10
1920	18.3	23	1984	10.2	8
1894	13.8	15	1906	10.0	8

Source: Kotwicki (1986, Table 4).

Table 7.6. *Estimated return periods of the Lake Eyre North floods at various water levels*

No.	Water level (m)	Return period (years)	No.	Water level (m)	Return period (years)
1	−12.5	5	5	−9.5	100
2	−11.7	10	6	−7.5	500
3	−11.0	20	7	−6.3	1000
4	−10.4	50	–	–	–

Source: Kotwicki (1986, p. 79).

- The level of Lake Torrens surveyed by the Hydraulic Engineer's surveyors was at 33.5 m above the sea level at Port Augusta.
- The suggested scheme to let the seawater into Lake Eyre was not practical. The slope of the canal would be less than 3 cm per km, and it was doubtful if the water would flow in such a canal.
- Digging the canal would cost more than £37 million ($3 billion in 2002 prices).

The Hon. R. C. Baker withdrew his motion following these answers.

In an article entitled "The Lake Eyre Basin: is an Inland Sea Feasible?", published on 18 May 1905 in *The Advertiser, Adelaide* (pp. 5 and 6), Sir Richard Baker, who was at the time President of the Senate, described the Scheme. The following brief description is based on his article.

Near the centre of the country there is a very large area that lies below sea level. If seawater is allowed into this depression, the evaporation from the created Inland Sea would alter the whole climate of Australia. The area of Lake Eyre is about 8900 km^2 and its shores are 11.9 m below sea level. However, the extent of the area below sea level could be as large as 26 000 km^2. Moreover, during comparatively recent geological times, the whole basin was well-watered and favourable vegetation supported gigantic herbivores.

The southern end of Lake Eyre is at a distance of 336 km from Port Augusta. An arm of the sea runs north for a considerable distance. The adjacent country is flat and not much above sea level for about 45 km. It is then only a distance of 13 km to the southern shores of Lake Torrens. This Lake is 192 km long and its shoreline is about 33.5 m above the low water mark at Port Augusta. The divide between the Lake Torrens and Lake Eyre catchments is at an elevation of 61 m. Therefore, a deep cutting would have to be excavated. Assuming that the required canal would be excavated to a depth 3 m below the low water mark at Port Augusta, there would be a fall of 8.9 m in 336 km along the bottom of the canal (a slope of 2.6 cm per km). It is evident that with such a small gradient, the velocity of water flow would be very small and a very wide canal would be required to obtain a reasonable flow into Lake Eyre.

It was estimated that a flow rate of 125 000 m^3 per minute (66×10^9 m^3 per annum) should be diverted. Almost one third of this volume would be required to compensate an assumed evaporation of about 2500 mm per annum from the Lake surface. A canal, which would convey this quantity of water at a velocity of 0.3 m per second, would require a cross-sectional area of 1800 m wide by 3.65 m deep. The Engineer-in-Chief of South Australia estimated that construction of the canal would require an excavation of 9.26×10^9 m^3 of materials. Because of differences in opinion amongst engineers and contractors, the probable cost of this excavation ranged between £151.5 million to £3 billion ($12 to 246 billion in 2002 prices). Sir Richard Baker concluded that the most discouraging part of the whole Scheme was that if Lake Eyre Basin was connected with the sea, there was no certainty that this enormous volume of water would increase the regional rainfall. He mentioned that the Red Sea has an enormous evaporation, but none of this evaporated water falls on or near its shores.

Gregory (1906) described the Scheme on pages 342–352 of his book entitled *The Dead Heart of Australia*. He argued that the size of the canal would depend on the amount of evaporation. He estimated the evaporation and dimensions of the canal as follows:

- Based on the data available at that time, he assumed an average annual evaporation rate of 2500 mm, or 7 mm per day and made his calculation based on a lower daily rate of 6.4 mm or 2335 mm per annum.
- Assuming that only 5200 km^2 of the Lake be flooded, the daily evaporation from this water surface, at the rate of 6.4 mm per day, would be about 33.3×10^6 m^3

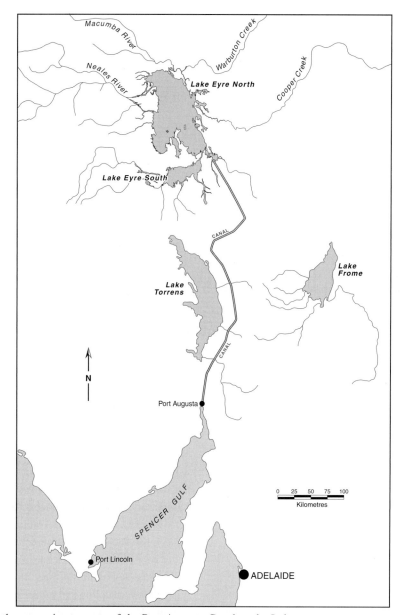

Figure 7.5 Lake Eyre and the approximate route of the Port Augusta Canal to the Lake.

(12×10^9 m^3 per annum). This meant that a canal carrying less than this daily volume would not maintain water in the Lake. Diversion of this volume of water required a canal that was 3 m deep and 300 m wide.

A ridge of hills divides the Lake Eyre Basin and Lake Torrens, and the lowest gap is about 53 m above sea level (Gregory, 1906). The shore of Lake Torrens is at about 33.5 m above sea level. Therefore, construction of the canal would require an enormous volume of excavation. As the cost of cutting a 15 m wide canal was estimated at £37 million, the cost of cutting this canal, which would be 20 times larger, would be about £740 million (approximately $70 billion in 2002 prices). He noted that this estimate was very approximate and was not based on a detailed costing. However, he concluded that the cost of flooding 5200 km^2 of Lake Eyre was extremely high and unaffordable. Moreover, because of salt deposition, the Lake would be gradually filled by salt.

Finally, Towner (1955) provides a description of the Lake Eyre Basin, full text of Sir Richard Baker's article in *The Advertiser* of 18 May 1905 and Appendix B of the Madigan (1946) book entitled *Crossing the Dead Heart*. This appendix entitled "The Supposed Deterioration of the Lake Eyre Country" rejects claims that the Lake Eyre country has suffered from decreased rainfall, sand drift and erosion.

It also rejects schemes for the flooding of Lake Eyre by seawater from Port Augusta, or as suggested by Ion Idriess (see the following section) by inland diversion of Queensland coastal rivers.

7.4 THE GREAT BOOMERANG SCHEME

During World War II, Idriess (1941) proposed the Great Boomerang[4] Scheme for the inland diversion of the Queensland coastal rivers. The "elbow" of the Boomerang rests on the central part of Australia known as the Dead Heart, one "blade" goes to the Queensland coast and the other through South Australia to the coast. He developed his scheme not as an engineer, backed up with data, measurements and cost estimates, but merely as a dreamer. He described the harsh environment of central Australia and compared it with the well-watered eastern and northern coastal areas of the country. He referred to the major floods of 1870, 1890 and 1904[5] (Idriess, 1941, pp. 211 and 212) and argued that once every 10 to 15 years[6] three major rivers (Georgina, Diamantina and Cooper Creek) discharge into Lake Eyre, as well as flooding of many other smaller lakes in the basin. Subsequently, the inland blossoms like a flowering garden and life returns to the catchment. Unfortunately, the water evaporates very rapidly and an impressive ecosystem dries up until the next major flooding. Idriess argued that if this is the situation after major floods why not permanently flood these lake systems and bring life and prosperity to the catchment on a continuing basis. To do so, a vast quantity of water is necessary. Idriess proposed that this could be achieved by diverting water from Queensland rivers that discharge to the Pacific Ocean and the Gulf of Carpentaria (Figure 7.6).

According to Idriess (1941, p. 217):

> We must lift or divert the floodwaters from the eastern Queensland coast and drop them back over the range into the headwater channels of the inland rivers: the Cooper and Diamantina in particular, the Georgina as the Plan develops, and later the Bullo, Paroo, and Warrego. We must divert the headwaters of the northern rivers which flow into the Gulf of Carpentaria, and by means of channel, cutting, or tunnel, lead them back into the head channels of the Diamantina and Georgina.

The harnessing of waterfalls for the generation of hydro-electric power was a major component of the proposed scheme. Indeed there are many waterfalls along the coastal ranges[7] such as Wallaman, Murray and Blencoe.[8] In the wet season, large volumes of water flow over these waterfalls, and Idriess proposed that hydro-electric power generators be installed at these sites. The power would then be available to pump water over the Great Dividing Range at sites where gravity flow through tunnels or along deep channels was not feasible (Idriess, 1941, pp. 229–231). As the harnessed water dropped down the west side of the Great Dividing Range, it could again be used for hydro-electric power generation. After the water was delivered into the upper reaches of the inland rivers, it could be used for irrigation as it flowed towards central Australia (Idriess, 1941, pp. 233–234). The remaining water would enter the north-eastern part of South Australia and would fill Lake Eyre and numerous other smaller lakes.

Idriess (1941, pp. 251–252) suggested the following outcomes for his scheme:

- **The expansion of industry** and increased employment during the construction phase.
- **The creation of a vast electrical industry**, ensuring an increased electric power for the whole nation. With the electrification of Queensland coastal towns, numerous industries would spring up because of cheap power.
- **Electrification** of the interior of Queensland which would pave the way for development and industrialisation in what now is purely a pastoral country.
- **Decentralisation**, which brings greater economic benefits and increased national security.
- **The elimination of droughts** in Queensland, one-half of South Australia, and north-west New South Wales.
- **The water and power enrichment** of a vast area of the continent stretching from the eastern and the northern coasts of Queensland to the South Australian coast.
- **The watering and irrigation** of south-west Queensland to the South Australian border and down through South Australia, to the north-west of New South Wales and beyond.

---

[4] Boomerang is a missile used by the Australian aborigines. Made of hard wood, it is roughly V-shaped with arms slightly skewed. The angle between the arms ranges from 90° to about 160°. The return-boomerang, which normally returns to the thrower, is about 30 cm to 75 cm in length and is used for sport and for hunting small birds. The non-return boomerang is straighter, heavier and ranging in length from 60 cm to 90 cm. It is used in war and for hunting large game.
[5] Only the major flood of 1890 is consistent with the data provided in Table 7.4.
[6] As described in Section 7.2.7, this occurs once every 8 years.
[7] These waterfalls are associated with the Great Escarpment in Queensland with a total length of 2800 km (Ollier and Stevens, 1989).
[8] See the end of this chapter for websites of a limited number of waterfalls.

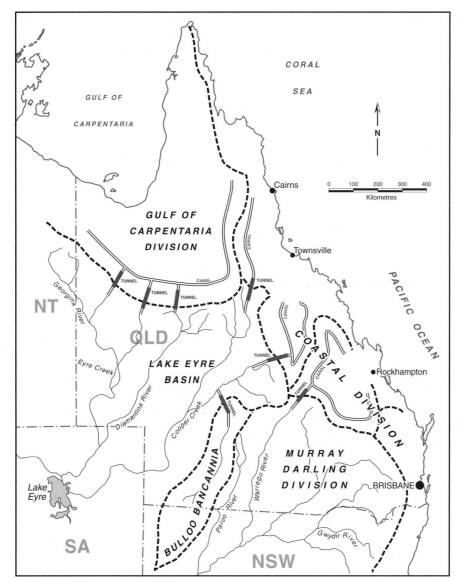

Figure 7.6 An approximate system of canals and tunnels for the inland diversion of the Queensland coastal rivers (Idriess, 1941).

- **The regeneration of the Dead Heart**, the end of the dust bowl, and the end of erosion over a vast area of central Australia.
- **The conversion of South Australia** into a State with a thousand lakes. South Australia would blossom into an entirely new State, where now she only has a very small and narrow strip of good land. The greater part of all her arid interior will become good land if abundant water is there to put life into it.
- **The gain to the entire Commonwealth**, a vast area of the continent, now comprising arid and even abandoned lands, would become well watered. It would be capable then of supporting many millions of people through its immeasurably increased productivity, and the industrialisation that will follow.

Concerning the cost of the scheme Idriess wrote (Idriess, 1941, pp. 250–251):

And the cost? Goodness knows. But it won't be anything like the cost of a war. And instead of killing millions of people and destroying cities, this expenditure will give a livelihood to millions of people – creating cities with hundred of thousands of homes for our industrial workers, and homesteads and farms.

Money and costs have taken on a new meaning. Money will in future, I believe, increasingly come to be spent at "long range". It will not be the amount to be spent that will be considered, but the snowball effect of the resulting benefits. The actual money that would have to be spent to set the Great Boomerang in motion might appear as vast as the scheme itself.

7.5 FLOODING LAKE EYRE WITH WATERS OF THE GREAT ARTESIAN BASIN

Idriess (1941, p. 250) described very briefly a proposal by S. E. Pearson from Queensland. His idea was to tap artesian waters of the Great Artesian Basin (GAB) by means of a great circle of bores and allow the water into Lake Eyre. He expected that the water in the Lake might possibly start the atmospheric processes that would bring rain to the area. Also, groundwater would be used for irrigation and at a later stage would help with forestation of the region. Lake Eyre has a vast area, therefore it would require a staggering number of bores to discharge water into the Lake and maintain it against evaporation. Idriess claimed that *this is the most practical scheme I have come across for the restoration of that land of desolation*. Then, he raised doubts about the capacity of the GAB for such a major undertaking.

7.6 CONCLUSIONS

This chapter has described the dramatic changes of the Lake Eyre Basin through the geological time and its cycle of dry and wet periods during the past 150 000 years. Because of high evaporation rates, the inflow exceeds evaporation once every 8 years resulting in the Lake filling. The three schemes described in this chapter (as well as the expanded Bradfield Scheme described in Chapter 6) were based on the erroneous assumption that the filling of Lake Eyre would increase the air moisture, which would then return to the catchment as rainfall.

The Port Augusta–Lake Eyre Canal Scheme proved impractical because a huge volume of excavation required to dig a very large canal capable of diverting enough water to the Lake to compensate for evaporation. The estimated cost of these earthworks proved prohibitive. Even if filling of the Lake with saline seawater was economically feasible, it could rapidly lose its efficiency because of a constant deposition of salt in the Lake.

The Great Boomerang Scheme proposed by Idriess was both idealistic and simplistic. He left many important issues unanswered. These include:

- Assessment of the volume of water available for inland diversion.
- An estimate of the volume of water that would evaporate as the diverted water flowed down the inland rivers towards central Australia.
- Engineering aspects of diversion such as: (1) what proportion of the water could be diverted by gravity and what proportion by pumping; (2) what would be the capacities and dimensions of reservoirs, tunnels and canals; (3) how much hydro-electric power could be produced by the scheme; (4) what proportion of the hydro-electric power would be required for pumping water and what proportion would be left for development of industry; and (5) how the electricity would be produced during the dry period.
- Suitability of inland soils for irrigated crop production.
- The environmental impacts of inland water diversion.
- The cost–benefit analysis of the Scheme.

The proposal to divert such large volumes of valuable freshwater into the Lake to be lost through evaporation seems irresponsible. Although the Great Boomerang Scheme cannot be implemented in its totality, some of its elements may have the potential to be implemented, such as the supply of water for irrigation to agricultural areas in close proximity to the western slopes of the Great Dividing Range, but certainly not for the filling of Lake Eyre.

Filling Lake Eyre with groundwater from the GAB is not feasible. This aquifer system with an annual recharge of about $1.1 \times 10^9 \, m^3$ is the main source of water supply of the pastoral, oil, gas and mining industries at the respective annual rates of $620 \times 10^6 \, m^3$, $26 \times 10^6 \, m^3$ and $13 \times 10^6 \, m^3$ (Habermehl, 2001). Moreover, approximately $40 \times 10^6 \, m^3$ of the aquifer's natural discharge feeds numerous springs and their associated ecosystems.

The climatological studies described in section 6.6.1 regarding the Bradfield Scheme, suggest that these schemes would have no measurable impact on the climate of the basin and would not increase local rainfall. This conclusion is supported by observations made during the major 1949–50 flood of Lake Eyre, which demonstrated that although the flood temporarily brought life back to the Lake, it had no impact on the climate of the Basin (Lake Eyre Committee, 1955, p. 29). Around the world there are many very dry areas close to large water bodies. These include coastal areas of the Persian Gulf, Gulf of Oman, Arabian Sea, Gulf of Aden, Red Sea, surrounding areas of the Dead Sea and the artificial Lake Nasser on the Nile River with a surface area of about $7000 \, km^2$, eastern coastal area of the Caspian Sea, Atacama Desert in Chile next to the Pacific Ocean, and many others. Therefore, the presence of a large evaporating water body is not a sufficient condition for promoting increased rainfall. Other conditions, such as the general atmospheric circulation pattern and topography, are required for any prospect of increased rainfall.

References

ABARE (1996). *Lake Eyre Basin: An Economic and Resource Profile of the South Australian Portion*. Canberra: Australian Bureau of Agricultural and Resource Economics.

Australian Water Resources Council (1987). *1985 Review of Australia's Water Resources and Water Use*. Canberra: Australian Government Publishing Service. V. **1**: Water Resources Data Set.

Bonython, C. W. and Fraser, A. S. eds. (1989). *The Great Filling of Lake Eyre in 1974*. Adelaide: Royal Geographical Society of Australasia (South Australian Branch) Inc.

Croke, J. C., Magee, J. W. and Wallensky, E. P. (1999). The role of the Australian monsoon in the western catchment of Lake Eyre, central Australia, during the last interglacial. *Quaternary International* **57/58**: 71–80.

DeVogel, S. B., Magee, J. W., Manley, W. F. and Miller, G. H. (2004). A GIS-based reconstruction of late Quaternary palaeohydrology: Lake Eyre, arid central Australia. *Palaeogeography, Palaeoclimatology, Palaeoecology* **204**: 1–13.

Drexel, J. F., Preiss, W. V. and Parker, A. J. eds. (1993). *The Geology of South Australia. V. 1, The Precambrian*. Adelaide: South Australian Geological Survey, Bulletin 54.

Drexel, J. F. and Preiss, W. V. eds. (1995). *The Geology of South Australia. V. 2, The Phanerozoic*. Adelaide: South Australian Geological Survey, Bulletin 54.

Gregory, J. W. (1906). *The Dead Heart of Australia: A Journey Around Lake Eyre in the Summer of 1901–1902, with some Account of the Lake Eyre Basin and the Flowing Wells of the Central Australia*. London: John Murray.

Habermehl, M. A. (2001). Hydrogeology and environmental geology of the Great Artesian Basin, Australia. In Gostin, V. A. ed. *Gondwana to Greenhouse: Australian Environmental Geoscience*. Geological Society of Australia Inc. Special Publication No. 21. pp. 127–143.

Idriess, I. L. (1941). *The Great Boomerang*. Angus and Robertson, Sydney.

Johnson, B. J., Miller, G. H., Fogel, M. L., Magee, J. W., Gagan, M. K. and Chivas, A. R. (1999). 65,000 years of vegetation change in central Australia and the Australian summer monsoon. *Science* **284**: 1150–1152.

Kotwicki, V. (1986). *Floods of Lake Eyre*. Adelaide: Engineering and Water Supply Department.

Lake Eyre Committee (1955). *Lake Eyre, South Australia: The Great Flooding of 1949–50*. Adelaide: Geographical Society of Australia, South Australian Branch.

Madigan, C. T. (1946). *Crossing the Dead Heart*. Melbourne: Georgian House.

Magee, J. W., Bowler, J. M., Miller, G. H. and Williams, D. L. G. (1995). Stratigraphy, sedimentology and palaeohydrology of Quaternary lacustrine deposits at Madigan Gulf, Lake Eyre, South Australia. *Palaeogeography, Palaeoclimatology, Palaeoecology* **113**: 3–42.

Magee, J. W. and Miller, G. F. (1998). Lake Eyre palaeohydrology from 60 ka to the present: beach ridges and glacial maximum aridity. *Palaeogeography, Palaeoclimatology, Palaeoecology* **144**: 307–329.

Magee, J. W., Miller, G. H., Spooner, N. A. and Questiaux, D. (2004). A continuous 150,000 yr monsoon record from Lake Eyre, Australia: Insolation-forcing implications and unexpected Holocene failure. *Geology* **32**(10): 885–888.

Miller, G. H., Mage, J. W., Johnson, B. J., Fogel, M. L., Spooner, N. A., McCulloch, M. T. and Ayliffe, L. K. (1999). Pleistocene extinction of *Genyornis newtoni*: human impact on Australian megafauna. *Science* **283**: 205–208.

National Land & Water Resources Audit (2001). *Australian Water Resources Assessment 2000*. Canberra: Natural Heritage Trust.

Ollier, C. D. and Stevens, N. C. (1989). The great escarpment in Queensland. In Le Maitre, R. W. Editor-in-Chief, *Pathways in Geology: Essays in Honour of Edwin Sherbon Hills*. Carlton, Victoria: Blackwell Scientific Publications, pp. 140–152.

Parliament of South Australia (1883). *Debates in the Houses of Legislature During the Third Session of the Tenth Parliament of South Australia, From May 31, 1883 to February 28, 1884*. Adelaide: W. K. Thomas & Co.

Parliament of South Australia (1883–84). Canal from Port Augusta to Lake Eyre. In *Proceedings of the Parliament of South Australia With Copies of Documents Ordered to be Printed, 1883–4. Volume IV, No. 88*. Adelaide: Government Printer.

Towner, E. T. (1955). Lake Eyre and its tributaries. *Queensland Geographical Journal* **43**: 65–94.

Wells, R. T. and Callen, R. A. eds. (1986). *The Lake Eyre Basin: Cainozoic Sediments, Fossil Vertebrates and Plants, Landforms, Silcretes and Climate Implications*. Sydney: Australian Sedimentologists Group, Field Guide Series No. 4, Geological Society of Australia.

Further Reading

Timbury, F. R. V. (1944). *The Battle for the Inland: The Case for the Bradfield and Idriess Plans*. Sydney: Angus and Robertson.

Websites For a Limited Number of Waterfalls in Queensland

http://www.gspeak.com.au/cardwell/waterfalls.html for Wallaman, Murray, Blencoe and Attie Creek waterfalls.[9]

http://www.walkabout.com.au/locations/QLDMillaaMillaa.shtml for Millaa Millaa Falls.

http://www.australianexplorer.com/photographs/qld_landscape_atherton_tablelands_waterfalls.htm for Millaa Millaa and Mungalli waterfalls.

http://www.barrierreefaustralia.com/IMAGEGALLERY/millstream-falls.htm for Millstream Falls.

http://www.skyrail.com.au/barronfalls.html for Barron Falls.

---

[9] All websites visited in July 2005.

8 The Goldfields Pipeline Scheme of Western Australia

8.1 INTRODUCTION

In 1826, the British set up a military settlement at King George Sound near Albany, in the south-west corner of Western Australia. Three years later, the first settlers arrived at the Swan River, and established the first huts and buildings at the site of Perth and at the nearby port of Fremantle. In February 1832, Western Australia officially became a Crown Colony and Captain James Stirling was appointed as the Governor. The population of the Colony gradually increased and by 1850 had reached about 6000. Western Australia was finally granted self-government in 1890 (Shaw, 1984, pp. 688 and 689).

In 1885, a small quantity of gold (10 ounces) was discovered at Kimberley in the far north of the Colony. By April 1886 this amount had grown to 400 ounces. The Kimberley Goldfields had limited reserves and did not last long (Ewers, 1935, p. 30). In late 1887, gold was also discovered at Yilgarn, north-east of Perth (Figure 8.1). This was followed by the discovery of gold at other locations around the Colony. At Coolgardie, Arthur Bailey and William Ford discovered gold in September 1892, and returned with 554 ounces (15.7 kg) of gold worth around £2000 (Evans, 2001, p. 138), or about $324 000 at 2002 gold prices.[1] In Kalgoorlie, Patrick Hannan, with two other Irish prospectors, discovered gold on 10 June 1893 (Blainey, 1993, p. 2).

The discovery of gold attracted thousands of people to Western Australia from the eastern Australian states and also from other countries. They arrived at Fremantle and Albany and set off for the goldfields, generally on foot. The population of the Colony, which was 29 700 in April 1881, increased to 49 700 in April 1891 and reached 184 100 in March 1901 (Fraser, 1906, p. 280). The Government was pressured to extend the eastern railway line to the township of Southern Cross to service the new goldfields. Construction commenced in 1891 and reached Southern Cross on 1 July 1894 (Ewers, 1935, p. 52). The line was then extended westward to Coolgardie and Kalgoorlie and was opened in January 1897 (Ewers, 1935, p. 47).

8.2 WATER SHORTAGE

The Goldfields region is in a low rainfall and high evaporation area of Western Australia (see Figures 3.2 and 3.3). Water supply for both domestic use and for the mining industry was a major problem in the early years. Water was obtained from:

(1) a very limited number of native wells;
(2) brackish groundwater;
(3) water carted from the west at a very high cost;
(4) water derived from reservoirs constructed on the goldfields, which were occasionally filled by rainwater; and
(5) water from condensers that distilled salty groundwater obtained in mine shafts and other sources.

The condensers were very simple. The boilers were 400 gallon[2] (1820 litre) tanks and the cooling system was made of galvanised iron pipes. In the early days, low quality water, hardly suitable for human consumption, was worth 2 shillings and 6 pence (2s 6d) per gallon, or $2.90 per litre at 2002 prices. Later, when condensed water became available, it was sold for 70 shillings per 1000 gallons (Anon, 1904, p. 2), or about 8 cents per litre at 2002 prices.

The nine reservoirs constructed between the townships of Southern Cross and Siberia, at a cost of £25 520 (approximately $2.7 million in 2002 prices) and with a total capacity of 54 500 m³, impounded only 45 400 m³ during 1895 (O'Connor, 1896, p. 19). However, 18 160 m³ of the impounded water was lost to evaporation and seepage and only 27 240 m³ was available for consumption. This water

---
[1] US$315 per ounce.
[2] One gallon = 4.55 litres.

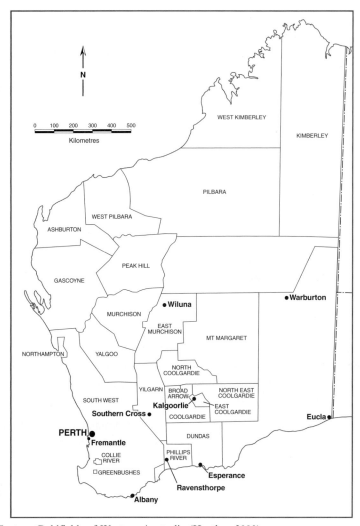

Figure 8.1 Mineral fields and Eastern Goldfields of Western Australia (Hartley, 2000).

was supplied at a cost of 10 shillings per 1000 gallons ($11.80 per m^3 at 2002 prices).

There were suggestions that deep groundwater could be a source of supply for the goldfields. However, expensive deep drillings, even one to a depth of 900 m through granite rocks, did not find any suitable supplies of fresh groundwater in the area (Evans, 2001, p. 147). It is now clear that the area consists of fractured to massive igneous and metamorphic rocks, containing only saline groundwater (Lau *et al.*, 1987).

Water was not only needed for domestic consumption and for gold processing, but also for the steam engines travelling along the railway system. Attempts were made to ease water shortages by dispatching water trains. One train could haul 30 wagons, carrying a total of 163.5 m^3. For the return journey the engine itself needed slightly more than half of this amount of water (Evans, 2001, pp. 147, 148). To address this problem, four large tanks were constructed along the railway line between Northam and Southern Cross, at Cunderdin, Kellerberrin, Merredin and Parkers Road. By 30 June 1895, they contained a useable supply of water.

Water shortages commonly caused health problems and led to disease outbreaks such as dysentery. Hundreds of people from Coolgardie died of typhoid through the drinking of stagnant water during a drought period.

8.3 PIPELINE PROPOSALS

It became evident that sufficiently large and reliable supplies of water were unavailable locally. By 1894, addressing the water shortages in the Goldfields had become a priority, and the Western Australian Government set up the Goldfields Water Supply Division within the Public Works Department. In the same year, there were two proposals for pumping water from the Avon Valley near Perth (Evans, 2001, pp. 156 and 157). Mr J.S. Talbot published the first proposal in the *West Australian* newspaper on 8 March 1894. He suggested that water could be brought to the Goldfields by means of a pipeline aided by pumping stations at Northam and along the railway line. He argued that the construction could be paid for by selling water at a shilling per gallon (about \$1 per litre at 2002 prices). The second proposal was similar and was put forward 5 months later by Mr John Maher. He applied to the government for the right to construct a pipeline from the vicinity of Northam and along the incomplete railway line to Coolgardie. Maher and his associates were confident that they could raise £2.5 million (about \$268 million in 2002 prices) on the London money market. They believed that they would be able to sell the water on the Goldfields for three shilling and six pence (3s 6d) per 1000 gallons (\$4.1 per m^3 at 2002 prices). Although neither of these proposals materialised, they helped raise a possible method of solving the water problem.

In October 1894 the Government introduced the *Water and Electric Works License Bill* into the Legislative Assembly. Its objective was to enable the Government or private enterprises to build a water pipeline to the Goldfields (Evans, 2001, p. 157).

8.3.1 Coolgardie Goldfields Water Supply Loan Bill

From the middle of October 1895, the Government Engineer-in-Chief, C.Y. O'Connor[3] gave his full attention to a pipeline scheme to bring water to the Goldfields. The Public Works Department, led by O'Connor, worked on collating technical information and cost estimates for the pumping of 4540 m^3, 22 700 m^3, and 45 400 m^3 of water per day. On 23 March 1896, the Premier of Western Australia, John Forrest[4] announced that the Government was determined that the Goldfields should be provided with an adequate water supply, by piping water from the coast (Anon, 1903). During the first half of 1896, O'Connor continued his exhaustive research in anticipation of legislation for the Goldfields Pipeline Scheme.

In July 1896, John Forrest tabled a comprehensive report from his Engineer-in-Chief (O'Connor, 1896), in both Houses of Parliament and introduced the *Coolgardie Goldfields Water Supply Loan Bill* to raise £2.5 million (\$268 million in 2002 prices) for the construction of the pipeline. During the second reading of the Bill on 21 July 1896, he made a long speech which described various aspects of the Scheme, including the following (Forrest, 1896): dryness of the interior of the Colony; water supply problems of the goldfield; likelihood of obtaining artesian water; technical features of the Scheme; practicality of the Scheme; adequacy of water supply from the Darling Range; capital cost of the Scheme; annual working expenses; price of water; time required for its construction; impacts of the Scheme on the inhabitants; welfare of the Colony; indebtedness of the Colony and the country; his opposition to undertaking the project by private enterprise; and the resultant increase in investment and export of gold. He concluded his speech by:

> I have proved that the work will pay; I have proved that it will reclaim the wilderness to a great extent, and spread comfort and plenty along its course. In conclusion I would like to emphasise this point that, not only are the mines languishing for the want of water, and the output of gold is being retarded, but our fellow-colonists who are trying to build up this country are also languishing, living in discomfort, without even the necessaries of life in regard to water. And when we remember that dirt and disease are fostered by the want of water, where health and cleanliness should prevail, surely a strong case is made out in favour of this scheme. I say sir, the scheme, which I have had the pleasure and honour of placing before the members of this House, and before my fellow-colonists, is a project worthy of an enterprising people. I believe if we carry on this great work, not only will the goldfields flourish, and not only shall we be relieved from our present anxiety in regard to the water question, but we shall also be repaid one hundred-fold. Future generations, I am quite certain, will think of us and bless us for our far-seeing patriotism, and it will be said of us, as Isaiah said of old: "They made a way in the wilderness, and rivers in the desert".

The second reading of the Bill was passed in the Legislative Assembly on 5 August 1896. Then, the Bill passed through the Legislative Council (Senate) on 3 September 1896 (Evans, 2001, p. 172), and Royal Assent was given to the Bill on 23 September 1896 (Anon, 1903, p. 14).

8.3.2 Technical Features of the Pipeline Scheme

The Scheme consisted of building a weir on the Helena River in the Darling Range and pumping water 530 km overland to Coolgardie (O'Connor, 1896). At this stage the intention was not to deliver water to Kalgoorlie. This section was added later by delivering water from Coolgardie to Kalgoorlie by gravity. Table 8.1 lists the major technical features of the Scheme. Eight pumping stations would deliver 22 700 m^3 of water per day. It was estimated that the Scheme would be completed in three years.

---

[3] See Appendix A for his short biography.
[4] See Appendix A for his short biography.

Table 8.1. *Technical features of the Goldfields Pipeline Scheme*

Item	Characteristics in			
	British system		Metric system	
	Unit	Amount	Unit	Amount
Length of pipeline	Miles	330	km	530
Velocity of water per second	Feet	2	m	0.61
Diameter of pipe	Inches	30	mm	762
Net height which water has to be pumped	Feet	1350	m	411.5
Head required to overcome friction	Feet/mile	3.5	m/km	0.66
Total head required to overcome friction	Feet	1155	m	352
Grand total head which pumps have to overcome	Feet	2505	m	763.5
Weight of water to be raised per 24 hours[a]	Tons	23 000	Tonnes	23 368

Note: [a] Including water for engines, evaporation and other losses.
Source: O'Connor (1896).

The population of the goldfields in 1896 was less than 40 000 but was expected to rapidly increase. Therefore, 9080 m^3 was allocated for domestic consumption by 100 000 people at a rate of 90 litres per person per day. The remaining 13 620 m^3 was allocated for the gold mining industry and railway requirements (Forrest, 1896, p. 16).

T. C. Hodgson, O'Connor's engineer, exhaustively examined seventeen watersheds of the Darling Range, which were within a moderate distance of the goldfields railway line. These included the Helena River, Talbot River, Dale River, Canning River and others. Several sites were surveyed on the Helena River and the most suitable was selected (Figure 8.2). The site was considered as ideal because the valley was very narrow at that point, with steep sides, and the bedrock was suitable for the foundation of the dam (O'Connor, 1896, pp. 17 and 18).

The riverbed at the proposed site was about 97.5 m above sea level. It was estimated that a 30.5 m concrete dam could be constructed at the site, with a length of 198 m at its top. At the base, the width of the dam was 36.6 m, with a width of 4.6 m at the top. The volume of the impounded water would be 21×10^6 m^3. The size of the drainage area was estimated at 1470 km^2. Assuming an annual rainfall of 510 mm over the catchment, the reservoir would be filled every year if only 3 percent of the rainfall run-off into the reservoir. The quality of water was also assessed as excellent (O'Connor, 1896, Appendix E, pp. 17 and 18).

Considering the topography between the dam site and Coolgardie, it was clear that water had to be pumped. O'Connor calculated that the total lift including friction was about 763.5 m. Eight pumping stations were required to pump 22 700 m^3 of water per day over a distance of 530 km at the required pressure. Twenty sets of pumps, with a total power of 6000 HP were needed (Anon, 1903, pp. 16 and 17). There were three sets of pumps at pumping stations 1 to 4 (Figure 8.3), and two sets at the other four stations. For all pumping stations, one pump was intended as a backup. The pipeline consisted of about 66 000 lengths of pipe, each 8.5 m long and 762 mm in diameter (Anon, 1903, p. 16 and Ayris, 2001, p. 24).

Pumping stations were located at the following sites (Anon, 1904, pp. 16 and 17):

- **Pumping Station No. 1** was located 400 m below the weir. The first stage was a lift of 128.3 m to Pumping Station No. 2, located at 2.6 km, where the water was received in a 2125 m^3 concrete tank.
- **Pumping Station No. 2** was capable of raising water a further 110 m into the first regulating concrete tank with a capacity of 2270 m^3 at Baker's Hill, at 329 m above sea level. From here the water gravitated to a second regulating concrete tank at Northam 28.8 km away. The Northam tank was 28.7 m lower than Baker's Hill. The water then gravitated to a large reservoir at Cunderdin with a capacity of 45 400 m^3 and located 125 km from the Helena reservoir.
- **Stations 3 to 7** pumped water to the eighth station at Dedari a distance of 349 km from Cunderdin, and at an elevation of 458.7 m. Each station was provided with a 4540 m^3 concrete tank.
- **Pumping Station 8** at Dedari, pumped water a distance of 19.2 km to the main service reservoir at Bulla Bulling with a capacity of 54 500 m^3. Bulla Bulling supplied a small service reservoir of 4540 m^3 on Toorak Hill, overlooking the town of Coolgardie with a mean elevation of 463.8 m. From Toorak Hill tank, water gravitated to a reservoir on Mt Charlotte to supply the town of Kalgoorlie.

A detailed longitudinal profile of the pipeline prepared in 1901 is available in Evans (2001, pp. 170 and 171). Further technical details such as the hydrology of the Helena River, its water quality, construction of the Mundaring Weir, the pipeline and its construction, and pumping stations and service reservoirs are available in Palmer (1905).

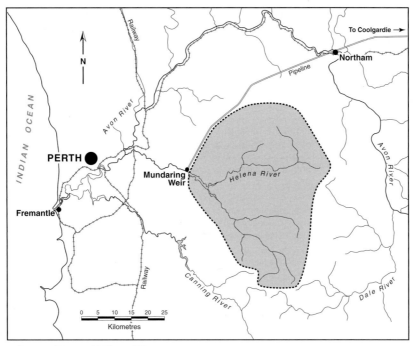

Figure 8.2 Helena River catchment (Ewers, 1935, p. 82).

O'Connor (1896) recommended that the Government arrange for a consultation with experts in England before proceeding with the project.

8.3.3 Critical Examination of the Scheme in London

O'Connor set off for London on 22 January 1897. In London, a Commission of English engineers examined details of O'Connor's proposal, such as the thickness and size of the pipe to be employed, whether it should be placed above or below ground, the number, position and power of the pumping stations, and other technical details. The Commission issued two reports. In the first or interim report, nine pumping stations were recommended. In the final report the Commission submitted an alternative arrangement with eight pumping stations (Palmer, 1905, p. 6). O'Connor then travelled to Germany and visited Krupps works in Essen and other engineering establishments. He returned to Albany on 16 September 1897 after being away for about 8 months (Evans, 2001, pp. 184 and 185). The following day, in a lengthy interview with the *West Australian*, O'Connor gave a detailed outline of the investigations conducted by the Commissioners into the Pipeline Scheme and said (Evans, 2001, pp. 184 and 185):

> While the Scheme, if carried out, will be the largest of its kind in the world, there is nothing in the nature of it, or in any of its details, which is in the least degree impractical or unprecedented. And with reasonable care and skill there is no reason to suppose otherwise than it will prove to be entirely satisfactory.

8.3.4 Cost Estimate of the Scheme

O'Connor (1896) estimated the cost for three daily capacities of the pipeline (Table 8.2). He demonstrated that the cost of delivery was dependent upon the cost of pumping which varied in an inverse ratio to the size of the pipe, and consequently to the total capital cost. However, considering the water requirements of the goldfields, he recommended that the 5 000 000 gallons (22 700 m^3) per day option be considered for implementation. His estimates of the capital and annual costs for this option are presented in Tables 8.3 and 8.4.

The estimated delivery cost per 1000 gallons, on the assumption of 5 000 000 gallons (22 700 m^3) sold daily, was 3s 5d to 6s 7d ($4.02 to $7.74 per m^3 at 2002 prices). This was the cheapest water supply method. As mentioned before, water supply from local reservoirs constructed by the Government had a cost of 10 shillings per 1000 gallons ($11.80 per m^3 at 2002 prices). As well, the estimated cost for supplying water by condensation was much more ($80 per m^3 at 2002 prices) than pumped water.

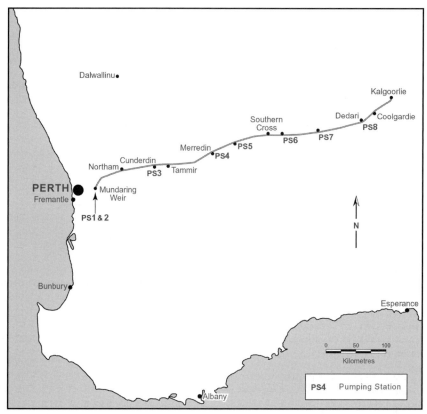

Figure 8.3 Goldfields pipeline and its pumping stations (Anon, 1904 and Evans, 2001, pp. 170 and 171).

Table 8.2. *Cost of water for various options*

Option	Daily capacity in Gallons	Daily capacity in m^3	Capital cost £m[a]	Capital cost $m[b]	Delivery cost per 1000 gallons[a]	Delivery cost per Cubic metre[b]
1	1 000 000	4540	0.7 to 1	75 to 107	5s 6d to 8s 6d	$6.47 to $9.41
2	5 000 000	22 700	2.2 to 2.7	236 to 289	3s 5d to 6s 7d	$4.02 to $7.74
3	10 000 000	45 400	3.5 to 4.6	375 to 492	3s to 5s	$3.53 to $5.88

Note: [a] 1896 prices; [b] equivalent 2002 prices.
Source: O'Connor (1896).

Construction of the pipeline required a loan of £2.5 million ($268 million in 2002 prices) to be repaid over 20 years at an annual rate of 6 percent (3 percent interest and 3 percent for a sink fund or depreciation of the Scheme). The first attempt to raise the necessary loan in the London money market failed. However, O'Connor and his assistants continued with their work of preparation for the pipeline in anticipation of the success of the loan. He was informed of its success on 17 January 1898. John Forrest instructed him as follows (Evans, 2001, pp. 190 and 191):

> You may now proceed to put the Coolgardie Scheme in hand at once, to proceed with the railway to the dam site, to build the dam, to call for tenders for pumps and pipes and generally push on with the great work.

A few months later, a French Engineer (Mr Bargigli) representing a French firm who had travelled throughout the Goldfields, offered to construct the pipeline for £2 272 000 which was £228 000 less than O'Connor's estimate (Ewers, 1935, pp. 62 and 63). Bargigli realised that his firm was not English and had no appeal to the people. He disappeared to form an English based company (The West Australian Goldfields Water Supply Limited). On his return, he twice approached John Forrest and defended his company's proposal, but the Premier resisted, because he

Table 8.3. *Capital cost for daily pumping of 5 000 000 gallons (22 700 m³)*

Item	Cost in 1896 (£)	Cost in 2002 ($)
Pumping engines	200 000	21 400 000
Main pipes at Fremantle (90 000 tonnes)	1 470 000	157 290 000
Carriage from Fremantle	140 000	14 980 000
Laying and jointing (including excavation and filling of trenches)	220 000	23 540 000
Reservoirs	300 000	32 100 000
Distributing mains (160 km of 305 mm diameter pipes)	170 000	18 190 000
Total	2 500 000	267 500 000

Source: O'Connor (1896).

Table 8.4. *Annual expenses for daily pumping of 5 000 000 gallons (22 700 m³)*

Item	Cost in 1896 (£)	Cost in 2002 ($)
Interest on capital cost at 3% plus 3% depreciation (6% in total)	150 000	16 050 000
Maintenance	45 000	4 815 000
Pumping	109 000	11 663 000
General and administration	16 000	1 712 000
Total	320 000	34 240 000

Source: O'Connor (1896).

was determined that the Government should do the work. The *Morning Herald* and the *West Australian* newspapers gave considerable publicity to this proposal. However, Bargigli was forced to withdraw when enquiries were made about his past illegal dealings.

8.3.5 Opposition to the Scheme and O'Connor

Although the pipeline project had the support of John Forrest, it still attracted considerable opposition and hostility. Some doubted the wisdom of the entire Scheme believing that, similar to the Kimberley Goldfields, the gold reserves could be limited and it would be unwise to consider provision of a permanent water supply scheme and that it would be impossible to convey water over the hills and over such a great distance. Others believed that the Scheme would cost considerably more than the estimate, and that construction would take a lot longer than estimated. Also, the competence of O'Connor was severely questioned. It was argued that his expertise was in the construction of railways and not in large water supply projects. O'Connor was also accused of dishonesty and helping himself to public money.

The most recent criticism of the Scheme came from the historian Geoffrey Blainey, who accused O'Connor of using low rainfall data to justify his project (Blainey, 1993, pp. 69–76). According to Blainey, it was possible to overcome the water shortage in the Goldfields by building an inexpensive dam, as in Broken Hill,[5] New South Wales. A detailed refutation of Blainey's accusation is available in Evans (2001, pp. 172–177).

8.3.6 Construction of the Scheme

Following the Premier's announcement that the Scheme would go ahead, tenders were invited for the supply of 530 km of pipes, each 8.5 m in length with diameter of 762 mm. The work on the construction of a railway from Mundaring to the dam site at the Helena River was planned to commence without delay and would be constructed by the Department of Public Works (Evans, 2001, p. 191).

Clearing the forest at the Helena Valley dam site near Mundaring and laying the railway lines commenced in early February 1898. In April 1898, excavation for the weir started and in August the railway to the construction site was completed, making transport of materials and personnel from Fremantle and Perth both rapid and comfortable.

Two key events happened in October 1898 (Evans, 2001, p. 193). The first was the passage of the *Coolgardie Goldfields Water Supply Construction Bill*. This was two years after the Scheme had been adopted by Parliament. The second was the signing of contracts for the supply of pipes with the Australian engineering firms Mephan Ferguson of Melbourne and G.Y.C. Hoskins of Sydney. The Australian tenders were considerably lower than those received from manufacturers in Europe and America.

Work at the dam site was well advanced in 1899. With the contracts signed and money allocated, there was no turning back. Steel plates imported from Germany and the United States for manufacturing the pipes began arriving at the new Fremantle port on 4 April 1899. Factories were erected at Maylands and Midland (both located in the north-west of Perth along the railway line) to manufacture the pipes (Ewers, 1935, p. 74).

The pipeline was constructed of 762 mm diameter steel pipes manufactured from imported steel sheets 8.5 m long

---

[5] See section D in Chapter 10.

and 1.2 m wide, using the locking-bar technique invented by Mephan Ferguson, a Melbourne engineer. Steel sheets were rolled into half pipe sections and jointed together along the longitudinal edge by an H-shaped locking-bar, which was pressed onto the edge of the rolled sheets (Figure 8.4). Factories worked two shifts of 8 hours making 150 to 160 pipes per day. Pipes were subject to a pressure test of 28 $kg cm^{-2}$. This was nearly twice the operating pressures (Palmer, 1905, p. 24). They were transported by rail and laid in trenches. Pipes were joined together with jointing rings. A ring 20 cm wide was slipped over the joints. Then, the space between the ring and the pipe was filled with lead and caulked by hand (Ewers, p. 76).

Two unforeseen events in 1899 seriously slowed progress (Evans, 2001, p. 199). The first was the discovery, during excavation at the dam site, of a fault line in the dam foundation. O'Connor inspected the site and decided not to shift the dam site elsewhere. Instead, he instructed that the fault and the loose rocks be drilled and excavated to a depth of 27.5 m below the riverbed. He also recommended installation of electric lights so that two shifts could work, by day and by night. The second incident that delayed construction of the pipeline and involved loss of life was the wreck of the ship *Carlisle Castle* in a gale south-west of Garden Island on 11 July 1899. The ship carrying locking-bars manufactured in England went down with all her cargo.

The repair work on the fault at the dam site was completed in January 1900. Pouring concrete for construction of the dam started soon after and continued day and night until completion in June 1902. A total of 77 508 casks of cement were used for construction of the dam (Palmer, 1905, p. 18). Some 19 767 casks were imported from Germany and the balance from England. The German cement was chiefly used in filling the deep excavation made to repair the fault in the bedrock.

In March 1900 a contract was let to Messrs James Simpson and Co. Limited of London to deliver and erect 20 complete sets of Worthington pumps, boilers, and accessories within 27 months of the signing of the contract (Anon, 1904, p. 18). The total cost was £241 750 (Evans, 2001, p. 203) equivalent to $26 million in 2002 prices. It was estimated that some 30 000 tonnes of local coal per annum from the Collie Coal Fields of Western Australia would be required to fuel the boilers (Anon, 1904, p. 35).

O'Connor decided to lay most of the pipes in trenches, because of the intense summer heat and winter cold, but also the daily temperature changes. This was contrary to the original advice given by the London Commissioners. By the middle of 1900, the first 48 km of trenching had been excavated (Evans, 2001, p. 203).

On 31 July 1900, Western Australia voted to enter the Australian Federation. This was a turning point, leading to the departure of John Forrest, O'Connor's main ally and supporter. On 1 January 1901 Queen Victoria proclaimed the Commonwealth of Australia. On 7 February 1901 John Forrest and his guest, the Federal Minister for Trade and Custom as well as a parliamentary delegation, inspected the dam site. This was the last of Forrest's public engagements in Western Australia before entering the new Federal Parliament.

A controversy over the joining of the pipe segments erupted in mid-1901. Traditionally this had been done by hand-caulkers packing the joints with lead. This was time-consuming and labour-intensive. A Victorian contractor, James Couston and his partner James Finlayson had invented an electric machine to simplify the caulking process and reduce the labour cost. The government purchased the patent for the sum of £7500 ($800 000 in 2002 prices). A number of machines were manufactured and Couston was contracted to supervise and train the caulking team.

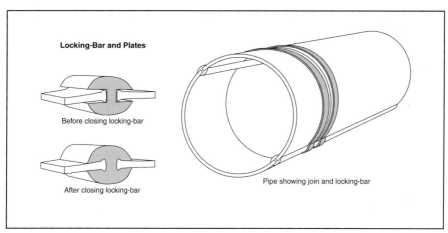

Figure 8.4 Locking-bar, plates and pipe (Ewers, 1935, pp. 65 and 75).

O'Connor came under pressure from the media for paying royalties to James Couston. O'Connor responded by pointing out that only £2700 had been paid to Couston as a deposit. The remaining £4800 would be paid when all joints were made to the satisfaction of the Department of Public Works. He also confirmed that approximately 24 km of pipeline had been caulked successfully by using the electric machines (Evans, 2001, p. 212).

Couston wrote to O'Connor on 23 December 1901, formally proposing that his firm (Couston, Finlayson and Porritt) be contracted to finish the work of pipe-laying. O'Connor accepted the proposal after consulting with his engineers. He wrote to the Minister of Public Works on 16 January 1902 recommending acceptance of Couston's proposal at a cost of £68 264 ($7 370 000 in 2002 prices). Although the Minister had the authority to make a decision, he tabled Couston's offer in the Legislative Assembly on 21 January 1902. The Parliamentary debate took place while O'Connor was in Adelaide on government business and concluded on 3 February 1902. The Parliament criticised not only the contract but also the Public Works Department and referred the matter to a Select Committee. The Committee, which consisted of prejudiced members, met from 4 to 17 February 1902 and conducted the inquiries while O'Connor was in Adelaide. On 9 February 1902, in the midst of the Select Committee's investigations, the *Sunday Times* accused O'Connor of corruption and claimed that in all his undertakings he mislead the State and concluded: *We need a court of justice in which to investigate O'Connor's relationship with the contractors.*

O'Connor returned to Western Australia on 17 February 1902 and was profoundly disturbed. In the evening of the same day, the Select Committee presented its report to Parliament. The Committee:

(1) exonerated Couston of any wrongdoing;
(2) investigated O'Connor's engineer T. C. Hodgson's land dealing of about 16 200 ha close to the site of the third pumping station at Cunderdin, laying him open to the charge that he had used his special knowledge of government works to his own advantage; and
(3) recommended that a Royal Commission investigate the issues.

The Legislative Assembly approved the appointment of a Royal Commission on 19 February 1902 to "*Enquire into the Report upon the Conduct and Completion of the Coolgardie Water Supply Scheme*". O'Connor realised that it was he who would be the focus on the Commission of Inquiry, and that his management and competency would be on trial. From this time forward, O'Connor became increasingly distressed. However, construction of the pipeline continued. Progress improved as the caulking team became more familiar with the equipment. Pumping stations 1, 2 and 3 were being erected and the pump manufacturers supervised pump installation. Water pumping tests were planned for the end of March 1902. The most difficult terrain in the pipeline's route over the Darling Range had been completed, and the remaining work became easier and faster. To protect the pipeline from the extreme daily and seasonal temperature variations, a special coating of asphalt and coal tar was applied hot and then sprinkled with sand (Evans, 2001, pp. 222 and 223).

O'Connor was confronted with bitter personal attacks in the press and Parliament. Lack of official support, scepticism, bureaucratic demands and an unsympathetic Royal Commission affected his mental health. In late February and early March 1902 the amount of personal criticisms he received overwhelmed him. He had no one in a senior position to turn to. O'Connor committed suicide in the early morning of Monday 10 March 1902 at South Fremantle, when the Scheme was less than a year away from completion.

The Royal Commission interviewed its first witnesses on 5 and 6 March 1902, and eventually held 47 meetings and interviewed 59 witnesses. It produced two interim reports; the first one on 2 April 1902 and the second one on 23 May of the same year. The first report recommended that (Evans, 2001, p. 237):

> It is desirable for the Government to invite tenders for those portions of the work remaining to be done east of Cunderdin, namely the pipe-laying, jointing, caulking and covering and such other works as may be suitable for public tender.

The second report was highly critical of T. C. Hodgson over his acquisition of land in the vicinity of the pipeline and pumping station at Cunderdin, borrowing money from the contractor (Couston) at advantageous interest and his self-serving advice to O'Connor. Hodgson was suspended from duty pending a disciplinary inquiry, which never took place. Hodgson resigned in August 1902 and lived as a landholder near Cunderdin (Evans, 2001, p. 237).

During the remaining months of 1902, work progressed under the guidance of the new Engineer-in-Chief, Charles Palmer, who previously served under O'Connor. Approximately 143 km of pipes were laid prior to O'Connor's death, and the remaining length of pipeline was completed by early 1903 (Evans, 2001, p. 239).

Pumping trials began in April 1902 with water being pumped to Pumping Station No. 2. Water reached Northam on 18 April and as each section of the pipeline was completed the water followed and reached Merredin on 22 August, Southern Cross on 30 October, and finally on 22 December water reached Coolgardie. By 16 January 1903, water was available to be supplied to the people of Kalgoorlie.[6]

Towards the end of the 1902, the Minister for Public Works, the Hon. C. H. Rason, in the course of a speech in Parliament, said (Ewers, 1935, p. 86):

> It has been said, not once but often, that we should never succeed in pumping the water from Mundaring to the goldfields; and that if we did succeed in pumping it, the water would not be fit to drink... The water has now reached Gilgai reservoir, which is 218.5 miles from Mundaring reservoir, and this (he held up a glass of water) is a sample of the water taken from the reservoir after being pumped 218.5 miles. I think no complaint will be found with it on the score of appearance, and I assure the House it is perfect to taste.

Construction of the pipeline was completed in January 1903. The whole construction, including the building of Mundaring Weir, took less than five years, at a total cost of £2 866 454 (about $307 million in 2002 prices), £366 454 more than O'Connor's original estimate (Tables 8.3 and 8.5). The consolidated revenue of Western Australia in the financial year of 1903–04 was £4 012 189 (Fraser, 1906, p. 455) or 1.4 times the cost of the Scheme and the population was about 200 000. These indicate that the Scheme was a major undertaking by the Western Australian Government.

8.3.7 Opening of the Scheme

Lady Forrest started the pumping machinery at Pumping Station No. 1 below the Mundaring Weir on Thursday 22 January 1903. Two days later on Saturday 24 January, in a summer heat of 41°C at Coolgardie, the Goldfields Water Supply Scheme was officially opened by John Forrest, Minister for Defence in the new Commonwealth Government and former Premier of Western Australia who had supported O'Connor in the design and construction of the pipeline. O'Connor's widow, Susan and his eldest son George Francis, who was a qualified engineer were present (Evans, 2001, pp. 239 and 240). The great ceremony of the day was at Kalgoorlie at 5 p.m. that afternoon when John Forrest opened the Mt Charlotte reservoir. The opening of the Scheme took place about 10 years after the discovery of gold by Patrick Hannan, and a little over 10 months after the death of O'Connor.

John Forrest in his opening address at Kalgoorlie said (Anon, 1903; Public Works Department, 1963):

> Ladies and Gentlemen,
>
> It is to me a great honour to be present here today to declare this Great Water Scheme open, and it is a source of much gratification that I should have been privileged, not only to introduce the necessary legislation for undertaking this great work, but also to declare it to be available for the use of the people.

After describing the history of the scheme and its benefits for the region, he praised O'Connor:

> In the midst of our rejoicing, however, my thoughts turn to the memory of the great builder of this work, the late Mr C. Y. O'Connor, C. M. G., who we all regret is not with us to receive the honour which is due to him, and our sorrow and regret is intensified when we remember how arduously and incessantly he laboured to complete this work, and how anxiously he looked forward to this day.

Then, he concluded his address by:

> I pray God that this river of pure water may give health, comfort and prosperity to all those who come within its life-giving influence, and that it may prove a benefit and a blessing to the Coolgardie Goldfields, and be far-reaching in its influence for good to all Australia. I now declare this great work open for the use of the people.

Appendix D provides a chronological table of the most important events in connection with the Scheme.

8.3.8 Controversy over the Originator of the Scheme

The controversy over who was the first person to put forward the idea of pumping water to the Goldfields has a long history. Perhaps the earliest recorded references are from Mr J. S. Talbot and Mr John Maher.

Table 8.5. *Final cost of the Goldfields Pipeline Scheme*

Item	Cost in 1903 (£)	Cost in 2002 prices ($)
Construction of the Mundaring Weir	249 000	26 643 000
Construction of the pipeline	1 796 988	192 277 700
Construction of the pumping stations	432 220	46 247 500
Construction of the reservoirs along the pipeline	58 500	6 259 500
Miscellaneous	329 746	35 282 800
Total	2 866 454	306 710 500

Source: Public Works Department (1963, p. 2).

---

[6] http://www.watercorporation.com.au/students/students_cyoconnor.cfm (visited in July 2005).

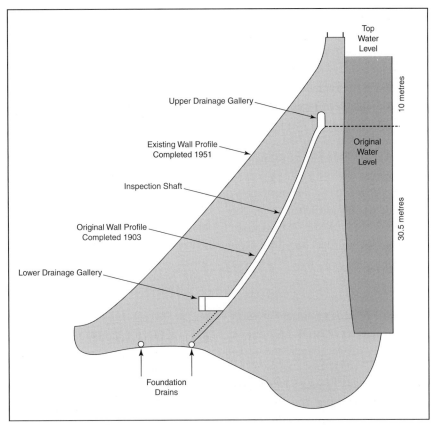

Figure 8.5 Cross-section of Mundaring Weir.[7]

Also, Mr N. W. Harper one-time manager of the Robinson Gold Mine, Kanowna claimed for many decades that he suggested the pumping scheme to the Premier John Forrest and persuaded him that such a scheme was practical. In an attempt to resolve the controversy, on 13 October 1952, the then Premier of Western Australia (Sir Ross McLarty), invited the Department of History, University of Western Australia to research the issue. The three academics that investigated the issue concluded that (Alexander et al., 1954):

> Broadly speaking, therefore, the Goldfields Water Scheme may well be described as politically Forrest's and technically O'Connor's, though neither man was responsible for the original idea of solving the Goldfields water problem by pumping water from the coast, credit for which may not safely be given to any one individual. It follows that the answers to the two questions submitted to the present writers in October 1952 by the then Premier of Western Australia must be:

Question 1. Who was responsible for placing before Sir John Forrest the plan for pumping water to the Goldfields by the method of pumping which was adopted in the Coolgardie Goldfields water supply scheme?

Answer: C. Y. O'Connor and his departmental officers.

Question 2. Who was chiefly responsible for persuading Sir John Forrest that the method adopted was practical and that it was within the financial means of the State?

Answer: The evidence does not permit the allocation of credit to any single person.

8.3.9 Expansion of the Scheme

Following the completion of the Scheme, numerous townships were connected to the pipeline and on 30 June 1905 the number of consumers reached 7960 with the average daily consumption of 1 400 000 gallons or 6370 m^3. Water was supplied at a cost ranging from 2s 6d per 1000 gallons in the western portion to 8s 4d at Kanowna in the east, with an average price of 5s 6d (Fraser, 1906, p. 1109). These prices are respectively equivalent to \$2.94, \$9.80 and \$6.47 per m^3 in 2002 prices.

Since the completion of the scheme, several major alterations and enlargements have been made to the scheme. In 1951, the height of Mundaring Weir was increased to 40.5 m (Figure 8.5). When the work was completed, the enlarged capacity of the reservoir was 69×10^6 m^3.

---

[7] http://www.watercorporation.com.au/dams/dams_mundaring.cfm (visited in July 2005).

The capacity has since been further increased to $77 \times 10^6 \, m^3$ by the erection of adjustable steel crest gates 1.2 m in height. In 1972, the Lower Helena Pipe-head Dam, 8 km below Mundaring Weir, was brought into operation, water being pumped from this source to augment the supply from the Mundaring Reservoir. The main pipeline between Mundaring and Kalgoorlie is 554 km long. It is for the most part 762 mm in diameter, but also has 1219 mm, 1067 mm and 914 mm sections (Australian Bureau of Statistics, 1985, p. 259).

The first branch line was built in 1907 to demonstrate to farmers that the pipeline could supply sufficient water for their stock and domestic needs (Public Works Department, 1963, p. 18). It was a big success and by 1947 about 1300 km of branch line had been laid, bringing water to more than 400 000 ha of farmland. In the same year, the Commonwealth Government agreed to assist the State Government to provide additional branches in agricultural districts. This resulted in the expansion of branches to more than 5000 km serving 1 659 000 ha of farmlands and associated townships. From 1963 to 1973, there were more extensions, which increased the area served with water to 3 160 000 ha (Ayris, 2001, p. 34). Table 8.6 shows the expansion of water supply by the Scheme for the period of 1960–61 to 1996–97.

The Scheme is currently known as the *Goldfields and Agricultural Areas Water Supply Scheme* (Figure 8.6). It supplies water to approximately 39 000 rural and town services. This water is used for household and commercial water supply, farm water supply for stock, and mineral processing. The cost of delivering water to Kalgoorlie is $3.71 per m^3, in 2005 prices. This includes operation, maintenance and capital cost.[8] However, the Water Corporation sells this water for $1 per m^3.

The original scheme was provided with eight steam-driven pumping stations. With the increased demand for water it became necessary to replace four of these with more powerful equipment. Between 1954 and 1960, the original pumping stations Nos. 1 and 2 at Mundaring Weir, No. 3 at Cunderdin and No. 4 at Merredin were replaced after more than 50 years of service with new electrical centrifugal pumps (Public Works Department, 1963, p. 19). Currently, there are 18 pumping stations with powerful electric pumps along the main pipeline. The original pumps were capable of delivering up to 22 700 m^3 of water daily. At Mundaring today, the pumps can deliver 134 000 m^3 per day (Ayris, 2001, p. 34).

8.4 CONCLUSIONS

The discovery of gold at Coolgardie in 1892 and Kalgoorlie in 1893 attracted thousands of people to the Western Australian Goldfields. The Goldfields region lies in a low rainfall and high evaporation area. Finding a reliable water supply for both domestic use and for the mining industry presented a major challenge. Initial water supplies were very restricted, expensive, and of poor quality. Water shortages also caused serious health problems such as dysentery and typhoid.

The Premier of Western Australia, John Forrest, tabled in July 1896 a comprehensive report from his Engineer-in-Chief, C. Y. O'Connor in both Houses of Parliament and introduced the *Coolgardie Goldfields Water Supply Loan Bill* to raise £2.5 million ($268 million in 2002 prices) for construction of a water pipeline. The Scheme consisted of building a weir on the Helena River in the Darling Range and pumping water 530 km overland to Coolgardie using eight pumping stations, delivering 22 700 m^3 of water per day at an estimated cost of about $4 per m^3 at 2002 prices. John Forrest instructed O'Connor to proceed with construction of the Scheme in January 1898. Western Australia voted to enter the Australian Federation on 31 July 1900. This was a turning point leading to the departure of John Forrest, O'Connor's main ally and supporter.

Although construction of the Scheme progressed well, O'Connor was personally attacked in the press and

Table 8.6. *Expansion of water supply by the Scheme from 1960–61 to 1996–97*

Item	Year				
	1960–61	1970–71	1980–81	1990–91	1996–97
Number of services connected	23 728	26 046	27 849	29 496	38 742
Length of water main (km)	5645	7261	7922	7981	8389
Water supplied ($10^6 \, m^3$)	11.5	16.0	18.8	28.9	28.8

Source: Australian Bureau of Statistics (1962, p. 220), (1973, p. 319), (1985, p. 259), (1992, pp. 12–15) and (1998, p. 192) for 1960–61 to 1996–97 data respectively.

---

[8] http://www.watercorporation.com.au/students/students_topics_ws_goldfields.cfm (visited in July 2005).

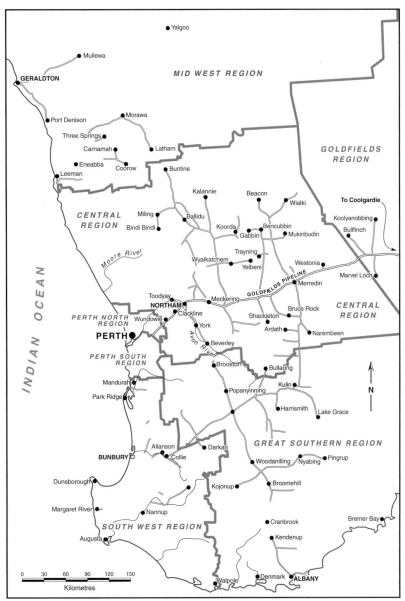

Figure 8.6 Major water supply regions and pipelines in south-west of Western Australia (Australian Bureau of Statistics, 1992, pp. 12–13).

Parliament. In February and early March 1902 these attacks reached an unbearable level, resulting in O'Connor's suicide on 10 March 1902 when the Scheme was less than a year away from completion. Construction of the pipeline continued and was completed in January 1903. The construction took about 5 years, at a total cost of £2 866 454 (about $307 million in 2002 prices), or £366 454 more than O'Connor's original estimate. The Goldfields Water Supply Scheme was officially opened by John Forrest, then Minister for Defence in the new Commonwealth Government and former Premier of Western Australia, who had supported and encouraged O'Connor in the design and construction of the pipeline, in a ceremony on Saturday 24 January 1903, at Coolgardie.

The Scheme provided a reliable supply of freshwater to Goldfields' residents and the mining industry. It brought comfort, health and prosperity to the region. Since its completion the Scheme has been expanded significantly. It is currently known as the *Goldfields and Agricultural Areas Water Supply Scheme*, and supplies water to approximately 39 000 rural and town services. This water is used for household and commercial water supply, farm water supply for stock, and mineral processing. The cost of delivering water to Kalgoorlie is $3.71 per m^3. This includes operation, maintenance and capital cost. However, the Water Corporation sells this water at a subsidised level of $1 per m^3.

The Scheme was a major undertaking for the Western Australian Government and, for its time,[9] was the largest of its kind in the world. John Forrest, in his speech on the 21 July 1896, during the second reading of the *Coolgardie Goldfields Water Supply Loan Bill* concluded:

> Future generations, I am quite certain, will think of us and bless us for our far-seeing patriotism.

References

Alexander, F., Crowlet, F. K. and Legge, J. D. (1954). *The Origins of the Eastern Goldfields Water Scheme in Western Australia*. Nedlands, W.A.: University of Western Australia Press.

Anon (1903). *The Coolgardie Water Supply: Souvenir of the Opening of the Works by Sir John and Lady Forrest at Mundaring, Coolgardie and Kalgoorlie, 22 and 24 January 1903*. Melbourne: D.W. Paterson, Printer and Bookbinder.

Anon (1904). *History of the Goldfields of Coolgardie and Kalgoorlie Water Supply Scheme*. Second Edition. London: Worthington Pump Co. Ltd.

Australian Bureau of Statistics (1962). *Official Year Book of Western Australia*. No. 3. Perth: Australian Bureau of Statistics, Western Australian Office.

Australian Bureau of Statistics (1973). *Western Australian Year Book*, No. 12. Perth: Australian Bureau of Statistics, Western Australian Office.

Australian Bureau of Statistics (1985). *Western Australian Year Book*, No. 23. Perth: Australian Bureau of Statistics, Western Australian Office.

Australian Bureau of Statistics (1992). *Western Australian Year Book*, No. 29. Perth: Australian Bureau of Statistics, Western Australian Office.

Australian Bureau of Statistics (1998). *Western Australian Year Book*, No. 34. Perth: Australian Bureau of Statistics, Western Australian Office.

Ayris, C. (2001). *C. Y. O'Connor: The Man for the Time*. Hamilton, W.A.: PK Print Pty Ltd.

Blainey, G. (1993). *The Golden Mile*. Sydney: Allen & Unwin Pty Ltd.

Evans, A. G. (2001). *C. Y. O'Connor: His Life and Legacy*. Crawley, WA: University of Western Australia Press.

Ewers, J. K. (1935). *The Story of the Pipe-Line*. Perth: Carroll's Ltd, Printers and Publishers.

Forrest, J. (1896). *Speech in the Legislative Assembly by the Honourable Sir John Forest on Moving the Second Reading of the Coolgardie Goldfields Water Supply Loan Bill on Tuesday 21st July 1896*. Perth: Government Printer.

Fraser, M. A. C. (1906). *Western Australia Year-Book for 1902–04*. Perth: Government Printer.

Hartley, R. (2000). *A Guide to Printed Sources for the History of the Eastern Goldfields Region of Western Australia*. Nedlands, W.A.: Centre for Western Australian History, The University of Western Australia, in Association with the University of Western Australia Press.

Lau, J. E., Commander, D. P. and Jacobson, G. (1987). *Hydrogeology of Australia*. Canberra: Australian Government Publishing Service.

O'Connor, C. Y. (1896). *Report on Proposed Water Supply (by Pumping) from Reservoirs in the Greenmount Ranges*. Perth: Government Printer.

Palmer, C. S. R. (1905). *Coolgardie Water Supply*. London: Institution of Civil Engineers.

Public Works Department (1963). *Goldfields Water Supply Scheme: Form of Proceedings at the Ceremony to be Held at Mt. Charlotte Reservoir on the 24th January 1963, at 3.30 p.m. to Commemorate the 60th Anniversary of the Official Opening of the Scheme at Coolgardie and Kalgoorlie on 24th January 1903 by the Rt. Hon. Sir John Forrest Minister for Defence*. Perth: Public Works Department, Western Australia.

Shaw, J. ed. (1984). *Collins Australian Encyclopedia*. Sydney: Collins.

---

[9] The Snowy Scheme described in Chapter 4 is the only other inter-basin water transfer scheme, which was undertaken by the Federal Government with a very high financial commitment.

9 Supplying Perth, Western Australia with Water: The Kimberley Pipeline Scheme

9.1 INTRODUCTION

Although Western Australia has significant volumes of surface and groundwater resources, the volume of water that can be economically and sustainably harvested is a fraction of the total available resources. Nearly 90 percent of the State's surface water is allocated to the environment and significant volumes of groundwater are reserved to protect groundwater dependent ecosystems (Water and Rivers Commission, 2000, p. 1). The sustainable surface water yield of the State is about $5.2 \times 10^9 \, m^3 \, yr^{-1}$, while the sustainable groundwater yield is estimated to be $6.3 \times 10^9 \, m^3 \, yr^{-1}$. These estimates are regardless of water quality and include fresh, marginal and saline waters.

Water withdrawal at the beginning of the twentieth century was very low. It increased gradually and reached $250 \times 10^6 \, m^3$ by 1960 when the State's population was about 750 000. Total withdrawal more than doubled between 1980 and 2000 (Figure 9.1) and is expected to double again by 2020 to around $3.6 \times 10^9 \, m^3$, reflecting the predicted increase in population to a total of more than 2.7 million. In 1999–2000, irrigated agriculture used 40 percent of the State's water withdrawal, followed by: mining 24 percent; households 13 percent; services 7 percent; gardening 5 percent; parks 4 percent; industry 4 percent; and stock water 3 percent (Government of Western Australia, 2003, pp. 7 and 8).

Water demand for irrigated agriculture is expected to increase rapidly from about $0.7 \times 10^9 \, m^3$ in 2000 to $1.6 \times 10^9 \, m^3$ in 2020. This is also the case for the mining industry, with a predicted increase in demand from $0.42 \times 10^9 \, m^3$ to $0.93 \times 10^9 \, m^3$ over the same period. However, growth in water demand for other sectors is forecast to be much slower (Government of Western Australia, 2003, Figure 6).

9.2 WATER CONSERVATION STRATEGY

The Western Australian Government released a Draft State Water Conservation Strategy for public comment in July 2002. From July to September 2002, 19 metropolitan and regional community water forums were held, leading to the State Water Symposium at State Parliament on 7–9 October 2002. The Symposium produced 22 recommendations as well as 40 wide-ranging conclusions. Following the process of public consultation, the Strategy was finalised and released in February 2003 (Government of Western Australia, 2003).

The State Water Strategy calls for strong community, government and industry partnership to ensure a sustainable water future for Western Australians. A prominent feature of the Strategy is that it recognises the regional diversity of the State and calls for tailor-made measures and different targets for different parts of the State. The objectives of the Strategy are:

(1) improving water use efficiency in all sectors including irrigated agriculture, mining and households;
(2) achieving significant advances in water reuse;
(3) fostering innovation and research;
(4) planning and developing new sources of water in a timely manner; and
(5) protecting the value of the State's water resources.

9.3 LONG-TERM WATER SUPPLY OPTIONS FOR PERTH

The Western Australian Water Resources Council (1988) studied the long-term options for water supply to Perth and the south of the State. The study is outdated, and current water supply policies have significantly changed with more focus on groundwater extraction and savings in water use applications. However, the work provides a useful

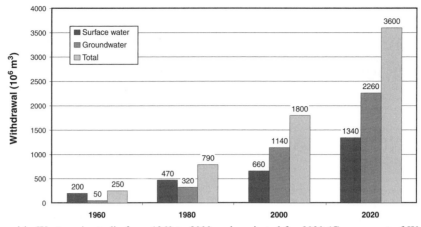

Figure 9.1 Water withdrawal in Western Australia from 1960 to 2000 and projected for 2020 (Government of Western Australia, 2003, Figure 4).

Table 9.1. *The 1988 water supply options for Perth, assuming moderate climate change*

Source	Yield[a] (10^6 m^3 yr^{-1})	Cost[d] ($ per m^3)	Comments
Additional Perth region sources			
Brackish water	37	1.00	Requires desalination
Forest thinning	29	0.05	Environmental impacts require further investigations
Reuse of wastewater[b]	25	0.07	Industrial
	15	0.2	Irrigation
	18	1.70	Aquifer recharge
Excess drainage	40	0.30–0.55	
Rainwater tanks	12	2.10	Potential health risks
Groundwater	Not estimated	Not estimated	Constrained by wetlands
Inter-regional transfers to Perth[c]			
South-west	810	0.53	Mainly surface water
Moore sub-region	110	0.54	Mostly marginal quality
Pilbara	210	4.90–5.10	Full yield uncertain
Kimberley (Fitzroy R.)	870	5.35	
Other source options			
Seawater desalination	>500	1.80	
Icebergs	>30	Unknown	Feasibility not proven
Bulk water transport by ship from the Kimberley	>300	3.30	Cheaper than piping water over very long distances

Note: [a] Yield by the year 2050 assuming moderate climate change; [b] Not for domestic use; [c] Surplus resources under medium regional demand scenario; [d] Cost in January 1988 dollars at 6 percent discount rate.
Source: Western Australian Water Resources Council (1988, Table 1).

background for the Kimberley Pipeline Scheme. The study examined the regional water situation for *high, medium* and *low* demand scenarios under conditions of a *stable climate*, and of *moderate* and *severe* rainfall decline. Table 9.1 shows the 1988 water supply options for Perth assuming *moderate* climate change.

Despite not being able to account for a major component (groundwater), the study indicated that sufficient water is available to meet Perth's needs until the mid twenty-first century. However, developments would become more expensive, and more importantly several options may have undesirable environmental impacts. In addition, there

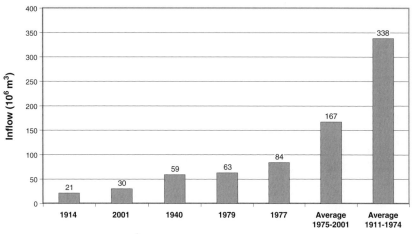

Figure 9.2 Annual inflow into dams of the Perth hills.[1]

would be strong public opposition to many of the options such as a major program of forest thinning or further damming of rivers in the south-west of the State. Drawing down the watertable in the northern Swan Coastal Plain close to Perth would have significant impact on its wetlands and would be vigorously opposed by the community. In contrast, there is scope for substantial reduction in water use in all sectors, and reuse of treated wastewater.

9.4 PERTH'S WATER SUPPLY OPTIONS

Climate change has contributed to a 10–20 percent reduction in rainfall in the south-west of the State since 1975 (Government of Western Australia, 2003, p. 3). This has led to a subsequent reduction of inflow into dams in the Perth hills, from $338 \times 10^6 \, m^3 \, yr^{-1}$ over the period of 1911–1974, to $167 \times 10^6 \, m^3 \, yr^{-1}$ over the period of 1975–2001 (Figure 9.2). The 2001 inflow was only $30 \times 10^6 \, m^3$, the second lowest inflow ever recorded. Inflow increased in 2002 to $100 \times 10^6 \, m^3$, and was $165 \times 10^6 \, m^3$ by the end of June 2004.

Results of a detailed study into the impact of the projected reduction in rainfall on the water yield of the Stirling Dam catchment, indicated that an 11 percent reduction in annual rainfall by the middle of the twenty-first century could result in a 31 percent reduction in annual water yield of the catchment (Berti et al., 2004). The predicted reduction in run-off from this catchment is likely to be representative of future responses of catchments in the high rainfall zone along the Darling Range to climate change. Reduction in catchment yields and inflows to the reservoirs has major impacts on the Integrated Water Supply Scheme (IWSS), which currently services more than 1.4 million of the 1.9 million people in Perth, Mandurah, Pinjarra, and towns and properties along the Goldfields Pipeline.

To illustrate the impacts of the long period of "below average" rainfall, Figure 9.3 shows the volume of water stored in the main dams that supply water for the IWSS at 30 June 2004, and compares them with the storage capacities of the dams. It indicates that the volume of water in dams ranged from about 42 percent for the Mundaring Weir down to 17 percent for the South Dandalup Dam, with an average of 25.5 percent for all dams.

There has been a 20 percent increase in per capita domestic water use in Perth over the past 20 years, from about $270 \, L \, d^{-1}$ to $330 \, L \, d^{-1}$ (Water Corporation, 2002). These figures do not include domestic pumping from shallow aquifers used by many households for garden watering. However, Perth's water supply received an extra boost in the 1960s from groundwater resources[2] mostly located in Perth's northern suburbs and adjacent pine plantations.

---

[1] http://www.watercorporation.com.au/dams/dams_streamflow.cfm (visited in July 2005).

[2] For the hydrogeology of the Swan Coastal Plain and Perth areas see:
  (a) Lowe, G. (1989). *Swan Coastal Plain Groundwater Management Conference, Proceedings.* Perth: Western Australian Water Resources Council. Publication No. 1/89.
  (b) Davidson, W. A. (1995). *Hydrogeology and Groundwater Resources of the Perth Region, Western Australia.* Perth: Geological Survey of Western Australia, Bulletin 142.
  (c) Water and Rivers Commission (1997). *Groundwater Allocation Plan: Swan Groundwater Area.* Perth: Water and Rivers Commission. Allocation and Planning Series Report No. WRAP 12.
  (d) *Perth Groundwater Atlas* at: http://www.wrc.wa.gov.au/infocentre/atlas/atlas_html (visited in July 2005).
  (e) *Hydrogeological Atlas of Western Australia* at: http://apostle.wrc.wa.gov.au/website/hydrogeological_atlas_wa/index.asp?form_id=1 (visited in July 2005).

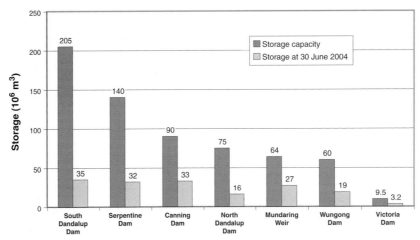

Figure 9.3 Storage capacities and volumes of water stored in major dams of the Integrated Water Supply Scheme, at 30 June 2004 (Water Corporation, 2004, p. 50).

Bores into the Yarragadee aquifer, north of Perth, can each supply $7.3 \times 10^6 \, \text{m}^3 \, \text{yr}^{-1}$ from a depth of 1000 m.

During the financial year 2003–04, the Water Corporation supplied $338 \times 10^6 \, \text{m}^3$ of water which included: $225.8 \times 10^6 \, \text{m}^3$ to Perth; $32.2 \times 10^6 \, \text{m}^3$ to North-West; $23.5 \times 10^6 \, \text{m}^3$ to South-West; and $17.5 \times 10^6 \, \text{m}^3$ to Mid-West (Water Corporation, 2004, p. 50). In order to supply Perth with $225.8 \times 10^6 \, \text{m}^3$ of water, $257.5 \times 10^6 \, \text{m}^3$ was extracted consisting of $100 \times 10^6 \, \text{m}^3$ from surface water and $157.5 \times 10^6 \, \text{m}^3$ from groundwater resources. Of the total volume of water extracted for water supply to Perth, internal transfers and system losses accounted for about $31.7 \times 10^6 \, \text{m}^3$.

Based on the estimated population growth of 1.7 percent per year, an additional $64 \times 10^6 \, \text{m}^3$ in new resources will be required by 2020 and $270 \times 10^6 \, \text{m}^3$ by 2050. Considering $338 \times 10^6 \, \text{m}^3$ was required in 2003–04, the Water Corporation will require more than $600 \times 10^6 \, \text{m}^3$ of water to supply Perth's metropolitan area by 2050. This estimate may increase if drier conditions prevail and the community is unable to further reduce its water use.[3]

Development of new sources for the Perth metropolitan area could include (Water Corporation, 2002):

- Extraction of more than $100 \times 10^6 \, \text{m}^3$ per annum from groundwater sources. The cost of water would be $0.30–$0.50 per m^3, but in the South-West it would be significantly greater, in the order of $0.70–$0.80 per m^3.
- Development of surface water resources. The cost of water would be about $0.35–$0.45 per m^3 or even more depending on the distance between the dams and Perth. If dams are a long distance from Perth, the extra length of pipelines, and extra pumping costs would increase the price of water.
- Reuse of treated wastewater. The three major wastewater treatment plants of Perth treat about $96 \times 10^6 \, \text{m}^3$ of wastewater and discharge it into the marine environment. This is an important potential water resource and in the future will be reused as much as possible.
- Desalination of seawater at a cost of $1.10 per m^3.
- Transfer of water from the Kimberley region at an estimated cost of more than $6.10 per m^3, and domestic rainwater tanks to supplement other options at a cost of $7.50 per m^3.

A more recent 2004 estimate[4] indicated that the population of the area served by IWSS would reach 2.1 million by 2025 and 2.6 million by 2050. This would require an additional supply of $130 \times 10^6 \, \text{m}^3 \, \text{yr}^{-1}$ by 2020, and $235 \times 10^6 \, \text{m}^3 \, \text{yr}^{-1}$ by 2050. Substantial further investment in new sources of supply will be necessary to restore the demand/supply balance. The key features of the Water Corporation's planning for water resource development up to 2014–15 are based on:

- Demand management to reduce water consumption to the level of $155 \, \text{m}^3 \, \text{yr}^{-1}$ per person by 2012.
- Desalination of seawater at a rate of $45 \times 10^6 \, \text{m}^3 \, \text{yr}^{-1}$ by using Kwinana Desalination Plant from the second half of 2006 (see Section 9.7).
- Water trading at a rate of $17 \times 10^6 \, \text{m}^3 \, \text{yr}^{-1}$ from the second half of 2006. This is associated with water saving by replacing open channels with pipelines in various south-west irrigation areas.

---

[3] http://www.ourwaterfuture.com.au/Factsheets/factsheet_allocatingwater.asp (visited in July 2005).
[4] Sarah Fisher, Senior Planning Engineer, Water Corporation, personal communication, December 2004.

- South-west Yarragadee groundwater extraction at a rate of $45 \times 10^6 \, \text{m}^3 \, \text{yr}^{-1}$, from December 2009. This is the most significant undeveloped water resource in the south-west of the State, which has been the subject of detailed investigations since early 2003.

In the longer term, the Water Corporation is actively progressing planning for a range of future options. The key initiatives being progressed include:

(1) smarter use of water;
(2) further water trading;
(3) additional desalination;
(4) additional surface water;
(5) further catchment management; and
(6) reuse of treated wastewater.

9.5 INTER-BASIN WATER TRANSFER FROM KIMBERLEY

Numerous proposals have been developed for transfer of water from the Kimberley region to Perth and Adelaide. These proposals and their related issues are described in the following sections, after a brief description of Kimberley's water resources.

9.5.1 Water in the Kimberley Region

The Kimberley Region is located in the northern part of Western Australia. Rainfall varies from about 350 mm along the southern border of the region to over 1400 mm in the north-western coastal area (Figure 3.2). The climate is tropical with a wet monsoonal or cyclonic season from November to March, and a long dry season from April to October. September and October are the driest months (Water and Rivers Commission, 1997). It has an estimated population of 31 200 and abundant but seasonal surface water resources (Table 9.2). The annual sustainable yield of its water resources is $3.2 \times 10^9 \, \text{m}^3$, while its water use is about $272 \times 10^6 \, \text{m}^3 \, \text{yr}^{-1}$ (Water and Rivers Commission, 2000, Figure 5.2), which indicates that only a small proportion of its water resources have been developed. The main regional development is the Ord River Irrigation Scheme (see Appendix F). In contrast, the Perth metropolitan area, has limited undeveloped freshwater resources and is faced with water shortages. As described before, rainfall and streamflow records in the south-western part of the State show a dramatic decline since the 1970s. The Kimberley Pipeline Scheme was proposed to supply additional water to the Perth metropolitan area from unused water resources of the Kimberley region.

Current Western Australian Government's policy is to use water resources of the Kimberley Region for local development only (see Appendices F and G). The Kimberley region's population is expected to expand to 43 000 by the year 2021 (Kimberley Development Commission, 2002, p. 7), and its regional economy is rapidly growing. This growth is based on a range of industries and activities including:

(1) a mining industry valued at $731 million;
(2) a tourism industry valued at $206 million;
(3) a pastoral industry covering 23 Mha, and exporting livestock to Asia and the Middle East;
(4) a cultured pearl industry valued at $145 million; and
(5) continued growth in the agriculture and horticulture industry, mainly in the Ord River Irrigation Area.

9.5.2 The Binnie & Partners Study of the Kimberley Pipeline

Binnie & Partners Pty Ltd (1988), in association with Wilson-Sayer-Core Pty Ltd, carried out a comprehensive evaluation

Table 9.2. *Characteristics of four major river basins of the Kimberley Region*

River basin	Catchment area (km^2)	Mean annual flow (10^9 m^3)	Main flow period	Maximum instantaneous recorded flow (m^3 s^{-1})
Fitzroy	88 900	14.2	Nov–May	12 200
Ord	46 100	3.9	Nov–April	30 800
Drysdale	12 200	2.2	Dec–May	NA
Prince Regent	3150	0.8	Dec–May	NA
Total	150 350	21.1	–	–

Source: Water and Rivers Commission (1997, Table 6).

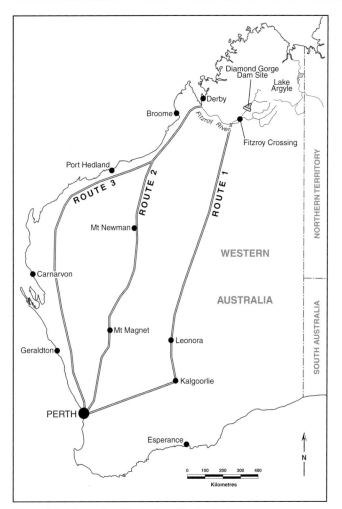

Figure 9.4 Alternative Kimberley–Perth pipeline routes (Binnie & Partners Pty Ltd, 1988, Figure 2).

of the costs and benefits of potential schemes for transferring Kimberley water to Perth. The study was undertaken for the Kimberley Regional Development Advisory Committee. Their assessments predicted that the overall water demand of the Perth–Mandurah region, which also supplies the Goldfields and Agricultural Areas Water Supply Systems (see Chapter 8), would exceed the region's water resources in the early twenty-first century. One supply option to meet the excess demand was to transfer water from the Kimberley region, which has large volumes of surplus water resources. They evaluated three alternative pipeline routes shown in Figure 9.4, taking into account major potential activities and geographical constraints. These routes are:

- **Route 1:** Inland route via Kalgoorlie to Perth.
- **Route 2:** Pilbara route via Mt Newman and Mt Magnet to Perth.
- **Route 3:** Coastal route via Broome, Port Hedland, Carnarvon and Geraldton to Perth.

The likely components of each pipeline are (Binnie & Partners Pty Ltd, 1988, pp. 53–55):

- Source works consisting of the Diamond Gorge Dam on the Fitzroy River (Figure 9.4) and associated intake works. The Diamond Gorge site appears to be the most suitable location for water diversion to the south for the following reasons: (1) the Fitzroy River is the closest river to the demand centres in the south; (2) it could provide about $670 \times 10^6 \, m^3$ of water per year which exceeds the expected demand; and (3) the divertible resources of the Fitzroy River are not committed. The cost of the Diamond Gorge Dam was estimated at about $80 to $100 million in 1988 prices ($125 to $156 million in 2002 prices).
- Pumping station to pump raw water from the intake structure to the treatment plant.
- Treatment plant in order to remove the sediment loads during flood periods.
- Pumping stations at intervals along the pipeline.
- Intermediate and terminal storage reservoirs to provide emergency reserve storage in the event of breakdown of the conveyance system.
- Electricity supply to pumping stations. Several options have been investigated: (1) high voltage above ground transmission lines from a power station in Perth; (2) diesel-fuelled power stations receiving fuel from fuel tankers, or less likely by fuel pipelines; (3) natural gas-fuelled power stations; (4) hydro-power from the Ord River Dam and transmission lines; (5) hydro-power from the Fitzroy River and transmission lines (but this would be less than the hydro-power generating capacity of the Ord River); (6) tidal power from the Kimberley region; and (7) solar power, which would be excessively expensive. The study adopted the power supply option (1) for cost estimation.

Economic Evaluation of Pipeline Options

The cost of water transfer from the Kimberley to the Perth region depends on the magnitude and nature of excess demands to be satisfied. Two scenarios of *medium* and *high* water demand to the year 2050 were considered. The *medium* and *high* demands are estimated to be 300 and $400 \times 10^6 \, m^3$ per annum respectively. The Diamond Gorge Dam on the Fitzroy River in the southern Kimberley region was considered as a possible source for all three routes.

The water conveyance systems in both cases were 1400 mm diameter high-pressure pipelines and associated pumping

Table 9.3. *Route lengths and unit costs of water delivery to Perth*

Options	Length (km)	Unit cost of water ($ per m^3) for:	
		Medium demand[a]	High demand[b]
Pipeline Route 1	1960	5.77	4.92
Pipeline Route 2	1840	5.35	4.53
Pipeline Route 3	2100	6.02	5.12

Note: [a] $300 \times 10^6 \, m^3$ per year; [b] $400 \times 10^6 \, m^3$ per year.
Source: Binnie & Partners Pty Ltd (1988, p. 4).

stations. The economic analysis assumed a 6 percent discount rate and a $0.10 per kWh energy cost, and gave the results shown in Table 9.3. This table indicates that:

(1) the cost of water delivered to Perth is the cheapest via Route 2, which could supply water at costs of $5.53 and $4.53 per m^3 for the medium and high demand rates respectively; and
(2) costs of water under the high demand option are lower than the medium demand for all three routes.

Evaluation of Alternative Routes

The three pipeline routes were evaluated based on the following criteria:

(1) the net 1988 present value of the benefits and the costs;
(2) estimation of the additional workforce and population; and
(3) the rank scores for a number of selected indicators of social and community benefits, which were quality of life, community stability, health and education.

Table 9.4 summarises the results of the evaluation. The main features of the evaluation were:

- Benefits for the mining industry are significant for all options particularly for Route 1.
- Benefits for the agricultural industry are significantly less than mining and the greatest agricultural benefits occur along Route 3.
- Industry benefits would be even lower than agriculture, with the highest benefits for Route 3.
- With the exception of tourism, the other sectors are largely dependent on the growth of the mining and the agricultural sectors due to their service nature.
- Route 1 would have the highest benefits under both *optimistic* and *pessimistic* State growth conditions.
- For all options, the costs exceed benefits with a maximum gap between benefits and costs of $8116 million for Route 3.
- Route 1 has the highest benefit/cost ratio of 0.55 under *optimistic* State growth conditions, and assuming a *high* level of benefits generated by the pipeline.
- The benefits to the quality of life and community stability would be higher for Routes 1 and 2, but substantially lower for Route 3.
- The aggregate score for social and community benefits are higher for Routes 1 and 2.
- The estimates for the additional workforce and population are higher for Route 1 compared to Routes 2 and 3.

If the economic and financial analysis were adopted as the main criterion for assessment of water supply by pipelines, then the Kimberley Pipeline Scheme would not proceed (Binnie & Partners Pty Ltd, 1988). However, justification of any route depends on the weight attached to the social and community benefits. On the basis of both economic and social benefits, Route 2, which has the lowest population size, scores lowest. Route 3 scores well due to its large population size as well as a significant share of established agricultural and mining activities. Yet at the same time, the regions along this route are more likely to experience continuing growth, irrespective of whether a pipeline is constructed or not. Route 1 provides the most economic and social benefits.

9.5.3 The Infrastructure Development Corporation Study of the Kimberley Pipeline

Following submission of the Binnie & Partners Pty Ltd (1988) report to the Kimberley Regional Development Advisory Committee, the Water Authority of Western Australia commissioned the Infrastructure Development Corporation (IDC) to undertake a preliminary feasibility study of establishing a water pipeline from the Kimberley Region to Perth. Its objectives were:

- To obtain an independent economic analysis reviewing works undertaken and completed previously on the Kimberley–Perth Pipeline Scheme.

Table 9.4. *Summary evaluation matrix for alternative pipeline routes for various criteria*

Route	Route options											
	Route 1				Route 2				Route 3			
Range of growth	Optimistic		Pessimistic		Optimistic		Pessimistic		Optimistic		Pessimistic	
Range of benefits	High	Low	High	Low	High	Low	High	Low	High	Low	High	Low
Criteria												
1.0 Economic and financial benefits (net present values in $M)												
1.1 Benefits												
Mining	3054	1108	1331	444	960	480	418	209	1997	999	870	435
Agriculture	105	53	46	23	92	46	40	20	231	77	101	34
Industrial	34	17	7	4	24	8	5	2	82	54	17	11
Other	202	121	89	53	120	72	53	32	322	193	141	85
Water revenue	500	500	740	740	500	500	740	740	500	500	740	740
Subtotal	3895	1709	2213	1264	1696	1106	1256	1003	3132	1823	1869	1305
1.2 Costs	Normal demand		High demand		Normal demand		High demand		Normal demand		High demand	
Water schemes	7027	7027	8808	8808	6675	6675	8333	8333	7510	7510	9421	9421
1.3 Gap between benefits and costs	3132	5318	6595	7544	4979	5569	7077	7330	4378	5687	7552	8116
1.4 Benefit/cost ratio	0.55	0.24	0.25	0.14	0.34	0.20	0.18	0.14	0.42	0.24	0.20	0.14
2.0 Social and community benefits												
2.1 Population figures for 1986	86 000				52 000				123 000			
2.2 Rank of the selected features												
Quality of life	118				111				87			
Community stability	103				107				68			
Health	175				175				182			
Education	93				94				82			
Subtotal	489				487				419			
2.3 Additional workforce (1000)	22	8	7	3	12	4	3	2	21	11	7	3
2.4 Additional population (1000)	50	17	16	6	33	12	8	4	40	21	13	7

Source: Binnie & Partners Pty Ltd (1988, Table 6.2).

- To obtain an independent report on the technical feasibility, economic and cash flow implications of the Scheme to assist with planning.

The IDC submitted a comprehensive report to the Water Authority in May 1990 (Infrastructure Development Corporation, 1990). The study examined the proposed water pipeline under two scenarios: the central inland Route 2, and the coastal Route 3 of Table 9.3 and Figure 9.4. The eastern inland route of the previous study was not examined in detail because of the logistic problems of construction and the much higher operating costs arising from the supply of energy for pumping.

The approach adopted was the design of a high pressure steel pipeline completely buried over its total length with a relatively small number of powerful pump stations using natural gas as the energy source. It was suggested that the pipeline with a minimum lifetime of 50 years could meet the increasing demand in the Perth region.

Based on principles and specifications commonly used in the oil and gas transmission industry, a high pressure (7000 kPa) welded steel pipeline, with possible diameters of 1400 mm or 1600 mm, was considered. The 7000 kPa operating pressure allowed greater spacing between pump stations than the conventional operating pressure of 2000 kPa, and only seven pumping stations were required

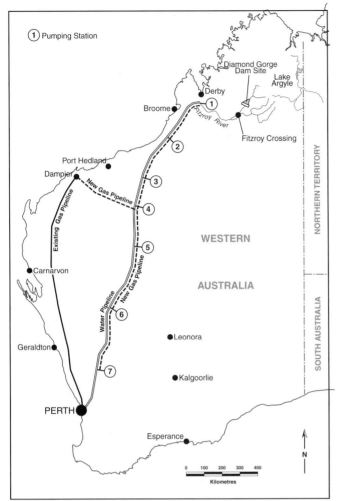

Figure 9.5 The central inland pipeline Route 2 and locations of the pumping stations for the 1400 mm pipeline option (Infrastructure Development Corporation, 1990, Figure 4.1).

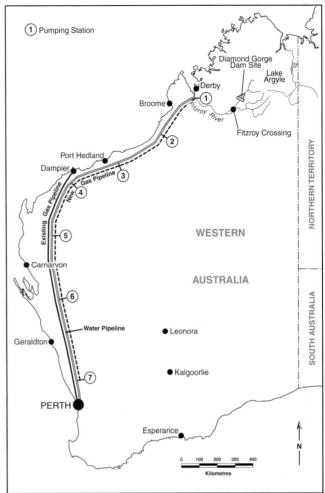

Figure 9.6 The coastal pipeline Route 3 and locations of the pumping stations for the 1400 mm pipeline option (Infrastructure Development Corporation, 1990, Figure 4.3).

instead of 20 for the conventional pressure. Pumps were required at Fitzroy River and at intermediate locations to maintain the required head pressures. The total pump energy requirements of the 1400 mm inland route and coastal route were estimated at 448 MW and 469 MW respectively. The central inland and coastal pipeline routes, and locations of the pumping stations for the 1400 mm pipeline options are shown on Figures 9.5 and 9.6 respectively.

Capital and Operating Cost Estimates

The cost estimates of the pipeline Scheme are summarised in Table 9.5 and indicates that the pipeline could be constructed at a lower cost than that estimated by Binnie & Partners Pty Ltd (1988).

The IDC estimates of the operating costs (Table 9.6) are significantly less than the earlier study of Binnie & Partners Pty Ltd (1988). The 1988 estimates of the operating costs were based on a "percentage of capital cost" calculation, while the IDC study estimated the operating costs by closer examination of the major operating cost categories. The IDC approach of using natural gas to fuel the pumping stations significantly reduced estimates of the energy cost.

Conclusions of the IDC Study

Some of the conclusions of the IDC study were:

- The Scheme is technically feasible.
- The capital cost would be 15 percent less than the 1988 study.
- Despite the project capital cost estimate of 15 percent less than the previous study, the cost of water ($3.45 per m^3) remained significantly higher than the cost of some of

Table 9.5. *Capital cost estimate of the pipeline Scheme*

Options	Pipeline characteristics:		1990 IDC estimate ($ million)	1988[a] estimate ($ million)
	Diameter (mm)	Length (km)		
Option 1: Inland Route 2	1400	1840	8448	10 407
Option 2: Inland Route 2	1600[b]	1840	9719	14 678
Option 3: Coastal Route 3	1400	2100	9852	11 722
Option 4: Coastal Route 3	1600[b]	2100	11 386	16 646

Note: [a] Binnie & Partners Pty Ltd (1988); [b] The 1600 mm pipe compares with the 1800 mm concrete lined of the 1988 study.
Source: Infrastructure Development Corporation (1990, Table 1.1).

Table 9.6. *Operating and energy costs for the Option 1, Inland Route 2 Pipeline (1840 km long with a diameter of 1400 mm)*

Cost items	1990 IDC estimate ($ million)	1988[a] estimate ($ million)
Operating costs	385	3452
Energy costs	1851	9958

Note: [a] Binnie & Partners Pty Ltd (1988).
Source: Infrastructure Development Corporation (1990, Table 1.2).

- the alternative water sources such as seawater desalination or surface water diversion from the south-west (see Table 9.1).
- The project does not appear to be capable of development as a viable business entity. However, the study acknowledged that the Western Australian Government has social, development and environmental responsibilities regarding long-term security of water supplies that extend beyond consideration of cost neutrality or profitability of public utilities.
- An inland 1400 mm diameter pipeline would be the most economical option for the Kimberley to Perth water pipeline (see Tables 9.5 and 9.6).
- Any significant development along the route of the pipeline would necessitate increased capital and operating costs to ensure continuity of supply for that development over the life of the project. This is unlikely to bring benefits in cost to the Perth consumers.

9.5.4 National Water Pipeline Scheme

Binnie & Partners Pty Ltd (1988, pp. 76–78) also provided a brief assessment of a proposal to supply both Perth and Adelaide with water from Lake Argyle on the Ord River. Figure 9.7 shows the layout of the pipeline. The approximate lengths of various sections of the pipelines are:

(1) from Ord Dam to WA/SA Junction 1150 km;
(2) WA/SA Junction to Adelaide 1600 km; and
(3) WA/SA Junction to Perth 1540 km.

Adelaide's demand was assumed to be the same as Perth's demand and the *normal* demand scenario was used.

The study estimated that, assuming the energy cost of $0.10 per kWh and the discount rate of 6 percent, the unit cost of water delivered to Perth and Adelaide would be $6.94 and $7.10 per m^3 respectively. These cost estimates indicate that the cost of piping water to Perth as part of a National Water Pipeline Scheme is greater than a direct piped supply from Fitzroy to Perth ($3.45 per m^3 according to the IDC 1990 study). It was suggested that Adelaide could obtain its water cheaper by other means including bulk transport by ship from Tasmania or by pipelines from the coastal area of New South Wales and Queensland (see sections 5.5 and 6.8) which are closer than the Kimberley Region.

9.5.5 Kimberley Aqueduct Scheme

Operation of the National Water Pipeline Scheme described above would have required high capital costs and consumption of considerable amounts of fossil fuel (oil or natural gas). In order to reduce capital costs, minimise the energy consumption and reduce greenhouse gas emissions, Ravine and Crawford (1990) designed a system for transfer of water by gravity via a covered aqueduct, which takes advantage of renewable resources as far as practicable.

The design capacity for the section of aqueduct between Lake Argyle and the WA/SA junction to Perth and Adelaide was $800 \times 10^6 \, m^3 \, yr^{-1}$. This capacity was assumed to be equally divided between the two branches from WA/SA

SUPPLYING PERTH, WESTERN AUSTRALIA WITH WATER: THE KIMBERLEY PIPELINE SCHEME 175

Figure 9.7 The National Water Pipeline Scheme (Binnie & Partners Pty Ltd, 1988, Figure 5.7).

Figure 9.8 Proposed Kimberley Aqueduct Scheme (Kimberley Pipeline Environmental Advisory Committee, 1990, Figure 2).

junction. Along a major part of the route, water would be transferred by gravity (Figure 9.8). However, a number of short sections required water to be pumped to higher elevations. For example, water would be pumped a distance of 175 km from an elevation of 75 m at the Lake Argyle intake, to an elevation of 425 m, using three pumping stations. The highest point of the route would be at the WA/SA junction where the elevation is 600 m. To pump water along these particular sections of the route, it was suggested that hydro-electric and tidal energy[5] in the Kimberley region, as well as the solar and wind energy from central Australia could be used. However, some diesel-fuelled backup would also be required.

The cost of the scheme was estimated at $5732 million with annual operating and maintenance costs of $93.5 million at 1990 prices. The price of water was estimated at between $2.48 and $2.99 per m^3 for discount rates of 6 percent and 8 percent respectively. These prices are much cheaper than those estimated for the pipeline schemes studied by Binnie & Partners (see Table 9.3) and $3.45 per m^3 of the IDC 1990 study, but still greater than desalination.

9.5.6 The environmental study of the Kimberley pipeline scheme

Early studies of the Kimberley pipeline scheme paid little attention to the environmental impacts of the proposals. In July 1989, the Minister for Water Resources, Hon Ernie Bridge, established a Kimberley Pipeline Environmental Advisory Committee whose terms of reference were to:

- Examine and assess previous studies on potential uses of Kimberley water within Western Australia and in other parts of the country.
- Formulate additional options and to recommend new studies if appropriate.
- Advise the Minister on long-term planning for the use of Kimberley water.
- Advise the Minister on the social and environmental impacts of exporting water from Kimberley within Western Australia and to other parts of Australia.

The Committee investigated the issues and submitted its report to the Minister on June 1990 (Kimberley Pipeline Environmental Advisory Committee, 1990). It considered that if the pipeline was to be constructed, it should predominantly use renewable energy sources for pumping. The Committee developed a conceptual plan for an aqueduct system (see Section 9.5.5). The concept appears to be workable from an engineering viewpoint but it needsto be fully costed to determine whether it would be economically attractive. Other advantages could include industrial development involving the renewable energy and assistance for settlements, tourism, mining and agriculture in inland areas of Australia. Some potential problems perceived were the growth of organisms in the aqueduct and the possible barrier presented to fauna and wildlife movement by an aqueduct.

The Committee also considered other options for the use of Kimberley water and concluded that there were substantial attractions in the use of the coastal route if the pipeline was to supply only Western Australia. Such a route would provide abundant water to existing settlements in the Kimberley, Pilbara and Gascoyne regions as a base for substantial decentralisation programmes. The social and political advantages of such a route could be justified on the need for northern development and decentralisation. The coastal option could also take advantage of abundant wind power and solar power along the route.

The Committee examined the possible use of Kimberley water for settlements, mining and agriculture along its route. The general conclusion was that opportunities do exist but was not substantial because of the remoteness of these areas from the markets. Nevertheless it was recognised that several isolated inland communities could benefit substantially from the Scheme. Also, the benefits to local communities, to national security, to decentralisation and to employment creation were perceived to be substantial.

The Committee concluded that the Scheme would provide significant potential benefits and with careful design and full environmental review, the impacts could probably be regarded as acceptable to most of the community. More detailed work is required to quantify costs and benefits. However, it was recommended that this should be deferred until a decision is made on the overall concept plan and possible routes for the pipeline.

9.5.7 GHD Studies of the Kimberley Pipeline

GHD (2002) reviewed the most favourable options of the Kimberly pipeline (see Figure 9.4, Routes 2 and 3),

---

[5] For a 48 MW tidal power proposal near Derby and its related environmental issues see:
(a) Halpern Glick Maunsell (1997). *Derby Tidal Power Project Doctors Creek Kimberley: Consultative Environmental Review*, for Derby Hydro Power Pty Ltd. and,
(b) Environment Protection Authority (2000). *Derby Power Project—Recommended Conditions*. Perth: EPA. Bulletin 984.

studied by Binnie & Partners Pty Ltd (1988), and the Infrastructure Corporation Development (1990). The report concluded that:

(1) there have not been any significant technological developments in the intervening period which would provide a breakthrough to reduce the cost of the project;
(2) construction costs have increased since 1990, but operating costs have remained about the same;
(3) the cost of the project would be at least $10 billion;
(4) the supply cost of water would be about $5.50 per m^3; and
(5) because of the comparatively high energy requirements and associated greenhouse gas emissions, environmental approval of the project would be difficult.

GHD (2004) evaluated the impact of the Kimberley pipeline from a sustainability perspective. The report has assessed the environmental, social, and economic issues associated with the project, and concluded that:

- Although technically feasible, the project is highly complex with significant uncertainties related to aboriginal heritage and environmental issues.
- The project would consume three times more energy than desalination. This would generate four and a half times more greenhouse gas emissions than desalination.
- The project would create significant ecological impacts particularly relating to Fitzroy River.
- Construction of the project would cost $11.7 billion, and its annual operating cost would be about $105 million.
- At an estimated cost of more than $6.10 per m^3 of delivered water the project remains economically unviable compared with desalination.

9.6 BULK WATER TRANSPORT BY SHIP FROM KIMBERLEY TO PERTH

Binnie & Partners Pty Ltd (1988, pp. 74–75) also investigated transport of water from the Kimberley region by supertankers over a distance of 2500 km to Perth. The main components of this scheme were:

- Diamond Gorge Dam on the Fitzroy River as the source of water supply.
- Diversion weir near Great Northern Highway, Fitzroy River road bridge.
- A 300 000 m^3 d^{-1} treatment plant for the removal of suspended sediments.
- A 300 000 m^3 d^{-1} pumping station and a 1400 mm pipeline to the loading port near Derby.
- Storage facility with a capacity of 300 000 m^3 d^{-1} at the loading port.
- 300 000 tonne supertankers for transportation of freshwater to Perth.
- Off-loading facilities in deep water off Rottnest Island, west of Perth.
- Storage facility and pumping station with a capacity of 300 000 m^3 per day at Rottnest Island.
- A submarine pipeline with a diameter of 1400 mm between Rottnest Island and Perth.
- An onshore pipeline with a diameter of 1400 mm to connect into the distribution system in Perth.

The cost of water supplied to Perth was estimated at $3.33 per m^3 in 1988 prices ($5.3 in 2003 values) which is lower than all Kimberley–Perth pipeline options. The disadvantage of this scheme is that it would not be able to supply water to the interior locations between the Kimberly region and Perth along the alternative pipeline routes.

9.7 SEAWATER DESALINATION FOR PERTH'S WATER SUPPLY

Desalination is a well-known process[6] for converting saline seawater or brackish water to drinkable freshwater for municipal water supply. Desalination is widely used in about 120 countries. At the beginning of 1998, a total of about 12 450 desalination units with capacities of 100 m^3 d^{-1} or larger had been installed or commissioned throughout the world. These plants had a total capacity of around 22.7 × 10^6 m^3 d^{-1}, consisting of 13.3 × 10^6 m^3 d^{-1} from seawater desalination and 9.4 × 10^6 m^3 d^{-1} from desalination of waters with lower salinities (Gleick, 2000, p. 96).

In 1998, the Water Corporation prepared a desalination strategy and in 2000 published information on various aspects of desalination and the possibilities for its application in Western Australia (Water Corporation, 2000). With the decreasing costs of seawater desalination and the increasing costs associated with other water sources, seawater desalination has become a viable source for Perth's potable water. In 2002, a proposal was developed for a desalination plant for Perth metropolitan area. The proposal

---

[6] The technologies include a number of methods such as: Multistage flash distillation (MSF); Reverse osmosis (RO); Electrodialysis (ED); Vapor compression (VC); Multi-effect distillation (ME); and Membrane softening (MS). For further information about desalination see Gleick, 2000, pp. 93–111.

Table 9.7. *Characteristics of the upgraded desalination project at the Kwinana Power Station site*

Characteristics	Units
Capacity	$45 \times 10^6\,\text{m}^3\,\text{yr}^{-1}$
Power requirements	24.1 MW
Characteristics of seawater intake:	
Average seawater intake	$300\,000\,\text{m}^3\,\text{d}^{-1}$
Length of pipeline	0.8 km
Diameter of pipeline	1500 mm
Characteristic of concentrated discharge:	
Volume	$180\,000\,\text{m}^3\,\text{d}^{-1}$ (weekly average)
Salinity	$65\,000\,\text{mg}\,\text{L}^{-1}$
Temperature	2°C above ambient
Characteristics of freshwater product pipeline:	
Capacity	$>150\,000\,\text{m}^3\,\text{d}^{-1}$
Length	10 km
Diameter	1000 mm
Destination	Thompson Reservoir

Source: Environmental Protection Authority (2004, Table 1).

involved construction and operation of a $30 \times 10^6\,\text{m}^3\,\text{yr}^{-1}$ Reverse Osmosis desalination plant at either Kwinana Power Station or at East Rockingham, both at less than 36 km south of Perth (Environmental Protection Authority, 2002). In May 2003, the Minister for Environment approved the proposal. The Water Corporation proposed, in September 2003, to upgrade the capacity from the approved $30 \times 10^6\,\text{m}^3\,\text{yr}^{-1}$ to $45 \times 10^6\,\text{m}^3\,\text{yr}^{-1}$. Table 9.7 summarises characteristics of the upgraded desalination project at the Kwinana Power Station site.

Following acceptance of these changes by the Environmental Protection Authority, the Western Australian Government approved the amended project on 29 July 2004.[7] The project will be built at a cost of $350 million and will be owned by the Water Corporation. It is expected to be completed in the second half of 2006 and will be the biggest desalination plant in the southern hemisphere. The cost of the water produced will be almost twice as much as current water supplies, at $1.11 per m³, and the average household water bill is expected to increase by $50 a year.

Inter-basin transfer of water was one of the key policy issues in the February 2005 Western Australian State election. The conservative Liberal–National Party coalition adopted the Kimberly–Perth water transfer as a central element of its policy platform. The Labor Party Government favoured the cheaper, already approved desalination option, and was re-elected.

9.8 CONCLUSIONS

This chapter has reviewed water resources and future water requirements of Western Australia and Perth metropolitan area. Also, it has described various proposals developed since 1988 for transfer of water from the Kimberley Region in the north of the State to Perth via pipelines, aqueduct or tanker from a port near Derby to overcome future water shortage. According to a 2004 consulting report, although transfer of water from Kimberly to Perth by pipeline is technically feasible, its construction and operation would have significant impacts on the Fitzroy River, and aboriginal heritage, and would require high energy consumption which would produce a large quantity of greenhouse gases. The scheme is expensive in terms of construction ($11.7 billion) and operation costs ($105 million per annum) and would supply water at a high price of above $6 per m³, which is almost six times higher than the cost of desalinised water. Therefore, construction of the pipeline schemes cannot be justified based on the economic criteria.

To overcome the above-mentioned problems, the Western Australian Government adopted a strategy which, among other things, would improve water use efficiency in all sectors, and would achieve significant advances in water reuse. Also groundwater resources of the south-west of the State will be used to overcome the shortfall. The strategy is capable of supplying future water requirements of the State in general and the Perth metropolitan area in particular. Moreover, with the implementation of the desalination plant for Perth in the second half of 2006, an additional $45 \times 10^6\,\text{m}^3$ of water will be available at a capital cost of $350 million, at a cost of $1.11 per m³.

Finally, as described in Appendices F and G, it is planned that water resources of the Kimberley region be used for local development, and that in the future there would be no excess water for diversion to other parts of the State or the country.

---

[7] http://www.watercorporation.com.au/media/media_detail.cfm?id=165 (visited in July 2005).

References

Berti, M.L., Bari, M.A., Charles, S.P. and Hauck, E.J. (2004). *Climate Change, Runoff and Risks to Water Supply in the South-West of Western Australia*. West Perth: Department of Environment.

Binnie & Partners Pty Ltd (1988). *Water for the South-West in the Twenty-First Century: Water from the Kimberleys*. Perth: Binnie & Partners Pty Ltd Consulting Engineers.

Environmental Protection Authority (2002). *Perth Metropolitan Desalination Proposal*. Perth: EPA. Bulletin 1070.

Environmental Protection Authority (2004). *Perth Metropolitan Desalination Proposal, Amendment of Implementation Conditions by Inquiry*. Perth: EPA. Bulletin 1137.

GHD (2002). *Development of a Water Supply Pipeline from Kimberleys to Perth: Independent Review of Feasibility Studies*. Adelaide: GHD Pty Ltd.

GHD (2004). *Kimberley Pipeline Project: Sustainability Review*. Adelaide: GHD Pty Ltd.

Gleick, P.H. (2000). *The World's Water 2000–2001: The Biennial Report on Freshwater Resources*. Washington D.C.: Island Press.

Government of Western Australia (2003). *Securing our Water Future: A State Water Strategy for Western Australia*. Perth: Government of Western Australia.

Infrastructure Development Corporation (1990). *Development of a Water Pipeline, Kimberleys to Perth: Preliminary Feasibility and Economic Appraisal Study*. Sydney: IDC.

Kimberley Development Commission (2002). *Annual Report for the Period Ended 30 June 2002*. Kununurra, WA: KDC.

Kimberley Pipeline Environmental Advisory Committee (1990). *Report of the Kimberley Pipeline Environmental Advisory Committee to the Hon Ernie Bridge, JP, MLA Minister for Agriculture, Water Resources and the North West*. Perth: KPEAC.

Ravine, P. and Crawford, T. (1990). The gravity aqueduct concept. In *Kimberley Pipeline Environmental Advisory Committee*. Perth: KPEAC, Appendix A, pp. 39–71.

Water and Rivers Commission (1997). *The State of the Northern Rivers*. East Perth: Water and Rivers Commission.

Water and Rivers Commission (2000). *Western Australia Water Assessment 2000: Water Availability and Use*. East Perth: Water and Rivers Commission.

Water Corporation (2000). *Desalination: A Viable Resource—A Strategic Review of Desalination Use in Western Australia*. Leederville, WA: Water Corporation.

Water Corporation (2002).[8] *Planning for Perth's Water Needs*. Leederville, WA: Water Corporation.

Water Corporation (2004). *New Solutions for a Changing World: Annual Report 2004*. Leederville, WA: Water Corporation.

Western Australian Water Resources Council (1988). *Water for the 21st Century: Supply Options for the Long Term Water Requirements of Southern Western Australia*. Leederville, WA: Western Australian Water Resources Council. Volume 1: Main report; Volume 2 (1989): Appendices.

Further Reading

A. Groundwater resources of the South-west, Western Australia: publications of the Department of Environment, South West Region, Bunbury, W.A., July 2003:[9]

Fact Sheet 1: Current investigations into groundwater in the South West.
Fact Sheet 2: Investigations into groundwater in the South West – a history.
Fact Sheet 3: The South West Yarragadee aquifer.
Fact Sheet 4: The Leederville aquifer.
Fact Sheet 5: The hydrology of the Blackwood River.
Fact Sheet 6: The 2003 drilling program on the Blackwood Plateau.
Fact Sheet 7: Geophysics investigations.
Fact Sheet 8: Aquifer water quality and salinity.
Fact Sheet 9: Recharge investigations.
Fact Sheet 10: Groundwater level trends in the South West.
Fact Sheet 11: Groundwater flow modelling.

B. Some of the articles and comments published in *The West Australian* regarding the Kimberley Pipeline Scheme's controversies:[10]

—Other states should share pipe, 18 July 1990.
—Governments sink Bridge's pipeline, 20 July 1990.
—Piping water from the north, 18 March 1991.
—Report rules Bridge water plan too dear, 21 March 1991.
—Pipeline of doubt, 18 April 1991.
—Lawrence backs $8b water pipe, 18 March 1992.
—Pipe could aid regions, 19 March 1992.
—Calamity awaits us, 23 March 1992.
—Kimberley water pipe plan a folly, 26 March 1992.
—Geologists on wrong trail, 4 April 1992.
—Ord alternatives are worth study, 8 April 1992.
—Hidden lake may be water saviour, 8 June 1992.
—Big water finds kept from public, 27 June 1992.
—Bridge denies he hid water details, 29 June 1992.
—Discovery to satisfy thirst for expansion, 30 June 1992.
—First a pipeline for us, 2 July 1992.
—Bridge adds to water confusion, 2 July 1992.
—Bridge: I knew all about lake, 3 July 1992.
—Underground water find erodes Kimberley pipeline plan, 4 July 1992.
—Water: debate policies, 7 July 1992.
—Pipeline plan under new fire, 8 July 1982.

---

[8] Also available at: http://www.watercorporation.com.au/publications/12/water_planning_pdf_doc.pdf (visited in July 2005).
[9] Information under this section is kindly provided by Philip Commander, from the Department of Environment, W.A.
[10] Information under this section is kindly provided by Philip Commander, from the Department of Environment, W.A.

10 Other Schemes in Australia

10.1 INTRODUCTION

This chapter covers a number of water transfer schemes in Australia. These are:

(1) inter-basin water transfers from Drainage Division IV to V (see Figure 3.12) for water supply of Adelaide and part of South Australia via pipelines from the River Murray;
(2) inter-basin water transfer in north-eastern Queensland from Drainage Division I to IX for irrigation water supply of the Mareeba–Dimbulah area;
(3) a number of water transfer schemes between (or within) catchments of Drainage Division I in Queensland for domestic and industrial water supply; and
(4) water supply of Broken Hill mines and township in New South Wales, partly by transfer of water from Drainage Division X to IV and mostly by diversion of water from Menindee Lakes within Drainage Division IV.

Numerous other water transfer schemes exist in various parts of Australia, which are not described in detail in this book. However, some of these schemes for water supply of Sydney and Melbourne, and hydro-power generation in Tasmania are described in Appendix H. These are supplemented by basic features of some other schemes.

SECTION A: RIVER MURRAY PIPELINES IN SOUTH AUSTRALIA

10.2 INTRODUCTION

South Australia covers an area of 984 377 km² or about 12.8 percent of the total area of Australia and was settled by Europeans in the 1830s. It is the driest State in Australia, with approximately 80 percent of the State receiving an average annual rainfall of less than 250 mm (see Figure 3.2). The wettest parts of the State are the Mount Lofty Ranges immediately east of Adelaide, the Flinders Ranges, and the southern coast.[1] Adelaide, the State capital has an average annual rainfall of 557 mm, while its annual evaporation rate[2] is approximately 1700 mm. South Australia had a population of 358 350 in 1901 (Shaw, 1984, p. 598), which increased to 1 527 000 by 30 June 2003 including 1 119 300 for Adelaide.[3] Since European settlement, numerous rivers have been dammed to supply the increasing population with adequate water (Table 10.1). However, the small size of river catchments in South Australia, and their low rainfall, have meant that dam capacities are significantly lower than in other states.[4]

The history of water resources development of South Australia is documented in Hammerton (1986). Because development of the State's surface and groundwater resources was not sufficient to satisfy demand, water resources planners had to rely on the water of the River Murray. The average annual flow of the River Murray to South Australia is $6.57 \times 10^9 \, m^3$ per annum (Murray–Darling Basin Commission, 1993). Under the terms of the Murray–Darling Basin Agreement, South Australia is entitled to $1.85 \times 10^9 \, m^3$ annually, except during extremely dry periods when the MDBC implements a period of special accounting (Australian

---

[1] http://www.bom.gov.au/cgi-bin/climate/cgi_bin_scripts/annual_rnfall.cgi (visited in July 2005).
[2] http://www.bom.gov.au/climate/map/evaporation/EVPANN.jpg (visited in July 2005).
[3] http://www.abs.gov.au/ausstats (visited in July 2005).
[4] For example, the capacity of some of the largest reservoirs in Australia are: Gordon (in TAS) $12.5 \times 10^9 \, m^3$, Eucumbene (in NSW) $4.80 \times 10^9 \, m^3$; and Dartmouth (in VIC) $4.0 \times 10^9 \, m^3$. Further information is available at: http://www.ancold.org.au/Table%203.PDF (visited in July 2005).

Table 10.1. *Major dams in South Australia in order of their capacity*

No.	Name of the dam	River[a]	Nearest city	Year	Type[b]	Height (m)	Length (m)	Capacity (10^6 m^3)	Area[c] (ha)
1	Mount Bold	Onkaparinga	Adelaide	1938	VA	58	232	46.18	305
2	South Para	South Para	Adelaide	1958	TE	48	284	45.33	444
3	Myponga	Myponga Creek	Adelaide	1962	VA	52	226	27.17	280
4	Little Para	Little Para	Salisbury	1977	ER	53	225	20.80	125
5	Kangaroo Creek	Torrens	Adelaide	1969	ER	63	131	19.16	103
6	Millbrook	Off-stream	Adelaide	1918	TE	38	288	16.50	178
7	Happy Valley	Off-stream	Adelaide	1896	TE	34	806	14.53	193
8	Tod River	Off-stream	Port Lincoln	1922	TE	32	351	11.30	134
9	Bundaleer	Off-stream	Port Pirie	1903	TE	38	334	6.37	8
10	Baroota	Baroota Creek	Port Pirie	1921	TE	37	301	6.14	63
11	Warren	South Para	Adelaide	1916	PG	26	150	4.79	100
12	Barossa	Off-stream	Adelaide	1902	VA	36	144	4.52	6
13	Hope Valley	Off-stream	Adelaide	1872	TE	22	765	3.63	?
14	Beetaloo	Crystal Brook	Port Pirie	1890	PG	37	180	3.18	33

Note: [a] Off-stream if the dam is not built on a watercourse; [b] TE (earthfill embankment), VA (concrete arch), ER (rockfill embankment), and PG (concrete gravity); [c] Reservoir area.
Source: Australian National Committee on Large Dams: National Register of Large Dams.[5]

Bureau of Statistics, 1999, p. 191). A major proportion of this allocation is used for irrigation of vineyards and orchards. Irrigation areas in South Australia are confined almost exclusively to the Murray Valley.

Currently, South Australia relies heavily on the River Murray to provide water for Adelaide and regional areas extending as far north as Woomera and as far south as Keith (Figure 10.1). Salinity of the River Murray water at Morgan is crucial for water supply of Adelaide and other population centres in South Australia. Its monthly average oscillated between 150–900 mg L^{-1} TDS over the period of 1940 to 1985 (Close, 1990). The objective is to maintain it below 500 mg L^{-1} TDS in order to be publicly acceptable. This target was exceeded 10 percent of the time between 1990 and 2000 and is estimated that it will exceed for 50 percent of the time by 2020 without intervention. Implementation of a Basinwide Salinity Management Strategy is expected to maintain the salinity at Morgan below 500 mg L^{-1} TDS 95 percent of the time.[6]

Water usage varies according to seasonal conditions, with the River supplying approximately 40 percent of demand in cool and wet conditions (as was the case in 2001–02) and up to 90 percent in drought years (SA Water, 2002). Water is supplied via six pipelines. Table 10.2 shows the main features of these pipelines. It indicates that the pipelines have a total daily capacity of about 1.2×10^6 m^3.

The following sections provide some information about the River Murray pipelines.

10.3 MORGAN–WHYALLA PIPELINES

By 1937, attention was focused on the development of Whyalla (Figure 10.1), which was being considered for the large-scale development of various industries. During the early years, Whyalla was supplied from local reservoirs and by water carting. Investigations indicated that there was little chance of securing significant supplies from the catchments in the region. This led to a scheme to pipe water from the River Murray to satisfy the increasing demand. The scheme was designed with a capacity of 9.55×10^6 m^3 per annum, consisting of 5.5×10^6 m^3 for Whyalla, and the balance for Port Pirie, Port Augusta and the rest of the northern area (Engineering and Water Supply Department, 1987, p. 10). Construction of the pipeline commenced in 1940 and was completed in 1944 at a cost of $5 million ($145 million in 2002 prices).

The pipeline follows a route from Morgan on the River Murray, through Port Augusta to Whyalla. The pipe diameter varies from 750 mm at Morgan to 525 mm at Whyalla and is laid above ground on concrete supports. Water is pumped through four electrical pumping stations over a distance of 92 km with a height lift of 475 m to a summit storage. From the summit storage water gravitates to Whyalla (SA Water, 1991). A branch pipeline of 177 km long was opened

---

[5] http://www.ancold.org.au/dam_register.html (visited in July 2005).
[6] http://www.dwlbc.sa.gov.au/murray/salinity/mdb_salinity.html (visited in July 2005).

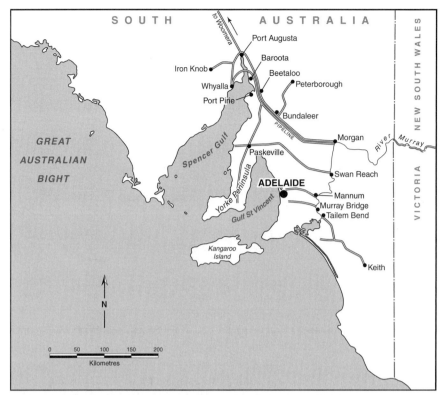

Figure 10.1 Pipelines serving metropolitan Adelaide and other areas of South Australia with River Murray water (Government of South Australia, 1989, Figure 1).

Table 10.2. *Main features of the River Murray pipelines for water supply of South Australia*

No.	Name of the pipeline	Date of completion	Length (km)	Diameter (mm)	Capacity ($m^3 d^{-1}$)
1 and 2	Morgan–Whyalla pipelines	1st: 1944	356	750 to 525	206 000
		2nd: 1966	281	1050 to 825	
3	Mannum–Adelaide	1954	60	1450 to 1150	380 000
4	Swan Reach–Paskeville	1969	183	900	80 000
5	Tailem Bend–Keith	1969	133	750 to 525	31 000
6	Murray Bridge–Onkaparinga	1973	48	1650 to 1350	514 000
Total	–	–	932	–	1 211 000

Source: SA Water (2002, p. 11) and other SA Water publications.

from Port Augusta to Woomera in 1949 (Crabb, 1968). Other branches serve Iron Knob, Peterborough and numerous other country towns and farming areas.

The rapid development that occurred in Whyalla, Port Augusta and other northern areas soon after World War II, indicated that future needs for water would exceed the capacity of the existing pipeline and it was decided to duplicate the pipeline. In the meantime, to meet demand until the duplicate pipeline was built, larger pumps were installed in the four pumping stations along the first pipeline. Construction of the second pipeline commenced in early 1962 and was completed in 1966. Its diameter ranges from 1050 mm at Morgan to 825 mm at Whyalla (SA Water, 1991). For most of its length, it follows the course of the first pipeline. However, 14 km of the pipeline was laid across the bed of Spencer Gulf, reducing the length of the pipeline by more than 70 km. The laying of the 825 mm submarine section of the pipeline was complex. Prior to installation, the 9 m pipe sections were assembled and welded into strings of 738 m each and one 387 m. They were then laid in a submarine trench varying in depth from 1.2 m to 9.1 m. Preparation for laying the pipeline across the Gulf took one year while actual laying the pipeline took only 8 days. As with the first pipeline, four pumping stations are used to lift water to a summit reservoir located at a distance of 92 km from Morgan. Then, water gravitates the remainder of the distance to Whyalla (SA Water, 1991). The Morgan–Whyalla pipelines are also linked to the Yorke Peninsula pipeline. A filtration plant at Morgan treats the River Murray water before entry into the pipelines.

10.4 MANNUM–ADELAIDE PIPELINE

The Mannum–Adelaide pipeline (Figure 10.1) was the first means of supplying River Murray water to metropolitan Adelaide. Although use of the Murray as a source of water for Adelaide was considered for many years, it was not practical until the barrages on the River mouth were constructed to prevent saline water intrusion in the lower reaches of the River. Construction of the Mannum–Adelaide pipeline was completed in 1954. As a result, Adelaide was the only Australian capital city to escape enforced water restrictions during the very dry 1954–55 summer. Three pumping stations (Table 10.3) lift the water to a summit storage. From this storage, water gravitates for about 30 km to a 136 000 m^3 terminal storage. The pipeline ranges in diameter from 1450 mm to 1150 mm and is laid above ground on concrete and steel supports. The River Murray water was often of very poor quality because of its turbidity. This problem was rectified with the commissioning of a filtration plant on 29 January 1980 at a cost of $14.5 million (SA Water 1993), which is about $42 million at 2002 prices.

10.5 SWAN REACH–PASKEVILLE PIPELINE

The main purpose of the Swan Reach–Paskeville pipeline (Figure 10.1) is to supplement supplies to the Barossa Valley,

Table 10.3. *Pumping stations of the Mannum–Adelaide pipeline*

Pumping station no.	Height lift (m)	Distance (km)
1	139	15
2	139	3
3	139	10

Source: (SA Water 1993).

Lower North and Yorke Peninsula areas.[7] This 900 mm diameter pipeline was commissioned in 1969. A water treatment plant at Swan Reach treats the water for its turbidity before being pumped into the pipeline.

10.6 TAILEM BEND–KEITH PIPELINE

The Tailem Bend–Keith pipeline (Figure 10.1) supplies water to 13 country towns and a large agricultural area.[8] The main trunk ranges in diameter from 750 mm to 525 mm. It feeds more than 800 km of branch main covering an area of 6470 km^2. Construction of the pipeline began in 1964 and was completed in 1969 with financial aid from the National Water Resources Development Program (Hammerton, 1986, p. 263). A water treatment plant at Tailem Bend treats the water before it is pumped into the pipeline.

10.7 MURRAY BRIDGE–ONKAPARINGA PIPELINE

During the late 1950s and early 1960s it became evident that a second pipeline would be needed for water supply of the Adelaide metropolitan area (SA Water, 1992). Studies were undertaken to identify the most suitable route and the pipeline's capacity. The chosen route extends from a point about 3 km north of Murray Bridge to the Onkaparinga River (Figure 10.2). Three pumping stations lift the water 418 m to the summit reservoir located at a distance of 38 km from the River Murray. Then, water discharges by gravity into the Onkaparinga River to supplement its natural flow. The river channel is used as a means of conveying water to the Mount Bold Reservoir. The pipeline is 48 km long, of which 23 km is laid below ground. It has a diameter of 1650 mm between the River Murray and Summit Reservoir and 1350 mm between Summit Reservoir and the Onkaparinga River, and its daily capacity is 514 000 m^3. Construction of this main source of water supply for Adelaide commenced in 1968 and the pipeline was officially commissioned in 1973 at an approximate cost of $23 million (about $147 million in 2002 prices).

---

[7] http://www.sawater.com.au/SAWater/Education/OurWaterSystems/Pipelines.htm (visited in July 2005).
[8] http://www.sawater.com.au/SAWater/Education/OurWaterSystems/Pipelines.htm (visited in July 2005).

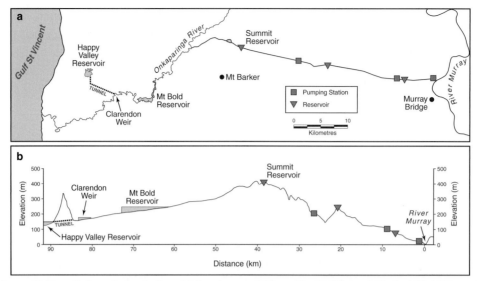

Figure 10.2 The Murray Bridge–Onkaparinga pipeline: (a) location; and (b) profile (Griffin and McCaskill, 1986, p. 60).

SECTION B: MAREEBA–DIMBULAH IRRIGATION SCHEME, QUEENSLAND

10.8 INTRODUCTION

The Mareeba Shire is located on the Atherton Tablelands in Queensland, south-west of Cairns. In 1996, it had a population of 18 000. The town of Mareeba is the administrative centre of the Shire with a 1996 population of 6873, while Dimbulah had a population of 431, and Kuranda 666 (Department of Natural Resources and Mines, 2002a, p. 29).

10.9 HISTORY OF THE SCHEME

A report presented to the Queensland Parliament in 1952 proposed a water conservation and irrigation scheme for the Mareeba–Dimbulah area (Figure 10.3).[9] The approved scheme included the following capital works:

(1) the construction of a concrete dam on the Barron River in the vicinity of Tinaroo Falls on the eastern side of the Great Dividing Range;
(2) provision of a supplementary weir known as Collins Weir on the Walsh River, west of the Great Dividing Range; and
(3) construction of main and distribution canals to convey water from the storages to various sections of the project (Commonwealth Bureau of Census and Statistics, 1960, p. 136).

The scheme envisaged 1100 irrigation farms with about 20 200 ha that could be irrigated to produce tobacco, mixed agricultural crops, and pasture (Commonwealth Bureau of Census and Statistics, 1972, p. 182). The soils (basaltic clay loams, sand and sandy loams of granitic origin, alluvial, heavy clays and others) and climate (warm summers and mild frost free winters) of the area were considered highly suitable for production of a wide range of crops.

The Tinaroo Falls Dam and the Mareeba–Dimbulah Irrigation Scheme was the first major water resources development in the Australian tropics. Tinaroo Falls Dam[10] was designed with a capacity of $407 \times 10^6 \, m^3$. Construction of the Dam started in 1953, the first water from the dam was released in 1958, and the storage first filled on 31 March 1963. Some of the technical details of the Dam are shown in Table 10.4.[11] Water is stored in the reservoir during the annual wet season (December, January, February and March), and is then available during the dry months for irrigation, power generation and urban uses.

In this Scheme, the Barron River waters (Drainage Division I), which naturally flow eastward into the Coral Sea, are diverted by gravity in an open channel across the Great Dividing Range to the Walsh/Mitchell River catchments (Drainage Division IX), which flow northward into the Gulf of Carpentaria.

---

[9] This scheme is currently known as the Mareeba–Dimbulah Water Supply Scheme (MDWSS).
[10] http://www.tinarooeec.qld.edu.au/Kids/tindam.html (visited in July 2005).
[11] Further technical and operational details of the Tinaroo Falls Dam, weirs and balancing storages are available in the Department of Natural Resources (2000).

OTHER SCHEMES IN AUSTRALIA

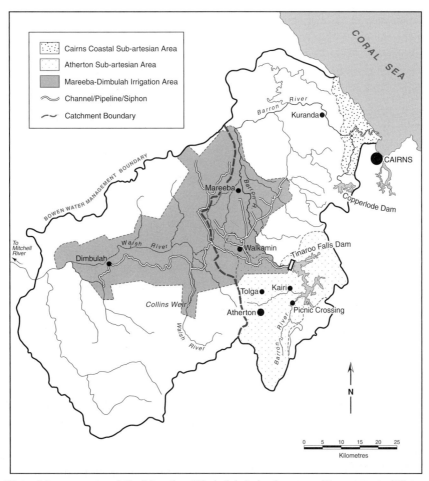

Figure 10.3 The Barron Water Management and the Mareeba–Dimbulah Irrigation areas (Department of Natural Resources and Mines, 2001a, Figure 1; and Gutteridge Haskins & Davey, 2001, Figure 2.1).

Table 10.4. *Some technical specifications of the Tinaroo Falls Dam*

Item	Unit
Period of construction	1953–58
Cost in 1958 prices	$12.6 million[a]
Catchment area	545 km^2
Reservoir area	3360 ha
Average annual rainfall	1300 mm
Storage capacity	407×10^6 m^3
Full supply level	670 m
Type of structure	Concrete gravity
Maximum base width	35.4 m
Total crest length	533.4 m
Height	45.1 m
Spillway dimensions	76.2 m × 3.7 m
Irrigation outlets	2 × 1500 mm
Barron River outlet	1 × 1500 mm

Note: [a] $133 million in 2002 prices.
Source: Tinaroo Environmental Education Centre.[12]

Construction of the channels that serve the irrigation area began in 1955 and was essentially completed by 1970.[13] The net cost of the Scheme up to 30 June 1971 was $32.7 million (about $243 million in 2002 prices), comprising of $12.5 million on the Tinaroo Falls Dam and $20.2 million on irrigation and other works (Commonwealth Bureau of Census and Statistics, 1972, p. 182).

The Mareeba–Dimbulah Irrigation Area (MDIA), with an elevation of 400 m to 600 m above mean sea level, spreads across the valleys of the Barron and Walsh rivers covering an area of about 820 km^2 (Gutteridge Haskins & Davey, 2001, p. 7). Farms are supplied with water from the Tinaroo Falls Dam in addition to the natural flows of local streams in the area. Water through the area is distributed by 195 km of main canals and pipelines consisting of 60.2 km unlined

---

[12] http://www.tinarooeec.qld.edu.au/Kids/tindam.html (visited in July 2005).
[13] http://www.tinarooeec.qld.edu.au/Kids/tindam.html (visited in July 2005).

earth canal, 59.5 km of pipeline, 52.6 km of concrete lined canal, 14.5 km of bench flume, 7.9 km of clay lined canal and 0.2 km of rock-cut canal (Gutteridge Haskins & Davey, 2001, Table 2.6). The capacity of the main canal leaving Tinaroo Falls Dam is $19.8\,m^3\,s^{-1}$ or $1.7 \times 10^6\,m^3\,d^{-1}$. A number of balancing storages ensure that supply levels are maintained throughout the system. Detailed plans and diagrams of the MDIA are available in Gutteridge Haskins & Davey (2001, pp. 10 and 11).

10.10 AGRICULTURAL DEVELOPMENT

Tobacco was the main crop produced on the irrigated farms, and about 3860 ha was planted on 52 farms in 1970–71. Of these plantings, 1940 ha were irrigated from the canal systems, 1890 ha by private pumping from regulated streams and the remaining 30 ha from unregulated streams (Commonwealth Bureau of Census and Statistics, 1972, p. 182).

The area of irrigated crop increased from approximately 5000 ha in 1979–80 to about 12 500 ha in 1983–84 and 22 000 ha in 1999–2000. The value of production was less than $25 million in 1979–80 ($84 million in 2002 prices), and increased to $115 million in 1998–99 (Department of Natural Resources and Mines, 2002a, p. 32) which is about 130 million in 2002 prices. Sprinklers are used to irrigate tobacco and vegetables. Trickle and mini-sprinkler systems are used in large orchards, while pasture is watered by conventional flood irrigation methods.[14]

With the downturn in the tobacco industry, sugarcane is now grown in large quantities. In 1992, the gross income for the Queensland tobacco industry was $46 million, and this dropped to $14 million in 2001 (Department of Natural Resources and Mines, 2002a, p. 41). Currently, sugarcane, tobacco, maize, sorghum, pumpkin, vegetables and others are grown in the area. Plantations of avocado, macadamia, mango, citrus and other fruits have also been established. The area is also recognised as the most suitable in Australia for coffee growing. Table 10.5 shows estimates of the area of irrigated crops within the MDIA.

10.11 WATER ALLOCATION AND WATER USE

The annual water requirement of the crops varies greatly. For example, tobacco requires approximately $4000\,m^3\,ha^{-1}$, while sugarcane requires 10 000 to $14\,100\,m^3\,ha^{-1}$ (Gutteridge Haskins & Davey, 2001, p. 1). The average irrigation application over the period of 1992–2002 was approximately $4.9 \times 10^3\,m^3\,ha^{-1}\,yr^{-1}$ (Department of Natural Resources and Mines, 2002a, p. 35). Approximately $161 \times 10^6\,m^3\,yr^{-1}$ is allocated for consumptive purposes (agriculture, urban and industrial) from Tinaroo Falls Dam. The water supply for the Scheme, although fully allocated, continues to be less than fully utilised. The maximum historical consumptive use of water has not exceeded 64 percent of the available allocation. In 1999–2000, the total allocation was $155 \times 10^6\,m^3$, while, the water use was about $75 \times 10^6\,m^3$ (Department of Natural Resources and Mines, 2002a, pp. 34 and 35).

Gutteridge Haskins & Davey (2001) investigated the efficiency of the irrigation water delivery system of the MDIA. They found that the average operational efficiencies in different sub-systems of the MDIA were ranging from 56 to 95 percent. Therefore some level of improvement is possible in lower efficiency sub-systems.

10.12 POWER GENERATION AND TOWN WATER SUPPLY

Water from the Tinaroo Falls Dam is also released into the Barron River to stabilise flow and provide an assured water supply to the Barron Gorge Hydro-electric Power Station at Kuranda (Figure 10.3). Engineering investigations into use of the river water for electricity generation began in 1906, but it was 1935 when the waters of the Barron River drove the first turbine. The small Barron Falls hydro-electric station of 3.8 MW capacity was commissioned in that year, and was the first underground power station in Australia. Before its closure in 1963, it supplied the Cairns area with electricity for 28 years. Construction of the present Barron Gorge Hydro-Electric Power Station began in 1960.[15] It was commissioned in September 1963 with a capacity of 60 MW, consisting of two units of 30 MW each.[16] An average of $270 \times 10^6\,m^3$ of water per year has been used by the power station, all of which is returned to the Barron River (Department of Natural Resources and Mines, 2002a, p. 35). The station is now a vital unit in the statewide generation and transmission of electricity. In addition, water is supplied to the towns of Mareeba, Dimbulah, Walkamin, and others.

---

[14] http://www.tinarooeec.qld.edu.au/Kids/tindam.html (visited in July 2005).

[15] http://www.tinarooeec.qld.edu.au/Kids/tindam.html (visited in July 2005).

[16] http://www.stanwell.com/frame.asp?ContentURL=/sites/locations.asp (visited in July 2005).

Table 10.5. *Area of irrigated crops within the Mareeba–Dimbulah Irrigation Area for 2001–02*

No.	Crops and their areas (ha)	Area (ha)
1	**Sugarcane**	10 000
2	**Tree crops:** Mango (2620 ha), avocado (1101 ha), macadamia (319 ha), cashews (220 ha), lychee (215 ha), longan (138 ha), citrus (72 ha), custard apple (24 ha), and peach (16 ha)	4725
3	**Pasture and seed:** Pasture bales (1499 ha), dollicos (260 ha), grass seed (230 ha), legume seed (96 ha), verano (40 ha), and forage sorghum (25 ha)	2150
4	**Crops:** Maize (451 ha), sorghum grain (115 ha), navy bean (105 ha), and cow pea (30 ha)	701
5	**Fruits and vegetables:** Paw-paw (685 ha), banana (625 ha), pumpkin (216 ha), pineapple (200 ha), grapes (106 ha), mixed vegetables (104 ha), potatoes (99 ha), sweet potato (17 ha), watermelon (11 ha), onions (6 ha), lettuce (5 ha), passion fruit (3 ha), and tomato (3 ha)	2080
6	**Others:** Tobacco (600 ha), coffee (174 ha), tea-tree (60 ha), native trees (41 ha), flowers (33 ha), turf (24 ha), and herbs (17 ha)	949
Total		20 605

Source: Department of Natural Resources and Mines (2002a, Table 3.3).

10.13 WATER QUALITY ISSUES OF THE TINAROO FALLS LAKE

Water quality is adversely affected if concentrations of phosphorous, nitrogen and chlorophyll rise to critical levels. Preliminary analysis of water quality data for the Tinaroo Falls Lake and its catchment has shown:[17]

- Significant occasional increases in nitrogen and phosphorus concentration exceeding current guidelines.
- Substantial increases in suspended sediments.
- Increases in nutrients and sediments following heavy rains.
- Excessive algal growth within the lake.
- Substantially reduced dissolved oxygen.

10.14 BARRON WATER RESOURCES PLAN

The *Water Act 2000* (State of Queensland, 2000), stipulated that only a single Water Resources Plan can apply to any specific area in Queensland. The process to prepare the Barron Water Allocation and Management Plan for surface water, and Atherton Sub-artesian Area Water Management Plan for groundwater began in April 1996 and November 1998 respectively. These were amalgamated for preparing the Barron Water Resources Plan. Following extensive investigations and consultations, numerous publications on the environmental, surface water, groundwater, social, indigenous, economic assessment and other issues, were produced (for references see the Department of Natural Resources and Mines, 2003, p. 45).

Some of the elements of the Water Resources Plan regarding surface water resources are (Department of Natural Resources and Mines, 2003):

(1) existing entitlements in the Mareeba–Dimbulah Irrigation Scheme to be fully used;
(2) conversion of area irrigation licences to volumetric licences at the rates of up to 6600 m^3 per year per hectare;
(3) in the Mareeba–Dimbulah Irrigation Scheme, existing irrigation entitlements will have a 90–95 percent monthly reliability and 75–80 percent annual reliability;
(4) continuation of arrangement for managing releases from Tinaroo Falls Dam for hydro-power generation; and
(5) if required an additional 4×10^6 m^3 of unallocated water to be made available from the lower Barron River to meet Cairns City's future water needs.

For groundwater, there are two proclaimed areas within the Barron Water Management area (Figure 10.3). These are the Atherton and the Cairns Coastal Sub-artesian areas located in the upper and lower reaches of the Barron catchment respectively (Department of Natural Resources and Mines, 2001a).

The Atherton Sub-artesian Area covers approximately 62 000 ha. Groundwater is used for irrigation, town water supply, industry, domestic purposes and aquaculture. The Atherton Basalt is the most extensively used aquifer in the area. Pumping rates are in the range of 15–20 L s^{-1} with maximum extraction rates of up to 40 L s^{-1}. Water supplies are typically of very good quality. The groundwater EC ranges from 45 to 350 µS cm^{-1}. In 1999, the estimated

---

[17] http://www.nrm.qld.gov.au/water/pdf/tinaroo_water.pdf (visited in July 2005).

annual groundwater use was 12×10^6 m^3, while the groundwater allocation was about 17×10^6 m^3. Preliminary investigations indicated that the groundwater system appears to have an annual sustainable yield of 14.5×10^6 m^3.

Cairns Coastal Sub-artesian Area consists of alluvial aquifers. Extraction rates are generally low and water is used for stock, and domestic water supply as well as irrigation of sugarcane and golf courses. Groundwater quality is generally good except near coastal areas where saltwater intrusion may be encountered. The 1999 licensed extraction volume was about 3.2×10^6 m^3.

For groundwater, the Barron Water Resources Plan (Department of Natural Resources and Mines, 2003, p. 42) provides:

(1) that no further entitlement be issued for the highly developed parts of the Atherton Sub-artesian Area; and
(2) the existing entitlements be maintained, except for those areas where licenced volumes exceed supply capacity.

10.15 POSSIBILITIES FOR FUTURE EXPANSION

Soil and crop suitability investigations indicated that potentially there is more than of 50 000 ha of soil suitable for irrigated cropping within the area, including the current cropped area of 21 000 ha. This suggests an additional 29 000 ha is suitable for irrigation. However, the best lands served by the existing canals and pipelines have already been developed (Department of Natural Resources and Mines, 2002a, pp. 34 and 35).

10.16 IMPACTS OF WATER RESOURCES DEVELOPMENT

Yu (2000) describes the flow regime changes of the Barron River for a 30-year period of pre-construction (July 1927 to June 1956) and post-impoundment (July 1960 to June 1990) of the Tinaroo Falls Dam at the Picnic Crossing and Mareeba stations, located upstream and downstream of the Dam respectively (Figure 10.3). Analysis of the maximum instantaneous discharges at Mareeba indicated that there has been a 13 percent decrease in the mean annual flood discharge since impoundment. Flow regulation of the Barron River via Tinaroo Falls Dam produced little or no channel change immediately downstream of the Dam because of the bedrock control and a general lack of sediment deposition within the first 20 km downstream of the Dam. However, a decrease in channel width and cross-sectional area had occurred far downstream from the Dam. For example, channel width at Mareeba reduced by 27 percent. This has occurred because, during prolonged periods of low flows, sand/gravel bars are stabilised by riparian vegetation encroachment.

The implications for three cases of future water resources management scenarios are (Department of Natural Resources and Mines, 2001b):

- **Scenario A** or **full utilisation** assumes that all existing entitlements are operated to their full in all years, compared with existing levels of use that are less than 70 percent.
- **Scenario B** or **existing applications** includes the water usage assumed in Scenario A, plus all outstanding licence applications for additional water such as Cairns town water supply from the Barron River.
- **Scenario C** or **potential development** includes the water usage assumed in the Scenarios A and B, plus a number of potential developments.

Under Scenarios A and B there is no need for any new major infrastructure, while under Scenario C provision is made for a new dam on the Walsh River. In each case, the hydrological, geomorphological and ecological implications were assessed. Details of these assessments are available at the Department of Natural Resources and Mines (2001b).

SECTION C: DOMESTIC AND INDUSTRIAL WATER SUPPLY IN NORTH QUEENSLAND

This section describes a number of water transfer schemes between (or within) catchments of Drainage Division I (Figure 3.12) in Queensland for domestic and industrial water supply. Although these schemes are not large-scale inter-basin water transfer schemes, they illustrate the importance of such transfers for mining and town water supply.

10.17 WATER SUPPLY FROM EUNGELLA DAM

Eungella Dam, located in the Bowen/Broken sub-catchments of the Burdekin River catchment, supplies water for industrial, urban and irrigation use. Constructed in 1969, the dam is 38.5 m high and 276 m long, has a capacity of 112.5×10^6 m^3 and a surface area of 848 ha. Originally it supplemented the irrigation water supply in the lower Burdekin River catchment, but currently supplies water for several townships and is used in conjunction with the Bowen River Weir. Water from the weir is pumped via a pipeline to supply a coal mine, power station and the Collinsville township. A separate pipeline feeds mining operations at Newlands and the township of Glenden. Two other pipelines from the Eungella Dam supply urban and mining activities at Goonyella and Moranbah. One pipeline belongs to BHP Billiton Mitsubishi Alliance (BMA) and the other to the Eungella Water Pipeline Company, which is a subsidiary of SunWater[18] (Department of Natural Resources and Mines, 2002b, p. 53). Irrigation water is also provided to riparian properties of the Bowen River to the junction with the Burdekin River. Some details of the above pipelines are in Table 10.6, while their locations are shown in Figure 10.4.

Table 10.6. *Some characteristics of the pipelines in the Burdekin River and coastal catchments*

Name	Year of construction	Capacity ($10^6 \, m^3 \, yr^{-1}$)	Length (km)	Diameter (mm)
Collinsville pipeline	1968	5.71	28	525–610
Newlands pipeline	1984	3.94	91	375–450
Eungella pipeline (BMA)	1972	5.8–6.2	157	400–550
Eungella Water Pipeline Pty Ltd.	–	15 to Goonyella 10 to Moranbah	123	711
Paluma pipeline	1958	14.5	69	475–600
Haughton pipeline	1987	10	35	900

10.18 PIPELINES FOR WATER SUPPLY OF TOWNSVILLE AND THURINGOWA

Water is supplied to Townsville and Thuringowa from the Ross River Dam[19] in the south of Townsville and from the Paluma Dam[20] in the upper Burdekin River catchment (Figure 10.4). These dams and their associated pipelines are managed by NQ Water[21] (formerly the Townsville Thuringowa Water Supply Board). Water is also supplied to the region by supplementing the Ross River Dam storage via the NQ Water Haughton Pipeline (Department of Natural Resources and Mines, 2002b, p. 53). Some details of these two pipelines are shown in Table 10.6. In 2003, NQ Water supplied $53.30 \times 10^6 \, m^3$ of water to its customers consisting of $34.46 \times 10^6 \, m^3$ to Townsville City, $15.07 \times 10^6 \, m^3$ to Thuringowa City, $2.06 \times 10^6 \, m^3$ to QLD Nickel and $1.71 \times 10^6 \, m^3$ to Sun Metals (NQ Water, 2003, p. 19).

10.19 PIPELINE TO BOWEN AREA

Water is imported to the Bowen area via a pipeline from the Proserpine River (Figure 10.4). The Peter Faust Dam[22] is located on the upper reaches of the Proserpine River and water is released from the dam to supply demands in the river system. The Bowen Shire Council's Proserpine River pumping station is located on the banks of the Proserpine River, approximately 4 km downstream of the dam. The pumping station has a gallery of three screens in the bed sands, and water for Bowen is drawn from this point. The water is pumped to the Bowen area and is used to supply residential, commercial and industrial requirements. The pipeline is 52 km long with a nominal diameter of 450 mm. Its construction was completed in 1991 and currently has a capacity of $180 \, L \, s^{-1}$ ($5.7 \times 10^6 \, m^3 \, yr^{-1}$). With the installation of booster pumps, its capacity may reach $360 \, L \, s^{-1}$ ($11.4 \times 10^6 \, m^3 \, yr^{-1}$). The Council has also separately committed to supply water to some agricultural users in the Bowen area in order to supplement their groundwater irrigation allocations.[23]

SECTION D: WATER SUPPLY TO THE BROKEN HILL MINES AND TOWNSHIP, NEW SOUTH WALES

10.20 INTRODUCTION

The Broken Hill area was discovered and named by the explorer Charles Sturt in 1844. It is located in the arid western region of NSW and has an average annual rainfall of about 253 mm. Its average daily maximum temperature ranges from 32.7°C in January to 15.1°C in July, while the

---

[18] http://www.sunwater.com.au (visited in July 2005).
[19] This dam, which was constructed in 1974, is 33 m high and 8670 m long, with a capacity of $417 \times 10^6 \, m^3$ and surface area of 8200 ha.
[20] This dam, which was constructed in 1959, is 20 m high and 318 m long, with a capacity of $12.34 \times 10^6 \, m^3$ and surface area of 800 ha.
[21] http://www.nqwater.com.au (visited in July 205).
[22] The Dam system is operated by SunWater.
[23] Gary Martin, Bowen Shire Council, personal communication (February 2004).

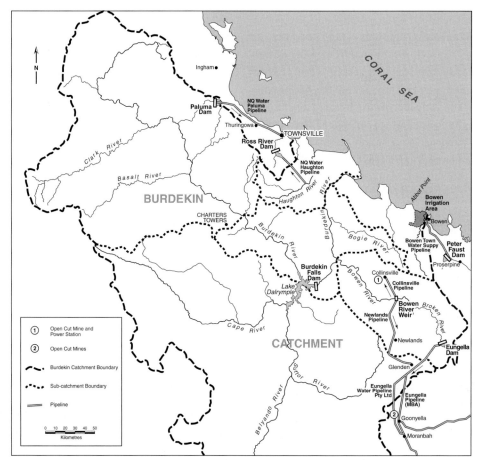

Figure 10.4 Locations of the water transfer pipelines in the Burdekin River and coastal catchments (Department of Natural Resources and Mines, 2002b, Figure 4.2).

average daily minimum temperature varies from 18.4°C in January to 5.3°C in July. Summer conditions are hot with an average of 35 days above 35°C and 6 days above 40°C (Australian Bureau of Meteorology website).[24]

Information on the history of the Broken Hill mining industry, which mined one of the world's richest deposits of lead, zinc and silver, is available in numerous publications including: Curtis (1908), Blainey (1968), Koenig (1983), and Solomon (1988).[25] The following is a brief description of its discovery, mainly from Koenig (1983).

Charles Rasp, accompanied by James Pool and David James, discovered silver in outcrops of the area in September 1883. Rasp registered his claim with the deputy-mining Registrar. His claim became the first mining lease in Broken Hill. A drought during the summer of 1883 made mining impossible, and it was not until October 1884 that a contract was let for the sinking of Rasp's shaft from which high grade silver ore was extracted. In 1885, another shaft, known as Knox's shaft, was sunk and a high grade deposit of lead, zinc and silver ore was discovered which ensured the future of Broken Hill. On 3 June 1885, a "Syndicate of Seven"[26] made

a decision to register the Broken Hill Proprietary Company Limited (BHP) with a capital of £320 000 ($29 million in 2002 prices). The Company was floated on 12 August 1885. Within a few years, the Broken Hill area became a booming mining town. Explorations indicated that the ore body was a massive arc shaped sulphide deposit more than 7 km long, outcropping at the centre with a maximum width of 90 m and plunging to a depth of about 1.5 km at its northern end (Figure 10.5).

Numerous mining companies made Broken Hill their home and struck their fortunes.[27] By October 1966, Broken Hill had produced 13.2 million tonnes of lead, 9.4 million tonnes of zinc and 693.4 million ounces (19 658 tonnes) of silver, with a value of $1.34 billion (Solomon, 1988, p. 52), which is about $12 billion in 2002 prices. The BHP Company

---

[24] http://www.bom.gov.au/climate/averages/tables/cw_047007.shtml (visited in July 2005).
[25] Also visit: http://au.geocities.com/bhsilvercity (visited in July 2005).
[26] Charles Rasp, George McCulloch, George Urquhart, George A. M. Lind, Philip Charley, David James and James Poole.
[27] http://www.visitbrokenhill.com.au/?id=history (visited in July 2005).

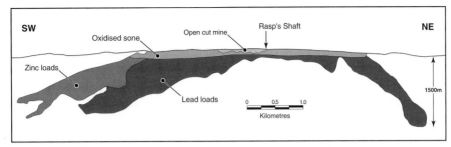

Figure 10.5 Cross-section of the Broken Hill ore body (Koenig, 1983, p. 6).

left Broken Hill in 1939. Towards the end of the twentieth century more companies gradually left, and by 2002 only one mine was operating which accounted for only 12 percent of the employment in Broken Hill (Moore et al., 2002, p. 25).

Broken Hill reached a maximum population of 35 000 in 1915. This declined to about 26 000 in 1921, increased to 33 000 in 1952, then declined to 27 000 in 1981 and 24 000 in 1982 (Solomon, 1988, p. 388). At 30 June 2001 it had a population of 21 100.

10.21 WATER SUPPLY

A short history of the Broken Hill water supply is described by Koenig (1983, pp. 27 and 28) and Solomon (1988, pp. 29–41). The following description is based on Hardy (1968), who provided a remarkably comprehensive history of the Broken Hill water supply.

10.21.1 Water Supply during the Early Years

From the early days of Broken Hill, supplying water proved to be a problem for mining activities as well as for the town population. The first attempt to organise a supply for the new mining centre was through the reservation of the Stephens Creek soaks. Here, a limited quantity of water could be obtained by sinking shallow wells in the sandy soils. This water was then transported to Broken Hill in 400 gallons[28] (1820 litres) square iron tanks on bullock drays. Water was sold to householders for from 2 to 4 pence per 4 gallon (18 litres) bucket, equivalent to $0.20–$0.40 per litre in 2002 prices. A public meeting in March 1886, urged the NSW Government to proceed with the construction of the White Leeds Dam (Figure 10.6), with no immediate result. By November 1887, the problem became serious and the dam was eventually completed in January 1888. Construction of the Imperial Dam (Figure 10.6) was completed about the middle of 1888, and these two reservoirs, together with the shallow wells at Stephens Creek, other Government wells, and some private storages, formed the main sources of water supply over the next years.

The year 1888 was very dry, with only 9 mm of rain (5 mm of this occurred in early February). Thirst was compounded by a typhoid epidemic and the great mining boom. Many came to Broken Hill by the newly constructed railway from South Australia and by the privately owned Silverton Tramway. New arrivals further reduced the amount of available water per head of population. As a relief measure, the State Government deepened its wells on the eastern side of the town which provided an additional 13 500 litres of water per day. To alleviate the problem, importation of water was suggested using the Silverton Tramway from the Government tank at Silverton. This was rejected because of the pressure that this would put on the water supply of Silverton. Another proposal was to rail water from the Rat Hole Tank located west of Silverton (Figure 10.6). This proposal also was rejected because the NSW Mines Department, which had the control of the tank, refused permission. Residents strongly protested as the conditions became intolerable.

10.21.2 Private Water Schemes

In the 1888 drought, parallel with persistent agitation for Government action, the private sector became interested in water supply of Broken Hill. In April 1888, the Broken Hill Water Supply Syndicate started to prepare plans and specifications for a weir on Stephens Creek. Water would be pumped through a 500 mm pipeline to a reservoir to be constructed in the town. A Bill was to be introduced in State Parliament to sanction the scheme. Three months later, surveys and levelling for reticulating the town were completed. However, rival interests blocked the company's Bill and work was stopped. In August 1888, the company was reformed under the name of Barrier Ranges and Broken Hill Water Supply Syndicate. The Syndicate developed a new and extended proposal to provide storage facilities on the

---

[28] One gallon is equal to 4.55 litres.

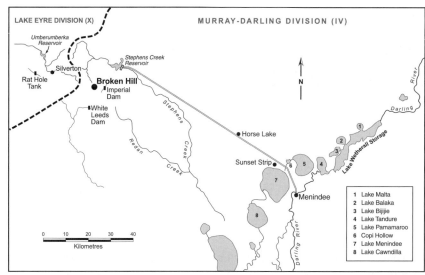

Figure 10.6 Broken Hill water supply systems (Broken Hill and Adelaide topographic map sheets at the scale on 1:1 000 000; and Maiden, 1989).

Stephens Creek with a capacity of $6 \times 10^6 \, m^3$, sufficient for a population of 20 000. By the end of the year, the Syndicate had a Bill before the NSW Parliament.

The Nolan's Stephens Creek Water Supply Company, locally known as the Nolan Brothers Scheme or Nolan and Lloyd's was floated in 1888. Its objective was to supply water from permanent waterholes several kilometres further downstream on Stephens Creek's proposed site of the Syndicates weir. By October 1888 a section of the plan was completed. Although the water was crystal clear, it could not be used for domestic purposes because it was brackish, but it was fit for stock water. Later, they installed a pump and pipeline from the Imperial Dam to their town reservoir, supplying people with more potable water. This water was bought from the Government and delivered by their own corps of water carriers.

A new contestant named Mr Harry Stockdale entered the field. He was the leader of a Sydney group whose plan was to bring water from the Darling River to Broken Hill. On 1 October 1889 an Act was passed by the NSW Parliament to enable the Stockdale Company to construct works and use water from the Menindee Lakes and the Darling River (Figure 10.6) to supply Broken Hill and District. Faced with this Act, the other two rivals (Broken Hill Water Supply Syndicate and Nolan's Company) decided to amalgamate and form the Broken Hill Water Supply Company. As a result, the newly formed company proceeded to implement plans for construction of a weir on Stephens Creek. This was supported by the Municipal Council without waiting for the authorisation of an Act of State Parliament. Nevertheless, the Council was confident that this would be forthcoming.

On 17 December 1890, the *Broken Hill Water Supply Act* was passed by the State Parliament. Construction of the reservoir on the Stephens Creek, which was started in April 1890, was completed in November 1891. No rain fell to create a run-off until June 1892, and it was necessary to resort to emergency measures in order to supply the population with desperately needed water. By the end of November 1891, the Municipal Council began negotiations for the transport of water by rail from South Australia. This provided a temporary relief, but the railed water never exceeded $225 \, m^3$ per day. With the fall of adequate rainfall in June 1892 and subsequent run-off into the Stephens Creek weir, the Broken Hill Water Supply Company was ready to commence its water supply business.

10.21.3 The Broken Hill Water Supply Company

The *Broken Hill Water Supply Act 1890* had some restrictive clauses, which included a maximum charge for the water. Moreover, at the end of 28 years, the whole of its works would revert to the Government without compensation. In spite of these limitations, the Company pushed on with the construction of the Stephens Creek Reservoir in 1891. The length of the earth dam was 140 m, its width at the base 73 m, and its height was 16.8 m[29] above the bedrock foundation level. By the end of 1891, construction of the reservoir and pumping stations were virtually completed. The whole plant was capable of delivering $273 \, m^3$ per hour. This capacity increased to $364 \, m^3$ per hour between 1892 and 1907.

---

[29] In 1909 the height was increased by about one metre.

Even before the completion of the scheme, doubts were being raised about its adequacy for the expanding water requirements of the town and mines. The issue was more about insufficient rainfall rather than insufficient storage.

In May 1891, evidence indicated that the Stockdale Company had difficulty in constructing its project, which was approved in October 1889. By October 1891, the Stockdale deadline had lapsed before construction of the project had commenced. The Stockdale interests then sought an amendment to their Bill to enable them to build the project. Amendment was passed on 1 April 1892. In May 1892 the Stockdale amended Bill failed to eventuate because they had failed to lodge a deposit of £10 000 ($1 million in 2002 prices) with the Colonial Treasurer within a month. In the meantime, the Broken Hill Water Supply Company continued to supply the population as well as the mining industry at an increasing rate.[30]

10.21.4 Cooperation between the Government and Mines

When the Stephens Creek Reservoir received its first intake of water in 1892, the water storage problem seemed to be solved. However, between 1892 and 1903 the reservoir was never more than half full because the average rainfall for these years was only 165 mm. During the 1896–97 drought, more water evaporated from the reservoir than flowed into it.

By September 1902 the water situation became serious and by June 1903 water became so scarce that a stoppage of the mines was imminent. By 23 June 1903 the Stephens Creek Reservoir had dried up completely, and as in 1892, water had to be railed from South Australia for both town water supply and mining requirements. The first train of water arrived on 14 July 1903 and deliveries soon reached 2290 m^3 per week. At the end of August rains came and continued to fall, and the Stephens Creek Reservoir filled and overflowed for 17 successive days.

On 23 June 1903, the Mayor of Broken Hill had sent an urgent telegram to the Premier of NSW requesting immediate construction of the Umberumberka Creek Dam. Three days later, the local office of the Public Works Department commenced work on the project. However, the partially constructed Umberumberka Creek Dam was washed away by floodwaters in September 1903.

The character of the Broken Hill mining industry was changing and requiring more water for its operations. Earlier methods of ore treatment had failed to extract a large proportion of some metals, particularly zinc. Therefore, huge mounds of tailings had accumulated, and were waiting to be re-processed by improved extraction methods. Improved flotation methods for extracting lead and zinc were developed, but the process required more water. Also, by 1907–1908 the population had risen to around 31 000. The Municipal Council, the mining companies and the State Government came together in an attempt to solve the problem of increasing water demand, and in December 1906 the Umberumberka Water Trust was formed based on an Act of the Parliament.[31] The Trust met for the first time on 17 April 1907, and plans and specifications for the new works were prepared. Construction commenced, and in a press statement issued in January 1908, it was promised that in eight months time water would be pumped into Broken Hill from the new dam. Unfortunately, in May 1908 the mining companies announced that they were unable to raise more than £100 000 ($10 million in 2002 prices) representing only half of their commitment. They advised the Trust to seek a Government loan for the balance. The Trust approached the Premier and the Minister for Public Works. However, the Government rejected the request. In September 1908, the citizens presented the Premier with a petition containing 10 000 signatures, which urged the Government to build the Umberumberka project. On the same day, a motion put to the NSW Parliament for construction of adequate water supply facilities for the domestic and mining purposes of Broken Hill was passed. At the end of 1909, the activities of the Trust were wound up pending the take over of the works by the Government.

10.21.5 Water Supply under the Public Works Department

The new proposal for the Umberumberka Creek consisted of a concrete dam with a crest of 205 m long, 25.3 m high above the Creek bed and 40.3 m above the bedrock. The spillway was 86 m wide. Other details were:

- Pumping machinery in duplicate, each unit capable of pumping 8200 m^3 per day.
- The concrete service reservoir with a capacity of 9100 m^3.
- 3 km of rising main and 27 km of gravitation main to Broken Hill.
- A system of distribution pipes from the 46 cm main to the mines.
- A reticulation system for the town consisting of 15 cm, 10 cm and 7.6 cm pipes.
- A total storage of 13.2×10^6 m^3.

---

[30] In 1889, water consumption was estimated at 11 litres per head per day. It increased gradually to 18 litres in 1910 and 60 litres per head per day in 1927.

[31] Broken Hill and Umberumberka Water Supply Act, No. 54, 1906.

- Out of the weekly supply of 49 185 m^3, it was estimated that the mines would require 35 000 m^3 and the town with its 31 000 inhabitants would need 14 185 m^3.

The total cost was estimated at £359 000 ($35 million in 2002 prices). No one now argued that the Scheme was unnecessary. The Stephens Creek Reservoir was subject to enormous evaporation,[32] and its storage capacity had already been reduced by siltation. Construction works on the Umberumberka Creek Dam commenced in February 1911. It was not until June 1914 that the work could be described as nearing completion. By October 1914, there was sufficient water in the dam for the supply to be turned on. By the middle of 1915 some 2250 house connections had been made, with an estimated 4750 houses to be connected. The total cost to that date was £425 000 ($29 million in 2002 prices).

In September 1915 a Bill was introduced regarding the administration of the completed scheme by the Government and terminating Broken Hill Water Supply Company's lease about 3 years before its expiry in 1918 without any compensation payment.[33] The Bill became law in December 1915 and under its terms the Chief of the Public Works Department became the administrative head of the Broken Hill Water Supply, with a manager and other officers acting for him in Broken Hill. At this point in time, the Department turned its attention to the newly acquired water works from the Broken Hill Water Supply Company including the Stephens Creek Reservoir, in which 2.4 m of silt had accumulated.[34]

During the 1930s the Broken Hill mining industry underwent another significant change. The central section of the ore body near the surface had been mined out. Mining activities on the southern and the northern extremities of the ore body then started to intensify, and the city gained a new lease of life. Additional parks, playing fields and tree-lined streets were established. Expansion of the mining activities and development of the recreational facilities required more water. Also sanitary facilities were inadequate. The financing of the undertaking was to be a joint effort of Government and mining companies. In addition, a Board was to be formed which would be responsible for the raising of revenue and management of all the works.

10.21.6 Water Supply under the Water Board

The *Broken Hill Water Supply and Sewerage Act* was passed on 3 November 1938 and the Broken Hill Water Board began to function on 1 January 1939. The Department of Public Works was to remain the constructing authority and the finished works were to be handed over to the Water Board. By the end of 1939, preliminary steps had been taken to initiate some of the new works authorised by the *Act*. However further action was deferred due to the closing down of the BHP mines and the commencement of World War II.

The first few years of the 1940s were very dry, and by 1944 Stephens Creek Reservoir was empty for the fourth year in succession. The storage in Umberumberka was so low that arrangements were made with the Railway Department to run eight trains of 13 tanks each day from the Darling River. This gave a supply of 22 750 m^3 per week to supplement any water that could be pumped from Umberumberka. The emergency rail lift commenced on 17 August 1944 and continued into 1945, keeping town and industry going but at an enormous cost. Rains at the end of February 1945 provided a small amount of run-off and sufficient water was impounded to terminate the emergency service temporarily. However, this relief was short lived, and the water trains were back in service by the beginning of May. The drought broke in the middle of January 1946. Umberumberka Reservoir overflowed and Stephens Creek Reservoir was half full.

In 1944, the mining companies appointed Mr J. R. Dridan, a water supply engineer with the Government of South Australia, to investigate various water supply proposals. He found that the two existing reservoirs had been seriously affected by siltation. Stephens Creek Reservoir, which had a capacity of 25×10^6 m^3 in 1909, was reduced to 21.4×10^6 m^3 by 1944. Also Umberumberka Reservoir, which had a capacity of 13.2×10^6 m^3 in 1916, only held 9.4×10^6 m^3 in 1944. The rate of loss here was much higher than that of Stephens Creek Reservoir. In addition to siltation, the annual rate of evaporation was high. At Umberumberka there was also considerable loss from seepage. Dridan assessed other proposals based on local sources and concluded that they would be too expensive. Moreover, no local scheme could be relied upon to continue for more than 30 years because of the inevitable siltation of storages.

J. R. Dridan examined proposals for supplies drawn from the River Murray and Darling River. These were:

- Water supply from the River Murray at Wentworth via a 50 cm diameter and 265 km long pipeline, with a cost of £2 068 000 ($102 million in 2002 prices).

---

[32] The average daily evaporation rate at the Stephens Creek Reservoir is about 7 mm, which is equivalent to 2555 mm per annum. http://www.bom.gov.au/climate/averages/tables/cw_047031.shtml (visited in July 2005).

[33] The reason was that the Company had charged the maximum rate for water and thus had compensated itself.

[34] By 1927, twelve years later, the silt level had risen to 5.2 m, more than double the siltation of the previous 23 years.

- Water supply from the Darling River at a point 70 km upstream of Wentworth via a 240 km pipeline, with a cost of £1 845 000 ($91 million in 2002 prices).
- Water supply from the Murray River via a pipeline from Morgan, South Australia.

All the above proposals were considered too expensive and impractical. Dridan's preferred option was the old proposal for a pipeline from the Darling River at Menindee by constructing a weir of approximately 7.2 m high upstream of Menindee. The cost was estimated at £897 500 ($44 million in 2002 prices), which included a sum of £120 000 for construction of the weir, construction of a 50 cm diameter and 110 km long pipeline between Menindee and the Stephens Creek Reservoir, following the railway line. This cost also included two pumping stations and a 4550 m^3 storage tank.

The Water Board considered the Dridan's report in 1945 and decided to adopt the Darling River proposal. The Board suggested that the mining companies and the Minister for Public Works should confer about ways and means of financing the scheme. In order to start the work immediately, the mining companies offered to advance a sum of £100 000 to the Board at 1.5 percent interest to be repaid when the negotiations were completed.

10.21.7 The Darling Pipeline

Construction of the Scheme commenced on 11 May 1946. At the end of the year only £53 000 of the temporary loan provided by the mining companies was spent, but the Board had finalised arrangements for the raising of £246 000 ($12 million in 2002 prices) to rapidly expand activities in the new year.

January 1947 brought a record summer demand for water, causing the Board to revise its ideas about the size of the installation necessary to bring water from the Darling River. The original estimate was for a maximum requirement of 3.6×10^6 m^3 per annum. By 1947 the mining industry was entering a boom period. Full employment and high wages had attracted new people to the city, creating a demand for new and better houses, surrounded by lawns and gardens. People now demanded an improved standard of living and water was the most vital ingredient.

A 50 cm diameter pipeline from the Darling River had been planned, capable of delivering 16 380 m^3 per day, but consumption had shown that a future peak demand of well above 18 200 m^3 per day might be expected. The Department of Public Works estimated that this would require a 60 cm pipeline at an extra cost of £215 000 above the original estimate. The mining companies endorsed the Board's decision to enlarge the Scheme and asked the Government to meet the additional costs.

The Darling Pipeline Scheme required a maximum lift of water of 179 m over a distance of 110 km. Construction work was hampered by post war shortages and other difficulties. By the end of 1950 only 16 km of the total length of the pipeline was completed. During 1949 and 1950 efforts were made to supplement and extend existing distribution facilities to meet the increased water demands.

By the end of 1951, drought conditions that had continued since 1948, had so depleted the storages that it once again became necessary to import water by rail and water trains began on 28 November 1951. By calling upon all available manpower, the construction of the pipeline from Menindee to Horse Lake (Figure 10.6) was completed by February 1952. At Horse Lake, a temporary installation allowed water to be pumped into the trains. This reduction in distance permitted increased deliveries of water. Finally, water was pumped from the Darling River via the pipeline to Stephens Creek Reservoir on 11 June 1952, and this brought to an end the dependence of Broken Hill on water trains in times of emergency, which had been practised for over 60 years.

10.21.8 The Menindee Lakes Storage Scheme

The pipeline was completed at a cost of over £2 million ($50 million in 2002 prices), but the only available storage on the Darling River was a 2.4 m high weir at Menindee built in 1941, which held back about 1.8×10^6 m^3 of water. At best this was no more than three to four months supply for Broken Hill. The effectiveness of the Scheme depended on the provision of a considerably larger storage on the Darling River.

The *Menindee Water Conservation Act* of 1949 empowered the Water Conservation and Irrigation Commission of NSW to construct dams, weirs, levee banks, regulators and other structures necessary to create the Menindee Lakes Storages Scheme. The objectives of the Scheme were (Maiden, 1989, p. 164):

- To provide water to the River Murray for the three states of New South Wales, Victoria and South Australia.
- To meet domestic, stock and irrigation requirements in the Darling River downstream of Menindee.
- To replenish annually the water supply for domestic and stock purposes in the Great Anabranch of the Darling River.
- To assure a supply of water from the Darling River to Broken Hill.

Turning the first sod of soil took place on 22 October 1949 and the major works were completed in 1960 (Water Conservation and Irrigation Commission of NSW, 1960). The entire Scheme was finally completed in December 1968 at a cost of $11.27 million[35] ($95 million in 2002 prices). The combined capacity of the lakes was $2.47 \times 10^9 \, m^3$ (Table 10.7). Completion of the Scheme increased the capacity of the natural lakes and allowed regulation of most flows. Water enters the lake system via Lake Wetherell and can pass downstream to Lake Pamamaroo, Lake Menindee and Lake Cawndilla (Figure 10.6). Releases are made from Lake Menindee and Lake Wetherell into the Darling River. Also, releases can be made from Lake Cawndilla to the Great Darling Anabranch.[36]

The Menindee Lakes are owned and operated by the NSW Government. The control and operation of the Lakes are included in the Murray–Darling Basin Agreement. The sharing of the available water resources is managed by the MDBC according to this Agreement. Rights to all stored waters reverts to the NSW Government whenever the total storage in the Lakes fall to less than $480 \times 10^6 \, m^3$ and until it rises again to exceed $640 \times 10^6 \, m^3$ (Department of Land and Water Conservation, 1988, p. 5).

The Menindee Lakes represent a significant natural, cultural and economic resource for Australia. The large wetland ecosystem supports a diverse range of native flora and fauna. However, a draft management plan identified significant information shortfalls regarding (Department of Land and Water Conservation, 1988):

(1) ecological processes within the lakes and immediately downstream;
(2) the impact of various flow regimes on the adjacent terrestrial ecosystems;
(3) hydrological processes;
(4) cultural heritage values;
(5) socio-economic values; and
(6) structural options.

The Menindee Lakes Ecologically Sustainable Development project was undertaken to address these shortfalls (Department of Land and Water Conservation, 2000).

10.21.9 Recent Water Supply to Broken Hill

Construction of the Menindee Lakes Storage Scheme guaranteed a more secure source of water supply for Broken Hill. On 15 December 2000 the *Australian Inland Energy Water Infrastructure Act 2000*[37] dissolved the Broken Hill Water Board and transferred its assets and liabilities to a new entity titled Australian Inland Energy Water Infrastructure.

In 2000–01, the water consumption of Broken Hill was $7.24 \times 10^6 \, m^3$, which included $1.64 \times 10^6 \, m^3$ for the mining industry. In 2001–02 it was reduced to $6.98 \times 10^6 \, m^3$ including $1.27 \times 10^6 \, m^3$ for the mining sector (Australian Inland, 2002, p. 35). Table 10.8 shows the volume of water pumped from various sources for the water supply of Broken Hill,

Table 10.7. *Capacity of the Menindee Lakes storages*

No.	Storage	Capacity ($10^6 \, m^3$)
1	Lake Wetherell	197.3
2	Lake Malta	4.9
3	Lake Balaka	24.7
4	Lake Bijijie	25.9
5	Lake Tandure	104.8
6	Lake Pamamaroo	382.2
7	Lake Menindee	924.8
8	Lake Cawndilla	702.8
9	Small lakes and depressions	98.6
Total		2466.0

Source: Water Conservation and Irrigation Commission of NSW (1960).

Table 10.8. *Water pumped from various sources over the period of 1999–2000 to 2001–02*

		Water pumped (in $10^3 \, m^3$) during:		
No.	Source	1999–2000	2000–01	2001–02
1	Darling River	3558	1249	5582
2	Stephens Creek Reservoir	2891	5851	260
3	Umberumberka Reservoir	805	370	1852
4	Imperial Dam	68	0	0
Total		7322	7470	7694

Source: Australian Inland (2002, p. 35).

---

[35] http://www.mdbc.gov.au/river_murray/river_murray_system/menindee/menindee.htm (visited in July 2005).
[36] http://www.mdbc.gov.au/river_murray/river_murray_system/menindee/design.htm (visited in July 2005).
[37] This is a NSW Parliament Act.

Menindee, Sunset Strip and Silverton. The 2001–02 water consumption of the last three locations were: 108 870 m^3 for the town of Menindee, 31 260 m^3 for the settlement of Sunset Strip around the shores of Lake Menindee, and 17 290 m^3 for the township of Silverton.

The major drought of 2002 affected the water supply of Broken Hill because of a shortage of water in the Menindee Lakes. On 20 December 2002 the Copi Hollow (Figure 10.6) pumping station, with a cost of one million dollars, went online. The pumping station is capable of delivering 42 000 m^3 of water per day from Copi Hollow into the Broken Hill pipeline.

10.22 CONCLUSIONS

This chapter has examined a number of water transfer schemes under four sections:

(A) inter-basin water transfer for water supply of Adelaide and part of South Australia;
(B) inter-basin water transfer for irrigation water supply of the Mareeba–Dimbulah Irrigation Area in Queensland;
(C) a number of water transfer schemes for domestic and industrial water supply in Queensland; and
(D) the water supply of Broken Hill mines and township in New South Wales.

South Australia, with a population of 1.5 million, relies heavily on the River Murray to provide water for Adelaide and a large regional area. Usage varies according to seasonal conditions, with the River supplying approximately 40 percent of demand in cool and wet conditions and up to 90 percent in drought years. Water is supplied via six pipelines constructed over the period of 1944 to 1973. Indeed, these pipelines are the lifelines of South Australia.

The salinity in the lower Murray poses significant water quality problems for Adelaide with concentrations exceeding 500 mg L^{-1} TDS target value in dry times. While a basinwide Salinity Management Strategy is being implemented to address this issue, the supply of lower salinity waters to South Australia has always had an attraction.

Through the Mareeba–Dimbulah Irrigation Scheme, the Barron River waters, which would naturally flow into the Coral Sea are diverted by gravity in an open channel across the Great Dividing Range to the Walsh/Mitchell River catchments, which flow into the Gulf of Carpentaria. In the financial year 1999–2000 this served to irrigate a crop area of 22 000 ha and produced more than $115 million of agricultural products. Without such inter-basin water transfer, irrigation could not be developed in this area where soils are generally highly productive and climate favourable.

A number of water transfer schemes between (or within) catchments of Drainage Division I in Queensland, for domestic and industrial water supply, have been outlined. Although these are not large-scale inter-basin water transfer schemes, they demonstrate the importance of such transfers for mining and town water supply.

From its early days, Broken Hill's water supply proved to be a problem, both for mining activities and the town's population. Following the completion of the White Leeds and Imperial dams in 1888, the area experienced a drought in the same year. The need for freshwater was further compounded by a typhoid epidemic. Several private sector proposals emerged to supply Broken Hill. These included a proposal to convey water along a 110 km pipeline from the Darling River to Broken Hill.

Stephens Creek weir was completed in 1891, and subsequently filled in June 1892. However, between 1892 and 1903 the reservoir was never more than half-full. As a result, construction of the Umberumberka Creek Dam commenced in June 1903, but the partially constructed Dam was washed away by floodwaters in September of the same year. Construction of the Umberumberka Creek Dam was recommenced in 1911 and completed in 1914.

The first few years of the 1940s were very dry, and by 1944 the Stephens Creek Reservoir was empty for the fourth year in succession. The length and severity of the drought provided a strong argument for an adequate water supply from the Darling River. The two existing reservoirs were found in 1944 to be seriously affected by siltation. This prompted examination of proposals for supplies drawn from the River Murray and Darling River. The preferred option was the 1889 proposal for a pipeline from the Darling River at Menindee. Work on the pipeline commenced on 11 May 1946, and was finally completed on 11 June 1952, marking an end to Broken Hill's dependence on water trains in times of emergency, a practice which had been employed for 60 years.

Construction of the major components of the Menindee Lakes Storage Scheme over the period of 1949–1960 guaranteed a more secure source of water supply for Broken Hill. However, the major drought of 2002 once again affected Broken Hill's water supply due to a shortage of water in the Menindee Lakes. Subsequently, the Copi Hollow pumping station, capable of delivering 42 000 m^3 of water per day into the Broken Hill Pipeline, went online on 20 December 2002.

References

Australian Bureau of Statistics (1999). *South Australian Year Book*. Adelaide: Australian Bureau of Statistics, South Australian Office.

Australian Inland (2002). *Australian Inland Annual Report 2001–2002*. Broken Hill, NSW: Australian Inland.

Blainey, G. (1968). *The Rise of Broken Hill*. South Melbourne: Macmillan of Australia.

Close, A. (1990). River salinity. In Mackay, N. and Eastburn, D. eds. *The Murray*. Canberra: Murray-Darling Basin Commission. pp. 127–144.

Commonwealth Bureau of Census and Statistics (1960). *Official Year Book of Queensland*. Brisbane: Government Printer.

Commonwealth Bureau of Census and Statistics (1972). *Queensland Year Book 1971 and 1972*. Brisbane: Government Printer.

Crabb, P. (1968). Water supplies in South Australia. *Geography* **53**: 282–293.

Curtis, L. S. (1908). *The History of Broken Hill: Its Rise and Progress*. Adelaide: Frearson's Printing House.

Department of Land and Water Conservation (1988). *The Menindee Lakes Storage: Draft Management Plan*. Sydney: Department of Land and Water Conservation and the Menindee Lakes Advisory Committee.

Department of Land and Water Conservation (2000). *Menindee Lakes Ecologically Sustainable Development Project*. Buronga, NSW: Department of Land and Water Conservation.

Department of Natural Resources and Mines (2000). *Interim Resource Operations Licence for Mareeba Dimbulah Water Supply Scheme Issued to SunWater*. Brisbane: Department of Natural Resources and Mines.

Department of Natural Resources and Mines (2001a).[38] *Groundwater Report: Barron Water Resources Plan*. Brisbane: Department of Natural Resources and Mines.

Department of Natural Resources and Mines (2001b). *Ecological Implications Report: Barron Water Resources Plan*. Brisbane: Department of Natural Resources and Mines.

Department of Natural Resources and Mines (2002a). *Water Supply Planing Study Report – Atherton Tableland/Cairns Region*. Brisbane: Department of Natural Resources and Mines.

Department of Natural Resources and Mines (2002b). *Water Supply Planning Study Report – Burdekin Basin Water Resources Plan*. Brisbane: Department of Natural Resources and Mines.

Department of Natural Resources and Mines (2003). *Consultation Report: Barron Water Resource Plan*. Brisbane: Department of Natural Resources and Mines.

Engineering and Water Supply Department (1987). *Water Resources Development 1836–1986: The Historical Background to the Water Resources Management Strategy*. Adelaide: Engineering and Water Supply Department.

Government of South Australia (1989). *South Australia Water Futures: 21 Options for the 21st Century*. Adelaide: Engineering and Water Supply Department.

Griffin, T. and McCaskill, M. eds. (1986). *Atlas of South Australia*. Adelaide: South Australian Government Printing Division.

Gutteridge Haskins & Davey (2001). *SWP Distribution System Efficiency Review: Report on Mareeba Dimbulah Irrigation Area*. Report prepared for the Queensland Department of Natural Resources. Brisbane: Gutteridge Haskins & Davey.

Hammerton, M. (1986). *Water South Australia: A History of the Engineering and Water Supply Department*. Netley, SA: Wakefield Press.

Hardy, B. (1968). *Water Carts to Pipelines: The History of the Broken Hill Water Supply*. Broken Hill: Broken Hill Water Board.

Koenig, K. (1983). *Broken Hill: 100 Years of Mining*. Sydney: New South Wales Department of Mineral Resources.

Maiden, S. (1989). *Menindee: First Town on the River Darling*. Red Cliffs, Victoria: The Sunnyland Press.

Moore, S., Clark, K. and Midgley, T. (2002). *The Local Socio-Economic Values of the Menindee Lakes Systems: A Report to the Menindee Lakes Ecologically Sustainable Development Project Steering Committee*. Buronga, NSW: Department of Land and Water Conservation.

Murray Darling Basin Commission (1993). *The Lower Murray: Morgan to the Mouth*. Canberra: Murray Darling Basin Commission.

NQ Water (2003). *NQ Water Annual Report 2002/03*. Townsville, QLD: NQ Water.

SA Water (1991). *Water Supply for Yorke Peninsula and the Northern Areas of South Australia*. Adelaide: SA Water. Information Bulletin No. 8.

SA Water (1992). *Adelaide's Water Supply – The Onkaparinga System and the Murray Bridge– Onkaparinga Pipeline*. Adelaide: SA Water. Information Bulletin No. 3.

SA Water (1993). *Adelaide's Water Supply – The Torrens System: The Little Para Reservoir and the Mannum–Adelaide Pipeline*. Adelaide: SA Water. Information Bulletin No. 2.

SA Water (2002). *Annual Report 2002*. Adelaide: SA Water.

Shaw, J. ed. (1984). *Collins Australian Encyclopedia*. Sydney: William Collins.

Solomon, R. J. (1988). *The Richest Lode: Broken Hill 1883–1988*. Sydney: Hale & Iremonger.

State of Queensland (2000).[39] *Water Act 2000: Act No. 34 of 2000*. Brisbane: Parliament of Queensland.

Water Conservation, and Irrigation Commission of NSW (1960). *The Menindee Lakes Storages: To Commemorate the Opening of the Menindee Lakes Storages by the Honourable R. J. Heffron Premier of New South Wales, 12 November 1960*. Sydney: Water Conservation and Irrigation Commission of NSW.

Yu, B. (2000). The hydrological and geomorphological impacts of the Tinaroo Falls Dam on the Barron River, North Queensland, Australia. In Brizga, S. and Finlayson, B. eds. *River Management: The Australian Experience*. Chichester: John Wiley & Sons. pp. 73–95.

---

[38] This publication is also available at: http://www.nrm.qld.gov.au/wrp/pdf/barron/barron_groundwater.pdf (visited in July 2005).

[39] This document is also available at: http://www.legislation.qld.gov.au/LEGISLTN/ACTS/2000/00AC034.pdf (visited in July 2005).

Part III
Inter-basin Water Transfer in Other Selected Countries

11 Inter-basin Water Transfer in the United States of America

SECTION A: OVERVIEW OF GEOGRAPHY, POPULATION, LAND AND WATER

11.1 GEOGRAPHY

The Unites States of America covers a total area of 9 372 570 km^2. To this, Alaska contributes 1 530 690 km^2 and Hawaii 16 760 km^2. Other areas outside the 48 contiguous States include Puerto Rico, American Samoa, Guam and the Virgin Islands. The east–west and north–south dimensions of the country are approximately 4517 km and 2572 km respectively. The United States has diverse physiographic features with broad plains, plateaus and mountain areas (Figure 11.1). The country's surface elevation ranges from 86 m below sea level in Death Valley, California to 4419 m at Mount Whitney, California, and 6198 m at Mount McKinley, Alaska (U.S. Census Bureau, 2003, Table 363). The approximate mean elevation of the country is 763 m, and its main physiographic regions are (Murphy, 1996):

- **The Coastal Plain** is located along the Atlantic Ocean and the Gulf of Mexico. Midway along the Coastal Plain, the Florida Peninsula separates the waters of the Atlantic from those of the Gulf of Mexico. West of the Peninsula, the Coastal Plain is wide, but towards the north-east it gradually narrows and in the neighbourhood of New York City the plain comes to an end.
- **The Appalachian Highlands** make up the dominant relief feature of the eastern United States, but they have relatively low elevation. Mount Mitchell in North Carolina, with an elevation of 2038 m, is the highest peak in eastern North America. The Appalachian Highlands form an effective watershed between the streams that flow to the Atlantic Ocean and those that flow westward to the Mississippi River.
- **The Central Lowlands** resemble a vast saucer, rising gradually to higher lands on all sides. Southward the land climbs to the Ozark Plateau, Ouachita Mountains and the Interior Low Plateau. On the eastern side it climbs towards the Appalachian Plateau and on the western side towards the Great Plains. The Central Lowlands constitute the nation's greatest agricultural resource and the richest single agricultural region in the world.
- **The Great Plains** stretch along the eastern edge of the Rocky Mountains. Its climate is generally semi-arid and it is noted for its dryland wheat farms and cattle ranches.
- **The Rocky Mountains** are the backbone of the North American continent. They form what is known as the Continental Divide, or the Great Divide. West of the Divide, streams find their way to the Pacific Ocean or into lakes and sinks. East of Divide, they flow to the Gulf of Mexico. The Rockies are higher than the Appalachians, with Mount Elbert, the highest peak at an elevation of 4399 m. The Rockies have three main divisions, the Northern, the Middle and the Southern.
- **The Intermontane Plateau** is located between the Rocky Mountains and the Pacific Mountains area. It includes the Great Basin, the Columbia Plateau and the Colorado Plateau. It contains the driest part of the United States and includes the Mojave Desert and Death Valley as well as the Sonoran Desert of south Arizona. The Great Basin occupies more than half of this area.
- **The Pacific Mountain Area** is a complicated assemblage of mountains and lowlands to the west of the Intermontane Plateau. The Pacific Coast Ranges extend along the western coast of the country. The Cascade and Sierra Nevada Ranges lie about 160 km further east and parallel to this range. Between the Sierra Nevada and Coast Ranges lie the Central Valley of California, while the Williamette and Puget Sound Lowland is

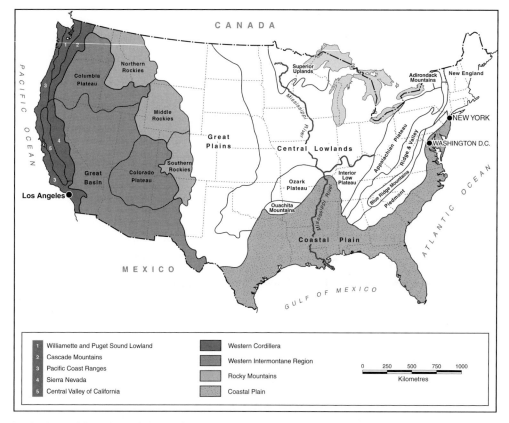

Figure 11.1 The main physiographic regions of the contiguous United States (Murphy, 1996, p. 500).

located further north, between the Cascade Mountains and the Pacific Coast Ranges.

11.2 POPULATION

The United States had a population of 226.5 million in 1980, and 288.4 million in 2002, representing an increase of about 62 million in 22 years or 2.8 million (one percent) per annum. Approximately 68 percent of the population lives in the eastern half of the country where a number of states have populations above 10 million (U.S. Census Bureau, 2003, Table 17). These are New York (19.2 million), Florida (16.7 million), Illinois (12.6 million), Pennsylvania (12.3 million), Ohio (11.4 million) and Michigan (10.1 million). In the western half of the United States, population ranges from 35.1 million for California, and 21.8 million for Texas, down to 0.5 million for Wyoming, which is the lowest in the country. In the western United States, many states have populations of less than 3 million (Figure 11.2).

The population of the United States is predicted to reach 349 million by the year 2025 and 420 million by the year 2050 (U.S. Census Bureau, 2003, Table 3).

11.3 PRECIPITATION AND CLIMATE

The average annual precipitation of the contiguous United States is approximately 762 mm and ranges from less than 100 mm in the desert areas of southern Arizona and California, to over 1500 mm in coastal areas of the Pacific north-west. In general, the average annual precipitation is high in the eastern states (1345 mm in Florida and Mississippi, 1270 mm in Georgia, Tennessee and North Carolina) while it is low in the central and western states (230 mm in Nevada, 330 mm in Utah and 355 mm in Arizona). Figure 11.3 shows the average annual precipitation of the country. Snowfalls represent a very important source of surface water through much of the country. Heavy snowfalls occur in the Rocky Mountains and also on the eastern shores of the Great Lakes.

The average annual potential evaporation rate of the United States ranges from less than 500 mm in the north–east and the north–west, to above 2000 mm in the desert areas of Arizona, California and Nevada.[1]

---

[1] Evaporation map of the USA is available at: http://www.shodor.org/~jingersoll/igems/Teacher1b/evaplv1teach.html (visited in July 2005).

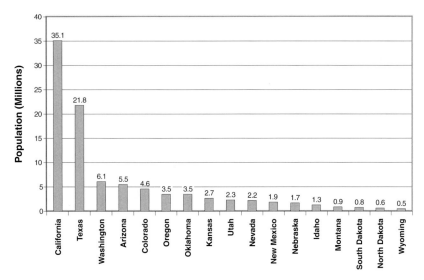

Figure 11.2 Population of the western states of the United States (U.S. Census Bureau, 2003, Table 17).

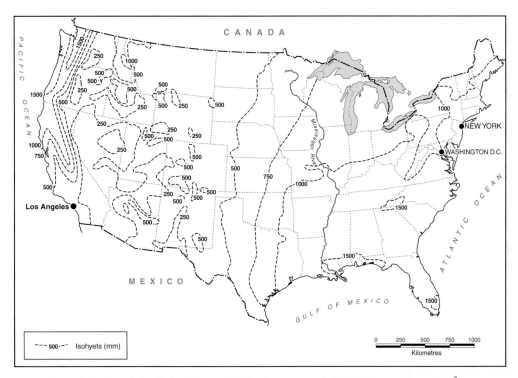

Figure 11.3 Average annual precipitation of the United States (Grolier Incorporated, 1996, Volume 27, p. 518).[2]

The United States has a wide variety of climatic types, ranging from sub-arctic regions on the higher mountain peaks to wet sub-tropical regions of the Gulf Coast and the southern tip of Florida, to the arid regions of the west. The arid and semi-arid west covers a large area of the country from the Great Plains to the Intermontane Plateau.

11.4 LAND USE

In 1997, about 767 Mha of land in the contiguous United States was under a variety of land uses. These included

[2] For the 2004 precipitation map of the United States visit the following website: http://www.cpc.noaa.gov/products/analysis_monitoring/regional_monitoring/us_12-month_precip.html (visited in July 2005).

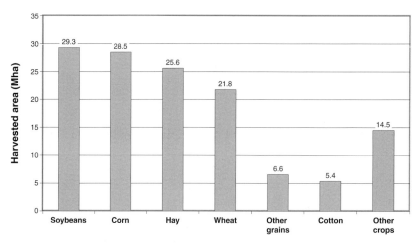

Figure 11.4 Extent of the harvested crop areas in the contiguous United States in 1999 (Economic Research Service, 2003a, Table 1.1.6 and Figure 1.1.3).

234 Mha as grassland pasture and rangelands, 224 Mha as forestland, 184 Mha as cropland, 84 Mha for urban, transport, recreation and defence, and 41 Mha for miscellaneous uses. The area of harvested crops in 1999 was about 131.7 Mha with the predominant crops being soybeans, corn and hay (Figure 11.4).

The area under irrigation in the United States is about 25 Mha.[3] Most of the states with large irrigated areas are in the dry western part of the country. California has the largest tract of irrigated land (4.1 Mha), followed by Nebraska (3.2 Mha) and Texas (2.6 Mha). Annually, $189 \times 10^9 \, m^3$ of water is withdrawn for irrigation. This represents an average of $7560 \, m^3$ per ha.

The estimated area of irrigated crops for 2000 was 22.5 Mha, consisting of corn 4.13 Mha, hay 3.90 Mha, orchard and vegetables 2.83 Mha, cotton 2.14 Mha, soybeans 2.10 Mha, sorghum, barley and wheat 1.98 Mha, rice 1.25 Mha, and other crops 4.17 Mha (Economic Research Service, 2003b, Chapter 2.1, p. 2).

11.5 WATER RESOURCES

11.5.1 Surface Water

The United States has a large number of rivers and is divided into 21 Water Resources Regions, 18 in the contiguous United States and one each in Alaska, Hawaii and the Caribbean. The largest river system is the Mississippi River, which drains an area of about $3.2 \times 10^6 \, km^2$, or 34 percent of the country between the Appalachian and Rocky Mountains (Figure 11.5).

In an average year, a volume of about $5800 \times 10^9 \, m^3$ falls as precipitation on the contiguous United States (U.S. Department of Agriculture, 1989, p. 181). About two-thirds of this volume or $3870 \times 10^9 \, m^3$, returns to the atmosphere through evaporation and transpiration. The remaining $1930 \times 10^9 \, m^3$ replenishes surface and groundwater resources. The average annual run-off of the 18 Water Resources Regions is about $1840 \times 10^9 \, m^3$ (Table 11.1), and the groundwater recharge is estimated to be $90 \times 10^9 \, m^3$.

The surface water resources of the United States are highly regulated. Reservoir construction peaked during the 1960s and has markedly slowed since then. Currently there are approximately 76 000 dams[4] higher than 1.8 m (6 feet) in the country. Because some reservoirs have multiple dams, this corresponds to about 68 000 separate reservoirs. The total volume of water stored in these reservoirs is about $520 \times 10^9 \, m^3$ or only 9 percent of the average annual precipitation (U.S. Geological Survey, 2002, p. 14). The two largest reservoirs are Lake Mead, created by the Hoover Dam (completed in 1936) in Nevada, and Lake Powell, created by the Glen Canyon Dam (completed in 1964) in Arizona, with capacities of $34.85 \times 10^9 \, m^3$ and $33.3 \times 10^9 \, m^3$ respectively.[5]

11.5.2 Groundwater

Groundwater is available nearly everywhere in the United States, but the conditions controlling its occurrence, quantity, quality, and development differ from one part of the country to another. It is found in unconsolidated sands and

---

[3] This includes irrigated crop areas (22.5 Mha), golf courses, parks and others (Hutson et al., 2004).
[4] Information about these dams is available at: http://crunch.tec.army.mil/nid/webpages/nid.cfm (visited in July 2005).
[5] U.S. Society on Dams at the URL http://www2.privatei.com/~uscold/uscold_s.html (visited in July 2005).

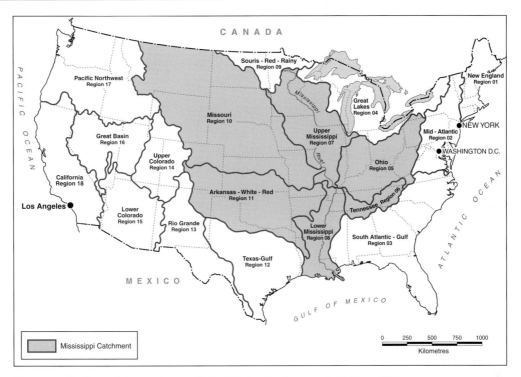

Figure 11.5 Water Resources Regions of the contiguous United States (Paulson *et al.*, 1993, p. 588).

Table 11.1. *Natural annual run-off for the average and dry year in the 18 Water Resources Regions of the contiguous United States*

	Natural annual run-off	
Water Resources Region	Average year (10^9 m^3)	Dry year (10^9 m^3)
1. New England	108.5	87.2
2. Mid-Atlantic	111.1	86.3
3. South Atlantic-Gulf	320.9	232.7
4. Great Lakes	103.9	82.5
5. Ohio	191.8	141.1
6. Tennessee	56.7	50.0
7. Upper Mississippi	104.9	85.1
8. Lower Mississippi	103.5	21.3
9. Souris-Red-Rainy	8.5	4.8
10. Missouri	84.9	64.5
11. Arkansas-White-Red	93.4	58.6
12. Texas-Gulf	49.2	27.1
13. Rio Grande	7.3	6.0
14. Upper Colorado	19.3	15.2
15. Lower Colorado	3.0	3.0
16. Great Basin	8.3	7.0
17. Pacific North-west	370.6	312.7
18. California	93.9	69.7
Total	1839.7	1354.8

Source: U.S. Department of Agriculture (1989, p. 256).

gravels, glacial deposits, semi-consolidated sands, sandstone, carbonate and fractured volcanic rocks. Detailed information concerning the hydrogeology of the country is available in Heath (1984), Back *et al.* (1988) and the *Ground Water Atlas*.[6]

Groundwater is the source of drinking water for approximately 140 million residents, or about 48 percent of the population. Groundwater also provides about 41.5 percent of the water used for irrigation at a rate of 78.6×10^9 m^3 per annum, or nearly 87 percent of the estimated annual recharge.

Excessive groundwater extraction threatens the sustainability of the resource in some areas. Impacts include storage depletion, land subsidence, saltwater intrusion in coastal aquifers, reduction in streamflow and loss of wetland and riparian habitats (Galloway *et al.*, 1999, 2003). Groundwater depletion has been a concern in the south-west and High Plains for many years, but increased demand has over-stressed aquifers in many other areas of the country as well. Groundwater depletion is occurring in the Atlantic Coastal Plain, west-central Florida, the Gulf Coastal Plain, the High Plains, Chicago–Milwaukee, the Pacific north-west, and the dry south-west.[7]

Heavy use of groundwater in coastal areas of southern California has lowered groundwater pressures. As a result, saltwater from the ocean has intruded into the coastal aquifers threatening local groundwater supplies.

[6] http://capp.water.usgs.gov/gwa/gwa.html (visited in July 2005).
[7] http://water.usgs.gov/pubs/fs/fs-103–03 (visited in July 2005).

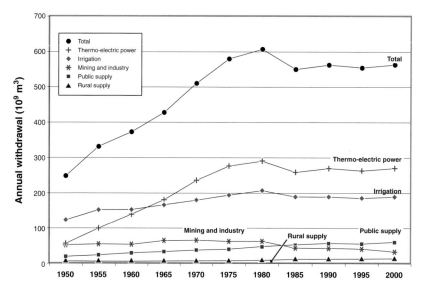

Figure 11.6 Water withdrawal in the United States from 1950 to 2000 (U.S. Census Bureau, 2003, Table 367 for the 1950–95 data; and Huston et al. 2004 for the 2000 data).

11.5.3 Water Quality Issues

Water quality issues in surface and groundwater include the levels of contaminants such as nitrogen, phosphorus, pesticides, herbicides, insecticides, endocrine disruptors, and other chemicals that might affect humans, aquatic life or the environment. Concern about water pollution resulted in the Federal Government passing the *Clean Water Act* in 1972. Since then, both private and public sectors have spent more than US$500 billion on water pollution control, much of which has been directed towards municipal and industrial point sources (U.S. Geological Survey, 1999, p. 2). Progress in cleaning up contamination from point sources has not been matched by control of contaminated discharge from non-point sources, including fertilisers and pesticides applied in agricultural and urban areas, and nutrients and pathogens from human and animal wastes. The U.S. Geological Survey (1999) provides an overview of the water quality issues as part of a National Water Quality Assessment Program, and is a source of extensive information on water quality.

11.5.4 Water Withdrawal and Use

In the United States water withdrawal steadily increased up to 1980, when it reached a peak value of about $610 \times 10^9\,\mathrm{m}^3$ or 10.5 percent of the average annual precipitation. It has since declined to $563 \times 10^9\,\mathrm{m}^3$ in 2000 (Figure 11.6). Table 11.2 shows the annual sectoral withdrawal and source of supply for the year 2000. It indicates that about 48 percent of water withdrawal was for the thermo-electric power utilities, followed by irrigation (34 percent). Surface water

Table 11.2. *The estimated water withdrawal and source of supply in the United States in 2000 (in $10^9\,m^3$)*

Withdrawal	Thermo-electric power utilities		270
	Irrigation		189
	Public supply		60
	Industrial		32
	Rural supply		12
	Total		563
Source of water	Surface water	Fresh	362
		Saline	84
		Total	446
	Groundwater	Fresh	115
		Saline	2
		Total	117
	Total		563

Source: Hutson et al. (2004, Table 1).

and groundwater contributed respectively $446 \times 10^9\,\mathrm{m}^3$ (79 percent) and $117 \times 10^9\,\mathrm{m}^3$ (21 percent). Freshwater withdrawal was $477 \times 10^9\,\mathrm{m}^3$ (85 percent), while the volume of saline water, mostly used by the power industry, was $86 \times 10^9\,\mathrm{m}^3$ (15 percent).

Although the annual volume of water withdrawal is high, consumptive use of water is much lower. For example in 1995,[8] total water consumption was $138 \times 10^9\,\mathrm{m}^3$ which is only about 25 percent of total water withdrawal (Solley et al., 1998). The main reason is that the thermo-electric power

---
[8] No data is available for 2000.

Table 11.3. *Irrigated area and irrigation water use in the United States in 2000*

State	Irrigated area (Mha)	Water withdrawal in $10^9 \, m^3$		
		Groundwater	Surface water	Total
1. California	4.1	16.1	26.0	42.1
2. Nebraska	3.2	10.3	1.9	12.2
3. Texas	2.6	9.0	3.0	12.0
4. Arkansas	1.8	9.0	2.0	11.0
5. Idaho	1.5	5.1	18.5	23.6
6. Colorado	1.4	3.0	12.8	15.8
7. Kansas	1.4	4.7	0.4	5.1
8. Oregon	0.9	1.1	7.3	8.4
9. Florida	0.8	3.0	2.9	5.9
10. Montana	0.7	0.1	10.9	11.0
11. Washington	0.6	1.0	3.2	4.2
12. Georgia	0.6	1.0	0.5	1.5
13. Mississippi	0.6	1.8	0.1	1.9
14. Utah	0.6	0.7	4.7	5.4
15. Missouri	0.5	1.9	0.1	2.0
16. Wyoming	0.5	0.6	5.7	6.3
17. Other States	3.2	10.2	10.7	20.9
Total	25.0	78.6	110.7	189.3

Source: Hutson *et al.* (2004, Table 7).

industry withdrew $262 \times 10^9 \, m^3$ of water but then returned 98 percent. Also, out of the $185 \times 10^9 \, m^3$ of water withdrawn for irrigation, consumptive use was only 61 percent. The remaining 39 percent consisted of 20 percent as return flow and 19 percent as conveyance losses.

Irrigation is the predominant consumptive user of water. It accounts for one-third of water withdrawals (Tables 11.2 and 11.3), and in 1995 was responsible for 85 percent of the Nation's consumptive water use (Solley *et al.*, 1998). California is the major user of irrigation water ($42.1 \times 10^9 \, m^3$) with 38 percent sourced from groundwater resources (Table 11.3). It is followed by Idaho and Colorado with respective irrigation water use of $23.6 \times 10^9 \, m^3$ and $15.8 \times 10^9 \, m^3$.

Based on the data in Table 11.2 for public and rural supplies, domestic water use in the United States was about 685 litres per capita per day in 2000.

Goklany (2002) analysed the trends in agricultural land and water use in the United States over the period of 1910 to 1998 and compared them with global trends. He concluded that the spectacular increase in agricultural productivity per unit area of land in both the United States and worldwide through much of the twentieth century was not matched by comparable increases in productivity per unit volume of water. Clearly, only improvements in water use efficiency will meet the increasing human and environmental needs.

11.6 FLOOD

Floods are the result of the interaction of many natural and human-induced factors, and include regional floods, flash floods, storm-surge floods, and dam-failure floods. In the United States, about 3800 towns and cities of more than 2500 inhabitants are on floodplains. Floods are the most chronic and costly natural hazard in the country, causing an average of 140 fatalities and US$5 billion in damage each year (O'Connor and Costa, 2003, p. 1). Perry (2000) has listed 32 major floods of the twentieth century, consisting of 20 regional floods, 4 flash floods, 3 storm-surge floods and 5 other types of floods. Many major floods occurred in the eastern states, but the western states such as California (1964–65, 1995 and 1996–97), Oregon (1903, 1964–65 and 1996), and Washington (1964–65 and 1980) have also been affected. In September 1900, the hurricane and storm-surge flood at Galveston, Texas, killed more than 6000 people, making it the worst natural disaster in the nation's history. The most destructive flood of the twentieth century was the regional flood of the Mississippi River in 1993 which caused 48 deaths and US$20 billion damage.

The problem of flood control in the United States assumed national importance in the twentieth century, because of the increasing frequency and intensity of floods in all of the major river valleys as a result of deforestation. After the devastating flood of April–May 1927 along the lower Mississippi River from Missouri to Louisiana, Congress passed the *Flood Control Act of 1928*, aimed at flood control along the Mississippi River.[9] In 1935, the Soil Conservation Service was established for the control and prevention of soil erosion and the reduction of flood hazard. Notable flood control measures were instituted in the 1930s. The most important measure was the establishment of the Tennessee Valley Authority in 1933 by the U.S. Congress for flood control, navigation, and replanting forests.[10] This was followed by the *Flood Control Act of 1936*,[11] and other legislation in the following decades.

Despite extensive flood control measures implemented by the state and federal agencies, such as the U.S. Bureau of Reclamation and the U.S. Army Corps of Engineers, flood damage in the United States increased during the twentieth century and reached approximately US$50 billion in the 1990s (Pielke, Jr. et al., 2002, p. 1 and Table 3.1). Speculative reasons for this increase include: climate change, population growth, land use change, and inadequate federal policies.

A detailed analysis of the flood damage in the country during the twentieth century is available in Pielke, Jr. et al. (2002). According to Pielke, Jr. et al. (2002, Figure 6.2) all states suffered flood damage during the period of 1955–78 and 1983–99.[12] The top 10 states in terms of total flood damage (in 1995 values) for these periods are Pennsylvania (US$12 billion), California (US$10.5 billion), Iowa (US$8.5 billion), Louisiana (US$8.4 billion), Missouri (US$6.2 billion), Texas (US$6 billion), Illinois (US$5 billion), Oregon (US$4.5 billion), North Dakota (US$4.4 billion), and New York (US$4 billion).

In the Central Valley of California, a complex system of reservoirs and levees has been built by local, state and federal authorities for flood control, power generation, and water supply for irrigation and other purposes. These include levees built in the Sacramento–San Joaquin Delta area, Shasta Dam, Friant Dam, Folsom Dam, and many others (see Section B of this chapter).

11.7 DROUGHT

Droughts occur throughout North America.[13] During the twentieth century, major droughts occurred in the 1930s (known as the Dust Bowl Drought),[14] 1950s, 1970s, and the 1980s. These droughts lasted five to seven years and covered large areas of continental United States. The Dust Bowl Drought came in three waves of 1934, 1936 and 1939–40, but some regions of the High Plains experienced drought conditions for up to eight years. At its peak in July 1934, this drought covered 70 percent of the country.

During the 1950s, the Great Plains and the south-western part of the country withstood a five-year drought, and in three of these years, drought conditions stretched coast to coast. The drought was first felt in the south-west in 1950 and spread to Oklahoma, Kansas and Nebraska by 1953. The drought persisted in the Great Plains, reaching a peak in 1956. The drought subsided in most areas with the spring rains of 1957.

Even the typically humid, north-eastern United States experienced a five-year drought in the 1960s that drained the water reservoirs of New York City down to 25 percent of capacity. The mid 1970s drought will probably be remembered as the one that impacted most on the water supplies of urban centres (Yevjevich et al., 1978, p. vii). Drought conditions in the summer of 1974 seriously damaged crops in the mid-west and the High Plains.

The three-year drought of 1987–89 covered 36 percent of the country at its peak. This drought began along the west coast and extended into the north-west. Its greatest impact was in the northern Great Plains. By 1988, the drought intensified over the northern Great Plains and spread across much of the eastern half of the country. It also encompassed the upper Mississippi River Basin, where low river levels caused major problems for barge navigation. The summer of 1988 is well remembered for the extensive forest fires that spread across western North America. The west coast of the United States experienced a six-year drought in the late 1980s and early 1990s, causing Californians to take aggressive water conservation measures.

The Federal Government spent US$3.3 billion (US$23 billion in 2004 values) responding to the 1953–1956 drought. During the 1976–77 drought, the federal drought response cost at least US$6.5 billion (US$21 billion in 2004 values), and about US$6 billion (US$9.5 billion in 2004 values) during the 1988–89 drought. These costs do not include

---

[9] http://www.mvd.usace.army.mil/MRC-History-Center/library/documents/1928doc.html (visited in July 2005).

[10] http://www.ourdocuments.gov/doc.php?doc=65 (visited in July 2005).

[11] http://www.ccrh.org/comm/cottage/primary/1936.htm (visited in July 2005).

[12] Because little data was collected for the period of 1979–1982, this period is not included in the analysis.

[13] Information provided in this section is based mainly on the information available at the URL http://www.ngdc.noaa.gov/paleo/drought/drght_history.html (visited in July 2005).

[14] Information regarding the Dust Bowl and its economic and social impacts is available in publications such as: Bonnifield (1979); Worster (1979); and Gregory (1989).

other costs such as losses in crop production and increases in food prices. For example, for the 1988 drought the estimated cost of losses in crop production was nearly US$20 billion and led to an increase in food prices of more than US$12 billion (National Drought Policy Commission, 2000, p. 1).

Palaeoclimatic records from historical documents, tree rings, archaeological remains, lake sediments and other sources indicate that:[15]

(1) droughts of similar duration and extent to the droughts of the 1930s and 1950s have occurred once or twice per century during the past 400 years;
(2) droughts of much longer duration than any drought experienced in twentieth century have occurred in the past 800 years, one lasting from about 1580–1600 and another that occurred in the last quarter of the thirteenth century; and
(3) based on palaeoclimatic records of the past 2000 years, droughts more severe than those of the 1930s and 1950s are likely to occur in the future, a possibility that might be exacerbated by global warming in the twenty-first century.

11.8 CLIMATE CHANGE IMPACTS

Impacts of climate change in North America on water resources, forests, agriculture, fisheries, human health, human settlements, infrastructures, tourism, and recreation are described in IPCC (2001, Chapter 15, pp. 735–800). IPCC found North America has warmed by about 0.7°C during the twentieth century. Climate modelling results suggest that North America could warm by 1–3°C during the twenty-first century for a low emission scenario, and as much as 3.5 to 7.7°C for high emissions (IPCC, 2001, p. 737). These changes are predicted to have severe impacts on water resources, ecosystems and agriculture of the United States, Canada and Mexico. Higher temperatures will decrease the extent of snowfall and will result in shorter snow accumulation periods. Where snow-melt is currently an important part of the hydrological regime, seasonal shifts in run-off are likely. A larger proportion of run-off will occur in winter, together with possible reductions in summer flows. Higher temperatures will also increase evaporation leading to reduced lake and reservoir water levels. The potential impact of climate change on water resources of various regions of the United States are shown in Table 11.4. It indicates lower snowfall for the north-west, less ice cover for the Great Lakes, and more floods and hurricanes for the Coastal Plain.

Adaptive responses to seasonal run-off changes could include altered management of constructed storages, increased reliance on conjunctive management of groundwater and surface water supplies, and voluntary water transfer between water users. However, it may not be possible to avoid adverse impacts on many aquatic ecosystems or to fully offset the impacts of reduced summer water availability for irrigation and other out-of-stream and in-stream water users. Moreover, where lower summer flows and higher water temperatures occur, water quality may be reduced, resulting in increased stress on aquatic ecosystems.

11.9 WATER TRANSFER PROJECTS IN THE UNITED STATES

The oldest water transfer facilities in the east and the west are the Portage Canal in Wisconsin, built in 1856 (Mooty and Jeffcoat, 1986, p. 5), and the Placer Ditch in Oregon, which was probably built about 1865 (Petsch, Jr., 1985, p. 6). Quinn (1968) was perhaps the first to study the inter-basin water transfer in western United States. According to his study, in 1965, some 146 water transfer projects existed in 15 western states. Major transfers occurred in four states of California, Texas, Washington and Colorado, with Colorado River Basin the main exporter of water, mostly to California. Quinn (1968) discussed a range of legal and political impediments to water transfer among different users and regions. He commented that although water transfer has been practiced in the west, no inter-basin water transfer crossed the state boundaries.

Petsch, Jr. (1985), and Mooty and Jeffcoat (1986) provide inventories of inter-basin water transfer for the western[16] and the eastern[17] states, respectively. These two inventories cover all 18 Water Resources Regions of the contiguous United States (Figure 11.5), and 204 sub-regions. They provide annual water transfer data between sub-regions and regions for the period of 1973 to 1982.

---

[15] http://www.usgcrp.gov/usgcrp/seminars/990120FO.html (visited in July 2005) and Woodhouse and Overpeck (1998).
[16] Western States included: Arizona, California, Colorado, Idaho, Iowa, Kansas, Missouri, Montana, Nebraska, Nevada, New Mexico, North Dakota, Oklahoma, Oregon, South Dakota, Texas, Utah, Washington, and Wyoming.
[17] Eastern states included: Alabama, Arkansas, Connecticut, Florida, Georgia, Illinois, Indiana, Kentucky, Louisiana, Maine, Maryland, Massachusetts, Michigan, Minnesota, Mississippi, North Carolina, New Hampshire, New Jersey, New York, Ohio, Pennsylvania, South Carolina, Tennessee, Vermont, Virginia, West Virginia, Wisconsin, and parts of Iowa, Missouri, South Dakota, and Texas.

Table 11.4. *Potential impacts of climate changes on water resources and aquatic species of the United States*

Region	Impacts
Pacific West Coast	– Less snowfall, more winter rainfall, earlier seasonal peak in run-off, increased autumn/winter flooding and decreased summer water supply. – Possible summer salinity increase in San Francisco Bay and Sacramento–San Joaquin Delta. – Benefits to warm-water fish species and damage to cold-water species.
Rocky Mountains	– Rise in snowline, earlier snow-melt, changes in seasonal streamflow, possible reduction in summer streamflow, and reduced summer soil moisture. – Stream temperature changes affecting species composition and increased isolation of cold-water stream fish.
South-west	– Possible declines in groundwater supplies due to reduced recharge. – Increased water temperature, causing further stress on aquatic species. – Increased frequency of intense precipitation events, which would increase risk of flash floods.
Mid-west	– Possible large declines in summer streamflow. – Increasing likelihood of severe droughts. – Possible increasing aridity in semi-arid zones. – Uncertain impacts on farm sector income, groundwater levels, streamflows, and water quality.
Great Lakes	– Possible precipitation decreases, coupled with reduced run-off and lake-level declines. – Reduced hydro-power production and reduced channel depths for shipping. – Decreases in lake ice extent. – Northward migration of fish species, and possible extermination of cold-water species.
North-east Region	– Decreased snow-cover amount and duration. – Possible large reductions in streamflow. – Accelerated coastal erosion, and saline intrusion into coastal aquifers. – Changes in magnitude, timing of ice freeze-up/break-up, with impacts on spring flooding. – Changes in fish species distributions and migration patterns.
Coastal Plain	– Heavily populated coastal plains at risk of flooding from extreme precipitation events and hurricanes. – Changes in estuary systems, wetland extent, and distribution of species.

Source: IPCC (2001, Figure 15.1).

In 1973, total water diversion between 204 sub-regions and 18 regions was $23 \times 10^9 \, m^3$, consisting of $7 \times 10^9 \, m^3$ in the east and $16 \times 10^9 \, m^3$ in the west. In 1982, these diversions were slightly lower at about $22.5 \times 10^9 \, m^3$ consisting of $7.2 \times 10^9 \, m^3$ in the east and $15.3 \times 10^9 \, m^3$ in the west. Exported water from the 18 Water Resources Region was $10.8 \times 10^9 \, m^3$ in 1973, and declined slightly to about $10.2 \times 10^9 \, m^3$ in 1982. Table 11.5 shows volumes of exported waters from these regions. It indicates that the Colorado River Basin had a contribution of 60 percent to the total water export between Water Resources Regions. It was followed by the Great Lakes Region (27 percent), and the Mid-Atlantic Region (12 percent).

The idea of inter-basin transfer from water surplus regions to water deficit areas remained popular up to the 1960s. Most of the water transfers in the west occur between sub-regions of California, and from the Colorado River Basin to its neighbouring states. Some of the projects developed for irrigation, urban water supplies and for flood control are described in Sections B and C of this chapter. Inter-basin water transfer projects are no longer favoured because of their high economic, environmental and social costs. Therefore, water requirements of a continuously increasing population can only be provided by water conservation, water trading, desalination and other measures.

Table 11.5. *Volumes of inter-basin exported waters from the 18 Water Resources Regions of the contiguous United States in 1982*

Water Resources Region	Export ($10^6 \, m^3$)
01. New England Region	–
02. Mid-Atlantic Region	1174
03. South Atlantic-Gulf Region	36
04. Great Lakes Region	2767
05. Ohio Region	2
06. Tennessee Region	5
07. Upper Mississippi Region	5
08. Lower Mississippi Region	30
09. Souris-Red-Rainy Region	–
10. Missouri Region	15
11. Arkansas-White-Red Region	7
12. Texas Gulf Region	51
13. Rio Grande Region	2
14. Upper Colorado Region	1012
15. Lower Colorado Region	5017
16. Great Basin Region	6
17. Pacific North-west Region	2
18. California Region	29
Total	10 160

Source: Petsch, Jr. (1985, Table 4); and Mooty and Jeffcoat (1986, Table 4).

11.10 AMBITIOUS PLANS FOR WATER TRANSFER

The 1950s drought and the 1963 U.S. Supreme Court's decision to restrict California's access to water from the Colorado River inspired development of numerous proposals for very large water transfer projects over great distances from areas of water surplus to areas of water deficiency. Howe and Easter (1971, pp. 11–17) summarised 20 proposals developed from 1950 to 1968. Seven of these ambitious proposals are shown in Table 11.6 and Figure 11.7. Four of them planned to tap surplus water of the Snake and Columbia rivers for conveyance to Lake Mead on the Colorado River and southern California. One undersea aqueduct was planned to take excess water from the rivers of northern California and Oregon for use in the Central Valley and central and southern coastal plains of California. The Beck Plan was proposed to divert water from the headwaters of the Missouri River across Nebraska, Colorado, Oklahoma, and Texas to New Mexico. The Texas Water Plan involved distribution of excess water from the rivers of east Texas by means of a coastal canal and the Trans-Texas distribution system (Geraghty *et al.*, 1973, description of Plate 16).

On a much larger scale, the North American Water and Power Alliance (NAWAPA) was proposed by Ralph M. Parsons Company (1964). The NAWAPA project (Figure 11.8) was designed to transfer water from the far north-western part of the North American continent (Alaska, Canada and Columbia River) to two provinces of Canada (Alberta and Saskatchewan), 22 states in the United States[18] and three states of Mexico (Baja California, Sonora and Chihuahua). It had an estimated cost of US$100 billion (US$600 billion in 2004 prices) requiring 10 years of planning and negotiations to satisfy state, federal and international political interests and to conclude treaties among the three countries. Construction was estimated to take a further twenty years. The NAWAPA proposal planned to deliver $31 \times 10^9 \, m^3$ of water to Canada, $165 \times 10^9 \, m^3$ to the United States ($120 \times 10^9 \, m^3$ for irrigation and $45 \times 10^9 \, m^3$ for domestic, industrial and other usages) and $24 \times 10^9 \, m^3$ to Mexico. The *Bulletin of Atomic Scientists* (v. 23, no. 7, September 1967, pp. 9–27) presents five articles which discuss the practical, political, economic and engineering aspects of the project and its predicted impacts on fish and fishing.

Canada objected to the transfer of its water resources (Howe and Easter, 1971, p. 11), arguing that these were not a continental resource as suggested by American officials; but were Canadian property to be used and managed by Canadians.[19]

Attitudes towards large-scale water transfer in the United States varied widely. Some saw it as the answer to a perceived crisis requiring immediate action, while others condemned it. The attitude is reflected in the following statement, which appeared in a 1964 Senate Sub-committee report (Howe and Easter, 1971, p. 9):

> It is a well established fact that a serious water problem exists in the western United States. It will grow steadily worse until it reaches alarming proportions in the years 1980 and 2000. The existing and proposed Federal developments will help to alleviate the situation but it is doubtful if such piecemeal developments will completely solve the water shortage problem. It therefore becomes imperative that new sources of water supply for the arid and semi-arid west be explored at an early date.

[18] **A. Water for irrigation:** Idaho, Montana, Oregon, Utah, Nevada, California, Arizona, New Mexico, Texas, Colorado, Nebraska, Kansas, Oklahoma, North Dakota and South Dakota. **B. Water for industry:** Wisconsin, Michigan, Illinois, Indiana, Ohio, Pennsylvania and New York.

[19] For description of other large-scale water export proposals from Canada to the United States and Canada's water export policy see sections 12.17 and 12.18.

Table 11.6. *Capacity of the proposed water diversion plans*

No.	Water diversion plan	Year of proposal	River basin(s) of: Source	River basin(s) of: Use	Annual capacity ($10^9 \, m^3$)
1	Snake-Colorado Project	1963	Snake River	Colorado River South Pacific Coastal Plain	3.0
2	Modified-Snake-Colorado Project	1965	Snake River Columbia River	Great Basin Snake River South Pacific Coastal Plain Colorado River	6.2
3	Sierra-Cascade Project	1965	Columbia River	Oregon valleys Central Valley South Pacific Coastal Plain	8.6
4	Western Water Project	1964	Columbia River	Colorado River Sacramento River South Pacific Coastal Plain	16.0
5	Beck Plan	1967	Missouri River	Texas High Plains	12.3
6	Texas Water Plan	1967	Mississippi River Texas rivers	High Plains of Texas New Mexico	13.6
7	Undersea Aqueduct System	1967	North Coast Pacific Rivers	Central Valley South Pacific Coastal Plain	13.6

Source: Howe and Easter (1971, Table 2).

In spite of some support for proposals for inter-basin water transfer, these proposals have faced significant resistance from environmentalists since the late 1960s. They raised issues related to inundation of fertile lands, destruction of fish and wildlife habitats and the lost value of recreational sites. Moreover, most water managers recognised that, firstly the cost of such projects is prohibitive and secondly, there are few river basins whose inhabitants would willingly give up their excess water to another geographic area, without compensation.

11.11 FEDERAL WATER PLAN FOR THE WEST (WATER 2025)

According to the U.S. Department of Interior (2003),[20] the western United States is faced with the following significant water issues:

- **Rapid population growth.** Explosive population growth is occurring in areas where water supplies are limited, and demand for water is increasing.

- **Shortage of water.** In some areas water supply will not be adequate to meet all demands (urban, agricultural and industrial requirements combined with increasing water demand for recreation, scenic value, fish and wildlife habitat) even in normal water years. Inevitable drought will magnify the impacts of water shortages.

- **Conflicts due to water shortages.** Water shortage creates conflicts among various water users (city residents, farmers, native Americans, fish and wildlife). The social, economic, and environmental consequences of a water supply crisis are severe.

- **Aging water facilities.** Most of the federal infrastructures that manage the finite usable water supply in the west are approaching 50–60 years of age, and some facilities are almost a century old (see Sections B and C of this chapter). Many of these facilities are not able to satisfy requirements in the twenty-first century and will need refurbishment or replacement.

[20] This document is also available at the URL http://www.doi.gov/water2025/Water2025.pdf (visited in July 2005).

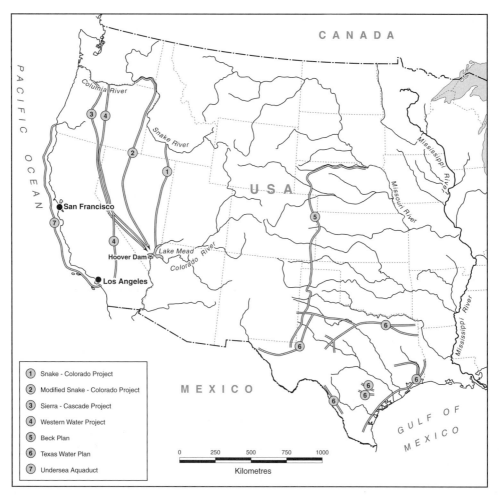

Figure 11.7 Location of some of the inter-basin water transfer proposals developed in the 1960s (Geraghty et al., 1973, Plate 16).

- **Crisis management is not effective.** Public and policy-level attention to water supply issues in drought conditions tends to disappear as soon as precipitation relieves the drought. But drought is only a magnifier of the larger problems associated with rapid population growth and environmental demands for water in areas where water supplies are already over allocated. Overall, existing crisis management is not an effective solution for addressing long-term, systematic water supply problems.

In order to address these issues, and to meet the challenge, the U.S. Department of Interior (2003) developed a water plan for 2025, which identifies four key tools to help prevent future conflicts and crises over water in the west. These tools are:

- **Conservation, efficiency and markets.** In many cases implementation of new water conservation and efficiency improvements through cooperative partnerships will result in an increased ability to meet otherwise conflicting demands for water. Most irrigation delivery systems were built during the first half of the twentieth century. These irrigation delivery systems can be modernised and retrofitted with new water management technologies. These include lining of the unlined canals, installation of automated control structures, and installation of remotely controlled pumping and canal facilities. Research has shown that for every US$1 spent on canal modernisation, an expected return of US$3 to US$5 in conserved water can be achieved. Also, for every US$1 spent on maintaining an existing canal lining, a return of up to US$10 in conserved water can be achieved.

Water banks and markets are essential to avoiding crises in critical areas of the west. These mechanisms should allow water to be shifted between competing water uses because they are based on recognition of the validity of existing water rights. Water banks also avoid

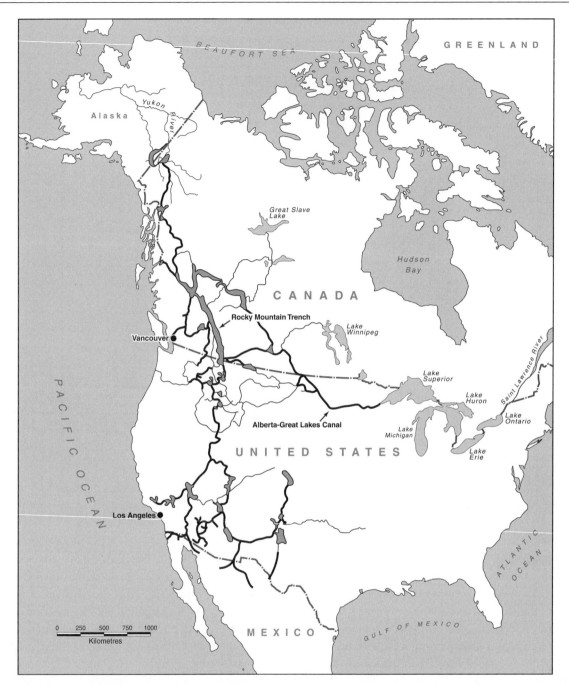

Figure 11.8 North American Water and Power Alliance System (Ralph M. Parsons Company, 1964, Figure 2).

or reduce the conflicts and crises. They can provide a mechanism for preserving irrigated agriculture and meeting other water supply needs. Also, permanent transfers of water can occur between willing buyers and sellers. Each transfer will take a few months at a cost of a few thousand dollars.

- **Collaboration.** Significant water supply issues must be addressed in advance of the crises. Collaborative processes that are based on recognition of the rights and interests of the stakeholders, fosters problem solving that maximises the opportunity for innovation and creativity. The integration of claims to water typically occurs in court proceedings. These proceedings can be complex and can take decades to complete. The U.S. Department of the Interior is committed to work with the states, tribes, and interested stakeholders to find ways to accelerate these proceedings in order to protect existing Federal and non-Federal rights.

A common element of many of the potential crises identified in Water 2025 is the need to provide water for people, cities, and farms in a manner that also attains the goals of the Federal *Endangered Species Act*. Success in meeting this challenge almost always requires a collaborative effort between stakeholders. The U.S. Department of Interior will partner states and local governments, tribes, water users and conservation groups to improve river systems.

- **Improved technology.** Wastewater, saline water and other low-quality sources of water can be purified to increase their utility. Water 2025's goal is to significantly aid technological advances and identify new supplies. Reducing desalination costs could enable the cost-effective treatment of brackish groundwater in traditionally water-short areas. The U.S. Department of Interior will facilitate implementation of desalination and advanced water treatment through improved inter-agency coordination of research and focused investment to areas that most need support.
- **Remove institutional barriers and increase inter-agency coordination.** In some areas of the west, Federal facilities have excess capacity during certain times of the year that could be used to satisfy unmet demands elsewhere. This excess capacity is, however, sometimes not available due to policy or legal constraints. In some cases, this additional capacity can be made available with appropriate changes in the policies of the Department of the Interior. In other cases, legislative action could be considered. The use of this excess capacity could minimise or avoid the need to build new infrastructure to satisfy additional water needs.

SECTION B: INTER-BASIN WATER TRANSFER IN CALIFORNIA

11.12 GEOGRAPHY AND POPULATION

The State of California covers an area of about 411 050 km^2. It has a varied geography and climate, which includes snow-capped mountains and dense forests in the north, fertile valleys, and scorching deserts in the south. Death Valley is the lowest and the hottest part of the nation. The average annual precipitation in California is about 584 mm. However, the average annual precipitation varies from more than 3500 mm in the north-western part of the State to less than 100 mm in the south-east. In the twentieth century alone, California experienced droughts in 1912–13,

Table 11.7. *Regional population of California in 2003*

Region	Population	Average annual population growth of 1990–2003 (in percent)
Southern coast	16 711 365	1.2
San Francisco Bay area	6 994 610	1.2
Inland Empire (Riverside and San Bernardino)	3 538 675	2.6
San Joaquin Valley	3 518 225	2.0
Sacramento Metropolitan area	1 932 625	2.2
Central coast	1 398 440	1.2
Rest of the State	1 497 565	1.3
Total	35 591 505	1.4

Source: Hanak and Simeti (2004, Table 2).

1918–20, 1922–24, 1929–34, 1947–50, 1959–61, 1976–77, and 1987–92.

California had a population of only 92 597 in 1850. Following the gold rush of 1849, its population increased rapidly and reached 1.5 million in 1900 (Grolier Incorporated, 1996, Vol. 5, p. 200b). Population increase continued and reached 10.6 million in 1950 and 35.6 million in 2003 (Table 11.7). It is estimated that California will have a population of about 48 million by 2030. Almost half the population increase is expected to occur in the water-scarce southern coastal region.

California has the strongest economy in the nation. The value of its gross domestic product in 2001 was US$1359.3 billion, or 13.4 percent of the total gross product of the United States (U.S. Census Bureau, 2003, Table 663). California's economy benefits from its natural resources. These resources have enabled the State to have a strong economy based on agriculture, manufacturing and tourism. The climate of the State provides it with a longer annual growing period than other states.

11.13 WATER SUPPLY AND DEMAND

The rapid increase in population of the State required development of water resources including transfer of water from one river basin to another. The predicted population growth of California means that demand for water will continue to increase with the potential for severe competition between users. Water is required to meet growing urban

Table 11.8. *California's water balance and water use summary for 1998, 2000 and 2001 (in $10^9 m^3$)*

Water balance and water use components	1998 Wet[a]	2000 Average[b]	2001 Dry[c]
A. Water balance components			
– Total supply (precipitation and imports)	415.6	240.2	179.5
– Evaporation, total use and outflows	408.4	247.3	197.1
Balance	7.2	−7.1	−17.6
B. Water use components (including reuse)			
– Urban uses[d]	9.6	11.0	10.6
– Agricultural uses	33.7	42.2	41.6
– Environmental water[e]	73.3	48.6	27.7
Total	116.6	101.8	79.9

Note: [a] 171 percent of average annual precipitation; [b] 97 percent of average annual precipitation; [c] 72 percent of average annual precipitation; [d] This includes industrial and commercial water uses; [e] Environmental water includes in-stream flows, wild and scenic flows, required Sacramento–San Joaquin Delta outflow, and managed wetlands water use.
Source: Department of Water Resources (2005b, Table 3.1).[21]

and industrial demands, agricultural production, and maintain streamflow for fish, recreation, water quality, salinity control, and navigation. In 2005, California had more than 190 reservoirs[22] with a total capacity of about $56 \times 10^9 m^3$ or 21 percent of the average annual precipitation of the State. In an average year, major sources of water supply and their contributions are (Department of Water Resources, 1998, Table ES3-1): groundwater ($15.4 \times 10^9 m^3$); local projects ($13.6 \times 10^9 m^3$), Central Valley Project ($8.6 \times 10^9 m^3$); Colorado River ($6.4 \times 10^9 m^3$);[23] and the State Water Project ($3.9 \times 10^9 m^3$). During dry years the contribution of surface water sources will be reduced while groundwater extraction will increase from about $15.4 \times 10^9 m^3$ to $19.5 \times 10^9 m^3$.

Table 11.8 shows the simplified water balance components and water use summary of the State for recent years. Year 1998 represents a recent wet year in California. Year 2000 is a representative average water year, and year 2001 provides a snapshot of a dry year. It shows that for the average year, urban (including industrial and commercial), agriculture, and environment had respective contributions of 11 percent, 41 percent and 48 percent of the total water use. In average years there is a net overdraft of water.

In estimating the water demand for 2030, the California Water Plan Update 2005,[24] has considered three plausible water demand scenarios. These scenarios do not include the impacts of climate change on water supplies and demands. Table 11.9 shows some factors and assumptions related to these scenarios.

The 2030 water demands have been estimated for urban, agricultural, and environmental sectors for the three scenarios. These estimates have been provided for each one of the ten Hydrological Regions[25] of California and for the State as a whole. Table 11.10 provides the estimated results for the State and compares them with the year 2000 water use data. It indicates a minor increase ($0.1 \times 10^9 m^3$) in water demand in 2030 under the Current Trends Scenario, slightly less under the Less Resource Intensive Scenario ($-0.4 \times 10^9 m^3$), and significantly more ($4.9 \times 10^9 m^3$) under the More Resource Intensive Scenario.

The pattern of a slight increase in water demand under the Current Trends Scenario may be surprising given projected population growth. This is because decline in agricultural water demand is slightly less than the sum of increase in urban and environmental demands. It should be noted that in each Scenario an additional $2.5 \times 10^9 m^3$ per year is required to eliminate groundwater overdraft over all the State. In general, California is expected to meet its water demands through the year 2030 if the State makes the right choice and investments (Department of Water Resources, 2005a).

Apart from water quantity, water quality is also of particular concern because in many areas surface water and

---

[21] http://www.waterplan.water.ca.gov/docs/cwpu2005/Vol_1/v1PRD.combined.pdf (visited in July 2005).
[22] http://cdec.water.ca.gov/misc/resinfo.html (visited in July 2005).
[23] This is above California's entitlement of $5.4 \times 10^9 m^3$ per annum (see description of the Colorado River Basin in Section 11.16).
[24] As above.
[25] These regions are: North Coast, San Francisco Bay, Central Coast, South Coast, Sacramento River, San Joaquin River, Tular Lake, Colorado River, North Lahontan, and South Lahontan.

Table 11.9. *Some of the factors and assumptions for the three California Water Plan 2005 scenarios for 2030*

Factor	Scenario 1 Current trends	Scenario 2 Less resource intensive	Scenario 3 More resource intensive
Population (in million)[a]	48.1	48.1	52.3
Commercial activity	Current trend	Increase in trend	Increase in trend
Industrial activity	Current trend	Increase in trend	Increase in trend
Crop area	Current trend	Level out at current area	Level out at current area
Crop unit water use	Current trend	Decrease in water use	Increase in water use
Environmental water flow	Current trend	High environmental flow	Year 2000 level of use
Naturally occurring conservation (NOC)	NOC trend	Higher than NOC	Lower than NOC
Urban water use efficiency	All cost effective BMPs[b]	All cost effective BMPs[b]	All cost effective BMPs[b]
Agricultural water use efficiency	All cost effective BMPs[b]	All cost effective BMPs[b]	All cost effective BMPs[b]
Per capita income	Current trend	Current trend	Current trend

Note: [a] Population for year 2000 is assumed to be 34.1 million; [b] Best Management Practices.
Source: Groves et al. (2005, Table 5).[26]

Table 11.10. *California's water use components in year 2000 and demand in 2030 for various scenarios (in $10^9 \, m^3$)*

Water use and demand components	Water use in year 2000	Demand in year 2030 for:		
		Current trends	Less resource intensive	More resource intensive
Urban	11.0	14.7	12.7	18.1
Agriculture	42.2	38.0	38.8	40.0
Environment	48.6	49.2	49.9	48.6
Total	101.8	101.9	101.4	106.7
Changes with respect to 2000	–	0.1	−0.4	4.9

Source: Groves et al. (2005, Tables 19 and 20).[27]

groundwater are being impaired by natural and human-made contaminants. These have effectively reduced the water supply that can be used. Contaminants degrade the environment, threaten human health, and increase water treatment costs.

11.14 WATER TRANSFER PROJECTS

In California, numerous water transfer facilities exist for the State's domestic, agricultural and industrial water supplies. These include: the Los Angeles Aqueducts; the Hetch Hetchy Aqueduct and the Mokelumne Aqueducts for water supply of San Francisco; the Central Valley Project; the State Water Project; water transfer from the Colorado River to Imperial Valley and Coachella Valley via the All-American Canal; and the Colorado River Aqueduct for water supply of Los Angeles. These facilities are described in the following.

11.14.1 Los Angeles Aqueducts

When the city of Los Angeles was founded in 1781, it depended on the Los Angeles River for its water supply. However, the population increased rapidly and reached about 85 000 in 1902. The daily water consumption was an extravagant 98 400 m³ or 1160 litres per head of population. It soon became clear that the local river would not be able to satisfy future demand. In 1903, introduction of a metering system reduced the daily water consumption to 750 litres

---
[26] http://www.waterplan.water.ca.gov/docs/cwpu2005/Vol_4/03-Data_and_Tools/V4PRD4-QUAN.PDF (visited in July 2005).
[27] As above.

Table 11.11. *Some features of the two Los Angeles Aqueducts*

Item	First Aqueduct	Second Aqueduct
Year completed	1913	1970
Construction duration	5 years	5 years
Capacity	13.7 m^3 s^{-1}	8.2 m^3 s^{-1}
Total cost	US$23 million (435 million in 2004 values)	US$89 million (430 million in 2004 values)
Average cost per km in 2004 values	US$1.21 million	US$1.95 million
Construction details	– Unlined channel: 38.6 km – Lined channel: 59.5 km – Concrete conduit: 157.7 km – Steel and concrete pipeline: 19.3 km – Tunnels 83.7 km	– Concrete conduit: 103 km – Steel pipeline: 111 km – Other facilities: 6.4 km
Total length	358.8 km	220.4 km

Source: Los Angeles Aqueducts Facts.[28]

per head. Estimates indicated that the population would increase to 390 000 by 1925. With the assumption of a daily water consumption of 570 litres per person, it was estimated that a daily water supply of more than 222 000 m^3 would be required, which could not be supplied by the Los Angeles River.[29]

In 1905–08 Los Angeles acquired water rights and riparian lands along the Owens River on the slopes of the Sierra Nevada to transfer water via an aqueduct. Construction of the Los Angeles Aqueduct commenced in 1908 and was completed in 1913 when the Los Angeles population reached 485 000. This project initially transferred about 185 × 10^6 m^3 of water per year (Howe and Easter, 1971, p. 6).

The Aqueduct was designed to deliver water to Los Angeles by gravity, and required no pumping along the route. Water carried by the aqueduct flows downhill from the Owens River, at an elevation of 1945 m, southward to the Los Angeles area. Using the gravity flow, engineers designed a number of hydro-electric power plants at suitable locations on the route (Kahrl, 1979, pp. 34 and 35).

In 1940 the Aqueduct was extended north to tap the waters of Mono Lake in the Sierra Mountains. In 1970, construction of a second aqueduct to Los Angeles was completed.[30] The two Aqueducts deliver an average of 594 × 10^6 m^3 of water per year to Los Angeles. Table 11.11 shows some features of the two aqueducts, while Figure 11.9 illustrates their locations.

Currently, there are three sources of water supply for Los Angeles: approximately 60 percent is supplied via the Los Angeles Aqueducts; 25 percent from the Metropolitan Water District (MWD) facilities,[31] which includes import of water via the Colorado River and California aqueducts; and 15 percent from the San Fernando groundwater basin. The groundwater basin acts as a vast underground reservoir where, during years of abundant rainfall, surface water is used to artificially recharge the aquifer so that it can be extracted during the dry years.[32]

During the 1977–78 drought, the City renewed efforts to strengthen its water conservation program. The result was one of the most aggressive and comprehensive on-going water conservation programs in California. The program educated residents in water conservation and reduced their water consumption. Also, a number of reclamation projects have been implemented to use reclaimed water for irrigating golf courses and parks.[33]

11.14.2 San Francisco Water Supplies

San Francisco was settled in 1776 by the Spanish. In 1846, after outbreak of the war with Mexico, the United States took over the area from Mexico. In 1848 gold was discovered near Sacramento and the ensuing gold rush rapidly transformed San Francisco into a booming community.

---

[28] http://wsoweb.ladwp.com/Aqueduct/historyoflaa/aqueductfacts.htm (visited in July 2005).
[29] http://wsoweb.ladwp.com/Aqueduct/historyoflaa (visited in July 2005).
[30] Further information regarding the history of the Los Angeles Aqueduct is available in Jones (1996), and Van Valen (1996).
[31] Established in 1928 by the cities in Los Angeles, Orange and San Bernardino counties.
[32] http://wsoweb.ladwp.com/Aqueduct/historyoflaa/waterquality.htm (visited in July 2005).
[33] http://wsoweb.ladwp.com/Aqueduct/historyoflaa/reclamation.htm (visited in July 2005).

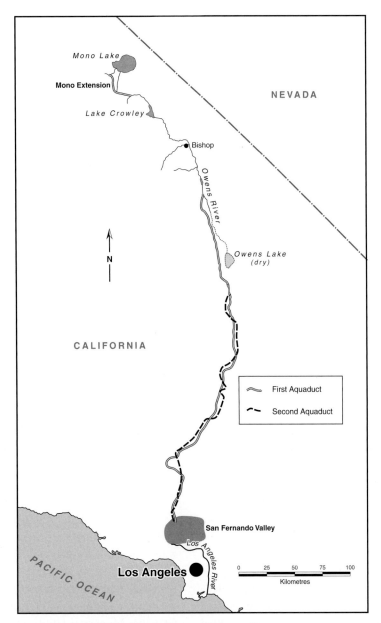

Figure 11.9 Los Angeles Aqueducts and Lake Mono Extension (Kahrl, 1979, p. 70).

By 1900, San Francisco had a population of more than 340 000. On 18 April 1906, a major earthquake caused a fire that raged for three days, destroying almost all of San Francisco's central business district and much of the residential area. However, it was rebuilt quickly.

Hetch Hetchy Aqueduct

In 1913, after many years of controversies, San Francisco acquired rights from the Federal Government to the Tuolumne River in the Yosemite National Park (Hundley, Jr, 2001, pp. 171–194). San Francisco's plan to build O'Shaughnessy[34] Dam in the Hetch Hetchy Valley was bitterly opposed by the famous environmentalist John Muir, founder of the Sierra Club[35] (Jones, 1965). However, Muir lost the battle in 1913, and San Francisco proceeded to build

---

[34] M. M. O'Shaughnessy was for twenty years (1912–32) the Chief of the San Francisco's Engineering Department, in charge of the Hetch Hetchy Project. Then, he became Consulting Engineer of the City's Public Utilities Commission.

[35] John Muir founded the Sierra Club in 1892 as a nonprofit conservation organisation. It is devoted to the study and protection of national scenic resources, particularly those of mountain regions. It is highly active in public environmental education. For further information visit http://www.sierraclub.org/history (visited in July 2005).

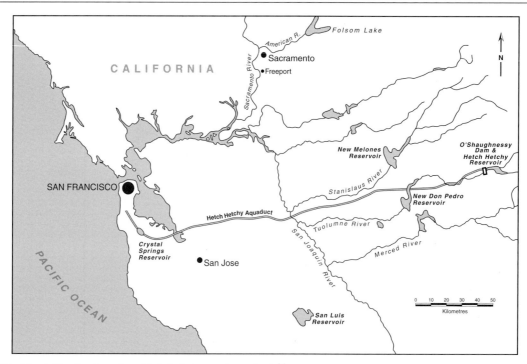

Figure 11.10 The Hetch Hetchy Aqueduct (Kahrl, 1979, pp. 70 and 71).[36]

the Hetch Hetchy Project. Work on the Hetch Hetchy Project began in 1914, eight years after the 1906 earthquake and fire. Water is carried from the Hetch Hetchy Reservoir, 249 km east of San Francisco, via tunnels and pipelines across the San Joaquin Valley (Figure 11.10). The first dam of the system (O'Shaughnessy Dam) was completed in 1923, and water was delivered to San Francisco in October 1934 through the Hetch Hetchy Aqueduct (O'Shaughnessy, 1934). The cost of the completed project was US$100 million (US$1.4 billion in 2004 values), which was US$23 million more than its initial estimate (Hundley, Jr, 2001, p. 193). In 1961, a decision was made to increase the Aqueduct's capacity. By the late 1970s San Francisco was importing nearly six times as much water as it did with the original project (Hundley, Jr, 2001, p. 194). Currently, the project supplies $440 \times 10^6 \, m^3$ of water per annum to San Francisco, which has a population of 2.4 million.

The Hetch Hetchy Project involved relatively simple engineering, but politically was very controversial and complex even in the early part of the twentieth century. The dispute over the impact of the Hetch Hetchy Project on the conservation of mountain areas continues today. Officials have long argued that the O'Shaughnessy Dam is essential for supplying San Francisco with high quality water as well as electricity. The Tuolumne River water, which comes from Yosemite National Park, is so pure that it does not need expensive filtration. However, environmentalists believe that this project was a monumental mistake, and San Francisco could manage without the Hetch Hetchy Reservoir, and they continue to request its removal.[37] Currently,[38] removal of the Dam is under investigation by the Department of Water Resources.

Mokelumne Aqueducts

The East Bay[39] Water Company was formed in 1916, and built the San Pablo and Upper San Leandro dams to supply water to San Francisco's East Bay Municipal District (Figure 11.11). A severe drought in 1918 emptied the reservoirs and caused widespread water shortages. In order to secure additional water from more distant sources, the East Bay Municipal Utility District (EBMUD) was formed in 1923. It was decided to import water from the Lower Mokelumne River in the Sierra Mountains. The Pardee Dam, with a height of 173 m and a reservoir capacity of $245 \times 10^6 \, m^3$, was constructed from 1925 to 1929. On 23 June 1929, water reached the East Bay via a 132.2 km aqueduct across the San Joaquin Valley, at a time when only a 21-day

---

[36] A more detailed map of the Aqueduct and its profile is available in O'Shaughnessy (1934).
[37] http://www.fresnobee.com/local/story/7946666p-8821920c.html (visited in July 2005).
[38] Early 2005.
[39] The East Bay refers to the area on the eastern shores of the San Francisco Bay which includes Richmond, Berkeley, Oakland, Fremont and others.

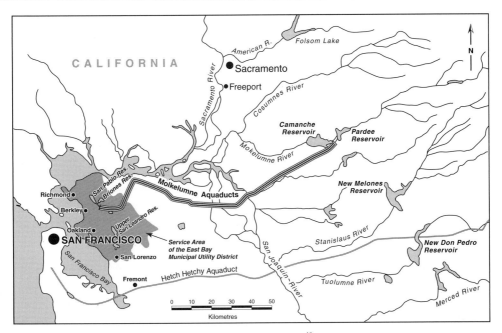

Figure 11.11 Mokelumne Aqueducts to the East Bay Municipal Utility District,[40] and the Hetch Hetchy Aqueduct.

supply of brackish water remained in local reservoirs (EBMUD, 2003).

The second and the third aqueducts were completed in 1949 and 1963 respectively. The Camanche and Briones reservoirs, with respective capacities of $515 \times 10^6 \, m^3$ and $74.6 \times 10^6 \, m^3$, were completed in 1964. The three aqueducts have respective diameters of 256.5 cm, 170.2 cm and 221 cm (EBMUD, 2001, pp. 2–10). On average, 95 percent of the water delivered to 1.4 million people in the EBMUD service area originates from the Mokelumne River Basin, and the remaining five percent is run-off within the service area (EBMUD, 2001, p. 1-1).

Table 11.12 shows that with the availability of water transferred from the Lower Mokelumne River, water use per head of population increased from $288 \, L \, d^{-1}$ in 1930 to an alarming rate of $757 \, L \, d^{-1}$ in 1970. By undertaking conservation measures, water consumption per head of population declined to $595 \, L \, d^{-1}$ in 2000, a reduction of $162 \, L \, d^{-1}$ or 21 percent.

The EBMUD's water supply system is vulnerable to a number of factors. The most important are the occurrences of drought, earthquake and flooding. In October 1993, following the 1976–77 and the 1986–92 droughts, EBMUD adopted a long-term Water Supply Management Program (WSMP), which identified actions and projects necessary to manage the water supply through to 2020. Key elements of the WSMP include: water conservation, water recycling, groundwater storage for drought years, desalination, and new storage and distribution facilities.[42]

Further changes to the water system were initiated in 2003 when EBMUD completed the draft environmental documentation for the Freeport Regional Water Project. This is a joint effort between EBMUD and Sacramento County, to draw water from the Sacramento River near the town of Freeport (Figure 11.11) and deliver up to $378\,500 \, m^3$ of water per day to the East Bay customers during drought, and up to $322\,000 \, m^3$ per day to Sacramento County. The EBMUD

Table 11.12. *Water use in the East Bay Municipal Utility District*

Year	Population	Total daily water use (m³)	Daily water use per head (L)
1930	460 000	132 500	288
1940	519 000	166 500	320
1950	851 000	412 600	485
1960	987 000	579 100	587
1970	1 100 000	832 700	757
1990	1 200 000	727 700	606
2000	1 400 000	833 000	595

Source: District History.[41]

---

[40] http://www.ebmud.com/water_&_environment/water_supply/system_maps/water_supply_system2003.pdf (visited in July 2005).
[41] http://www.ebmud.com/about_ebmud/overview/district_history/default.htm (visited in July 2005).
[42] http://www.ebmud.com/water_&_environment/water_supply/current_projects/default.htm (visited in July 2005) and EBMUD (2001).

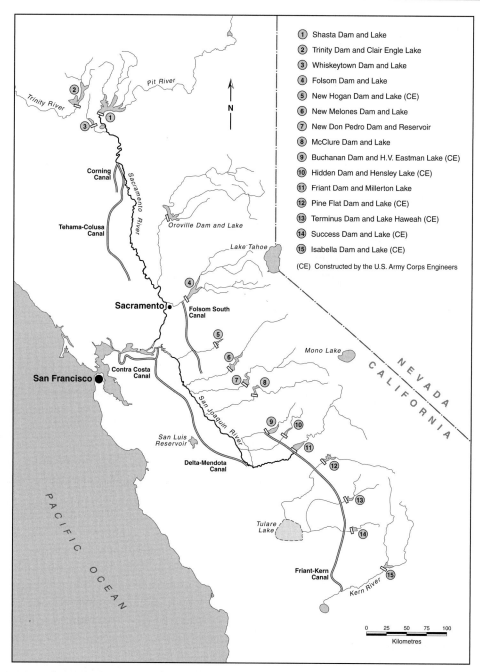

Figure 11.12 Some of the hydraulic structures of the Central Valley Project (based on the USBR's Central Valley Project map no. 214-208-5133 and Kahrl, 1979, pp. 70 and 71).

water supply system will also be connected to San Francisco's Hetch Hetchy system to provide back-up water supplies for emergencies (EBMUD, 2003).

11.14.3 Central Valley Project

California's Central Valley (Figure 11.12) is one of the major agricultural regions of the United States. It consists of two major watersheds, the Sacramento River in the north and the San Joaquin River in the south. The two rivers join together at the Sacramento–San Joaquin Delta (Delta) and eventually discharge into San Francisco Bay. Johnson *et al.* (1993) describe in detail the physical features, native people, history of settlement, economic development and environmental degradation (including loss of native vegetation and wildlife, salinity problems, and land subsidence) of the

Central Valley. Information on the Delta, including hydrology, water quality, soils, agriculture, and flood control structures, is available from Department of Water Resources (1993).

De Roos (1948) outlines the early history of the Central Valley Project (CVP) when the project consisted of three dams, five canals, and three hydro-electric power plants. A recent and detailed history of the CVP and its divisions is available in Autobee et al. (unpublished). The following summary is mainly based on this document.[43]

The idea of transferring excess water from the Sacramento River to the San Joaquin Valley was first proposed in the 1870s. However, after decades of planning and debate, no progress had been made. In 1919, Colonel Robert Bradford Marshall put forward a plan to build storage reservoirs along the Sacramento River system to transfer water to San Joaquin Valley via two large canals on both sides of the Sacramento River. This plan earned him the nickname of *The Father of the Central Valley Project*. At that time the government of California had become interested in a comprehensive water plan for the State. Between 1920 and 1932 approximately fourteen reports detailed water flow, drought, flood control and irrigation issues in California. In 1933, the California Legislature authorised the future Central Valley Project as a State project. However, because the State government was unable to finance the project, assistance was requested from the Federal Government. President Roosevelt finally approved the Central Valley Project on 2 December 1935. The *Rivers and Harbors Act* of 1937 authorised the CVP and construction of the project started in the late 1930s and continued through 1940s and 1950s. By the end of the 1960s it was almost completed. The CVP is owned and managed by a conglomeration of various Federal and State government agencies.[44]

The 1960s marked the end of the era of large dam building in the United States. In the 1970s environmental concerns started to gain widespread support. President Richard Nixon signed the *Endangered Species Act* in 1973. The CVP felt the consequences of the *Act* because of its impacts on migratory salmon. In 1969, the population of winter-run Chinook salmon was about 118 000 at Red Bluff Diversion Dam, located about 70 km below Shasta Dam on the Sacramento River (Figure 11.12). By 1990, the population had dropped to less than 5 percent of the 1969 level.

The *Central Valley Project Improvement Act* (CVPIA) of 1992 moved the CVP in a new direction. The CVPIA reallocated $985 \times 10^6 \, m^3$ ($740 \times 10^6 \, m^3$ in dry years) of the $8.6 \times 10^9 \, m^3$ of annual CVP water diversion to in-stream uses to regenerate salmon runs. This was proclaimed a victory by environmentalists, but was viewed as a disaster by the farmers.

In spite of the social, environmental and political controversies, CVP remains an impressive accomplishment. It contains about 75 percent of the irrigated land in California and ranks first among the Reclamation projects in the value of flood damage prevented between 1950 and 1991 when the CVP prevented more than an estimated US$5 billion in flood damage. The value of CVP annual farm production exceeds the total value of all gold mined in California since 1848.

The CVP is the largest water storage and delivery system in California. The main features of the CVP are:[45]

- Twenty-one dams and reservoirs, eleven power plants and 800 km of major canals, conduits, tunnels and related facilities.
- Management of $11 \times 10^9 \, m^3$ of water per year.
- Diversion of about $8.6 \times 10^9 \, m^3$ of water per year for agriculture, urban and wildlife.
- Provision of about $6.2 \times 10^9 \, m^3$ of water per year to farmers to irrigate 1.2 Mha of land.
- Provision of about $740 \times 10^6 \, m^3$ of water for municipal and industrial use, enough to supply about one million households with their water needs each year.
- Generation of enough electricity to meet the needs of about two million people.
- Dedication of $985 \times 10^6 \, m^3$ of water per year to fish and wildlife and their habitat and $505 \times 10^6 \, m^3$ to State and Federal wildlife refuges and wetlands.

Technical features of 21 dams and their associated reservoirs, 11 hydro-electric power plants, 10 canals, 5 tunnels, 2 diversion dams, 7 pumping stations, and 4 conduits are available in Autobee et al. (unpublished) and also at the websites provided in Appendix I (Table I.1).

Table 11.13 provides some technical features of five major dams and reservoirs of the CVP, while technical features of five power plants and four canals are provided in Appendix I (Tables I.2 and I.3).

Within the CVP there are several projects constructed by the U.S. Army Corps of Engineers that have been fully or partially integrated into the CVP (see Appendix I, Table I.4).

---

[43] The introductory chapter of this document by Eric A. Stene is available at the URL http://www.usbr.gov/history/cvpintro.htm (visited in July 2005).
[44] These included: the U.S. Bureau of Reclamation; U.S. Army Corps of Engineers; and the California Department of Water Resources.
[45] http://www.usbr.gov/dataweb/html/cvp.html (visited in July 2005).

Table 11.13. *Selected technical features of five major dams and reservoirs of the Central Valley Project, in order of capacity*

No.	Name	Type	Year of completion	Height (m)	Capacity ($10^6 \, m^3$)	Surface area (km^2)
1	Shasta Dam and Lake	Curved concrete gravity with earthfill wings	1945	183.5	5613	120.4
2	Trinity Dam and Clair Engle Lake	Zoned earthfill	1961	164.0	3020	66.9
3	New Melones Dam and Lake	Rockfill with earth core	1979	190.5	2960	50.6
4	Folsom Dam and Lake	Concrete gravity with earthfill wings	1965	103.6	1246	46.3
5	Friant Dam and Millerton Lake	Concrete gravity	1944	97.2	641	18.6

Source: Autobee *et al.* (Unpublished, Appendix).

11.14.4 State Water Project

The California Department of Water Resources (DWR) was created in July 1956 to plan and guide the development of the State's water resources. Planning for the State Water Project (SWP) began when it became evident that local developments and the Central Valley Project could not keep pace with the water requirements of the State's rapidly growing population. The following description of the Project is based mainly on the Department of Water Resources (1999).

Work on the SWP started in May 1957 with an initial emergency funding of US$25.2 million (US$168 million in 2004 prices). Although the Legislature continued to allocate funds to build units of the SWP, it was not until 1959 that the *California Water Resources Development Bond Act* was passed. The *Act* authorised funding of US$1.75 billion (US$11.25 billion in 2004 prices) in general obligation bonds to finance the Project.

The SWP is a water storage and delivery system of reservoirs, aqueducts, tunnels, power plants and pumping plants stretching from north of Sacramento to Perris Dam and Lake in the south (Figure 11.13). Its main purpose is to store water and distribute it to 29 urban and agricultural water agencies in northern California, San Francisco Bay Area, San Joaquin Valley, Central Coast, and southern California. These 29 agencies have long-term contracts for water entitlements through to the year 2035 with a maximum annual entitlement of about $5.15 \times 10^9 \, m^3$. Of the contracted water supply, 70 percent goes to urban users and 30 percent goes to agricultural users.

The SWP conveys an annual average of $3 \times 10^9 \, m^3$. It consists of 32 storage facilities; 8 hydro-electric power plants, 17 pumping plants; and 1065 km of canals, pipelines and tunnels (Table 11.14). It is the State's fourth largest energy supplier and the single largest user of power. Selected technical features of the SWP's major components are presented in Appendix J. Other features and technical details of all components of the Project (locations, dimensions, capacities, plans, cross-sections, photos, etc.) are available in the Department of Water Resources (1999).

The San Luis Unit (Sisk Dam, San Luis Reservoir, etc.) located on the west side of the San Joaquin Valley, is the result of a partnership between the State of California and the Federal Government. The State financed and owns 55 percent of the Unit, while the remaining 45 percent belongs to the Federal Government. San Luis Reservoir is the nation's largest off-stream man-made lake (Autobee *et al.*, Unpublished). Water in the Lake originates from the Sacramento–San Joaquin Delta, which is pumped by the Harvey O. Banks Pumping Plant and transferred south to the San Luis Reservoir via the California Aqueduct (Figure 11.13).

While the SWP was built primarily for water supply, the project and its facilities also provide the people of California with many other benefits which include: flood control, recreation, fish and wildlife enhancement, power, and salinity control. With respect to salinity control, the SWP in coordination with CVP is operated to limit salinity intrusion into the Sacramento–San Joaquin Delta and Suisun Marsh. This is accomplished by supplementing freshwater outflow to San Francisco Bay and limiting water export from the Delta during specific times of the year.

The SWP made its first deliveries in 1962 to Alameda County. In 1965, deliveries were made to the Santa Clara Valley, where imported supplies were used to solve a land subsidence problem caused by long-term water extraction from a local groundwater basin. In 1968, service was extended into the central and southern San Joaquin Valley, and by 1972, southern California began receiving their first deliveries. On 24 November 1987, after more than 25 years of negotiations and Congressional approval, Department of Water Resources and U.S. Bureau of Reclamation authorities signed the *Coordinated Operation Agreement*, which

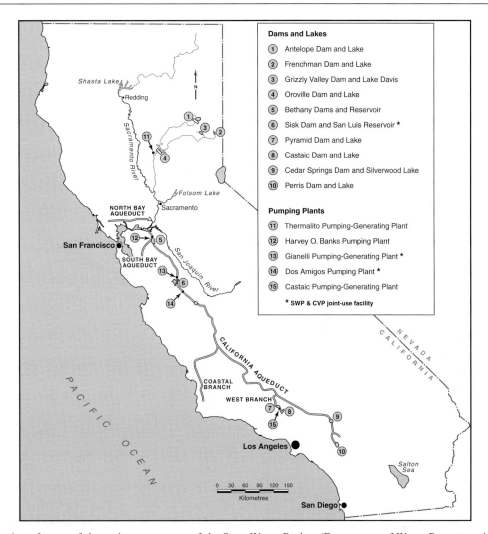

Figure 11.13 Location of some of the main components of the State Water Project (Department of Water Resources, 1999).

Table 11.14. *Main features of the State Water Project*

Feature	Unit
Number of storage facilities	32
Total reservoir storage	$7.2 \times 10^9 \, m^3$
Largest reservoir capacity (Oroville)	$4.3 \times 10^9 \, m^3$
Highest dam (Oroville)	234.7 m
Highest dam crest elevation	1763 m
Length of canals and pipelines	1065 km
Number of hydro-electric power plants	8
Number of coal-fired plants	1
Average annual energy generation	7.6 billion kWh
Average annual energy use	12.2 billion kWh
Number of pumping units	17
Highest pump lift (Edmonston Pumping Plant)	587 m

Source: Department of Water Resources (1999, p. 29).

initiated a new era in cooperation between the SWP and the CVP.

In December 1994, representatives of the DWR and the SWP contractors signed the *Monterey Agreement* that refined the way DWR administers its long-term water contracts. Major issues covered by the agreement include:

(1) adjustment in water allocations among contractors in years of short supply;
(2) future transfers of entitlements from agricultural contractors to urban users;
(3) financial restructuring to establish a SWP-operating reserve and water rate management; and
(4) added operational flexibility such as storage of SWP water in non-SWP surface storage facilities for later use, expanded rules for carryover in SWP conservation reservoirs, and no limits for groundwater storage of SWP water outside a contractor's service area for later use within the service area.

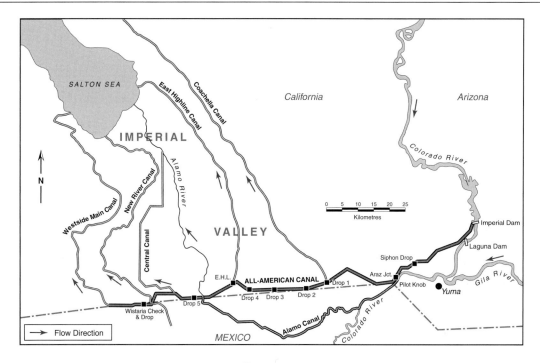

Figure 11.14 The Imperial Valley and the All-American Canal.[46]

11.14.5 Water Transfer from the Colorado River to California

Colorado River[47] water is transferred to Imperial Valley and Coachella Valley via the All-American Canal and its branch, the Coachella Canal (Figures 11.14 and 11.15), and to Los Angeles and San Diego via the Colorado River Aqueduct (Figure 11.15). These facilities are described in the following sections. However, it should be noted that water transfer from the Colorado River to the Imperial Valley and the Coachella Valley are not considered as inter-basin water transfer, because these areas are part of the Colorado River Basin. These will be described for their importance in development of a very dry area of the State and also for major water-saving activities in these areas. The objective of water saving is to better manage California's share of the Colorado water and to assist water supply of Los Angeles and San Diego by transferring the saved water to these population centres via the Colorado River Aqueduct.

The Imperial Valley

The Imperial Valley in southern California, lies to the west of the Colorado River (Figures 11.14 and 11.15). The Valley is surrounded on the north, east and west by mountains and the entire valley floor is below sea level. The United States–Mexican border forms the southern boundary of the Valley. Winters are mild and dry with maximum temperatures in the range of 18–24°C. Summers are extremely hot with daily maximum temperatures of 40–46°C. The average annual rainfall is only about 75 mm. The Alamo and New rivers flow from Mexico into the Salton Sea. These rivers follow old, shallow riverbeds along which the Colorado River once flowed before its normal channel became silted up and overflowed its banks in a south-westerly direction (National Research Council, 1992, pp. 235–236).

Development of the Imperial Valley has always been dependent on the availability of water. The quest to bring water from the Colorado River to irrigate land in Imperial Valley began in the 1850s. However, it was not until 1901 that the California Development Company established by Charles Rockwood and George Chaffy built the Alamo Canal to deliver water by gravity from the Colorado River to the southern end of the Imperial Valley. Promotion for settlement in the Imperial Valley began in 1901 by the California Development Company, which renamed the Valley from *Valley of the Death* to *Imperial Valley*. Within 8 months, there were two towns, 2000 settlers and about 40 000 ha of land ready for harvesting (Reisner, 1987, pp. 127–128). Between 1901 and 1904 about 7000 people

---

[46] http://www.iid.com/water/works-allamerican-map.html (visited in July 2005).
[47] For a brief description of the Colorado River Basin, its water resources and salinity, see Section 11.16 and 11.17 of this Chapter.

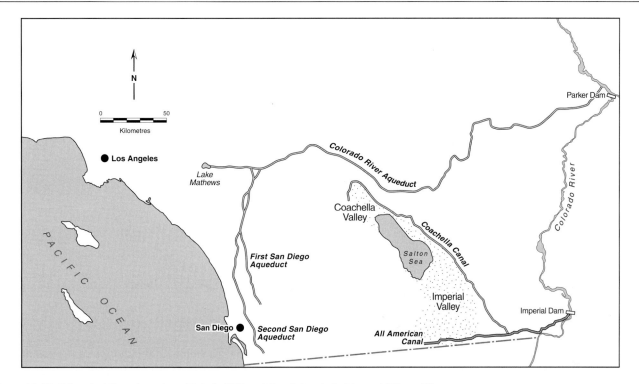

Figure 11.15 Colorado River Aqueduct (Kahrl, 1979, p. 25 and *Reader's Digest*, 1998, p. 89).

settled in the Valley. By 1915, 121 000 ha were under cultivation. In 1999 Imperial Valley had 231 600 ha of harvested crops including vegetables, alfalfa, wheat, fruit, and nuts, and 325 000 head of cattle in feedlots. The total value of its crops and livestock was about US$1.05 billion. Imperial Valley had a population of 142 100 in 1999.

All-American Canal

The Imperial Irrigation District (IID) was formed in 1911. Because its main canal (Alamo Canal) and levees were located in Mexico, the IID had little security for its water supplies. The IID recognised the need for construction of the All-American Canal along the US–Mexico international border. The U.S. Bureau of Reclamation constructed the All-American Canal during the 1930s. In 1940 the first water was delivered to Imperial Valley. By 1942, the All-American Canal was the sole water source for Imperial Valley residents and farmlands.[48]

California has the right to divert $5.4 \times 10^9 \, m^3$ of Colorado River water per year. The IID diverts $3.2 \times 10^9 \, m^3$ of this volume (National Research Council, 1992, p. 239) via the All-American Canal and its branch to the Imperial Valley and the Coachella Valley. Technical features of the All-American Canal are listed in Table 11.15. It delivers water to nine population centres and more than 230 000 ha of irrigated land through the Imperial Valley. Irrigation water is delivered via 30 000 km of irrigation canals and ditches.[49]

The IID has about 1700 km of surface drains and 10 regulating reservoirs with a total storage capacity of $4 \times 10^6 \, m^3$.[50]

Water use by the IID has been very inefficient partly due to water losses from unlined canals. In 1984, the State Water Resources Control Board concluded that the IID's use of water was unreasonable under Californian Law and constituted waste. The USBR issued a report in 1985 that identified measures that could conserve $436 \times 10^6 \, m^3$ of water per year (National Research Council, 1992, p. 240).

The IID and the Los Angeles Metropolitan Water District (MWD) signed the *Water Conservation Agreement* in 1989 in order to transfer annually about $131 \times 10^6 \, m^3$ of conserved irrigation water from the IID to MWD for a period of 35 years. The total capital cost was estimated at US$97.8 million plus US$23 million in indirect costs (National Research Council, 1992, pp. 241–242). This program is currently in effect and the agreed volume of water is transferred annually to the MWD.[51]

The All-American Canal was originally unlined. However, the USBR concluded that an estimated $86 \times 10^6 \, m^3$ per year

---

[48] http://www.iid.com/water/works-allamerican.html (visited in July 2005).
[49] http://commserv.ucdavis.edu/CEImperial/overview.htm (visited in July 2005).
[50] http://www.iid.com/water/works-reservoirs.html (visited in July 2005).
[51] Tina Shields, IID (personal communication, May 2004).

Table 11.15. *Technical features of the All-American Canal*

No.	Components		Dimensions
1	Length		130 km
2	Width		46–61 m
3	Depth		2.1–6.1 m
4	Total drop		53.3 m
5	Capacity	From Imperial Dam to Siphon Drop Power Plant	430 m^3 s^{-1}
		From Siphon Drop Power Plant to Pilot Knob	430 m^3 s^{-1}
		From Pilot Knob to Drop No.1	288 m^3 s^{-1}
6	Diversions	Yuma Project at Siphon Drop and other outlets on the canal upstream from Siphon Drop	56.6 m^3 s^{-1}
		Coachella Main Canal at Drop No. 1, about 32 km west of Yuma	70.8 m^3 s^{-1}
7	Cost in 1944 values		US$25 million
8	Cost in 2004 values		US$268 million

Source: Website of the All-American Canal.[52]

of water was lost due to seepage along a 37 km section of the Canal running through the sand dunes from Pilot Knob to Drop 3 (Figure 11.14). In 1998, a decision was made to concrete line part of the Canal and its Coachella branch at a cost of US$235 million. The concrete lining of the Canal is expected to be completed in 2006.[53]

In October 2003, the IID Board of Directors approved an agreement with the San Diego County Water Authority (SDCWA), which had been under negotiation since 1998. The objective of this long-term agreement (45 years with a possible 30 years extension) is to transfer annually 247×10^6 m^3 of conserved water from the Imperial Valley to San Diego.[54] The IID would conserve water through on-farm and other projects. Water transfer would increase at a rate of 24.7×10^6 m^3 per annum to reach its maximum in 10 years. Implementation of the agreement has the potential to benefit California in several ways including:[55]

(1) reduce reliance on State Water Project deliveries; and
(2) relieve pressure on the Sacramento–San Joaquin Delta.

Moreover, the agricultural water saving projects in the IID, and implementation of the MWD and SDCWA agreements will reduce by an equivalent amount the level of Colorado River water use in California and will assist the State to resolve disputes with the other Colorado River Basin states, which requested California to reduce its water intake to its entitlement of 5.4×10^9 m^3 (Colorado River Board of California, 2000, p. 37, and the URL provided in the footnote).[56]

The first hydro-electric power plants on the All-American Canal were completed at Drops 3 and 4 in 1941 and at Drop 2 in 1953. The Pilot Knob went into operation in 1957. The Drop 5 installation was completed in 1982 and the Drop 1 and East Highline Turnout plants were commissioned in 1984.[57] Further information regarding the Imperial Irrigation District is available at the IID's website.[58]

Coachella Canal

The original 197 km Coachella Canal was a feature of the Coachella Division of the All-American Canal System. The Coachella Canal delivers Colorado River water from the All-American Canal to irrigate 32 000 ha of agricultural land in the Coachella Valley (Figure 11.15). Originally the first 138 km were unlined and the remaining 59 km were concrete lined. The capacity of the original canal was 70.8 m^3 s^{-1} at the turnout from the All-American Canal and this decreased through successive reaches to 36.8 m^3 s^{-1} at the beginning of the last concrete lined section,[59] due to seepage and evaporation losses.

After completion of the canal in 1948, seepage losses became evident along the unlined canal. The first 78.5 km of

---

[52] http://www.iid.com/water/works-allamerican.html (visited in July 2005).
[53] http://www.iid.com/water/works-allamerican.html (visited in July 2005).
[54] http://www.iid.com/water/agmt/index.shtml (visited in July 2005).
[55] http://www.iid.com/water/qa.html (visited in July 2005).
[56] http://www.iid.com/water/qa.html (visited in July 2005).
[57] http://www.iid.com/water/works-allamerican.html (visited in July 2005).
[58] http://www.iid.com (visited in July 2005).
[59] http://www.usbr.gov/dataweb/html/crbcspccu.html (visited in July 2005).

the unlined section traversed coarse sandy soils where the most severe seepage occurred. The average annual seepage rate was estimated at $208 \times 10^6 \, m^3$, one-third of the annual diversion of $614 \times 10^6 \, m^3$. It was decided to replace the initial 78.5 km of the unlined section of the canal with a new concrete-lined canal. It was projected that seepage losses would be reduced significantly, resulting in an annual saving of $163 \times 10^6 \, m^3$ of water. Construction of the relocated portion of the Coachella Canal started in 1979 and was completed in 1982.[60]

The Colorado River Aqueduct

Construction of the Colorado River Aqueduct (Figure 11.15) was commenced by the MWD in 1933 and delivery of Colorado River water to Greater Los Angeles commenced in 1941.

The Colorado River Aqueduct is an important source of water supply for south-western California. Annually, it imports about $1.5 \times 10^9 \, m^3$ of Colorado River water to this region.[61] The Aqueduct is 389 km long and crosses the harsh Colorado Desert. There are five pumping stations along the route lifting water about 493 m. The Aqueduct starts at Lake Havasu created by the Parker Dam and ends in Lake Mathews. From Lake Mathews, water is transported and distributed to Greater Los Angeles. Approximately 45 km east of Lake Mathews, two aqueducts transport water to the San Diego region (Figure 11.15).

11.15 MAJOR MANAGEMENT PROGRAMS AND STRATEGIES

Numerous programs and strategies have been developed to meet the water needs of California's increasing population and at the same time satisfy the environmental requirements. The California Water Plan, CALFED-Bay Delta Program, water conservation, water trading, water recycling, conjunctive management of surface and groundwater resources, water desalinisation, precipitation enhancement, and ecosystem restoration are all components of the State's resource management strategies, and are described briefly in the following sections. Information regarding other resource management strategies[62] is available in the Department of Water Resources (2005c).

11.15.1 California Water Plan

The California Water Plan is a continuing strategic plan for managing the State's water needs. The first plan was published in 1957 by the DWR and has been updated seven times between 1966 and 1998.[63] A 1991 amendment to the *California Water Code* directed the DWR to update the plan every five years. By statute, the California Water Plan cannot mandate actions nor authorise spending for its recommendations. The latest Water Plan update is for 2005. The following summary of California Water Plan 2005 is based on the information available on the website of the California Water Plan Update 2005.[64]

The California Water Plan Update 2005 addresses the State's changing water management and better reflects the roles of the State and Federal Governments and the growing role of regional and local agencies in California water management. It goes beyond trying to forecast and quantify a simple gap between supply and demand. It is a roadmap for meeting the State's water demand through the year 2030. Update 2005 charts a Framework for Action that will help sustain California's water resources use and management to ensure that water is available where and when it is needed. Its new features include a strategic plan with vision, mission, goals and recommendations.

Vision

The vision statement of Update 2005 describes the desired future for California water resources and management and serves as a foundation for water planning:

> California's water resources management preserves and enhances public health and the standard of living for

---

[60] http://www.usbr.gov/dataweb/html/crbcspccu.html (visited in July 2005).

[61] http://hoover.sandi.net/course/STRAND/WtrProj/Period6/documents/colorado6.html (visited in July 2005).

[62] Agricultural lands stewardship, conveyance, drinking water treatment and distribution, economic incentives (loans, grants and water pricing), floodplain management, groundwater/aquifer remediation, matching water quality to use, pollution prevention, recharge area protection, surface storage, system re-operation, urban land use management, urban run-off management, water-dependent recreation, watershed management, and others.

[63] The titles and publication dates of these updates are: (1) Implementation of the California Water Plan (1966), (2) Water for California: The California Water Plan, Outlook in 1970, (3) The California Water Plan: Outlook in 1974, (4) The California Water Plan: Projected Use and Available Water Supplies to 2010 (1983), (5) California Water Plan: Looking to the Future, Bulletin 160-87 (1987), (6) California Water Plan Update: Bulletin 160-93 (1994), and (7) California Water Plan Update: Bulletin 160-98 (1998).

[64] http://www.waterplan.water.ca.gov/cwpu2005/index.cfm#vol1 (visited in July 2005). California Water Plan Update 2005 is organised in five volumes: (1) Strategic Plan; (2) Resource Management Strategies; (3) Regional Reports; (4) Reference Guide; and (5) Technical Guide.

Californians; strengthens economic growth, business vitality, and the agricultural industry; and restores and protects California's unique environmental diversity.

Mission

The mission statement describes the water plan's unique purpose and its overarching reason for existence. It identifies what it should do and why and for whom it does it:

> To develop a strategic plan that guides State, local, and regional entities in planning, developing, and managing adequate, reliable, secure, affordable, and sustainable water of suitable quality for all beneficial uses.

Goals

The following goals are the desired outcome of Update 2005 over its planning horizon to 2030. The goals are founded on the statewide vision. Meeting the goals requires coordination among the State, Federal, and local governments and agencies.

- State government supports good water planning and management through leadership, oversight, and public funding.
- Regional efforts play a central role in California water planning and management.
- Water planning and urban development protect, preserve, and enhance environmental and agricultural resources.
- Natural resources and land use planners make informed water management decisions.
- Water decisions are equitable across all communities.

Recommendations

The California Water Plan Update 2005 has made 14 recommendations and for each one has provided an action plan, intended outcomes, resource assumptions, implementation challenges and performance measures.[65] These recommendations are directed at decision makers throughout the State, the executives and legislative branches of State Government, and DWR and other State agencies. These recommendations are:

- **1 – Diversify regional water portfolios.** California needs to invest in reliable, high quality, sustainable and affordable water conservation, efficient water management and development of water supplies to protect public health, and to maintain and improve California's economy, environment and standard of living.
- **2 – Promote and implement integrated regional water management.** State government must provide incentives and assist regional and local agencies and governments and private utilities to: (1) prepare integrated resource and drought contingency plans on a watershed basis; (2) diversify their regional resource management strategies; and (3) empower them to implement their plans.
- **3 – Improve water quality.** State government must lead an effort with local agencies and governments to inventory, evaluate, and propose management strategies to remediate the causes and effects of contaminants on surface and groundwater quality.
- **4 – Maintain and improve aging water infrastructure.** California needs to rehabilitate and maintain its aging water infrastructure, especially drinking water and sewage treatment facilities, operated by State, Federal and local entities.
- **5 – Implement the CALFED[66] Program.** State government must continue to provide leadership for the CALFED-Bay-Delta Program (see the following section) to ensure continued and balanced progress on greater water supply reliability, water quality, ecosystem restoration and levee system integrity.
- **6 – Provide effective State government leadership, assistance, and oversight.** State government needs to take the lead in water planning and management activities that: (1) regions cannot accomplish on their own; (2) the State can do more efficiently; (3) involve interregional, interstate, or international issues; or (4) have broad benefits.
- **7 – Clarify State, Federal and local roles and responsibilities.** California needs to define and articulate the respective roles, authorities, and responsibilities of State, Federal, and local agencies and governments responsible for water.
- **8 – Develop funding strategies and clarify role of public investments.** California needs to develop broad and realistic funding strategies that define the role of public investments for water and other water-related resource needs over the next quarter of a century.
- **9 – Invest in new water technology.** State government should invest in research and development to help local agencies and governments implement promising water technologies more cost effectively.
- **10 – Adapt to global climate change impacts.** State government should help predict and prepare for the

---

[65] http://www.waterplan.water.ca.gov/docs/cwpu2005/Vol_1/v1PRD.combined.pdf (visited in July 2005).
[66] **CAL**ifornia **FED**eral.

effects of global climate change on the State's water resources and water management systems.

- **11 – Improve water data management and scientific understanding.** DWR and other State agencies should improve data, analytical tools, and information management needed to prepare, evaluate, and implement regional integrated resource plans and programs in cooperation with Federal, tribal, local and research entities.
- **12 – Protect public trust resources.** DWR and other State agencies should explicitly consider public trust values in the planning and allocation of water resources and protect public trust uses whenever feasible.
- **13 – Increase tribal participation and access to funding.** DWR and other State agencies should invite, encourage, and assist tribal government representatives to participate in statewide, regional, and local water planning processes and to access State funding for water projects.
- **14 – Ensure environmental justice across all communities.** DWR and other State agencies should encourage and assist representatives from disadvantaged communities and vulnerable populations, and the local agencies and private utilities serving them, to participate in statewide, regional, and local water planning processes and to get equal access to State funding for water projects.

11.15.2 CALFED-Bay Delta Program

The Bay–Delta (San Francisco Bay–Sacramento San Joaquin Delta) is one of California's unique and valuable resources. The Delta covers an area of about 3000 km^2, and is interlaced with hundreds of kilometres of waterways (Department of Water Resources, 1993). A large part of the Delta is between 3 and 5 m below sea level and relies on more than 1700 km of levees for protection against flooding. The Delta is the major collection and distribution point for the SWP and the CVP that serves over 20 million people and large tracts of farmland in central and southern California. Also, the Delta is a rich agricultural region. Before construction of Shasta Dam in 1943 on the headwaters of the Sacramento River, intrusion of saline water from the Pacific Ocean was widespread in the Delta during the dry seasons. Since then, the extent of the area affected by salinity has decreased markedly because of the release of fresh water. The Delta is the habitat for 230 species of birds, 45 species of mammals, 52 species of fish, 25 species of reptiles and amphibians, and 150 species of flowering plants. Anadromous fish species include salmon, striped bass, steelhead trout,[67] American shad and sturgeon.

By the mid-1980s the Delta was facing numerous major problems, including an inadequate levee system, decreasing water supply, pollution by selenium, mercury and pesticides, saltwater intrusion, loss of wildlife, and degradation of wetlands (Hundley, Jr., 2001, pp. 398–405).

In June 1994, State and Federal agencies[68] signed the *Framework Agreement* to develop a long-term plan to:

(1) restore the health of the San Francisco Bay/Sacramento–San Joaquin Delta ecosystem by assuring availability of sufficient water to meet fishery protection and restoration needs;
(2) provide more reliable water supplies for water users of northern and southern California;
(3) improve water quality (reducing intrusion of saltwater in the San Francisco Bay, reducing discharge of organic materials, and pollutants);
(4) reducing vulnerability of the water supply system with respect to major earthquakes and flooding; and
(5) improve levee stability.

In September 1996, the *Bay–Delta Act* provided Federal funding for the Bay–Delta Program, and in November of the same year *Proposition 204* provided State funding for the Program.[69]

In August 2000, implementation of the CALFED Bay–Delta Program began and a *Record of Decision* for a 30-year plan to address ecosystem health and water supply reliability problems in the Bay–Delta was issued (California Bay–Delta Authority, 2003, pp. 4 and 5). The Program addresses four resource management objectives:

(1) water supply reliability,
(2) water quality,

---

[67] Salmon and steelhead trout evolved with the California's landscape. They reach maturity at sea and return to the rivers to spawn. Due to construction of dams and reduction of flow in the river systems they lost their spawning environment. Subsequently, their population has declined by 80 to 90 percent of their 1940s level (Johnson *et al.*, 1993, p. 98).

[68] **A. State agencies:** California Bay–Delta Authority, California State Parks, Department of Water Resources, Department of Fish and Game, The Reclamation Board, Delta Protection Commission, Department of Conservation, San Francisco Bay Conservation and Development Commission, California Environmental Protection Agency, State Water Resources Control Board, Department of Health Services, and Department of Food and Agriculture.
B. Federal agencies: Bureau of Reclamation, Fish and Wildlife Service, Geological Survey, Bureau of Land Management, Environmental Protection Agency, Army Corps of Engineers, Department of Agriculture, Natural Resources Conservation Service, Forest Service, National Marine Fisheries Service, and Western Area Power Administration.

[69] http://calwater.ca.gov (visited in July 2005).

(3) ecosystem restoration, and
(4) levee system integrity.

Table 11.16 shows a summary of the major components of the CALFED Bay–Delta Plan. In January 2003 the California Bay–Delta Authority was created to oversee implementation and coordination of the Program. In November 2003, the State Government provided funding for three years of Program implementations.

Over the three-year period of 2001–2003, nearly US$1.89 billion[70] was invested in water supply, water quality, and ecosystem restoration programs. Some of the achievements are:

- In 2003, water trading totalling $617 \times 10^6 \, m^3$ has supplemented existing water supplies. The transactions moved water from willing sellers to areas of need, while protecting other water users, local economies and the environment.
- Over the three-year period of 2001–2003, about $1100 \times 10^6 \, m^3$ of water became available to be used for fish protection measures, better protecting the Delta without reducing deliveries to cities and farms.
- US$131 million was invested in 104 groundwater projects with a potential yield of about $260 \times 10^6 \, m^3$.
- Studies were continued on four potential surface storage projects: north of the Delta ($2344 \times 10^6 \, m^3$), Upper San Joaquin (310 000 to 860 000 m^3), enlargement of Shasta Dam reservoir (370 000 m^3), and Delta storage (310 000 m^3).
- Investments were made in numerous agricultural and urban water conservation and water recycling projects.

It is expected that implementation of the first seven years (Stage 1) of this 30-year Program will be completed in 2007.

11.15.3 Water Trading

The development of a water market has become a key component of California's water policy.[71] The 1987–92 drought sharply diminished surface water supplies for many urban and agricultural users, as well as the environment (Rosegrant, 1995). Increased water conservation was necessary in much of California, including mandatory water rationing for urban users. The prolonged drought prompted urban water agencies to develop drought emergency plans to address water supply shortages. This led to the development of the California State Emergency Drought Water Bank, which brokered large-scale water trading in 1991.

A number of other events changed the water trading climate (Hanak, 2002):

- As mentioned previously,[72] in 1992 the U.S. Congress passed the CVPIA to increase environmental flows. The CVPIA contained provisions to facilitate water marketing and introduced a mechanism for the CVP to purchase additional water for environmental processes.
- In 1994, contractors of the SWP concluded negotiations for the *Monterey Agreement*,[73] which included a number of measures to make it easier for contractors to transfer water to one another.
- In 2000, State and Federal authorities launched the *Environmental Water Account* (EWA), a program of water purchases for the environment under the CALFED program to restore health of the fisheries of the Bay–Delta system.

The water market enables the historical water rights holders (mainly farmers) to transfer water to other users willing to pay for it. Potential buyers include urban and industrial users, other farmers with higher-value crops and limited supplies, and environmental programs to support fish and wildlife habitats. Municipal agencies are the principal buyers of long-term and permanent contracts, which account for approximately 20 percent of all sales (Hanak, 2003, pp. v and vi).

Figure 11.16 indicates that water trading has increased steadily between 1985 and 2001. Over this period a total of $14 \times 10^9 \, m^3$ of water was traded. The share of water transfers by type of market was (Hanak 2002, Figure 2): direct government purchase (31 percent); within CVP (28 percent); within the Colorado River Region (16 percent); open market (15 percent); and within SWP (10 percent). The peak in water trading in 1991 reflects the severe drought, while decreases in 1995 and 1998 reflect wetter than normal conditions. It is important to note that the 2000 and 2001 levels of water trading represent about 3 percent of total water supplied in these two years for municipal, industrial and agricultural purposes (see Table 11.8).

Agriculture is the leading source of supply for water trading. In most years, agricultural water users provide

---
[70] This consisted of 61.3 percent by the State Government, 8.8 percent by the Federal Government, 6.8 percent by water users, 5.8 percent by local authorities and 17.3 percent by non-federal matching funds (California Bay–Delta Authority, 2003, p. 53).
[71] Frederick (2001) provides an overview of water rights and water market in the United States. Also, a number of articles are available in March/April issue of *Southwest Hydrology* (2004).
[72] See section 11.14.3.
[73] See section 11.14.4.

Table 11.16. *Major components of the CALFED Bay–Delta Plan*

Program	Goals
Water supply reliability	– Assist local partners in developing $615-1230 \times 10^6 \, m^3$ of groundwater storage. – Pursue planning and other actions at State and Federal level to expand surface storage capacity by up to $4.3 \times 10^9 \, m^3$. – Optimise water conveyance facilities in the Delta and in other locations to maximise flexibility, protect water quality and fish species, and increase water supply reliability. – Invest in local projects that boost water use efficiency through annual water conservation and recycling competitive grants/loan program. – Streamline water trading approval process and development and effective water transfer market that protects water rights, the environment and local economies.
Water quality	– Develop and implement source control and drainage management programs. – Invest in treatment technology. – Implement aggressive measures to improve Delta water quality and water quality science.
Ecosystem restoration	– Conduct grant program to fund local projects in habitat restoration, fish passage, invasive species management, and environmental water quality. – Recover at risk native species and their habitat. – Augment streamflow in upstream areas to benefit native fish and invest in fish passage improvements through dam removal and improved fish ladders. – Provide local and technological assistance to assess watershed conditions and develop plans to address watershed problems. – Manage Environmental Water Account to acquire water from willing sellers to protect fish species without reducing water supply reliability.
Levee system integrity	– Maintain and strengthen Delta levees, provide protection to Delta resources and drinking water quality. – Develop best management practices for beneficial reuse of dredged material. – Improve Delta Emergency Management Plan and develop Risk Management Strategy to identify risks to Delta levees, evaluate consequences and recommend actions.
Science	– Establish Independent Science Board to integrate world class science into program implementation. – Implement comprehensive monitoring and research programs. Develop performance measures to evaluate program accomplishments.
Oversight and coordination	– Develop and implement program tracking system to ensure accountability and assess program progress. – Submit annual report to the Legislature and Congress to assure balanced progress in meeting program goals. – Establish a public advisory council and ensure public involvement in program implementation. – Address environmental justice and tribal needs associated with program implementation.

Source: California Bay–Delta Authority (2003, pp. 5 and 6).

at least 90 percent of supply (Hanak, 2002). The main source regions are the Central Valley (served by the CVP and SWP) and the areas irrigated by the Colorado River water. In most years the Central Valley has provided about 75 percent of the total volume transferred. Approximately 25 percent of the non-environmental volume of water purchased is from parties in the same county, 50 percent from parties in the same region and the remaining 25 percent is inter-regional trading. Over the period of 1988–2001, approximately $2.31 \times 10^9 \, m^3$ of water was purchased for the environment.

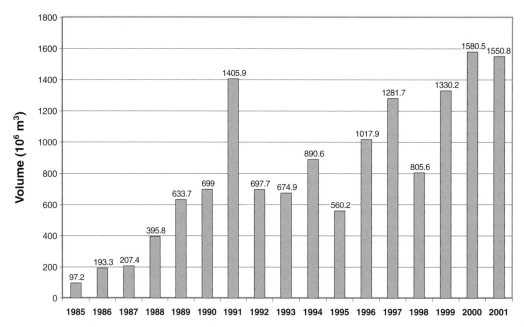

Figure 11.16 Water trading in California over the period of 1985 to 2001 (source of data: Hanak, 2002, Appendix Table 1, and Hanak, 2003, Table A.1).

Some 70 percent of this volume came from San Joaquin Valley, 28 percent from Sacramento Valley and only 2 percent from the San Francisco Bay area (Hanak, 2002, Appendix Table 4).

Water traded by farmers comes from various sources such as (Rosegrant, 1995):

- **Fallowing.** Water saved by withholding irrigation water from fields for an entire irrigation season can be temporarily transferred to another use.
- **Shift in cropping patterns.** Shifts from water-intensive crops to crops which use less water can save water for longer-term trading.
- **Groundwater substitution.** If a farmer grows the same crop by using groundwater instead of surface water, then he/she can transfer his/her surface water entitlement to another user. However, these transfers should be within the sustainability limits of the groundwater resources.
- **Conservation.** Water conserved through a reduction in crop consumption, canal lining, and use of improved irrigation technology, can also be traded on a longer-term basis.

Hanak (2003) points out that although water transfer has resolved some water supply problems, it has created economic and social problems in some areas due to fallowing farmland and fear of uncontrolled groundwater mining. The extent of the problem is reflected in the fact that by late 2002, 22 of the 58 State's counties adopted export restrictions, which includes permit requirements to export groundwater or to extract groundwater in substitution for transferred surface water.

Water Trading Regulations and Procedures

Water trading is one of a suite of measures that can help California meets its water supply requirements. Some trades are as simple as reducing one user's contract of water to allow an increase by a neighbour. Others can be as complex as reallocating water from reservoirs in northern California for use in the south of the State. State law supports voluntary water trading, and directs State agencies to encourage and facilitate voluntary transfer in a manner that protects existing water uses and provides technical assistance to interested parties to implement water conservation measures which will make additional water available for trading.

Considering the complexities of water rights and water trading issues, and to assist parties undertaking water transfers, the State Water Resources Control Board (1999)[74] describes basic principles of the California water right (riparian rights, appropriative rights, pre-1914 and post-1914 appropriative rights, and others) laws related to water transfer, and defines different types of surface water and groundwater transfers and procedures.

---

[74] This publication and other information regarding water rights are available at: http://www.waterrights.ca.gov (visited in July 2005).

To start the process of water trading, two important decisions need to be made. The first decision is whether the trading in question relates to water that originates within the watershed or is imported into the watershed. The next decision is whether the water to be transferred is surface water or groundwater.[75] No water rights are granted to the party receiving the water. The original water right holder retains all the water rights.

It is important that water market transactions take place within a reasonably short period of time. An online facility called *On Tap*[76] has been developed and launched on 28 December 2000 (Young and Hunn, 2001) to assist water trading proponents to understand approval requirements, undertake better planning to secure needed permits, preparing applications, expediting agency review and permit issuance. The website is administered and managed jointly by the DWR, USBR and the SWRCB. *On Tap* consists of two major components (Young and Hunn, 2001):

- **Transaction guide**, directs proponents through a series of questions, returning relevant information about application requirements specific to their proposed transactions. The intent of the guide is to ensure that a proponent understands all pertinent details for a proposed transfer prior to submitting a request for review to DWR, USBR or SWRCB. Case-specific feedback will be provided based on information from the applicant regarding, but not limited to the: (1) transaction participant (seller, buyer, intermediary); (2) underlying water right; (3) method proposed to make the water available for transfer; (4) destination of the water proposed for transfer; and (5) duration of the transfer.
- **Transaction database**, intends to allow the public to have direct access to available data on specific water transfers. This database contains data as far back as 1945, when limited data were collected, to the most recent completed transfers. The target audience for the database includes consultants, researchers and public policy advocates.

11.15.4 Water Use Efficiency

In order to increase water use efficiency, the Department of Water Resources established the Office of Water Use Efficiency (OWUE) in 2001 as an expansion and reorganisation of DWR's water conservation program that had been in place for two decades. The expansion was necessary to respond to the implementation of CALFED water use efficiency program, as well as the greater need for water conservation statewide (Alemi, 2003). The expanded mission not only includes water saving, but also improving water quality, water recycling and desalination. OWUE provides technical and financial assistance in cooperation with local authorities and other agencies. Its services include (Alemi, 2003):

- Assisting agricultural water supply agencies to prepare agricultural water management plans.
- Reviewing and evaluating the plans, and helping implement efficient water management practices and drainage management measures.
- Assisting urban water agencies to prepare urban water management plans.
- Reviewing and evaluating urban water management plans.
- Serving the public's needs for information on water use and water conservation.
- Managing the California Irrigation Management Information System (CIMIS), which provides near real-time weather and evapotranspiration data[77] to the irrigators and general public.
- Providing financial assistance to local agencies for water use efficiency, water desalination, and water recycling.

Agriculture is an important element of California's economy, and generated US$27.6 billion in 2001. However, as shown in Tables 11.3 and 11.8, it uses a large proportion of surface water and groundwater resources of the State. Because these resources are limited, many farmers and irrigation districts have implemented state-of-the-art design, delivery, and management practices to increase production efficiency and conserve water. One indicator of agricultural water use efficiency improvement is that agricultural production per unit of applied water for 32 important crops increased by 38 percent between 1980 and 2000 (Department of Water Resources, 2005, Chapter 3). Over the same period the inflation-adjusted gross crop revenue per unit of applied water increased by 11 percent. Table 11.17 shows changes in irrigation methods between 1990 and 2000. It indicates that the irrigated area remained unchanged during this 10-year period. However, the area irrigated by gravity methods (furrow and flood) declined by 16 percent (650 000 ha), while the area irrigated by drip method increased by 11 percent (450 000 ha), and the area irrigated by sprinkler increased by 5 percent (200 000 ha).

---

[75] For the other decisions to be made see Figure 1, Water Transfer Decision Tree in the State Water Resources Control Board (1999).
[76] Available at: http://ontap.ca.gov (visited in July 2005).
[77] http://wwwcimis.water.ca.gov/cimis/welcome.jsp (visited in July 2005).

Table 11.17. *Changes in irrigation methods in California between 1990 and 2000*

Irrigation method	Year 1990		Year 2000		Changes from 1990 to 2000 (Percentage)
	Area (Mha)	Percentage of total	Area (Mha)	Percentage of total	
Gravity (furrow, flood)	2.63	67	1.98	51	−16
Sprinkler	0.93	24	1.13	29	5
Drip	0.32	9	0.77	20	11
Total	3.88	100	3.88	100	–

Source: Department of Water Resources (2005c, Chapter 3, Table 1).

The CALFED Record of Decisions estimated that efficiency improvement would result in water saving ranging from $148 \times 10^6 \, m^3$ to $695 \times 10^6 \, m^3$ by 2030. Estimates also included $1.97 \times 10^9 \, m^3$ reduction in applied water (recoverable losses) being reused that provides environmental and crop production benefits. Also, there is significant water saving in the Colorado River Region. Lining of the All-American Canal and Coachella Canal, and increasing water use efficiency in the IID and Coachella Irrigation District, has resulted in substantial savings (see sections on All-American Canal and Coachella Canal in this chapter). In spite of these improvements, more efforts are underway to further increase agricultural water use efficiency.

The urban water use of California was $11 \times 10^9 \, m^3$ in the year 2000 (see Table 11.8). This was the result of progress made toward improved water use efficiency. For example, the Los Angeles Department of Water and Power in their Urban Water Management Plan Update 2002–03 reported "*water conservation continues to play an important part in keeping the City's water use equivalent to levels seen 20 years ago*". Overall, in various regions of the State, an increase in population has not resulted in a proportionate increase in urban water use (Department of Water Resources, 2005c, Chapter 22). These improvements are partly due to implementation of water use efficiency practices that have been institutionalised through the California Urban Water Conservation Council's Memorandum of Understanding (MoU). This involves active participation of urban water agencies, environmental interests, and the business community. They planned and implemented a set of 14 urban Best Management Practices.[78] As of 1 September 2003, there were 309 signatories to the Urban MoU, representing 80 percent of all urban water supplied in California (Department of Water Resources, 2005c, Chapter 22). One example of their activity is that 2.3 million water efficient toilets have been retrofitted statewide in the past 12 years. The total number of toilets installed before 1992 that still need to be replaced is about 10 million. A recent state-sponsored study indicates achievable savings of $2.5–2.8 \times 10^9 \, m^3$ per year by the year 2030. Overall, urban water use efficiency can be a very cost-effective strategy, with the cost of most measures ranging from US$47 to US$575 per $1000 \, m^3$, depending on the program.

Funds dedicated to water use efficiency have fallen below commitments made in 2000 through the CALFED Record of Decisions that called for a State and Federal investment of US$1.5 billion to US$2 billion during Stage 1 from 2000–07 (Department of Water Resources, 2005c, Chapter 22). For example, in 2002 and 2003, investments lagged projected expenditure by US$4 million and US$235 million respectively.

The OWUE agricultural and urban grants to support water conservation projects in 2001 and 2002 were US$12.9 million and US$9.8 million respectively. An additional funding of US$120 million became available for the following three years for water efficiency projects (Alemi, 2003). Further information regarding agricultural and urban water saving is available in the Department of Water Resources (2005c, Chapters 3 and 22), and the OWUE's website.[79]

11.15.5 Water Recycling and Reclamation

Recycled water is defined as the wastewater that has been treated to a quality that is suitable to use again. "Water reclamation", "Wastewater reclamation and reuse", and "Greywater mining" are other equivalent terms. Since the 1890s, Californians have been reusing municipal wastewater for agriculture and farm irrigation (Recycled

---

[78] These included: Public information programs; school education programs; residential ultra low flush toilet replacement programs; use of high efficient washing machines; large landscape irrigation water conservation; and pricing (Department of Water Resources, 2005c, Chapter 22).

[79] http://www.owue.water.ca.gov (visited in July 2005).

Water Task Force,[80] 2003, pp. 5–7). It is likely, at that time, that wastewater was untreated. By 1910, at least 35 communities were using wastewater for farm irrigation, 11 without treatment and 24 with treatment. In 1970, approximately $215 \times 10^6 \, m^3$ of water was recycled. In 2000, this amount increased to $495 \times 10^6 \, m^3$ supplied by 234 wastewater treatment plants. By 2003, the amount of treated municipal wastewater in California was estimated to be $6.2 \times 10^9 \, m^3$, and 10 percent of this volume or $620 \times 10^6 \, m^3$ was reused. Major recycled water users consisted of agricultural irrigation (48 percent), irrigation of golf courses, school yards, parks and others (20 percent), groundwater recharge (15 percent), environmental uses (6 percent), industrial uses (5 percent) and others (6 percent).

Aquifers have been recharged with recycled water. The most notable use of recycled water for this purpose is recharge in the Montebello Groundwater Project in East Los Angeles, which has occurred since 1962. In coastal areas, excessive groundwater pumping has resulted in seawater intrusion. Highly treated recycled water has been used to recharge the aquifers to create a hydraulic barrier to seawater intrusion in Orange County since 1976, and a recently completed project now operates along the coast in Los Angeles County.

By 2030, there will be $8 \times 10^9 \, m^3$ of wastewater in California and it is estimated that $1.8 \times 10^9 \, m^3$ could be recycled. However, to achieve that potential, Californians will have to invest nearly US$6 to US$9 billion for additional infrastructure to produce and deliver the recycled water.[81] The actual cost will depend on the quality of the wastewater, the treatment level to meet recycled water intended use, and the availability of a distribution network. Uses such as irrigation near the treatment plant will benefit from lower treatment and distribution cost. Irrigation of a wide range of agricultural crops can even benefit from the nutrients present in the recycled water by lowering the need for applied fertilisers. However, the use of recycled water for irrigation without adequate soil and water management may cause accumulation of salts or specific ions in soil and groundwater (Department of Water Resources, 2005c, Chapter 16).

The Recycled Water Task Force (2003) made numerous recommendations, which include:

- **Funding.** State funding for water recycling facilities and infrastructure should be increased.
- **Community involvement.** Local agencies should engage the public in an active dialogue and participation in planning water recycling projects.
- **Leadership.** State government should take a leadership role in encouraging recycled water use and improve consistency of policy within branches of the State government.
- **Education.** The State should develop comprehensive education curricula for public schools, and institutions of higher education.
- **Health and safety regulation.** The Department of Health Services should involve stakeholders in a review of various factors to identify any needs for enhancing existing local and State health regulation associated with the use of recycled water.
- **Uniform analytical method for economic analysis.** A uniform and economically valid procedural framework should be developed to determine the economic benefits and costs of water recycling projects.
- **Research funding.** The State should expand funding sources for research on recycled water issues.

11.15.6 Conjunctive Management of Surface and Groundwater

Groundwater is one of California's greatest natural resources. Currently, there are 431 groundwater basins underlying about 40 percent of the surface area of the State. In 1995, an estimated 13 million Californians, or nearly 40 percent of the State's population, were supplied with groundwater. Hydrologic regions with significant annual groundwater extraction are (Department of Water Resources, 2003a, Table 12 and Figure 23): Tulare Lake ($5.35 \times 10^9 \, m^3$); Sacramento River ($3.30 \times 10^9 \, m^3$); San Joaquin River ($2.71 \times 10^9 \, m^3$); South Coast ($1.45 \times 10^9 \, m^3$); and Central Coast ($1.29 \times 10^9 \, m^3$). It is estimated that the overdraft is between 1.2 and $2.5 \times 10^9 \, m^3$ annually with the most overdraft occurring in the Tulare Lake, San Joaquin River and Central Coast hydrologic regions.

Conjunctive management of surface and groundwater resources takes advantage of surface storage facilities to temporarily store stormwater and the ability of aquifers to serve as long-term storage.[82] The main components of a conjunctive management plan are:

(1) recharging the aquifer when surface water is available;
(2) extracting groundwater in dry years when surface water is scarce; and

---

[80] The Task Force consisted of 40 members representing Federal, State and local governmental and private sector entities, environmental organisations, University of California and public interest groups.
[81] The estimated cost to build the capacity to yield $1000 \, m^3$ of recycled water per annum is about US$5000, making this water expensive.
[82] Description of the conjunctive management of surface and groundwater in this section is based on the Department of Water Resources (2005c, Chapter 4).

(3) having a monitoring program to allow water managers to respond to changes in groundwater, surface water, or environmental conditions that could violate management objectives or impact other water users.

Conjunctive management improves water supply reliability, reduces groundwater overdraft and land subsidence, reduces water loss due to evaporation, protects water quality, and improves environmental conditions.

Currently, conjunctive management is carried out at regional and local scales. Two examples illustrate these activities. In southern California, including Kern County, conjunctive management has substantially increased average year water deliveries. Over the years, artificial recharge in these areas has increased the water in aquifer storage by approximately $8.6 \times 10^9 \, \text{m}^3$. Santa Clara Valley Water District releases local supplies and imported water into more than 20 local creeks for artificial in-stream recharge and into more than 70 recharge ponds to recharge a total of about $194 \times 10^6 \, \text{m}^3$ annually. To assist further conjunctive use of surface and groundwater, DWR's Conjunctive Water Management Program awarded over US$130 million in grants and loans for project funding and study throughout California in fiscal years 2001 and 2002.

Conservative estimates indicate that conjunctive management has the potential to increase average annual water deliveries throughout the State by about $0.62 \times 10^9 \, \text{m}^3$. Project costs range from US$8 to US$485 per $1000 \, \text{m}^3$. This wide range of costs is due to many factors including complexity of the projects, regional differences in construction costs, availability and quality of recharge supply, intended use of water, and treatment requirements.

11.15.7 Water Desalination

The possibility of using desalination as a method of producing additional freshwater in California, has been investigated by a task force consisting of representatives from 27 organisations (Desalination Task Force, 2003). The following description is mainly based on this publication.

In the late 1980s, during a period of extended drought, several localities either considered or actually built relatively small desalination facilities along the Californian coast. But with the end of drought, the high cost of desalinated water could not be justified. For example, in Santa Barbara the desalination facility built in response to the drought was decommissioned. However, by the late 1990s, desalination received renewed interest. On the coast, new facilities were considered in response to anticipated growth, the potential impacts of another drought and the prospect of potential reductions in imported water. As of 2002, six plants with a cumulative capacity of $86.3 \times 10^6 \, \text{m}^3 \, \text{yr}^{-1}$ were in operation, four plants were in design and construction with a capacity of $39.2 \times 10^6 \, \text{m}^3 \, \text{yr}^{-1}$, and nine plants were planned or projected with a capacity of $300 \times 10^6 \, \text{m}^3 \, \text{yr}^{-1}$ (Department of Water Resources, 2005c, Chapter 6). Out of the total desalination capacity of $425.5 \times 10^6 \, \text{m}^3 \, \text{yr}^{-1}$ in operation, design, construction, planned or projected, about $192 \times 10^6 \, \text{m}^3 \, \text{yr}^{-1}$ or 45 percent is based on groundwater desalination. Although not all of designed, planned or projected plants are likely to be constructed this decade, it is assumed that they, or an equivalent number, will be operational by 2030.

Recent technological advances have significantly reduced the cost of desalinated water to levels that are comparable, and in some instances competitive with other alternatives for acquiring new water supplies. Desalination technologies are becoming more efficient, less energy demanding and less expensive. As an indication, the cost of brackish groundwater desalination is about US$200–400 per $1000 \, \text{m}^3$, while it is about US$650–1600 per $1000 \, \text{m}^3$ for seawater.

While most estimates predict that seawater desalination will contribute less than 10 percent of the total water supply requirements of California, some suggest that with the unlimited supply of seawater, this percentage could be even higher if the environmental and energy issues associated with seawater desalination can be adequately addressed. Some of these critical issues are:

- **Energy consumption and cost.** Seawater desalination requires large amounts of electricity. New reverse osmosis technologies are reducing the amount of electricity needed to produce potable water. Also, desalination of seawater requires nearly four times more energy than brackish groundwater. Therefore, where brackish water is available, it may be economically preferable to use that rather than seawater. Estimates indicate that energy generation capacity would not be a constraint to implementation of the planned seawater desalination facilities for 2010. At a normal price of 8 to 11 cents per kWh, energy constitutes 40 to 45 percent of the operating cost. To reduce the cost of energy some suggest co-location of desalination facilities with existing power plants and then purchase the power directly at wholesale rates.
- **Ecological impacts of feed-water intake and brine discharge.** Both impingement and entrainment can cause ecological impacts. Impingement is the process of trapping organisms on screens covering feed-water intakes. Entrainment refers to the organisms in the

water column that pass through intake screens and then through the feed-water system. The impacts vary considerably with the volume and velocity of feed-water intake, and the use of mitigation measures developed to minimise impacts. Co-location of desalination plants with coastal power plants and using the power plant's cooling water as the feed-water for the desalinisation facility would eliminate or minimise these ecological impacts. Another option is using wells for water intakes. This would eliminate major ecological concerns related to feed-water intake, and would lower pre-treatment operation and maintenance costs.

The major by-product of the desalination process is brine, which contains concentrated salts removed from feed-water. The primary concern with brine discharges is the potential effect of increased salinity on habitat and organisms surrounding the outfall. While some organisms are accustomed to variability in salinity, others are extremely sensitive. Engineering approaches to mitigate the impacts of brine discharges include diffusers, which encourage dilution and prevent one large discharge point, standpipes, which allow better mixing based on density and temperature gradients, and mixing brine with other wastewaters discharging to the sea.

- **Linkages between local and regional planning, growth forecasts and water supply demand.** As desalinated water becomes part of the water supply portfolio, it is increasingly linked to the process of assessing existing and future water supplies and water demand in concert with land use planning and growth forecasting.
- **Land use and infrastructure compatibility.** Desalination facilities need to be compatible with adjacent land use and supported by adequate infrastructure.
- **Realistic assessment of economic costs.** Project analysis needs to include all of the direct and indirect benefit and cost components. If there is State and Federal government funding involved, then the analysis should also cover the economic costs to them and the taxpayers.
- **Distribution and public health.** Water systems may need to mix desalinated water with other freshwater supplies prior to distribution. These waters may be chemically dissimilar and may not be compatible resulting in quality problems that will impact consumer acceptance. Desalinised water may be highly corrosive and may require treatment and pH adjustment to minimise degradation of distribution system.

The discharge of wastes, including municipal wastewater and stormwater, and also the presence of toxic algal blooms, may affect water intakes for desalination facilities. In addition, certain constituents in ocean water such as boron may not be adequately removed during desalination.

Based on the findings of the Desalination Task Force (2003), 29 general and specific recommendations covering issues such as energy and environment, planning, and funding, are provided by the Department of Water Resources (2003b). The overarching recommendation is that desalination projects should be evaluated on a case-by-case basis since each facility is essentially unique, given local water supply and reliability needs, environmental conditions, project objectives, and proposed technology.

11.15.8 Precipitation Enhancement

As described before (see Section 11.8), climate change predictions indicate that California will be faced with higher temperatures and changed precipitation patterns including reduced snowfall. These predicted changes would mean less surface run-off flowing into water storage facilities. Under favourable conditions, cloud seeding can be used to alleviate this problem. It consists of injecting special substances (mostly silver or potassium iodide) into the clouds to assist formation of snowflakes and raindrops. The State law concerning water rights treats water gained from cloud seeding as natural supply. The following description is based on the Department of Water Resources (2005c, Chapter 14).

The first serious cloud seeding program in California began in 1948 on Bishop Creek catchment in the Owens River Basin for California Electric Power Company. Cloud seeding has been practised continuously in several river basins of California since the early 1950s. Most projects are located along the central and southern Sierra Nevada, and some others in the coastal ranges. The number of operating projects has tended to increase during droughts (up to 20 in 1991), but has levelled off to about 12 or 13 in recent years. The total area covered by these projects is around $33\,700\,km^2$. One conservative estimate indicates that the combined effect of these projects is the generation of $370-490 \times 10^6\,m^3$ of extra water per year, which represents about a 4 percent increase in average annual run-off. A similar volume of water could be generated in other river basins, particularly in the Sacramento River Basin, in watersheds that are not seeded now.

The costs of cloud seeding are about US$16 per $1000\,m^3$ of rain produced. Therefore, generating an extra $370-490 \times 10^6\,m^3$ per year would require about

US$5.9–7.8 million per annum, plus an initial investment of US$1.5–2 million in planning and environmental studies.

11.15.9 Ecosystem Restoration

California's ecosystems, particularly aquatic ecosystems, have been significantly modified over the last two hundred years.[83] Construction and operation of dams and diversion structures, poor water quality (including pollution by point and non-point sources, reduction in dissolved oxygen levels, and changes in temperature regime), flood control, urbanisation, and introduction of exotic species, have all contributed to a decline in ecosystem health.

Ecosystem restoration is the activity of improving the condition of California's modified natural landscapes and biotic communities. Ecosystem restoration includes in-stream flow changes, habitat restoration, physical modification to water bodies, control of waste discharges into waterways, control of exotic species, removal of barriers to fish migration, and other activities.

The State's ecosystems cannot be restored to their pristine state, nor is that restoration desirable. Instead, ecosystem restoration focuses on the rehabilitation of ecosystems so that they supply important elements of their original structure and function in a sustainable manner. Ecosystem restoration is one of the water management strategies in the California Water Plan, particularly because it is linked with improvement of water supply reliability and water quality. Currently, major river restoration projects are underway in different parts of the State, including the San Joaquin, Sacramento, and Trinity rivers. Also the CALFED Bay–Delta Program serves as an example of integrated resource management, by improving water supply reliability while simultaneously restoring ecosystems.

Detailed statewide ecosystem needs and their cost do not exist. However, preliminary estimates indicate that the cost of ecosystem restoration associated with the CALFED Bay–Delta program (US$150 million per year), and restorations in other areas of the State to year 2030 could be within the range of US$7.5–11.3 billion.

SECTION C: INTER-BASIN WATER TRANSFER FROM THE COLORADO RIVER

11.16 COLORADO RIVER BASIN

The Colorado River is about 2250 km long, originating as small streams fed by snow-melt in the mountains of Colorado, Wyoming and New Mexico, and discharging to the Gulf of Mexico. It drains about 632 000 km^2 of land across seven states (Colorado, Wyoming, Utah, New Mexico, Arizona, Nevada and California) before entering Mexico (Figure 11.17). The Colorado River Basin is divided into the Upper and Lower Basins. Colorado, New Mexico, Utah and Wyoming are generally known as the Upper Basin states, while Arizona, California and Nevada are generally referred to as the Lower Basin states. The Colorado River is the primary source of domestic water supply for some 27 million people in the seven Colorado River Basin states. It also provides irrigation water for more than 1.4 Mha of farmland within the Basin and large tracts of lands outside the Basin.

The long history of negotiations, disputes, court actions and even a military expedition for the settlement of water-sharing disputes between the seven states is documented in numerous publications, including Reisner (1987) and Fradkin (1996).

The headwaters of the Colorado River and its major tributaries lie in the high mountains where precipitation averages 1000 to 1500 mm per year. However, most of its course crosses the semi-arid Colorado Plateau and desert, where average annual precipitation may be as low as 60 mm (Mueller and Moody, 1984).

Historically, run-off within the catchment was sporadic with most occurring as floods in the spring and early summer, and very little flow during the rest of the year. The annual flow varied from 6.75×10^9 m^3 to 29.65×10^9 m^3 (Barton, 1981). This highly variable flow was a problem. Most of the agricultural land in the area and major cities including Denver, Los Angeles, Phoenix and San Diego, depended on the Colorado River or its tributaries for their water supply. Ten major storage dams including the Glen Canyon Dam in the Upper Basin and the Hoover Dam in the Lower Basin have been built on the river to regulate its flow. These dams have a reported storage capacity of about 76×10^9 m^3 (Barton, 1981). However, construction of these reservoirs has caused environmental impacts such as sediment trapping, increased evaporation, salt concentration, thermal stratification of water in the reservoirs, and changes in aquatic species (El-Ashry and Gibbons, 1990).

In 1922, negotiation of the *Colorado River Compact* took place in New Mexico. Delegations from the seven states arbitrarily divided the river basin into the Upper and Lower Basins. They also divided the 21.6×10^9 m^3 average annual

---

[83] Description of this section is based on the Department of Water Resources (2005c, Chapter 9).

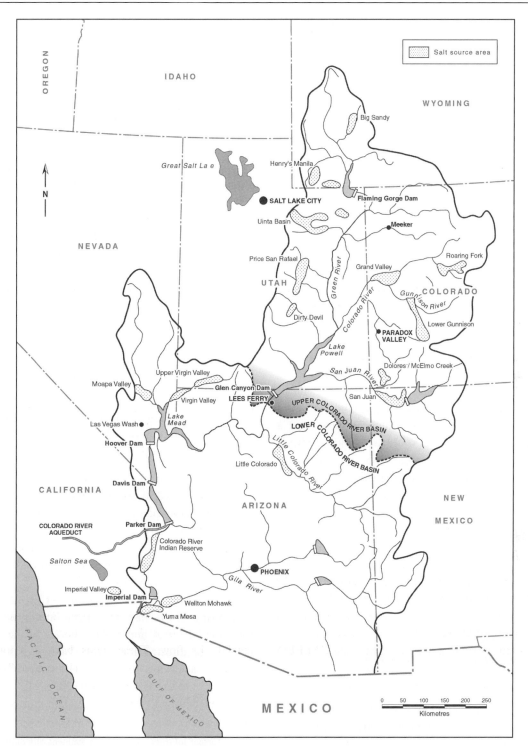

Figure 11.17 Colorado River Basin and locations of irrigation salt source areas (Hedlund, 1984; and Colorado River Basin Salinity Control Forum, 1993, Figure 4.1).

flow of the river as follows: $9.3 \times 10^9 \, m^3 \, yr^{-1}$ for each of the Upper and the Lower Basins; $1.8 \times 10^9 \, m^3 \, yr^{-1}$ reserved for Mexico; and $1.2 \times 10^9 \, m^3 \, yr^{-1}$ apportioned as a bonus to the Lower Basin (Reisner, 1987).

In 1928, the U.S. Congress divided $9.3 \times 10^9 \, m^3$ of the Lower Basin's annual share of the Colorado River between California ($5.4 \times 10^9 \, m^3$), Arizona ($3.5 \times 10^9 \, m^3$) and Nevada ($0.4 \times 10^9 \, m^3$). The *California Seven Party Agreement* of

1931 helped resolve the long-standing conflict between Californian agricultural and municipal interests over distribution of California's entitlement of the Colorado River waters. The seven principal claimants (Palo Verde Irrigation District, Yuma Project, Imperial Irrigation District, Coachella Valley Irrigation District, Los Angeles Metropolitan Water District, and the City and County of San Diego) reached consensus on the amount of water to be allocated to each entity on an annual basis.

The *Upper Colorado River Basin Compact* of 1948 apportioned the Upper Basin's $9.3 \times 10^9 \, m^3 \, yr^{-1}$ between Colorado (51.75 percent), New Mexico (11.25 percent), Utah (23 percent), and Wyoming (14 percent). The *Colorado River Basin Project Act* of 1968 authorised construction of a number of water development projects in both the Upper and the Lower Basins, including the Central Arizona Project.[84]

During the 1990s, California's average annual withdrawal from Colorado River ranged from $5.6 \times 10^9 \, m^3$ to $6.2 \times 10^9 \, m^3$, which was above its allocated share of $5.4 \times 10^9 \, m^3$. Excess water came from unused water allocated to Arizona and Nevada. However, the water requirements of Arizona and Nevada have increased rapidly. Under pressure from Colorado River Basin states, California developed a draft plan for its Colorado River water use (Colorado River Board of California, 2000). The objective of the plan was to cut the State's over-use of the Colorado River water over a period of 15 years.

11.16.1 Colorado River Salinity

Salinity has long been recognised as one of the major problems of the Colorado River. At its headwaters in the mountains of Colorado, the River has a TDS of less than $30 \, mg \, L^{-1}$. The salinity progressively increases downstream as a result of two processes:

(1) salt loading, or addition of soluble salt from natural sources as well as return of irrigation drainage waters; and
(2) salt concentration, caused by a reduction in the volume of river water due to evaporation, transpiration and withdrawal.

Without salinity control, the TDS was projected to increase to about $970 \, mg \, L^{-1}$ at the Imperial Dam by the year 2010. Implementation of the salinity control programs was predicted to keep TDS slightly below $880 \, mg \, L^{-1}$ (Colorado River Salinity Program Coordinator, 1993, Figure 2).

In 1944, the United States and Mexico signed a treaty requiring the United States to deliver $1.85 \times 10^9 \, m^3$ per year of Colorado River water to Mexico (U.S. Bureau of Reclamation, 1992). This treaty did not address the salinity of the delivered water. In 1961, the TDS of the water delivered to Mexico increased sharply from about $700-900 \, mg \, L^{-1}$ to about $1340 \, mg \, L^{-1}$. Through the 1960s and early 1970s, the United States and Mexico pursued a series of temporary solutions to the salinity problem. In 1974 the Federal *Colorado River Basin Salinity Control Act* approved, among other measures, construction of the Yuma Desalting Plant for desalination of drainage waters from the Wellton–Mohawk Irrigation and Drainage District in Arizona. The desalinised water was expected to return to the Colorado River for delivery to Mexico as partial fulfilment of the 1944 Treaty.[85] However, after its completion in 1992 at a cost of US$256 million, it was operated only for a short period from May 1992 to January 1993 at up to one-third of its capacity.[86] It was shut down for a number of reasons which included:

(1) in the years immediately after completion of the Plant, and before a drought which commenced in 2000, there was sufficient water in the system to satisfy water requirements and meet the salinity standard;
(2) implementation of other salinity control measures had helped reduce salinity levels; and
(3) the high operating cost of the Plant.

Although over the period of 2001–05, about US$50 million has been allocated to make the Plant operational within the promised timeframe of 24–30 months, there is no evidence to suggest that any of these funds have been effectively used.[87] In the meantime, Lake Mead continues to lose about 30 cm of its waters per year in order to fulfil United States obligations towards Mexico.

The *Colorado River Basin Salinity Control Program* is a partnership effort between agricultural producers, water users, and Federal and State agencies in the seven states. Efforts are directed towards reduction of salinity in the Basin by achieving the following numeric TDS criteria: $723 \, mg \, L^{-1}$ below Hoover Dam; $747 \, mg \, L^{-1}$ below Parker Dam; and $879 \, mg \, L^{-1}$ at Imperial Dam. Further information

---

[84] For further information on apportionment of Colorado River water visit: http://www.crwua.org/Colorado_river/lor.htm and http://www.usbr.gov/lc/region/g1000/lawofrvr.html (both visited in July 2005).
[85] http://www.usbr.gov/dataweb/html/yumadesalt.html#general (visited in July 2005).
[86] http://www.hcn.org/sevlets/hcn.Article?article_id=97 and http://ag.arizona.edu/AZWATER/awr/mayjune04/policy.html (both visited in July 2005).
[87] http://www.cap-az.com/breifings/05criticalissues/OpYumaDesaltingPlant2–05.pdf (visited in July 2005).

Figure 11.18 Inter-basin water transfers from Colorado River Basin to its neighbouring states (Fradkin, 1996).

regarding salinity in the basin and achievements of the salinity control programs are available in Colorado River Basin Salinity Control Forum (2002), U.S. Department of the Interior (2001),[88] and numerous links at the USBR's website.[89]

11.17 WATER TRANSFER PROJECTS

The Colorado River Basin is a major source of inter-basin water transfer to the states of Colorado, New Mexico, California, and Utah (Figure 11.18). The Colorado–Big Thompson Project, Fryingpan–Arkansas Project, and a number of small projects divert water to the State of Colorado. The San Juan–Chama Project diverts water to New Mexico, while the Central Utah Project that includes the Strawberry Valley Project, which was developed earlier, diverts water to Utah.

Water transfers from the Colorado River Basin to California via the All-American Canal and the Colorado River Aqueduct has been described before (see Section 11.14.5). In the following sections, the other above-mentioned projects will be described.

11.17.1 Water Transfer Projects to the State of Colorado

The State of Colorado covers an area of 269 600 km^2 with elevations ranging from about 1000 m to 4400 m. Its land area encompasses parts of three physiographic regions of the western United States: the Great Plains, the Rocky Mountains, and the Colorado Plateau (Figure 11.1). It achieved statehood in 1876, on the 100th anniversary of the Declaration of Independence. The average annual precipitation of the State ranges from less than 400 mm to more than 1500 mm. It had a population of 2.9 million in 1980 and reached 4.5 million in 2002 and had a growth rate of 4.8 percent over the period of April 2000 to July 2002 (U.S. Census Bureau, 2003, Table 17 and Figure 1.1).

[88] This publication is also available at http://www.usbr.gov/uc/progact/salinity/pdfs/PR20.pdf (visited in July 2005).
[89] http://www.usbr.gov/uc/progact/salinity (visited in July 2005).

Table 11.18. *Total annual water storage and usage in the seven River Basin Divisions of the State of Colorado from 1998 to 2002 (in $10^9 m^3$)*

Year	Storage	Irrigation area (Mha)	Water use	Municipal	Industrial	Others[a]	Total
1998	4.22	1.23	14.20	1.04	0.31	1.98	21.75
1999	4.94	1.24	13.45	1.03	0.26	2.07	21.74
2000	3.63	1.21	12.80	1.17	0.23	1.77	19.60
2001	3.57	1.23	12.25	1.16	0.24	1.74	18.96
2002	1.96	0.94	8.11	1.13	0.21	1.21	12.62

Note: [a] Mainly stock, fishery, and recreation.
Source: Cumulative yearly statistics of the Colorado Division of Water Resources.[90]

Estimates suggest the State will reach a population of about 7.2 million by 2030.[91]

Colorado is divided into seven divisions based on its major river basins:[92] Division 1, South Platte River Basin; Division 2, Arkansas River Basin; Division 3, Rio Grande Basin; Division 4, Gunnison River Basin; Division 5, Colorado River Basin; Division 6, Yampa/White River Basin; and Division 7, San Juan Dolores River Basin. Table 11.18 shows the total water storage and usage in the seven River Basin Divisions of the State from 1998 to 2002. A significant drop in water storage and usage was evident in 2002, due to the worst drought in the recorded history of the State.[93] Irrigation has been a major water user (64 percent) at an average high rate of over $10 000 m^3$ per ha. Domestic consumption was about 700 litres per day and per head of population in 2002.

The inter-basin water transfer projects from Colorado River Basin to the State of Colorado are described in the following three sections, followed by descriptions of the "Colorado Statewide Water Supply Initiative" and "Drought and Water Supply Assessment".

Colorado–Big Thompson Project

Tyler (1992) provides a detailed history of the Colorado–Big Thompson (C-BT) Project with numerous maps and photographs, while Autobee (unpublished) presents a concise description of the Project as part of the Historic Reclamation Projects studies. The following short history and description of the Project is mainly based on the second reference.

In 1889, the State of Colorado undertook a survey to investigate the feasibility of drilling a tunnel from Monarch Lake[94] on a Colorado River tributary, through the Continental Divide, to St Vrain Creek, a tributary of South Platte River. In 1904, the United States Reclamation Service (USRS)[95] produced a report, which suggested raising the elevation of Grand Lake by about 6 m through construction of a dam at the Lake's outlet. However, no action resulted from these and a number of other investigations.

By mid-1930s, progress toward the C-BT project was underway. On 21 January 1935, US$150 000 (US$2 million in 2004 values) was allocated to the US Bureau of Reclamation in order to survey the Grand Lake–Big Thompson proposal. On 8 February 1936, USBR delivered a preliminary report, which included a request for additional funding to complete the Grand Lake–Big Thompson Trans-mountain Project. The proposal came under attack by the National Park Service, conservation groups, residents of the Western Slope and grassroots organisations, because of its impacts on the Rocky Mountain National Park,[96] wildlife and water rights issues. On 18 July 1936, the USBR changed the project's name to Colorado–Big Thompson Project. The argument over C-BT lingered for another year, both in the media, and in the executive and legislative branches of government. Construction of the project was contingent on

---

[90] http://water.state.co.us/pubs/cumulative/CYS_rpt_2002.pdf (visited in July 2005).
[91] See Table 4-2 at the URL http://cwcb.state.co.us/owc/Drought_Water/pdf/Chapter%204.pdf (visited in July 2005).
[92] Information regarding these river basins is available at the URL http://cdss.state.co.us (visited in July 2005).
[93] http://water.state.co.us/pubs/annualreport/annlrpt_2002.PDF (visited in July 2005).
[94] An artificial lake located about 29 km north-east of Granby on the Arapahoe Creek, which had an average annual flow of $75.6 \times 10^6 m^3$ over the period of 1935–71. The maximum height of the dam is 5.5 m, surface area of the lake is 65 ha, and normal average storage capacity of the lake is $1.2 \times 10^6 m^3$, which could be increased to $2.3 \times 10^6 m^3$ (Brit Storey, USBR, Denver, personal communication, August 2004).
[95] Predecessor of the USBR (see **Abbreviations**).
[96] The Rocky Mountain National Park was established on 26 January 1915. It covers an area of $1080 km^2$. The highest peak in the Park is about 4345 m above mean sea level (http://www.rocky.mountain.national-park.com/info.htm, visited in July 2005).

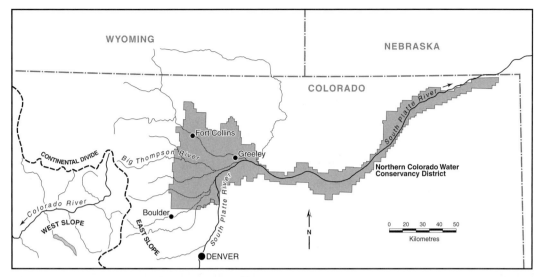

Figure 11.19 Northern Colorado Water Conservancy District.[97]

the development of a conservancy district contract with the Federal Government. The Northern Colorado Water Conservancy District (NCWCD) was established in September 1937 and reached an agreement with the US Bureau of Reclamation for construction of the Project and repayment of Federal funds over a period of 40 years. President Roosevelt finally approved the C-BT Project on 21 December 1937 and paved the way for construction of the Project.

Development of the Project was faced with a series of obstacles. These included protests over preservation of Rocky Mountain National Park, dealing with water rights issues, labour disputes during construction of the project, and shortages of materials and manpower caused by World War II. To protect the Rocky Mountain National Park, the Bureau of Reclamation agreed to abstain from construction within the Park boundaries by constructing a tunnel underneath the Continental Divide. In addition, the Park received both electricity and a guaranteed supply of water from C-BT.

The C-BT Project, one of the most successful Bureau of Reclamation projects, spans 240 km east to west and 105 km from north to south. It captures run-off from the headwaters of the Colorado River on the Western Slope of the Continental Divide. The water is then transferred via a tunnel to reservoirs and distribution facilities on the East Slope of the Divide (Figures 11.19 and 11.20). Diverted waters supplement run-off of the South Platte River, which flows from Colorado to Nebraska. The agreement with Western Slope water users assured them that diverted water to the Eastern Slope would not impinge on their existing water rights.

The Project collects snow-melt in four reservoirs (Windy Gap Reservoir, Willow Creek Reservoir, Lake Granby, and Shadow Mountain Reservoir). Water is transported west to east through the Alva B. Adams Tunnel.[98] Construction of the Green Mountain Dam, which included a hydro-electric plant, commenced in 1938 and was completed in 1943 when the Green Mountain Power plant started to supply power to the city of Denver.

Construction of the Adams Tunnel commenced on 15 June 1940. The 3 m diameter, and 21 km long tunnel located 1158 m below the Continental Divide was excavated from both ends. On 10 June 1944, excavation was completed and concrete lining of the tunnel was finished three years later in 1947. It has a capacity of $15.6 \, m^3 s^{-1}$. East of the Tunnel, the diverted water drops about 884 m as it flows through a series of canals, power plants and regulating reservoirs for delivery to farms and urban areas.

Construction of the Granby Dam, which is the centrepiece of the Project, commenced in December 1941. This dam and its four dikes collect water from the Colorado River and its tributaries and stores it for pumping into Shadow Mountain and Grand Lakes. Lake Granby's additional water comes from Willow Creek, a westerly tributary entering the Colorado River below Granby Dam. Willow Creek Dam was constructed between 1951 and 1953. Water collected in the Willow Creek Reservoir is lifted 53.3 m by pumps into Lake Granby.

---

[97] Simplified from the URL http://www.ncwcd.org/images/maps/Main_C-BT.jpg (visited in July 2005).
[98] Named after the U.S. Democrat Senator from Colorado who played a key role in convincing Congress to fund and construct the Project.

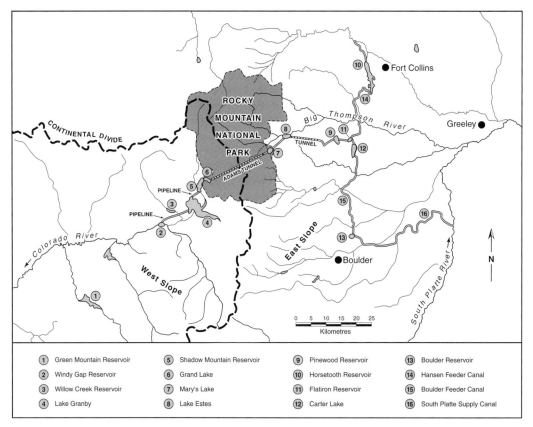

Figure 11.20 The Colorado-Big Thompson Project's reservoirs, tunnels and canals (Northern Colorado Water Conservancy District).[99]

Table 11.19. *Selected features of six dams and reservoirs of the Colorado–Big Thompson Project, in order of reservoir capacity*

Name of the dam	Height (m)	Crest length (m)	Name of the reservoir	Capacity ($10^6 \, m^3$)
Granby	68.0	262	Lake Granby	665.9
Green Mountain	80.5	351	Green Mountain	190.7
Horsetooth	33.8	561	Horsetooth	187.3
Carter Lake	57.9	376	Carter Lake	138.4
Shadow Mountain	11.3	938	Shadow Mountain Lake	22.7
Willow Creek	29.0	335	Willow Creek	13.1

Source: Autobee (unpublished).

Almost all components of the project were completed and opened on 11 August 1956. The only exception was the Big Thompson Power Plant, which was completed in 1959. After the CVP, C-BT is the most productive undertaking launched by the USBR. It has assisted agricultural development, generated hydro-power, supplied municipal and industrial water, and allowed urban and agricultural growth in Colorado.

The Project consists of 12 man-made reservoirs, with a total capacity of about $1230 \times 10^6 \, m^3$, one natural lake (Grand Lake), 56.3 km of tunnels, 153 km of canals, 18 pumping plants, 11 power plants and 1126 km of transmission lines.[100]

Table 11.19 shows some features of six major dams and reservoirs of the project. It supplies water to 30 cities and towns and 250 000 ha of irrigated land.

The USBR operates most West Slope storage and diversion facilities as well as power plants. It also manages similar works on the East Slope above the supply canal from Horsetooth Reservoirs to Carter Lake and Boulder

---

[99] http://www.ncwcd.org/project_features/cbt_maps.asp and http://www.ncwcd.org/images/maps/Main_C-BT.jpg (both visited in July 2005).

[100] http://www.ncwcd.org/project_features/cbt_by_the_numbers.asp (visited in July 2005).

Reservoir. The NCWCD operates and maintains Lake Granby, Willow Creek Reservoir, and other diversion features supplying its service area.

The C-BT Project was designed to deliver an average annual volume of $382.4 \times 10^6 \, m^3$ (Zimbelman and Werner, 2001). However, deliveries ranged from 50 to 100 percent of the designed capacity. The average for the first 43 years of operation of the project being slightly above 72 percent or $276 \times 10^6 \, m^3$. The District's policies and procedures allow water to be transferred between water users annually or permanently. Of the average annual deliveries, 67–75 percent can be transferred between users on an annual basis.

As municipal areas within the District have grown and expanded, much of the water supplies to support this growth were transferred from agricultural users. In 1957, municipal and industrial allocations were about 15 percent of all water deliveries. In 1999, this amount had grown to 54 percent. This is a viable transfer because much of the land annexed by municipalities had been agricultural land. It makes sense for the water previously used by agriculture to be transferred to the relevant municipality now occupying that land (Zimbelman and Werner, 2001).

Small Projects

Based on the 2000 Census, and the Colorado State Engineer's records, the Front Range of Colorado (the East Slope of the Continental Divide, excluding the North Platte and Rio Grande basins) has 90 percent of the State's population, but only 16 percent of the State's water (Winchester, 2001). Because the Front Range and the eastern plains of Colorado State are in a semi-arid environment, 30 diversion projects[101] have been constructed to divert water from the Colorado River and its tributaries (Gunnison and San Juan rivers)[102] to the east of the Continental Divide, to satisfy the region's demand for water. Winchester (2001) describes these diversions and provides their average annual diversion capacities. Over the period of 1990–99 these 30 projects diverted an annual average of $713 \times 10^6 \, m^3$ of water from the Colorado River Basin to the South Platte River, Arkansas River and Rio Grande basins. Among these projects, Colorado–Big Thompson, San Juan–Chama, and Fryingpan–Arkansas Projects delivered about 62 percent (35 percent, 17 percent and 10 percent respectively) and the other 27 projects delivered the remaining 38 percent of the average annual diversion.

Thaemert and Faucett (2001) believe that inter-basin water diversions along the Front Range have facilitated development of irrigated agriculture and urbanisation by providing reliable sources of water supply. However, because of the expansion of the urban areas, water use has been shifting from agricultural use towards municipal water supply. This trend has been accompanied with changes in seasonal water use patterns, higher values for water, and increasing pressure for supply of environmental water requirements. These require further investment in infrastructure to improve water supply reliability and ease operations to satisfy various demands.

Fryingpan–Arkansas Project

The Fryingpan–Arkansas (Fry–Ark) Project[103] is a multi-purpose inter-basin water transfer and delivery system from the Colorado River. Initial studies by the USBR began in 1936. Intensive investigations started in 1941, resulting in planning reports in 1947 and 1948, followed by a special report in 1949 and official recommendations in 1951. A revised planning report "Fryingpan–Arkansas Project" in 1953 led to congressional approval of the Project. The Project was authorised for construction in 1962, and construction of the Project began in 1964 with the building of the Ruedi Dam and Reservoir. The authorisation was later amended in 1978.

The Fry–Ark Project transfers water from Fryingpan River and Hunter Creek on the western slope of the Continental Divide to the Arkansas River on the eastern slope of the Divide (Figure 11.21). The diverted water, together with available water supplies in the Arkansas River Basin, provides an average annual water supply of $99 \times 10^6 \, m^3$ for both municipal use and supplemental irrigation water in the Arkansas Valley.

There are five dams and reservoirs in the Project (Table 11.20). Ruedi Dam and Reservoir is on the western slope, while the other four are on the eastern slope, with the Pueblo Dam and Reservoir on the Arkansas River being the largest reservoir of the Project.

Seventeen diversion structures on the western slope are used to divert water into the collection system, which includes nine tunnels with a combined length of 43 km. The system collects $85.4 \times 10^6 \, m^3$ of water from melting snow in the high mountains at elevations of about 3000 m. These waters flow into the inlet portal of the Charles H. Boustead Tunnel. The 8.7 km long, and 3.2 m diameter

---

[101] These include Colorado–Big Thompson, Fryingpan–Arkansas, and San Juan–Chama Projects described in this chapter.
[102] For locations, see Figure 11.17.
[103] Description of the Project provided here is based on the URL http://www.usbr.gov/dataweb/html/fryark.html (visited in July 2005).

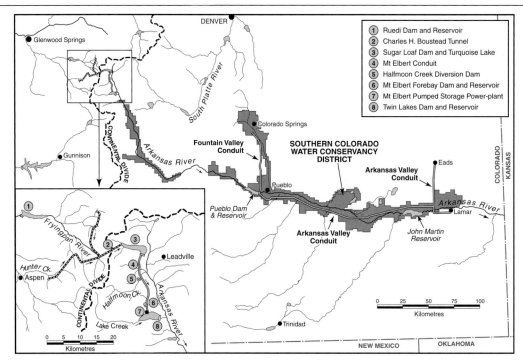

Figure 11.21 Fryingpan–Arkansas Project (based on the USBR map number 382-706-2646 dated August 1984).

Table 11.20. *Selected features of five dams and reservoirs of the Fryingpan–Arkansas Project, in order of reservoir capacity*

Name of the dam	Construction period	Type	Height (m)	Crest length (m)	Name of the reservoir	Capacity[b] (10^6 m^3)
Pueblo Dam	1970–75[a]	Concrete	76.2	3118	Pueblo Reservoir	441.2
Twin Lakes Dam	1978	Earthfill	N.A.	960	Twin Lakes Reservoir	173.8
Sugar Loaf Dam	1965–68	Earthfill	41.1	616	Turquoise Lake	159.6
Ruedi Dam	1964–68	Earthfill	86.9	318	Ruedi Reservoir	126.3
Mt Elbert Forebay Dam	1977–78	Earthfill	28.0	793	Mt Elbert Forebay Reservoir	13.7

Note: [a] Modified over the period of 1998–2000; [b] Data from the Fryingpan–Arkansas Project website.[104]
Source: U.S. Bureau of Reclamation's website.[105]

horseshoe-shaped tunnel has a capacity of 26.8 m^{3}s^{-1} and conveys the water through the Continental Divide to Turquoise Lake.

Turquoise Lake (3)[106] provides a storage facility for regulation of Fry–Ark Project water flowing from the Boustead Tunnel (2). The Mt Elbert Conduit (4) conveys water from Turquoise Lake to the Mt Elbert Forebay Reservoir (6). The Halfmoon Diversion Dam (5) intercepts flows of Halfmoon Creek for diversion to the Mt Elbert Conduit (4). Water from Mt Elbert Forebay Reservoir (6) is used to generate hydro-electric power in the Mt Elbert Pumped-Storage Power-plant (7). From the Power plant, water exits into Twin Lakes (8). From Twin Lakes, water is released to Lake Creek and the Arkansas River for water users upstream and downstream of Pueblo Reservoir.

Water from the Fry–Ark Project, which is used for municipal, industrial, and irrigation purposes, became available in September 1975. Initial deliveries of water to the Fountain Valley Conduit occurred in the mid-1980s. The Conduit begins at the Pueblo Dam and ends south of Colorado Springs. It is 72.4 km long with a diameter ranging from 1.07 m to 0.36 m. The Conduit has five pumping plants, and conveys approximately 24.8 × 10^6 m^3 water annually for municipal and industrial purposes. Colorado Springs and communities along the Conduit depend heavily on the diverted water. Supplemental irrigation water delivered to

[104] http://www.usbr.gov/dataweb/html/fryark.html (visited in July 2005).
[105] http://www.usbr.gov/dataweb/dams (visited in July 2005).
[106] See Figure 11.21 for location of numbers provided in brackets.

Table 11.21. *Summary of the Colorado Statewide Water Supply Initiative tasks*

Phase 1: Develop project objectives and initiate public participant
Task 1: Project kick-off and public information and basin roundtable participant involvement activities
Task 2: Data collection (studies, plans, reports and databases) and review
Task 3: Final scoping activities
Phase 2: Conduct statewide and basin inventories
Task 4: Statewide physical environment (climate, topography, geology, surface water, groundwater, and water quality)
Task 5: Statewide institutional setting (regulatory and management agencies, laws, funding mechanisms, etc.)
Task 6: Statewide demography, economy, social setting, and environmental issues
Task 7: Basin descriptions (physical, water supply, demography, economic, social and institutional settings)
Task 8: Water supplies available for development in each basin (hydrology, drought, and available supplies)
Task 9: Estimate of existing and projected water demands
Phase 3: Alternatives identification and evaluation
Task 10: Catalogue basin options (conservation, rehabilitation of existing facilities, development of new sources, etc.)
Task 11: Overview and analysis of water supply options and demands at the State level
Phase 4: Develop implementation strategies
Task 12: Develop implementation strategies that will enhance implementation of the selected projects
Task 13: Identify implementation processes (planning, authorisation requirements, legislations, funding, etc.)
Phase 5: Documentation and management
Task 14: Develop an electronic database to provide a single-source documentation for implementation of the SWSI
Task 15: Study documentation to compile project information and task memoranda into a final report
Task 16: Management of project scope, schedule, budget, staffing and related activities by Colorado Water Conservation Board

Source: Colorado Water Conservation Board: Statewide Water Supply Initiative.[107]

the Arkansas Valley since 1975, has enabled farmers to increase their agricultural productivity over an area of 113 600 ha, with the major crops being alfalfa, corn, sorghum and sugar beet.

The Fry–Ark Project also generates electricity at the Mt Elbert Pumped-Storage Power Plant with an installed capacity of 200 MW.[108] This electricity is delivered to the transmission system near Leadville. Recreation facilities have also been developed throughout the Project area (Ruedi Reservoir, Turquoise Lake, Twin Lakes, Pueblo Reservoir, etc.), and water has been provided for the Leadville Fish Hatchery. Up to 1995, the Project had also provided an estimated accumulated US$16.8 million in flood control benefits.

Colorado Statewide Water Supply Initiative

The overall objective of the Colorado Statewide Water Supply Initiative (SWSI) released on 28 May 2003 is to help Colorado maintain adequate water supplies for its citizens and the environment for the next 30 years.[109] It consists of 5 phases and 16 tasks (Table 11.21).

The SWSI is not intended to take the place of local water planning initiatives. Rather, it is a "forum" to develop a common understanding of existing and future water supply requirements throughout Colorado, and possible means of meeting these needs. The Colorado Water Conservation Board (CWCB), through the SWSI, will support and identify solutions to water supply needs. Examples of potential solutions include: conservation, and rehabilitation of existing water supply facilities; enlargement and/or more efficient use of existing facilities; and development of new water supply projects. To achieve these goals, the SWSI will assess the existing water supplies and projected demands up to 30 years into the future for each one of the seven Water Divisions of the State, and will develop a range of potential options to meet demands.

Drought and Water Supply Assessment

The major drought of 2000–03 galvanised the CWCB to undertake a project to better understand the challenges and

---

[107] http://cwcb.state.co.us/SWSI/SWSI_SOW.pdf (visited in July 2005).
[108] http://www.usbr.gov/power/data/sites/mtelbert/mtelbert.html (visited in July 2005).
[109] Description of the SWSI provided here is based on the URL http://cwcb.state.co.us/SWSI/SWSI_SOW.pdf (visited in July 2005).

Table 11.22. *Summary of effective drought and water conservation measures for municipal and agricultural water use for the State of Colorado*

	Municipal	Agricultural
Drought	(1) Public education and involvement (2) Lawn and garden watering restrictions (3) Fines and tiered rates for water use	(1) Water conservation (2) Cooperative agreements
Water conservation	(1) Public education and involvement (2) Water metering (3) Distribution system leak detection (4) Fines and tiered rates for water use	(1) Alternative irrigation practices (includes alternative crops and planting strategies) (2) Lining ditches and canals (3) Conjunctive use of surface and groundwater and recycled water (4) Water metering

Source: Drought & Water Supply Assessment (2004, Table 17.1).[110]

issues that Colorado's water users are facing regarding drought and water supply. The outcome, published in 2004, is a comprehensive document entitled "Drought and Water Supply Assessment".[111] The major objective that has been identified is to improve water availability and reliability for municipal, agricultural, and other water use sectors. Water users indicated that meaningful and effective drought and water conservation measures would need to include the components listed in Table 11.22.

To achieve water availability and reliability objectives, the document has recommended 14 key tasks to be performed by the CWCB. These tasks include:

- Evaluate, improve, and coordinate the role and relationship of the CWCB public information and education efforts with those being conducted by local water authorities, users, and suppliers.
- Revise and update the CWCB long-term and strategic plans to ensure performance of the identified implementation tasks and activities.
- Examine the CWCB internal budget and organisational structure to determine how to best achieve the desired objectives.
- Evaluate means to fund public education, infrastructure construction and maintenance and technical assistance programs.
- Continue to support development and implementation of the SWSI as it relates to the identification of areas with critical water management issues, water development projects, water supply and demand imbalances, and infrastructure needs.
- Integrate the results of this project, and other relevant projects, into the SWSI, Federal Water Plan for the West (see Section 11.11), and other State and regional water planning efforts.

11.17.2 Water Transfer to the State of New Mexico: San Juan–Chama Project

The San Juan–Chama Project diverts water from San Juan River, a tributary of the Colorado River, into the Rio Chama, a tributary of the Rio Grande in New Mexico (Figure 11.22). The State of New Mexico covers about 314 900 km^2 with elevations ranging from 859 m in the southeast to 4011 m in the north. It entered the Union on 6 January 1912 as the 47th state. Much of the State is semi-arid to arid. Its average annual precipitation ranges from 200 mm in the south to more than 700 mm in the mountainous areas. Its major river is the Rio Grande, which flows north to south through the centre of the State and provides water for many settlements in its valley. The *New Mexico Water Resources Atlas*[112] provides information about water resources of the State. New Mexico had a population of 1.82 million in 2000, and is expected to reach 2.38 million by the year 2020.

---

[110] http://cwcb.state.co.us/owc/Drought_Water/index_DWSA.html (visited in July 2005).
[111] This document is available at the URL: http://cwcb.state.co.us/owc/Drought_Water/index_DWSA.html (visited in July 2005).
[112] http://www.ose.state.nm.us/water-info/NMWaterPlanning/nmwateratlas.pdf (visited in July 2005).

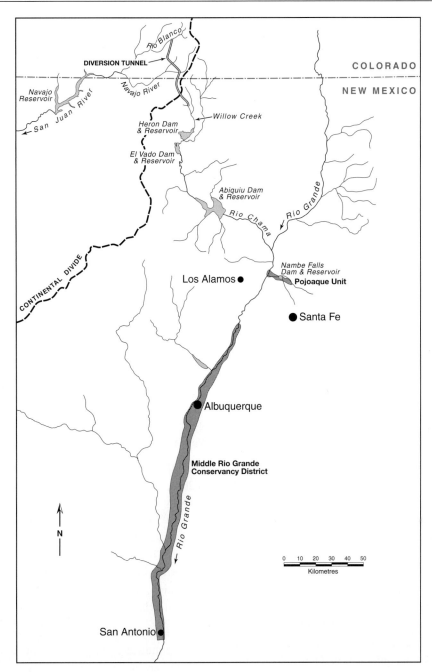

Figure 11.22 Main features of San Juan Chama Project (based on the USBR map of the Project dated January 1964).

Studies investigating the diversion of water from San Juan River into the Rio Chama began in 1901 and continued after World War I.[113] A 1933–34 investigation of Rio Grande water, known as the Bunger Survey, resulted in a proposed project called the San Juan–Chama Diversion Project which aimed to deliver water to Albuquerque across the Continental Divide. This survey was resumed in 1936 as part of the Rio Grande Joint Investigations and formed the basis for the Rio Grande Compact, which was approved by Congress on 31 May 1939. The Compact constituted an agreement between the states of Colorado, New Mexico and Texas on the appropriation of Rio Grande Water. In 1948, the *Upper Colorado River Basin Compact* established New Mexico's water right at about $1.05 \times 10^9 \, m^3$ per year (see Section 11.16).

The San Juan Technical Committee was appointed in 1950. It prepared a survey report in May 1950, and presented

---

[113] The history of the Project described here is based on the URL http://www.usbr.gov/dataweb/html/sjcdraft.html (visited in July 2005).

Table 11.23. *Selected features of two dams and reservoirs of the San Juan–Chama Project, in order of reservoir capacity*

Name of the dam	Construction period	Type	Height (m)	Crest length (m)	Name of the reservoir	Capacity (10^6 m^3)
Heron	1967–71	Earthfill	82.0	372	Heron	506.0
Nambe Falls	1974–76	Concrete	45.7	98	Nambe Falls	2.5

Source: U.S. Bureau of Reclamation's Dataweb.[114]

progress reports in 1951 and 1952. The 1952 report suggested possible annual diversion at three different levels of 325×10^6 m^3, 290×10^6 m^3 and 201×10^6 m^3. In 1955, a feasibility study was prepared by the USBR. In April 1958, New Mexico Senators introduced the *Navajo Irrigation–San Juan Chama Diversion Bill* to Congress. The *Bill* went through several hearings and amendments from 1958 to 1961. Finally, the US Senate approved the *Bill* on 29 May 1962 to initiate the first stage of the San Juan–Chama Project, and President Kennedy signed it on 13 June 1962. On 25 June 1963, the Middle Rio Grande Conservancy District and the City of Albuquerque signed a contract with the Federal Government for respective repayment of US$30.9 million and US$3.4 million of the Federal funds (US$189 million and US$20.8 million in 2004 prices).

The landscape in the Project area varies from forested mountain ranges to low and flat desert valleys. The Project consists of a system of diversion structures and tunnels for inter-basin water transfer.[115] The Project diverts 137×10^6 m^3 of water from the upper tributaries of the San Juan River. The primary objectives of the Project are to supply water to the middle Rio Grande Valley for municipal, agricultural and industrial uses.

The Project takes water from the Rio Blanco and the Navajo River, which are the upper tributaries of the San Juan River. The collection and diversion elements (diversion dams, tunnels and siphons) were constructed over the period of 1964–70, and deliver water to a tributary of Willow Creek in the Rio Grande Basin, and then to Heron Reservoir. Under various laws, only imported San Juan–Chama Project water is allowed to be stored in Heron Reservoir. All local waters are released back into the Creek below the Dam. The outlet works for El Vado Dam, located downstream of Heron Dam, were enlarged in 1965–66 to about 187 m^{3}s^{-1}, so that water released from Heron Reservoir could be passed unimpeded through El Vado Reservoir.

The Nambe Falls Dam is a pre-stressed concrete thin arch structure with thrust block and embankment wing. Its reservoir stores excess flows from Rio Nambe during high run-off periods and provides supplemental irrigation water for the Pojoaque Valley Irrigation District formed in October 1969, and also provides water supplies for a number of the surrounding communities.

The Project consists of two storage dams, three diversion dams, three tunnels and two siphons. Table 11.23 lists some features of the two dams and associated reservoirs of the Project.

Major beneficiaries of the Project are the City of Albuquerque (59.5×10^6 m^3), Santa Fe (6.9×10^6 m^3), and Los Alamos (1.5×10^6 m^3). Supplemental irrigation water (25.8×10^6 m^3) is provided for 36 300 ha in the Middle Rio Grande Conservancy District and 1.3×10^6 m^3 for 1120 ha in the Pojoaque Valley Irrigation District. The Project also provides facilities at Heron and Nambe Falls dams for fish and wildlife conservation and for recreation.

New Mexico State Water Plan

The Office of the State Engineer (2004, p. A-1) identified that in 2000, total withdrawal of water from streams and aquifers of New Mexico was more than 5.2×10^9 m^3, of which 3.2×10^9 m^3 was consumed. At present, even during periods of average water supply, demand for water exceeds the supply. This problem becomes acute during drought years and with increasing demand associated with increasing population. The State's water resources are faced with the following challenges:

(1) in many areas total water use exceeds total entitlements;
(2) groundwater is a primary source of drinking water for 90 percent of the State's population, and is also often used for agricultural and industrial purposes, however groundwater supplies are being depleted in some areas and are subject to contamination; and
(3) environmental flows required by the Federal *Endangered Species Act* have not been fully quantified and were not previously taken into account when water supplies were fully allocated to other uses.

[114] http://www.usbr.gov/dataweb/dams/index.html (visited in July 2005).
[115] Description of the project provided here is based on the URL http://www.usbr.gov/dataweb/html/sjuanchama.html (visited in July 2005).

The State Water Plan (Office of the State Engineer, 2003) articulates policies and strategies that will guide and assist the State in better management of its water resources (in terms of quantity and quality) in the early decades of the twenty-first century. The statewide common priorities and objectives of the Water Plan include:

(1) ensuring that water is available for the future economic vitality of the State;
(2) ensuring a safe and adequate drinking water supply for New Mexico;
(3) developing water resources (including desalination of brackish and saline waters) to expand the available supply;
(4) promoting water conservation measures, increasing water use efficiency in all sectors (agricultural, domestic and industrial), and water recycling;
(5) promoting drought planning;
(6) protecting, maintaining and enhancing the quality of the State's waters;
(7) providing for fish and wildlife habitat preservation and maintenance and for river restoration; and
(8) protecting pre-existing water rights.

11.17.3 Water Transfer to the State of Utah

The State of Utah covers about 220 000 km^2 and contains a variety of landforms consisting of high mountains, desert, forests, and fertile valleys. It was first settled by Mormons in 1847, and had a population of more than 2.3 million in 2002 (U.S. Census Bureau, 2003, Table 17). It is expected that the State's population will increase to 3.2 million by 2020 and 5 million by 2050 (Division of Water Resources, 2001, Figure 8).

The average annual precipitation of the State is about 330 mm, ranging from 127 mm over Great Salt Lake Desert to 380 mm in most populated areas, and more than 1000 mm in mountainous regions. The Colorado, Green and the San Juan rivers drain most of the east Utah (Figure 11.17), while the Great Basin drains the western half of the State. The State's two largest natural lakes, Great Salt Lake and Utah Lake, are remnants of the ancient Lake Bonneville, which at one time occupied much of Utah's Great Basin.[116] The largest artificial lake of the State is Lake Powell (Figure 11.17) created by construction of Glen Canyon Dam[117] on the Colorado River over the period of 1957–64.

Early Mormon settlers first started irrigation in Utah's Salt Lake Valley, and since then the irrigated area has expanded to 0.6 Mha, using 5.4×10^9 m^3 of surface water and groundwater (Table 11.3), representing an average water use of 9000 m^3 par ha. As the population increased and the irrigation projects developed, authorities turned to the Federal Government for expertise and funding. The early Federal projects include (Murray and Johnston, 2001) the Strawberry Valley Project (1906–18), Uintah Indian Irrigation Project (1906–22), Provo River Project (1938–41), and Moon Lake Project (1935–41). The early Federal projects served the people for a time, but as demand for water increased, the idea of the Central Utah Project (CUP) developed. In the following sections the Strawberry Valley Project and the Central Utah Project are described. These are followed by a short description of the Utah State Water Plan.

Strawberry Valley Project

The Strawberry Valley Project is the forerunner of the Central Utah Project. The following description of the project is based on Stene (1995).[118]

Utah Valley desperately needed a reliable source of water long before 1900. Around this time, Utah State Senator, Henry Gardner, visited Strawberry Valley, and a proposal was developed to construct a reservoir to store water and to transfer it westward through the Wasatch Divide between the Colorado River Basin and the Great Basin. Further investigations followed in 1902 when an engineer conducted preliminary surveys and concluded that the project would require more money than the citizens of the Valley could afford. Therefore, Utah Valley residents requested a USRS[119] investigation of the project's feasibility. Preliminary investigations by the USRS were carried out in 1903 and 1904. On 2 October 1905 a board of engineers determined that the project was feasible, and construction of the Strawberry Valley Project was authorised on 6 March 1906.

Water from the Strawberry Reservoir flows through Strawberry Tunnel into Six Water Creek, then travels to the Diamond Fork River, and flows into Spanish Fork River (Figure 11.23). The Spanish Fork Diversion Dam diverts the

---

[116] http://www.ugs.state.ut.us/online/PI-39/pi39pg01.htm (visited in July 2005).
[117] Technical details of the dam and its reservoir are (http://www2.privatei.com/~uscold/uscold_s.html and http://www.usbr.gov/dataweb/dams/az10307.htm, both visited in July 2005): type concrete arch; height 216.4 m; crest length 475.5 m; crest width 7.6 m; base width 91.5 m; reservoir capacity 33.3×10^9 m^3.
[118] Stene, Eric A. (1995). Research on Historic Reclamation Projects: The Strawberry Valley Project (second draft). Denver, Colorado. USBR. Available at the URL http://www.usbr.gov/dataweb/html/strawh.html (visited in July 2005).
[119] See **Abbreviations**.

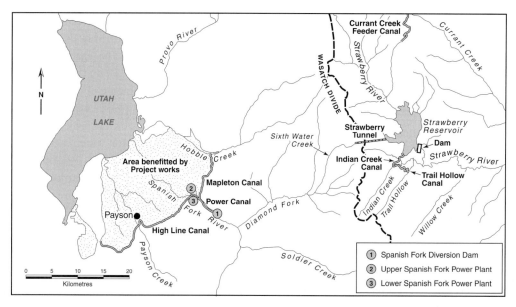

Figure 11.23 Simplified map of the Strawberry Valley Project (based on the USBR map number 27-400-139, dated December 1966).

water into the Power Canal, feeding the Upper and Lower Spanish Fork power plants. The High Line and the Mapleton canals, and associated laterals, transport water across the farmlands, south-east of Utah Lake.

Construction of the Project commenced with excavation of the 6.1 km Strawberry Tunnel in late August 1906 and was completed in June 1912, followed by completion of its lining in December 1912. It had a capacity of $17 \, m^3 \, s^{-1}$, was 2.13 m wide and 2.74 m high with an arched ceiling.

Construction of the Power Canal for supplying water to the hydro-electric power plant on the Spanish Fork River, commenced on 1 May 1907 and the first water flowed through the canal on 13 December 1908. Construction of Upper Spanish Fork Power Plant was completed on 10 January 1909. Work crews began excavation for the Spanish Fork Diversion Dam in October 1907, and completed the structure on 1 July 1908.

Construction of the Strawberry Dam commenced on 18 June 1911 and completed on 20 September 1913. It is an earthfill dam 21.9 m high with a crest length of 149.4 m, and a capacity of $1.4 \times 10^9 \, m^3$. The High Line Canal was constructed over the period of December 1914 to June 1917. This was followed by construction of the Mapleton Canal in 1918.

After receiving control of the Project on 6 October 1926, the Strawberry Water Users Association (SWUA) made some additions to the facilities. For example, in the 1930s, it built the Currant Creek Feeder Canal (Figure 11.23), which transfers water from Currant Creek into the Strawberry River, and the Lower Spanish Fork Power Plant on the Power Canal was constructed in 1937.

Central Utah Project

The Central Utah Project (Figure 11.24) is the largest water resources development program ever undertaken in the State of Utah. Two detailed reports (Division of Water Resources, 1997, 1999) provide information on all aspects of the project, including water resources, water supply and use, agriculture, water quality, fisheries, management, planning and economics. The CUP water has been used for municipal, industrial and irrigation purposes, as well as hydro-electric power, fish, wildlife, conservation and recreation. The Project is also designed to improve flood control and to assist water quality improvement.

The concept of a project for Central Utah was envisioned when a reconnaissance investigation of the newly conceived Colorado River-Great Basin Project was conducted by the USBR from 1939 to 1943 (Murray and Johnston, 2001). The project plan called for an annual diversion of $1.23 \times 10^9 \, m^3$ of water from Green River to the Great Basin. To satisfy future water requirements in Central Utah, the possibility of expanding the existing Strawberry Valley Project was considered. Reconnaissance investigations for additional water for the Strawberry Valley Project started in the spring of 1945, and the name Central Utah Project was given to an expanded version of the plan. Results of the investigations were contained in a planning interim report of September 1945, suggesting transfer of $710 \times 10^6 \, m^3$ of water from Colorado River Basin to the Bonneville Basin.

A Central Utah Project Office was established in 1946 and feasibility investigations were carried out over the following

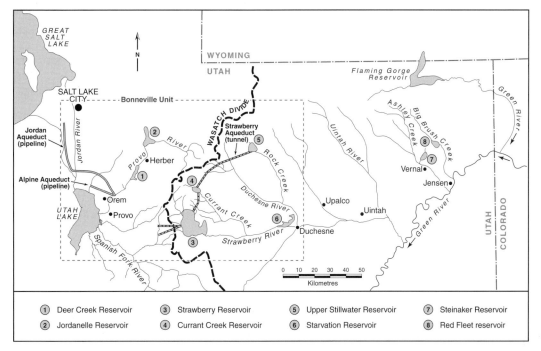

Figure 11.24 Main features of the Central Utah Project (Murray and Johnston, 2001, Figure 3 and Central Utah Water Conservancy District, 2003).

years. Results of these investigations were compiled in a feasibility report released in 1951, which greatly reduced the volume of water to be diverted to $175 \times 10^6 \, m^3$ per annum. In 1956, the Congress authorised construction of the Project.

On 2 March 1964 the Central Utah Water Conservancy District (CUWCD) was established. A repayment contract between the USBR and the CUWCD for construction of the CUP was executed on 28 December 1965. In 1967, the US Congress approved US$3.5 million (US$19.6 million in 2004 prices) to initiate construction of the Project (Murray and Johnston, 2001).

The Sierra Club opposed the water transfer required for the Project (Murray and Johnston, 2001). Following the enactment of the *National Environmental Policy Act* of 1969, signed into the law on 1 January 1970, USBR began to work on an Environmental Impact Statement (EIS) for the Project. In August 1973, USBR issued the Bonneville Unit Final EIS and committed to prepare a site-specific EIS for each of the remaining Units before initiating their construction. USBR published the Bonneville Unit Municipal and Industrial System final EIS report on 25 October 1979. Also, USBR had to comply with the *Endangered Species Act* of 1973 in order to protect endangered species within the Project area. In 1988, USBR prepared the Supplement to the Definite Plan report to address refinements made to the Bonneville Unit since 1964. In the meantime construction was progressing slowly because of both Federal environmental laws and inadequate Federal funding. The slow progress prompted State and local officials to request Congress to make unprecedented changes to the way Federal water projects are planned and constructed. In response to these requests the Congress enacted the *Central Utah Project Completion Act (CUPCA)* on 30 October 1992. To implement CUPCA, Congress established a partnership arrangement between the Department of the Interior, CUWCD, the Utah Reclamation Mitigation and Conservation Commission, and the Ute Indian Tribe.

Out of the five authorised units of the CUP, only the Vernal and Jensen Units have been completed (Murray and Johnston, 2001). The Upalco Unit has been indefinitely postponed, the Uintah unit has been replaced,[120] and the Bonneville Unit is currently (2005) under construction.

Description of the original units of the Central Utah Project is available at the Central Utah Project Overview website.[121] The following short description of the two completed units (Jensen and Vernal) is based on this website,

---

[120] Information regarding the Replacement Project which is under construction is available at the URL: http://www.cuwcd.com/cupca/projects/ubrp/index.htm (visited in July 2005).

[121] http://www.usbr.gov/dataweb/html/cupoverview.html (visited in July 2005).

Table 11.24. *Characteristics of the major dams and reservoirs of the Central Utah Project*

Name	Year of completion	Type	Height (m)	Crest length (m)	Capacity ($10^6 \, m^3$)
Strawberry	1913	Earthfill	21.9	149	1365.0[a]
Jordanelle	1993	Earthfill	91.5	1128	387.0[b]
Starvation	1970	Earthfill	64.0	936	206.6[a]
Deer Creek	1941	Earthfill	71.6	398	188.2[b]
Steinaker	1962	Earthfill	49.4	609	41.1[a]
Upper Stillwater	1987	Concrete	89.0	808	39.5[a]
Red Fleet	1980	Earthfill	49.1	1010	32.3[a]
Currant Creek	1975	Earthfill	39.6	488	19.3[a]

Note: [a] Division of Water Resources (1999, Table 6.1); [b] Division of Water Resources (1997, Table 6.1).
Source: Central Utah Project Overview.[122]

while description of the Bonneville Unit is based on the Bonneville Unit website,[123] and other sources.

Vernal Unit

The Vernal Unit completed in 1962 provides supplemental irrigation water to about 6000 ha of land in Ashley Valley by storing the high spring flows of Ashley Creek for late season use. Flows from Ashley Creek are diverted via a canal to Steinaker Reservoir. This off-stream reservoir has a total capacity of $47 \times 10^6 \, m^3$. The Vernal Unit also furnishes municipal water for Vernal and other communities. Recreation and fishing facilities have been provided at Steinaker Reservoir.

Jensen Unit

The Jensen Unit provides annually about $22.2 \times 10^6 \, m^3$ of municipal and industrial water and $5.7 \times 10^6 \, m^3$ of irrigation water to lands in the vicinity of Jensen. The Red Fleet Reservoir on Big Brush Creek, the Unit's major feature, has a total capacity of $32.3 \times 10^6 \, m^3$. The reservoir stores early spring run-off and surplus flows on Big Brush Creek for subsequent municipal, industrial and irrigation use. Recreation, fish and wildlife, and flood control benefits are also provided.

Bonneville Unit

The Bonneville Unit is the largest and most complex of the authorised units of the CUP. It has been designed to serve the needs of a growing population in the Unit area. The CUPCA provided direction for the completion of CUP and modified some of the Bonneville Unit's systems. Notable changes occurred to the irrigation and drainage, and Diamond Fork Systems. The CUPCA also authorised new components to improve water management and wildlife habitats.

Starvation Reservoir on the Strawberry River, completed in 1970, has a capacity of $206.6 \times 10^6 \, m^3$. It stores flows of the Strawberry River, and diverted waters of the Duchense River via Knight Diversion Dam (completed in July 1968, located about 8 km upstream from the town of Duchense) and Starvation Feeder Conduit completed in November 1968. The Starvation Collection System provides flood control facility and supplies water for irrigation, municipal and industrial use.

The Strawberry Aqueduct, which is about 60 km long, collects flows from Rock Creek and tributaries of the Duchesne River and Currant Creek. It delivers the water to Strawberry Reservoir. The Upper Stillwater and Currant Creek reservoirs serve as regulating reservoirs. Table 11.24 shows some of the characteristics of the major dams and reservoir of the CUP.

Since the inception of the CUP, the social and political climate of Utah and the United States changed significantly, resulting in major financial and environmental challenges for the Project. When the CUP was initiated, the project was primarily for agricultural development. Since then, the objectives have changed because of increasing attention to environmental issues. These changes are reflected in the enactment of CUPCA, which added water conservation measures, water use efficiency, conjunctive use of surface and groundwater resources, recovery of endangered species and consideration of water quality issues (Murray and Johnston, 2001). The CUWCD has been allocated with a credit of US$50 million in Federal funds for implementation of conservation measures (Division of Water Resources, 2001, p. 25). This money is distributed to projects

---

[122] http://www.usbr.gov/dataweb/html/cupoverview.html (visited in July 2005).
[123] http://www.usbr.gov/dataweb/html/bonneville.html (visited in July 2005).

that meet certain criteria, including a requirement of a 35 percent cost sharing from local sources. The CUWCD expects to conserve $61.2 \times 10^6 \, m^3$ of water per annum by 2013, by implementing various water conservation projects.

Utah State Water Plan

The Division of Water Resources (2001, Tables 6 and 7) predicted that agricultural water use will decline to $4.93 \times 10^9 \, m^3$ by 2020 and $4.68 \times 10^9 \, m^3$ by 2050 because of a reduction in the extent of irrigated land. For municipal and industrial water use, estimates indicate that if the present trend in water use continues, water requirements of these sectors will increase from their 1998 level of $1.12 \times 10^9 \, m^3$ to $1.63 \times 10^9 \, m^3$ in 2020 and $2.41 \times 10^9 \, m^3$ in 2050. These estimates for the future water requirements exceed the current level of the State's developed water resources. Although the State has the potential to further develop its surface water resources by about $0.98 \times 10^9 \, m^3$, which includes $0.5 \times 10^9 \, m^3$ from Colorado River, their development is not straightforward because of the environmental and economic considerations. To overcome the problem, the State has developed a water plan with a major water conservation component (Division of Water Resources, 2001). Elements of this plan include: conservation of agricultural, municipal, and industrial water; conjunctive use of surface and groundwater supplies; water trading between users; and reuse of wastewater. With respect to the municipal and industrial water use, the goal is to reduce per capita water demand from public community systems by 12.5 percent by 2020 and 25 percent before the year 2050. The State's water plan also includes development of Bear River, construction of Lake Powell Pipeline, and continuation of cloud seeding projects. Cloud seeding projects were undertaken for the first time in the State from 1951 to 1955. In water year 2000, there were four active cloud seeding projects in the State, which resulted in a 13 percent increase in average annual run-off in the project areas compared with the historic records. This was achieved at a cost of US$0.8 per $1000 \, m^3$ (Division of Water Resources, 2001, pp. 48–49).

11.18 CONCLUSIONS

This chapter has demonstrated the extent of water export from the 18 Water Resources Regions of the United States, and water transfer between their sub-regions. It has focused on water resources issues and water transfer in the drier western United States and described major water transfer projects in California and from the Colorado River Basin to eastern Colorado, New Mexico, California and Utah. Although it has not covered all inter-basin water transfer projects in the western United States, it has clearly indicated the enormous engineering efforts and financial resources, which were devoted to the design and construction of these projects during the twentieth century and particularly after the World War II. These projects have provided western states with irrigation, municipal, and industrial water supplies as well as hydro-power, flood control, and recreational facilities. The Sierra Club first raised environmental objections to some of these projects, such as the Hetch Hetchy Aqueduct Project in the early twentieth century. Increasing environmental awareness in the last decades of the twentieth century has resulted in an end to most new inter-basin water transfer projects. This means that greater emphasis is being placed on water conservation, water reuse and on new sources within basins.

Because of the expected significant increase in population of the western states of the United States in the next two or three decades, and potential impacts of climate change, as well as protection of the environment, western states have already developed or are developing long-term water management plans. Their current efforts are oriented towards better management of the existing developed water resources, rather than development of new ones, particularly transfer of water from one basin to another by building new dams, tunnels, aqueducts and other facilities. Current water management policies encourage and facilitate activities such as: public education and involvement; increasing water use efficiency in agricultural, domestic and industrial sectors; reuse of treated wastewater; conjunctive use of surface and groundwater resources; water trading; water pricing; desalination of brackish and saline water; and implementation of cloud seeding projects. These activities are technically and financially supported by the Federal Government under the umbrella of the *Federal Water Plan for the West (Water 2025)* described in this chapter. These initiatives have seen some dramatic changes in water use, especially in agriculture where the production per unit of applied water for 32 important crops increased by 38 percent between 1980 and 2000. As well, in some of the large cities, per capita water use has declined by about 20 percent.

Water trading, in spite of its advantages, has created economic and social problems in some water exporting areas of California. The extent of the problem has been widespread and by late 2002, 22 of the 58 counties in California have adopted export restrictions. This indicates that water trading is not quite the panacea that economists had envisaged.

References

Alemi, M. (2003). Office of Water Use Efficiency: the services we offer. *Water Conservation News*. January 2003, pp. 1–2.

Autobee, R. (unpublished). *Colorado–Big Thompson Project: Historic Reclamation Projects Book*. Denver, Colorado: U.S. Bureau of Reclamation.

Autobee, R., Simonds, W. J., Stene, E. A. and Whynot, W. E. (unpublished). *Central Valley Project: Historic Reclamation Projects Book*. Denver, Colorado: U.S. Bureau of Reclamation.

Back, W., Rosenshein, J. S. and Seaber, P. R. eds. (1988). *Hydrogeology*. Boulder, Colorado: The Geological Society of America.

Barton, R. V. (1981). Computer simulation of the Colorado River for long-term operation studies. In Framji, K. K. ed. *State-of-the-Art Irrigation Drainage and Flood Control*. New Delhi: International Commission on Irrigation and Drainage, V. II, pp. 67–92.

Bonnifield, P. (1979). *The Dust Bowl: Men, Dirt, and Depression*. Albuquerque: University of New Mexico Press.

California Bay–Delta Authority (2003). *California Bay–Delta Authority: 2003 Annual Report*. Sacramento, California: California Bay–Delta Authority.

Central Utah Water Conservancy District (2003). *Annual Report*. Orem, Utah: CUWCD, Operation and Maintenance Department.

Colorado River Basin Salinity Control Forum (1993). *1993 Review: Water Quality Standards for Salinity, Colorado River System*. Bountiful, Utah: Colorado River Basin Salinity Control Forum.

Colorado River Basin Salinity Control Forum (2002). *2002 Review: Water Quality Standards for Salinity, Colorado River System*. Bountiful, Utah: Colorado River Basin Salinity Control Forum.

Colorado River Board of California (2000). *Draft California's Colorado River Water Use Plan*. Glendale, California: Colorado River Board of California.

Colorado River Salinity Program Coordinator (1993). *Salinity Update*. Denver, Colorado: U.S. Bureau of Reclamation.

Department of Water Resources (1993). *Sacramento San Joaquin Delta Atlas*. Sacramento: DWR.

Department of Water Resources (1998). *California Water Plan Update: Bulletin 160–98*. Sacramento: DWR.

Department of Water Resources (1999). *California State Water Project Atlas*. Sacramento: DWR.

Department of Water Resources (2003a). *California's Groundwater: Bulletin 118 Update 2003*. Sacramento: DWR.

Department of Water Resources (2003b). *Water Desalination: Findings and Recommendations*. Sacramento: DWR.

Department of Water Resources (2005a).[124] *California Water Plan: Highlights*. Sacramento: DWR.

Department of Water Resources (2005b).[125] *California Water Plan Update 2005: Volume 1 – Strategic Plan*. Sacramento: DWR.

Department of Water Resources (2005c).[126] *California Water Plan Update 2005: Volume 2 – Resource Management Strategies*. Sacramento: DWR.

De Roos, R. (1948).[127] *The Thirsty Land: The Story of the Central Valley Project*. Stanford, California: Stanford University Press.

Desalination Task Force (2003). *Desalination Task Force Report: Draft*. Sacramento: Department of Water Resources.

Division of Water Resources (1997). *Utah State Water Plan: Utah Lake Basin*. Salt Lake City: Division of Water Resources, Department of Natural Resources.

Division of Water Resources (1999). *Utah State Water Plan: Uintah Basin*. Salt Lake City: Division of Water Resources, Department of Natural Resources.

Division of Water Resources (2001). *Utah's Water Resources Planning for the Future*. Salt Lake City: Division of Water Resources, Department of Natural Resources.

EBMUD (2001). *Urban Water Management Plan 2000*. Oakland, California. East Bay Municipal Utility District.

EBMUD (2003). *East Bay Municipal Utility District: Annual Report 2003*. Oakland, California. East Bay Municipal Utility District.

Economic Research Service (2003a).[128] *Chapter 1.1: Land Use*. Economic Research Service, United States Department of Agriculture.

Economic Research Service (2003b).[129] *Chapter 2.1: Water Use and Pricing in Agriculture*. Economic Research Service, United States Department of Agriculture.

El-Ashry, M. T. and Gibbons, D. C. (1990). Adverse impacts and alternatives to large dams and interbasin transfers in the Colorado River Basin, USA. In UNESCO. *The Impact of Large Water Projects on the Environment*. Paris: UNESCO, pp. 137–144.

Fradkin, P. L. (1996). *A River No More: The Colorado River and the West*. Berkeley, California: University of California Press.

Frederick, K. (2001). Water marketing: obstacles and opportunities. *Forum for Applied Research and Public Policy* **16**(1): 54–62.

Galloway, D. L., Jones, D. R. and Ingebritsen, S. E. eds. (1999). *Land Subsidence in the United States*. Circular 1182. Reston, Virginia: U.S. Geological Survey.

Galloway, D. L., Alley, W. M., Barlow, P. M., Reilly, T. E. and Tucci, P. (2003). *Evolving Issues and Practices in Managing Ground-Water Resources: Case Studies on the Role of Science*. Circular 1247. Reston, Virginia: U.S. Geological Survey.

Geraghty, J. J., Miller, D. W., van der Leeden, F. and Trois, F. L. (1973). *Water Atlas of the United States*. Third Edition. Port Washington: Water Information Center.

Goklany, I. M. (2002). Comparing 20th century trends in U.S. and global agricultural water and land use. *Water International* **27**(3): 321–329.

Gregory, J. N. (1989). *American Exodus: The Dust Bowl Migration and Okie Culture in California*. New York: Oxford University Press.

Grolier Incorporated (1996). *The Encyclopedia Americana*: International Edition. Danbury, Connecticut: Grolier Incorporated.

Hanak, E. (2002). *California's Water Market by the Numbers*. San Francisco: Public Policy Institute of California.

Hanak, E. (2003). *Who Should be Allowed to Sell Water in California? Third-Party Issues and the Water Market*. San Francisco: Public Policy Institute of California.

Hanak, E. and Simeti, A. (2004). *Water Supply and Growth in California: A Survey of City and County Land-Use Planners*. San Francisco: Public Policy Institute of California.

Heath, R. C. (1984). *Ground-Water Regions of the United States*. U.S. Geological Survey Water Supply Paper 2242. Washington D.C.: U.S. Government Printing Office.

Hedlund, J. D. (1984). USDA planning process for Colorado River Basin Salinity Control. In French, R. H. ed. *Salinity in Watercourses and Reservoirs*. Proceedings of the 1983 International Symposium on State-of-the-Art Control of Salinity, Salt Lake City, Utah, 13–15 July 1983. Boston: Butterworth, pp. 63–77.

Howe, C. W. and Easter, K. W. (1971). *Interbasin Transfer of Water: Economic Issues and Impacts*. Baltimore: The John Hopkins Press.

Humlum, J. (1969). *Water Development and Water Planning in the Southwestern United States*. Kulturgeografisk Institute, Aarhus University, Denmark.

Hundley, Jr., N. (2001). *The Great Thirst: Californians and Water, A History*. Berkeley: University of California Press.

Hutson, S. S., Barber, N. L., Kenny, J. F., Linsey, K. S., Lumia, D. S. and Maupin, M. A. (2004).[130] *Estimated Use of Water in the United States in 2000*. U.S. Geological Survey Circular 1268.

---

[124] Available at the URL: http://www.waterplan.water.ca.gov/docs/cwpu2005/highlights/Highlights-web.pdf (visited in July 2005).

[125] Available at the URL: http://www.waterplan.water.ca.gov/docs/cwpu2005/Vol_1/v1PRD.combined.pdf (visited in July 2005).

[126] Available at the URL: http://www.waterplan.water.ca.gov/docs/cwpu2005/Vol_2/_V2PRDFULL.pdf (visited in July 2005).

[127] This book is reprinted in 2000 by Beard Books, Washington D.C.

[128] Available at the URL: http://www.ers.usda.gov/publications/arei/ah722/arei1_1/arei1_1landuse.pdf (visited in July 2005).

[129] Available at the URL: http://www.ers.usda.gov/publications/arei/ah722/arei2_1/AREIch2_1Watuse_pricing.pdf (visited in July 2005).

[130] Available at the URL: http://water.usgs.gov/pubs/circ/2004/circ1268/index.html (visited in July 2005).

IPCC (2001). *Climate Change 2001: Impacts, Adaptation and Vulnerability*. Contribution of Working Group II to the Third Assessment Report of the Intergovernmental Panel on Climate Change. Cambridge: Cambridge University Press.

Johnson, S., Haslam, G. and Dawson, R. (1993). *The Great Central Valley: California's Heartland*. Berkeley: University of California Press.

Jones, H. R. (1965). *John Muir and the Sierra Club: The Battle for Yosemite*. San Francisco: Sierra Club.

Jones, W. K. (1996). Los Angeles Aqueduct: a search for water. In Shumsky, N. L. ed. *The Physical City: Public Space and the Infrastructure*. New York: Garland Publishing, pp. 369–385.

Kahrl, W. L. ed. (1979). *The California Water Atlas*. California: Governor's Office of Planning and Research.

Mooty, W. S. and Jeffcoat, H. H. (1986). *Inventory of Interbasin Transfers of Water in the Eastern United States*. Denver, Colorado: U.S. Geological Survey. Open-File Report 86–148.

Mueller, D. K. and Moody, C. D. (1984). Historical trends in concentration and load of major ions in the Colorado River System. In French, R. H. ed. *Salinity in Watercourses and Reservoirs*. Proceedings of the 1983 International Symposium on State-of-the-Art Control of Salinity. Salt Lake City, Utah, 13–15 July 1983. Boston: Butterworth, pp. 181–192.

Murphy, R. E. (1996). Face of the land. In *The Encyclopedia Americana, International Edition*. Danbury, Connecticut: Grolier Incorporated. V. 27, pp. 500–515.

Murray, R. R. and Johnston, R. (2001). History of the Central Utah Project: a Federal perspective. In Schaack, J. and Anderson, S. S. eds. *Transbasin Water Transfers*. Proceedings of the 2001 USCID Water Management Conference, Denver, Colorado, 27–30 June 2001. Denver, Colorado: U.S. Committee on Irrigation and Drainage, pp. 87–102.

National Drought Policy Commission (2000). *Preparing for Drought in the 21st Century*. Washington D.C.: National Drought Policy Commission.

National Research Council (1992). *Water Transfers in the West: Efficiency, Equity, and the Environment*. Washington D.C.: National Academy Press.

O'Connor, J. E. and Costa, J. E. (2003). *Large Floods in the United States: Where they Happen and Why*. Circular 1245. Reston, Virginia: U.S. Geological Survey.

Office of the State Engineer (2003).[131] *New Mexico State Water Plan*. Santa Fe, New Mexico: New Mexico Office of the State Engineer and the Interstate Stream Commission.

Office of the State Engineer (2004).[132] *2003 New Mexico State Water Plan, Appendix A: Water Resources Issues*. Santa Fe, New Mexico: New Mexico Office of the State Engineer and the Interstate Stream Commission.

O'Shaughnessy, M. M. (1934). *Hetch Hetchy: Its Origin and History*. San Francisco: Recorder Printing and Publishing Company.

Paulson, R. W., Chase, E. B., Williams, J. S. and Moody, D. W., Compilers (1993). *National Water Summary 1990–91: Hydrologic Events and Stream Water Quality*. U.S. Geological Survey, Water Supply Paper 2400. Washington D.C.: U.S. Government Printing Office.

Perry, C. A. (2000).[133] Significant floods in the United States During the 20th century. Lawrence, Kansas: U.S. Geological Survey. Fact Sheet 024-00.

Petsch, Jr., H. E. (1985). *Inventory of Interbasin Transfers of Water in the Western Conterminous United States*. Denver, Colorado: U.S. Geological Survey. Open-File Report 85-166.

Pielke, Jr., R. A., Downton, M. W. and Miller, J. Z. B. (2002). *Flood Damage in the United States, 1926–2000: A Reanalysis of National Weather Service Estimates*. Boulder, Colorado: National Center for Atmospheric Research.

Quinn, F. (1968). Water transfer: must the American west be won again? *Geographical Review* **58**: 108–132.

Ralph M. Parsons Company (1964). *NAWAPA, North American Water and Power Alliance: Water and Power Plan*. Los Angeles: Ralph M. Parsons Company.

Reader's Digest (1998). *Reader's Digest Atlas of America: Our Nation in Maps, Facts, and Pictures*. New York: The Reader's Digest Association.

Recycled Water Task Force (2003). *Water Recycling 2030: Recommendations of California's Recycled Water Task Force*. Sacramento: Department of Water Resources.

Reisner, M. (1987). *Cadillac Desert: The American West and Its Disappearing Water*. New York: Penguin Books.

Rosegrant, M. W. (1995). Water transfers in California: potentials and constraints. *Water International* **20**: 72–87.

Solley, W. B., Pierce, R. R. and Perlman, H. A. (1998). *Estimated Use of Water in the United States in 1995*. Circular 1200. Reston, Virginia: U.S. Geological Survey.

State Water Resources Control Board (1999). *A Guide to Water Transfers*. Sacramento: State Water Resources Control Board, Division of Water Rights.

Thaemert, D. K. and Faucett, A. H. (2001). Evolution of transmountain water diversions in Colorado. In Schaack, J. and Anderson, S. S. eds. *Transbasin Water Transfers*. Proceedings of the 2001 USCID Water Management Conference, Denver, Colorado, 27–30 June 2001. Denver, Colorado: U.S. Committee on Irrigation and Drainage pp. 139–147.

Tyler, D. (1992). *The Last Water Hole in the West: The Colorado–Big Thompson Project and the Northern Colorado Water Conservancy District*. Niwot, Colorado: The University Press of Colorado.

U.S. Bureau of Reclamation (1992). *Title I Program Colorado River Basin Salinity Control Act*. Report to the Secretary of the Interior and the Congress. Nevada: U.S. Bureau of Reclamation, Lower Region.

U.S. Census Bureau (2003). *Statistical Abstract of the United States*. Washington D.C.: U.S. Department of Commerce.

U.S. Department of Agriculture (1989). *The Second RCA Appraisal, Soil Water and Related Resources on Nonfederal Land in the United States, Analysis of Conditions and Trends*. Washington D.C.: U.S. Government Printing Office.

U.S. Department of the Interior (2001). *Quality of Water: Colorado River Basin*. Progress Report No. 20. Denver: U.S. Bureau of Reclamation.

U.S. Department of Interior (2003). *Water 2025: Preventing Crises and Conflict in the West*. Washington D.C.: U.S. Department of Interior.

U.S. Geological Survey (1999). *The Quality of Our Nation's Waters: Nutrients and Pesticides*. Circular 1225. Reston, Virginia: U.S. Geological Survey.

U.S. Geological Survey (2002). *Report to Congress: Concepts for National Assessment of Water Availability and Use*. Circular 1223. Reston, Virginia: U.S. Geological Survey.

Van Valen, N. (1996). A neglected aspect of Owens River Aqueduct story: the inception of the Los Angeles municipal electric system. In Shumsky, N. L. ed. *The Physical City: Public Space and the Infrastructure*. New York: Garland Publishing, pp. 387–411.

Winchester, J. N. (2001). A historical view: Transmountain diversion development in Colorado. In Schaack, J. and Anderson, S. S. eds. *Transbasin Water Transfers*. Proceedings of the 2001 USCID Water Management Conference, Denver, Colorado, 27–30 June, 2001. Denver, Colorado: U.S. Committee on Irrigation and Drainage, pp. 479–497.

Woodhouse, C. A. and Overpeck, J. T. (1998). 2000 years of drought variability in the central United States. *Bulletin of the American Meteorological Society* **79**(12): 2693–2714.

Worster, D. (1979). *Dust Bowl: The Southern Plains in the 1930s*. New York: Oxford University Press.

Yevjevich, V., Hall, W. A. and Salas, J. D. eds. (1978). *Drought Research Needs*. Proceedings of the Conference on Drought Research Needs, held at Colorado State University, Fort Collins, Colorado, 12–15 December, 1977. Fort Collins: Water Resources Publications.

---

[131] Available at the URL http://www.seo.state.nm.us/water-info/NMWaterPlanning/2003StateWaterPlan.pdf (visited in July 2005).

[132] Available at the URL http://www.seo.state.nm.us/water-info/NMWaterPlanning/A_Water_Resources_Issues1.pdf (visited in July 2005).

[133] Available at the URL: http://ks.water.usgs.gov/Kansas/pubs/factsheets/fs.024–00.pdf (visited in July 2005).

Young, G. and Hunn, R. (2001). *On Tap:* An interactive web site to facilitate water transfer in California. In Schaack, J. and Anderson, S. S. eds. *Transbasin Water Transfers*. Proceedings of the 2001 USCID Water Management Conference, Denver, Colorado, 27–30 June, 2001. Denver, Colorado: U.S. Committee on Irrigation and Drainage, pp. 287–297.

Zimbelman, D. D. and Werner, B. R. (2001). Water management in the Northern Colorado Water Conservation District. In Schaack, J. and Anderson, S. S. eds. *Transbasin Water Transfers*. Proceedings of the 2001 USCID Water Management Conference, Denver, Colorado, 27–30 June 2001. Denver, Colorado: U.S. Committee on Irrigation and Drainage, pp. 331–339.

12 Inter-basin Water Transfer in Canada

SECTION A: OVERVIEW OF GEOGRAPHY, POPULATION, LAND AND WATER

12.1 GEOGRAPHY

Canada covers an area of about 9 984 670 km², consisting of 9 093 507 km² of land and 891 163 km² of freshwater lakes. It is the second largest country in the world after Russia and covers more than half of the North American continent. It extends from the Atlantic Ocean to the Pacific Ocean and from the northern boundary of the United States to the Polar Regions. Canada is a federation of 10 provinces and three territories.[1] Seven physiographic regions are distinguishable in Canada (Figure 12.1). These regions are (Bone, 2002):

- **Canadian Shield** is the largest region and covers about half the total area of Canada. It forms an incomplete ring around Hudson Bay and consists of Precambrian volcanic rocks of 3500 million to 600 million years old. Originally, it was the site of high mountain ranges, which have since been eroded over geological time. It consists mainly of a rugged, rolling upland.
- **Cordillera** is a complex region of mountains, plateaus, and valleys. It is a young geological structure, which was formed 40 to 80 million years ago. The Rocky Mountains are the best-known mountain ranges of Cordillera, which have elevations between 3000 m to 4000 m.
- **Interior Plains** is a northward extension of the Great Plains of the United States. The oldest sedimentary rocks of the region were formed about 500 million years ago. Since then, other sedimentary layers have been deposited on top of them.
- **Arctic Lands** is centred in the Canadian Arctic Archipelago and occupies over 25 percent of the area of Canada. It is a complex composite of coastal plains, plateaus and mountains. Across these lands, the ground is permanently frozen to great depths.
- **Hudson Bay Lowland** is the youngest of the physiographic regions in Canada. It was formed when the ice sheet no longer blocked the Atlantic Ocean from entering what is now Hudson Bay.
- **Appalachian Uplands** is the northern end of a long belt of old mountains running north-easterly along most of the Atlantic seaboard of North America. It covers only about two percent of Canada's landmass.
- **Great Lakes–St Lawrence Lowlands** covers less than two percent of the area of Canada, but it is the most important manufacturing and the most intensively settled region in the country.

12.2 POPULATION

Canada had a population of 24.3 million in 1981, and reached 32.1 million on 1 January 2005.[2] Approximately 90 percent of the Canadian population lives within 250 km of the southern Canadian border with the United States, while vast areas of the country are sparsely inhabited. The highly populated provinces are: Ontario (12.4 million); Quebec (7.5 million); British Columbia (4.2 million); Alberta (3.2 million); Manitoba (1.2 million); and

---

[1] British Colombia, Alberta, Saskatchewan, Manitoba, Ontario, Quebec, Newfoundland, Nova Scotia, New Brunswick, Prince Edward Island, Yukon Territory, Northwest Territories and Nunavut Territory. Description of geography, population, history, economy, social services and education of each province and territory is available in Watkins (1993) and Bone (2002).
[2] http://www.statcan.ca/start.html (visited in July 2005).

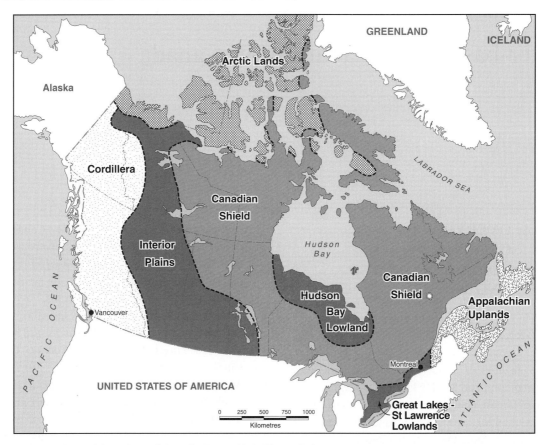

Figure 12.1 Main physiographic regions of Canada (Bone, 2002, Figure 2.1).

Saskatchewan (1.0 million). From July 2003 to July 2004, Canada's population grew by 0.9 percent. The slowing of population growth is related to a continuing decline in natural increase (birth minus death), but also due to a decrease in net international migration. Canada's population is projected to reach 34.4 million by 2016,[3] and 36.2 million by 2026.[4]

12.3 ECONOMY

The Gross Domestic Product of Canada in 1999 was CA$880.3 billion (Statistics Canada, 2001, Table 9.1, p. 329). Since 1961, there has been a shift away from resource-based industries to service-based industries. In 1961, the agriculture, forest, metal and mineral products industries accounted for 63 percent of total exports. By 2000, these industries accounted for only 25 percent of total exports. In contrast, the fuel and energy industries increased from 4 percent to 11 percent, and the transport equipment industry increased its share from 2 to 23 percent (Statistics Canada, 2004, p. 56).

12.4 CLIMATE AND PRECIPITATION

Over the past 900 000 years, the Canadian region has experienced a succession of glacial periods, each lasting about 100 000 years, interspersed with brief warm spells less than 20 000 years long. The most recent glaciation began about 100 000 years ago (Statistics Canada, 2001, p. 19). Some 18 000 years ago, the ice sheet reached its maximum expanse, covering nearly all of Canada's landmass with a maximum thickness of approximately 5 km. About 15 000 years ago, the climate began to warm, causing these ice sheets to retreat. Seven thousand years ago, the last remnants of these ice sheets were in the Rocky Mountains and in the uplands of the Canadian Shield in northern Quebec–Labrador and on Baffin Island (Bone, 2002, pp. 48–49).

The physical geography of Canada has a marked influence on its current climate. Air moving off the Pacific Ocean to

[3] http://www.statcan.ca/english/Pgdb/demo23b.htm (visited in July 2005).
[4] http://www.statcan.ca/english/Pgdb/demo23c.htm (visited in July 2005).

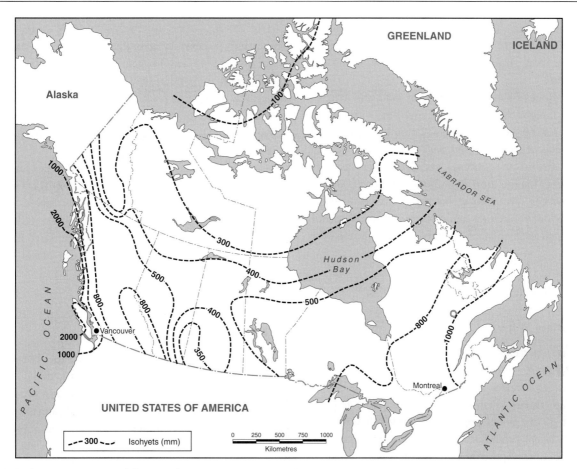

Figure 12.2 Average annual precipitation of Canada (Bone, 2002, Figure 2.5).

the shores of British Columbia gives a narrow coastal band with the most pleasant and moderate climate in Canada. This is partly because the surrounding mountains effectively block most outbreaks of cold arctic air from the interior. The mountains of Cordillera force the westerly air stream to rise, producing copious rain along the Pacific coast.

In Canada, the coldest month is January when the minimum temperature ranges from $-15°C$ in the south to $-35°C$ in the Arctic Archipelago, while the warmest month is July with the average temperature ranging from $5°C$ in the extreme north to $20°C$ in the south (Bone, 2002, Figures 2.3 and 2.4).

Annually, on average some 545 mm or $5450 \times 10^9 \, m^3$ of precipitation falls on Canada, mainly in the form of rain and snow (Statistics Canada, 2004, Table A.3). Surface run-off is about 332 mm, evaporation 176 mm and groundwater recharge 37 mm per annum. Precipitation in Canada is not uniformly distributed (Figure 12.2). It ranges from only 100 mm in the high Arctic in the form of snow, to more than 2500 mm along the Pacific Coast.

12.5 LAND COVER AND USE

In Canada, 266.6 Mha of land is barren or sparsely vegetated, 155.5 Mha is covered by snow, ice or water bodies, 70.3 Mha is shrubland, 10.7 Mha is grassland, and urban areas covers 8.2 Mha (Statistics Canada, 2004, Table A.1).

Forest covers 417.6 Mha or about 45 percent of the country, stretching from the Atlantic to the Pacific coasts and to the Arctic tree line.[5] Approximately 234.5 Mha or 56 percent of the forested lands are considered commercial forests, with 1 Mha of this being harvested each year.

Large-scale agriculture in the Prairie Provinces (Alberta, Saskatchewan and Manitoba) was introduced with the influx of European settlers at the beginning of the twentieth century. In 1996, farmlands covered 68.1 Mha across Canada, which included 26.6 Mha in Saskatchewan, 21 Mha in Alberta, 7.7 Mha in Manitoba, 5.6 Mha in Ontario,

---

[5] http://atlas.gc.ca/site/english/maps/environment/forest/useforest/proforlanduse/ (visited in July 2005).

3.5 Mha in Quebec and 2.5 Mha in British Columbia (Statistics Canada, 1997). The remaining farmland was in Nova Scotia, New Brunswick, Prince Edward Island and Newfoundland.

In 2000, about one million hectares of land was irrigated (Coote and Gregorich, 2000, pp. 17 and 19), with Alberta accounting for 60 percent, British Columbia 13 percent, Ontario/Quebec 12 percent and Saskatchewan 11 percent. Canada is one of the leading world exporters of food products. Wheat is the most important single crop, and the Prairie Provinces form one of the greatest wheat-growing areas of the world. Other agricultural products include barley, oats, oilseeds, tobacco vegetables, potatoes and corn.

12.6 WATER RESOURCES

12.6.1 Surface Water

Canada has 7 percent of the world's renewable supply of freshwater and 20 percent of the world's total freshwater resources, which includes waters captured in glaciers and the polar ice cap (Environment Canada, 2003a, p. 1), and only 0.5 percent of the global population. Although Canada is rich in large rivers and lakes, water is not evenly distributed in space and time. Approximately two thirds of Canada's freshwater flows north toward the Arctic Ocean and Hudson Bay. As a result, most regions of the country have experienced water-related problems, such as drought, flood and water quality.

Canada has five ocean-drainage basins (Biswas, 2003),[6] Arctic Ocean, Hudson Bay, Atlantic Ocean, Pacific Ocean and Gulf of Mexico.[7] The country can be divided into 23 river basins (Figure 12.3). Annually, about $3316 \times 10^9 \, m^3$ of water flows in the Canadian river basins, representing 60 percent of the average annual precipitation (Table 12.1). This means that approximately $104\,000 \, m^3$ of surface water is available per head of population. The two major rivers in terms of flow are the Mackenzie and St Lawrence rivers with average annual flows of $325 \times 10^9 \, m^3$ and $318 \times 10^9 \, m^3$, respectively.[8] The St Lawrence River is Canada's most important river, providing a seaway for ships from the Great Lakes to the Atlantic Ocean.

There are about two million lakes in Canada, covering approximately 7.6 percent of the Canadian landmass (Statistics Canada, 2000, p. 27). The Great Lakes (Table 12.2) are the largest group of freshwater lakes in the world. Approximately 36 percent of their total surface area is in Canada. Lake Superior is the largest and the deepest of the five lakes.

During the period following World War II, the rapid increase in population, manufacturing activity, and agricultural production led to significant deterioration in water quality in many parts of the Great Lakes Basin. Information regarding issues such as hydrology, water level, water quality, flora and fauna, and ecosystem management of the Great Lakes is available at the URLs provided in the footnote.[9]

Table 12.3 shows the main characteristics of the five largest lakes lying entirely inside Canada. The Great Bear Lake is the largest lake and the fourth largest in North and South America. Only Lake Superior, Lake Huron and Lake Michigan are larger. The Great Slave Lake is the second largest and the deepest lake of Canada with a depth of 614 m.

12.6.2 Groundwater

Although groundwater has a contribution of about 7 percent in total water supply of Canada, it is a major source of municipal and domestic water supply. It also provides vital water supplies for agriculture and major industries involved in manufacturing, mining and petroleum production. In 1996, about 30.3 percent (8.9 million) of Canada's population relied on groundwater for their domestic water supply with approximately two-thirds of these users living in rural areas. Dependence on groundwater for domestic water supply in the provinces ranges from 23.1 percent for Alberta to 100 percent for Prince Edward Island (Table 12.4). Many important aquifers in Canada are composed of thick glacial deposits of sand and gravel. Groundwater is also available in sandstones (e.g. Prince Edward Island) and carbonate rocks (e.g. Winnipeg). Some features of Canada's groundwater regions are summarised in Table 12.5. The average recharge rate of the aquifers is about $370 \times 10^9 \, m^3$ per annum,[10] or 7 percent of the average annual precipitations.

---

[6] Also see http://atlas.gc.ca/site/english/maps/peopleandsociety/nunavut/land/drainagebasins (visited in July 2005).

[7] This drainage basin is located in the extreme south of Alberta and Saskatchewan and is drained by the Mississippi River System into the Gulf of Mexico.

[8] http://www.ec.gc.ca/water/images/nature/prop/a2f3e.htm (visited in April 2006).

[9] http://www.great-lakes.net/envt/water/levels/hydro.html and http://www.lre.usace.army.mil/greatlakes/hh/greatlakeswaterlevels/historicdata/greatlakeshydrographs/ (both visited in July 2005).

[10] http://earthtrends.wri.org/text/water-resources/country-profile-33.html (visited in July 2005).

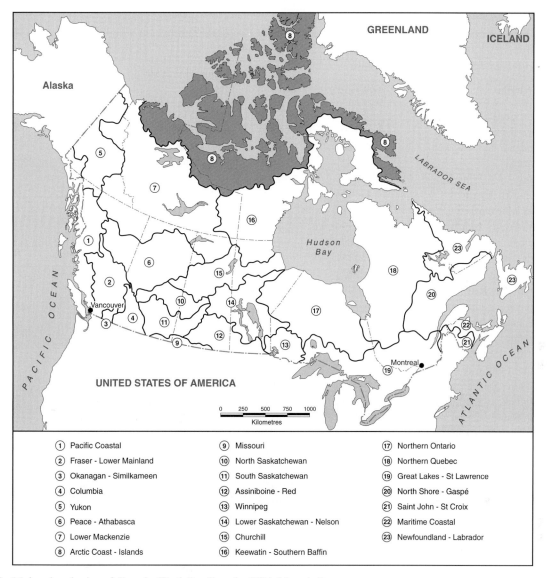

Figure 12.3 Major river basins of Canada (Statistics Canada, 2004, Map A.4).

12.6.3 Water Quality Issues

Despite significant progress with pollution control, contamination of water resources by industrial, agricultural and municipal pollutants is still a major concern in Canada. The National Water Research Institute (2001) describes 15 threats to sources of drinking water and aquatic ecosystem health in Canada. These threats include waterborne pathogens, pesticides, heavy metals, endocrine disrupting substances, municipal wastewater effluents, industrial point source discharges, landfills and waste disposals, agricultural and forestry land use impacts, and impacts of dams, diversions and climate changes.

In 1996, the National Pollutant Release Inventory documented that 13 027 tonnes of industrial pollutants were released into the waterways. The river systems into which the largest amounts of noxious substances were released were the Saint John River in New Brunswick and the St Lawrence River in Quebec and Ontario. These two rivers received 19 percent and 15 percent of all pollutants respectively (Statistics Canada, 2001, p. 57). Pollutants reaching coastal waters can also harm the environment and affect some commercial activities. For example, a number of shellfish-harvesting areas on the Atlantic and Pacific coasts have been closed due to pollution (Statistics Canada, 2001, p. 57).

Over the past 40 years, the number of farms in Canada has declined; however, those that remain have become larger and more productive. This transformation was made possible by greater mechanisation, better crop varieties, innovative farming practices and the use of fertilisers and pesticides. Intensification of agriculture has increased the risk of

Table 12.1. *Characteristics of major river basins of Canada*

No.	River basin	Area (km^2)	Streamflow		Precipitation	
			Rate (m^3 s^{-1})	Annual volume (10^9 m^3)	Rate (mm yr^{-1})	Annual volume (10^9 m^3)
1	Pacific Coastal	334 452	16 390	516.9	1 354	451
2	Fraser–Lower Mainland	233 105	3972	125.3	670	156
3	Okanagan–Similkameen	15 603	74	2.3	466	7
4	Columbia	87 321	2009	63.4	776	68
5	Yukon	332 906	2506	79.0	346	115
6	Peace–Athabasca	485 146	2903	91.5	497	241
7	Lower Mackenzie	1 330 481	7337	231.4	365	486
8	Arctic Coast–Islands	1 764 279	8744	275.8	189	333
9	Missouri	27 097	12	0.4	390	11
10	North Saskatchewan	150 151	234	7.4	443	67
11	South Saskatchewan	177 623	239	7.5	419	74
12	Assiniboine–Red	190 705	50	1.6	450	86
13	Winnipeg	107 654	758	23.9	683	74
14	L. Saskatchewan–Nelson	360 883	1911	60.3	508	183
15	Churchill	313 572	701	22.1	480	151
16	Keewatin–Southern Baffin	939 568	5383	169.8	330	310
17	Northern Ontario	691 811	5995	189.1	674	466
18	Northern Quebec	940 194	16 830	530.8	698	656
19	Great Lakes–St Lawrence	582 945	7197	227.0	957	556
20	North Shore–Gaspé	369 094	8159	257.3	994	367
21	Saint John–St Croix	41 904	779	24.6	1 147	48
22	Maritime Coastal	122 058	3628	114.4	1 251	153
23	Newfoundland–Labrador	380 355	9324	294.0	1 030	392
Canada		9 978 904	105 135	3315.5	545	5451

Source: Statistics Canada (2004, Table A.3).

Table 12.2. *Selected characteristics of the Great Lakes*

Name	Levation (m)	Length (km)	Breadth (km)	Maximum depth (m)	Total area (km^2)	Area in Canada (km^2)
Lake Superior	184	563	257	406	82 100	28 700
Lake Michigan	176	494	190	282	57 800	–
Lake Huron	177	332	295	229	59 600	36 000
Lake Erie	174	388	92	64	25 700	12 800
Lake Ontario	75	311	85	244	18 960	10 000
Total					244 160	87 500

Source: Statistics Canada (2001, Table 1.5).

contamination of surface and groundwater by pollutants such as nutrients (nitrogen and phosphorus) and agricultural chemicals (Chambers *et al.*, 2002; and Coote and Gregorich, 2000, Chapters 5 to 7).

Discharge of untreated municipal waste is now uncommon. Approximately 89 percent of Canadians are served by wastewater treatment plants, compared with 72 percent in the United States, 92 percent in Germany, 95 percent in Australia, and 98 percent in the Netherlands (OECD, 1999, Figure 6).

Groundwater contaminants across Canada come from various sources (Crowe *et al.*, 2003). Analyses of

Table 12.3. *The five largest internal lakes of Canada*

Name	Location	Elevation (m)	Area (km^2)	Maximum depth (m)
Great Bear Lake	Northwest Territory	156	31 328	413
Great Slave Lake	Northwest Territory	156	28 568	614
Lake Winnipeg	Manitoba	217	24 387	18
Lake Athabasca	Saskatchewan	213	7935	120
Reindeer Lake	Saskatchewan	337	6650	219

Source: Statistics Canada (2000, Table 3.3.2).

Table 12.4. *Reliance of Canadian population on groundwater for their domestic water supply*

Province and territory	Percentage of reliance	Province and territory	Percentage of reliance
Alberta	23.1	Newfoundland and Labrador	33.9
Quebec	27.7	Saskatchewan	42.8
Northwest Territory and Nunavut	28.1	Nova Scotia	45.8
British Columbia	28.5	Yukon	47.9
Ontario	28.5	New Brunswick	66.5
Manitoba	30.2	Prince Edward Island	100.0
Canada			30.3

Source: Website of the Environment Canada.[11]

groundwater from rural wells commonly exhibit nitrate, bacteria and/or pesticides. Manure or pesticide spreading on the land surface is particularly a problem if undertaken close to a well that is improperly constructed or located. The threat to groundwater from urban sources of contamination will increase as urban areas expand into rural areas traditionally serviced by wells. Groundwater contamination by chemical leaks or spills from gasoline stations, dry cleaning stores, petroleum refineries, chemical plants, waste disposal facilities and industrial sites has attracted considerable attention as they can pose a significant risk to human health at very low concentrations. Groundwater contamination by the petroleum industry stems from the legacy of over a century of exploration, development and refining, and aging production facilities. The mining industry is another source of groundwater contamination. There are 90 active mines and over 10 000 abandoned mines across Canada. The waste rock and tailing at these sites can release high concentrations of acid, sulphate and heavy metals several orders of magnitude above the acceptable levels.

Water quality monitoring is an important issue in identification and management of water contamination. In this respect, de Rosemond *et al.* (2003) describe issues such as current federal, provincial and territorial water quality monitoring activities, design of water quality monitoring program and networks, integration of water quality monitoring practices, and recommendations regarding further steps towards the development of a Canada wide water quality monitoring system.

12.6.4 Water Withdrawal and Use

Water withdrawal has been increasing over the past few decades. For example, between 1972 and 1996, Canada's rate of water withdrawal increased by almost 90 percent from $24 \times 10^9 \, m^3 \, yr^{-1}$ to $46 \times 10^9 \, m^3 \, yr^{-1}$, while the population increased by only 33.6 percent,[12] illustrating a major change in water consumption. Figure 12.4 shows water withdrawal, return and consumption from 1981 to 1996, and Table 12.6 shows the same parameters for the four main categories of water users (power generation, industry and mining, agriculture, and residential and commercial) over the same period. It shows that the major part of water withdrawal was returned to the source of water supply. In 1996, irrigation consumed 73 percent of water withdrawal,

---

[11] http://www.ec.gc.ca/water/images/nature/grdwtr/a5f6e.htm (visited in July 2005).
[12] http://www.ec.gc.ca/water/en/info/pubs/FS/e_FSA6.htm (visited in July 2005).

Table 12.5. *Major groundwater regions of Canada*

Region	Geology	Aquifer properties		
		Transmissivity ($m^2 d^{-1}$)	Recharge rate (mm yr^{-1})	Well yield (L s^{-1})
Western mountain ranges	Mountains with thin soils over fractured rocks of Precambrian to Cenozoic age, underlain by alluvial and glacial deposits of Pleistocene age.	0.5–500	3–50	0.8–17
Western glaciated plains	Hills and relatively undissected plains underlain by glacial deposits over consolidated sedimentary rocks of Mesozoic and Cenozoic age.	25–2500	5–200	3.3–33
Precambrian shield	A hilly terrain underlain by glacial deposits over complexly folded to flat-lying metamorphic rocks of Precambrian age. Along the southwest side of James Bay, the bedrock consists of relatively flat-lying consolidated sedimentary rocks mostly of Palaeozoic age.	10–500	10–300	1.7–17
Northeastern Appalachians	Hilly to mountainous area underlain by glacial deposits over folded metamorphic rocks of Palaeozoic age complexly intruded by igneous rocks.	50–500	30–300	1.7–33
St Lawrence Lowlands	A hilly area underlain by glacial deposits over flat-lying consolidated sedimentary rocks of Palaeozoic age.	100–2000	5–300	3.3–33
Permafrost region	A topographically diverse area. Commonly underlain by unconsolidated deposits, partly of glacial origin, overlying fractured rocks of Precambrian to Cenozoic age. Hydrogeological conditions are dominated by continuous permafrost in the northern part of the region and discontinuous permafrost in the southern part.	10–2000	5–100	0.2–33

Source: Back *et al.* (1988, pp. 20–23).

residential and commercial 11 percent, industry and mining 10 percent, and power generation less than 2 percent (Table 12.6).

Canadian households used on average 326 L h^{-1} d^{-1} in 1996 (Burke *et al.*, 2001). This was significantly higher than the household water consumption of Germany (128 L h^{-1} d^{-1}), the Netherlands (130 L h^{-1} d^{-1}), the United Kingdom (149 L h^{-1} d^{-1}), and Australia (268 L h^{-1} d^{-1}) in the same year (OECD, 1999, Table 1). One of the key factors explaining the high rate of consumption is the absence of

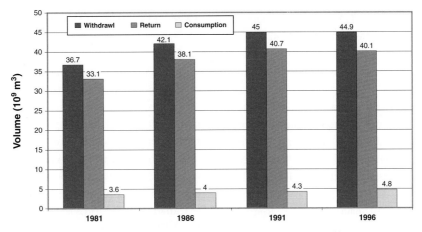

Figure 12.4 Water withdrawal, return and consumption in Canada for the period of 1981–96.[13]

Table 12.6. *Water withdrawal and consumption of Canada for the period of 1981–1996 (in $10^9 m^3$)*

Year	Population (million)	Power generation			Industry and mining			Agriculture			Residential and commercial			Total		
		W	R	C	W	R	C	W	R	C	W	R	C	W	R	C
1981	24.34	18.2	18.1	0.1	11.7	10.9	0.7	3.1	0.7	2.4	3.8	3.4	0.40	36.7	33.1	3.6
1986	25.31	25.0	24.7	0.3	9.8	9.2	0.6	3.6	0.8	2.8	3.7	3.3	0.38	42.1	38.1	4.0
1991	27.30	28.3	28.2	0.1	8.9	8.2	0.7	4.0	0.9	3.1	3.8	3.4	0.43	45.0	40.7	4.3
1996	28.85	28.7	28.2	0.5	8.2	7.4	0.8	4.1	1.1	3.0	3.9	3.5	0.44	44.9	40.1	4.8

Note: W = Withdrawal, R = Return, C = Consumption.
Source: Website of the Statistics Canada.[14]

appropriate pricing mechanisms. In 1999, 44 percent of Canadian residences served by municipal water systems were not metered. In addition, 55 percent of Canadians faced residential water use charges that discouraged water conservation.

Scharf *et al.* (2002) provide a detailed analysis of industrial water use based on the 1996 survey. They reported that between 1991 and 1996, manufacturing establishments reduced their total water use by 8 percent, through improved efficiencies and increased recycling.

12.7 FLOOD

In Canada, flood risk varies with location and season. In early spring, rapidly melting snow packs and ice jams pose a risk of flooding on rivers in almost every part of the country. In summer however, thunderstorms are the most frequent cause of floods in all inland areas of southern Canada (Francis and Hengeveld, 1998, p. 16). Annually, about CA$3–5 billion is spent on flood prevention, storm and other drainage networks.[15] Because of the economic losses caused by flooding and the cost of prevention, flooding is considered a significant hazard in Canada.

Major floods occurred in various parts of Canada[16] in 1798, 1826, 1865, 1883, 1928, 1937, 1948, 1950, 1954, 1973, 1974, 1979, 1980, 1986, 1987, 1989, 1993, 1996 and 1997. The 1996 flooding in the Saguenay River Valley of Quebec and the 1997 flooding of Manitoba's Red River Valley were two of the costliest natural disasters in Canadian history. In the Saguenay River Valley, 290 mm of rainfall fell in less than

---

[13] http://www40.statcan.ca/l01/cst01/envir05.htm (visited in July 2005).
[14] http://www40.statcan.ca/l01/cst01/envir05.htm (visited in July 2005).
[15] http://www.utoronto.ca/env/nh/pt2ch2-3-2.htm (visited in July 2005).
[16] http://www.utoronto.ca/env/nh/tab2-11.htm (visited in July 2005).

36 hours and caused an estimated CA$1 billion in damages in the region, and at least 10 deaths. A total of 15 825 people had to leave their homes. At least 20 major bridges were heavily damaged and more than 50 towns and villages were inundated.

The Red River flows north from North Dakota in the United States to Manitoba in Canada and discharges into Lake Winnipeg. Its flooding is a significant recurring natural hazard in southern Manitoba. About 70 percent of Manitoba's 1.2 million citizens live in the Red River Valley, particularly in the Winnipeg area. Twenty large floods with discharge rates of 1950–6400 m^3 s^{-1} at Redwood Bridge have occurred since 1800.[17] The 1997 Red River flood was caused by a combination of factors, a wet autumn that left soil saturated, heavier than normal snowfall during the winter, and an early April blizzard that dropped another 0.5–0.7 m of snow. During this flood, approximately 28 000 people were evacuated from flood-risk areas.

Overall, the risk of flooding in Canada is predicted to increase as a result of a warmer climate. This increase would come mainly from rainstorm floods, with heavier rainfall expected to come from more thunderstorms, and from fewer but larger rainstorms associated with large-scale weather systems (Francis and Hengeveld, 1998, p. 17).

12.8 DROUGHT

During the past two centuries, at least 40 long duration droughts have occurred in western Canada.[18] More recently, severe drought occurred in Canada in the 1930s, 1961, 1973, 1977–80, 1983–85, 1988, and again in 2001 with major impacts in the Prairie Provinces. In contrast, droughts in eastern Canada are usually shorter, smaller in area, less frequent and less intense. The most drought prone region is the Palliser Triangle in the southern portions of Alberta, Saskatchewan and Manitoba. The drought of 2001 was widespread and affected Canada from British Columbia, through the Prairies, into the Great Lakes–St Lawrence region and even the Atlantic Provinces (Table 12.7), with significant economic costs (Figure 12.5). Drought causes economic losses, particularly to agriculture, forestry and hydro-power generation, and also impacts the environment by increased bushfires, decline in surface and groundwater supplies, decline in water quality, and reduction in the extent of wetlands.

12.9 HYDRO-POWER GENERATION

In 2000, Canada had 849 dams[19] with crests higher than 10 metres in operation or under construction, with 50 percent of them constructed over the three decades of 1950s, 1960s and 1970s. The vast majority (70 percent) of these dams were constructed solely for hydro-power generation. Of the remainder, 7 percent were constructed for water supply, and 6 percent for irrigation in the Prairie Provinces. The rest (17 percent) serve a variety of purposes including flood control, navigation, and recreation.

Canada is the world's largest producer of hydro-electricity. In 2001, it produced 335 000 GWh, followed by China (277 000 GWh), Brazil (268 000 GWh), United States (223 000 GWh), and Russia (176 000 GWh). In 2001, total electricity production in Canada was 588 000 GWh (Statistics Canada, 2004). Of this total, hydro-power accounted for 57 percent, nuclear 13 percent,[20] and the balance was produced by oil, natural gas, coal, liquefied petroleum gas, and biomass. In the same year, the value of Canadian energy exports amounted to CA$55.1 billion or 14 percent of all exports.

Canada's principal hydro-electric generating stations and their installed capacities are (Bone, 2002, p. 82):

- LG-2 on the La Grande River, Quebec (5328 MW).
- Churchill Falls on the Churchill River, Labrador (5225 MW).
- Gordon M. Shrum on the Peace River, British Columbia (2416 MW).
- LG-4 on the La Grande River, Quebec (1650 MW).
- Kettle Rapids on the Nelson River, Manitoba (1255 MW).

12.10 CLIMATE CHANGE IMPACTS

Temperature records indicate that Canada's average annual temperature increased by approximately 0.9°C between 1948 and 2000 (Environment Canada, 2002a). Furthermore, during the last two decades of the twentieth century, Canada experienced a record number of very warm years.

---

[17] http://gsc.nrcan.gc.ca/floods/redriver/table1_e.php (visited in July 2005).
[18] http://www.nwri.ca/threats2full/ch3-1-e.html (visited in July 2005).
[19] http://www.nwri.ca/threats2full/ch2-1-e.html and http://www.nwri.ca/threats2full/ch2-2-e.html (both visited in July 2005).
[20] In 2004, Canada had 17 nuclear rectors: 15 in Ontario, one in Quebec and one in New Brunswick (Statistics Canada, 2004, Table 1.6).

Table 12.7. *Some impacts of the 2001 drought across Canada*

Region	Conditions in 2001
British Columbia	– Driest winter on record, with precipitation half of the historic average across the coast and southern interior. – Snow packs in southern regions were at or below historic low.
Prairies	– Saskatoon in Saskatchewan was 30 percent drier than 110-year record. – Many areas experienced lowest precipitation in historic record. – Parts of the Palliser Triangle experienced a second or third consecutive drought year.
Great Lakes–St Lawrence basin	– Driest summer in 54 years. – Southern Ontario experienced the driest 8 weeks on record. – Montreal experienced driest April on record, and set summer record with 35 consecutive days without measurable precipitation.
Atlantic	– Third driest summer on record. – Large regions experienced only 25 percent of normal rainfall in July, and August was the driest on record. – July, with 5 mm of rain, was the driest month ever recorded in Charlottetown (Prince Edward Island).

Source: Lemmen and Warren (2004, Table 1, p. 35).

For example, in 1988 and 1998 spring temperatures were respectively 2.0°C and 3.2°C above the average.[21] Canada is already feeling the effects of climate change in the form of (Environment Canada, 2002b):

(1) increasing number and intensity of heat waves with related health problems;
(2) declining water levels in the Great Lakes;
(3) changes in fish migration and melting of the polar ice cap;
(4) insect infestations in British Colombia's forests;
(5) hotter summers and higher levels of smog in major urban centres; and
(6) more extreme weather events such as drought on the Prairies, and flooding in Manitoba and Quebec.

The expected impacts of climate changes in various parts of Canada are:[22]

- **British Columbia and Yukon Territory.** (1) Increased spring flood damage in coastal areas and throughout the interior; (2) summer droughts along the south coast; (3) sea level rise of up to 30 cm on the north coast of British Columbia and up to 50 cm on the north Yukon coast by 2050; (4) increased winter precipitation, permafrost degradation and glacier retreat; and (5) changes in fish and wildlife habitat.
- **Prairies.** (1) Increased air temperatures and decreased soil moisture; (2) increase in frequency and length of droughts; (3) decrease in average crop yields by approximately 10 to 30 percent; (4) increased demand for water; and (5) drying up of semi-permanent and seasonal wetlands.
- **Ontario.** (1) Average annual warming of 3–8°C by the end of the twenty-first century leading to fewer weeks of snow, a longer growing season, less moisture in the soil, and an increase in the frequency and severity of drought; (2) likely increase in the frequency and severity of forest fires; and (3) decline in water levels in the Great Lakes.
- **Quebec.** (1) Increase in average annual temperature of 1–4°C in the south and 2–6°C in the north;

---
[21] http://www.msc.ec.gc.ca/ccrm/bulletin/national_e.cfm (visited in July 2005).
[22] http://www.climatechange.gc.ca/english/publications/ccs (visited in July 2005)

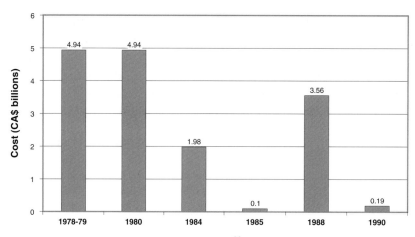

Figure 12.5 Cost estimates for recent droughts in Canada, in 2000 values.[23]

(2) stable or slight decrease in precipitation in the south and an increase of 10 to 20 percent in the north; (3) lower water levels in the St Lawrence River affecting shipping, and the river environment; and (4) longer growing season and the extension of agriculture further north.

- **Atlantic region.** This region has not followed the national warming trend of the twentieth century, and in fact a slight cooling trend has been experienced over the second half of the past century. This region is particularly vulnerable to rising sea level.

Lemmen and Warren (2004) present an overview of climate change impacts and adaptation in Canada. Their publication covers:

(1) water resources;
(2) agriculture;
(3) forestry;
(4) fisheries;
(5) coastal zone;
(6) transport; and
(7) human health and well being.

It concludes that climate change impacts and the ability to adapt to those impacts will differ among sectors and also across regions of Canada. Although some negative impacts are expected for all sectors, these impacts appear more important in the areas of water resources and health.[24]

Water quality would be affected by increased surface water temperatures, decreased concentrations in dissolved oxygen, and higher concentrations of nutrients (Lemmen and Warren, 2004, Box 2, p. 40). Other impacts would include saltwater intrusion into coastal aquifers and river mouths due to a rise in sea level and increased water demand, increased health risks from water-borne diseases caused by increased flooding and sewer overflows, increased water turbidity and sediment loads from increased landslides and surface erosion, and degradation of stream habitat.

Canada, as a signatory of the *Kyoto Protocol*,[25] has set a target to reduce greenhouse gas emissions to an average level of 6 percent below the 1990 level during the first commitment period of 2008–12. This represents a reduction of 240 Mt of greenhouse gas emissions. To achieve this objective it has developed the *Climate Change Plan for Canada* (Environment Canada, 2002b).

12.11 MANAGEMENT OF WATER RESOURCES

Canadian provinces and territories are responsible for the management of their water resources. The *Canada Water Act* (proclaimed on 30 September 1970) provides the framework for cooperation between federal, provincial and territorial governments in the conservation, development and utilisation of Canada's water resources.

---

[23] http://www.utoronto.ca/env/nato/proceedings-april2002/mp-drought.pdf (visited in July 2005).
[24] Further information regarding potential impacts of climate change on Canada's water resources is available in Lemmen and Warren (2004).
[25] Text of the Kyoto Protocol is available at the URL http://unfccc.int/resource/docs/convkp/kpeng.pdf (visited in July 2005).

Governments have developed a substantial range of policies and regulations to enhance the reliability of water supplies and to protect aquatic ecosystems. Generally, jurisdictions are moving towards integrated ecosystem and watershed management approaches that ensure environmental, economic, and social considerations are factored into the decision-making process (Environment Canada, 2003a). Moreover, Canadian governments have created institutions to focus on water issues that have implications for more than one province or territory. These include the Prairie Provinces Water Board, the Mackenzie River Basin Board, the Canadian Council of Ministers of the Environment, and the Federation of Canadian Municipalities. Also, a number of international institutions such as the International Joint Commission and the Council of Great Lakes Governors have been established by the Canadian and the United States governments for better management of their shared lakes and river systems. Some of the international institutions are described in the following sections. These are followed by a description of water conservation and wastewater reclamation.

12.11.1 International Joint Commission

The International Joint Commission (IJC) was established in 1912 under the *Boundary Waters Treaty* of 1909.[26] Its objective is to prevent and resolve disputes between Canada and the United States in relation to the use and quality of boundary waters, and to advise both countries on related questions.[27] Its main activities include:

- **Great Lakes Water Quality.** The objective is to restore and maintain the chemical, physical, and biological integrity of the waters of the Great Lakes Basin. The 12th Biennial Report to the governments of the United States and Canada highlights the key issues and contains specific recommendations relating to impacts and threats on the aquatic environment (International Joint Commission, 2004). The report indicates that the number of toxic chemical releases have declined over the past decades. However, natural habitat continues to be lost as urban areas expand, and new aquatic alien species continue to be introduced into the lakes at a rate of one per eight months.
- **Great Lakes Water Levels.** The Lake Ontario–St Lawrence River Study Board is undertaking a comprehensive five-year study for the IJC to assess the current criteria used for regulating water levels of Lake Ontario and of the St Lawrence River.

- **Watershed Initiative.** The IJC continues to support the establishment of ecosystem-focused watershed boards, in accordance with a 1998 request from the United States and Canadian governments. Core elements of the concept include recognising local expertise and initiatives, and coordinating among numerous organisations within the watershed. The IJC's fundamental interest in promoting the watershed board concept is to more effectively meet its mandate of preventing and resolving transboundary water disputes.

12.11.2 The Council of Great Lakes Governors

The Council of Great Lakes Governors is a partnership of the Governors of the eight Great Lakes States: Illinois, Indiana, Michigan, Minnesota, New York, Ohio, Pennsylvania and Wisconsin.[28] In 1983, the region's Governors joined forces to create the Council and tackle the severe environmental and economic challenges facing the citizens of their States. In more recent years, the Premiers of Ontario and Quebec have joined the Council. The Council's mission is to encourage and facilitate environmentally responsible economic growth of the industrial heartland of North America. This is accomplished by establishing a cooperative effort between the public and private sectors among the member States and Provinces in the United States and Canada. Through the Council, Governors work collectively to ensure that the entire Great Lakes region is both economically sound and environmentally conscious in addressing current problems and challenges of the future.

Current projects of the Council include:[29]

- **Great Lakes Water Management Initiative.** Through this initiative the Council is taking the lead in protecting waters of the Great Lakes. The Council assist the Governors and Premiers in coordinating activities under the *Great Lakes Charter* of 1985.[30] The Council also coordinates the authority granted to the Governors under the United States Federal *Water Resources Development Act* of 1986.[31] In order to update

---

[26] http://www.ijc.org/rel/agree/water.html (visited in July 2005).
[27] http://www.ijc.org (visited in July 2005).
[28] http://www.cglg.org (visited in April 2006).
[29] http://www.cglg.org/projects/index.asp (visited in April 2006).
[30] http://www.cglg.org/projects/water/docs/GreatLakesCharter.pdf (visited in April 2006).
[31] http://www.cglg.org/projects/water/docs/WRDA86−Amended 2000.pdf (visited in April 2006).

the regional water management system and ensure that the Great Lakes are protected, the Governors and Premiers signed the *Great Lakes Charter Annex* in 2001.[32]

- **Great Lakes Priorities Initiative.** The Council has established nine priorities to guide the restoration and protection of the Great Lakes: (1) ensure the sustainable use of water resources; (2) promote programs to protect human health against pollution; (3) control pollution from diffuse sources; (4) continue to reduce the introduction of toxics into the Great Lakes ecosystem; (5) stop the introduction and spread of non-native invasive aquatic species; (6) enhance fish and wildlife habitats; (7) restore the environmental health of the areas of concern; (8) standardise and enhance methods by which information is collected, recorded and shared within the region; and (9) adopt sustainable use practices that protect environmental resources and may enhance the recreational and commercial value of the Great Lakes.
- **Great Lakes International Trade Initiative.** The mission of this initiative is to offer responsive and comprehensive services to small- and medium-sized companies from the Great Lakes region seeking to expand product and services sales.
- **Aquatic Invasive Species Task Force.** The goal of this Task Force is to stop further introduction and spread of aquatic invasive species into the Great Lakes, which is one of the nine priority initiatives described above.

12.11.3 Water Conservation

The perception that Canada has more than $100\,000\,\text{m}^3$ of fresh surface water per head of population has led to excessive use of these resources. However, there are some limitations with respect to freshwater resources. These limitations in population centres include:[33]

(1) growing populations and subsequent increase in water demand are concentrated in the expanding metropolitan areas, and are forcing regulators and policy makers to find ways to stretch available supplies even further;
(2) increasing pollution of surface and groundwater is further reducing the supplies of readily available, clean water;
(3) reduction in water use will help ecosystems; and
(4) an increase in water demand will increase the cost of both supplying and treating more freshwater, and also treating and disposing of the wastewater.

To overcome these problems, in May 1994, the Canadian Council of Ministers of the Environment approved a national action plan to encourage municipal water use efficiency.[34] The goal of this action plan was to achieve more efficient use of water in Canadian municipalities, which would save money and energy, delay or reduce expansion of existing water and wastewater systems, and conserve water. Development of this action plan included the principles of:

- **User pays on basis of volume.** Consumer shall pay for water and wastewater services on the basis of measured actual use.
- **Full cost pricing.** Municipalities shall move towards water and wastewater rate structure that reflect the full costs of delivery and treatment.
- **An informed public.** The public shall be informed of the real cost of water use and the savings that can be achieved through water efficiency, and of actions they can take to reduce usage.

Municipalities across Canada are beginning to take action to manage the demand for water, instead of seeking new sources of supply. Demand management incorporating water efficient applications, is rapidly gaining popularity as a low cost, effective way to get more service out of existing systems. The wide range of water efficiency initiatives currently being undertaken can be grouped under four principal categories.[35]

- **Structural.** Water metering,[36] water recycling systems, wastewater reuse, flow control devices, distribution system pressure reduction, use of water saving devices, drought resistance landscaping, efficient sprinkling/irrigation technology, new process technologies, and improvements to treatment plants.
- **Operational.** Leak detection and repair, water use restrictions, and elimination of combined sanitary/storm sewers to reduce loadings of sewage treatment plants.
- **Economic.** Rate structure,[37] pricing policies, incentives through rebates and tax credits, and sanctions (fines).

---

[32] http://www.cglg.org/projects/water/docs/GreatLakesCharter Annex.pdf (visited in April 2006).
[33] http://www.ec.gc.ca/water/en/info/pubs/FS/e_FSA6.htm (visited in July 2005).
[34] http://www.ec.gc.ca/water/en/info/pubs/action/e_action.htm (visited in July 2005).
[35] http://www.ec.gc.ca/water/en/info/pubs/FS/e_FSA6.htm (visited in July 2005).
[36] In 1999, only about 56 percent of Canada's urban population was metered (http://www.ec.gc.ca/water/en/info/pubs/FS/e_FSA6.htm visited in July 2005).
[37] In 1999, about 43 percent of the population was under a flat rate structure, and their water use was 70 percent higher than those with volume-based rates (http://www.ec.gc.ca/water/en/info/pubs/FS/e_FSA6.htm visited in July 2005).

- **Socio-political.** Public education, information transfer and training, and regulatory measures (legislation, standards and others).

Agriculture is the largest water consumer in Canada (see Section 12.6.4). To limit competition with other water users, irrigators, governments, and researchers have cooperated to introduce greater efficiency in the way irrigation water is stored, conveyed and applied in the field (Coote and Gregorich, 2000, p. 19). For example:

- Irrigation headwaters, main canals, and whole distribution systems have been renovated to minimise water loss.
- Irrigators are encouraged to switch from less efficient gravity systems to more efficient sprinkler systems or to high efficiency drip or trickle systems.
- Governments and industry are conducting research and setting up demonstration projects to test new irrigation technology.
- Water meters are being used at the district and farm levels to measure water use and charge for water, based on consumption.

12.11.4 Wastewater Reclamation

Marsalek *et al.* (2002) provide an overview of the major issues related to reclamation and reuse of wastewater in Canada. The following provides a brief summary of some of the issues related to this topic.

In Canada, wastewater reclamation is practiced on a relatively small scale, and mostly in isolated cases. However, interest in wastewater reclamation and reuse is growing due to:

(1) the steady increase in water demand while supplies are limited;
(2) opportunities to save on future expansion of the water supply infrastructures;
(3) the need to reduce or eliminate wastewater discharge to sensitive receiving waters; and
(4) opportunities for inexpensive provision of water services in isolated places.

The provinces of British Columbia and Alberta have the most experience in wastewater reclamation, and have developed regulatory guidance documents. Other provinces may allow individual wastewater reuse projects on an experimental basis, but do not yet have written regulatory guidance for routine applications of reuse. Municipalities in these provinces are typically reusing treated wastewater to irrigate urban parklands, golf courses and agricultural non-food crops.

In spite of the fact that wastewater reclamation is growing in popularity, the most significant concerns are those related to health risks. This includes the potential risks of using water that might contain traces of chemicals such as endocrine disruptors, pharmaceutical chemicals, organic industrial chemicals and heavy metals. Another important issue is evaluation of the impact of reclaimed water on irrigated crops and soils.

SECTION B: INTER-BASIN WATER TRANSFER PROJECTS

12.12 INTRODUCTION

In Canada, more streamflows are diverted out of their basin of origin than any other country in the world. For example, the average rate of inter-basin water transfer flow in Canada is $4424\,m^3\,s^{-1}$ (Quin, 2004), which is more than six times greater than the United States with a transfer rate of about $713\,m^3\,s^{-1}$. There are 62 diversion projects developed across nine provinces of Canada (Table 12.8). If all of the diverted waters in Canada were concentrated into a "hypothetical river", it would be the third largest river of the country, after the Mackenzie and St Lawrence rivers (Quinn, 2004). These diversions have the following characteristics:

(1) diverted flow does not return to stream of origin within 25 km from the point of withdrawal; and
(2) average annual diverted flow is more than $1\,m^3\,s^{-1}$.

These characteristics exclude localised and smaller withdrawals operated by numerous municipalities, power plants, and individual irrigators.

Among the inter-basin water transfer projects, hydropower dominates in the number and scale of diversions. Irrigation, flood control and municipal uses only assume importance regionally or locally. Selected characteristics of 18 major diversion projects with individual flow rates of above $25\,m^3\,s^{-1}$ are provided in Table 12.9, while Figure 12.6 shows their locations. The three largest inter-basin water transfer projects are numbers 12, 14, and 6 with the respective diversion rates of $845\,m^3\,s^{-1}$, $790\,m^3\,s^{-1}$, and $775\,m^3\,s^{-1}$.

Many regions of Canada have natural conditions that are highly favourable for inter-basin water transfer. Unlike the United States, Australia and many other countries,

Table 12.8. *Inter-basin water transfer projects in Canada by province, in order of average flow rates*

Province	Number of projects	Average flow rates ($m^3 s^{-1}$)	Major use
Quebec	9	1851	Hydro-power
Manitoba	7	784	Hydro-power
Newfoundland	5	716	Hydro-power
Ontario	9	555	Hydro-power
British Columbia	10	334	Hydro-power
Alberta	10	126	Irrigation
Saskatchewan	5	33	Hydro-power
Nova Scotia	6	23	Hydro-power
New Brunswick	1	2	Municipal
Total	62	4424	–

Source: Quinn (2004).

Canadian water diversions have not resorted to high pumping lifts, or long canals and pipelines. Instead, they have the advantage of using short cuts between adjacent water bodies and gravity flows for water diversion (Day and Quinn, 1992, p. 10). This is mainly due to the dense network of interconnected lakes, rivers, and watercourses in the country. Another advantage has been that until recently these diversion projects have not been the subject of conflicts between different provincial, federal and local governments (Day and Quinn, 1992, p. 18). Unlike other countries, water diversion in Canada is not from wetter to drier regions, nor from less populated to more populated regions. They are mostly for hydro-electric power generation.

12.13 EXAMPLES OF WATER TRANSFER PROJECTS

Day and Quinn (1992) provide a detailed description of five projects, from the earliest to most recently implemented. These also happen to be from smallest to largest in scale. They provide details of each project's historical development, technical aspects, biophysical, social, and economical impacts. The following descriptions are primarily based on this reference.

12.13.1 The Long Lake and Ogoki Diversions

These diversions[38] are separate projects but often considered together because both divert waters that originally drained northward into Lake Superior. These projects were developed to generate hydro-electric power for Canada's defence industries during World War II, and in the case of Long Lake to transport pulpwood. Through an Exchange of Note in 1940 and a clause in the Niagara River Water Diversion Treaty of 1950, the United States agreed to Canada deriving exclusive power benefits at Niagara from $143 m^3 s^{-1}$ of the Long Lake and Ogoki flow contributions (Quinn and Edstrom, 2000).

The Long Lake Diversion (Figure 12.7), which was completed in 1939, transfers water from the Kenogami River in the James Bay drainage system into Lake Superior. It redirects Kenogami River flow south into the Aguasabon River, which discharges into Lake Superior. The Kenogami Control Dam was constructed north of Long Lake to reverse northward Kenogami River discharges. At the south end of the lake, a diversion canal of 8.5 km long was excavated through the watershed boundary. Further downstream, the South Regulating Dam was erected to control southward diversion flows, averaging $45 m^3 s^{-1}$, into the Great Lakes. Construction of the Hays Lake Dam and the Aguasabon Generating Station was completed in 1948. At that time, the Long Lake Diversion began inter-basin water transfer, and power generation, locally as well as in the Ste Mary's River in Sault Ste. Marie, the Niagara River near Niagara Falls, and St Lawrence River.

The Ogoki Diversion project rationale was to divert the north-eastward flowing Ogoki River, a tributary of the Albany River, southward through Lake Nipigon into the Great Lakes system (Figure 12.7) providing an average $113 m^3 s^{-1}$ flow increment for hydro-power generating stations. The project involved construction of a diversion

---
[38] Projects Nos. 8 and 9 in Table 12.9 and Figure 12.6.

Table 12.9. *Major inter-basin water diversion projects in Canada*

No.[a]	Diversion project and jurisdiction	Diversion rate ($m^3 s^{-1}$)	Major use[b]	Date	No.[a]	Diversion project and jurisdiction	Diversion rate ($m^3 s^{-1}$)	Major use	Date
1	Nechako–Kemano (B.C.)	115	H.P.	1952	10	Little Abitibi–Abitibi (Ont.)	40	H.P.	1963
2	Bridge–Seton Lake (B.C.)	92	H.P.	1934	11	Welland Canal, Lake Erie–Lake Ontario (Ont.)	239	H.P. and navigation	1829
3	Cheakamus–Squamish (B.C.)	37	H.P.	1957	12	Eastmain Opinaca–La Grande (Que.)	845	H.P.	1980
4	Coquitlam–Buntzen Lake (B.C.)	28	H.P.	1912	13	Fregate–La Grande (Que.)	31	H.P.	1982
5	Tazin Lake–Charlot (Sask.)	28	H.P.	1958	14	Caniapiscau–La Grande (Que.)	790	H.P.	1983
6	Churchill–Nelson (Man.)	775	H.P.	1976	15[c]	Julian Unknown–Churchill (Nfld.)	196	H.P.	1971
7	Lake St Joseph–Root (Ont.)	86	H.P.	1958	16[c]	Naskaupi–Churchill (Nfld.)	200	H.P.	1971
8	Ogoki–Nipigon (Ont.)	113	H.P.	1943	17[c]	Kanairiktok–Churchill (Nfld.)	130	H.P.	1971
9	Long Lake–Lake Superior (Ont.)	45	H.P. and pulpwood transport	1939	18	Victoria, White Bear, Grey and Salmon–Northwest Brook (Nfld.)	185	H.P.	1969

Note: B.C.: British Columbia; Sask.: Saskatchewan; Man.: Manitoba; Ont.: Ontario; Que.: Quebec; Nfld.: Newfoundland. [a] For locations see Figure 12.6; [b] H.P. = Hydro-power; [c] These are components of the Churchill Falls projects.
Source: Day and Quinn (1992, Table 3) and Quinn (2004).

dam on the Ogoki River to raise its water level by 12 m. At this level, a 400 m diversion channel was excavated and a dam was constructed to regulate southerly flows. The project became operational in 1943.

Theses wartime projects continue to pay off handsomely. From 1943 to 1974, net economic benefits from hydro-electric generation by the Ogoki–Long Lake diversions exceeded CA$220 million in 1974 dollars (CA$840 million in 2004 values).

12.13.2 The Nechako–Kemano Diversion Project

In 1948, the Province of British Columbia invited the Aluminium Company of Canada (ALCAN) to explore the feasibility of developing hydro-electric capacity as a basis for an aluminium smelting industry. This led to a significantly different type of diversion. Instead of using a large volume of water and low vertical drop, a comparatively small flow falls through a high head to generate electricity. In 1950, a licence was granted to ALCAN for diversion of up to $269\,m^3 s^{-1}$. The Nechako–Kemano Diversion Project (Figure 12.8) was financed by the private sector without contribution from public funds.

The Nechako–Kemano Diversion Project[39] was completed in 1954 and was the largest corporate project undertaken in Canada at that time. The Kenny Dam on the Nechako River was completed in October 1952 and created a 900 km² reservoir which extended to the west end

---

[39] Project No. 1 in Table 12.9 and on Figure 12.6.

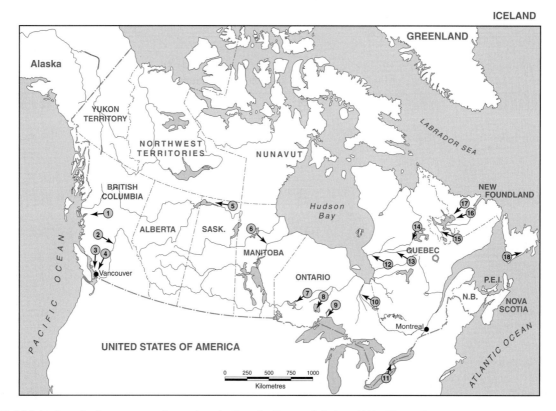

Figure 12.6 Major inter-basin water transfer projects in Canada (Day and Quinn, 1992, Figure 2).

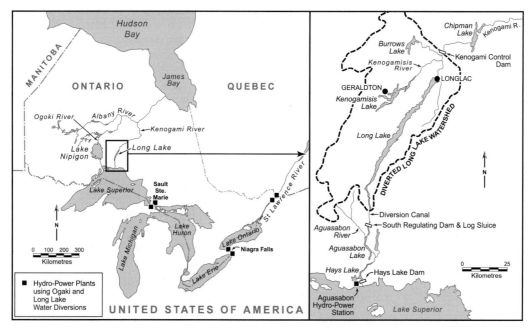

Figure 12.7 The Long Lake and Ogoki Diversions (Day and Quinn, 1992, Figures 9 and 10).

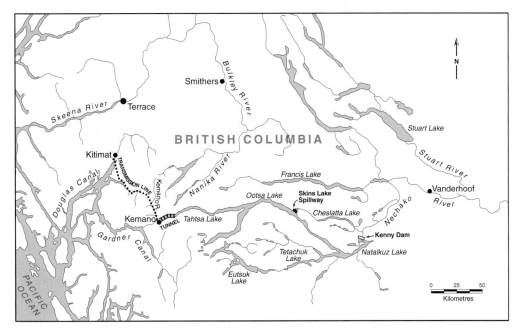

Figure 12.8 The Nechako–Kemano Diversion Project (Day and Quinn, 1992, Figure 12).

of Tahtsa Lake (Figure 12.8). A spillway at Skins Lake, on the north side of Ootsa Lake, is used to control the level of water in the reservoir. From the west end of the Tahtsa Lake, a 16 km tunnel transports a 115 m^3 s^{-1} average flow through the mountains and provides a 792 m drop[40] to the 896 MW Kemano Power Plant. Water from the Power Plant is then discharged to the Kemano River and ultimately via the Gardner Channel and Douglas Channel to the Pacific Ocean. Electricity from Kemano Power Plant is transported to the Kitimat aluminium smelter. In 1979, the ALCAN electricity system was linked with the British Columbia energy grid.

12.13.3 The Churchill–Nelson Diversion Project

The Nelson River is 656 km long and descends 217 m through various falls before entering Hudson Bay. In 1960, the Kelsey Hydro-power Plant on the Nelson River was completed. This was followed by construction of the Kettle Hydro-power Plant. In 1966, the Province of Manitoba decided to divert the Churchill River into the Nelson River and to regulate Lake Winnipeg outflow to meet its growing demands for electricity. Subsequently, Manitoba Hydro was licensed to undertake the diversion in 1972.

The Churchill–Nelson Diversion Project[41] consisted of three major components (Figure 12.9):

(1) a control dam was constructed at Missi Falls, which raised the Southern Indian Lake level by 3 m;

(2) a channel was excavated from South Bay of Southern Indian Lake, allowing an average flow of 775 m^3 s^{-1}, or approximately 35 times the normal flow of the Rat River and seven times the flow of the Burntwood River; and

(3) construction of a dam to control discharge of Lake Winnipeg at Jenpeg.[42]

The Diversion Project was in operation by 1977.

As a result of this diversion project, about 213 680 ha of land was flooded and thousands of indigenous[43] and non-indigenous people were affected. Also, the water regimes of both the Churchill and the Nelson river systems have been dramatically changed. For example, the mean annual flow rate of the Churchill River at Missi Falls decreased from 1011 m^3 s^{-1} to 251 m^3 s^{-1}, and at the town of Churchill, flow rate decreased from 1131 m^3 s^{-1} to 371 m^3 s^{-1}. At the same time, the mean annual flow of the Lower Nelson River at the Split Lake outlet increased approximately from 2265 m^3 s^{-1} to 3030 m^3 s^{-1} (Day and Quinn, 1992).

Table 12.10 shows selected characteristics of the five hydro-power plants built on the Nelson River. These plants

---

[40] Nearly 16 times the height of Niagara Falls which has a drop of 49 m in Canada and 51 m in the United States.
[41] Project No. 6 on Figure 12.6 and in Table 12.9.
[42] The Lake Winnipeg long-term mean outflow is about 2016 m^3 s^{-1} with winter outflow normally being reduced to 700 m^3 s^{-1} due to ice blockage.
[43] Hertlein (1999) describes the impacts of the diversion on the indigenous people.

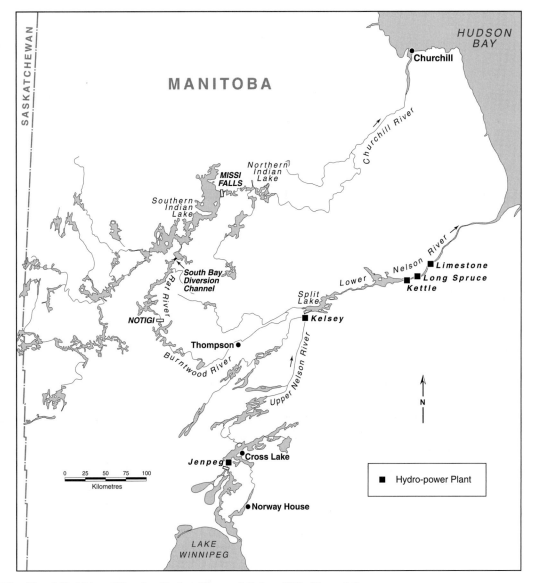

Figure 12.9 The Churchill–Nelson Diversion Project (Day and Quinn, 1992, Figure 14).

produced 86.3 percent of the total generating capacity of Manitoba Hydro in the year ending 31 March 1998. The 1997–98 year proved a profitable endeavour with gross revenue for Manitoba Hydro of CA$1.04 billion, of which about CA$250 million was from exports to the United States (Hertlein, 1999).

12.13.4 La Grande (James Bay) Diversion Project

The La Grande Diversion Project[44] was initially proposed in 1971, with an estimated cost of CA$2 billion, to satisfy future demands for hydro-electric power by Quebec. During the next two years different options were examined in detail, and a decision was finally made to use the massive potential of La Grande River Basin, which had an average discharge rate of $1710 \, m^3 \, s^{-1}$.

The La Grande River drops approximately 500 m over its 850 km length from east to west before reaching James Bay (Figure 12.10). To maximise the output from hydro-electric power plants, water was diverted into La Grande River from adjacent river basins. From the south, $845 \, m^3 \, s^{-1}$ from the Eastmain River (87 percent of the flow measured at the mouth) was redirected into the LG-2 reservoir. Also $790 \, m^3 \, s^{-1}$ of the Caniapiscau River flow was diverted into the LG-4 reservoir, to eventually flow through four hydro-power plants (LG-1 to LG-4). A third smaller

---

[44] Projects Nos. 12, 13 and 14 in Table 12.9 and on Figure 12.6.

diversion transfers $31\,m^3\,s^{-1}$ from Frigate Dam on the Sakami River into the LG-3 reservoir. Collectively, these diversions nearly doubled the natural flow of the La Grande River.

When the decision was made in May 1972 to proceed with the La Grande project, the cost estimate was CA$5.8 billion involving 4 power plants with a total capacity of 8330 MW. As detailed studies were undertaken, the cost was revised to CA$14.6 billion. Ultimately a number of power plants (LG-1, LG-2, LG-3, LG-4, Laforge-1, Laforge-2 and others)[45] with a combined installed capacity of about 15 550 MW were constructed in the La Grande River Basin and commissioned over the period of 1981–1996 at a cost of more than CA$20 billion (Quinn, 2004).

12.14 GREAT LAKES BASIN DIVERSIONS

Quinn and Edstrom (2000) describe exchange of water between the Great Lakes Basin and its neighbouring basins, as well as between the sub-basins of the Great Lakes (Table 12.11 and Figure 12.11). The two previously described Ogoki and the Long Lake diversions are the major importing water projects into the Great Lakes Basin, while the Chicago diversion is the major exporting one.

The Chicago diversion originated in 1848 as the Illinois–Michigan Canal, providing a navigation link between the Chicago River, draining into the Great Lakes and the Illinois River of the Mississippi River system. In 1900 the Chicago River flow was reversed causing Lake Michigan to drain southward into the Mississippi River. The diversion project caused dispute between various Great Lakes states and Canada.[46] Subsequently, the United States Supreme Court gradually reduced the diversion rate from $290\,m^3\,s^{-1}$ to $91\,m^3\,s^{-1}$ (Quinn and Edstrom, 2000).

The Welland Canal was originally built in 1829, providing a navigable route between Lakes Erie and Ontario, and linking with Canadian ports on the St Lawrence River. The Canal was reconstructed several times, and now is operated as an integral part of the deep-draft international St Lawrence Seaway. It diverts $260\,m^3\,s^{-1}$ of water from Lake Erie at Port Colborne at the north-east corner of the Lake in Ontario, and bypasses the Niagara River and Falls to reach Lake Ontario at Port Welland (Quinn and Edstrom, 2000).

The net balance of all diversions has been an increase in water supply to the Great Lakes Basin (see Table 12.11). The Ogoki and Long Lake diversions have increased the average outflow from Lake Superior by about $160\,m^3\,s^{-1}$, while the average outflow from Lake Michigan has decreased by $90\,m^3\,s^{-1}$ because of the Chicago diversion. In terms of cumulative effects, the average outflows from lakes Huron, Erie and Ontario have increased by about $68\,m^3\,s^{-1}$ (Quinn and Edstrom, 2000). The regulation plans in operation on the Great Lakes have been designed to accommodate changes in flow rates due to diversions.

12.15 IMPACTS OF THE DIVERSION PROJECTS

Development of water diversion projects, and hydroelectric power generation, have contributed to the

Table 12.10. *Selected characteristics of the five power plants on the Nelson River*

No.	Name	Year of completion	Capacity (MW)	Operating head (m)
1	Kelsey	1960	224	17.1
2	Kettle	1971	1272	30.0
3	Long Spruce	1977	980	26.0
4	Jenpeg	1977	126	7.3
5	Limestone	1990	1330	27.6
Total		–	3932	–

Source: Hertlein (1999).[47]

---

[45] http://www.hydroquebec.com/generation/hydroelectric/la_grande/index.html (visited in July 2005).

[46] The Chicago diversion was initially opposed by the Mississippi River States, which didn't want Chicago's sewage effluent, as well as by most Great Lakes States and Canada, which didn't want to lose flow for their own hydro-power generation, navigation and other interests (Frank Quinn, personal communication, April 2006).

[47] http://www.dams.org/docs/kbase/contrib/soc205.pdf (visited in July 2005).

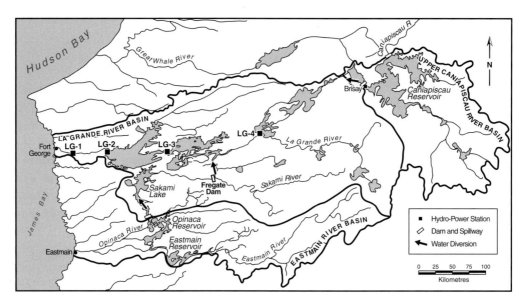

Figure 12.10 La Grande Diversion Project (Day and Quinn, 1992, Figure 17).

Table 12.11. *Existing diversions in the Great Lakes Basin*

Diversion project	Water transfer From	To	Date of original project	Average annual flow $(10^6 \text{m}^3)^{(a)}$
A. Exchange between the Great Lakes Basin and its neighbouring basins				
Ogoki	Albany River	L. Superior	1943	+3564.0
Long Lake	Kenogami River	L. Superior	1939	+1419.0
Chicago	L. Michigan	Illinois River	1848	−270.0
Forestport	L. Ontario Basin	Hudson River Basin	1825	−44.0
Portage Canal	Mississippi Basin	L. Michigan Basin	1860	+31.5
Ohio and Erie Canal	Ohio River Basin	L. Erie Basin	1847	+9.5
Pleasant Prairie	L. Michigan Basin	Illinois River	1990	−3.2
Akron	L. Erie Basin	Ohio River Basin	1998	−0.3
B. Exchanges between sub-basins of the Great Lakes Basin				
Welland Canal	L. Erie	L. Ontario	1829	8200.0
New York State Barge Canal	Niagara River	L. Ontario	1825	631.0
Detroit	L. Huron	Detroit River	1975	126.0
London	L. Huron and L. Erie	Thames River and L. St Clair	1967	94.0
Raisin River	St Lawrence River	Raisin River and St Lawrence River	1968	22.0
Haldimand	L. Ontario Basin	L. Erie Basin	1997	3.2

Note: (a) + inflow; (b) − outflow.
Source: Quinn and Edstrom (2000, Table 1).

industrialisation and economic development of Canada. However, these developments, in particular the early ones, took place at a time when integrated catchment management was not yet a widely accepted strategy. Therefore, impacts of the projects on issues such as erosion, fish, forests, water quality, and displacement of indigenous people were largely ignored. In fact, provincial governments approved the projects without any input from other

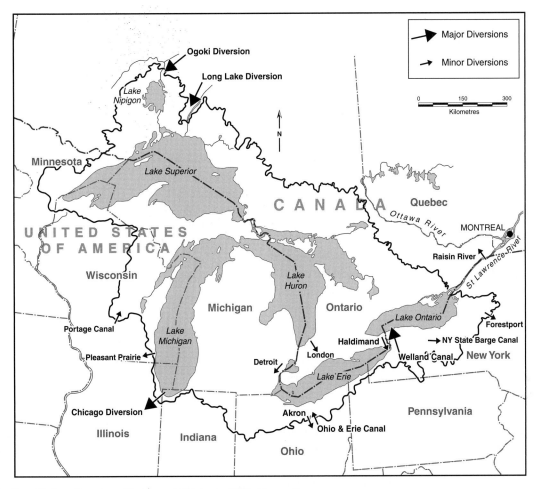

Figure 12.11 Great Lakes diversions (Quinn and Edstrom, 2000, Figure 1).

affected interests. Day and Quinn (1992) provide an analysis of these impacts as well as their economic outcomes for each of the five projects described earlier in this chapter. The following is a short description of some of the impacts:

- Ecological and social concerns were not generally considered at the time of project development, and little documentation was prepared before and during the project construction.
- Diversions induced erosion in reservoirs, channels and receiving bodies. Turbidity degraded water quality, and impaired habitats for fish.
- No provision was made to provide adequate flow in the diverted rivers for fish spawning, fish migration and other fish habitat requirements.
- Diversions have been operated with little concern for the affected public, who depended on the river for their livelihood. Water was occasionally released down its original channel, without warning the downstream residents.
- Failure to clear trees from reservoirs, diversion channels and receiving water bodies produced debris, which may take hundreds of years to decompose. Submerged standing trees cause navigation and shoreline access hazards. Furthermore, drowned vegetation constitutes a hazard for commercial fishing.
- Parkland and forests were flooded by man-made reservoirs.
- Fluoride emission from the Kitimat smelter damaged Kitimat Valley vegetation and caused health problems. Discharge of the smelter effluents also impacted the aquatic environment.
- Indigenous communities have been displaced, paid marginal compensation, and their life style has been permanently changed.

It should be noted that since completion of the projects, attempts have been made to rectify the above problems.

12.16 LEARNING FROM CANADIAN EXPERIENCE

Day and Quinn (1992, pp. 171–183) describe various aspects of Canadian experience with water diversion plans. The following is based on this publication.

Canada's rivers have been dammed and diverted, mainly for electricity generation. As a result, Canada has become the world's leading producer of hydro-electricity. However, not all Canadians are proud of the past attitude with which hydro-electric utilities and governments handled environmental issues, and the native people affected by major hydro-electric related water diversions. Some of the issues and the lessons that can be learned are as follows.

Planning and assessment. Provincial utilities and governments constructed most diversions with little considerations for human, biophysical and ecological imapcts. Indeed, the driving force to protect the environment and indigenous people came from the impacted groups, rather than the provincial or federal governments. Little effort was made to consult those affected and to canvas their perceptions and attitudes about development options. To improve the situation, governments need to provide choices for public consideration, and institute alternative demand management such as higher prices, increased efficiency, and other means for conserving water and electricity. A wider range of alternatives might help to reduce or defer as long as possible, the environmental and social disturbances and public debt associated with large diversion projects.

Rivalries between jurisdictions, closed decision-making processes, and a lack of cooperation among federal, provincial, regional, and local authorities, were at the heart of the problems. Diversions were planned and executed at the provincial level, with only modest and ineffective federal intervention. There was a clear need for the federal government to take a more proactive role within the framework of the Canadian constitution.

Biophysical changes. Diversions have impacted forests, agricultural lands, wildlife, fisheries and increased erosion and water turbidity. Only small amounts of money have been spent on the mitigation of these impacts, or on compensation to those who have been affected. Questions relating to the sustainability of the interacting environmental components, and the human populations dependent on them, have also been ignored.

Economic issues. Manitoba in the 1970s and Quebec in the 1980s rapidly expanded their hydro-electric capacity. Construction costs rose well above the initial estimates, and the market for electric power failed to develop at the same growth rate. Rising interest rates and a falling Canadian dollar made things worse. With a large energy surplus available, it was impossible to promote energy conservation.

Politics. A lack of effective control by provincial governments over their resource development agencies contributed substantially to the problems. This was because major water projects were not subject to thorough public assessment of their biophysical and socio-economic impacts before political commitments had been made to build them. The political push for development dominated at every level of government.

Native communities. A disturbing aspect of developments was the secrecy with which planning was undertaken. Initially, those to be directly affected were not consulted, nor were they informed of what would happen to their environment. Moreover, small native communities, those most vulnerable to changes in their water regime, were unable to influence the basic nature of the development. This confrontational approach was in stark contrast with the currently recognised principle of negotiation and consultation with stakeholders.

12.17 LARGE-SCALE WATER EXPORT PROPOSALS

Since the late 1950s numerous large-scale proposals have been developed to export water to the United States, particularly to the drier western states where demand exceeds available resources. The North American Water and Power Alliance (NAWAPA) proposal is described in section 11.10. Bryan (1973), Scott *et al.* (1986), and de Silva (1997) have discussed a number of other proposals, which are presented in the following sections.

12.17.1 The Central North American Water Project

The Central North American Water Project (CeNAWP) was put forward as an alternative to NAWAPA (Tinney, 1967). It proposed to divert Canadian western waters to Great Bear Lake, Great Slave Lake, Lake Athabasca, Lake Winnipeg, Lake Nipigon and Lake Superior. Water from Lake Winnipeg would be transferred to the Missouri River, in

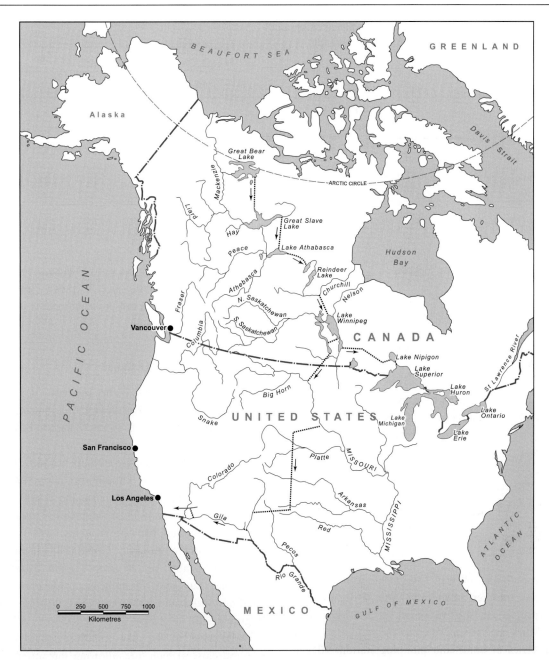

Figure 12.12 The Central North American Water Project (Bryan, 1973, Figure 9.3; and Tinney, 1967).

the United States (Figure 12.12), and then, via a canal, to the Rio Grande. Finally, water from the Rio Grande would be diverted to the Gila River in the Southern Colorado River Basin and then to southern California. Tinney (1967) claimed that his proposal did not require any reservoirs because it relied entirely on approximately 130 000 km^2 of existing lakes, which are so closely spaced within Canada, that only a few hundred kilometres of linking canals would be needed. It would generate hydro-power at three major points: a drop from Churchill River to Lake Winnipeg; a drop from Lake Winnipeg to Lake Superior; and drops throughout the Great Lakes and along the St Lawrence River. The generated power would more than offset the required power in northern Canada for pumping. Tinney (1967) does not provide any cost estimates for the CeNAWP proposal. However, he believed that the proposal would be cheaper to construct (without taking into account other economic, environmental and social costs), and most likely only one-third the cost of the NAWAPA proposal.

Table 12.12. *Estimated costs of water for a number of diversions*

Water diversion		Cost of water per 1000 m³ in CA$	
From	To	In 1966 values	In 2004 values
Lake Winnipeg	Canadian Prairies	8	48
Lake Winnipeg	US Midwest	16	96
Lake Winnipeg	Lake Superior	4	24
Lake Winnipeg	Texas	28	165
Churchill River	Lake Winnipeg	2	12
Lake Athabasca	Lake Winnipeg	4	24

Source: Kuiper (1966).

12.17.2 The Kuiper Diversion Scheme

Kuiper (1966) described future water needs in western Canada and water availability in the Yukon, Mackenzie, Churchill, and Nelson river systems, and proposed a diversion scheme, which is very similar to CeNAWP. The Kuiper Diversion scheme would divert water from Mackenzie drainage basin into rivers across western Canada to Lake Winnipeg. From there, water could be diverted east to the Great Lakes, or south to the United States. Using a simple, yet inaccurate procedure, Kuiper (1966) provided rough estimates of the cost of water for a number of diversions (Table 12.12). His estimates appear grossly below the real costs.

12.17.3 The Western States Water Augmentation Concept

The Western States Water Augmentation Concept (WSWAC) is another elaborate alternative to the NAWAPA project. It was proposed in 1968 to divert part of the Mackenzie River water to the southern United States by a series of dams and pumping stations (Bryan, 1973, pp. 162–164). The WSWAC is a two-part proposal. The first half proposes to divert water from the Liard Basin south to the Rocky Mountain Trench. Then, water would be transferred through tunnels and canals into the Fraser, Columbia or Kootenay rivers to the United States. The second half of the proposal suggests transferring waters from the Athabasca and Saskatchewan rivers through the Qu'Appelle River to Lake Winnipeg and then south to the United States (Figure 12.13). The proposal would yield about 47×10^9 m³ of water per year, and would cost approximately CA$75 billion (CA$450 billion in 2004 values) to construct, in addition to costs within the United States. Of the CA$75 billion, CA$12 billion was for the collection systems, CA$51 billion for water distribution systems, and CA$12 billion for hydro-electric power generation. The proposal appeared to be less environmentally damaging than the NAWAPA proposal.

12.17.4 The GRAND Canal

Kierans (1965) proposed the concept of the GRAND (Great Recycling and Northern Development) Canal. The proposal was intended to convert James Bay, which has an average inflow of about $9000 \, \text{m}^3 \, \text{s}^{-1}$, to a source of freshwater that could be used to transfer 36×10^9 m³ of water per annum to the Lake Huron (Day, 1985, Table 3). This project expected to stabilise the levels of the Great Lakes[48] and assist transfer of water to dry areas of Canada, United States and Mexico (Kierans, 1987). The project cost was estimated at about CA$100 billion in 1984 values (Day, 1985, Table 3). The proposal was based on turning the James Bay into an immense freshwater reservoir by building a dike across the Bay where it meets Hudson Bay (Figure 12.14).

Sluice gates in the dike enclosure would open at low tide and close at high tide, allowing salt water to flow into Hudson Bay while retaining freshwater from local rivers discharging into James Bay. It was assumed that within a few years, James Bay would become a freshwater lake (de Silva, 1997). Other components of the concept are (Kierans, 1987):

- The transfer system from James Bay to the Great Lakes would include intake systems, aqueducts, reservoirs, pumping plants to lift water from James Bay (at sea

---

[48] Years of higher or lower than average precipitation result in levels which are respectively very high or very low. High levels cause heavy erosion. Low levels deteriorate water quality, and impair shipping and hydro-power production.

INTER-BASIN WATER TRANSFER IN CANADA 287

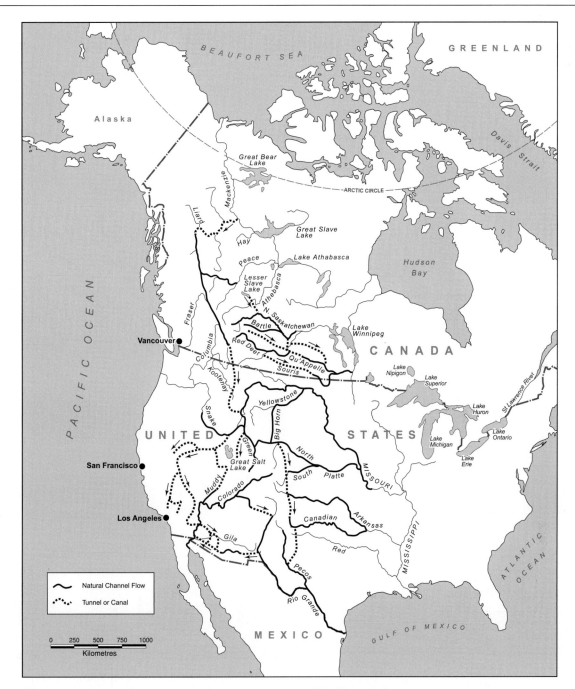

Figure 12.13 The Western States Water Augmentation Concept (Bryan, 1973, Figure 9.4).

level) to Lake Huron (at about 177 m), energy supply systems, and inflow and outflow control facilities.
- The Great Lakes new inflow/outflow stabilisation control system.
- The water transfer system from Lake Superior to the water deficient Canadian Prairies. This transfer requires reservoirs, aqueducts, pumping plants, energy supply facilities and distribution systems.
- The water transfer system from Lake Michigan to water deficient areas in western United States, that would require numerous intermediate reservoirs, aqueducts, pumping plants, energy supply facilities and distribution systems.

In the 1980s, the proposal had the support of engineering companies in Canada including Bechtel Canada, a subsidiary

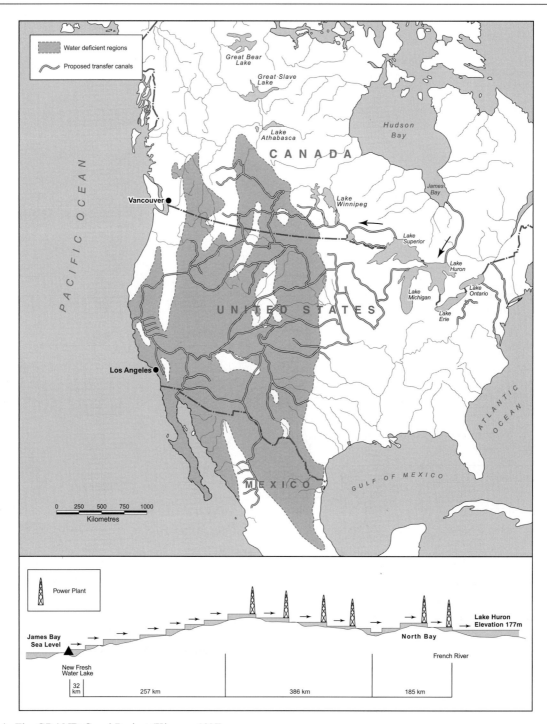

Figure 12.14 The GRAND Canal Project (Kierans, 1987).

of the giant United States engineering company Bechtel, and from Simon Reisman who was later appointed as the head of the Canadian negotiating team for the Free Trade Agreement with the United States. The proposal also had support from the U.S. Army Corps of Engineers. One of the arguments in support of the proposal was that construction of the project would create 150 000 direct jobs in Canada, and at least as many again to supply goods and services (Crane, 1988). However, the proposal had numerous serious economic, environmental and social problems.

Gamble (1987) examined the proposal from a policy, economic and environmental perspective. He argued that this project with a capital cost of CA$80 billion to CA$130

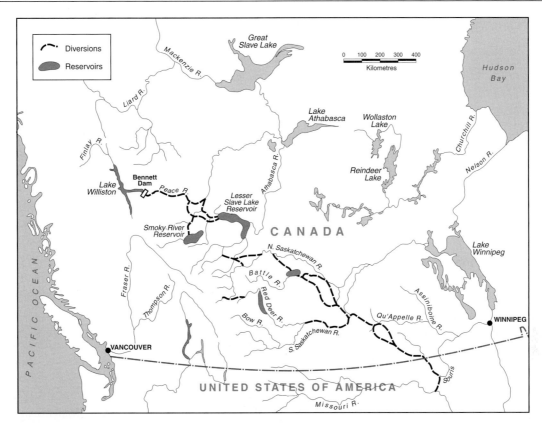

Figure 12.15 The Magnum Diversion Scheme (Bryan, 1973, Figure 9.5).

billion (CA$145 billion to CA$235 billion in 2004 values), could not be undertaken on a "user pay" basis and required unjustified subsidy. Moreover, it would have significant environmental impacts, social costs, and policy implications for both Canada and the United States.

Milko (1987) described that changes in nutrient content and freshwater circulation out of Hudson Bay could potentially affect productivity downstream on the Labrador Shelf, and changes in productivity and ice pack within Hudson Bay could be detrimental to fish and marine mammals.

12.17.5 The Magnum Diversion Scheme

The Magnum scheme was developed in the late 1960s (Bryan, 1973, pp. 164 and 165), to divert water from Lake Williston to the Peace River below Bennett Dam into a canal leading to Lesser Slave Lake (Figure 12.15). Subsidiary storage would be established on the Smokey River for diversion by another canal to Lesser Slave Lake. The storage capacity of Lesser Slave Lake would be increased to allow the level to rise 6 m to 12 m, and would receive water from the Athabasca River, as well as from the Peace and Smokey rivers. Subsequently, water would be transferred to the North and the South Saskatchewan rivers, Qu'Appelle River, and Souris River to the Missouri River, and from there to demand areas in the United States. The plan expected to divert $31 \times 10^9 \, m^3$ of water per year to the United States at an unspecified cost (Day, 1985, Table 3). This proposal involved considerable water diversion, which could have serious ecological impacts on the river systems.

12.17.6 Saskatchewan-Nelson Basin

The Saskatchewan–Nelson Basin (SNB) proposal was developed in the 1970s to link a number of lakes and rivers across western Canada (de Silva, 1997). Water from the Mackenzie River Basin would be diverted into the Wollaston Lake, Reindeer Lake and finally into the Churchill River system. From the Churchill River system, the Saskatchewan River would be used as storage until the water could be delivered into Cedar Lake. From there, water would be delivered south through canals into Lake Manitoba. This would then be pumped into the Assiniboine River and would be diverted into the Souris River. Water from Souris River could then be delivered into the Garrison River in North Dakota for diversion to demand areas in the western United States.

12.17.7 The North Thompson Project

The objective of this scheme was to export $1.2 \times 10^9 \, m^3$ of water a year from the North Thompson River in British Columbia to the United States (de Silva, 1997). It consisted of diverting water from the Columbia Drainage Basin into the John Day Dam Reservoir. The water would then be pumped to Shasta Lake in northern California and from there would be transferred to southern California through the Sacramento River and California Aqueduct (see Figure 11.13). A number of dams necessary for the project have already been built for other purposes.

12.18 WATER EXPORT POLICY

For many decades, the stated policy of both the federal and provincial governments has been one of "no water export". In December 1965, the Parliamentary Secretary to the Ministry of Northern Affairs gave a speech at the United States Chamber of Commerce's National Water Conference in Washington D.C. and said (Howe and Easter, 1971, p. 11):

- First, Canada does not agree that its water is a "continental" resource as often suggested by American officials. In Ottawa's opinion it is Canadian property to be used as Canada alone decides.
- Second, Canada alone will survey its water, find out exactly how much it owns and how much it is likely to need for a rapidly expanding population.
- Third, when these unknown facts are clarified and if a water surplus is revealed, Canada will discuss the possible sale of water with the United States.
- Fourth, we might wish to export water not for money but in return for access to your market.

In 1965, the Government of Alberta rejected the NAWAPA proposal as a foreign plot and promoted its own plan for diversion of rivers southward within the Province. But in the following decade this plan was rejected as it became a source of conflict between north-central and southern Alberta. Subsequently, Alberta abandoned the plan in favour of improved river basin planning (Day and Quinn, 1992, p. 172).

In August 1984, the Federal Minister of the Environment stated that (Scott *et al.*, 1986, p. 194):

> Canada's position to oppose the export of water hasn't changed. ... We reject the contention that water is available for export. This will be a very important commodity for Canadians in the decades ahead. We therefore reject any such notion whether it comes from provinces, municipalities or regions in the north. Our position on that is clear and consistent.

Scott *et al.* (1986) presented important background information on water export, which included: historical information relevant to Canadian water exports; physical factors that affect water export; environmental impacts; Canadian costs and benefits; international law, Canadian constitutional law, and legal aspects in the United States; and the Canadian water export policy. They argued that neither a simple *permissive* nor a *prohibitive* policy is good enough because they are too inflexible to deal with all possible situations that might arise in the various Canadian regions. Instead, they proposed that water export projects should be assessed on a project-by-project basis by a government review process using cost–benefit analysis. However, strong objection existed, and continues to exist, towards large-scale export of Canadian waters to the United States, because these projects have high environmental, social and economic costs. For example, the average cost of large-scale water export would be US$1500 per $1000 \, m^3$, while the United States would likely pay US$130 for the same volume (de Silva, 1997), so that most proposals would rely heavily on government subsidies.

12.18.1 International Agreements and Export of Canada's Waters

Canada is a major exporter of bottled water to the United States. In 1999, the United States bottled water sale was approximately 18 billion litres, of which 8 percent (1.4 billion litres) was imported primarily from Canada (51 percent) and France (33 percent). Water traded as bottled or value-added water is covered by international trade rules like any other goods (Gleick *et al.*, 2002).

Negotiations for the *Free Trade Agreement* (FTA) between Canada and the United States were initiated in 1986, completed in 1988 and went into effect on 1 January 1989. This agreement was followed by the *North American Free Trade Agreements* (NAFTA)[49] signed between the United States, Canada and Mexico in December 1992. It was

---

[49] Information regarding FTA and NAFTA is available in numerous publications including: Johnson and Schachter (1988); Fatemi and Salvatore (1994); and Baer and Weintraub (1994).

approved by the United States House of Representatives and Senate in November 1993, and took effect on 1 January 1994. Although Canada exports bottled water under international trade laws, Canadians are strongly opposed to bulk export of their freshwater to the United States. A July 1988 Gallup Poll showed that 69 percent of Canadians disapprove the export of water to the United States (Holm, 1988, p. 15).

While the Canadian government argued that the FTA applied only to bottled water and not natural water in all its form, experts were highly sceptical and believed that the GATT[50] and FTA would open the door to large-scale water export to the United States. Holm (1988) provided views of experts in the fields of economics, international law, constitutional law, international trade, the GATT, the environment, natural resources and public policy, about this controversial issue. She argued that despite the claims of the Mulroney government, the FTA is not restricted to the export of bottled water and that there are no provisions that give Canada the right to permanently prohibit the bulk export of water.[51]

To reassure public opinion, the United States, Canadian and Mexican governments have frequently claimed that under the FTA or NAFTA, Canada is not obliged to bulk export its waters. For example, in 1993, the three NAFTA parties signed the following joint declaration[52] to provide explicit protection for water resources in their natural state and the rights of the country of origin under NAFTA:

> Unless water, in any form, has entered into commerce and becomes a good or product, it is not covered by the provisions of any trade agreement, including the NAFTA. And nothing in the NAFTA would obligate any NAFTA party to either exploit its water for commercial use, or to begin exporting water in any form. Water in its natural state in lakes, rivers, reservoirs, aquifers, water basins and the like is not a good or product, is not traded, and therefore is not and never has been subject to the terms of any trade agreement.
>
> International rights and obligations respecting water in its natural state are contained in separate treaties and agreements negotiated for that purpose. Examples are the United States–Canada Boundary Waters Treaty of 1909 and the 1944 Waters Treaty between Mexico and the United States.

Although this joint declaration was the clearest exposition of the intent to protect natural waters from bulk withdrawals for international trade, some analysts argued that it had limited weight in international law, was not legally binding and established no legal obligations (Gleick et al., 2002).

While British Columbia and Alberta already had legislation that prohibited the removal of water, including for export, Ontario was finalising regulations to accomplish the same goal, with other provinces having similar policies. On 10 February 1999, the government of Canada launched a strategy to prohibit the bulk removal[53] of Canadian water, including water for export. The strategy responded to Canadian concerns about the security of their freshwater resources. The strategy recognised that provinces have the primary responsibility for water management, and that the government of Canada has certain legislative authorities in the areas of navigation, fisheries, federal land, and shared water resources with the United States under the *Boundary Waters Treaty*. The strategy also respected Canada's trade obligations as it focused on water in its natural state (e.g. in rivers and lakes). The Federal Government claimed that water in its natural state is not a good or product, and is therefore not subject to international trade agreements. Some of the outcomes of the Federal Government strategy to prohibit bulk water removal include:[54]

- In February 1999, Canada and the United States agreed on a joint reference to the International Joint Commission (IJC) to study the effects of water consumption, diversion and removal, including export from boundary waters.
- In August 1999, the IJC submitted an interim report to the Canadian and the United States governments. The key recommendation called for an immediate

---

[50] The General Agreement on Tariffs and Trade (GATT) for goods was established in 1947. On the 1 January 1995, the World Trade Organization (WTO) was established as the successor to the GATT (World Trade Organization, 1999).

[51] The Federal Government and its legal and policy advisors disagreed with views presented by Holm (1988) and proceeded to submit amendments to the International Boundary Waters Treaty Act to prohibit bulk removal of water from boundary waters. The amendments went into effect after approval by the parliament in 2002 (personal communication, Frank Quinn, April 2006).

[52] http://www.scics.gc.ca/cinfo99/83067000_e.html#statement (visited in July 2005).

[53] The removal and transfer of water out of its basin of origin by man-made diversions (e.g. canals), tanker ships or trucks, and pipelines. Such removals have the potential (directly or cumulatively) to harm the health of a drainage basin. Small-scale removal, such as water in small portable containers, is not considered bulk.

[54] http://www.ec.gc.ca/water/en/manage/removal/e_backgr.htm (visited in July 2005).

moratorium on bulk removal of waters from the Great Lakes.
- In November 1999, the Canadian Council of Ministers of the Environment considered the proposed Canada-wide Accord, with all jurisdictions[55] agreeing on a common objective, namely to prohibit bulk water removal from major drainage basins in Canada.
- In March 2000, IJC submitted its final report entitled "Protection of the Waters of the Great Lakes".[56] It recommended that the governments should not permit any new proposals for removal of water from the Great Lakes Basin to proceed unless the proponents can demonstrate that the removal would not endanger the integrity of the ecosystem of the Great Lakes Basin. Application of the IJC recommendations would effectively prevent any large-scale or long-distance removal of water from the Basin.
- In December 2002, amendments to the *Boundary Waters Treaty* and related regulations came into force. The main purpose of the amendments is to prohibit the removal of boundary waters from their water basins, principally the Great Lakes. In addition, water related projects in boundary waters, such as dams, dikes or other obstructions that affect the level or flow of waters on the United States side of the boundary would require licenses from the Canadian Minister of Foreign Affairs. These changes will strengthen Canada's implementation of the *Boundary Water Treaty*.

Currently, all provinces have in place, or are developing legislation, regulations or policies prohibiting the bulk removal of water. This provides solid assurance that bulk removals and export will not proceed any time in the near future.

Current public opinion and government policy are against bulk water transfer to the United States. Although the United States is implementing a major water management plan to satisfy its water requirements over the next few decades by undertaking water conservation and other measures (see section 11.11), some Canadians believe that the United States will sooner or later run out of water and this will inevitably put pressure on Canada to share its water resources.

12.19 CONCLUSIONS

Canada enjoys more than $100\,000\,m^3$ of fresh surface water per head of population. This, combined with inappropriate water pricing, has historically led to profligate use of water by industry, irrigators, and households, which in turn has caused significant impacts on the environment and native communities. In more recent years, Canada has developed numerous policies promoting water saving, water pricing and wastewater reuse, with the aim of reducing water withdrawal and consumption. Future water resources planning and management will continue to promote these initiatives, for sound social, economic and environmental reasons.

Average annual temperature of the country has increased more than 0.9°C since 1982, and it is expected that global warming could reduce the snow cover and change the flow regimes of river systems. These issues must also be addressed in future water resources planning and management. Another policy consideration is that Canada is a signatory of the Kyoto Agreement, and therefore supports a reduction of greenhouse gas emissions. This favours the continuing use of large-scale hydro-electric power generation, so that energy conservation is also linked with water issue.

Hydro-electric power schemes dominate inter-basin water transfer projects, both in the number and scale of diversions. Irrigation, flood control and municipal uses only assume importance regionally or locally. Canada has become the world's leading producer of hydro-electricity, and in 2001 it provided 57 percent of country's electricity. Those facilities export significant amounts of electricity to the United States.

Development of water diversion projects, and hydro-electric power generation, have contributed to the industrialisation and economic development of Canada. However, many of these developments took place at a time when integrated catchment management and public participation were not widely accepted strategies. The impacts of these projects on issues such as erosion, fish, forests, water quality, and displacement of indigenous people, were largely ignored. Provincial governments approved many of these projects without any input from the Federal Government or stakeholders. In recent years there have been considerable changes, and such developments now have to carefully consider all social, economic and environmental impacts and benefits.

---

[55] Alberta, British Columbia, Manitoba, and Saskatchewan reserved their position pending further consideration. Quebec declared that it will set out its position in its own communiqué (http://www.scics.gc.ca/cinfo99/83067000_e.html visited in July 2005).

[56] http://www.ijc.org/php/publications/html/finalreport.html (visited in July 2005).

The export of water to the United States remains controversial. The large-scale proposals developed in the 1960s were the subject of intense debate. There was, and continues to be, strong public opinion and government policy against such projects. The Federal Government has declared that water in its natural state is not a good or product. Therefore, under the FTA and NAFTA agreements, Canada has no obligation to trade its water resources.

References

Back, W., Rosenshein, J. S. and Seaber, P. R. eds. (1988). *Hydrogeology*. Boulder, Colorado: Geological Society of America, Inc.

Baer, M. D. and Weintraub, S. eds. (1994). *The NAFTA Debate: Grappling with Unconventional Trade Issues*. Boulder, Colorado: Lynne Rienner Publishers.

Biswas, A. ed. (2003). *Water Resources of North America*. Berlin: Springer.

Bone, R. M. (2002). *The Regional Geography of Canada*, Second Edition. Toronto, Canada: Oxford University Press.

Bryan, R. (1973). *Much is Taken, Much Remains*. North Scituate, Massachusetts: Duxbury Press.

Burke, D., Leigh, L. and Sexton, V. (2001). *Municipal Water Pricing, 1991–1999*. Ottawa: Environmental Economics Branch, Environment Canada.

Chambers, P. A., Dupont, J., Schaefer, K. A. and Bielak, A. T. (2002). *Effects of Agricultural Activities on Water Quality*. Winnipeg: Canadian Council of Ministers of the Environment. CCME Linking Water Science to Policy Workshop Series. Report No. 1.

Coote, D. R. and Gregorich, L. J. eds. (2000). *The Health of our Water: Toward Sustainable Agriculture in Canada*. Ottawa: Research Branch, Agriculture and Agri-Food Canada.

Crane, D. (1988). The pressure to sell our water. In Holm, W. ed. *Water and Free Trade: The Mulroney Government's Agenda for Canada's Most Precious Resource*. Toronto: James Lorimer & Company, pp. 21–27.

Crowe, A. S., Schaefer, K. A., Kohut, A., Shikaze, S. G. and Ptacek, C. J. (2003). *Groundwater Quality*. Winnipeg: Canadian Council of Ministers of the Environment. CCME Linking Water Science to Policy Workshop Series. Report No. 2.

Day, J. C. (1985). *Canadian Interbasin Diversions*. Burnaby, B. C.: Natural Resources Management Program, Simon Fraser University.

Day, J. C. and Quinn, F. (1992). *Water Diversion and Export: Learning from Canadian Experience*. Waterloo: Department of Geography, University of Waterloo. Publication Series No. 36.

de Rosemond, S., Kent, R., Murray, J. and Swain, L. (2003). *Experts Workshop on Water Quality Monitoring: The Current State of Science and Practice*. Winnipeg: Canadian Council of Ministers of the Environment. CCME Linking Water Science to Policy Workshop Series. Report No. 5.

de Silva, A. (1997).[57] The sale of Canadian water to the United States: a review of proposals, agreements and policies regarding large scale interbasin exports.

Environment Canada (2002a). *The Meteorological Service of Canada: Annual Report 2000–2001*. Ottawa: Environment Canada.

Environment Canada (2002b). *Climate Change Plan for Canada*. Ottawa: Environment Canada.

Environment Canada (2003a). *Water and Canada: Preserving a Legacy for People and the Environment*. Ottawa: Environment Canada.

Environment Canada (2003b). *The Canada Water Act Annual Report 2001–2002*. Ottawa: Environment Canada.

Fatemi, K. and Salvatore, D. eds. (1994). *The North American Free Trade Agreement*. Oxford, U.K.: Elsevier Science.

Francis, D. and Hengeveld, H. (1998). *Extreme Weather and Climate Change*. Ontario: Atmospheric Environment Service.

Gamble, D. J. (1987). The Grand Canal Scheme: some observations on research and policy implications. In Nicholaichuk, W. and Quinn, F. *Proceedings of the Symposium on Interbasin Transfer of Water: Impacts and Research Needs for Canada*. Saskatoon, Saskatchewan, 9 and 10 November 1987. Saskatoon: National Hydrology Research Centre, pp. 71–83.

Gleick, P. H., Wolff, G., Chalecki, L. and Reyes, R. (2002). Globalization and international trade of water. In Gleick P. H. *The World's Water 2002–2003: The Biennial Report on Freshwater Resources*. Washington: Island Press, pp. 33–56.

Great Lakes Commission (2003). *A Year of Progress: 2003 Annual Report*. Ann Arbor, Michigan: Great Lakes Commission.

Hertlein, L. (1999).[58] Lake Winnipeg Regulation: Churchill–Nelson River Diversion Project in the Cree on Northern Manitoba, Canada.

Holm, W. ed. (1988). *Water and Free Trade: The Mulroney Government's Agenda for Canada's Most Precious Resource*. Toronto: James Lorimer & Company.

Howe, C. W. and Easter, K. W. (1971). *Interbasin Transfer of Water: Economic Issues and Impacts*. Baltimore: The John Hopkins Press.

International Joint Commission (2004). *Twelfth Biennial Report on Great Lakes Water Quality*. Detroit, Michigan: IJC.

Johnson, J. R. and Schachter, J. S. (1988). *The Free Trade Agreement: A Comprehensive Guide*. Aurora, Ontario: Canada Law Book Inc.

Kierans, T. W. (1965). The GRAND Canal concept. *Engineering Journal* **48**(12): 39–42.

Kierans, T. (1987). Recycled water from the north: the alternative to interbasin water transfer. In Nicholaichuk, W. and Quinn, F. *Proceedings of the Symposium on Interbasin Transfer of Water: Impacts and Research Needs for Canada*. Saskatoon, Saskatchewan, 9 and 10 November 1987. Saskatoon: National Hydrology Research Centre, pp. 59–70.

Kuiper, E. (1966). Canadian water export. *Engineering Journal, Engineering Institute of Canada* **49**(7): 13–18.

Lemmen, D. S. and Warren, F. J. eds. (2004). *Climate Change Impacts and Adaptation: A Canadian Perspective*. Ottawa: Natural Resources Canada.

Marsalek, J., Schaefer, K., Exall, K., Brannen, L. and Aidun, B. (2002). *Water Reuse and Recycling*. Winnipeg: Canadian Council of Ministers of the Environment. CCME Linking Water Science to Policy Workshop Series. Report No. 3.

Milko, R. J. (1987). The GRAND Canal: potential ecological impacts to the North and research needs. In Nicholaichuk, W. and Quinn, F. *Proceedings of the Symposium on Interbasin Transfer of Water: Impacts and Research Needs for Canada*. Saskatoon, Saskatchewan, 9 and 10 November 1987. Saskatoon: National Hydrology Research Centre, pp. 85–99.

National Water Research Institute (2001). *Threats to Sources of Drinking Water and Aquatic Ecosystem Health in Canada*. Burlington, Ontario: NWRI Scientific Assessment Report Series No. 1.

OECD (1999). *The Price of Water: Trends in OECD Countries*. Paris: Organisation for Economic Co-operation and Development.

Quinn, F. and Edstrom, J. (2000). Great Lakes diversions and other removals. *Canadian Water Resources Journal* **25**(2): 125–151.

Quinn, F. (2004). Interbasin water diversions in Canada. A report to the International Commission on Irrigation and Drainage (ICID). Ottawa: Environment Canada.

Scharf, D., Burke, D. W., Villeneuve, M. and Leight, L. (2002). *Industrial Water Use 1996*. Ottawa: Environmental Economics Branch, Environment Canada.

Scott, A., Olynyk, J. and Renzetti, S. (1986). The design of water-export policy. In Whalley, J. *Canada's Resource Industries*. Toronto: University of Toronto Press. pp. 161–246.

Statistics Canada (1997). *Historical Overview of Canadian Agriculture*. Catalogue no. 93-358-XPB. Ottawa: Statistics Canada, Agriculture Division.

---

[57] Available at the URL: http://www.environmentprobe.org/enviroprobe/pubs/ev-540.htm (visited in July 2005).

[58] Available at the URL: http://www.dams.org/docs/kbase/contrib/soc205.pdf (visited in July 2005).

Statistics Canada (2000). *Human Activity and the Environment 2000*. Ottawa: Statistics Canada, Dissemination Division.

Statistics Canada (2001). *Canada Year Book 2001*. Ottawa: Statistics Canada, Dissemination Division.

Statistics Canada (2004). *Human Activity and the Environment: Annual Statistics 2004*. Ottawa: Statistics Canada, Dissemination Division.

Tinney, E. R. (1967). Engineering aspects. *Bulletin of the Atomic Scientists* **23**(7): 21–25.

Watkins, M. ed. (1993). *Handbook of the Modern World: Canada*. New York: Facts on File.

World Trade Organization (1999). *WTO Agreements Series: (No.1) Agreement Establishing the WTO*. Geneva, Switzerland: WTO.

13 Inter-basin Water Transfer in China

SECTION A: OVERVIEW OF GEOGRAPHY, POPULATION, LAND AND WATER

13.1 GEOGRAPHY

The People's Republic of China (PRC) was established in 1949. Its mainland covers an area of 9 560 980 km² and has a varied topography with highlands in the west and plains in the east. This physical feature of gradual descent from the west to the east causes all the major rivers to run eastwards. China has the following physiographic features (China Handbook Editorial Committee, 2002):

- **Mountain ranges** run in different directions across the length and breadth of the country (Figure 13.1). These are: (1) the west−east ranges consisting of the Tianshan, Yanshan, Kunlun, Qinling and Nanling Mountains; (2) the north-east to south-west ranges consisting of the Greater Hinggan, Taihang, Changbai, Wuyi and Taiwan Mountains; (3) the north-west to south-east ranges consisting of the Qilian and Altai Mountains; and (4) the north−south ranges consisting of the Helan, Liupan and Hengduan Mountains. The Himalayas lie on the south-western border of China and run in a west−east direction. Of the world's 14 mountain peaks exceeding 8000 m, nine are in China or on its borders.
- **Plateaus** which vary in both height and physical features cover about 25 percent of China's total area. They are located mainly in the western and central parts of the country. The major plateaus are: (1) the Tibet Plateau in western China, which is the highest plateau in the world with an elevation of 4000−5000 m; (2) the Inner Mongolia Plateau in northern China, which is the second largest plateau in the country and stands at an average elevation of 1000−2000 m above sea level; (3) the Loess Plateau, bounded by the Qinling Mountains in the south and the Taihang Mountains in the east, which rises 800−2000 m above sea level; and (4) the Yunnan−Guizhou Plateau in south-western China, which is about 2000 m high in Yunnan Province and 1000 m high in Guizhou Province.
- **Basins** as large as hundreds of thousands of square kilometres are found in the west, notably the Tarim, Junggar, Turpan, Chaidam and Sichuan basins. Smaller sized basins are found in the east.
- **Plains** lie mainly in the north-east and eastern seaboard regions, covering 1.12 million km² or a little more than 10 percent of the country's total area. They have gentle terrain, fertile soil, mild climate, and provide a base for China's major agriculture and industry. The main plains are: (1) the North-east Plain, which measures 1000 km from north to south and 400 km from west to east in its widest part; (2) the North China Plain, also called the Huang-Huai-Hai Plain, which is the second largest plain, covering an area of about 300 000 km², and is the product of the alluvial deposits from the Huanghe (Yellow), Huaihe and Haihe rivers. It is an important industrial and agricultural region, with the largest cultivated area in China; and (3) the Middle-Lower Yangtze Plain, which consists of a series of plains of varying width on both sides of the Yangtze River, covering 200 000 km² and mostly with an elevation of less than 50 m above sea level.
- **Deserts** cover 1 095 000 km² or approximately 11 percent of the country's total area, of which sand deserts account for 637 000 km² and the Gobi and stone deserts 458 000 km². The major deserts are the Taklimakan, Gurbantunggut, Badainjaran, Tengger and Muus Deserts. Most of these deserts are found in Xinjiang, Qinghai, Gansu, and Inner Mongolia Provinces in north-western and northern China (Figure 13.2). The deserts in Xinjiang constitute about 60 percent of the Country's total.

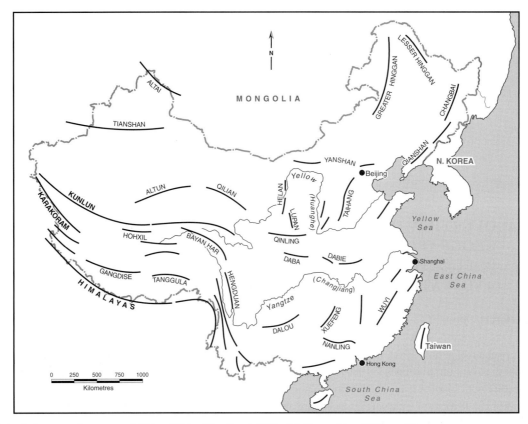

Figure 13.1 Principal mountain ranges of China (China Handbook Editorial Committee, 2002, p. 18).

13.2 POPULATION

China is the most populated country in the world. It supported a population of 555 million in 1950, and this had increased to about 1.28 billion by 2000, including 800 million in rural areas. It is expected that the country's population will reach 1.48 billion by 2025 (World Resources Institute, 2000, p. 296). The annual rate of population growth was as high as 2.8 percent over the period of 1969–71, and since then has gradually declined to 0.7 percent in 2000 as a result of the one child per family policy adopted in 1979. China's population is unevenly distributed. The population density ranges from more than 2000 persons per km² in Shanghai to one person per km² in Tibet (James, 1989, p. 46). The eastern seaboard has a population density of 300–400 persons per km², while western China is only sparsely populated. The nation's average population density is approximately 135 persons per km². China is rapidly transforming from a rural agrarian society to an industrialised urban society. Currently, its level of urbanisation is less than 40 percent, but if the current trend continues, the level of urbanisation will reach 60 percent by 2050. This will have a major impact on the land, water and energy requirements of the country.

Mainland China is divided into 22 provinces, five autonomous regions and three independent municipalities[1] (Figure 13.2). These municipalities and their populations are Shanghai (20 million), Beijing (15 million) and Tianjin (10 million).

13.3 ECONOMY

At the time of its establishment in 1949, the PRC had a poor, undeveloped and ravaged economy as a result of two decades of war. The economic situation has improved significantly since then, and in recent decades has been shifting from an agricultural to a manufacturing economy, and from a centrally planned to a socialist market economy. This has been supported by foreign investments and low cost labour, which makes its products highly competitive on the international market. The average annual growth rate of the country's GDP during the 1970s, 1980s and 1990s were 6.6 percent, 9.4 percent and 10.1 percent respectively. In absolute terms, its GDP (in 1995 values) increased from

---

[1] Descriptions of these administrative divisions is available in China Handbook Editorial Committee (2002, pp. 131–257).

INTER-BASIN WATER TRANSFER IN CHINA

Figure 13.2 Administrative divisions of China.

US$63 billion in 1960 to US$164 billion in 1980 and US$1042 billion in 2000 (World Bank, 2003). It is estimated that China's GDP will continue to increase, with a more than four fold increase projected between 2000 and 2025. The per capita GDP, which was US$94 in 1960, increased to US$167 in 1980, US$825 in 2000, and is expected to continue growing strongly.

13.4 CLIMATE AND PRECIPITATION

Zhang and Lin (1992) provide a detailed description of China's climate, which is affected by the summer monsoon. Every year from October to April cold winds sweep southward across China from Siberia and Mongolia. As a result, winters in China are dry and cold. January is the coldest and July is the warmest month for all parts of China. For example, the average monthly minimum and maximum temperatures for January and July for Beijing are −5.4°C and 29.3°C, and for Mohe in the extreme north-east of China are −33.3°C and 22.6°C.

Precipitation in China, which falls as rain and snow, is unevenly distributed across the country and varies considerably from year to year, causing major drought and flood problems. The coastal area in the south-east enjoys plentiful rainfall and a humid climate with an annual precipitation of 1000–2000 mm, while in some parts of the north-west, the annual rainfall is below 25 mm. If the isohyet of 400 mm is taken as the boundary line, China can be divided into two parts (Figure 13.3). East of the line is the humid area and to the west is the arid zone. A large part of the arid zone receives less than 200 mm annually. South-eastern China is under the influence of monsoon from May to August during which a one month rainfall can make up to between 25 and 50 percent of total annual rainfall.

The long-term average annual precipitation of China is 648 mm, which is lower than the global land average of 800 mm and less than the Asian average of 740 mm (Qian, 1994, p. 7). The potential evaporation is high in the arid areas of the north-west, exceeding 2500 mm per annum, while in the mountain regions of north-east and central China it is less than 800 mm.

13.5 LAND COVER AND USE

A large proportion of the 956 Mha landmass of China consists of mountains, high plateaus, deserts, and other lands

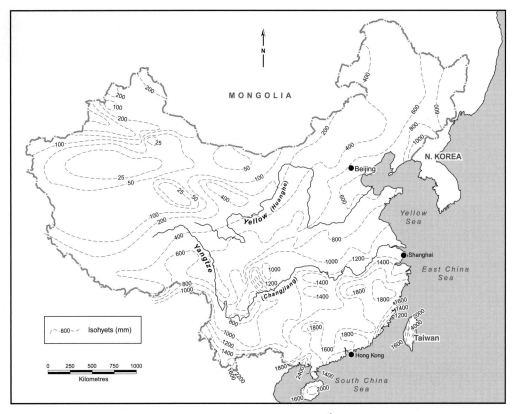

Figure 13.3 China's average annual precipitation (Zhang and Lin, 1992, Figure 4.1).[2]

that are not suitable for cultivation. In 2001, 143.6 Mha was arable, and 11.7 Mha was covered by permanent crops (FAO, 2003). These arable lands are located mostly in eastern China. The area of arable land per head of population was 0.18 ha in 1949 and has decreased to 0.1 ha in 2001. This is about one third of the world average[3] and reflects the most striking feature of Chinese agriculture – the challenge of feeding the world's largest population with a limited area of arable land.

Eight thousand years ago forests covered about 52 percent of China (497 Mha), but much of these have been gradually destroyed and reduced to about 133 Mha in 1995 (World Resources Institute, 2000, Tables FG.1 and FG.2). Although deforestation has occurred for thousands of years, the rate has accelerated over the past 300 years. But nothing could match the destruction of the past fifty years when state-run logging firms cut down vital forests at the headwaters of major rivers (Ma, 2004, p. ix). The main causes of deforestation were:

(1) clearance of land for farming and settlements;
(2) provision of fuel for heating and cooking; and
(3) supply of timber for building houses, boats and ships.

Elvin (2004, Chapters 3 and 4) describes the history of China's deforestation and its impacts, such as an increase in the sediment loads of the rivers, silt deposition on the flood plains, expansion of river deltas, loss of habitats for birds and animals, and extinction of species such as rhinoceroses, elephants, and tigers. Deforestation had serious impacts on the country's environment, particularly on river water quality which was affected by increased turbidity and high sediment loads.

13.6 IRRIGATION

Framji et al. (1981, pp. 222–237) have documented the historical development of irrigation and drainage in China over the last 4000 years, and have described the main features of more recent projects. The maximum irrigable area of the country has been estimated at between 64 and 67 Mha (Fuggle and Smith, 2000, p. 17). In 1949, 16 Mha of land was irrigated, and this has increased to 54.8 Mha by 2001 (FAO, 2003). This significant expansion required major development of the surface and groundwater resources of the country. Initially, yields of the irrigated farms were

---

[2] Also visit http://www.iiasa.ac.at/Research/LUC/ChinaFood/data/maps/precip/pre_0_m.htm (visited in July 2005).
[3] See China, Country Position Paper at the URL http://www.icid.org/index_e.html (visited in July 2005).

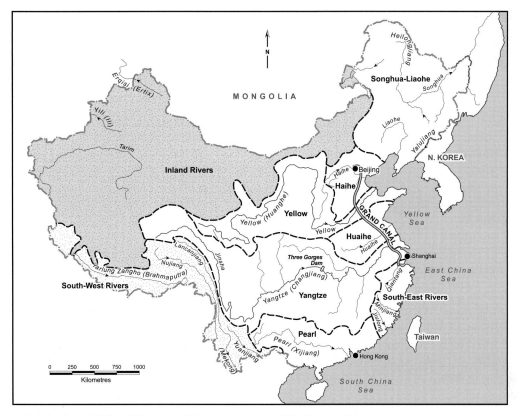

Figure 13.4 Major river basins of China (Ministry of Water Resources, 2002, Figure 11).

very low and the average annual cereal consumption was only about 210 kg per capita. By the end of the twentieth century, the expansion of the irrigated area and improvements in farming practices resulted in increased cereal production to the extent that the average annual cereal consumption was 400 kg per capita.[4] Expansion of the irrigated area has led to significant problems. For example, it was estimated that in the early 1990s, 6.7 Mha were affected by irrigation-induced salinisation (Ghassemi et al., 1995, Chapter 3).

13.7 WATER RESOURCES

13.7.1 Surface Water

On average, $6190 \times 10^9 \, m^3$ of water per year falls on China as precipitations. China has a large number of rivers with a total length of 420 000 km (Ministry of Water Resources, 2004a). There are more than 50 000 rivers, each with a drainage area of over 100 km^2 including more than 1580 rivers with a basin area of over 1000 km^2 and 79 over 10 000 km^2 (James, 1989, p. 73). Most of the rivers are situated in the wetter eastern climatic zone, flowing eastward into the sea, with the major ones including the Yangtze, Yellow,[5] Heilong, Pearl, Liaohe, Huaihe and Haihe. In the drier north-western part of the country there are only a small number of internally draining rivers.

China is divided into nine major river basins (Figure 13.4). The average annual volume of run-off from China's rivers is approximately $2711 \times 10^9 \, m^3$ with an average annual depth 284 mm (Table 13.1), which is equivalent to 44 percent of the precipitation. This means that the total amount of evaporation is approximately $3478 \times 10^9 \, m^3$, which is equivalent to a nationwide water depth of 364 mm (Ministry of Water Resources, 2004a). The Yangtze River Basin, with an average annual run-off of $951 \times 10^9 \, m^3$ or 35 percent of the country's run-off, is the most important river basin in China.

In China there are about 2300 lakes each with a water surface area larger than 1 km^2 (Qian, 1994, pp. 14 and 15). Among them, there are more than 10 large lakes with a water surface area larger than 1000 km^2. Lakes fall into two main categories of freshwater and saline. The total water

---

[4] See China, Country Position Paper at the URL http://www.icid.org/index_e.html (visited in July 2005).
[5] A number of articles analysing various aspects of the Yellow River's water resources are available in *Water International* (2004), Volume 29, No. 4; Zhu et al. (2004); and JBICI (2004).

Table 13.1. *Run-off in major river basins of China*

River basin	Run-off depth (mm)	Run-off volume (10^9 m^3)	Run-off as percentage of national total
1. Songhua-Liaohe	132	165	6.1
2. Haihe	90	29	1.1
3. Huaihe	225	74	2.7
4. Yellow (Huanghe)	83	66	2.4
5. Yangtze (Changjiang)	525	951	35.1
6. Pearl (Xijiang)	807	469	17.3
7. South-east Rivers	1066	256	9.4
8. South-west Rivers	687	585	21.6
9. Inland Rivers	35	116	4.3
Total	284	2711	100.0

Source: Ministry of Water Resources (2004a, Table 1).

surface area of all lakes is 72 000 km^2, and the total storage capacity is 709×10^9 m^3. The freshwater capacity is 32 percent of the total lakes storage, and covers an area of about 35 000 km^2. Salt lakes are found mainly in the drier areas of the country such as the Tibet Plateau, Xinjiang and Inner Mongolia provinces.

In addition to its natural rivers and lakes, China has many canals. The Grand Canal (Figure 13.4), which was first built around AD 470, was an enormous project for its time and was used for the transport of rice and other goods from production areas in the south to Beijing in the north. It is 1794 km long, and is one of the oldest and longest man-made waterways in the world.[6] The other outstanding example is the web of canals in the Yangtze Delta.

The clearing of forests has caused a massive increase in soil erosion, with 3.6×10^6 km^2, or 37 percent of the total area of China being affected (Ministry of water Resources, 2003a, p. 11). The Yellow River is famous for its large sediment load and is now perched up to 10 m above the floodplain because of silt deposition along its lower course. A major source of this sediment is the Loess Plateau, which generates about 3700 tonnes of sediments per square kilometer per annum.[7] The high sediment load of the Yellow River results in serious problems with sedimentation of reservoirs and the river channel, and has affected the severity of flooding.

Dams and Reservoirs

Before 1949, China had 22 large dams with only one exceeding 100 m in height. There had been no strategic planning for dam construction. Since then, development of water resources has been a high priority. Over the period of 1949–99 more than 84 900 reservoirs were constructed with a total capacity of 470×10^9 m^3, including 397 large reservoirs with a total capacity of 327×10^9 m^3 (Fuggle and Smith, 2000, pp. 6–12). China intends to continue constructing large dams over a large part of the twenty-first century for hydro-power generation, flood control, irrigation and domestic water supply.[8] However, international agencies have urged China to develop detailed plans of how social and environmental impacts caused by these new dams can be minimised. China has identified 149 sites for economically feasible large dams, which could provide 224 550 MW of hydro-power and 1700 sites for smaller dams that could provide a further 45 000 MW of hydro-electric energy.

Among the major dams currently under construction is the controversial Three Gorges Dam on the Yangtze River (Figure 13.4), designed for flood control and hydro-power generation. The National People's Congress approved the project in 1992 and its construction began in 1994. This 181 m high gravity dam will have a capacity of 39.3×10^9 m^3, and an installed hydro-electric capacity of 18 200 MW.[9] To provide the equivalent amount of power using coal-fired power stations would require 40 million tonnes of coal

---

[6] http://www.mwr.gov.cn/english1/20040802/38049.asp (visited in July 2005).

[7] Ministry of Water Resources at the URL http://www.mwr.gov.cn/english1/20040827/39304.asp (visited in July 2005).

[8] http://www.mwr.gov.cn/english1/20000922/37422.asp (visited in July 2005).

[9] In 1998, China's total installed electricity generating capacity was 270 000 MW including 44 600 MW (16.5 percent) by hydro-power (Fuggle and Smith, 2000, p. 23). Approximately 78 percent was produced by coal, 4.5 percent by oil and 1 percent by nuclear facilities.

per year. The project cost, based on 1993 values was about US$11 billion,[10] including US$4.8 billion for resettlement of 1.2 million displaced people (Heggelund, 2004, pp. 20 and 21). The project is expected to be completed by 2009. Further information about the Three Gorges Dam project is available in various publications including Barber and Ryder (1993) and Heggelund (2004).

Another controversial dam construction program is underway along the Mekong River. So far China has completed two high dams on the river, two others are under construction and two more are planned (Osborne, 2004). The main objectives are the production of 14 000 MW of hydro-power for internal consumption in China and for export to Thailand, and to boost regional development of the country. These dams will have serious environmental and social impacts in China and the lower Mekong countries, particularly Cambodia and Vietnam. Indigenous communities in the Lower Mekong rely on the river for subsistence.

13.7.2 Groundwater

Groundwater resources of China occur mainly in the following geological formations (Institute of Hydrogeology and Engineering Geology, 1988):

- **Unconsolidated sediments.** In the vast plains in the eastern part of China and in the Inland Rivers Basin, groundwater occurs in the unconsolidated sediments. Aquifers consist mainly of sands and gravel and often appear in multi-layered formations. In the Huang-Huai-Hai Plain, Quaternary sediments are quite well developed. The thickness of sediments varies from 200 to 600 m and reaches a maximum of more than 1000 m in some depressions.
- **Karstic formations.** There are appreciable differences in the distribution of karstic formations between the areas north and south of the Qinling-Kunlun Mountains (Figure 13.1). In the northern area they occur mainly in dolomitised Cambrian and Ordovician rocks which are moderately karstified. In the southern areas, karstic formations occur abundantly in the Upper Paleozoic and Lower Mesozoic carbonate rocks.
- **Fractured rocks.** The exposed bedrock in mountainous areas makes up about two-thirds of the total territory of China. The main rock types include magmatic, metamorphic and clastic rocks. Among them granites are rather widespread, accounting for about one-third of the total exposed area of bedrock over the whole country. Aquifers are mainly of the fractured rock type and locally of the porous type.

Groundwater resources of China are estimated at about $830 \times 10^9 \, m^3$, consisting of $677 \times 10^9 \, m^3$ in mountain areas and $188 \times 10^9 \, m^3$ in the plains, with $35 \times 10^9 \, m^3$ overlapping volume between them (Table 13.2). Groundwater resources in plain areas are mainly in the Inland Rivers Basin, followed by Songhua-Liaohe, Huaihe, Yangtze, Haihe and Yellow River Basins.

Over extraction of groundwater is a significant problem in some parts of China. There are 69 overdraft areas in North China covering 55 000 km^2, with a total overdraft volume of $9 \times 10^9 \, m^3$ per annum (Ministry of Water Resources, 2003a, p. 11). Numerous cities in the North China Plain including Beijing extract large volumes of groundwater for their drinking and other water supply requirements. This has resulted in significant drops in water levels (Han, 1997; Xia et al., 2003). For example, groundwater levels around the Beijing metropolitan area declined by more than 30 m between 1959 and 1999. Shanghai also has been extracting groundwater for its water supply for many decades. As a result of groundwater overdraft, 46 cities in China, including Beijing and Shanghai, are faced with land subsidence, causing damage to buildings, infrastructures, and creating problem for drainage of floodwaters. Groundwater extraction has also decreased base flows in some river system such as the Haihe River, where flows have decreased by nearly 40 percent over the last 40 years.

13.7.3 Total Water Resources

Table 13.3 lists total water resources of China, including both surface (Table 13.1) and groundwater resources (Table 13.2). Approximately 88 percent of the groundwater resource constitutes base flow of the river systems, thus overlapping with surface water resources. The net water resources of the country are $2812 \times 10^9 \, m^3$ or approximately 2200 m^3 per head of population.[11] The per capita volume of water resources in the Haihe, Huaihe and Yellow river basins are lowest at 290 m^3, 478 m^3 and 633 m^3 respectively (Ministry of Water Resources, 2004a), while other river basins such as Yangtze, South-west Rivers, and Pearl are significantly better off.

13.7.4 Water Withdrawal and Use

Water use varies considerably within river basins of China. While the overall use of available water resources in China

---

[10] Considering the inflation, the project is expected to cost US$24 billion in 2009 (Heggelund, 2004, p. 21).
[11] The World Bank definition of a country in water stress is when the availability of water is less than 1000 m^3 per head of population.

Table 13.2. *Groundwater resources of China by river basin (in $10^9 m^3$)*

River basin	Mountain area	Plain area	Overlapping volume	Net volume
1. Songhua-Liaohe	32	33	2	63
2. Haihe	13	18	4	27
3. Huaihe	11	30	1	40
4. Yellow (Huanghe)	29	16	4	41
5. Yangtze (Changjiang)	222	26	2	246
6. Pearl (Xijiang)	103	9	1	111
7. South-east Rivers	56	5	–	61
8. South-west Rivers	154	–	–	154
9. Inland Rivers	57	51	21	87
Total	677	188	35	830

Source: Ministry of Water Resources (2004a, Table 2).

Table 13.3. *Total annual water resources of China by river basin*

River basin	Surface water ($10^9 m^3$)	Groundwater ($10^9 m^3$)	Overlapping volume ($10^9 m^3$)	Net total ($10^9 m^3$)	Water yield ($10^3 m^3/km^2$)
1. Songhua-Liaohe	165	63	35	193	155.6
2. Haihe	29	27	13	43	132.4
3. Huaihe	74	40	17	97	289.5
4. Yellow (Huanghe)	66	41	32	75	93.0
5. Yangtze (Changjiang)	951	246	236	961	534.4
6. Pearl (Xijiang)	469	111	109	471	816.0
7. South-east Rivers	256	61	58	259	1080.8
8. South-west Rivers	585	154	157	582	687.5
9. Inland Rivers	116	87	72	131	38.6
Total	2711	830	729	2812	294.0

Source: Ministry of Water Resources (2004a, Table 3).

is 19.5 percent, in some river basins such as the Haihe and Huaihe, water use has reached very high levels of 93 percent and 63 percent respectively (Table 13.4). In 2000, total water withdrawal was almost $550 \times 10^9 m^3$, including $440 \times 10^9 m^3$ of surface water (80 percent) and $107 \times 10^9 m^3$ of groundwater (19.5 percent). The balance was provided by recycled and collected rainwater.

Out of a total water withdrawal from surface and groundwater supplies of $549.7 \times 10^9 m^3$ in 2000, $373.6 \times 10^9 m^3$ was for agriculture,[12] or 68 percent of the total (Table 13.5). This was followed by $114.2 \times 10^9 m^3$ for industry (20.8 percent) and $61.9 \times 10^9 m^3$ for domestic usage (11.2 percent). The national water consumption in 2000 was $298.5 \times 10^9 m^3$, accounting for 54 percent of the total volume of water withdrawn. The ratio of water consumption over water withdrawal was 68 percent for irrigation, 24 percent for industries, 24 percent for urban domestic use and 88 percent for rural domestic use (Ministry of Water Resources, 2004a).

The trends in China's water withdrawal for the agricultural, industrial, and domestic sectors for the period of 1949 to 2000 are plotted in Figure 13.5. It shows that agriculture has been the major water user, and also demonstrates that the total water withdrawal increased by more than five-fold from 1949 to 1997 (from $103 \times 10^9 m^3$ to $572 \times 10^9 m^3$), while population increased by slightly more than two-fold over the same period. The increase in water withdrawal for the industrial and domestic sectors over the period of 1949 to 2000 were 48- and 103-fold respectively. The per capita water withdrawal, which was

---

[12] Approximately 90 percent of the agricultural water use was for irrigation.

Table 13.4. *Water withdrawal and degree of use in China by river basin in 2000*

River basin	Water withdrawal ($10^9 m^3$)				Water resources ($10^9 m^3$)	Degree of use[b] (%)
	Surface water	Groundwater	Others[a]	Total		
1. Songhua Liaohe	30.8	25.8	0.0	56.6	193	30.9
2. Haihe	12.8	27.0	0.2	40.0	43	93.0
3. Huaihe	42.1	18.9	0.2	61.2	97	63.1
4. Yellow (Huanghe)	25.1	13.5	0.2	38.8	75	51.9
5. Yangtze (Changjiang)	159.4	8.2	0.7	168.3	961	17.5
6. Pearl (Xijiang)	80.5	4.2	0.4	85.1	471	18.1
7. South-east Rivers	30.8	0.9	0.2	31.9	259	12.3
8. South-west Rivers	9.9	0.3	0.1	10.3	582	1.8
9. Inland Rivers	49.0	8.4	0.1	57.5	131	43.9
Total	440.4	107.2	2.1	549.7	2812	19.5

Note: [a] Recycled and collected rainwater; [b] Ratio of total water withdrawal to water resources.
Source: Ministry of Water Resources (2004a, Table 4).

Table 13.5. *Water withdrawal and consumption for major sectors of the economy in 2000*

River basin	Water withdrawal ($10^9 m^3$)				Consumption	
	Agriculture	Industry	Domestic	Total	Volume ($10^9 m^3$)	Rate (%)
1. Songhua-Liaohe	40.8	10.5	5.3	56.6	31.0	55
2. Haihe	28.6	6.2	5.2	40.0	27.9	70
3. Huaihe	44.7	9.3	7.2	61.2	39.3	64
4. Yellow (Huanghe)	29.8	5.5	3.5	38.8	2.0	57
5. Yangtze (Changjiang)	93.3	53.3	21.5	168.3	76.2	32
6. Pearl (Xijiang)	54.9	18.3	11.9	85.1	39.3	46
7. South-east Rivers	18.4	9.0	4.5	31.9	16.5	52
8. South-west Rivers	8.8	0.6	0.9	10.3	7.1	69
9. Inland Rivers	54.0	1.5	1.9	57.4	39.0	68
Total	373.6	114.2	61.9	549.7	298.3	54

Source: Ministry of Water Resources (2004a, Table 5).

$187 m^3$ in 1949, increased to $430 m^3$ in 2000. These increases are the result of population growth, economic development and improvement in living standard.

13.7.5 Water Quality

Surface water and groundwater pollution, due to population growth, rapid urbanisation and industrial development, has become a national issue over the last few decades with serious social, political and economic implications. Untreated domestic and industrial wastewaters have been discharged into waterways. Also, diffuse release of nutrients and agricultural chemicals from farmland is another important increasing source of contamination. In 2000, the total discharge of sewage and wastewater was $62 \times 10^9 m^3$, of which only 24 percent was treated. As a result, 75 percent of China's lakes are now polluted. A survey of drinking water for 118 cities indicated that groundwater has been polluted to some degree in 97 percent of the cities, and 64 percent have a serious groundwater pollution problem (Ministry of Water Resources, 2004a).

Water quality in China is assessed based on a classification of Grade-I to Grade-V, with Grade-I the best and Grade-V the worst. In 2002, water quality monitoring at over 3200 sites along 1300 rivers throughout the country indicated that Grade-I water accounted for 6 percent of river lengths, Grade-II for 33 percent, Grade-III for 26 percent, Grade-IV for 12 percent, Grate-V for 6 percent, and poorer than grade-V for 17 percent (Ministry of Water Resources, 2002, pp. 26 and 27). Overall, 35 percent of river lengths had a rating of

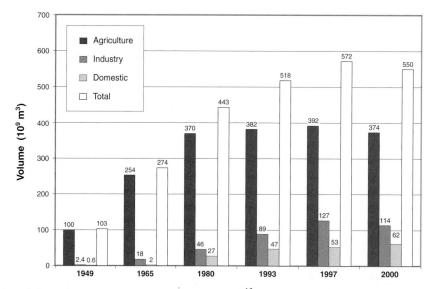

Figure 13.5 Water withdrawal for various sectors from 1949 to 2000 (ICID[13] for 1949 and 1965 data; Fuggle and Smith, 2000, Table 2.5 for 1980, 1993 and 1997 data; and Ministry of Water Resources, 2004a, Table 5 for 2000 data).

Grade-IV or worse, indicating high levels of pollution. The least polluted region was the South-west River Basin with only 7 percent of river length having a rating of grade-IV or worse, while the three most polluted river basins were the Haihe, Huaihe and Yellow, with ratings of grade-IV or worse for 61, 61, and 59 percent of river lengths respectively. The Haihe River Basin is the most polluted river basin in the country.

In China, untreated sewerage and urban wastewaters have been traditionally used by the agricultural sector to save water and reduce the need for fertilisers. In the North China Plain, 80 percent of irrigation water used on farms near big cities is untreated urban wastewater. This has caused serious groundwater pollution by nitrate (Tang et al., 2004; Chen et al., 2004). Further information regarding water quality issues in China is available in a number articles published in *Water International* (2004, volume 29, Number 3, pp. 269–306).

13.8 FLOOD

Zhang and Lin (1992, Chapter 11) provide an analysis of flood and drought records in China over the past 500 years (Research Institute of Meteorological Sciences, 1982). They concluded that a major flood or drought occurs every five years. Approximately two-thirds of China is subject to flooding (Qian, 1994, Chapter 3). A north-east to south-west diagonal line from the north of Heilongjiang Province (Figure 13.2) to the west of Yunnan Province divides China into two different regions. In the eastern region, flood disasters are primarily caused by severe rainstorms and by coastal storm surges. During the monsoon season from May through to August, just a few big rainstorms can contribute up to 50 percent of annual rainfall. The severity of storm floods caused by high intensity rainfall is rarely seen in other countries. Maximum one-hour rainfall of 400 mm, six-hour of 830 mm and 24-hour of 1672 mm are recorded in this region. Such high intensity rainstorms over a large area often form extremely large flood peaks and result in serious flooding. In the western region, glaciers occupy 58 700 km^2. The largest glaciers in Tibet make up 47 percent of the country's total, Xinjiang accounts for 44 percent, and the remaining are in Qinghai, Gansu and other provinces. Many rivers are frozen for a long period in winter. Floods caused by rapid snowmelt and icemelt can have disastrous effects. Rainstorm floods do also occur in this region, but they are generally more localized and with lower intensity.

Qian (1994, Chapter 3) describes flood disasters and control measures for the major river basins in China including the Yellow, Yangtze, Huaihe, Haihe, and Liaohe. Major floods occurred in 1996 and 1998 (Fuggle and Smith, 2000, p. 21). Estimates suggested that the nationwide flood of 1996 affected more than 5 million houses, caused 4400 deaths, impacted 31 Mha of farmland, and caused economic losses of US$27 billion. The 1998 flood in the Yangtze and

---

[13] International Commission on Irrigation and Drainage (ICID), China Country profile at the URL: http://www.icid.org/index_e.html (visited in July 2005).

Heilongjiang river basins affected 7.3 million houses, caused 3656 deaths, impacted 25 Mha of farmland and resulted in losses of US$30 billion. Further information regarding historical floods in China is available in Jiang *et al.* (2002a, 2002b), while Cheng (2005) and GHD (2005) describe the issue of flood management strategies.

13.9 DROUGHT

Droughts lasting from just a few months to several consecutive years frequently occur in China. During the 2155 years from 206 BC to AD 1949, hundreds of droughts occurred in various parts of China (Qian, 1994, pp. 60 and 61). Within the 400 years from the sixteenth to the nineteenth century, there were occasions (1640, 1671, 1679, 1721, 1785, 1835, 1856 and 1877) when very large portions of the country were affected by drought. The 1640 drought lasted for 4–7 years in the affected areas. The 1920 drought extended to five provinces (Shaanxi, Henan, Hebei, Shandong, and Shanxi), 20 million people were affected by famine, and 500 000 people died. A major drought struck all provinces in the Yellow River Basin in 1929, affecting 34 million people. In 1972, a severe drought covered seven provinces of north China. Droughts also occurred between 1978 and 1986, 1997, 2000 and 2002. The severe drought of 2000 hit over 20 provinces, more than 40 Mha of farmland were affected and water shortages occurred in the major cities of north China (Ministry of Water Resources, 2003b, p. 41).

A major drought affected southern China in 2004.[14] In the Guangxi Province, which had its worst drought in 50 years, about 1100 reservoirs dried up and hydro-power generation reduced dramatically. This drought also damaged crops such as rice, sugarcane and vegetables in the neighbouring Guangdong Province. The drought also spread to the northern provinces and authorities declared that 3.3 Mha of crops had been damaged in Guangxi, Guangdong, Hunan and Anhui Provinces.

13.10 CLIMATE CHANGE IMPACTS

Ying (2000) analysed climate change impacts on the water resources of six river basins representing the various climatic regions of China. Up to the year 2030, average annual temperatures in China will increase by 0.9°C to 1.2°C with changes in the south less pronounced than in the north. The average annual precipitation will rise very slightly, but the increase could be as high as 4 percent in the Songhua and Liaohe river basins, while in other regions, such as the Huaihe River Basin, the increase will be almost negligible (0.05 percent). The average annual run-off would decrease by about 10.5 percent in the Huaihe River Basin, which is already suffering water shortages. This decrease is mainly attributed to an evaporation increase of 3.5 percent in the Basin. In the Liaohe and Songhua river basins, the average annual run-off is expected to increase by 6.9 percent and 6.2 percent respectively, which could aggravate their flood problems (Ying, 2000). Agricultural water demand is expected to decrease in south China, and increase in north China (Tao *et al.*, 2003). Crop production in the North China Plain and north-east China could therefore face serious water related challenges in the coming decades. Further information regarding climate change and its impact on water resources in China is available in Gao and Huang (2001), Xia and Zhang (2005), Ye *et al.* (2005), Yang *et al.* (2003), Yang *et al.* (2005), Fu and Chen (2005), and Wang and Xia (2005).

13.11 SUSTAINABLE WATER RESOURCES DEVELOPMENT

China's environment has been severely degraded because of its rapid economic growth, massive expansion of industry, increasing demand for natural resources and forest clearing. Many rivers are badly polluted, groundwater resources are being used at unsustainable levels, air pollution and acid rain are widespread problems, and large areas of forests have been logged. China's long-term economic prosperity cannot be maintained unless there is concerted effort to protect the environment and use natural resources at more sustainable levels. Numerous publications such as World Bank (1997), Asian Development Bank (2000), Edmonds (2000), Smil (2004), Economy (2004), and Ma (2004) describe these issues in detail.

Water resources are at particular risk, and China is faced with numerous challenges including: providing water supplies for rapidly expanding urban areas, growing industries and modernising agricultural sector; food supply security; raising people's living standard; providing flood control and drought relief; regional development; poverty alleviation; supply of electricity requirements through hydro-power generation; and ensuring environmental protection.

---

[14] http://www.chinadaily.com.cn/english/doc/2004-11/04/content_388648.htm and http://www.china.org.cn/english/2004/Dec/113911.htm (both visited in July 2005).

The Chinese Government is attempting to face these challenges by modernising its procedures, legislations, and institutions, and implementing a sustainable water resources development strategy (Ministry of Water Resources, 2003b).

Until the late 1980s the main focus of China's water resources policy was the construction and management of dams and other structures for flood control, irrigation, water supply, and drainage facilities. The first national Water Law was approved in 1988. This was followed by a large number of environmental management laws including the 1989 Environmental Protection, the 1991 Water and Soil Conservation, the 1994 Water Pollution Control, and the 1997 Flood Control (Fuggle and Smith, 2000, pp. 32–52).

The 1988 Water Law was the fundamental law for water management in China. Its aims were to coordinate and standardise all activities for development, utilisation and protection of water resources, harnessing of rivers, and management of water-related disasters. It consisted of seven chapters and 53 articles.[15] In brief it (Fuggle and Smith, 2000, p. 39):

(1) declared that surface and groundwater resources are the property of the State;
(2) established that the State shall take effective measures to protect and preserve water resources and the ecological environment;
(3) proposed a system of centralised and integrated water resources management;
(4) required preparation of basin wide and regional plans;
(5) addressed problems of groundwater extraction;
(6) identified domestic, irrigation, industrial, navigation water use priorities, and specified needs for multi-purpose development;
(7) established a water permit system for water diversion and extraction;
(8) required water charges for water use; and
(9) payed considerable attention to flood prevention measures.

In terms of the institutional reform, seven River Basin Commissions have been set up under the overall structure of the Ministry of Water Resources for the Yangtze, Yellow, Huaihe, Haihe, Pearl, Songhua-Liaohe, and Taihu (Lake).[16]

The National People's Congress approved the revised Water Law of China on the 29 August 2002. The law became effective on 1 October 2002. The 2002 Water Law is expected to change water resources management from a traditional exploitative approach to a modern and sustainable one. In particular, it should speed up development of water conservation measures, help reduce water pollution, and support economic and social development of the country (Ministry of Water Resources, 2002, pp. 30 and 31). The new Water Law was developed based on more than 10 years experience with the 1988 Water Law and on water resources legislation in other countries. The 2002 Water Law will (Ministry of Water Resources, 2002, pp. 30 and 31; Shao et al., 2003):

(1) intensify the integrated management of water resources;
(2) establish river basin management agencies;
(3) promote water saving and water use efficiency;
(4) optimise allocation of water resources;
(5) stipulate state-owned water rights;
(6) implement water pricing according to local conditions and level of water use; and
(7) stress the relationships between water resources, economy, society and the environmental protection in the development and use of water resources.

13.12 WATER CONSERVATION

Inadequate water resources management in China has resulted in serious water shortage, water pollution and environmental degradation. To address these issues, authorities have launched a series of national campaigns to urge people to save water. In the irrigated areas, water consumption dropped from $8000\,m^3\,ha^{-1}$ in 1993 to $6500\,m^3\,ha^{-1}$ in 2000. The Chinese Government established the National Water Saving Office in 1980 (Ministry of Water Resources, 2003b, p. 23).

The publication of the first guideline on the development of water-efficient technology marks a significant step forward in China's effort to tackle its water shortage. The guideline, jointly released by five government departments,[17] consists of five chapters and 146 articles. It is an attempt to cap the growth of water use for agricultural purposes and reduce the amount of water use in industrial production and domestic use between 2005 and 2010, through the broad application of water-efficient technology.[18]

---

[15] The chapter titles were: (1) General provisions; (2) Development and utilisation; (3) Protection of water resources, river basins and water related projects; (4) Management of water utilities; (5) Prevention and fight against floods and inundations; (6) Legal responsibilities; and (7) Supplementary provisions.
[16] The Yellow River Basin Commission was originally set up in 1918, the Yangtze in 1922 and the Pearl in 1929.
[17] The National Development and Reform Commission, Ministry of Science and Technology, Ministry of Water Resources, Ministry of Construction, and Ministry of Agriculture.
[18] China Daily at: http://www.chinadaily.com.cn/english/doc/2005-05/25/content_445465.htm (visited in July 2005).

Water pricing is another measure promoted by the government. Most water consumers were not charged for their water use prior to 1985. Because of this, very few corporations or farms were compelled to invest in water treatment and water recycling technology. Over the past 20 years, prices have risen very slowly and remain among the world's lowest. In general, water is purchased at about 40 percent below cost.[19] Although authorities have raised water prices to encourage water saving in recent years, it has so far failed to deliver the desired results, because current pricing does not cover the real cost of water, particularly for the big users. An adequate pricing mechanism is required to generate the economic incentives needed to persuade all users to conserve water.

SECTION B: INTER-BASIN WATER TRANSFER PROJECTS

13.13 INTRODUCTION

Although China has an average annual water resources of 2812×10^9 m^3 or approximately 2200 m^3 of water per head of population, its water resources are unevenly distributed in space and time. Before 1949, many small-scale water diversion projects were undertaken. In 1961, Jiangsu Province constructed a major diversion project through installation of a pumping station on the lower Yangtze River and using the Grand Canal to transfer water northward (Liu, 2000). Since then, other water diversion projects (mostly inter-basin) have been developed or are under development to supply the increasing demands of urbanisation, industries and the agricultural sector (Shao et al., 2003). Figure 13.6 and Table 13.6 show the locations and characteristics of these projects.

The heavily populated and polluted Haihe, Huaihe and Yellow (Huanghe) river basins have been faced with serious shortages of water in recent decades for their domestic, industrial and agricultural water supplies. Due to a combination of excessive water withdrawal and drought, these rivers have, on a several occasions, dried up. For example, in 1997, the Yellow River failed to flow to the sea for 226 days, with the dry stretch reaching roughly 700 km inland (Liu, 2000). To overcome the problem of water shortage, the South to North Water Transfer Project (projects 12, 13 and 14 in Table 13.6) has been developed. The following sections describe this project and its implications. A detailed description of water transfer from the Yellow River to Taiyuan City (project no. 10 in Table 13.6) is available in Xie et al. (1999).

13.14 SOUTH TO NORTH WATER TRANSFER PROJECT

The Huang-Huai-Hai area, which encompasses the Huanghe (Yellow), Huaihe and Haihe river basins, is a major political, economic, and cultural centre in China. It supports 35 percent of China's population, generates 35 percent of the GDP, contains 36 percent of the cultivated land and 37 percent of grain production (Ministry of Water Resources, 2004b), but, has only about 7.5 percent of the country's water resources (Table 13.3).

Approximately 80 percent of the annual precipitation in the Huang-Huai-Hai area normally falls during the period of July to October, and dry years occur frequently. Historically, there was a dry period of 11 successive years from 1922 to 1932 in the Yellow River Basin (Ministry of Water Resources, 2004b).

In 2000, the total water withdrawal in the area was 140×10^9 m^3. Some 74 percent of water withdrawal was for agricultural activities, with 15 percent for industry and 11 percent for domestic use (Table 13.5). Surface water supplied 57.2 percent of water requirements, groundwater 42.2 percent and others (rainwater and recycled water) 0.4 percent (Table 13.4). The annual rate of groundwater overdraft in the area has reached 6×10^9 m^3 which is 66 percent of the country's total volume of overdraft. Subsequently, groundwater levels have dropped from their initial depth of 2–3 m to 10–31 m (Ministry of Water Resources, 2004b), and land has subsided in numerous cities including Beijing and Tianjin, damaging buildings and infrastructure.

Analysis of water supply and demand indicate that currently there is a shortage of $14-21 \times 10^9$ m^3. It is estimated that this shortfall will reach $21-28 \times 10^9$ m^3 by 2010 and $32-40 \times 10^9$ m^3 by 2030 (Ministry of Water Resources, 2004b). In the Haihe River Basin, the water shortage will be $10-12 \times 10^9$ m^3 by 2010.[20] To address this shortfall it is planned to increase water saving measures and implement the South to North Water Transfer Project (SNWTP). This Project is a strategic and ambitious

---

[19] http://news.mongabay.com/2005/0531-tina_butler.html (visited in July 2005).

[20] A detailed analysis of water related issues (surface water, groundwater, water withdrawal, water pollution, water demand, water reuse, water resources management, etc.) in the Hunag-Huai-Hai area is available in the World Bank (2001).

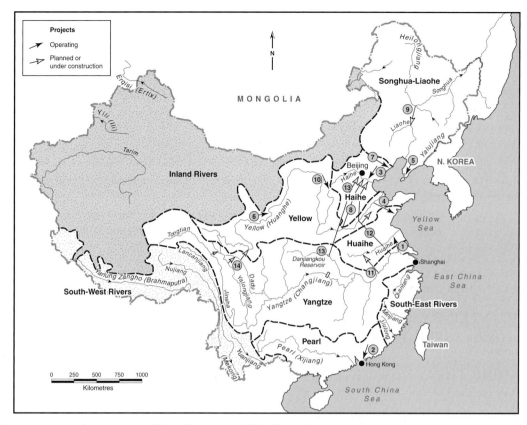

Figure 13.6 Major water transfer projects in China (Shao *et al.*, 2003, Figure 3).

approach to resolve water shortage problems in the North China Plain and overcome the continuous drying up of the Yellow River at its mouth (Liu and Zheng, 2002).

Chairman Mao Zedong first proposed the SNWTP in the early 1950s when he inspected the Yellow River Basin in 1953 and suggested transferring water from the Yangtze River to the Yellow River (Liu and Zheng, 2002). In the late 1950s a number of investigations were undertaken in the upper reaches of the Yangtze River and the southwestern rivers on the feasibility of transferring water to the upper catchment of the Yellow River. These resulted in numerous preliminary proposals. In 1972, a severe drought covered seven provinces of North China and prompted Premier Zhou Enlai to suggest a scheme to divert lower Yangtze River water to the North China Plain. Economic development and population growth have greatly increased water consumption in the North China Plain since the 1980s and have accelerated the development of plans for the SNWTP and its three sub-projects. These are: Eastern Route, Middle Route and the Western Route Schemes (Figure 13.7). The State Council approved construction of the Eastern Route in March 1983. But the project was suspended because of impasses in negotiations between Jiangsu and the northern provinces and municipalities of Shandong, Hebei, Beijing and Tianjin. In particular, Jiangsu Province wanted to keep more of the diverted water than was allocated under the approved plan (Liu, 2000).

The State Council approved the overall plan for the South to North Water Transfer Project submitted by the State Development and Reform Commission and the Ministry of Water Resources on 23 December 2002 (Ministry of Water Resources, 2002, pp. 31 and 32). Approval was for construction of the Eastern, Middle and Western Routes Schemes. These Schemes will divert water from lower, middle and upper reaches of the Yangtze River Basin respectively to meet the increasing water requirements of the North China Plain. By 2050, the annual total water transfer will be about $45 \times 10^9 \, m^3$ (Shao *et al.*, 2003). The Eastern and the Middle Route Schemes will be built first. The opening ceremony for the construction work took place in Beijing and at the construction sites in Jiangsu and Shandong Provinces on 27 December 2002. The cost of the Eastern and the Middle Routes is estimated at US$19 billion, while the total cost of all three Routes is estimated at about US$56 billion. At full development, the Eastern and the Middle Routes will divert $28 \times 10^9 \, m^3$ of water from the Yangtze River, representing less than three percent of its average annual flow of $951 \times 10^9 \, m^3$.

Table 13.6. *Selected characteristics of the water transfer projects in China*

No.	Transfer From (river)	To (river/city)	Diverted volume (10^9 m^3 yr^{-1})	Length of transfer (km)	Number of reservoirs	Number of pump stations	Year of completion
1	Yangtze	Huaihe	–	400	–	21	1961
2	Dongjiang	Hong Kong	0.62	83	2	8	1964
3	Luanhe	Tianjin	2.00	286	4	0	1982
4	Yellow	Tsingdao	0.64	262	1	3	1986
5	Biluhe	Dalin	0.13	150	2	5	1982
6	Datonghe	Yongden	0.40	70	–	0	1980
7	Qinglong	Qinhuangdao	0.17	63	1	0	1989
8	Yellow	Baiyangdian	1.25	779	–	–	Planned
9	Songhua	Liaohe	4.40	656	7	2	Planned
10	Yellow	Taiyum	1.40	453	1	5	2005
11	Yangtze	Huaihe	–	269	–	2	Planned
12[a]	Yangtze	Yellow	15.00	1150	10	23	Under construction
13[b]	Yangtze	Huaihe, Haihe	13.00	1267	40	0	Under construction
14[c]	Yangtze	Yellow	17.00	700	7	10	Planned

Note: [a] The Eastern Route Scheme; [b] The Middle Route Scheme; [c] The Western Route Scheme.
Source: Shao *et al.* (2003, Table 3).

While implementation of this expensive project is expected to alleviate water shortage problems in the North China Plain, its success will depend on implementation of better management of surface and groundwater resources, water conservation, adequate water pricing, pollution control and wastewater reuse (World Bank, 2001).

13.14.1 The Eastern Route Scheme

The Eastern Route will use the existing Grand Canal as the main transfer channel, with four natural lakes (Hongze, Luoma, Nansi and Dongping) and several planned reservoirs. Water would be pumped in several stages to the Yellow River crossing (Figure 13.8). The crossing will be through two 8.7 km long and 8.3 m diameter tunnels beneath the riverbed (China Internet Information Centre, 2002). This will be the most technically challenging part of the Route. After the river crossing, water flows by gravity, mainly along the Grand Canal, and will finally reach Tianjin City (Liu and Zheng, 2002). The water transfer capacity of the Scheme will be 15×10^9 m^3 per annum, and will supply water to Jiangsu, Anhui, Shandong, and Hebi Provinces and Tianjin Municipality (Figure 13.2).

The total length of the main canal will be 1150 km. Since the water level of the Dongping Lake (Figures 13.7 and 13.8) is higher than that of the Yangtze River, water will have to be pumped in stages through a series of pumping stations (Figure 13.8). The designed inflow capacity to Hongze Lake is 1000 m^3 s^{-1}, and the northern flow will gradually decrease to 180 m^3 s^{-1} at Tianjin. The capacity of reservoirs along the transfer route is 8.7×10^9 m^3, consisting of 7.5×10^9 m^3 in the south and 1.2×10^9 m^3 in the north of the Yellow River (Liu and Zheng, 2002). The estimated cost of the three phases of the Scheme is about US$9 billion. Compared with the Middle Route, construction of the Eastern Route is relatively simple, its operation is less risky and the management more flexible. However, poor water quality along the Route poses a problem. Although the water at the source is good quality, it deteriorates northward, especially in the lakes and channel segments north of the Yellow River, where lowland regions are polluted by wastes from various sources (Liu and Zheng, 2002).

Construction of the Eastern Route commenced on 27 December 2002, and will be built in three phases (US Embassy, 2003):

- **Phase I** is currently under construction and is expected to be completed by 2008. It consists of upgrading and extending the Grand Canal and other existing infrastructures in Jiangsu Province, so that water can be pumped from Jiangdu City on the Yangtze River as far north as Dezhou City in northern Shandong Province.
- **Phase II** will enable the water to flow underneath the Yellow River to Tianjin and Beijing by 2012.

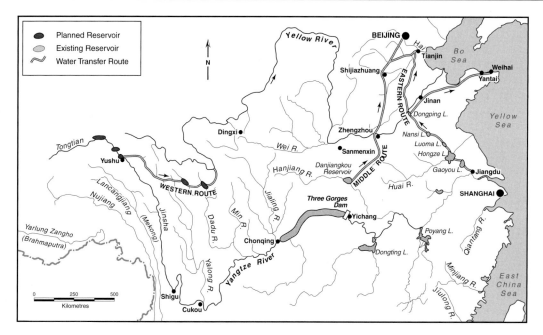

Figure 13.7 The Eastern, the Middle and the Western Routs (Liu and Zheng, 2002, Figure 1).[21]

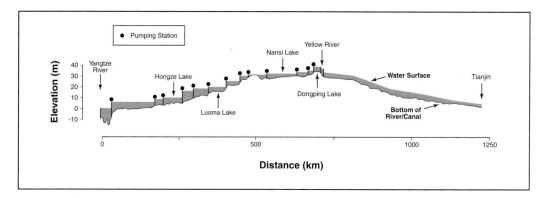

Figure 13.8 Profile of the Eastern Route (Berkoff, 2003, Figure 4).

- **Phase III** consists of constructing a 690 km branch line which will transfer water from Dongping Lake eastward to Jinan, the capital city of Shandong Province and onward to Yantai and Weihai on the shores of the Bo Sea (Figure 13.7).

13.14.2 The Middle Route Scheme

The Middle Route will divert water from Danjiangkou Reservoir[22] on the Hanjiang River (a major tributary of the Yangtze River) into a large canal that will deliver the water to Hubei, Henan, and Hebi Provinces and ultimately to Beijing and Tianjin Municipalities (Figure 13.7). The average annual capacity of the Scheme will be $13 \times 10^9 \, m^3$. The project requires the existing Danjiangkou Dam to be elevated to 176.6 m above sea level from its present level of 162 m, and the normal reservoir level raised to 170 m from the present level of 157 m. This will increase its capacity to $29 \times 10^9 \, m^3$. The total length of the diversion canal will be 1267 km with a 142 km branch to Tianjin, and will have a maximum design capacity of $800 \, m^3 \, s^{-1}$ (Shao et al., 2003). The water will pass under the Yellow River through a 7.2 km long tunnel (Berkoff, 2003). This scheme will supply water to the middle and western parts of the Huang-Huai-Hai Plain.

The limited supply of water from the Hanjiang River (Figure 13.7) means that the Scheme cannot meet all water

---

[21] Also visit the URL http://www.cawra.org/PDFtext/SouthToNorth WaterTransfer.pdf (visited in July 2005).

[22] The Danjiangkou Dam was built in 1974. The reservoir has an annual natural inflow of $41 \times 10^9 \, m^3$ from a drainage area of 65 217 km^2, and a current capacity of $17.5 \times 10^9 \, m^3$ (Liu and Zheng, 2002).

requirements of the supply area. It will mainly provide water for over 20 large- and medium-sized cities such as Beijing, Tianjin and others. It will also supply agricultural and environmental water requirements in its supply area. Construction of the Scheme started in 2003 and is expected to be completed by 2012 at a cost of US$10 billion, consisting of US$7 billion for construction and US$3 billion for resettlement and other costs (Berkoff, 2003).

Unlike the Eastern Route, an important advantage of the Middle Route is that the transferred water flows northward entirely by gravity (Figure 13.9). However, rainfall in the Hanjiang River Basin is highly variable and during low rainfall periods, similar to those experienced in 1965–66 and 1991–95, there will not be enough water available for transfers to occur (Smil, 2004, p. 167). In the long-term, building a connection between the Danjiangkou and the Three Gorges Reservoirs on the Yangtze River could solve this problem. Such a scheme would require pumping facilities from the Three Gorges Dam Reservoir to the Danjiangkou Reservoir.

Construction of the Middle Route will be affected by several geological problems, mainly the slope stability of swelling clay and rock, and land subsidence in coal mining areas (Wang and Ma, 1999). Approximately 160 km of the Route have problems related to swelling soils and unstable rock formations. Some 51 km of the transfer channel pass through seven coal mine areas, and land subsidence and ground fissures are expected along these parts of the Route. Other problems include accumulation of floodwater on the western side of the channel, and seismic activity. All these problems are being addressed in the planning of both the construction and operation phases of the project.

13.14.3 The Western Route Scheme

The Western Route will divert water from three major headwater tributaries of the Yangtze (Tongtian, Yalong and Dadu rivers) to the Yellow River (Figures 13.6 and 13.7). It consists of three sub-routes of tunnels through mountain ranges of the eastern Tibet, western Qinghai, and Sichuan provinces (Figure 13.2). Water diversion from the Tongtian River to the Yalong River will be by gravity, requiring a 302 m high dam on the Tongtian River. The tunnel from Tongtian River to Yalong River will be 158 km long. Diversion from the Yalong River to the Dadu River will be by gravity requiring a 175 m high dam on the Changxu reach of Yalong River and a tunnel with a length of 131 km. Diversion from the Dadu River to the Yellow River will require pumping and a 296 m high dam on the Dadu River. Then, water will be lifted 458 m to Jiaqu River, a tributary of the Yellow River, through a 28.5 km tunnel (China Internet Information Centre, 2002). Dams and tunnels will have to be built in areas that are mountainous and not easily accessible, with elevations ranging from 3000 m to 5000 m. Much of the area is frozen for large parts of the year, it has a very complex geological structure, and is prone to frequent earthquakes (China Internet Information Centre, 2002).

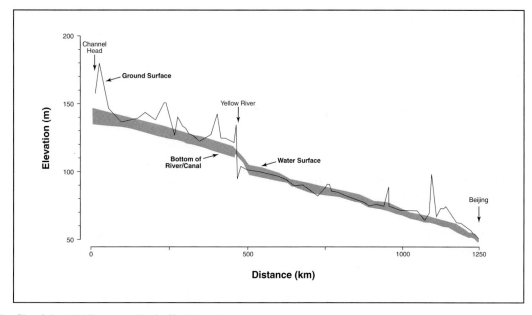

Figure 13.9 Profile of the Middle Route (Berkoff, 2003, Figure 3).

The Western Route will divert $17 \times 10^9 \, \text{m}^3$ per annum (Shao et al., 2003). Waters will be delivered to Qinghai, Gansu, Shaanxi and Shanxi and the Inner Mongolia Provinces (Figure 13.2). Compared with the Eastern and Middle Routes, the Western Route is expected to provide the best quality water and there is little need for displacement of people. However, the Western Route will face much greater engineering and construction challenges. Preliminary cost estimates exceed US$ 37 billion. The difficult geological and other conditions of the Route and its high cost, mean that it is unlikely that construction of the Western Route will begin before 2020, and there is a possibility that it may not be constructed at all (Liu and Zheng, 2002).

13.14.4 Finances for the Project

The SNWTP will be built as part of the infrastructure of the national economy, with the three routes running through a number of provinces that have their own administrative powers and economic interests. Construction of the main infrastructure, including reservoirs and main diversion channels, will be financed through the establishment of a construction fund to cover the construction, interest and maintenance costs. Each province wishing to purchase water rights will have to contribute to the fund and provinces receiving the most water will have to pay the greatest share (Shao et al., 2003). Provinces will be able to recover these costs by charging water users for the volume of water withdraw. The order of priority for water allocations will be:

(1) domestic water use for urban centres;
(2) industrial water use;
(3) agricultural requirements; and
(4) instream flow and ecological needs.

It is planned that the system will have enough flexibility to transfer water from one sector to another if required. During the transitional period from a planned economy to market economy in China, a quasi-market type of management will be adopted for the process of pricing and the transfer of water rights (Shao et al., 2003).

Although the Central Government is playing the key role in the planning, construction and management of the SNWTP, it will supply only 30 percent (US$4.5 billion) of the capital cost for construction of the first phase of the Eastern and the Middle Routes with a recently estimated total cost of US$15 billion. Another 30 percent (US$4.5 billion) is expected to come from water charges[23] in provinces that will benefit from the project. The final 40 percent (US$6 billion) will come from bank loans.[24]

On 25 June 2005, the Vice Minister of Water Resources announced that the price of water transferred to Beijing through the SNWTP would be US$0.86 per m^3.[25]

13.14.5 Is the Project Justified?

Authorities in China and the World Bank (2001) concluded that the SNWTP is justified. In contrast, the World Wildlife Fund[26] (2001) has argued that it is not worth the cost because improved water management would deliver more water as well as considerable environmental benefits. The World Wildlife Fund estimated that approximately $44-55 \times 10^9 \, \text{m}^3$ could be saved by investing in pollution control and water saving measures. This could be achieved by reducing leakage from urban distribution systems, improving the collection and treatment of wastewater for downstream reuse, treatment of industrial wastewater for recycling, discouraging irrigated agriculture in arid regions and reallocating this water to municipal and industrial uses, improving irrigation practice, encouraging export of grain from regions where rainfall provides at least part of the crop water requirments, and the use of desalination technology. Berkoff (2003) argued that the alternatives suggested by the World Wildlife Fund (2001) would be unlikely to provide a satisfactory solution because:

(1) the contribution from improved efficiency is likely to be much less than what was claimed;
(2) reallocating water from irrigation to municipal and industrial uses raises significant socio-political problems; and
(3) desalination remains a possibility only for some priority uses in coastal cities, and can be ruled out for large-scale use.

Berkoff (2003) assessed the economic analysis of the two above reports and concluded that little confidence can be placed in either. He placed the project in a broader regional

---

[23] Charges will take into account depreciation, maintenance, management and operation costs (Liu, 2000).
[24] A Consortium of seven Chinese Banks led by the National Development Bank is providing loan for construction of the Eastern and the Middle Routes because they have huge amount of saving from Chinese people. On 29 March 2005, they signed an agreement with the legal representatives from four units undertaking the construction works. Although the World Bank has some involvement in the related projects such as wastewater treatment, soil conservation and water supply, so far there has been no indication that it has been involved in financing the principal construction works of the Eastern and the Middle Routes (Professor Shao, personal communication, July 2005).
[25] Professor Shao, personal communication, August 2005.
[26] Currently Worldwide Fund for Nature.

and agricultural development context, and concluded that, irrespective of the economic viability of the scheme, the pace and scale of socio-economic change in China are without precedent. The adjustment problems on the North China Plain are greatly exacerbated by water scarcity. Also, reallocation of water from irrigation to municipal and industrial uses, or to the environment, is socially divisive and in some instances physically impractical. The SNWTP would greatly alleviate socio-economic and environmental problems being faced in the region. Therefore, it is the political and pragmatic arguments rather than those based on economic profitability or environmental impact that justify the Government's decision to proceed with the project.

13.14.6 Environmental Impacts

Possible environmental impacts of the SNWTP have been investigated for more than 20 years. Biswas *et al.* (1983) presented an early analysis of the impacts of the Eastern and the Middle Routes on the natural environment of both the water exporting and importing regions. These included the impact on the surface water, groundwater, water balance, aquatic life, saltwater intrusion, and land salinisation. More recent reports and publications are in the Chinese language. The SNWTP will transfer water from a region with water surplus to a region with a water deficit, and this can yield important environmental benefits to the areas receiving the extra water, such as alleviating the drying up of the lower Yellow River during drought periods, diluting pollution, improving riverine ecology, and ameliorating groundwater depletion and associated land subsidence.

Between the Eastern and the Western Routes, the Eastern Route will have the least environmental impacts. The potential impacts are (US Embassy, 2003):

- Although the annual volume of water diversion for the Eastern and the Middle Routes is less than three percent of the average annual flow of the Yangtze River, it will still affect downstream ecosystems of this river. There will be less sediment delivery needed to maintain riparian and coastal wetlands. Pollutants also will be marginally less diluted, increasing their concentrations in the lower Yangtze River.
- As water flows north along the Eastern Route, it will be impounded in four major lakes along the Route. The introduction of chemicals and biota from different ecosystems may affect these lakes and their surrounding ecosystems. On the other hand, more abundant water for wetlands adjacent to the lakes will help these ecosystems.
- Agricultural run-off, sewage, factory waste, river transport pollution, and intensive aquaculture already heavily pollute the existing waterways along the Eastern Route. The worst pollution is in the Haihe River Basin (see Section 13.7.5), where virtually every river and stream is polluted. Water moving through watersheds in the Eastern Route will dilute the high concentrations of pollutants. This is considered to be a significant benefit of the project. However, the Eastern Route requires a huge investment in the construction of treatment facilities. At least 13 major new facilities are planned to be built. Shandong Province alone will spend over US$1 billion on wastewater treatment plants. Local authorities along the Route will also be required to enforce water pollution regulations more strictly.

Diversion of a substantial amount of water from the Hanjiang River to the Middle Route Scheme will reduce flows in the downstream section of the river and may lead to the worsening of the existing eutrophication outbreaks, which occurred in 1992, 1998 and 2000 (Shao *et al.*, 2003).

The rapid progradation of the Yangtze River estuary began about 2000 years ago because of an increase in riverine sediments (Yang *et al.*, 2001). It has grown on average, at a rate of $5\,km^2$ per year, and the shoreline at different locations has migrated seaward at a rate of about $17-50\,m$ per year. The tidal flat at the forward end of the Yangtze River estuary could be radically changed due to the reduction in riverine sediment supply caused by the operation of the Three Gorges Dam and the SNWTP.

Saltwater intrusion in the estuary of the Yangtze River has natural daily and seasonal variations. Daily variations are related to diurnal tidal effects. Seasonal variations are due to seasonally dependent river flow rate. The period of low salinity is in the rainy season and the high river flow from May to October, with minimum salinity in July. High salinities occur in the dry season from November to April, with the highest salinity in February (Biswas *et al.*, 1983, Chapter 25). Saltwater intrusion can extend about $100\,km$ upstream during low flow periods. Operation of the SNWTP and the Three Gorges Dam will lead to a slightly longer saltwater intrusion up the Yangtze River during the dry season, but the effect is predicted to be negligible for the rest of the year (Shao *et al.*, 2003). Saltwater intrusion may affect the Yangtze Delta fisheries, urban water supply and the intake of the Eastern Route itself. To mitigate these effects, engineers and researchers have recommended that water withdrawal to the SNWTP should not occur when the Yangtze flow rate falls from its average annual flow rate of about 30 000 m³ s⁻¹ to below the critical level of 8000 m³ s⁻¹ (Liu, 2000).

In the 1980s, there was some concern about the parasitic disease schistosomiasis. It was suggested that the SNWTP might allow the parasitic host, a water snail, to migrate northward into the waterways and irrigation areas of the North China Plain. However, experiments have shown that the water snail has a northern migration limit of 33° 15′ N due to cooler water temperatures in winter (Biswas et al, 1983, p. 164). Increased soil salinisation also does not appear to be a problem. Improvements for flood protection have increased drainage capacity from fields and channels. Also, the drilling of more than 1.6 million tube wells in the North China Plain has resulted in a fall in the watertable, virtually eliminating the possibility of soil salinisation (Liu, 2000).

13.14.7 Resettlement and Social Issue

Construction of the SNWTP will require resettlement of about 275 000 people, mostly related to the Middle Route. Increasing the height of the Danjiangkou Dam from 162 m to 176.6 m above sea level requires relocation of 225 000 residents in the reservoir area. Construction of the Eastern Route Canal would displace a further 50 000 people (Berkoff, 2003). Resettlement of the displaced people would cost about US$3 billion. Compensation payment is required, not only for the displaced people, but also for the agricultural, industrial, mining, and commercial enterprises that are affected. Although each displaced resident is expected to receive US$5000 in compensation, which is roughly equivalent to six years wages for a central China farmer, the Danjiangkou resettlement is likely to become the most controversial aspect of the entire SNWTP (US Embassy, 2003).

Elsewhere in Asia, there has been significant criticism of major water infrastructure projects (Kaosa-ard, 2003). These criticisms have centred on the fact that the impacts of major water infrastructure projects fell inequitably on rural poor. The benefits of these projects are claimed to flow to the rich, widening the gap between rich and poor. In addition, resettlement of rural communities to urban fringes has increased social problems. In the Mekong River Basin, six factors have been identified, which unless addressed, will cause poverty to increase (Kaosa-ard, 2003, pp. 103–105):

- Continuing deterioration of natural resources on which poor depend.
- Conflict between national laws and customary rights.
- Spending of public resources on large infrastructure projects rather than on social investment.
- Institutional failures, and conflicting jurisdictions and agendas between government agencies.
- Reluctance to empower community participation in development and natural resources management.
- The internal inability to transfer the opportunities of globalisation to the poor and its relation to property rights.

While these factors may be relevant in the lower Mekong countries, it remains to be seen whether China's socialist market economy will permit the same problems to persist.

13.15 ACTION PLAN FOR THE NORTH CHINA PLAIN

Although the supply of more water to the North China Plain will alleviate water shortages in the area, it will not be a long-term solution. The World Bank (2001) has proposed an Action Plan covering various aspects of water resources. The recommendations include structural changes, policy changes and institutional changes, which if applied in the proposed timeframe and with a sufficient degree of political and financial commitment, will ensure that limited water resources do not impede China's development. These recommendations can be summarised as (World Bank, 2001, Chapter 11):

Water scarcity and water resources. The Action Plan recommended two measures to reduce water consumption to a feasible level:

(1) increasing water prices for all water users so that revenues reflect the full costs of supply; and
(2) undertaking a series of measures for increasing water use efficiency in all sectors.

However, demand management alone will not be able to reduce consumption to meet the current supply level. The Action Plan therefore recommended augmenting supplies by construction of the SNWTP and by increasing reuse of treated wastewater.

Agriculture. The changes occurring in China's economy will continue to reduce the relative importance of the agricultural sector's contribution to GDP. However, it is predicted that agriculture itself will continue to grow over the coming decades and will undergo a shift from grain production to higher value crops. The Action Plan recognised the important social and political (including food security) implications of maintaining adequate water supplies to rural areas for agriculture. At the same time, the cost to the government of supplying large volumes of water to low value-adding agricultural activities is no longer sustainable. Thus the Action Plan recommended development of innovative models for management of irrigation districts including formation of user associations, water supply companies and self-financed irrigation and drainage districts. It also recommended water saving measures such as:

(1) improvement of canal and on-farm irrigation and drainage systems;
(2) agronomic measures; and
(3) innovative irrigation management practices.

Water pollution. Surface and groundwater quality has been declining over the last three decades to an alarming level. The Action Plan recognised that water pollution is not only a threat to public health and the environment, but also diminishes the total resource available to water users. The Action Plan recommended structural and non-structural investments and programs to enhance existing government efforts to improve water quality. Structural options to reduce industrial pollution include industrial wastewater treatment, cleaner production technology, and reuse of treated wastewater. Pollution from urban sources should be addressed by construction of wastewater treatment plants. Pollution from livestock is another major problem and the Action Plan recommended appropriately designed stabilisation ponds. The non-structural part of the Action Plan includes review of water quality and development of standard methods for laboratory analytical methods, review of monitoring programs, review of regulatory systems, inclusion of environmental impact assessment processes for small industry planning, and review of the pollution permit allocation system.

Groundwater. The alarming decline in watertables and piezometric heads in many parts of the North China Plain is symptomatic of excessive withdrawals, which in turn are due to excessive pollution of surface water resources, general water scarcity, low or non-existent prices for groundwater, and ineffective or non-existent regulatory mechanisms to control withdrawals. The key actions recommended by the Action Plan are:

(1) definition of groundwater management units with determination of sustainable yields;
(2) preparation of groundwater management plans;
(3) allocation licensing linked to sustainable yield;
(4) licensing of well construction drillers;
(5) development of a national groundwater database; and
(6) preparation and implementation of a groundwater pollution strategy, including provision in selected cities for recharging aquifers by spreading of treated wastewater effluents or floodwaters on permeable surface areas, and for injection of adequately treated effluents into the aquifer systems.

Wastewater reuse. As the price for water increases, the value of wastewater will also increase, permitting the development of wastewater reuse schemes. Treated municipal wastewaters represent a very valuable source of supplementary water for industrial and agricultural sectors. Therefore, the Action Plan has recommended that options for reuse of wastewater for various sectors be investigated. The accompanying development of appropriate regulatory, institutional and legal mechanisms will be required to facilitate water supply and to protect public health.

Flood control. Economic losses caused by flood damage can be very high despite the massive efforts and resources allocated to flood protection. The Action Plan proposed a series of strategic projects in parallel with existing recommendations by Chinese planners. These projects are in the areas of:

(1) construction of additional reservoirs;
(2) road construction to serve as auxiliary dikes and to enable escape from flooded areas;
(3) risk assessment;
(4) upgrading of flood forecasting and warning systems; and
(5) development of models for flood damage assessment.

Institutional arrangements. Recommendations regarding the institutional arrangements include:

(1) establishment of high level River Basin Coordinating Councils (RBCNs) in each of the basins of the North China Plain, controlled by boards with balanced representation from both central government and provincial and local government agencies;
(2) giving the RBCNs authority for determining water resources allocations and developing water resource policies for optimal overall utilisation of water resources;
(3) reorienting the existing River Basin Commissions to serve as the working arms for the RBCNs;
(4) establishing provincial water resources coordinating committees for the management of water resources at the local levels;
(5) ensuring that only one agency is responsible for issuing permits for water extraction;

(6) transferring management of irrigation districts to local institutions; and

(7) arranging for enlargement of the scope of the permit system in municipalities that receive expensive imported water, to cover water supply, water use, and wastewater treatment and reuse, to ensure optimal use of both local and imported water.

13.16 CONCLUSIONS

The People's Republic of China, the most populated country of the world, has an average annual renewable water resources of $2812 \times 10^9 \, m^3$ or about $2200 \, m^3$ per head of population. These resources are not evenly distributed. While the southern River Basins are well watered, the northern areas are short of water. For example, in the Haihe River Basin the per capita water available is only $290 \, m^3$. Moreover, rapid transformation of China's society from a rural to an urban industrialised society, and the shift in its economy from a centrally planned to a socialist market economy, has forced the country to face some serious challenges in the supply of water for the agricultural, industrial and domestic sectors. Other major challenges include the need for improved flood control, and the increasing demand for hydro-electric power generation. Past mistakes in management of the country's water resources, particularly in the heavily populated area of the North China Plain, has resulted in the drying up of major river systems during drought years, overdraft of groundwater resources and contamination of water resources by domestic, industrial and agricultural wastes.

Current and future water resources development in China includes construction of a significant number of dams and major inter-basin water transfer projects. The most important one is the South to North Water Transfer Project, which consists of three (Eastern, Middle and Western) Routes, that would transfer water from the Yangtze River Basin to the North China Plain, where surface and groundwater resources are under stress and heavily polluted. Completion of the Eastern and the Middle Routes, which are currently under construction, will alleviate water shortages and water quality problems, but construction of the Western Route has been postponed because of its complex geotechnical problems and high cost. The long-term success of the project will depend on implementation of water saving, water pricing and pollution prevention measures, and the treatment of the 275 000 people displaced by the project.

Climate change in China is expected to increase the average annual temperatures by 0.9°C to 1.2°C by the year 2030. The average annual precipitation is predicted to rise slightly in the Songhua River Basin, while the increase would be negligible for the Huaihe River Basin. The predicted increase in evaporation means that demand for water is expected to increase, while less water may be available in the river basins already under stress.

Currently, China is continuing its program of dam building to achieve its economic and social development objectives. This program is expected to continue through a large part of the twenty-first century. China intends to develop and manage its water resources within a framework of sustainable development. This requires that the implementation of the dam building program and inter-basin water transfer projects are accompanied by better management of surface and groundwater resources, implementation of water saving measures, adequate water pricing, pollution control and environmental protection measures. Recent policy developments, legislations and institutional changes in the areas of water resources and environmental protection, are indications that the right steps towards the sustainable development are being taken. However, China is facing immense challenges and it is unlikely that it can achieve its sustainable development objectives in the near future.

References

Asian Development Bank (2000). *Reform of Environmental and Land Legislation in the People's Republic of China*. Manila: Asian Development Bank.

Barber, M. and Ryder, G. eds. (1993). *Damming the Three Gorges: What Dam Builders Don't Want you to Know*. Second Edition. London: Earthscane Publications.

Berkoff, J. (2003). China: The South–North Water Transfer Project – is it justified? *Water Policy* **5**: 1–28.

Biswas, A. K., Zuo, D., Nickum, J. E. and Liu, C. (1983). *Long-Distance Water Transfer: A Chinese Case Study and International Experiences*. Dublin: Tycooly International Publishing Limited.

Chen, J., Tang, C., Shen, Y., Sakura, Y. and Fukushima, Y. (2004). Nitrate pollution of groundwater in a wastewater irrigated field in Hebei Province, China. In Steenvoorden, J. and Endreny, T. eds. *Wastewater Re-use and Groundwater Quality*. Publication No. 285. Wallingford: International Association of Hydrological Sciences, pp. 23–27.

Cheng, X. (2005). Changes of flood control situations and adjustment of flood management strategies in China. *Water International* **30**(1): 108–113.

China Handbook Editorial Committee (2002). *Geography of China*. Honolulu, Hawaii: University Press of the Pacific.

China Internet Information Centre (2002).[27] Water Conservancy: South-to-North Water Transfer Project.

Economy, E. C. (2004). *The River Runs Black: The Environmental Challenge to China's Future*. Ithaca: Cornell University Press.

---

[27] http://us.tom.com/english/1815.htm (visited in July 2005).

Edmonds, R. L. ed. (2000). *Managing the Chinese Environment*. New York: Oxford University Press.

Elvin, M. (2004). *The Retreat of the Elephants: An Environmental History of China*. New Haven: Yale University Press.

FAO (2003). *Production Yearbook 2002*. Volume 56. Rome: FAO.

Framji, K. K., Garg, B. C. and Luthra, S. D. L. (1981). *Irrigation and Drainage in the World: A Global Review*. Third Edition. New Delhi: International Commission on Irrigation and Drainage. Volume 1.

Fu, G. and Chen, S. (2005). Geostatistical analysis of observed streamflow and its response to precipitation and temperature changes in the Yellow River. In Franks, S. *et al.* eds. *Regional Hydrological Impacts of Climatic Chang: Hydroclimatic Variability*. Publication No. 295. Wallingford: International Association of Hydrological Sciences, pp. 238–245.

Fuggle, R. and Smith, W. T. (2000).[28] *Experience with Dams in Water and Energy Resource Development in the People's Republic of China*. Cape Town: World Commission on Dams.

Gao, G. and Huang, C. (2001). Climate change and its impact on water resources in North China. *Advances in Atmospheric Sciences* **18**(5): 718–732.

Ghassemi, F., Jakeman, A. J. and Nix, H. A. (1995). *Salinisation of Land and Water Resources: Human Causes, Extent, Management and Case Studies*. Sydney: University of NSW Press.

GHD (2005).[29] *National Flood Management Strategy Study: Inception Report*. Melbourne: GHD Pty Ltd.

Han, Z. (1997). Groundwater for urban water supply in northern China. In Chilton, J. *et al.* eds. *Groundwater in the Urban Environment*. Volume 1: Problems, Processes and Management. Rotterdam: A. A. Balkema, pp. 331–334.

Heggelund, G. (2004). *Environment and Resettlement Politics in China: The Three Gorges Project*. Hants, England: Ashgate Publishing Limited.

Institute of Hydrogeology and Engineering Geology (1988). *Hydrogeological Map of China (1:4,000,000)*. Beijing: China Cartographic Publishing House.

James, C. V. ed. (1989). *Information China: The Comprehensive and Authoritarian Reference Source of New China*, Volume 1. Oxford: Pergamon Press.

JBICI (2004).[30] *Issues and Challenges for Water Resources in North China: Case of the Yellow River Basin*. Research Paper No. 28. Tokyo: Japan Bank for International Cooperation Institute.

Jiang, T. Chen, J., Ke, C., Gemmer, M., Metzler, M. and King, L. (2002a). Visualisation of historical flood and drought information (1100–1940) for the middle reaches of the Yangtze Rive valley. In Wu, B. *et al.* ed. *Flood Defence, 2002*. New York: Science Press. pp. 802–808.

Jiang, T., Wollesen, D., King, L., Wang, R. and Chen, J. (2002b). Chinese historical flood/drought data as an indicator to reconstruct for El Niño and La Niña Events (1470–1990). In Wu, B. *et al.* ed. *Flood Defence, 2002*. New York: Science Press, pp. 809–814.

Kaosa-ard, M. (2003). Poverty and globalisation. In Kaosa-ard, M. and Done, J. eds. *Social Challenges for the Mekong Region*. Bangkok: White Lotus, pp. 81–108.

Liu, C. (2000). Environmental issues and the South–North Water Transfer Scheme. In Edmonds, R. L. ed. *Managing the Chinese Environment*. New York: Oxford University Press, pp. 175–186.

Liu, C. and Zheng, H. (2002). South-to-north water transfer schemes for China. *Water Resources Development* **18**(3): 453–471.

Ma, J. (2004). *China's Water Crisis*. Norwalk, CT, USA: EastBridge.

Ministry of Water Resources (2002). *Water Resources Bulletin*. Beijing: Ministry of Water Resources.

Ministry of Water Resources (2003a).[31] *2003 Annual Report: Ministry of Water Resources of the People's Republic of China*. Beijing: Ministry of Water Resources.

Ministry of Water Resources (2003b).[32] *China Country Report on Sustainable Development: Water Resources*. Beijing: Ministry of Water Resources.

Ministry of Water Resources (2004a).[33] *Water Resources in China*. Beijing: Ministry of Water Resources.

Ministry of Water Resources (2004b).[34] *South-to-North Water Transfer Project*. Beijing: Ministry of Water Resources.

Osborne, M. (2004).[35] *River at Risk: The Mekong and the Water Politics on China and Southeast Asia*. Sydney: Lowy Institute for International Policy.

Qian, Z. ed. (1994). *Water Resources Development in China*. Beijing: China Water & Power Press.

Research Institute of Meteorological Sciences (1982). *Atlas of Drought and Flood in China in the Past 500 years*. Beijing: Atlas Press.

Shao, X., Wang, H. and Wang, Z. (2003). Interbasin transfer projects and their implications: a China case study. *International Journal of River Basin Management* **1**(1): 5–14.

Smil, V. (2004). *China's Past, China's Future: Energy, Food, Environment*. New York: RoutledgeCurzon.

Tang, C., Chen, J. and Shen, Y. (2004). Long-term effect of wastewater irrigation on nitrate in groundwater in the North China Plain. In Steenvoorden, J. and Endreny, T. eds. *Wastewater Re-use and Groundwater Quality*. Publication No. 285. Wallingford: International Association of Hydrological Sciences, pp. 34–40.

Tao, F., Yokozawa, M., Hayashi, Y. and Lin, E. (2003). Future climate change, the agricultural water cycle, and agricultural production in China. *Agriculture, Ecosystems and Environment* **95**: 203–215.

US Embassy (2003).[36] Update on China's South–North Water Transfer Project. Beijing: US Embassy.

Wang, L. and Ma, C. (1999). A study on the environmental geology of the Middle Route Project of the South–North water transfer. *Engineering Geology* **51**: 153–165.

Wang, M. and Xia, J. (2005). Impact of climate fluctuation and land-cover changes on runoff in the Yellow River basin. In Wagener, T. *et al.* eds. *Regional Hydrological Impacts of Climatic Change: Impact Assessment and Decision Making*. Publication No. 295. Wallingford: International Association of Hydrological Sciences, pp. 183–188.

World Bank (1997). *Clear Water, Blue Skies: China's Environment in the New Century*. Washington D.C.: World Bank.

World Bank (2001). China: Agenda for Water Sector Strategy for North China. Report No. 22040-CHA (4 volumes).[37] Volume 1: Summary Report.[38] Washington D.C.: World Bank.

World Bank (2003). *World Development Indicators*. Washington D.C.: World Bank. CD-ROM.

World Resources Institute (2000). *World Resources 2000–2001: People and Ecosystems, The Fraying Web of Life*. Oxford, UK: Elsevier Science.

World Wildlife Fund (2001). *The Proposed South–North Water Transfer Scheme in China; Need, Justification and Cost*. Draft Report. Beijing: World Wildlife Fund.

---

[28] Also available at the URL: http://www.dams.org/docs/kbase/studies/cscnmain.pdf (visited in July 2005).

[29] Also available at the URL: http://www.mwr.gov.cn/english1/20050223/20050223110722NEICZQ.pdf (visited in July 2005).

[30] Also available at the URL: http://www.jbic.go.jp/english/research/report/paper/pdf/rp28_e.pdf (visited in July 2005).

[31] Also available at the URL: http://www.mwr.gov.cn/english1/pdf/Annual2003.pdf (visited in July 2005).

[32] Also available at the URL: http://www.mwr.gov.cn/english1/pdf/china2003.pdf (visited in July 2003).

[33] Available at the URL: http://www.mwr.gov.cn/english1/20040802/38161.asp (visited in July 2005).

[34] Available at the URL: http://www.mwr.gov.cn/english1/20040827/39304.asp (visited in July 2005).

[35] Also available at the URL: http://www.lowyinstitute.org (visited in July 2005).

[36] http://www.usembassy-china.org.cn/sandt/ptr/SNWT-East-Route-prt.htm (visited in July 2005).

[37] Volume I, Summary Report; Volume 2, The Main Report; Volumes 3 and 4, The Statistical and GIS Maps Annexes.

[38] Available at: http://lnweb18.worldbank.org/eap/eap.nsf/Attachments/WaterSectorReport/$File/Vol[v]3A4a1.pdf (visited in July 2005).

Xia, J., Wong, H. and Ip, W. C. (2003). Water problems and sustainability in North China. In Franks, S. *et al.* eds. *Water Resources Systems: Water Availability and Global Change*. Publication No. 280. Wallingford: International Association of Hydrological Sciences, pp. 12–22.

Xia, J. and Zhang, L. (2005). Climate change and water resources security in North China. In Wagener, T. *et al.* eds. *Regional Hydrological Impacts of Climatic Change: Impact Assessment and Decision Making*. Publication No. 295. Wallingford: International Association of Hydrological Sciences, pp. 167–173.

Xie, Q., Guo, X. and Ludwig, H. F. (1999). The Wanjiazhai Water Transfer Project, China: an environmentally integrated water transfer system. *The Environmentalist* 19: 39–60.

Yang, S. L., Ding, P. X. and Chen, S. L. (2001). Changes in progradation rate of the tidal flats at the mouth of the Changjiang (Yangtze) River, China. *Geomorphology* 38: 167–180.

Yang, D., Li, C., Musiake, K. and Kusuda, T. (2003). Analysis of water resources in the Yellow River basin in the last century. In Franks, S. *et al.* eds. *Water Resources Systems: Water Availability and Global Change*. Publication No. 280. Wallingford: International Association of Hydrological Sciences, pp. 70–78.

Yang, D., Ni, G., Kanae, S., Li, C. and Kusuda, T. (2005). Water Resources variability from the past to future in the Yellow River, China. In Wagener, T. *et al.* eds. *Regional Hydrological Impacts of Climatic Change: Impact Assessment and Decision Making*. Publication No. 295. Wallingford: International Association of Hydrological Sciences, pp. 174–182.

Ye, B., Li, C., Yang, D., Ding, Y. and Shen, Y. (2005). Precipitation trends and their impact of the discharge of China's four largest rivers, 1951–1998. In Franks, S. *et al.* eds. *Regional Hydrological Impacts of Climatic Change: Hydroclimatic Variability*. Publication No. 296. Wallingford: International Association of Hydrological Sciences, pp. 228–237.

Ying, A. (2000). Impact of global climate change on China's water resources. *Environmental Monitoring and Assessment* 61(1): 187–191.

Zhang, J. and Lin, Z. (1992). *Climate of China*. New York: John Wiley & Sons.

Zhu, Z., Giordano, M., Cai, X. and Molden, D. (2004). The Yellow River Basin: water accounting, water accounts, and current issues. *Water International* 29(1): 2–10.

14 India: The National River-Linking Project

SECTION A: OVERVIEW OF GEOGRAPHY, POPULATION, LAND AND WATER

14.1 GEOGRAPHY

India is a vast country covering an area of about 3 287 726 km². It measures 2980 km from east to west and 3220 km from north to south. India has the following principal physiographic regions (Figure 14.1).

- **The Himalaya Mountains** with Mt Everest the world's highest peak at 8848 m, run for more than 2400 km along the northern frontiers of India. Numerous peaks of Himalaya with elevations of above 7700 m are within India.
- **The Indo-Gangetic Plains** in the south and parallel to the Himalaya Mountains, comprise a belt of flat alluvial lowlands. The average elevation is less than 150 m above mean sea level. These plains are watered by three distinct river systems: the Indus, the Ganga, and the Brahmaputra. The Indus River drains through Pakistan into the Arabian Sea, while the other two have their outfall in the Bay of Bengal.
- **The Great Indian Desert** is an arid region in the northwestern part of the country and contains large tracts of fertile soils. The Indira Gandhi Canal diverts tributaries of the Indus River to these areas.
- **The Deccan Plateau** in the south of the Indo-Gangetic Plains, is a vast triangular tableland occupying most of the Indian Peninsula. It is a generally uneven plateau naturally divided into regions by low mountain ranges and deep valleys. The elevation of the plateau ranges mainly from about 305 to 915 m, with a highest peak of 2134 m.
- **The coastal Mountain Belts** on the eastern and western sides of the triangular Deccan Plateau converge in south India. These ranges, the Eastern and Western Ghats have elevations of about 475 and 914 m above mean sea level respectively. The Western Ghats rise sharply to elevations of 1000 to 1300 m, with some peaks as high as 2438 m. The Eastern Ghats are less prominent and rise to elevations of 800–1000 m. There are extensive deltas along the eastern Ghats. These flat and fertile deltas have some of the largest irrigation systems in the country. The deltas of the west-flowing rivers are much less extensive and the western coastal belt is comparatively narrow.

14.2 POPULATION

India is a union of 35 states and territories most of which are shown in Figure 14.2. The country had a population of 238 million in 1900, which increased to 1.03 billion by 1 March 2001 (Research, Reference and Training Division, 2005). It is the second most populous country in the world after China. Only 28 percent of the population is in urban areas. The remaining 72 percent live in nearly 600 000 villages. Population density was around 72 per km² in 1900 and had increased to 314 per km² in 2001. The approximate population of the five mostly populated states in 2001 were Uttar Pradesh (166 million), Maharashtra (97 million), Bihar (83 million), West Bengal (80 million) and Andhra Pradesh (76 million). India's population is expected to grow and reach 1.4 billion by 2025 and 1.8 billion by 2050, with the predicted gradual rural population decline to 50 percent by 2050.

14.3 ECONOMY

The Indian economy is still predominately agricultural and about 25 percent of the national income is derived from agriculture and associated activities, employing some 65–70 percent of the workforce (Research, Reference and

Figure 14.1 Major physiographic features of mainland India.

Training Division, 2005, p. 60). The Indian economy has grown at rates of 5.8 percent and 5.4 percent over the periods of 1981–90 and 1991–2000 respectively. Its GDP (in 1995 values) has increased from US$80 billion in 1960 to US$257 in 1990, and to US$493 in 2001 (World Bank, 2003). The per capita GDP of US$186 in 1960 increased to US$323 in 1990 and to US$460 in 2000.

14.4 CLIMATE AND PRECIPITATION

Because of its size, peninsularity, topography and geographical position, climatic conditions in India are diverse ranging from tropical wet to arid and semi-arid (Pant and Rupa Kumar, 1997). The west coast, especially in Kerala, Karnataka and Goa, has a wet tropical climate with high temperatures and rainfalls. Both wet and dry tropical climates are found along much of the east coast and in the interior of the northern peninsula. The entire Indo-Gangetic Plains have a humid sub-tropical climate. There is also a humid sub-tropical region in the extreme north-east, where rainfall is the highest. Arid and semi-arid climate mainly occur in the west and north-west of India in Rajasthan, Gujarat, Punjab, Haryana and Maharashtra.

India has two monsoons, the south-west or summer monsoon from June to September and the north-east monsoon from October to December. The country depends heavily on the 90 to 105 days of summer monsoon rainfall for its water supply. These rainfalls account for 85 percent of

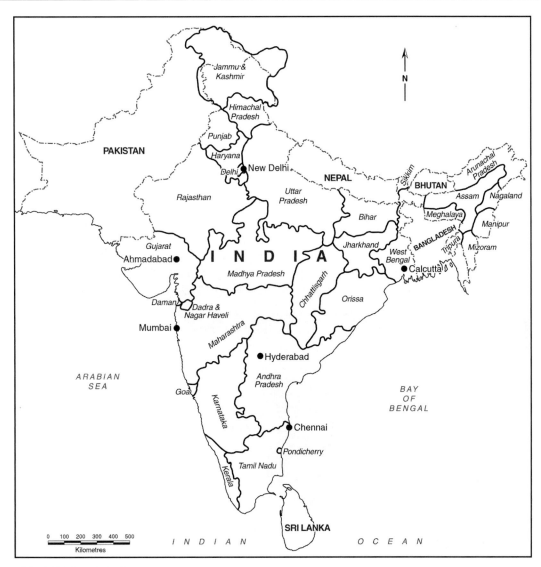

Figure 14.2 Political divisions in India.

the country's precipitation and furnish water for agriculture, water supply and hydro-power production. Excess run-off is stored in reservoirs for use in dry seasons. Although considerable areas of the country are irrigated, nearly 80 percent of the cropped areas are still dependent on seasonal rainfall.

India's average annual precipitation is estimated to be about 1217 mm or $4000 \times 10^9 \, m^3$. Approximately 42 percent of precipitation is lost to evaporation, with the balance being surface run-off and aquifer recharge. Rainfall is unevenly distributed in space and time, with the heaviest rains occurring over the north-eastern states and along the west coast (Figure 14.3). Cherrapunjee, in the State of Meghalaya has an average annual rainfall of nearly 10 000 mm a year while some districts in western Rajasthan receive barely 100 mm a year.

14.5 IRRIGATION

While irrigation in India has been practiced for thousands of years, the total area irrigated was relatively small until British influence in the nineteenth century (Framji et al., 1982, pp. 516–597). From 1800–36, some of the pre-existing canals in the north and the south were remodelled. After 1836, a number of large-scale irrigation projects were designed and constructed by British Army engineers including the Upper Ganga Canal in Uttar Pradesh, the Upper Bari Doab Canal in Punjab and the Godavari Delta System in Andhra Pradesh. The Upper Ganga Canal, with a capacity of $269 \, m^3 \, s^{-1}$, was opened in 1854. The system had 914 km of main canals and 3846 km of distributaries, irrigating an area of 639 000 ha. The Upper Bari Doab Canal was completed in 1859. However, after partitioning of the

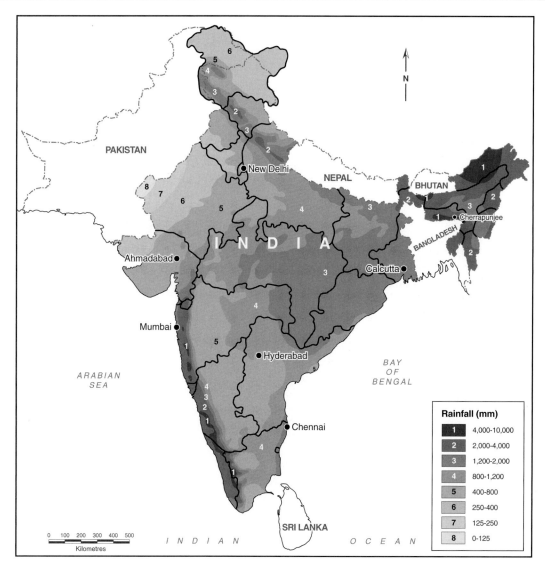

Figure 14.3 Average annual rainfall of India (Government of India, 1990, p. 5).

country in 1947, a substantial part of the area irrigated by this canal was included in the newly created State of Pakistan. The Godavari Delta System was completed in 1890 and had 805 km of canals and 3219 km of distributaries. In 1961–62 the area irrigated by this system was 558 466 ha.

Irrigation in India has undergone continuous development. The total irrigated area, which was about 1 Mha in 1850, reached 11.7 Mha in 1900, 42.1 Mha in 1987 and 57.2 Mha in 2002 (FAO, 2004). The ultimate irrigation potential expected to be about 148 Mha (Ministry of Water Resources, 2005, p. 16). Irrigation has played a major role in food self-sufficiency of the country. Food grain production, which was about 50 Mt in the 1950s, has increased to more than 208 Mt in 1999–2000 and is expected to reach 350 Mt by 2025 to satisfy the food requirements of 1.4 billion people.

The efficiency of irrigation water use has been estimated at about 40 percent for canal irrigation and 60 percent for groundwater schemes (Rangachari *et al.*, 2000, p. 178). Various studies of canal-irrigated areas indicate that less than one percent of the cost of farming is due to water charges. In addition, across the country, irrigation water charges are based on the cropped area and the crop type rather than being volumetric (Rangachari *et al.*, 2000, Annexes, p. 96). This water pricing regime has resulted in excessive water application, waterlogging of 2.5 Mha and salinity problems in 3.3 Mha of irrigated lands, totaling about 10 percent of current irrigation areas. Adoption of a volumetric charge policy for irrigation water combined with a reasonable price structure will encourage increased water use efficiency and the water saved could be either allocated to other users or to irrigate new lands.

Table 14.1. *Surface water resources of India*

No.	River	Average annual run-off ($10^9 \, m^3$)
1	Indus	73.3
2	Ganga	525.0
3	Brahmaputra	585.6
4	Brahmani, Baitarni and Subarnarekha	40.9
5	Mahanadi	66.9
6	East flowing rivers between Mahanadi and Godavari	22.5
7	Godavari	110.5
8	Krishna	78.1
9	Pennar	6.3
10	Cauvery	21.4
11	East flowing rivers between Cauvery and Kanyakumari	16.5
12	Rajasthan Desert	Negligible
13	West flowing rivers of Kutch, Saurashtra and Luni	15.1
14	Sabarmati and Mahi	14.8
15	Narmada and Tapi	60.5
16	West flowing rivers from Tapi to Tadri	87.4
17	West flowing rivers from Tadri to Kanyakumari	113.5
18	Minor river basins draining into Bangladesh and Burma	31.0
Total		1869.3

Source: Website of the Ministry of Water Resources.[1]

14.6 WATER RESOURCES

14.6.1 Surface Water

Surface run-off constitutes $1869 \times 10^9 \, m^3$ (Table 14.1) out of the average annual precipitation of about $4000 \times 10^9 \, m^3$. The 2001 surface water availability was approximately $1815 \, m^3$ per capita. Some of India's rivers are amongst the mighty rivers of the world such as Brahmaputra, Ganga and the Indus with average annual flows of about 585, 525 and $73 \times 10^9 \, m^3$, respectively. The Brahmaputra rises in Tibet and runs eastward parallel to the Himalayas. In India, it flows westward for approximately 916 km before entering Bangladesh. Its catchment area in India is about $194\,400 \, km^2$, smaller than the catchments of the Ganga and the Indus. The Ganga River is 2525 km long and has a catchment area of $860\,000 \, km^2$, covering virtually the entire northern India. The Ganga River has a large number of tributaries. Some of them are of Himalaya origin, while the other tributaries such as Chambal and Banas originate in the Deccan Plateau. The Indus River has its origin in the Tibet Plateau. It is 2800 km long and less than 40 percent of its catchment is in India. The main tributaries of the Indus River are the Jhelum, Chenab, Ravi, Beas and Sutlej. Other important Indian rivers include the Mahanadi, Godavari, Krishna, Sabarmati and Narmada (Table 14.1 and Figure 14.4).

India had only 42 large dams in 1900, consisting of 28 dams higher than 15 m and 14 with heights of 10–15 m. By May 1994 there were 4291 dams higher than 10 m and 2342 dams higher than 15 m. Half of the large dams were constructed during the period of 1970 to 1989 (Table 14.2). The concentration of major dams is 1529 dams in Maharashtra, 1093 in Madhya Pradesh, 537 in Gujarat and the remaining 1132 dams in all other states. Reservoirs created by construction of dams have a total capacity of $250 \times 10^9 \, m^3$ (Rangachari *et al.*, 2000, pp. 9–10). The five river basins with the highest reservoir capacities are Ganga ($53 \times 10^9 \, m^3$), Krishna ($42.3 \times 10^9 \, m^3$), Godavari ($30.2 \times 10^9 \, m^3$), Narmada ($23.3 \times 10^9 \, m^3$) and Indus ($16.3 \times 10^9 \, m^3$).

A significant number of dams were built for irrigation purposes, while only some were constructed mainly for hydro-power generation. Some were also built to satisfy water requirements of mega-cities (Bangalore, Mumbai, Chennai, Delhi, Hyderabad and others), large towns and industries (steel, fertiliser, textile and others).

---

[1] http://wrmin.nic.in (visited in January 2006).

Figure 14.4 River Basins of India (Mohile, 2005, Figure 2.1, and the website provided in the footnote).[2]

Table 14.2. *Large dams in India up to May 1994*

	Number of large dams		
Period	>15 m high	10–15 high	Total
Up to 1900	28	14	42
1901–50	118	133	251
1951–70	418	277	695
1971–89	1187	1069	2256
1990–94	56	60	116
Details not available	74	162	236
Under construction	461	234	695
Total	2342	1949	4291

Source: Rangachari *et al.* (2000, Table 1.1).

Hydro-power Generation

The first major hydro-power station with a capacity of 4.5 MW was commissioned in 1902 in the State of Karnataka. The pace of electricity development was slow until independence in 1947, when the total installed capacity was 1362 MW, including 508 MW of hydro-power. Power generation has increased rapidly since independence and reached 89 000 MW by March 1998 which included about 21 900 MW of hydro-power (Rangachari *et al.*, 2000, p. 8). India's hydro-power potential has been estimated at about 148 600 MW. Sources of hydro-power potential are within

[2] http://www.wrmin.nic.in/riverbasin/allindia.htm (visited in January 2006)

the river basins of the Brahmaputra (66 000 MW), Indus (33 840 MW), Ganga (20 700 MW), east flowing rivers of south India (14 500 MW), west flowing rivers of south India (9420 MW), and central rivers (4140 MW).

14.6.2 Groundwater

Groundwater has been used for irrigation in India for many centuries, but mainly from shallow dug-wells. In the mid 1930s, the Uttar Pradesh Government installed tube-well schemes tapping deep aquifers. Since then, similar groundwater schemes have been developed in other states. Groundwater resources exist in a wide variety of formations from Precambrian crystalline and sedimentary rocks to Quaternary alluvium. The average annual groundwater recharge is about $432 \times 10^9 \, m^3$ or 11 percent of precipitations. The annual sustainable yield of aquifer systems ($432 \times 10^9 \, m^3$) includes $171 \times 10^9 \, m^3$ in Ganga, $41 \times 10^9 \, m^3$ in Godavari, $27 \times 10^9 \, m^3$ in Brahmaputra and $27 \times 10^9 \, m^3$ in Indus river basins and the balance ($166 \times 10^9 \, m^3$) in other river basins. Groundwater resources play a major role in annual irrigation, domestic and industrial water supply of the country.

At the end of the VIth Five Year Plan (1980–85), there were 8.7 million irrigation dug-wells, 3.36 million private tube-wells and 46 000 state tube-wells in India. The Target for the VIIth Five Year Plan (1986–90) was an additional 1.25 million irrigation dug-wells, 1.41 million private tube-wells and 25 000 state tube-wells (Ghosh and Phadtare, 1990).

Although watertable rise and associated waterlogging and salinisation are serious issues in some command areas, in many arid zones, overdraft and associated water quality problems are increasingly emerging. Decline in groundwater levels exclude the poor from access to groundwater due to the cost of increasing well depth, and would reduce base flows in streams impacting on surface water availability, riverine ecosystems and water quality. Groundwater overdraft in western and peninsular India has taken the form of depletion.[3] Declining watertable and groundwater mining in Gujarat and Rajasthan has meant rising fluoride contamination of drinking water supplies. In coastal aquifers this has resulted in seawater intrusion into the aquifers and deteriorating water quality.[4]

14.6.3 Water Use

Irrigation is the major water user in the country. In 1990, out of a total water use of $510 \times 10^9 \, m^3$, approximately $439 \times 10^9 \, m^3$ or 86 percent was used for irrigation (Navalawala, 1992). Surface water contributed 71 percent ($360 \times 10^9 \, m^3$) to the water supply, with the remainder ($150 \times 10^9 \, m^3$) from groundwater. In 2000, total water use in India was estimated to be $750 \times 10^9 \, m^3$ with 84 percent being used for irrigation. The remaining 16 percent was shared between domestic, industry, energy and other sectors (Table 14.3). Total annual water use is expected to rise to $1050 \times 10^9 \, m^3$ by 2025 but with a major decline in irrigated water use being offset by a significant increase in industrial use. By 2050, the annual water use could be as high as $1300 \times 10^9 \, m^3$.

Table 14.3. *The annual water use in 2000 and its projections for 2025 and 2050*

Category	Water use in 2000 ($10^9 \, m^3$)	(percent)	Water use in 2025 ($10^9 \, m^3$)	(percent)	Water use in 2050 ($10^9 \, m^3$)
Irrigation	630	84.0	770	73.3	–
Domestic	33	4.4	52	5.0	–
Industry	30	4.0	120	11.4	–
Energy	27	3.6	71	6.8	–
Others	30	4.0	37	3.5	–
Total	750	100.0	1050	100.0	1300

Source: Singh and Khurana (2001).

14.6.4 Water Quality

Although Indians consider rivers as holy, streams are highly polluted due to point sources (disposal of untreated sewage from large and medium population centres and industries) and diffuse agricultural sources (run-off from agricultural lands carrying sediment, fertilisers, herbicides, and pesticides). In India, faecal contamination is the primary water quality issue in rivers, especially where human and animal wastes are not adequately collected and treated. Although this occurs in both rural and urban areas, the situation is more critical in fast growing cities.[5] Industries which discharge large volumes of effluents containing organic matter (pulp and paper products and food residues) and chemicals (heavy metals and toxic organic compounds) affect the use of water for drinking and impact on riverine ecologies. As an example, the Damodar River

---

[3] http://www.india-seminar.com/2000/486/486%20moench.htm (visited in January 2006).
[4] http://www.iwmi.cgiar.org/rthemes/ground/intro.htm (visited in January 2006).
[5] http://wrmin.nic.in/problems/pb_faced.htm (visited in January 2006).

in the West Bengal is heavily polluted by heavy metals from electroplating, tanning and metal-based industries.

Shallow groundwater resources are also contaminated by pathogenic bacteria, nitrate, heavy metals, chemical compounds and salt. Pit latrines, septic tanks, inadequately constructed landfills, leaching of mine tailings, excessive applications of fertilisers for agricultural developments and heavy groundwater extraction in coastal areas have contributed to groundwater contamination.

In West Bengal groundwater contamination by naturally occurring arsenic was first recognised in 1983, but not appreciated internationally until the mid 1990s. It is estimated that more than 5 million people are drinking water with arsenic concentrations greater than the accepted standard of 50 µg L^{-1} (Smedley, 2003). Arsenic concentrates in hair, nails, skin and liver tissue of the affected people and causes a wide rage of symptoms including depigmentation, keratosis and gangrene. Arsenic occurs mainly in the groundwater of the valley-fill sequence deposited during the Holocene marine transgression. Concentrations are highest in the upper 50 m of sedimentary sequence. Below 100 m, arsenic concentration reduces, and below 200 m the chance of drilling an arsenic-safe well approaches 99 percent (Ravenscroft *et al.*, 2005).

Water quality is a major concern in India and the government established the *Water Quality Assessment Authority* in May 2000 to address the problem. The Authority as a first step set up Water Quality Review Committees at state levels to review monitoring practices and to highlight important issues for consideration.

14.7 FLOOD

Monsoon floods occur frequently in India. Pant and Rupa Kumar (1997, Table 6.5) list 19 major floods from 1874 to 1994. Floods in 1917 covered up to 53 percent of the country while those in 1971 inundated 21 percent. Major river basins such as Brahmaputra and Ganga are frequently affected by floods with serious social and economic damages and widespread suffering in the states of Assam, Bihar, West Bengal and Uttar Pradesh. The 2002 flood in the state of Bihar affected 11 million people and took more than 140 lives. Of the country's total area of approximately 329 Mha, 40 Mha is prone to floods. Only about 32 Mha of this can be provided with a reasonable degree of flood protection (Research, Reference and Training Division, 2005, p. 706). On average, floods affect an area of around 7.5 Mha each year.[6] Up to March 2002, an area of 16.4 Mha had some degree of flood protection against flood through construction of embankments, drainage channels, town protection works and by raising villages. In order to mitigate flood damage, the Central Water Commission has set up a nationwide Flood Forecasting and Warning System on inter-state river basins (Research, Reference and Training Division, 2005, p. 707). Although dams have been effective in flood control, it is suspected that they have been merely exporting floods outside the traditional flood prone areas (Rangachari *et al.*, 2000, p. 196).

14.8 DROUGHT

Droughts also occur frequently in India. From 1873 to 1987, 21 large-scale droughts occurred. In 1899, up to 73 percent of the country was drought affected, while in 1986 drought covered 10 percent (Pant and Rupa Kumar, 1997, Table 6.5). The worst drought of the twentieth century in 1918 affected 68 percent of India. The second worst drought in 1987–88 covered 64 percent of the country and seriously affected the states of Punjab, Haryana, Rajasthan, Uttar Pradesh, Gujarat, Madhya Pradesh, Orissa, Maharashtra, Andhra Pradesh, Karnataka and Himachal Pradesh. Environmentalists believe that this was not just a meteorological drought but was also a consequence of deforestation, mismanagement of water resources and desertification, all of which have been taking place over decades (Bandyopadhyay, 1988). A drought in 2000–01 affected more than 145 million people in eight states,[7] including Gujarat, Madhya Pradesh, Orissa, and Rajasthan, with some states in their second or third consecutive year of drought.

14.9 CLIMATE CHANGE IMPACTS

Pant and Rupa Kumar (1997, pp. 182–186) examined changes in surface air temperature over India from 1901 to 1987 and showed that there has been an increasing trend of maximum and minimum temperatures of 0.6°C and 0.1°C per 100 years, respectively. Singh *et al.* (2005) analysed rainfall data over the period of 1820s–2001 and demonstrated that rainfall had a decreasing trend in the following river basins: Narmada (from 1950); Sabarmati (from 1960);

---

[6] http://wrmin.nic.in/policy/nwp2002.pdf (visited in January 2006). See also Rangachari *et al.* (2000, pp. 26–28).
[7] http://www.rainwaterharvesting.org/Crisis/Drought.htm (visited in January 2006).

Mahanadi (from 1962); Mahi (from 1964); Godavari (from 1964); and Tapi (from 1965). In other river basins there has been an increasing trend in rainfall: Cauvery (from 1929); Krishna (from 1953); Indus (from 1954); Brahmaputra (from 1988); and Ganga (from 1993). These trends in temperature and rainfall across India have been attributed to global warming. Climate change is predicted to have major impacts on India's water resources, agriculture, forestry, ecology and economy.

Climate model simulations for the Indian sub-continent predict an increase in average annual minimum and maximum surface air temperatures of 1.0°C and 0.7°C respectively by the 2040s relative to 1980s.[8] Climate change is likely to worsen existing water availability problems. For any given region, the combined effects of predicted lower rainfall, higher temperature and evaporation should have direct consequences on regional water balances. These could lead to less run-off and groundwater recharge, substantially changing the availability of freshwater in the river basins. Furthermore, potential changes in temperature and precipitation might have a dramatic impact on soil moisture and aridity levels. Climate model simulations indicate, by the year 2050, the average annual run-off of the Brahmaputra River will decline by 14 percent[9] and the average annual run-off of the Himalayan tributaries of the Ganga River will also decline because of the recession of glaciers.

14.10 IMPACTS OF DAM BUILDING

A number of significant benefits have resulted from the construction of large dams and development of India's water resources since its independence. These include the development of irrigated food production, hydro-power generation, flood control, and municipal and industrial water supplies. In recent years there has been fierce controversy over the net benefits of large dams. Opponents argue that:

(1) large dams have caused substantial environmental and social costs;
(2) major resources have been allocated to a large number of projects;
(3) constructions have been delayed and costs increased because of both delays and initial underestimation of costs;
(4) projected benefits have failed to materialise;
(5) inequities have arisen in distribution of benefits;
(6) significant numbers of people, particularly the poor, have been displaced, and
(7) rehabilitation of displaced communities has been poor.

Rangachari *et al.* (2000, Chapters 4 and 5) provide information on the environmental, financial, economic, and social impacts of large dams. A brief account of some of the salient issues follows.

Environmental impacts. The identified impacts due to dam construction were:

(1) premature cutting of trees in the area to be submerged;
(2) harvesting of forests and vegetation in catchment area for cooking fuel by construction workers during project construction and by local communities after completion of project; and
(3) dust pollution, disturbing wildlife and destruction of vegetation due to mining/quarrying for construction materials and lack of planning for rehabilitation of mining sites.

Impacts due to dam filling were:

(1) submergence and destruction of flora and fauna in large tracts of forests, grasslands and wetlands;
(2) submergence of productive agricultural and grazing lands;
(3) spread of vector borne diseases such as malaria;
(4) changes to the downstream river ecosystems due to changes in natural flow regime; and
(5) broken ecological continuity for those species of fish whose passage up river to their breeding ground is blocked by dams.

Once the dams had filled, it was found that there were:

(1) changes in ecological conditions of the rivers (temperature, oxygen levels and chemical and physical characteristics of the water) due to dam discharge;
(2) changes in coastal and marine ecology due to changes in water flow regimes and their silt contents;
(3) development of waterlogging and land salinisation in irrigation areas due to leakage from unlined canals, excessive water use, and lack of drainage facilities;
(4) much higher rates (up to eight times) of reservoir siltation than anticipated; and
(5) overestimation of water availability.

Escalation of project costs. The marked difference between actual and original estimated costs of the large irrigation development and dam construction projects

---
[8] http://www.teriin.org/climate/impacts.htm (visited in January 2006).
[9] http://www.teriin.org/climate/impacts.htm (visited in January 2006).

Table 14.4. *Escalation of project costs*

Year	Report	Cost over-runs
1973	Report of the Expert Committee on Rise in Cost of Irrigation and Multipurpose projects	Revised estimates of 64 major projects demonstrated costs to be on average 108 percent higher than approved estimates, and 32 projects showed escalation exceeding 100 percent.
1978	Estimates Committee, Ministry of Agriculture and Irrigation	In the IVth Plan (1969–74), expenditure exceeded outlay by 19.4 percent, while physical targets in area irrigated showed a shortfall of 51.4 percent.
1979	Indian National Committee on Large Dams	Average cost escalation of 41 dams was 254 percent with only six dams showing escalation of less than 100 percent.
1983	Public Accounts Committee	Cost overruns of 159 projects averaged 232 percent, and 32 projects showed over-runs of 500 percent.
1983	Desai Committee Report	During Vth Plan (1974–78) revised estimates of all irrigation projects were 3.2 times the original cost.

Source: Rangachari *et al.* (2000, pp. 50–51).

is well documented. Table 14.4 summarises some of the major cost over-runs identified in various Indian national reports.

Adverse social impacts. Among the most significant adverse social impacts of dams are those that result from forced displacement of people from their homes, fields, towns and regions. A survey of 140 large to medium dams in India indicated displacement of 4 387 625 people or approximately 31 340 persons per dam (Rangachari *et al.*, 2000, p. 116). While this sample was not representative of all of India's dams, it is clear that millions of people have been displaced by dam construction. Most of these tend to be poor with meagre resources and limited resilience. Many displaced gravitate to the large cities where they subsist on the fringes in poverty. The adverse social impacts include:

(1) loss of cultural heritage sites and monuments;
(2) loss of home, loss of familiar social and geographical surrounds;
(3) loss of preferred or familiar sources of livelihood;
(4) trauma, uncertainties and insecurities;
(5) impacts on physical health;
(6) impacts on living standards;
(7) social alienation from, and conflicts with host communities; and
(8) increased fringe urbanisation.

In most dam construction projects, families are eligible for some form of compensation. Though many rehabilitation packages attempt to compensate for loss of individual property and livelihood, very rarely is there an attempt to compensate for the loss of common property resources such as:

(1) free access to water and other resources of the river, including riverbed land and fish; and
(2) grasslands, forests, wetlands and a host of natural resources, from which they derived not only subsistence but also income.

Equity issues.

(1) Members of scheduled castes make up approximately 24.5 percent of the population of Indian and tribal people are a further 8 percent. Data available for a limited number of dams have indicated that nearly 15 percent of the displaced populations were members of scheduled castes and 47 percent were tribal peoples.
(2) The irrigation benefits of dams accrue to the downstream populations and benefits are not shared with the upstream residents.
(3) Between 1980 and 2000, approximately 1.3 Mha of forests were submerged. This has had a devastating impact on the poorest segments of the society, especially tribal people, who heavily depend on the forest for their subsistence needs.
(4) Recipients of hydro-electricity are beneficiaries of dams, whether they live in urban or rural areas. In contrast, those who are too poor or isolated without access to electricity are the losers.
(5) Only in rare cases, displaced people were given an adequate amount of good agricultural land as compensation. There were positive economic and social benefits from this for the displaced people.

14.11 NATIONAL WATER POLICY

India has a federal system of government where ownership of water is vested in the state governments. The Indian National Water Resources Council adopted the *National Water Policy* in September 1987. Since then, a number of issues and challenges have emerged in the development and management of water resources and the *National Water Policy* was revised and updated in 2002. It recognises that water is a scarce and precious national resource and lays down the broad principles that govern the management of the country's water resources. Some of the policies are as follows.[10]

Information System. A well-developed information system for water related data in its entirety at the national/state level is a prime requisite for resource planning. A standardised national information system should be established with a network of data banks and databases, integrating and strengthening the existing central and state level agencies.

Water Resources Planning.

(1) The amount of water resources available for use should be maximised;
(2) non-conventional methods for use of water, such as inter-basin transfers, artificial recharge of aquifers and desalination of brackish or seawater, as well as traditional water conservation practices like rainwater harvesting, need to be encouraged;
(3) sustainable water resources development and management will have to be planned for hydrological units such as drainage basins or sub-basins taking into account groundwater;
(4) development and management will need to incorporate quantity and quality aspects as well as environmental considerations; and
(5) water should be made available to areas of water deficit by transfer from areas which are water rich.

Institutional Mechanisms. In order to sustainably plan, develop and manage water resources, existing institutions will have to be appropriately reoriented/reorganised and even created wherever necessary. Appropriate river basin organisations should be established for the planned development and management of river basins.

Water Allocation Priorities. Water allocation priorities for planning and operating water systems should be, broadly: first drinking water, second irrigation, third hydro-power, fourth ecology, fifth industries, sixth navigation, and then other uses.[11]

Drinking Water. Adequate safe drinking water should be provided to the entire population both in urban and rural areas.

Project Planning. In the planning, implementation and operation of a water resources project, the preservation of the quality of the environment and the ecological balance should be a primary consideration. Special efforts should be made to investigate and formulate projects either in, or for benefit of, areas inhabited by tribal or other specially disadvantaged groups. The drainage system should form an integral part of any irrigation project right from the planning stage, and the involvement and participation of beneficiaries and other stakeholders should be encouraged.

Groundwater Development. Extraction of groundwater should be so regulated as not to exceed the recharge rate. Detrimental environmental consequences of over extraction of groundwater need to be prevented. Groundwater recharge projects should be developed and implemented for improving both the quality and availability of groundwater resources. The integrated development of surface and groundwater resources and their conjunctive use should be an integral part of the project development and implementation.

Water Charges. Charges for various water uses should be fixed in such a way that they cover at least the operation and maintenance costs of providing the service initially, and a part of the capital costs subsequently. These rates should be linked directly to the quality of service provided. The subsidy on water rates to the disadvantaged and poorer sections of society should be well targeted and transparent.[12]

Water Quality. Effluents should be treated to acceptable levels and standards before discharge into the natural streams. Minimum flow should be ensured in perennial streams to maintain the ecology and for social considerations. The principle of polluter pays should be followed in management of polluted water.

Water Conservation. Water use efficiency should be optimised and an awareness of water as a scarce resource should be fostered. Conservation consciousness should be promoted through education, regulation, incentives and disincentives.

---

[10] http://wrmin.nic.in/policy/nwp2002.pdf (visited in January 2006).
[11] This ranking of priorities is in contrast to those in Australia where, in some jurisdictions, meeting the needs of the environment has the highest priority.
[12] It is interesting to note that the environmental costs of water use are not being addressed through the water charge.

14.12 INTER-STATE WATER DISPUTES

Most of the major rivers in India cross state boundaries. Often water disputes arise among the basin states over the use, distribution or control of the waters. The *Inter-State Water Dispute Act* was enacted in 1956 to adjudicate disputes over waters of inter-state rivers. The *Act* was amended in March 2002 and came into force in August 2002. Amendments included the time frame for the establishment of the Inter-State Water Disputes Tribunal and also prescribed the time limit for tribunals to hand down their decisions (Ministry of Water Resources, 2005, pp. 56 and 57). In the amended *Act*, the National Government will have to constitute a Water Dispute Tribunal within a year of the date of receipt of a request from any state government. The decision of the Tribunal shall have the force of decree of the Supreme Court. To date, the Indian Government has set up five tribunals.[13] These are:

(1) Godavari Water Disputes Tribunal in April 1969;
(2) Krishna Water Disputes Tribunal in April 1969;
(3) Narmada Water Disputes Tribunal in October 1969;
(4) Ravi and Beas Waters Tribunal in April 1986; and
(5) Cauvery Water Disputes Tribunal in June 1990.

SECTION B: THE NATIONAL RIVER-LINKING PROJECT

14.13 INTRODUCTION

India has a rapidly increasing population, which requires more food production and hence more land to be developed for irrigated crops. It is also faced with serious flood and drought problems that may be exacerbated by climate change and imbalance in water availability between different parts of the country. The Indian Government considers large-scale inter-basin water transfer from water-rich to water deficit river basins as one of the ways to address these problems. It is envisaged that reservoirs in water-rich river basins such as Brahmaputra, Ganga, Mahanadi, Godavari and west flowing rivers could store large volumes of water to be transferred to water deficit areas to develop irrigation, and provide additional domestic and industrial water supply and generate hydro-power.

Currently, inter-basin water transfer helps supply water to major Indian cities. Delhi, on the banks of the Yamuna River, now depends on long distance water transfer of water from many river basins. The Bhakra Dam on the Sutlej River in the Indus Basin, as well as Ramganga Dam on the Ganga system, already supply part of Delhi's needs. The Tehri, Renuka and Kishau dams are expected to meet additional future demands of Delhi. Hyderabad, Chennai, Mumbai and a number of other cities are similarly dependent on inter-basin water transfer (Rangachari *et al.*, 2000, p. 231).

In the following sections some of the existing inter-basin water transfer projects, major river-link proposals suggested in 1970s, and the current National River-Linking Project are described.

14.14 EXISTING PROJECTS

Existing inter-basin water transfer projects in India includes:[14]

- **Kurnool Cuddapah Canal.** This project was started by a private company in 1863. It transfers water from Kurnook on the Tungabhadra River (tributary of Krishna River) southeastward to Cuddapah on the Pennar River. The canal is 304 km long with a capacity at its head of $85\,m^3\,s^{-1}$ ($2.7 \times 10^9\,m^3$ per annum), and is used to irrigate 52 750 ha. The scheme was taken over by the Government of India in 1882.
- **Periyar Project.** The project is one of the most notable engineering achievements of the nineteenth century. The cross-boundary project transfers water from Periyar River catchment in Kerala to Vaigai River catchment in Tamil Nadu. A masonry gravity dam 47.3 m high was constructed across the west flowing Periyar River and a 1740 m long tunnel with a discharge capacity of about $41\,m^3\,s^{-1}$ ($1.3 \times 10^9\,m^3$ per annum) was built to convey water eastward to Vaigai catchment. The project, commissioned in 1895, initially provided irrigation water to 57 900 ha, which has since been expanded to 81 000 ha. The project also generates 140 MW of hydro-power.
- **Parambikulam Aliyar Project.** This project was built during the 1960s, following the 1958 agreement between Kerala and Tamil Nadu. It is a complex multi-purpose project of seven streams consisting of five west and two east flowing rivers. These rivers were dammed and their reservoirs linked by tunnels. The project enables transfer of water eastward from Parambikulam (Chalakudi) catchment in Kerala to

---

[13] http://wrmin.nic.in/cooperation/disputes.htm (visited in January 2006).
[14] Description of these projects is based on the URL: http://www.sdnpbd.org/river_basin/whatis/whatis_history.htm (visited in January 2006).

Aliyar (Bharathapuzha) and Cauvery catchments in Tamil Nadu and water is ultimately delivered to drought prone areas of Coimbatore district of Tamil Nadu and the Chittur area of Kerala states. Water is used to irrigate 162 000 ha. It also generates 185 MW of hydro-power at four power stations.

- **Telugu Ganga Project.** This project connects the Srisailam Reservoir on the Krishna River in Andhra Pradesh with the Poondi Reservoir near Chennai in Tamil Nadu. The length of the main canal connecting two reservoirs is 434 km. The main objectives of this project were to:[15] convey $425 \times 10^6 \, m^3$ of Krishna water to supply Chennai metropolitan area; divert $821 \times 10^6 \, m^3$ of Krishna water to irrigate 111 290 ha in Kurnool and Cuddapah districts of Andhra Pradesh; and to divert $890 \times 10^6 \, m^3$ of Pennar waters in Andhra Pradesh to irrigate 123 434 ha.

- **Ravi-Beas-Sutlej links and the Indira Gandhi Canal Project.** Under the Indus Waters Treaty of 1960 (see section 2.9.1), waters of three eastern rivers were allocated to India. As the land benefited in India lies mostly to the east and south of these rivers, the rivers had to be linked and the water conveyed to canal systems. Ravi waters are diverted eastward to the Beas River and from there to Sutlej River via link canals (Gulhati, 1973, Figure 6). The Sutlej River supplies the 645 km long Indira Gandhi Canal along the border with Pakistan. It has a capacity of $524 \, m^3 \, s^{-1}$ and delivers $9.36 \times 10^9 \, m^3$ of $10.6 \times 10^9 \, m^3$ of water per annum allocated to Rajasthan from Ravi and Beas rivers.[16] Diverted waters have been used for development of irrigation in the Thar Desert. Construction of the Indira Gandhi Canal commenced in 1958 and completed in 1986.

14.15 RIVER-LINKING PROPOSALS OF THE 1970S

Two proposals for a National Water Grid were developed in the 1970s and attracted considerable attention. These were:[17]

- **The 1972 proposal developed by K. L. Rao.** This proposal linked the Ganga River in the State of Bihar to Cauvery River in the Tamil Nadu in the south (Figure 14.5) using 2640 km of canals. It involved the withdrawal of $1680 \, m^3 \, s^{-1}$ of the flood flows from the Ganga for about 150 days in a year and pumping approximately $1400 \, m^3 \, s^{-1}$ of this over a head of 549 m for delivery to the Peninsular region. It was proposed to use the remaining $280 \, m^3 \, s^{-1}$ in the Ganga Basin itself. The scheme required approximately 5000–7000 MW to lift the water and aimed to bring an additional 4 Mha under irrigation. The estimated cost of the proposal in 2002 values was around 1500 billion Rupees or US$33 billion. The Central Water Commission examined the proposal and found it to be economically prohibitive and the proposal was not pursued.

- **The 1977 Himalayan and Garland Canals proposed by Captain Dastur.** The proposal consisted of two canals[18] (Figure 14.6). The first was a 4200 km long, 300 m wide canal at the foot of Himalayan slopes, running from Chenab in the west to Brahmaputra and beyond in the east to be fed by the Himalayan river waters stored in 50 reservoirs and 40 reservoirs beyond Brahmaputra. The total annual capacity was $247 \times 10^9 \, m^3$. The second was 9300 km long, 300 m wide Garland Canal in the central and southern parts of the country. The Garland Canal was intended to have about 200 reservoirs with a total storage capacity of $497 \times 10^9 \, m^3$. The Himalayan and Garland canals were to be inter-connected at two places by pipelines. It was estimated that all the surplus waters in the country would be used to irrigate approximately 219 Mha. The estimated cost of the proposal in 2002 values was about 700 000 billion Rupees[19] or US$15 500 billion. Two committees of senior engineers, government officials, and scientists from universities and research institutions examined the proposal and concluded that it was technically and economically unsound.

14.16 THE NATIONAL RIVER-LINKING PROJECT

In 1980, the Ministry of Water Resources devised a National Perspective Plan (NPP) for water resources development by transferring water from water rich to water deficit basins through the linking of rivers.[20] The Plan has two main components: the Peninsular Rivers Development; and the Himalayan Rivers Development. A feature of the NPP

---

[15] http://nwda.gov.in/writereaddata/sublink2images/193.pdf (visited in January 2006).
[16] http://www.rajirrigation.gov.in/4ignp.htm (visited in January 2006).
[17] Description of these projects is based on: http://nwda.gov.in/index2.asp?sublinkid=47 (visited in January 2006).
[18] http://www.sdnpbd.org/river_basin/whatis/whatis_history.htm (visited in January 2006).
[19] http://nwda.gov.in/index2.asp?sublinkid=47 (visited in January 2006).
[20] http://wrmin.nic.in/interbasin/riverlink.htm (visited in January 2006).

Figure 14.5 Rao's River Linking Proposal.[21]

is that the transfer of water from water rich basins to water deficit basins would essentially be by gravity. Any pumping required would not exceed lifts of 120 m. The National Water Development Agency (NWDA) was set up in July 1982 to carry out studies for optimum use of the water resources of the Peninsular river systems and to prepare feasibility reports. In 1990, NWDA was also entrusted with the task of the Himalayan Rivers Development component of the Plan.[22] The functions of NWDA includes:[23]

- Promotion of scientific development for optimum use of water resources in India.
- Carrying out detailed surveys and investigations of possible storage reservoir sites and inter-connecting links in order to establish feasibility of the proposal of Peninsular and Himalayan Rivers Development components of the NPP.
- Undertaking detailed studies on the amount of water in various Peninsular and Himalayan river systems, which can be transferred to other basins/states after meeting reasonable future needs of source basin/states.
- Preparation of feasibility reports of various components of the scheme relating to Peninsular and Himalayan rivers.

[21] http://nwda.gov.in/writereaddata/sublinkimages/14.jpg (visited in January 2006).
[22] http://nwda.gov.in/indexab.asp (visited in January 2006).
[23] http://nwda.gov.in/indexab.asp (visited in January 2006).

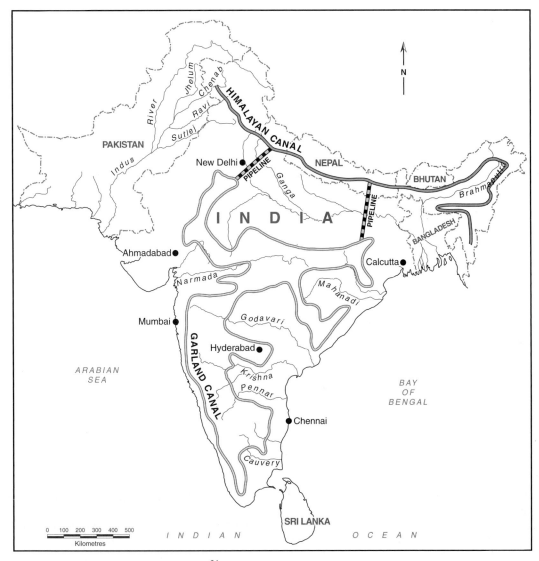

Figure 14.6 Captain Dastur's Garland Canal proposal.[24]

- Undertaking other actions that the society may consider necessary to achieve the above objectives.

The NWDA originally planned to complete the Peninsular Link Project by 2035 and the Himalayan Link Project by 2043. In September 2002, a petition was filed in the Supreme Court, quoting a speech of the President of India on 14 August 2002. In it he mentioned the need for the networking of rivers and prayed for appropriate directions. The Supreme Court made a series of orders from 31 October 2002 to 8 April 2005 that included:[25]

- The planned timetable for construction of the project was found to be unacceptable. In response a Task Force was set up in December 2002 and set out the timetable for commencement of construction works by 2007 and their completion by the end of the 2016.

- The preparation of Detailed Project Reports must include detailed Environmental Impact Assessments, Environmental Management Plans and Rehabilitation and Resettlement Plans for affected communities. In response, the Ministry of Water Resources set up a Committee of Experts of environmentalists, social scientists and others in December 2004. This committee will be involved in the consultative process. The Terms of Reference and other information about the Committee are available at the URL provided.[26]

[24] http://nwda.gov.in/writereaddata/sublinkimages/15.jpg (visited in January 2006).
[25] http://nwda.gov.in/psearchdetailmain.asp?pageid=97&linkpos=1 (visited in January 2006).
[26] http://nwda.gov.in/psearchdetailmain.asp?pageid=96&linkpos=1 (visited in January 2006).

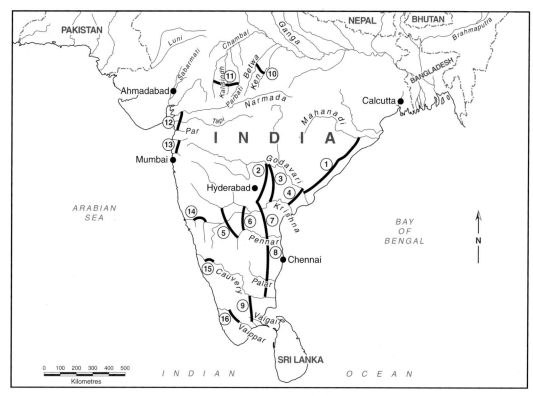

Figure 14.7 The Peninsular component of the National River-Linking Project.[27]

- The feasibility reports must be placed on a website soon after their completions so that the environmentalists and concerned individuals can provide comments that must be considered. In response, most of the feasibility reports prepared for the Peninsular Development component have become available at the URL provided.[28]

The work of the Task Force set up in response to the Supreme Court's orders included[29] the suggestion of methods for arriving at speedy consensus amongst the States for the sharing and transfer of water and the provision of guidance on standards for appraisal of individual projects in terms of economic viability, socio-economic impacts, environmental impacts and preparation of resettlement. In addition the Task Force had to prioritise project components, to prepare and implement Detailed Project Reports, and develop a suitable organisational structure to implement the project and funding. The Task Force held 15 meetings between 2002 and 2004 and submitted two reports to the Government about its activities. The Task Force activities ended on 31 December 2004 and a Special Cell has been created under the Ministry of Water Resources to look after the remaining work of the Task Force and to take further action.

14.16.1 Peninsular Rivers Development Component

The Peninsular Rivers Development is divided into four major parts (Ministry of Water Resources, 2005):

(1) linking of the Mahanadi, Godavari, Krishna, Pennar, and Cauvery rivers (Figure 14.7);
(2) linking of the west flowing rivers north of Mumbai;
(3) linking of the Ken–Betwa and Parbati–Chambal rivers; and
(4) diversion of the west flowing rivers of Kerala and Karnataka to water deficit areas east of the Western Ghats.

The NWDA has completed water balance studies of all 137 catchments/sub-catchments and 52 diversion points, 58 reservoir studies, topographic studies of link canals and almost all pre-feasibility studies. The present status of the feasibility reports for these links is shown in Table 14.5.

---

[27] http://nwda.gov.in/writereaddata/sublinkimages/13.jpg (visited in January 2006).
[28] http://nwda.gov.in/index2.asp?sublinkid=62 (visited in January 2006).
[29] http://nwda.gov.in/psearchdetailmain.asp?pageid=88&linkpos=1 (visited in January 2006).

Table 14.5. *Status of feasibility reports of the Peninsular Rivers Development Component*

No.	Name of the link project	Status of feasibility report[a]
1	Mahanadi (Manibhadra)–Godavari (Dowlaiswaram) link	Completed
2	Godavari (Inchampalli)–Krishna (Nagarjunasagar) link	Completed
3	Godavari (Inchampalli)–Krishna (Pulichintala) link	Completed
4	Godavari (Polavaram)–Krishna (Vijayawada) link	Completed
5	Krishna (Almatti)–Pennar link	Completed
6	Krishna (Srisailam)–Pennar link	Completed
7	Krishna (Nagarjunasagar)–Pennar (Somasila) link	Completed
8	Pennar (Somasila)–Palar–Cauvery (Grand Anicut) link	Completed
9	Cauvery (Kattalai)–Vaigai–Gundar link	Completed
10	Ken–Betwa link	Completed
11	Parbati–Kalisindh–Chambal link	Completed
12	Par–Tapi–Narmada link	Completed
13	Damanganga–Pinjal link	Completed
14	Bedti–Varda link	[b]
15	Netravati–Hemavati link	[b]
16	Pamba–Achankovil–Vaippar link	Completed

Note: [a] Completed feasibility reports are available at the URL http://nwda.gov.in/index2.asp?sublinkid=62 (visited in January 2006). [b] The field survey and investigations for preparation of feasibility report of these links will be conducted after obtaining clearance from the Government of Karnathaka.
Source: Ministry of Water Resources (2005, p. 12).

14.16.2 Himalayan Rivers Development Component

Studies of Himalayan Rivers Development Components started during 1991–92 (Ministry of Water Resources, 2005). These envisage construction of reservoirs on the principal tributaries of the Ganga and the Brahmaputra in India, Nepal and Bhutan, along with linking canal systems to transfer excess waters of the western tributaries of the Ganga to the west (Luni and Sabarmati), as well as linking of the eastern tributaries of Ganga, and main Brahmaputra and its tributaries with Mahanadi (Figure 14.8).

The NDWA has completed water balance studies at 19 diversion points, topographic studies of 16 reservoirs and pre-feasibility report of 14 links. The status of the feasibility reports of the links is shown in Table 14.6. To date (January 2006) no feasibility reports of the link projects for this component have been released.

14.16.3 Benefits of the Project

Preliminary assessment indicates that implementation of the National River-Linking Project would provide between 200 and 300×10^9 m³ of additional water per year by transferring water from water rich to water deficit basins.[30] An additional 25 Mha of land could be irrigated by surface water, and 10 Mha by increased use of groundwater, raising the ultimate irrigation potential from 113 to 148 Mha (Ministry of Water Resources, 2005, p. 16). The Scheme would also generate 34 000 MW of hydro-power. Other projected benefits include flood control, navigation improvement, fisheries, and water quality control.

The Himalayan component of the Project is planned to benefit the states of Uttar Pradesh, Haryana, Rajasthan, Gujarat, Assam, West Bengal, Bihar, and Orissa, and enrich the peninsular component with waters of the Brahmaputra. The peninsular component is aimed at benefiting Andhra Pradesh, Orissa, Karnataka, Tamil Nadu, Kerala, Madhya Pradesh, Rajasthan, Maharashtra and Gujarat (Ministry of Water Resources, 2005, p. 16).

Food production is projected to be almost doubled from 220 Mt to 430 Mt. Other identified benefits include creation of numerous construction jobs as well as new jobs in manufacturing, transportation, agriculture, fisheries and tourism. Creation of on-farm and off-farm jobs is designed to prevent the exodus of rural populations.

14.16.4 Timetable and Cost of the Project

Planning of the Project is intended to be completed by 2006 and its construction by 2016. The tentative estimated cost of

---
[30] http://www.sdnpbd.org/river_basin/progress_so_far.htm (visited in January 2006).

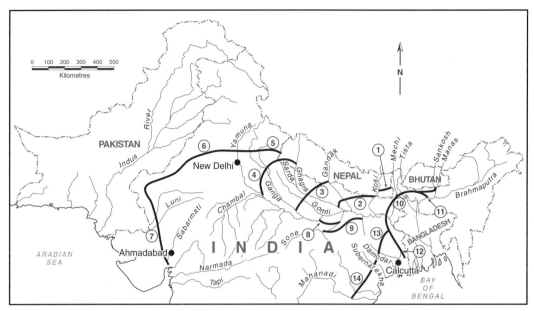

Figure 14.8 The Himalayan component of the National River Linking Project.[31]

the Project is approximately US$112 to US$200 billion,[32] representing 23 to 40 percent of the country's GDP in 2001 (US$493 billion). The scale of this water project is beyond current world experience. The haste to complete such a complex mega-project by 2016 seems unreasonable, given the plethora of economic, social, environmental and political issues that still need to be investigated and considered before a final decision can be made (Ray, Unpublished, Chapter 5).

The released feasibility reports have only short chapters on the environmental impacts of the project. However each link project requires a detailed *Environmental Impact Assessment* and *Environmental Management Plan*. These reports are critical for the Indian Government in order to persuade State Governments and the Indian community about the soundness of the Scheme and its impacts; and to obtain loans from the World Bank.

14.16.5 Impacts of the Project

The Project raises a vast array of uncertainties over adverse environmental, economic and social impacts not just in India, but also in Bangladesh.[33] In 1974, India completed the Farakka Barrage on the Ganga River close to the border with Bangladesh, diverting crucial dry season flows into Indian irrigation fields. This was a source of considerable tension between two countries and resulted in the 1996 treaty, which commits India not to reduce transboundary flows further. The National River-Linking Project is also opposed by NGOs in Nepal, which have launched vigorous campaigns against it (Ray, unpublished, Chapter 5). The Himalayan component of the project, which is the major source of excess water, is faced with numerous problems, the main ones of which are summarised below (Ray, unpublished, Chapter 5).

Geological risks. Construction of a large number of dams in this geologically unstable and seismically active area has great risks of triggering earthquakes and consequent dam failure.

Reduction in sediment loads. The Ganga and Brahmaputra rivers together carry on average sediment loads of $1.7-2.4 \times 10^9$ tonnes per year to the $20\,000\,km^2$ delta of these rivers in Bangladesh and India. Construction of a large number of dams in these two vast river basins will trap sediments, reducing sediment supply to the delta. This could have serious environmental impacts on fishery, wildlife, forests, and on stability of the coastlines, and social costs on those that depend on these areas for their livelihoods.

Environmental and social impacts. Substantial deforestation, mismanagement of land and water resources, and sea level rise in the Ganga–Brahmaputra Delta have already

---

[31] http://nwda.gov.in/writereaddata/sublinkimages/12.jpg (visited in January 2006).
[32] http://www.sdnpbd.org/river_basin/whatis/whatis_cost_project.htm (visited in January 2006).
[33] Probable impacts on Bangladesh include reduction in availability of surface water and groundwater, increase in water pollution and deterioration of water quality, changes in river morphology, increase in surface water, groundwater and soil salinity, change in flood regime impacting on the agricultural practices, destruction of the ecosystems, and reduction in fish production (http://www.sdnpbd.org/river_basin/bangladesh/bangladesh_probable_impact_on_bd.htm visited in January 2006).

Table 14.6. *Status of feasibility studies of the Himalayan Components of the Linking Project*

No.	Name of the link project	Status of feasibility report
1	Kosi–Mechi link	In progress
2	Kosi–Ghagra link	In progress
3	Gandak–Ganga link	In progress
4	Ghagra–Yamuna link	Completed
5	Sarda–Yamuna link	Completed
6	Yamuna–Rajasthan link	In progress
7	Rajasthan–Sabarmati link	In progress
8	Chunar–Sone Barrage link	In progress
9	Sone Dam–Southern tributaries of Ganga link	In progress
10	Manas–Sankosh–Tista–Ganga link	In progress
11	Jogighopa–Tista–Frakka link	In progress
12	Farakka–Sunderbans link	In progress
13	Ganga (Farakka)–Damodar–Subernarekha link	In progress
14	Subernarekha–Mahanadi link	In progress

Source: Ministry of Water Resources (2005, p. 16).

had severe effects on its ecological balance. It has been estimated that within the next 15–20 years 15 percent of the habitat area could be lost, displacing more than 30 000 people. Diversion of the Himalayan rivers will aggravate the situation with serious impacts on fisheries, mangroves and livelihood of the local population who depend on these resources.

Construction of the Project within India will inundate an estimated 8000 km^2 of land and will destroy habitats of native flora and fauna. It could leave up to 3 million people homeless.[34] Opponents of the project argue that in the past, the Central Water Commission has manipulated water resources and related data to influence outcomes of feasibility reports. There is no guarantee to believe that these practices have not been used again in the Project's feasibility reports (Ray, unpublished, Chapter 5).

14.16.6 Financial Perspectives

To finance the project, a number of options have been considered (Ray, unpublished, Chapter 7). The Task Force has suggested imposing additional taxes. This mechanism is not considered realistic because most farmers currently cannot even afford the modest payment for irrigation water, let alone any additional taxes. It has been suggested that the private sector could finance the project. However, this option is out of the question because the private sector currently does not have the necessary resources. International agencies such as the World Bank and the Asian Development Bank are capable of fully or partially funding the project. But they would demand an open, transparent and publicly verifiable environmental, social and economic impact assessment and financial capability evaluation. Such assessments would explore alternative and perhaps cheaper and environmentally less damaging options to achieve similar outcomes.

In spite of the above speculations regarding financing of the project the feasibility reports will have to be followed by more detailed studies in the form of Detailed Project Reports (DPR) to address various aspects of the Project including its financial requirements. Therefore, the exact requirement of funds can be firmed up only after preparation of DPRs for all the links. So far (January 2006), preparation of a DPR for only one of the links (the Ken-Betwa Link) has commenced and is expected to be completed in 2008. It is only after the DPRs are completed and the financial dimensions are clearly delineated, it would be possible to consider various options for financing.[35]

14.16.7 Feasibility Reports of the Peninsular Rivers Development Component

The preparation of feasibility reports has aimed mainly to facilitate firming up of the proposals and discussions among the concerned states. These discussions are intended to reach broad agreements on the amount of diversions, the uses of water, and how costs and benefits will be shared. Preliminary

---

[34] http://www.sdnpbd.org/river_basin/persons_behind/canappan1.htm (visited in January 2006).
[35] Bhandari, N. K., NWDA's Chief Engineer (personal communication, January 2006).

feasibility reports have been prepared progressively and submitted to the relevant states. These have been revised and released on the internet following comment by the states. To date (January 2006), 14 completed feasibility reports for the Peninsular component of the Project (see Table 14.5) have been released on the website of the NWDA.[36] Each feasibility report contains an executive summary, location map, salient features, introduction, physical features, interstate aspects, survey and investigations, hydrology, design features, reservoirs, irrigation planning, command area development, construction programme, manpower, environmental and ecological issues, cost estimate, and financial aspects.

Table 14.7 shows some features of the Peninsular Link projects. It indicates that the volume of diversion will be more than $70 \times 10^9 \, m^3$. This water will be used for irrigating 3.4 Mha and supplying domestic and industrial water to major population centres such as Chennai and Mumbai. As well, a large volume of water will be released to other rivers. The hydro-power generating capacity of the projects is 3874 MW. Water transfer between rivers will be mostly by gravity and to a small extent by pumping via 4440 km of lined canals and 153 km of tunnels. The only major project that needs pumping is Project No. 2 where a large volume of water ($16.4 \times 10^9 \, m^3$ per year) will need to be lifted by 107 m in four stages requiring 1981 MW. This energy requirement is more than double of the 975 MW expected installed capacity of this component of the Project.

All 14 projects in Table 14.7 would cost about US$19.2 billion, with the benefit/cost ratios range between 1.01 and 3.93. Approximately 1320 villages will be totally or partially affected by the reservoirs and canals requiring resettlement of about 264 000 people. The expected construction time for various projects varies between 5 and 12 years depending on their complexities.

Impacts of the Peninsular Rivers Development Component

Each one of the 14 released feasibility reports contains a chapter assessing the impacts of the link projects. Assessed issues for reservoirs include: submerged area, land acquisition, affected people, resettlement and rehabilitation, reservoir sedimentation, seismicity, aquatic life, flora and fauna, and public health. For the link canals and the irrigation command areas, issues include: land acquisition, rehabilitation and resettlement of the affected people, groundwater regime, flood control, pollution and industrial development, aquatic life, public health, and waterlogging and salinity.

Feasibility reports recognise that construction of reservoirs will cause loss of habitat for wildlife and endangered species such as the Leopard (*Panthere pardus*), the Leopard cats (*Felis bengalensis*), Indian Wolf (*Canis lupus*), the Great Indian Bustard (*Choriotis nigriceps*) and Black Buck (*Antelope ceivicapra*). However, they believe that large forests are available in the surrounding regions for the migration of wildlife and endangered species. Reports also argue that creation of permanent waterbodies will be beneficial for wildlife existing in the surrounding forests.

Watertables in command areas are expected to rise due to application of irrigation water and could result in waterlogging and salinisation. Construction of surface or subsurface drains has not been considered because of their costs. Instead, conjunctive use of surface and groundwater as well as tree planting have been recommended to prevent waterlogging and land salinisation. Little emphasis have been given to improve irrigation efficiencies to lower accessions to watertables.

It has been claimed that despite previous experience, the Project will have *tremendous socioeconomic* benefits and will generate employment during and after its construction. Moreover, it has been claimed that it will not have any serious adverse impacts.[37] The only admitted serious problem is resettlement of about 264 000 people (Table 14.7). To overcome this, resettlement and rehabilitation plans have been developed for each link project considering requirements of resettled people. It has been proposed to resettle people in groups of 50 to 100 families in separate colonies with all basic facilities such as housing, land, road, water supply, electricity, school, health services, shopping centre, post office, police, and places of worship. It is claimed that settlements will be located in appropriate places so that the affected communities should be able to live in harmony with nature. It does not appear that the views of the affected communities have been sought in these feasibility reports.

Comments of the State Governments

The Peninsular Rivers Development Component of the National River-Linking Project cut across current water sharing agreements between the states of Andhra Pradesh, Chhattisgarh, Orissa, Maharashtra, Tamil Nadu, Gujarat, Madhya Pradesh, Rajasthan and Kerala. These states have commented on the feasibility reports. Some of their

---

[36] http://nwda.gov.in/index2.asp?sublinkid=62 (visited in January 2006).

[37] These claims are in contradiction with the past Indian experience regarding building new dams and development of surface water resources (see section 14.10).

Table 14.7. *Some features of the Peninsular Rivers Development Component*

Link project No.[a]	Transfer details and use (10^6 m^3)					Installed hydro-power capacity (MW)	Link (km)		Capital cost[b] (US$M)	Benefit/ cost ratio	Number of displaced people	Construction time (Years)
	Intake	Irrigation	Domestic and industrial	Losses	End delivery to another river		Canal	Tunnel				
1	12 165	3790	802	1073	6500	445	822	6.2	3900	1.63	21 170	8
2	16 426	1427	237	562	14 200	975	290	9.2	5800	2.25	11 220	8
3	4370	3665	413	292	–	1002	300	12.5	1120	3.41	25 000	10
4	5325	1402	162	260	3501	720	174	–	330	1.22	144 812	12
5	1980	1714	56	210	–	13	552	35.7	1500	1.20	1333	10
6	2310	–	–	215	2095	17	204[c]	–	18	2.50	–	5
7	12 146	3264	124	332	8426	90	392	1.3	1400	1.86	5148	8
8	8565	3048	1105[d]	557	3855	–	529	–	1500	3.93	–	10
9	2252	1952	185	115	–	–	256	–	595	3.14	–	6
10	1020	312	12	37	659	72	231	2.0	442	1.87	8550	9
11	1360	631	13	40	676	–	244	20.6	665	1.67	27 077	8
12	1350	1200	70	80	–	33	395	5.5	1340	1.08	14 832	7
13	577	–	577[e]	–	–	–	–	42.6	285	1.38	4530	9
14[f]	–	–	–	–	–	–	–	–	–	–	–	–
15[f]	–	–	–	–	–	–	–	–	–	–	–	–
16	634	634	–	–	–	507	51	17.0	310	1.01	297	8
Total	70 480	23 039	3756	3773	39 912	3874	4440	152.6	19 205	–	263 969	–

Note: [a] For the link project names and locations see Table 14.5 and Figure 14.7; [b] Costs (mostly in 2003–04 values) in Rupees have been converted to US$ assuming 45 Rupees for 1 US$; [c] Including 180 km of natural streams; [d] Including 813 × 10^6 m^3 for water supply of Chennai; [e] For water supply of Mumbai; [f] Feasibility report is not available.

Source: Feasibility reports at the URL http://nwda.gov.in/index2.asp?sublinkid=62 (visited in January 2006).

comments on various river link projects are summarised in the following.[38]

Project 1: Mahanadi (Manibhadra)–Godavari (Dowlaiswaram) Link

- The Government of Andhra Pradesh recognises that the Peninsular rivers are seasonal. Linking these rivers is therefore not attractive unless linked with perennial Himalayan rivers.
- The State of Orissa did not accept the validity of the NWDA studies on the availability of surplus water at Manibhadra. It believes that most of the Peninsular River Development Component projects are integrated with the Mahanadi–Godavari Link, which depends on the availability of surplus water in the Mahanadi and Godavari rivers. Therefore, the availability of surplus water at Manibhadra should be first substantiated. Moreover, the Subernarekha–Mahanadi link, part of the Himalayan River Development Component (see Figure 14.8 and Table 14.6), should be constructed first before diverting water from Mahanadi.
- The Government of Chhattisgarh did not accept the feasibility report, because it underestimates the State's water requirements and the extent of the proposed irrigated area.

Projects 2 and 3: Godavari (Inchampalli)–Krishna (Nagarjunasagar) link and Godavari (Inchampalli)–Krishna (Pulichintala) Link

- The Government of Andhra Pradesh claims that following the Godavari Water Dispute Tribunal Award, all waters proposed for inter-basin transfer from the Godavari Basin belongs to Andhra Pradesh.
- The Government of Chhattisgarh does not accept the proposed links. It believes their impacts far outweigh their limited benefits for the State. The NWDA water requirements assessment of the newly formed State of Chhattisgarh is low. New industries and power plants are being commissioned and a new capital is also to be developed. Work on these new developments will increase the State's water requirements. As well, the area of Chhattisgarh State inundated by the Inchampalli reservoir will be 5033 ha, rather than the projected 2140 ha. This submergence includes 3079 ha of reserve forest and 50 ha of the National Indravati Tiger Reserve Sanctuary and wildlife.

Project 4: Godavari (Polavaram)–Krishna (Vijayawada) Link

- The Government of Chhattisgarh found this proposal unacceptable. It considers that the NWDA study report has not properly assessed irrigation in the sub-basins to be benefited by the surplus water. The study should have considered the transfer of $2265 \times 10^6 \, m^3$ identified in the Godavari Water Dispute Tribunal Award rather than the $5525 \times 10^6 \, m^3$ proposed in the Feasibility Report. The benefits to the State are meagre compared with the cost of the valuable 3605 ha of reserve forest area and 13 villages to be submerged.
- The Government of Orissa requested a detailed study of land submergence in its state as a consequence of the Polavaram dam construction. It stresses that environmental impact assessment should include rehabilitation and resettlement of affected communities.

Project 8: Pennar (Somasila)–Palar–Cauvery (Grand Anicut) Link

- The Government of Kerala found that the feasibility report did not contain a provision for any share of water for the upper riparian states of Cauvery Basin. Instead, the proposed link appears to unfairly benefit only one basin state (Tamil Nadu). Kerala requested a proportionate share for the upper riparian states to be included in the feasibility report. Moreover, because of the claims by the Cauvery River Basin states of Kerala, Tamil Nadu, Karnataka and Pondicherry, no link project should be undertaken in the Basin until final adjudication of the case by the Cauvery Water Disputes Tribunal.

Project 9: Cauvery (Kattalai)–Vaigai–Gundar Link

- This is the last component of the Mahanadi–Godavari–Krishna–Pennar–Cauvery–Vaigai–Gundar Peninsular River links. The Government of Kerala has commented that the feasibility study has substantially decreased estimates of the ultimate requirements of the State by 2050. The Cauvery River has its catchment spread over three states of Kerala, Karnataka and Tamil Nadu as well as the Union Territory of Pondicherry. Any link project that would potentially

---

[38] http://nwda.gov.in/index2.asp?sublinkid-62 (visited in January 2006).

benefit one particular state and overlook the upper riparian states is unacceptable. The sharing of Cauvery water among the basin states is under dispute so the Central Government of India has set up the Cauvery Water Disputes Tribunal to resolve the dispute. Therefore, diversion of Cauvery waters to other basins cannot be undertaken until the final adjudication by the Tribunal.

Project 11: Parbati–Kalisindh–Chambal Link

- The Government of Rajasthan has objected to construction of seven dams on the upper reaches of Chambal River in Madhya Pradesh that would affect water availability in Rajasthan. An agreement already exists between the two states regarding equal sharing of the Chambal River's water. This link project would violate the agreement. As a consequence, a new agreement on sharing cost and benefits would have to be reached between Madhya Pradesh and Rajasthan before the link project could be constructed.

Project 13: Damanganga–Pinjal Link

- The Government of Gujarat commented that this project, paradoxically, diverts water from a surplus basin to another surplus basin and is therefore not in line with the National Water Policy[39] which emphasises the diversion from surplus to deficit basins. The objective is to supply Mumbai annually with $577 \times 10^6 \, m^3$ of domestic and industrial water. This could be achieved by diverting water from large, high rainfall catchments close to Mumbai. The Government of Gujarat therefore requested reexamination of the project and suggested linking the Damanganga River with the Sabarmati River in order to overcome acute shortage of water in drought-prone regions of Gujarat. The Government of Maharashtra considered Gujarat's request unacceptable, because the National Water Policy gives highest priority to domestic water supply. Maharashtra wishes to divert Damanganga water to Mumbai for domestic purposes.

Project 16: Pamba–Achankovil–Vaippar Link

- While the project is acceptable to the Government of Tamil Nadu, the Government of Kerala commented that the feasibility report is based on the incorrect presumption that surplus water is available in the Pamba and Achankovil rivers in Kerala, so that $634 \times 10^6 \, m^3$ can be diverted annually to Tamil Nadu. As well, the NWDA has not investigated the ecological and environmental impacts of the link project on the Vembanad Lake and has overlooked the water requirements of a planned 1500 MW power production plant. A 1999 detailed investigation of six rivers draining into Vembanad Lake demonstrated that all six rivers, particularly the Pamba River, would face deficits. The estimated deficit for the Pamba and Achankovil rivers by 2051 will be $3537 \times 10^6 \, m^3$ and $495 \times 10^6 \, m^3$ respectively. The Government of Kerala maintains that the export of more water from Kerala to Tamil Nadu is not feasible.

The above gives a snapshot of the complexity of the issues and the strong objections of the concerned states to the Peninsular Rivers Development Component. A consensus has to be reached among them before this component can be implemented. Even if a consensus is reached, separate inter-state agreements will still have to be reached between all the concerned states for sharing of the surplus water. As well, existing inter-state agreements for various river basins will have to be reviewed and modified (Feasibility Report of the Link Project 1, Chapter 3, pp. 3–4).

14.16.8 Alternatives to the National River-Linking Project

Opponents believe that the project is already expensive and its cost will escalate as has happened in numerous water projects in the past. The project will inundate large areas of valuable forests, damage the environment and impact on the wildlife and endangered species. They argue that the State and Central Governments can ease water shortages in India by undertaking a range of measures.[40] India has numerous uncompleted water projects worth billions of dollars that should be completed before undertaking any new projects. Efforts should be made to harvest local rainwater by storing it in tens of thousands of reservoirs or tanks that governments have abandoned. Governments could do more to develop alternative strategies such as increasing water use efficiency in irrigated farms, in domestic water uses, and in industries. The price of water is too low and is heavily subsidised.

---

[39] See section 14.11.
[40] http://www.sdnpbd.org/river_basin/persons_behind/canappan1.htm (visited in January 2006).

Efforts to discourage overuse, wastage and pollution should be increased.

14.16.9 Assessment of the Project by the International Water Management Institute

Because of the number of controversies surrounding the project, and strong opposition to it by academia, environmental groups, NGOs, and media, the International Water Management Institute (IWMI) has commenced a three-year research programme in 2005. This project, in collaboration with a number of institutions in India,[41] aims to study various alternatives and to develop a coherent plan of action to meet India's water challenges up to 2050. The research program is entitled *Strategic Analyses of India's National River-Linking Project*. The ultimate goals are to promote a balanced, analytical and informed national discourse on India's Water Future, and assess various aspects of the National River-Linking Project. It has four specific objectives:[42]

- To build scenarios of what India would be like (economy, society, demography, habitat, and environment) in 2025 and 2050, and to predict their implications for its water future?
- To analyse whether the National River-Linking Project is an adequate, cost effective, and sustainable in social, ecological, and political terms towards meeting the water challenge.
- To bring together a number of institutional and policy interventions into a National Water Sector Perspective Plan.
- To identify best practices to implement the National River-Linking Project as well as the National Water Sector Perspective Plan.

The research programme will have three phases:

- **Phase I (India's Water Future to 2025 and 2050).** This phase will last for 9 months and research will focus on building a sharp prognosis of India's water future for 2025 and 2050. It includes activities such as: (1) estimates of river basins demographic projection; (2) impact assessment of demographic growth on domestic water demand; (3) scenarios of regional (river basins) patterns of changing food consumption and their impacts on agriculture water demand; (4) national and regional (river basins) economic growth patterns and implications for domestic and industrial water demand; (5) national food security; (6) potential for improving water productivity at different river basins; (7) assessment of present state of the river ecologies and ecosystems; (8) environmental impacts of growing intensification of groundwater irrigation; and (9) scenarios of water supply and demand for 2025 and 2050.
- **Phase II (Strategic Analysis of National River-Linking Project).** A comprehensive financial, economic and social cost benefit analysis of the National River-Linking Project will be carried out in this 15-month phase. In addition, three important link projects will be chosen for detailed study.
- **Phase III (National Water Sector Perspective Plan).** This 12-month phase will explore alternate options if the National River-Linking Project fails to be implemented. IWMI will study various alternatives and will develop a coherent plan of action to meet India's water challenge up to 2050.

14.17 CONCLUSIONS

In this chapter, some of the complex issues surrounding inter-basin water transfer in India, which is rapidly developing, have been considered. India has a federal system of government where ownership of water is vested in the state governments. This ensures that inter-basin transfer of water between states will seldom be straightforward. The distribution of water in India is spatially and temporally variable with heavy reliance on monsoon rains that fall in 90 to 105 days per year. Because of this, water harvesting and redistribution to irrigation have been used for thousands of years. The British colonial period saw major expansion in water transfer and irrigation schemes which has intensified since independence in 1947.

India's population of 1.03 billion in 2000 is expected to reach 1.8 billion by 2050 so that the present average annual surface run-off per head of population of approximately 1800 m^3 will fall to around 1000 m^3, suggesting that the country as a whole could then be considered water stressed. More than 4000 large dams have been built since its independence to feed its population, supply population centres,

---

[41] (1) Institute of Rural Management, Anand, Gujarat; (2) Gujarat Institute of Development Research, Ahmedabad; (3) Central Water Commission, New Delhi; and (4) National Water Development Agency, New Delhi.
[42] http://www.iwmi.cgiar.org/iwmi-tata/files/PDF/CP_NRLP_B.pdf (visited in January 2006).

generate hydro-power, control floods and manage through periodic severe droughts.

Around 84 percent of India's annual water use of $750 \times 10^9 \, m^3$ in 2000 was used for irrigation. Irrigation has relatively low water use efficiency, approximately 40 percent in canal irrigation and 60 percent for groundwater irrigation schemes. Currently, approximately 10 percent of India's irrigated area is affected by salinity or waterlogging. The plans to massively expand the area irrigated poses additional problems since the drainage necessary to control waterlogging and salinity is seen as too expensive to be included. Water pricing for irrigation is calculated on the land area under irrigation and on crop type rather than on the volume of water used. It has been estimated that in canal-irrigated areas, less than 1 percent of the cost of farming is due to the cost of water. In addition, many farmers currently cannot even afford the present meagre prices for irrigation water. There appear to be potentially vast water savings possible if water conservation and realistic water pricing schemes can be implemented.

The annual water use in India is expected to rise to $1300 \times 10^9 \, m^3$ by 2050, a rise of 73 percent over use in 2000. Such a large expansion in water requirements, coupled with the projected impacts of global warming, poses major problems. The *National Water Policy*, revised and updated in 2002, recognises that water is a scarce and precious national resource and lays down the broad principles that govern the management of the country's water resources. The policy sees the provision of adequate safe drinking water in both urban and rural areas as a national goal with the treatment of discharge waters to acceptable standards as a necessary measure. It emphasises that the preservation of the quality of the environment and the ecological balance should be a primary consideration and that groundwater extraction should not exceed the recharge rate. The Policy requires that water use charges should at least cover the supply systems operation and maintenance costs. Its priorities for water allocation planning and operations are, broadly, first drinking water, second irrigation, third hydro-power, fourth ecology, fifth secondary and tertiary industries, sixth navigation and then other uses. While it seeks to raise awareness on water use efficiency and to promote water conservation education, its goal is also to maximise the use of the country's water resources through non-conventional means such as inter-basin water transfer. Like many national water policies elsewhere in the world there are tensions and potential conflicts within and between the principles enunciated in the *National Water Policy*.

In inter-basin water transfer, India has been planning the massive National River-Linking Project since 1980 to link seasonal Peninsular rivers with perennial Himalayan rivers. The current ambitious objectives are to finalise planning by late 2006 and complete the project by 2016. The Project faces major objections within India because of its huge cost (US$112 to US$200 billion), social and environmental impacts and disputed benefits. Outside India, Bangladesh, Bhutan and Nepal oppose the Himalayan Rivers Development Component of the Project because of its impacts on water availability, environments, and their economic and social well being.

To date (January 2006), 14 feasibility reports for the Peninsular Rivers Development Component have been released, but no feasibility report for the Himalayan Rivers Development Component is available. The response from the affected states has been less than enthusiastic. Linking seasonal Peninsular rivers is viewed unfavourably unless they are linked with perennial Himalayan rivers.

Peninsular states identified as having excess water, are currently opposed to project proposals to export their surpluses. They argue that their future water requirements have been underestimated and current agreed shares of inter-state rivers have been ignored. Tamil Nadu is the only state happy with the project, because it is planned to be a recipient of water from eastern Peninsular rivers as well as from the westerly flowing rivers of Kerala. There is no agreement between the states involved in the Peninsular Rivers Development Component. Implementation of the component cannot proceed without a consensus. Even after such a consensus is reached, separate inter-state agreements will have to be drawn among all the states involved for sharing surplus water. Existing inter-state agreements for various river basins will also have to be reviewed and modified accordingly. Fulfilling these legal requirements is complex and will require years of negotiations.

Opponents of the project believe that the project is already expensive and, if past experience is a guide, these costs will escalate. They point out that the project will inundate large areas of valuable forests, damage the environment and seriously affect wildlife and endangered species. Importantly they see the displacement of hundreds of thousands of poor and indigenous communities as a major impediment. Instead, they argue that the states and Central Governments can ease water shortages in India by first finishing the numerous uncompleted water projects worth billion of dollars. Second, serious efforts should be made to harvest local rainwater by storing it in tens of thousands of reservoirs or tanks that the government has abandoned. Third, they believe that governments could do much more to develop alternative strategies like increasing water use efficiency in irrigated farms, and in domestic and industrial water use. Fourth, they believe that the cost of water is both too low and heavily subsidised and encourages over use and wastage. Finally, they consider that the vast sums of money involved would be better spent on other priorities such as health, education and poverty and inequity alleviation.

In an attempt to address the controversies surrounding the project, the International Water Management Institute (IWMI) commenced a three-year study in 2005. The objective is to investigate potential alternatives and to develop a coherent plan of action to meet India's water challenge up to 2050. There is of course no guarantee that either the proponents or opponents of the project will accept IWMI's findings, should they be adverse to their causes. At this stage, it would seem that it is highly unlikely that construction of the National River-Linking Project will commence in 2007 and be completed by 2016.

References

Bandyopadhyay, J. (1988). *The Indian Drought 1987–88: The Ecological Causes of Water Crises*. Penang, Malaysia: Third World Network for Third World Science Movement.

FAO (2004). *Production Yearbook 2003*. Volume 57. Rome: FAO.

Framji, K. K., Garg, B. C. and Luthra, S. D. L. (1982). *Irrigation and Drainage in the World: A Global Review*. New Delhi: International Commission on Irrigation and Drainage. Volume II, pp. 493–1159.

Ghosh, G. and Phadtare, P. N. (1990). Policy issues regarding groundwater management in India. In *Proceedings of the International Conference on Groundwater Resources Management*. Asian Institute of Technology, Bangkok, 5–7 November 1990. Bangkok: Division of Water Resources Engineering, Asian Institute of Technology, pp. 433–457.

Government of India (1990). *An Atlas of India*. Delhi: Oxford University Press.

Gulhati, N. D. (1973). *Indus Waters Treaty: An Exercise in International Mediation*. Bombay: Allied Publishers.

Ministry of Water Resources (2005).[43] *Annual Report 2004–2005*. New Delhi: Ministry of Water Resources.

Mohile, A. D. (2005). Integration in bits and parts: a case study for India. In Biswas, A. K., Varis, O. and Tortajada, C. eds. *Integrated Water Resources Management in South and South-East Asia*. New Delhi: Oxford University Press. pp. 39–66.

Navalawala, B. N. (1992). Indian perspective in water resources planning. In *Pre-Seminar Proceedings of Seminar on Irrigation Water Management*. Gandhinagar: Water Management Forum. Volume 1, pp. 18–42.

Pant, G. B. and Rupa Kumar, K. (1997). *Climate of South Asia*. Chichester: John Wiley and Sons.

Rangachari, R., Sengupta, N., Iyer, R. R., Banerji, P. and Singh, S. (2000).[44] *Large Dams: India's Experience*. Final Report, November 2000. Cape Town, South Africa: World Commission on Dams.

Ravenscroft, P., Burgess, W. G., Ahmed, K. M., Burren, M. and Perrin, J. (2005). Arsenic in groundwater of the Bengal Basin, Bangladesh: distribution, field relations, and hydrogeological setting. *Hydrogeology Journal* **13**(5–6): 727–751.

Ray, B. (unpublished). *Water: The looming crisis in India and Regional-Environs, with Reference to River-Linking Project*.[45]

Research, Reference and Training Division (2005). *India 2005: A Reference Annual*. New Delhi: Publications Division, Ministry of Information and Broadcasting.

Singh, N., Sontakke, N. A., Singh, H. N. and Pandey, A. K. (2005). Recent trend in spatiotemporal variation of rainfall over India: an investigation into basin-scale rainfall fluctuations. In Franks, S. et al. eds. *Regional Hydrological Impacts of Climatic Change: Hydroclimatic Variability*. Publication No. 296. Wallingford: International Association of Hydrological Sciences, pp. 273–282.

Smedley, P. L. (2003). Arsenic in groundwater – south and east Asia. In Welch, A. H. and Stollenwerk, K. G. eds. *Arsenic in Ground Water*. Boston: Kluwer Academic Publishers, pp. 179–209.

World Bank (2003). *World Development Indicators*. Washington D.C.: World Bank. CD-ROM.

---

[43] Available at the URL http://wrmin.nic.in/publication/ar2005/ar2004–05.pdf (visited in January 2006).

[44] Also available at the URLs http://www.dams.org/docs/kbase/studies/csinmain.pdf and http://www.dams.org/docs/kbase/studies/csinanx.pdf (both visited in January 2006).

[45] This book is expected to be published in English and Bengali in 2007.

15 Inter-basin Water Transfer, Successes, Failures and the Future

15.1 INTRODUCTION

World population is projected to reach 9.3 billion by 2050, about 50 percent higher than the 2000 population of 6.1 billion. While global freshwater resources are adequate for that population and its water-dependent support systems, the uneven spatial distribution of freshwater means that by 2050 nearly two thirds of the world's population will live in water-stressed countries. Water shortages and extreme events are expected to be exacerbated by the impacts of global warming and land use changes. The problems that individual governments face in both limiting and supplying demand, in finding adequate sources of freshwater and in protecting communities and ecosystems are complex and difficult. As shown in this book, inevitably, decisions involve a trade-off between competing elements. There are no simple, universal solutions because of context-specific factors and location-dependent political considerations. None the less some lessons do emerge from analysis of the past inter-basin water transfer projects.

World water resources were developed rapidly over the past century to satisfy increasing demands. Large dams and numerous inter-basin water transfer projects have been constructed in all continents. It has been estimated that globally about 47 000 high dams have been built for town water supply, irrigation, flood control and hydro-power generation. The total investment in large dams is estimated at more than US$2000 billion. Global water withdrawal has increased approximately seven-fold from about $579 \times 10^9 \mathrm{m}^3$ in 1900 to $3917 \times 10^9 \mathrm{m}^3$ in 2000 compared with only a four-fold increase in population from 1.5 billion to 6.1 billion in the same period. A large proportion of this water has been used for irrigation. Increasing abstraction has been accompanied by significant community and environmental impacts.

In the preceding chapters, inter-basin water transfer projects either operating or proposed for five contrasting countries have been described (for a tabulated summary see Appendix K). The selected case studies span from the driest inhabited continent to a country with the highest per capita annual freshwater flow; from a country with a centrally controlled economy to others with an essentially free economy; from rapidly developing economies to developed economies; and from countries where power generation is the main beneficiary of developed water resources to others where irrigated agriculture is the main user. Lessons from these countries may help others to avoid mistakes of the past.

Table 15.1 summarises population data, average annual run-off and withdrawals in the selected countries. Australia has the lowest average annual run-off because of its dry climate. Its annual water withdrawal is proportionately low due to its relatively limited population. China, however, with its rapidly developing economy, has experienced a tremendous growth rate in water withdrawals with an over five-fold increase between 1949 and 2000 (from $103 \times 10^9 \mathrm{m}^3$ to $550 \times 10^9 \mathrm{m}^3$), while its population increased by more than two fold (from 555 million to 1280 million). The demand for inter-basin water transfer to fuel its rapid economic development and to more evenly distribute water is obvious. In contrast, in the United States water withdrawal over a similar period has increased by slightly more than two-fold (from about $248 \times 10^9 \mathrm{m}^3$ to $563 \times 10^6 \mathrm{m}^3$), while the population increased by less than two-fold (from 152.3 million to 281.4 million). Its water withdrawal is mostly used in thermal electric power generation and irrigation. India, though less populated than China, has a much higher annual withdrawal of $750 \times 10^9 \mathrm{m}^3$ amounting to a per capita water withdrawal of $730 \mathrm{m}^3$, suggesting profligate use of water. Canada has the world's highest per capita annual run-off of $104 000 \mathrm{m}^3$. Its water withdrawal is mostly used for hydro-power generation and then returns to streams.

Table 15.1. *Summary of the water withdrawal of the countries studied in this book*

Country	Population (million)		Average annual run-off		Annual water withdrawal		Main use of withdrawal water
	Current (in year)	Projected (for year)	Total (10^9 m^3)	Per capita (m^3)	Total (10^9 m^3)	Per capita (m^3)	
Australia	20 (2003)	23–31 (2050)	387	19 300	24	1200	Irrigation
United States	288 (2002)	349 (2025)	1840	6400	563	1955	Thermo-electric power and irrigation
Canada	32 (2005)	36 (2026)	3316	104 000	45	1400	Hydro-power generation
China	1280 (2000)	1480 (2025)	2711	2120	550	430	Irrigation
India	1030 (2001)	1400 (2025)	1869	1815	750	730	Irrigation

15.2 BENEFITS OF INTER-BASIN WATER TRANSFER PROJECTS

The main benefits of inter-basin water transfer projects in all countries examined have been to supply water for domestic, irrigation, industrial and mining activities, hydro-power generation, and flood control. The benefits flowing from the projects studied here are discussed in the following sections.

15.2.1 Australia

Goldfields Pipeline Scheme

The discovery of gold at Coolgardie and Kalgoorlie in 1892 and 1893 respectively, attracted thousands of people to the Western Australian goldfields. The goldfields region lies in a low rainfall and high evaporation area. Finding a reliable water supply for both domestic use and for the mining industry, presented a major challenge. Existing water supplies were very restricted, expensive, of poor quality, and caused serious health problems including outbreaks of dysentery and typhoid. To address these problems, the first inter-basin water transfer project in Australia, the Goldfields Pipeline Scheme, was commissioned in 1903. It enabled 22 700 m^3 of water per day to be pumped from a weir on the Helena River in the Darling Range 530 km overland to Coolgardie using eight pumping stations. At the time, the scheme was the largest of its kind in the world. The reliable supply of freshwater to Goldfields dramatically improved health and comfort and was responsible for wealth generation in the region by enabling the continued development of the mining industry. Strenuous opposition arising as a result of the cost and management of the project led to the tragic suicide of the scheme's Engineer-in-Chief just prior to its completion.

The Snowy Mountains Hydro-electric Scheme

The scheme is the largest engineering project ever undertaken in Australia. It diverts more than 1×10^9 m^3 of water from the southward flowing Snowy River to the Murray and Murrumbidgee Rivers in Victoria and New South Wales, generating hydro-electricity and supplying major irrigation schemes. Construction commenced in 1949 and took 25 years to complete at a total cost of US$7 billion in current values. Although water is a State responsibility in Australia, the Snowy Scheme was initiated by the Federal Government as part of the post World War II development of the country. Its completion boosted economic activities in inland areas of New South Wales and Victoria through irrigation, facilitated settlement of Returned Soldiers, created a renewable energy source[1] and greatly increased the technical skills base and competence of the country.

Shoalhaven Diversion Scheme

The past history of water supply in Sydney, the capital and major city of the state of New South Wales, reveals an on-going series of domestic and industrial water supply crises, due in large part to mostly El Niño Southern Oscillation related droughts. The Shoalhaven Diversion Scheme was designed to transfer water to meet Sydney's ever increasing needs and to generate hydro-electric power. Construction was completed in 1977 at a cost of US$375 million in 2003 values. The scheme transfers water from Shoalhaven River catchment to Sydney water supply systems and plays a major role in Sydney's water supply, particularly during drought periods. The estimated annual diversion capacity of the Scheme is approximately 284×10^6 m^3 and generates 240 MW of hydro-power.

---

[1] The scheme has a current installed hydro-electric generating capacity of 3756 MW.

River Murray Pipelines

South Australia relies heavily on inter-basin water transfer from the River Murray to provide water for Adelaide's population of more than one million. Water use varies according to seasonal conditions, with the River Murray supplying approximately 40 percent of demand in cool and wet conditions and up to 90 percent in drought years. Water is supplied via six pipelines commissioned over the period of 1944 to 1973, with a total capacity of $1.2 \times 10^6 \, m^3$ per day. Without this supply, Adelaide would not be able to exist in its current form.

Thomson Diversion Scheme

The Thomson River is a major tributary of the La Trobe River system in Victoria. The Thomson Diversion Scheme for domestic and industrial water supply to Melbourne, the State's capital, was completed in 1984 and consists of the Thomson Dam with a reservoir capacity of $1.1 \times 10^9 \, m^3$ and a diversion tunnel to the Yarra River. It has a diversion capacity of $148 \times 10^6 \, m^3$ per annum, and makes a substantial and essential contribution to the city's water demand.

Hydro-Power Projects in Tasmania

Water resources of the island State of Tasmania have been developed since the late nineteenth century for hydro-power generation. The first hydro-electric power station was commissioned in 1895. In 2004, mainland Tasmania had 29 hydro-power stations with a total installed capacity of 2265 MW. Most of Tasmania's hydro-electric power stations are located along river systems. However, inter-basin water transfers have assisted a limited number of schemes. Three schemes, the Great Lake, Mersey–Forth and Gordon use inter-basin water transfer for hydro-power generation and have a total water diversion capacity of more than $1.5 \times 10^9 \, m^3$ per annum. Their associated hydro-power stations have a combined installed capacity of 1120 MW and generate a significant proportion of the State's hydro-power.

Mareeba–Dimbulah Irrigation Scheme

This scheme was the first major water resource development in the Australian tropics. Tinaroo Falls Dam was designed with a capacity of $407 \times 10^6 \, m^3$ and the water from the dam was first released in 1958. The Barron River waters, which naturally flow eastward into the Coral Sea, are diverted by gravity across the Great Dividing Range to the Walsh/Mitchell River catchments, which flow northward into the Gulf of Carpentaria. The scheme irrigates 22 000 ha, originally mostly tobacco, and generates 60 MW of hydro-power.

Other Schemes

Numerous other inter-basin water transfer schemes are in operation in Australia. These schemes are for water supplies to population centres such as Adelaide, Perth and Darwin, which rely wholly or partly on inter-basin water transfer projects.

15.2.2 United States

Inter-basin water transfer has been widely used in the United States since 1856 to supply water scarce regions or promote new developments, particularly irrigated agriculture in drier regions of western states. The volume of water diversion steadily increased to the extent that in 1982, the volume of water diversion between sub-regions was about $22.5 \times 10^9 \, m^3$ including $10.2 \times 10^9 \, m^3$ 204 exported from the 18 Water Resources Regions. In California, numerous water transfer schemes are in place for a wide range of uses including domestic, agricultural and industrial water supplies, hydro-power generation, flood control, recreation, management of saline water intrusion into the San Francisco Bay and Sacramento–San Joaquin Delta, and restoration of fluvial ecosystems. These schemes include the Los Angeles Aqueducts, the Hetch Hetchy Aqueduct, and the Mokelumne Aqueducts for water supply of San Francisco, the Central Valley Project, the State Water Project, and the Colorado River Aqueduct for water supply of Los Angeles and San Diego. Water is also transferred from the Colorado River Basin to the states of Colorado, New Mexico and Utah. These projects and their benefits are summarised below.

Los Angeles Aqueducts

The rapid population expansion in Los Angeles in the nineteenth century meant that demand quickly outstripped local supply. The Los Angeles Aqueduct, completed in 1913, was designed to deliver water to Los Angeles by gravity. Water carried by the aqueduct flows downhill from the Owens River to the Los Angeles area. In 1940 the Aqueduct was extended north to tap the waters of Mono Lake in the Sierra Mountains and a second aqueduct to Los Angeles was completed in 1970. The two Aqueducts deliver

$594 \times 10^6 \, \text{m}^3$ of water per year to Los Angeles which is absolutely vital to the city.

Hetch Hetchy Aqueduct

Water was first delivered to San Francisco through the Hetch Hetchy Aqueduct in October 1934, and in 1961 it was decided to increase the aqueduct's capacity. By the late 1970s demand had grown at such a rate that San Francisco was importing nearly six times as much water as it did through the original scheme. Currently, the Hetch Hetchy supplies $440 \times 10^6 \, \text{m}^3$ of water per annum to San Francisco.

Mokelumne Aqueducts

The East Bay Municipal Utility District was formed in 1923 to supply water to San Francisco's East Bay Municipal District. The Lower Mokelumne River in the Sierra Mountains was identified as a reliable water source and the Pardee Dam was completed there in 1929. This enabled water to be diverted to East Bay via a 132.2 km aqueduct across the San Joaquin Valley. The second and the third aqueducts were completed in 1949 and 1963 respectively.

The Colorado River Aqueduct

The delivery of water to Greater Los Angeles through the Colorado River Aqueduct commenced in 1941. It currently imports about $1.5 \times 10^9 \, \text{m}^3$ of Colorado River water annually. The Aqueduct draws water from Lake Havasu, created by the Parker Dam, and ends in Lake Mathews from where water is transported and distributed to Greater Los Angeles. Approximately 45 km east of Lake Mathews, two aqueducts also transport water to the San Diego region. The Colorado River Aqueduct is clearly an important source of water supply for south-western California.

Central Valley Project (CVP)

The CVP is the largest water storage and delivery system in California. It was engineered by the US Bureau of Reclamation from the late 1930s to the end of 1960s with several components constructed by the US Army Corps of Engineers. This scheme manages $11 \times 10^9 \, \text{m}^3$ of water per year of which about $8.6 \times 10^9 \, \text{m}^3$ per year are diverted for agriculture, urban and wildlife. Of the latter, about $6.2 \times 10^9 \, \text{m}^3$ of water per year is supplied to farmers to irrigate 1.2 Mha of land. A further $740 \times 10^6 \, \text{m}^3$ is supplied for municipal and industrial use, enough to supply about one million households with their water needs each year. The scheme also generates enough electricity to meet the energy needs of two million people. Of the remaining diversion, $985 \times 10^6 \, \text{m}^3$ of water per year is dedicated to fish and wildlife and their habitats and $505 \times 10^6 \, \text{m}^3$ to State and Federal wildlife refuges and wetlands. The CVP is the first ranked US Bureau of Reclamation project in terms of the value of its flood damage prevention. It has been estimated that between 1950 and 1991 it prevented more than US$5 billion in flood damage. The annual value of farm production in the CVP exceeds the total value of all gold mined in California since 1848.

State Water Project (SWP)

Planning for the SWP began when it became evident that local developments and the Central Valley Project in California could not keep pace with the water requirements of the State's rapidly growing population. Deliveries from the SWP were first made in 1962 and deliveries to southern California commenced in 1972. The SWP is a water storage and delivery system of reservoirs, aqueducts, tunnels, power plants and pumping plants. Its main purpose is to store water and distribute it to 29 urban and agricultural water agencies in northern California, San Francisco Bay Area, San Joaquin Valley, Central Coast, and southern California. These 29 agencies have long-term contracts for water entitlements through to the year 2035 with a maximum annual entitlement of about $5.15 \times 10^9 \, \text{m}^3$. Of the contracted water supply, 70 percent goes to urban users and 30 percent to agricultural users. The SWP is the State's fourth largest energy supplier and the single largest user of power. While the SWP was built primarily for water supply, the project and its facilities also provide the people of California with many additional benefits which include flood control, recreation, fish and wildlife enhancement, power generation and control of saline seawater intrusion into the Sacramento–San Joaquin Delta.

Water Transfer from the Colorado River

The Colorado River Basin is a major source of inter-basin water transfer to the states of Colorado, New Mexico, California, and Utah. The Colorado–Big Thompson Project, Fryingpan–Arkansas Project, and a number of smaller projects divert water to the State of Colorado. The San Juan–Chama Project diverts water to New Mexico, while the Central Utah Project that includes the Strawberry Valley Project diverts water to Utah. These projects have assisted agricultural development in arid regions of these states. They also generate hydro-power, supply municipal and industrial water, and aid fish and wildlife conservation.

15.2.3 Canada

In Canada, hydro-electric power schemes dominate inter-basin water transfer projects, both in the number and scale of diversions. Irrigation, flood control and municipal uses only assume importance regionally or locally. Some of the benefits of Canada's inter-basin water transfer projects are described in the following sections.

The Long Lake and Ogoki Diversions

These diversions were developed to generate hydro-electric power for Canada's defence industries during the World War II. The Long Lake Diversion began inter-basin water transfer and power generation in 1948. It transfers water from the Kenogami River within the James Bay drainage system into Lake Superior at an average rate of $45 \, m^3 \, s^{-1}$. The rationale of the Ogoki Diversion project was to divert the north-eastward flowing Ogoki River southward through Lake Nipigon into the Great Lakes system providing an average $113 \, m^3 \, s^{-1}$ flow increment for hydro-power generating stations.

The Nechako–Kemano Diversion Project

In 1948, the Province of British Columbia invited the Aluminium Company of Canada (ALCAN) to explore the feasibility of developing hydro-electric power as a basis for an aluminium smelting industry. This led to a significantly different type of diversion. Instead of using the usual large volume of water over a relatively low vertical drop, this scheme employed a comparatively small flow volume falling through a high head to generate electricity. ALCAN was licensed in 1950 to divert up to $269 \, m^3 \, s^{-1}$ of water. The Nechako–Kemano Diversion Project was completed in 1954, and was the largest corporate project that had been undertaken in Canada at that time.

The Churchill–Nelson Diversion Project

This project was completed in 1976 and has been in operation since 1977. It diverts an average flow of $775 \, m^3 \, s^{-1}$, or approximately 35 times the normal flow of the Rat River. The project has an installed hydro-power generating capacity of 3932 MW.

La Grande (James Bay) Diversion Project

The project was initially proposed in 1971 to satisfy Quebec's future demands for hydro-electric power. Options were examined in detail over the next two years before it was decided to use the massive potential of the La Grande River Basin, with an average discharge rate of $1710 \, m^3 \, s^{-1}$. The La Grande River drops approximately 500 m over its 850 km length from east to west before reaching James Bay. To maximise the output from hydro-electric power plants, water was diverted into La Grande River from adjacent catchments. Collectively, these diversions nearly doubled the natural flow of the La Grande River. Currently, the project has a massive installed hydro-power capacity of 15 550 MW.

Great Lakes Basin Diversions

Water has been exchanged for some time between the Great Lakes Basin and its neighbouring basins, as well as between the sub-basins of the Great Lakes. Diversions have been used for navigation, hydro-power generation and control of the Great Lakes water levels.

15.2.4 China

The rapid economic development of China over the past decades in combination with its population increase and significant improvements in living standards has increased pressures on its water resources. The heavily populated and polluted Haihe, Huaihe and Huanghe (Yellow) river basins are facing serious shortages of water for domestic, industrial and agricultural water supplies. The Huang-Huai-Hai area, a major political, economic, and cultural centre in China supports 35 percent of China's population, generates 35 percent of the GDP, contains 36 percent of the cultivated land and 37 percent of grain production, but has only about 7.5 percent of the country's water resources. The South to North Water Transfer Project has been planned to address these water shortages. It consists of three sub-projects. These are Eastern Route, Middle Route and the Western Route Schemes. These schemes will divert water from lower, middle and upper reaches of the Yangtze River Basin respectively to meet the increasing water requirements of the North China Plain. Projected benefits of the scheme include improving domestic and industrial water supply, alleviating the drying up of the lower Yellow River during drought periods, diluting pollution, improving riverine ecology, and ameliorating groundwater depletion and associated land subsidence.

15.2.5 India

India has a rapidly increasing population and growing economy, which requires more water, higher food production, and hence additional land development for irrigated crops. It has also experienced serious flood and drought problems. These may be exacerbated by climate change as well as the uneven distribution of water availability across the country. The Indian Government considers large-scale inter-basin water transfer from water rich to water deficit river basins as a means to address these problems.

In 1980, the Ministry of Water Resources devised a National Perspective Plan (NPP) for water resources development by transferring water from water rich to water deficit basins through the linking of river networks. The National River-Linking Project has two main components, the Peninsular Rivers Development and the Himalayan Rivers Development. A feature of the NPP is that the transfer of water from water rich basins to deficit basins would essentially be by gravity. The proposal ultimately aims to augment water supplies to major population centres in the Peninsular region and to develop irrigation in drier states. Preliminary assessment indicates that the National River-Linking Project would provide between 200 and 300×10^9 m^3 of additional water per year to water deficit basins. It is estimated that an additional 25 Mha of land could be irrigated by surface water, and a further 10 Mha by increased use of groundwater. These would raise the ultimate irrigation potential of the country from 113 to 148 Mha and would also generate 34 000 MW of hydro-power. Other projected benefits include flood control, navigation improvement, fisheries, and water quality control. The National River-Linking Project however faces significant opposition from the states involved in the plan, environmental groups, academics and others.

15.3 IMPACTS OF INTER-BASIN WATER TRANSFER PROJECTS

Inter-basin water transfer projects, in spite of their significant benefits, have impacted river ecosystems, increased erosion in water receiving river systems, caused silting of dams, and in some cases resulted in major displacement of local populations affecting their livelihoods. The following sections summarise the impacts of some of the major projects in the studied countries.

15.3.1 Australia

The Snowy Mountains Hydro-electric Scheme was highly subsidised and farmers received water at a very low cost with little incentive for improving water use efficiency. The scheme has contributed to the development of irrigated land salinity across the lower Murray and Murrumbidgee valleys. A large area of the Alpine region, now a National Park, has been inundated and communities relocated due to construction of reservoirs. Furthermore, the legendary Snowy River has become severely degraded as a result of diversion of 99 percent of its headwaters. Under pressure from the community and environmentalists, the New South Wales, Victoria and Federal Governments have agreed to increase the average annual flow rate of the Snowy River at Dalgety to 15 percent and ultimately to 28 percent of its pre-scheme rates with the water being sourced from increased irrigation efficiency. A 2006 plan to privatise the managing agency, the *Snowy Hydro Ltd*, raised concerns about the long-term commitment to the return of flows to the Snowy River, and resulted in abandonment of the plan by The Federal and State (New South Wales and Victoria) governments in early June 2006.

The Shoalhaven Diversion Scheme impacted the natural environment of the Shoalhaven River catchment. Assessment of the Shoalhaven Catchment with respect to streamflow, aquatic plants and animals, bed and bank erosion, and other issues has indicated that more than half of its sub-catchments are under high environmental stress. The main issues are streambed and bank erosion and changes to fish habitat due to construction of dams and weirs. The proposal to increase storage volumes in the Shoalhaven to meet Sydney's water needs following the 2000–05 drought was met with strenuous opposition, resulting in abandonment of the proposal, at least for the time being.

The daily transfer of a large volume of water from River Murray to Adelaide via River Murray Pipelines has impacted the environment of the Lower River Murray, particularly during the dry periods. The impacts include reduction and more frequent cessation of flow at the River mouth and sediment deposition.

Assessment of the environmental impacts of the Thomson Diversion Scheme for water supply of Melbourne indicated that no unique or rare species have been threatened by construction and operation of the project. However, local habitat has been destroyed in the area inundated by the reservoir, and a migratory fish, the eel, has been excluded from the upper portion of the Thomson River.

More than 1100 km^2 of river valleys, wetlands and pre-existing lakes have been inundated through the development of hydro-electricity in Tasmania. Construction of dams and weirs has affected 21 creeks, 25 rivers and 7 estuaries. A total length of at least 1200 km of natural creeks and rivers has been impacted by diversions and alterations of flow regimes. These hydrological changes have had implications for sediment movement, stream channel instability, siltation, instream biota, fish passage, riparian habitats, and water quality. In particular, the Mersey River has been seriously affected by construction of the Rowallan and Parangana dams and water diversion to the Forth River, with the average annual discharge at the Parangana Dam reduced to about 15 percent of the natural flow at this point.

Analyses of the flow regime changes of the Barron River in Queensland for periods of pre and past construction of the Tinaroo Falls Dam as the major source of water supply for the Mareeba–Dimbulah Irrigation Scheme, indicated that there has been a 13 percent decrease in the mean annual flood discharge since impoundment. Flow regulation of the Barron River via Tinaroo Falls Dam produced little or no channel change immediately downstream of the Dam because of the bedrock control within the first 20 km downstream of the Dam. However, a decrease in channel width and cross-sectional area had occurred far downstream from the Dam with the channel width at Mareeba reduced by 27 percent.

15.3.2 United States

Construction of a large number of dams for inter-basin water transfer projects and long distance water transfer for irrigation and other uses have adversely impacted the environment. Impacts include inundation of sensitive catchment areas such as the Sierra and Colorado mountains, increased river erosion, development of the irrigated land salinisation, significant changes in flow regimes and their temporal variations, resulting in changes in riverine ecosystems and destruction of native fish habitat. For example in 1969, the population of Chinook Salmon was about 118 000 at Red Bluff Diversion Dam located about 70 km below Shasta Dam on the Sacramento River. By 1990 the population had dropped to less than 5 percent of the 1969 numbers.

The famous environmentalist John Muir, founder of the Sierra Club, raised attention to the potential major environmental impacts of the Hetch Hetchy Aqueduct for water supply of San Francisco. Muir's efforts to prevent construction of the dam failed in 1913 and in the following year work on the Aqueduct was commenced. However, the dispute over the impacts of the project on the conservation of mountain areas continues to this day. The Sierra Club also opposed the Central Utah Project and has succeeded in having some of the components largely modified.

Inter-basin water transfer projects in the United States constructed before the late 1960s were developed mostly for irrigation and municipal water supply and were largely bereft of any detailed assessment of the environmental or social consequences. The 1960s marked the end of the era of large dam building in the United States. Environmental concerns started to gain widespread support in the 1970s. Following the enactment of the *National Environmental Policy Act* of 1969, and *Endangered Species Act* of 1973 water transfer projects had to comply with stringent new requirements. In California, the *Central Valley Project Improvement Act (CVPIA)* of 1992 moved the project in a new direction. The CVPIA reallocated $985 \times 10^6 \, m^3$ ($740 \times 10^6 \, m^3$ in dry years) of the annual CVP water diversion of $8.6 \times 10^9 \, m^3$ in order to supply the needs of fish and wildlife and their habitat. Major components of the State Water Plan and CALFED Bay–Delta Program now concentrate on ecosystem restoration.

15.3.3 Canada

Canada's rivers have been dammed and diverted, mainly for electricity generation. As a result, Canada has become the world's leading producer of hydro-electricity. However, these developments, particularly the early ones, have displaced indigenous communities and had serious adverse impacts on erosion, fish, forests and water quality. For example, as a result of the Churchill–Nelson Diversion Project about 213 680 ha of land was flooded, thousands of indigenous and non-indigenous people were affected. Furthermore, the water regimes of both the Churchill and the Nelson river systems were dramatically changed. As is the case with many other water infrastructure projects, some developments failed to achieve their expected economic goals. These problems were due to the following issues.

Inadequate Planning

Provincial utilities and governments constructed most diversions with little considerations for human, biophysical and ecological impacts. Little effort was made to consult those affected and to canvas their perceptions and attitudes with regard to development options. Rivalries between jurisdictions, closed decision-making processes, and a lack of cooperation among federal, provincial, regional, and local authorities, were at the heart of the problems.

Politics

A lack of effective control by provincial governments over their resource development agencies contributed substantially to the problems. This was because major water projects were not subject to thorough public assessment of their biophysical and socio-economic impacts before political commitments had been made to their construction.

Economic Issues

Manitoba in the 1970s and Quebec in the 1980s rapidly expanded their hydro-electric capacity. Construction costs rose well above the initial estimates, and the market for electric power failed to develop at the same rate. Rising interest rates and a falling Canadian dollar exacerbated the problems.

Native Communities

A disturbing aspect of developments was the secrecy with which planning was undertaken. Initially, communities directly affected were not consulted, nor were they informed of the planned changes to their environment. Small native communities, those most vulnerable to changes in their water regime, were unable to influence decisions or the basic nature of the development.

15.3.4 China

Some of the predicted impacts of the South to North Water Transfer Project (SNWTP) in China are:

- Although the annual volume of water diversion for the Eastern and the Middle Routes is less than 3 percent of the average annual flow of the Yangtze River, it will still affect downstream ecosystems of this river. There will be less sediment delivery needed to maintain riparian and coastal wetlands. Pollutants also will be marginally less diluted, increasing their concentrations in the lower Yangtze River.
- As water flows north along the Eastern Route, it will be impounded in four major lakes along the route. The introduction of chemicals and biota from different ecosystems may affect these lakes and their ecosystems. On the other hand, more abundant water for wetlands adjacent to the lakes will help these ecosystems.
- Diversion of a substantial amount of water from the Hanjiang River to the Middle Route Scheme will reduce flows in the downstream section of the river and may lead to the worsening of the existing eutrophication outbreaks.
- Saltwater intrusion in the estuary of the Yangtze River can extend about 100 km upstream during low flow periods. Operation of the SNWTP and the Three Gorges Dam will lead to a slightly longer saltwater intrusion up the Yangtze River during the dry season, but the effect is predicted to be negligible for the rest of the year. Saltwater intrusion may affect the Yangtze Delta fisheries, urban water supply and the intake of the Eastern Route Scheme.
- Construction of the SNWTP will require resettlement of about 275 000 people, mostly related to the Middle Route.

15.3.5 India

The National River-Linking Project raises a vast array of uncertainties over its adverse environmental, economic and social impacts not just in India, but also in Bangladesh. In 1974, India completed the Farakka Barrage on the Ganga River close to the border with Bangladesh, diverting crucial dry season flows into Indian irrigation fields. This was a source of considerable tension between the two countries and resulted in the 1996 treaty, which commits India not to reduce transboundary flows further. The National River-Linking Project is also opposed by NGOs in Nepal, which have launched vigorous campaigns against it. The Himalayan component of the project, which is the major source of excess water, is faced with numerous problems, the main ones are as follows.

Geological Risks

Construction of a large number of dams in this geologically unstable and seismically active area has great risks of triggering earthquakes and consequent dam failure.

Reduction in Sediment Loads

The Ganga and Brahmaputra rivers together carry on average sediment loads of $1.7-2.4 \times 10^9$ tonnes per year to the $20\,000\,km^2$ delta of these rivers in Bangladesh and India. Construction of a large number of dams in these two vast river basins will trap sediments, reducing sediment supply to the delta. This could have serious environmental impacts on fisheries, wildlife, forests, the productivity of floodplains and on the stability of the coastlines. The social costs on those that depend on these areas for their livelihoods, particularly the poor, are severe.

Environmental and Social Impacts

Substantial deforestation, mismanagement of land and water resources, and sea level rise in the Ganga–Brahmaputra Delta have already had severe effects on its ecological balance. It has been estimated that within the next 15 to 20 years 15 percent of the habitat area could be lost, displacing more than 30 000 people. Diversion of the Himalayan rivers will aggravate the situation with serious impacts on fisheries, mangroves and livelihood of the local population who depend on these resources.

Construction of the National River-Linking Project will inundate an estimated 8000 km^2 of land within India and will destroy habitats of native flora and fauna. It could potentially leave up to 3 million people homeless. Opponents of the project argue that in the past, the Central Water Commission has manipulated water resources and related data to influence outcomes of feasibility reports. There is no guarantee that the project's feasibility reports will be free of such deceitful practices.

15.4 MEGA-SCALE WATER TRANSFER PROPOSALS

Numerous mega-scale water transfer projects have been proposed in Australia, North America and India. The main goal of these proposals was to transfer water from water surplus to water deficit areas. In the case of Canada the objective of the private-sector[2] proponents was to export water to the Unites States and Mexico. Appendix K includes a summary of these proposals, while detailed descriptions are provided in the relevant sections of the book.

15.5 NECESSARY KNOWLEDGE FOR INTER-BASIN WATER TRANSFER

A number of key lessons can be drawn from the examples considered in the preceding chapters. Inter-basin water transfer proposals require detailed investigations of topography, geology, hydrology, potential climate change impacts, as well as engineering and geotechnical issues. They also require comprehensive environmental, social, and economic assessment of both the impacts and benefits of projects in relation to water exporting and importing basins. Environmental assessment of such projects includes impacts of water diversion on river flow and its temporal variation, riverine ecosystems and wetlands, water quality, erosion and sedimentation, groundwater levels, and quality. In particular, it is important to adequately assess environmental flow requirements of the river systems affected by the project. As part of the project assessment, a critical examination of the long-term social impacts is necessary and should include strategies to minimise the social costs. Major issues include threats to indigenous people, loss of cultural properties, loss of traditional livelihoods, and involuntary resettlement.

Detailed and realistic cost–benefit analysis are necessary to ensure that any proposal for inter-basin transfer is economically feasible. The capital cost for construction of dams, tunnels, aqueducts, pipelines, pumping stations, access roads and hydro-power plants, as well as the annual operating costs of the project including loan repayments, operation and maintenance, energy and administration, require assessment and clear identification of underlying assumptions. The proposal could be considered economically viable if the revenue generated by sale of water and electricity covers all costs, including environmental and social, without any subsidy. It is, however, fundamentally important to investigate all realistic alternatives to inter-basin water transfer such as increasing water use efficiency, water recycling, water trading, water pricing and dam building within the basin before considering inter-basin water transfer.

Meeting the challenges of sustainable catchment management requires an integrated approach involving physical, social and economical aspects of the project. Moreover, decisions and project development should be democratic with participation of all stakeholders involved, particularly communities that may be relocated.

Planning for the inter-basin water transfer projects also requires detailed monitoring facilities and data analysis. These include:

(1) monitoring of the factors influencing reservoirs and catchments such as climatic, sediment loads, water quality parameters, erosion, impacts on wildlife species, changes in flora and fauna, public health issues, and changes in economic and social status; and

(2) monitoring of irrigation systems developed as a result of the inter-basin water transfer project including climate factors, stream discharge above and below the irrigation project at various points, nutrient content of discharge water, watertable depth in project area and

---

[2] Neither the Canadian nor the United States governments ever endorsed any of these schemes (Frank Quinn, personal communication, April 2006).

downstream, groundwater quality, soil physical and chemical properties, erosion and sedimentation, condition of distribution and drainage canals, and incidence of disease.

It is critically important to have substantial baseline information, particularly in areas with highly variable climate and streamflows, before any decisions can be reached.

15.6 INTER-BASIN WATER TRANSFER, WATER CONSERVATION AND NEW SOURCES OF SUPPLY

Current water resources management policies in the studied developed countries (Australia, United States and Canada) place environmental sustainability as the top priority. These, therefore, discourage inter-basin water transfer projects because of their high costs and adverse environmental and social impacts. Instead, they espouse a range of water conservation and reuse measures and only if necessary and as a last resort, the development of local sources of water supply. In China and India with rapidly expanding economies, supplying water requirements of their growing population and industry, generation of hydro-power and other pragmatic arguments, appear to have priority over the economic profitability of the water projects or their environmental impacts or sustainability. However, more recent policy shifts have tended to encourage some degree of water conservation to satisfy their water requirements.

Water conservation measures that can be used as a substitute for inter-basin water transfer projects and new sources of water supply include the following.

Increasing water use efficiency. Irrigation is the major global water user. Lining of unlined irrigation canals, transport of water by pipeline instead of canals, use of sprinklers and drip irrigation methods instead of flood and furrow irrigation, can substantially increase its efficiency. In urban areas, leakage in distribution systems can be reduced and water use efficiency can be increased by use of water efficient toilets, washing machines and dishwashers. Furthermore, gardens can be redesigned in order to use water efficient native species. Industries can also increase their water use efficiency by recycling their water and replacing use of high quality water with treated wastewater.

Water trading. Over the past two decades, the development of a water market has been promoted internationally as a means of increasing the economic efficiency of water use. The water market enables the water right holders (mainly farmers) to transfer their entitlements to other water users willing to pay for it. Potential buyers include: urban and industrial users, other farmers with higher-value crops and limited supplies, and environmental programmes to support fish and wildlife habitat. Farmers' tradable water could come potentially from various savings such as fallowing irrigated lands, shifts in cropping pattern from water intensive crops to less water demanding crops, water conservation measures (canal lining, improved irrigation technology and others), and use of groundwater instead of surface water. Water trading can be for a short period (one season), long term (a defined number of years) or permanent. A major concern in short term water trading is that the transaction costs and time for approvals of trade may prohibit exchanges in some jurisdictions.

Wastewater reclamation. Wastewater can be treated and used for aquifer recharge, irrigation, flushing toilets and industrial uses; however the use of treated wastewater for human consumption is not recommended or viewed favourably by some societies.[3] The most significant concerns are those related to health risks. This includes the potential risks of using water that might contain traces of chemicals such as endocrine disruptors, pharmaceutical chemicals, organic industrial chemicals and heavy metals. These concerns became more serious in closed basins where concentration of chemicals could reach an alarming rate over prolonged periods. The level of wastewater treatment depends on the type of designated use. Treated wastewater for aquifer recharge should have a very high quality. In case of irrigation, its quality should be compatible with soil characteristics, since there is the potential for increasing sodicity and soil structural instability, while for toilet flushing and industrial use it could have a much lower quality.

Conjunctive use of surface and groundwater. This measure takes advantage of surface storage facilities to temporarily store stormwater and the ability of aquifers to serve as long-term, low-cost storage. The main components of a conjunctive management plan are:

(1) recharging the aquifer when surface water is available;
(2) extracting groundwater in dry periods when surface water is scarce; and

---

[3] In Singapore, only one percent of the daily potable water requirements of $1.12 \times 10^6 \, \text{m}^3$ is treated wastewater. Dual-membrane (microfiltration and reverse osmosis) and ultraviolet technologies have been used, in addition to conventional water treatment processes (http://en.wikipedia.org/wiki/Water_resources_of_Singapore visited in December 2005).

(3) having a monitoring program to allow water managers to respond to changes in groundwater level and quality, surface water, or environmental conditions that could violate management objectives or impact other water users.

Conjunctive management improves water supply reliability and reduces groundwater overdraft and water loss due to evaporation.

Water metering and pricing. Correct pricing of water to cover the true costs of supply has been long recommended as an effective tool for controlling demand. Experience in numerous countries including the United States, Canada and Australia has indicated that water metering and a tiered price structure[4] will prevent wastage of water resources and will reduce the needs for development of new sources of water supply for a relatively long period of time. In India, where the price of irrigation water is not volumetrically based, there are few incentives for improved efficiency.

Cloud seeding. Under favourable conditions, cloud seeding has been shown to cost effectively increase precipitation and run-off in river catchments in the United States and the State of Tasmania in Australia. Research is in progress to test the efficacy of cloud seeding for increasing precipitation over Australia's Snowy Mountains area.

Water desalination. Desalination of brackish or saline waters provides opportunities to increase water supplies. Recent technological advances have significantly reduced the cost of desalinated water to the level that is comparable and in some instances competitive with other alternatives for acquiring new water supplies. Desalination technologies are becoming more efficient, less energy demanding and less expensive. However, their associated problems include energy consumption and cost, greenhouse gas emission due to production of required electricity, ecological impacts of feed-water intake and brine disposal.

Building new dams. If hydrological, environmental and other conditions are favourable, building new dams within the catchment where water is in demand is less damaging than the inter-basin water transfer.

15.7 INTER-BASIN WATER TRANSFER AND CROSS JURISDICTIONAL AGREEMENTS

Globally, there are 261 international river basins covering 45 percent of the land surface of the earth (excluding Antarctica). Overall, 145 nations have part of their territories within these international river basins. Disagreements over the sharing of water within these basins have been the cause of many international disputes and resulted in more than 3600 agreements between AD 800 and 1985. Similar conflicts also exist between states, provinces and territories within countries with the federal systems.

Inter-basin water transfer in multi-jurisdictional river basins creates numerous conflicts among jurisdictions and a wide range of stakeholders regarding issues such as water rights, water allocation, protection of the environment, loss of flora and fauna, resettlement, compensation and others. The extent of these problems depends on the extent of water transfer between jurisdictions. Their resolution requires an agreement between parties involved. Selected examples of broad agreements are as follows.

Colorado River Compact. The Colorado River drains seven states (Colorado, Wyoming, Utah, New Mexico, Arizona, Nevada and California) in the United States before entering Mexico. The 1922 negotiation of the *Colorado River Compact* divided the river basin into the Upper and the Lower Basins. Colorado, New Mexico, Utah and Wyoming constitute the Upper Basin States, while Arizona, California and Nevada are the Lower Basin States. Delegations from the seven states also divided the $21.6 \times 10^9 \, m^3$ average annual flow of the river between the Upper and the Lower Basins. In 1928, the United States Congress divided $9.3 \times 10^9 \, m^3$ of the Lower Basin's annual share of the Colorado River between California ($5.4 \times 10^9 \, m^3$), Arizona ($3.5 \times 10^9 \, m^3$), and Nevada ($0.4 \times 10^9 \, m^3$). The *Upper Colorado River Basin Compact* of 1948 apportioned the $9.3 \times 10^9 \, m^3$ between Colorado (51.75 percent), New Mexico (11.25 percent), Utah (23 percent) and Wyoming (14 percent). Based on these agreements, water has been diverted from the Colorado River to the States of Colorado, New Mexico, Utah and California via numerous inter-basin water transfer schemes.

During the 1990s, California's average withdrawal from the Colorado River ranged from $5.6 \times 10^9 \, m^3$ to $6.2 \times 10^9 \, m^3$, which are above its allocated share of $5.4 \times 10^9 \, m^3$. Excess water came from unused water allocated to Arizona and Nevada. However, the water requirements of Arizona and Nevada have increased rapidly. Under pressure from other Colorado River Basin states, California is required to cut its over-use of Colorado River Water over a period of 15 years.

The California Seven Party Agreement of 1931 helped resolve the long-standing conflict between California

---

[4] Price structure where users pay higher prices for higher water consumptions. For example US\$1 per m^3 for the first $200 \, m^3 \, yr^{-1}$, US\$1.5 per m^3 for $200-400 \, m^3 \, yr^{-1}$ and US\$2 per m^3 for consumptions between $400-600 \, m^3 \, yr^{-1}$.

agricultural and municipal interests over distribution of California's entitlement of the Colorado River waters. The six principal claimants (Palo Verde Irrigation District, Yuma Project, Imperial Irrigation District, Coachella Valley Irrigation District, Los Angeles Metropolitan Water District, and the City and County of San Diego) reached consensus on the amount of water to be allocated to each entity on an annual basis.

Snowy Mountains Hydro-electric Agreement 1957. The proposal for construction of the Snowy Scheme and inter-basin water transfer of Snowy river waters to the Murray and Murrumbidgee rivers was a source of dispute between the Australian Federal Government and the states of New South Wales and Victoria. The Federal Government passed the *Snowy Mountains Hydro-electric Power Act* in 1949 by using its defence power as defined under section 51 of the Australian Constitution without any agreement with the affected state governments. Under the *Act*, the Snowy Mountains Hydro-electric Authority was empowered to construct, maintain, operate, protect, manage and control works for collection, diversion and storage of water, and for generation of electricity in the Snowy Mountains area. Construction works was launched on 17 October 1949. While construction of the scheme was in progress, the *Snowy Mountains Hydro-electric Agreement* of 1957 was finally signed by the three concerned parties in September 1957 and became formally effective on 2 January 1959.

Kerala–Tamil Nadu Agreement. In India, an agreement was reached in 1958 between the states of Kerala and Tamil Nadu for inter-basin water transfer between river catchments of the two states and construction of the Parambikulam Aliyar Project. The project uses the transferred water for development of irrigation in drought prone areas of Tamil Nadu and hydro-power generation.

15.8 RECOMMENDATIONS OF THE WORLD COMMISSION ON DAMS

Many water resources development projects, including those involving inter-basin water transfer have been criticised for their negative environmental impacts on rivers, watersheds and aquatic ecosystems, poor economic performance and lack of equity in the distribution of benefits, lack of development of alternative options for water supply, and shortfall in achieving physical targets, cost recovery and economic profitability. In order to lower the chance of repeating these past mistakes, the World Commission on Dams has recommended the following strategic priorities and related policy principles for decision-making:

- **Gaining public acceptance.** Key decisions need to be accepted by all groups of affected people.
- **Comprehensive options assessment.** The selection should be based on comprehensive and participatory assessment of a full range of policy, institutional and technical options.
- **Addressing existing dams.** Opportunities exist to optimise benefits from many existing dams, and addressing outstanding social, economic and environmental issues.
- **Sustaining rivers and livelihoods.** Rivers, watersheds and aquatic ecosystems are the basis for life and livelihood of local communities. Dams transform landscapes and create risks of irreversible impacts. Understanding, protecting and restoring ecosystems at river basin level is essential to foster equitable human development and welfare of all species.
- **Recognising entitlements and sharing benefits.** Negotiations are required with adversely affected people to reach a mutually agreed and legally enforceable mitigation, resettlement and development provisions.
- **Ensuring compliance.** Compliance with applicable regulations and guidelines, and project-specific negotiated agreements, should be secured at all critical stages in project planning and implementation. Also, a set of mutually reinforcing incentives and mechanisms is required for social, environmental and technical measures.
- **Sharing rivers for peace, development and security.** Storage and diversion of water on transboundary rivers have been the source of considerable tension both between countries and within countries. Specific interventions for diverting water require constructive regional cooperation and peaceful collaboration.

15.9 CONCLUDING COMMENTS

Without question water resources developments have resulted in remarkable global economic prosperity, significant rise in health and living standards and major increases in food and fibre production. However, the world is faced with spatially uneven shortages of water for satisfying long-term agricultural, domestic, industrial, and environmental requirements. This is mostly due to a predicted 50 percent increase in the world's population by the middle of twenty-first century, particularly in Asia and the Middle East.

As countries develop, increasing demand per head of population also plays an important role in the projected increase in water requirements. Global warming and global change are expected to aggravate the situation.

In the past, dam building and inter-basin water transfer have been largely used to meet demands. The environmental and social impacts of these developments means that these options are no longer favoured by communities in developed countries. The Sierra Club first raised environmental objections to the Hetch Hetchy Aqueduct Project for water supply to San Francisco but failed to prevent its construction in 1913. Inter-basin water transfer from water surplus regions to water deficit areas remained popular until the late 1960s. The numerous examples of inter-basin water transfer projects in this book from Australia, United States, and Canada have clearly indicated the enormous engineering efforts and financial resources devoted to the construction of these projects, particularly after World War II. These projects have provided water deficit areas of these countries with irrigation, municipal, mining and industrial water supplies as well as hydro-power, flood control, and recreational facilities.

Almost all the projects considered here were designed and constructed at a time when only engineers and politicians were involved in the decision-making process and when integrated catchment management and public participation were not considered. The cost–benefit analyses of these projects were inadequate and their impacts on the environment (flora, faunas, water quality and quantity) and the displacement of affected people were largely ignored. Increasing environmental and social awareness in the last decades of the twentieth century has led to the understanding that the past projects had high economic, environmental and social costs, particularly for native and disadvantaged communities. Awareness resulted in an end to most new inter-basin water transfer proposals in developed countries. Currently, in countries such as Australia, United States and Canada, greater emphasis is being placed on alternative options such as increasing water use efficiency in all sectors of the economy (agricultural, domestic and industrial), water metering, adequate water pricing, water trading, reuse of treated wastewaters for agricultural and industrial use, conjunctive use of surface and groundwater resources, desalination of brackish and saline waters, and the use of precipitation enhancement techniques.

In China and India, with their rapidly developing economies, the situation is different. China is continuing its programme of dam building to achieve its economic and social development objectives. This programme is expected to continue through a large part of the twenty-first century. However, it now aims to develop and manage its water resources within a framework of sustainable development. This requires that the implementation of the dam building programme and inter-basin water transfer projects are accompanied by better management of surface and groundwater resources, implementation of water saving measures, adequate water pricing, pollution control and environmental protection measures. Recent policy developments, legislations and institutional changes in the areas of water resources and environmental protection, are indications that steps towards sustainable development are being taken. However, China is facing immense challenges and it is unlikely that it can achieve its sustainable development objectives in the near future.

In India, the water demands of urban mega-cities and the desire to expand irrigation to meet food and fibre needs has seen the development of plans for major inter-basin transfer of water. While requirements are in place to consider the social, economic and environmental impacts of the National River-Linking Project, the short deadlines for completion appear to place serious limitations on the thoroughness of the assessments.

Based on the experience of the countries presented in this book, a number of points regarding water conservation measures and development of new sources of water supply as alternatives to inter-basin water transfer are worth emphasising:

- Irrigation is the largest water user in the world. Lining of the irrigation canals and the use of more efficient sprinklers and drip irrigation methods could save very large volumes of water in the United States, research has shown that for every US$1 spent on canal modernisation, an expected return of US$3 to US$5 in conserved water can be achieved.
- Water metering combined with an adequate pricing structure is a strong tool in the reduction of water consumption and preventing water wastage.
- Urban water use efficiency can be a very cost effective strategy. In California, a recent study indicates that water savings of $2.5-2.8 \times 10^9 \, m^3$ per year by the year 2030 is achievable, with the cost of most measures ranging from US$47 to US$575 per $1000 \, m^3$, depending on the programme.
- Educating the public and involving them in water management and decision-making are essential elements in the success of water saving measures.
- Wastewater reclamation is gaining popularity. Treated wastewater appears suitable for various uses. However, in agricultural areas, compatibility of water quality with

- soil characteristics should be investigated. The use of treated wastewater for human consumption is not recommended.
- Technological advances in recent decades have made desalination of saline and brackish waters highly competitive with other water supply measures for domestic consumption. As generation of greenhouse gases related to electricity requirements of these facilities is a source of concern, research and developments are continuing to reduce their electricity consumptions. Another concern is the environmental impacts of the brine disposal produced by desalination facilities which is seldom adequately addressed.
- Water trading, in spite of its advantages, has created economic and social problems in some water exporting areas of California. The extent of the problem has been widespread and by late 2002, 22 of the 58 counties in California have adopted export restrictions.
- In favourable conditions, cloud seeding has proved to be a cost effective way of increasing atmospheric precipitations. In California, the costs of cloud seeding are about US$16 per 1000 m^3 of rainwater. In Utah, cloud seeding projects resulted in a 13 percent increase in average annual run-off in the project areas compared with the historic records. This was achieved at a cost of approximately US$1 per 1000 m^3.
- If dam building within the catchment could have any chance to increase water supply of the demand area, its details should be investigated as a preferred option compared with the inter-basin water transfer.

In numerous cases, in both developed and developing countries, alternatives to inter-basin water transfer may exist. All these options require thorough exploration, adequate assessment of their economic, social and environmental costs and a committed participatory process. Inter-basin water transfer is an option when all other methods of satisfying demand are inadequate. In that case, the recommendations of the World Commission on Dams are then particularly apposite.

Part IV
Appendices

A Some of the Australian Pioneers of Inter-basin Water Transfer

A.1 BRADFIELD, JOHN JOB CREW (1867–1943)

John Job Bradfield,[1] known as "Jack", was born in Sandgate near Brisbane on 26 December 1867. He was the ninth of ten children born to Maria and John Edward Bradfield, a brick maker and a Crimean War veteran. His parents migrated to Australia with their first three children and settled in Ipswich, Queensland in 1857. He was educated at Ipswich Grammar School and won one of the three Queensland Government Scholarships to study engineering at the University of Sydney. In March 1889 he passed his final engineering exams, presented his final-year project (design of a bridge) and obtained his Bachelor of Engineering with Honours.

Bradfield started work in the office of the Chief Engineer of Queensland Railways. His initial assignment was the design of a footbridge over the Roma Street railway station in Brisbane.

Within the year, as a result of the worsening economy, all new railway works in Queensland were shut down. In 1891, Bradfield was out of a job and returned to Sydney to work for the Public Works Department. While working, Bradfield continued his academic studies. In 1896, he graduated with the degree of Master of Engineering from the University of Sydney and was awarded University's Gold Medal. Bradfield used the problems posed by his work as the basis for his thesis. These problems involved calculations for the four gigantic 45.9 m trusses used on the Kempsey Bridge over the Macleay River. These trusses at the time were a Southern Hemisphere record.

In December 1901, the Sydney water supply failed, and in May 1902 restrictions were imposed throughout the metropolitan area because of the continuing drought. In March 1902, a Royal Commission was appointed to investigate Sydney's water supply. The Commission instructed government agencies to examine the whole of the Cataract, Nepean and Cordeaux catchments for suitable dam sites. After the surveys were completed, Bradfield was instructed to prepare detailed design and cost estimates for the Cataract Dam, which was constructed over the period of 1902–07 (see section H.2 in Appendix H). Bradfield was also involved in investigations of the availability of water supply for the nine sites proposed for the Federal capital city. He also played a role in the design of Burrinjuck Dam, the first large dam in New South Wales as part of a major water conservation and irrigation scheme. It was constructed on the Murrumbidgee River over the period of 1907–12, and was used for irrigation of crop and pasture in the Coleambally, Hay and Murrumbidgee irrigation areas.

Bradfield's professional life was intimately connected with the design and construction of bridges and in particular the Sydney Harbour Bridge. The history of building a bridge over Sydney Harbour goes back to 1815 when the architect Francis Green put forward his proposal. The first drawing of a proposed bridge appeared in 1857. In 1900, competitive bridge designs were sought and 24 designs were submitted. All submissions were considered and rejected by leading engineers. Later, proposals suggested for subways beneath Sydney Harbour, but Bradfield

---

[1] *Source of photograph:* National Library of Australia (nla.pic-an23278893).

opposed this idea because his preferred option was a bridge over the Harbour. In 1912, he was appointed Chief Engineer for the Sydney Metropolitan Railways Construction and Harbour Bridge. He was sent to Europe and North America in 1914 to study modern practices in underground railway and long span bridge construction. A worldwide tender was called for the Harbour Bridge on 8 December 1921, and on 16 March 1922 Bradfield was sent again to Europe and North America to talk to prospective tenderers. Towards the end of 1923, he submitted details of his design for the Sydney Harbour Bridge and City Railway to the University of Sydney as a thesis for the degree of Doctorate in Engineering. His thesis was accepted and he was awarded a Doctorate in 1924 at the age of 57. Twenty tenders were received on the closing date, 16 January 1924. The British firm Dorman Long and Co. was selected and the contract was signed on 24 March 1924. Ralph Freeman, Dorman Long's consultant, carried out design, calculations and preparation of plans in London. Bradfield was sent to London in 1924 with three engineers to check calculations and design details.

Construction of the Bridge was slow to start. Before works could commence, 438 houses had to be bought and demolished. On 26 March 1925 the foundation stone of the southern abutment tower of the Bridge was laid by the Minister for Public Works and Railways, and construction commenced under Bradfield's supervision. The bridge took 7 years to complete and was opened on 19 March 1932. At the time, it was the largest of its type in the world (the maximum height of the arch above mean high tide is 135 m, the length of the span 503 m, and the total length of the Bridge including approaches is almost 3.9 km). Completion of the Bridge was a major source of satisfaction for Bradfield whose dream finally came true. His name was given to the major highway that crosses the Bridge. Controversy regarding the design of the Bridge started in 1928 and continued for the following years. Both Bradfield and Freeman claimed that they were the sole designer of the Bridge. The controversy was never finally resolved, but in 1933, the Director of Public Works stated that Bradfield was the designer of the Bridge and that "*no other person by any stretch of imagination can claim that distinction*" (for details see Raxworthy, 1989, pp. 99–104; Spearritt, 1982, pp. 46–47; and Ellyard and Wraxworthy, 1982, pp. 136–140). Perhaps it could be said that Bradfield and Freeman, by working together, produced one of the most iconic bridges of the world.

Bradfield retired from the New South Wales Public Service in 1933, after completion of the Sydney Harbour Bridge. He set up a consulting engineering business with his son Bill Bradfield in Sydney. He was appointed consulting engineer for the construction of the Story and St Lucia bridges in Brisbane and awarded an honorary doctorate from the University of Queensland in 1935. In 1938 he drew up his controversial plan to divert water from north Queensland rivers across the Great Dividing Range into the dry centre of the State. Over the period of 1942–43 he was deputy Chancellor of Sydney University. After a short illness, he died in Sydney on 23 September 1943 at the age of 76.

Further Reading

Ellyard, D. and Wraxworthy, R. (1982). *The Proud Arch: The History of the Sydney Harbour Bridge*. Sydney: Bay Books.

Raxworthy, R. (1989). *The Unreasonable Man: The Life and Works of J. J. C. Bradfield*. Sydney: Hale & Iremonger.

Spearritt, P. (1982). *The Sydney Harbour Bridge*. Sydney: George Allen & Unwin.

A.2 CHIFLEY, JOSEPH BENEDICT "BEN" (1885–1951)

Ben Chifley[2] was born on 22 September 1885 in Bathurst, New South Wales, where his father Patrick Chifley was a blacksmith. After leaving school he worked briefly in a store and then in a tannery. On 13 September 1903, at the age of 18, he joined the New South Wales Railways as shop-boy in Bathurst. His studious habits helped his advancement. After a series of promotions, he gained a locomotive driver's certificate in 1912. On 6 June 1914 he married Elizabeth Mckenzie, daughter of a Bathurst railway man. He took part in the anti-conscription campaign of 1916–17. In August 1917 there was a statewide railway strike. Chifley was one of the leaders of the Bathurst strikers. When the strike was over, the NSW Railways administration gave promotions to workers who had not supported the strike, and downgraded or even dismissed strike supporters. Chifley was dismissed but promptly appealed and was reinstated with reduced seniority. Later attempts to secure full reinstatement failed and Chifley and his fellow strikers had to wait for a Labor victory in the 1925 State election for reinstatement. Chifley received a medallion for his service to the Union, but he wanted something more than a medallion, and was determined to enter Federal politics. Years later he acknowledged the strike and its aftermath as a turning-point in his career: "*I should not be a Member of this Parliament today if some tolerance had been extended to the men who took part in the strike of 1917*".

After World War I, he felt the time was right for him to try for Parliament. However, he was defeated in Labor Party pre-selection for the State seat of Bathurst in 1922 and 1924, and for the Macquarie Federal seat in 1925. He finally won the Macquarie seat in 1928 and entered Federal Parliament. One of the tasks which he enjoyed in Federal Parliament, and was of much benefit to him later, was the membership of the Joint Committee of Public Accounts, a parliamentary body which gave him an opportunity to familiarise himself with the processes of public finances of State and Commonwealth governments. In March 1931, he became Minister for Defence in the Labor Government of James Scullin.

[2] *Source of photograph:* http://psephos.adam-carr.net/countries/a/australia/gallery/chifley.jpg (visited in August 2005).

For years, Chifley had studied banking laws and practice, and he was gradually recognised as an able administrator with a sound knowledge of finance. In 1936, he was appointed a member of the Royal Commission on Banking, whose report appeared in July 1937. In October 1937, he was appointed to the Capital Issues Advisory Board. In February 1941, Chifley became a member of a Board of Inquiry into Hire Purchase and Cash Order Systems. All these duties gave him a wealth of experience and insight into financial matters.

John Curtin became Labor Party Prime Minister of Australia in October 1941. His main task was to guide Australia through World War II. He appointed Chifley as his Treasurer and in December 1942, he was given a second portfolio as Minister for Post-War Reconstruction. Chifley wanted to secure full employment in peace as well as war, and he believed that uniform taxation across Australia was necessary. He acted under the Commonwealth wartime defence power without States' acquiescence, and uniform taxation was made law for the duration of the war plus one year. Until this time, taxation was the responsibility of individual States in the Australian Federated system. In his view *National rights must take precedent over the State rights. The rights of the sovereign people are paramount to the sovereign right of the States.* The States challenged the legislation in the High Court and much to Chifley's surprise and pleasure the court ruled in favour of the Federal legislation.[3]

In March 1945, while still Treasurer in John Curtin's Cabinet, he turned his attention to reforming the banking system. He introduced the *Commonwealth Bank Act* of 1945 and *Banking Act* 1945 to the House of Representatives. His reforms aimed at greater Commonwealth control in order to avoid the chaos and misery of the 1930s depression. His legislations tightened Commonwealth control over general banking, gave the Federal Government greater power to direct the Commonwealth Bank in policy matters, and guaranteed the safe return of deposits in all banks. It also compelled public bodies, such as states and local governments to bank with the Commonwealth Bank. After a long debate that spread over five months and a furious campaign waged by the private banks, legislations passed all stages and were assented in August 1945.[4]

The Prime Minister John Curtin died on 5 July 1945. After some indecision, Chifley contested the Labor leadership and became the sixteenth Prime Minister of Australia on 13 July 1945. One of his first duties was the announcement of the end of the war in a broadcast to the nation. In his broadcast he declared that "*there is much to be done*" and asked "*the state governments and all sections of the community*" to cooperate "*in facing the tasks and solving the problems that are ahead*".

Chifley's main task was to bring the nation back to a peacetime footing. The government had to assimilate returned servicemen and women into the workforce. Industry was encouraged to develop by favourable financial treatment. A War Service Land Settlement Scheme was developed to place servicemen on the land. Large projects such as the Snowy Mountains Hydro-electric Scheme also helped provide employment and infrastructure. Government loans enabled many people to begin in business. The Commonwealth Reconstruction Training Scheme provided free tuition and living allowances for many who wished to complete university courses or learn a trade. A program of assisted migration was also set up. The program was so successful that the initial target of 70 000 migrants per year was more than doubled within five years. The Constitution was amended to grant the Commonwealth power to introduce social services such as maternity allowance, child endowment, pharmaceutical, sickness and hospital benefits. In external affairs, the government worked with Asian countries in regional associations and supported the Indonesian independence against Dutch rule.

On 16 August 1947, only three days after the High Court declared part of the banking system reform unconstitutional, Chifley decided to nationalise the banks because he believed that this measure would ensure full employment and economic stability. However, the High Court invalidated bank nationalisation in August 1948.

Chifley was faced with various problems in implementing his policies. For example, doctors defeated his plans for a national

---

[3] After the war, Chifley announced at the 1946 Premiers' Conference that the uniform taxation was to remain, labelling the Premiers' complaints as "*nonsense*".

[4] Banks appealed to the High Court against the section that obliged public bodies to trade with the Commonwealth Bank, and on 13 August 1947 this section was declared unconstitutional.

health scheme. Troubles also came from the unions. During the war, Communists won many key union positions, and the Communist Party began to use the unions as a base to cripple Australian post-war industry. Strikes were called repeatedly. The nation lost millions of pounds in lost man-hours and major cities were blacked out because of power shortages. In July 1949, despite opposition in his own ranks, he sent Army troops to the coalfields to end a miners' strike which had severely restricted coal supplies.

In spite of many successes, Chiefly entered the 1949 election year with the attention of voters fixed on issues such as petrol rationing, banking, decline in housing construction, inflation, fear of socialism and the Communist "menace". All these issues were widely publicised by the Opposition and the public was receptive. Chifley lost the election in December 1949, and Robert Menzies became Prime Minister of Australia, but Labor retained its majority in the Senate. In 1950, Chifley was elected as the Leader of the Labor Party in Opposition. Soon after he had his first heart attack. In 1950, Menzies secured a double dissolution of the House of Representative and the Senate. Chifley passed away on 13 June 1951, at the age of 65, as a result of a second heart attack during a week of Commonwealth Jubilee celebrations. On 17 June 1951, he was given a State funeral in Bathurst. Later, the Labor Party erected a monument over his grave.

Ben Chifley was one of the Australia's greatest and most popular Prime Ministers. He was a man of considerable charm, and remarkable simplicity. Chifley won wide respect for his integrity, intelligence and capacity for hard work. Despite his lack of formal education, he developed a deep understanding of financial matters and was widely regarded as a particularly able Treasurer. Many policies that his Labor government introduced after World War II were to define the "modern Australia". These included a mass immigration scheme, improved social services, the beginning of a free health service, a commitment to full employment, the opening up of access to university education, establishment of the Australian National University in Canberra in 1946 as Australia's only research-oriented university, a greater engagement with Asia, the promotion of an Australian sentiment, and a commitment to civil liberties. His other achievements include the foundation of the Snowy Mountains Hydro-electric Authority, Trans-Australian Airlines, and the nationalisation of Qantas and overseas telecommunications. His most controversial act was using troops to crush the coal miners' strike in 1949.

Further Reading

Bennett, S. (1973). *J. B. Chifley*. Melbourne: Oxford University Press.

Crisp, L. F. (1961). *Ben Chifley*. London: Longmans.

Day, D. (2001). *Chifley*. Sydney: Harper Collins Publishers.

A.3 FORREST, SIR JOHN (1847–1918)

John Forrest's Scottish parents, William and Margaret Forrest, arrived in 1842 in the colony of Western Australia, which had only been established 13 years earlier. William Forrest was a determined and energetic man and for some time he worked for a prosperous settler. He then established a flour mill near the mouth of Preston River, and shortly afterwards built a house at Picton near Bunbury, approximately 150 km south of Perth. John, the third son of the family, was born at Preston Point on 22 August 1847. The total white population of Western Australia at this time was less than 5000 and about half of them lived around Perth. At the age of 12, John was sent to Perth to attend Bishop's School, the only boy's college in the colony of Western Australia. He was a hard-working student and did well in mathematics. At the age of 17, after leaving the Bishop's School, he worked as an apprentice with a surveyor in Bunbury. A year later he entered the colony's Survey Department as a junior surveyor.

Forrest[5] gradually gained experience and rose to higher positions. He also won a reputation as an explorer, based mainly on three expeditions into the interior of Western and South Australia (Figure A.1). In 1869, he travelled north-west from Perth to well beyond Lake Barlee. On his return to Perth he reported that there was no good grazing land in the area he had explored but there was some interesting evidence of minerals. He suggested that it might be useful to send geologists to the area. The Governor did not act on this suggestion. The three hills, which he had named (Mt Margaret, Mt Leonora and Mt Malcolm), became the sites of gold discoveries more than 30 years later. In 1870, he explored from Perth to Adelaide

[5] *Source of photograph:* http://en.wikipedia.org/wiki/Premier_of_Western_Australia (visited in August 2005).

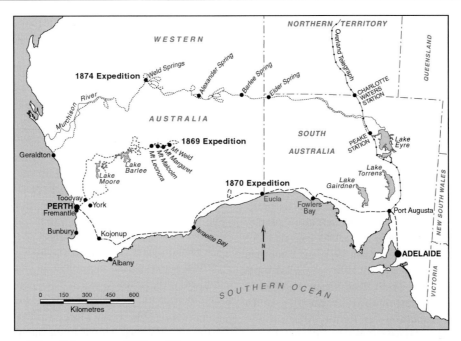

Figure A.1 Forrest's expeditions (Mossenson, 1960, p. 17).

around the Great Australian Bight and in 1874, he crossed the western half of the continent from Geraldton and the Murchison River in Western Australia to Peake Station in the north-west of Lake Eyre in South Australia. Although his expeditions did not make discoveries of great importance, they were notable for their endurance in some of the harshest regions in Australia. After the 1874 expedition he returned to the Survey Department and in 1883 became Surveyor-General and Commissioner of Crown Lands.

When Western Australia became self-governing in 1890, Forrest was elected to parliament as the Member of Bunbury. He became the first Premier of the colony on 29 December 1890, a position he held for 10 years until 15 February 1901. Forrest was respected for his loyalty to the British Empire, efficient administration, progressive social legislation and public works policies, and was Knighted by the Queen Victoria. The ten years of his administration were highly fruitful in Western Australian history. Gold discoveries attracted people to the State on a much larger scale than ever before. He developed railways, harbours and telegraph lines, opened up the agricultural districts, and assisted the goldfields. He encouraged and supported his Chief-Engineer Charles O'Connor in planning and constructing the Goldfields Water Supply Scheme.

Forrest represented Western Australia at meetings of the Commonwealth Council during the 1890s and took part in Commonwealth Conventions, in which he was particularly concerned with protecting the rights of the less populous colonies. He was elected to represent the Swan electorate in the first Commonwealth Parliament in 1901. He served as a Federal Minister for Defence, Treasurer, and External Affairs in five Federal Cabinets. Forrest came close to being Prime Minister four times (in 1908, 1913, 1917 and 1918) during his 18 year career in Federal politics. For a few months from March until June 1907 he was the Acting Prime Minister. He was associated with the formation of the Australian Army and the construction of the east–west transcontinental railway which was opened in 1917. The culmination of his lifetime service came on 9 February 1918, when it was announced by King George V that he had been made Lord Forrest, Baron of Bunbury. Although he was not in good health and previously had two operations on his temple for a cancerous growth, he decided to go to London to take his seat in the House of Lords and for further medical treatment. He left Albany with his wife and a nurse on 30 July 1918, and celebrated his 71st birthday at sea on 22 August whilst sailing north up the west coast of Africa. He died at sea in great agony on 3 September 1918. He was buried in Sierra Leone, but his body was later brought to Western Australia for reburial.

Further Reading

Crowley, F. K. (1968). *Sir John Forrest*. St. Lucia: University of Queensland Press.

McEwans, M. (1985). *Great Australian Explorers*. Sydney: Bay Books, pp. 358–363.

Mossenson, D. (1960). *John Forrest*. Melbourne: Oxford University Press.

A.4 HUDSON, SIR WILLIAM (1896–1978)

The son of Dr James Hudson, William Hudson[6] was born in the little town of Nelson on the South Island of New Zealand on 27 April 1896. He had six brothers and two sisters. While his brothers became doctors or farmers, he dreamt of becoming an engineer who built bridges, dams and railways. He was educated at Nelson College, and in 1914, at the age of 18, he went to London to study engineering at University College, London University. World War I interrupted his studies. After completing the first two years of the course, he enlisted for active duty. He served with the London Regiment in France as an infantry lieutenant until wounded in the leg in 1917. After convalescence in New Zealand, he returned to London in 1918, and completed his Bachelor degree in Engineering in 1920 with First Class Honours and also won the Archibald Head Medal. He enrolled in a postgraduate course in hydro-electric engineering at the University of Grenoble in south-eastern France. After a period of service with Sir W. G. Armstrong-Whitworth and Co. in England, he returned to New Zealand, where he joined the Public Works Department. Initially he was employed on railway construction works, and then as assistant engineer on the Mangahao hydro-electric scheme outside Wellington from 1922 to 1924. From 1924 to 1927 he worked on the construction of the 70 m high Arapuni Dam in New Zealand. He moved then to Australia and for three years (1928–30) worked for the New South Wales Department of Public Works as the supervisory engineer on the construction of the Nepean Dam, one of the storage reservoirs for Sydney's water supply system (see section H.2 in Appendix H). Hudson married Eileen Trotter in 1926 at the age of 30. They had two daughters, Margaret and Anne.

Over the period of 1931–37, he worked as a member of the staff of Sir Alexander Gibb & Partners in England. He was the engineer-in-chief of the Galloway hydro-electric scheme in Scotland, a project involving construction of seven storage dams. He returned to Australia in 1937 and joined the Sydney Metropolitan Water, Sewerage and Drainage Board as resident engineer on the construction of the Woronora Dam in south-west Sydney. During World War II he was resident engineer for the construction of the Captain Cook Dock in Sydney. By 1948, he was appointed the Board's engineer-in-chief.

A suitable appointee was being sought in 1949 for the position of Commissioner of the Snowy Mountains Hydro-electric Authority The Federal Minister for Public Works and Housing, Mr Nelson Lemmon, took Hudson to meet the Prime Minister, Ben Chifley. At the age of 53, he was appointed as Commissioner of the Authority by the Governor General, Sir William McKell.

Hudson's skills were not just in engineering, he was an extraordinary administrator and father-figure to some 100 000 workers from more than 30 countries. He successfully worked with his own staff, consultants, contractors and the Commonwealth, as well as the State authorities. With tremendous capacity for work, which characterised the whole of his career, he dedicated himself to the Snowy Scheme, inspiring all those people who had a part in this outstanding development. His position as the Commissioner was extended twice after he had reached the retiring age of 65. He fulfilled his duties over a long period of 18 years until his retirement on 26 April 1967 at the age of 70. This was about 7 years before completion of the Scheme in 1974. Hudson advocated using the expertise of the Snowy Mountains Authority for the development of an inland diversion scheme for northern Queensland rivers. He was unsuccessful in persuading the Federal Government to carry out such a project.

Hudson received numerous recognitions for his achievements. In 1955 he was Knighted; in 1957 he received the Australian Engineering Award; in 1959 he received the Kernot Memorial Medal for distinguished engineering achievement in Australia; in January 1961 he was elected a Fellow of University College, London University; late in 1961 he was made an honorary member of the Institute of Mining and Metallurgy; in March 1962 he became an honorary member of the Institution of Engineers, Australia; in April 1962 he received an Honorary Doctorate of Law from the Australian National University; and in 1964 was elected Fellow of the Royal Society, London. He was also president of the National Safety Council of Australia, and Chairman of the Road Safety Council of New South Wales (1968–71).

Hudson lived in Canberra during his retirement and worked as a consultant for Sir Alexander Gibb & Partners over the period of 1969–74. He died in Canberra on 12 September 1978 at the age of 82. He loved the Snowy Mountains area and it was his wish to be buried in Cooma (located about 100 km south of Canberra) where he had lived with his family and worked for

[6] *Source of photograph:* http://www.phm.gov.au/hsc/snowy/img/003229.jpg (visited in August 2005).

18 years. Some weeks before his death, he agreed to have his name used for fund raising to build a 50 bed Nursing Home in Cooma to be named *Sir William Hudson Memorial Centre*. Construction of the centre commenced on 27 September 1980, and was officially opened by Lady Hudson on 1 May 1982 as a reminder of his great contribution to Australia.

Further Reading

Australian Dictionary of Biography (1996). Melbourne University Press. Vol. 14, pp. 510–512.

Websites

http://nzedge.com/heroes/hudson.html (visited in August 2005)

http://www.nla.gov.au/ms/findaids/9025.html (visited in August 2005)

A.5 IDRIESS, ION LLEWELLYN (1889–1979)

Walter Owen Idriess[7] was born on 3 March 1862 in Dolgelly, North Wales. He was an officer in Her Majesty's Navy and travelled a number of times to Sydney between 1883 and 1887, before migrating to Australia and marrying his wife Juliette in Sydney on 1 December 1888. Ion Idriess, known as "Jack", was born in Waverly, Sydney on 20 September 1889. He grew up in a family that wandered around New South Wales before finally settling in Broken Hill, where his father took up the office of Sheriff and Mines Inspector in 1899. Ion was educated at Lismore, Tamworth, and Broken Hill. In November 1907, after completing his final examination for qualification as an

assayer, he got typhoid fever. He was lucky to survive but his mother died from the same illness on 13 January 1908. He was employed in an assay office in Broken Hill, but stayed for only a short while in the position. He subsequently worked as seaman, opal miner, tin prospector, drover, rabbit poisoner, timber cutter, and dived for pearls. In 1914, he enlisted for World War I, serving throughout its duration with the Fifth Light Horse at the Gallipoli, Sinai and Palestine campaigns against the Turks, and was wounded three times. After the war he travelled widely in north and inland Australia, New Guinea and the Pacific. In 1928, he finally settled down in Sydney and in 1931 married Eta Gibson, who migrated from England to Australia in 1929.

His first book *Madam's Island* was published in 1927. His other books include *Prospecting for Gold* (1931), *Lasseter's Last Ride* (1931), *Flynn of the Inland* (1932), *The Desert Column* (1932), *Drums of Mer* (1933), *Man Tracks* (1935), *The Cattle King* (1936), *Forty Fathoms Deep* (1939), *Headhunters of the Coral Sea* (1940), *In Crocodile Land* (1946), *The Great Boomerang* (1948), *The Wild White Man of Badu* (1950), *The Red Chief* (1953), *The Silver City* (1956), and *Challenge of the North* (1969).

His narrative and descriptive works are vigorously written and show great knowledge of the scenes of his travels. His books were immensely popular. His love of Australia and Australian folklore was evident in all his writing. He was an extraordinarily fast writer and was capable of writing three books in a year. Between 1927 and 1969, he published 56 books with some of them, such as *Flynn of the Inland* and *Lasseter's Last Ride*, running to 40 and 50 editions. Overall, more than three million copies of his books were sold, making him one of Australia's most popular writers.

Idriess was also a visionary about the future of the Australian inland and its need for water. In his book *The Great Boomerang* he put forward his plan for watering the dry interior by inland diversion of Queensland's coastal rivers.

Idriess had a love–hate relationship with his wife. In 1954, when his daughters Judy and Wendy were living in England, Eta left him to live with a young man who she had a relationship with for eight years. In December 1964, Idriess had a stroke, which prevented him from writing as before, and finally in 1969 he published his last book *Challenge of the North*. In late 1978, he was admitted to a nursing home in Mona Vale, New South Wales. He died there on 6 June 1979 at the age of 89. Eta refused to attend his funeral. He was survived by his estranged wife Eta, and two daughters. His awards include Fellowship of the Royal Geographical Society of Australia (1944), and the Order of the British Empire (1967).

---

[7] *Source of photograph:* National Library of Australia (nla.pic-an12776350).

Further Reading

Eley, B. (1995). *Ion Idriess*. Sydney: IMPRINT.

A.6 MENZIES, SIR ROBERT GORDON (1894–1978)

Robert Menzies[8] was born on 20 December 1894 in the small town of Jeparit in Victoria, about 400 km north-west of Melbourne. His father was a storekeeper at Jeparit. Robert Menzies was educated at Jeparit and Ballarat. He had a natural thirst for learning, an excellent memory, and was determined to be a barrister. He studied at Grenville College in Ballarat after winning a State Scholarship. At this time his father decided to enter State politics. He stood for the Victorian Legislative Assembly and was elected and moved with his family to Melbourne. Robert Menzies won a State Scholarship to study at Wesley College in Melbourne. He studied three years at Wesley College to prepare himself for Melbourne University. In 1912, at the age of 18, he started a four-year course at Melbourne University after winning a State Scholarship. In the summer of 1916, he passed his final examinations and became a graduate in law with First Class Honours.

He entered an outstandingly successful career as a barrister in the Victorian High Court in 1918, dealing with issues related to Constitutional laws. He married Pattie, daughter of Senator J. W. Leckie on 27 September 1920. He decided to enter the Victorian Parliament in 1928. He stood for one of the seats in the Legislative Council (Upper House) for Yarra. His first attempt failed, but a few months later at a by-election, he was elected to the Legislative Council. Shortly after, Menzies decided to resign and offer himself as a candidate for the Victorian Legislative Assembly (Lower House) at the election in December 1929. He was successful and when the new parliament met, he took his seat for the constituency of Nunawading. However, The United Australia Party (UAP) was defeated and he sat in opposition for two and a half years.

The United Australia Party in Victoria was led by a group of elderly politicians. Menzies founded the Society of Young Nationalists of Victoria. The Society progressed rapidly and their ideas so impressed people that in the 1932 Victorian general election the majority of UAP elected to Parliament were Young Nationalists. Menzies held various Cabinet positions, such as the Attorney General, Solicitor-General and Minister of Railways. Menzies also became Acting Premier of Victoria from February to June 1934. In October 1934 he entered the Commonwealth Parliament as a Member of the House of Representative for Kooyong, and was immediately appointed Attorney General and Minister for Industry in the Cabinet of Joseph Lyons, positions held until March 1939.

In April 1939 Menzies, at the age of 45, became Prime Minister for the first time. At this stage of his career he lacked patience in dealing with colleagues. In 1941, as the military situation deteriorated in Europe, many government members became dissatisfied with Menzies' leadership of Australia's war effort, and in August 1941 he resigned the position of Prime Minister, passing it first to Sir Arthur Fadden from the Country/United Australian Coalition, and then in October 1941 to John Curtin of the Australian Labor Party who ran the country until his death on the 5 July 1945.

In late 1944, Menzies succeeded in changing the name of the United Australia Party to the Liberal Party. On 13 July 1945, Ben Chifley from the Labor Party became Prime Minister of Australia. In the Parliament, Menzies continuously attacked the Labor Government of Ben Chifley and the Labor Party's policies in the post war years, and particularly nationalisation of the banks. Menzies led the Liberal Party to victory in the December 1949 election and became Prime Minister for the second time. He had changed very much since 1939, and was no longer as arrogant as before. He had learned much political wisdom during the years in opposition and became more patient and more experienced in handling politicians and running the Government. He retained the position of Prime Minister until his retirement 17 years later in 1966, and became Australia's longest serving Prime Minister.

[8] *Source of photograph:* http://en.wikipedia.org/wiki/Robert_Menzies (visited in August 2005).

Menzies achievements included social stability, economic growth, gradual expansion of social and health services, raising the standard of the Public Service, assisting the State universities in the expansion of tertiary education, encouragement of scientific research, and development of Canberra as the national capital. Although Menzies opposed the *Snowy Mountains Hydro-electric Power Act* in July 1949 as the leader of Opposition, after he led the Liberal Party to victory in December 1949 he realised the importance of the Scheme, supported the project, and facilitated its construction.

Menzies was Knighted of the Thistle in 1963 and upon the death of Sir Winston Churchill in 1965 was appointed Lord Warden of the Cinque Ports. He was Chancellor of Melbourne University from 1967 to 1972. His publications include two volumes of memoirs, *Afternoon Light* and *The Measure of the Year*, published in 1967 and 1970 respectively. His publications are listed below. In 1971, Menzies suffered a severe stroke, which incapacitated him physically and put limits on his remaining public appearances. He died in Melbourne on 15 May 1978, and was given a State funeral.

Further Reading

Bunting, J. (1988). *R.G. Menzies: A Portrait*. Sydney: Allen & Unwin.

Hasluck, P. (1980). *Sir Robert Menzies*. Melbourne University Press.

Joske, P. (1978). *Sir Robert Menzies, 1894–1978: A New Informal Memoir*. London: Angus & Robertson.

Menzies, R.G. (1941). *To the People of Britain at War*. London: Longmans Green and Co.

Menzies, R.G. (1943). *The Forgotten People and Other Studies in Democracy*. Sydney: Angus & Robertson.

Menzies, R.G. (1958). *Speech is of Time: Selected Speeches and Writings*. London: Cassell & Co. Ltd.

Menzies, R.G. (1967). *Afternoon Light: Some Memoires of Men and Events*. Melbourne: Cassell Australia Ltd.

Menzies, R.G. (1967). *Central Power in the Australian Commonwealth: An Examination of the Growth of Commonwealth Power in the Australian Federation*. London: Cassell & Co. Ltd.

Menzies, R.G. (1970). The *Measure of the Year*. Melbourne: Cassell Australia Ltd.

Seth, R. (1960). *Robert Gordon Menzies*. London: Cassell & Co. Ltd.

A.7 O'CONNOR, CHARLES YELVERTON (1843–1902)

Charles Yelverton O'Connor[9] was born in Gravelmount, Ireland in 1843. He studied engineering at Dublin University, and worked for a while as a railway engineer in Ireland. In 1865, at the age of 22, he migrated to New Zealand to avoid a long and devastating famine in Ireland. In New Zealand he worked as a civil engineer for the province of Canterbury. He constructed the first road over New Zealand's Southern Alps to the newly discovered goldfield at Hokitika. In succeeding years he planned roads, railways, bridges and harbours all over New Zealand. In March 1874, he married Susan Ness of Christchurch, the eldest daughter of the Government architect. He was appointed Under Secretary of the Public Works Department in 1883. Eight years later in 1891, with the New Zealand economy in a deep recession and having been denied the position of Chief Engineer, O'Connor decided to migrate to the colony of Western Australia, with a small population of only about 48 000. Conditions were harsh and there were fears that the Government would go broke.

In Perth he was appointed Engineer-in-Chief of Public Works and Acting Manager of the Railways. The planning and construction of a suitable harbour for the colony of Western Australia was his first major undertaking. He proposed the Swan River mouth as the site for this development and although there was considerable opposition to his plan, it was accepted and the work on the Fremantle Harbour project started in 1892 and opened on 4 May 1897. In the 1890s, he was also responsible for several railway developments such as the rebuilding of the line through the Darling Range and the extension of the route to Southern Cross. Perhaps his most important project was the Goldfields Pipeline Scheme, which consisted of building the Mundaring Reservoir on the Helena River and construction of the pipeline to Coolgardie and Kalgoorlie. This project attracted much criticism over

[9] *Source of photograph:* http://www.fremantleports.com.au/system/graphics/full/1896_FPA_CY_OConnor_F.jpg (visited in August 2005).

techniques and cost. Although the scheme stands as a tribute to his vision, O'Connor was not destined to see it completed. While he had the support of the government, there were strong views from the government's oppositions regarding the practicality of the pipeline project and rumours about corruption in his administration. These controversies played on O'Connor. On the morning of 10 March 1902, O'Connor left home for his usual early morning horse riding. Generally he was accompanied by one of his three daughters, but this morning he left early to be alone. When he did not return for breakfast a search began. He had committed suicide and his body was found lying on a beach about 5 km from Fremantle. A farewell note to his family dated 10 March 1902 was found on his desk at home and read (Evans, 2001, p. 229):

> The position has become impossible. Anxious important work to do and three commissions of inquiry to attend to. We may not have done as well as possible in the past but we will necessarily be too hampered to do well in the immediate future. I feel that my brain is suffering and I am in great fear of what effect all this worry may have upon me. I have lost control of my thoughts. The Coolgardie Scheme is all right and I could finish it if I got a chance and protection from misrepresentation but there is no hope for that now and it's better that it should be given to some entirely new man to do who will be untrammelled by prior responsibility.

The news spread rapidly and the country was shocked. Even his critics felt that they had gone too far. Telegrams and letters of sympathy poured into the State from all over Australia. The Governor of Western Australia, Captain Sir Arthur Lawley, wrote (Evans, 2001, p. 228):

> When I consider what a long and valuable service he had rendered in the development on Western Australia, with what zeal and patience he devoted to the carrying out of the great works, in the initiation of which he was instrumental and with which his name will always be associated long after they have been brought to successful issue, I realise how deplorable, from a rational point of view, is this tragedy.

Sir John Forrest was in Melbourne when he learned of his trusted Chief-Engineer's death. He wrote in his cable:

> I mourn with the people of Western Australia the loss of one who has left behind a high and honourable record of splendid public service and I mourn the loss, also, of a dear and valuable friend.

On 12 March 1902, two days after his suicide, O'Connor was buried in the Fremantle Cemetery. Subsequent investigations indicated that there was no truth in any of the accusations against him. He owned no property and had no savings. His family would have been destitute without government support. In September 1902, the government passed an Act to provide an annuity for his widow amounting to £250 per annum ($22 500 in 2002 prices).

On 24 January 1903, about 10 months after his suicide, Sir John Forrest opened the pipeline project. Today there is a museum at the Mundaring Weir site commemorating the gold discoveries in Western Australia, the subsequent building of the pipeline, and C. Y. O'Connor who was responsible for the planning, design and construction of the Goldfields Pipeline Scheme.

Further Reading

Ayris, C. (2001). *C. Y. O'Connor: The Man for the Time*. Hamilton, Western Australia: PK Print Pty Ltd.

Evans, A. G. (2001). *C. Y. O'Connor: His Life and Legacy*. University of Western Australia Press.

B Construction Timetable of the Snowy Mountains Hydro-electric Scheme

No.	Name of the structure	Construction period		1950–54					1955–59					1960–64					1965–69					1970–74				
		From:	To:	0	1	2	3	4	5	6	7	8	9	0	1	2	3	4	5	6	7	8	9	0	1	2	3	4
1	Guthega Power Station	Nov. 1951	Apr. 1955																								
2	Guthega Pressure Tunnel	Dec. 1951	Apr. 1955																								
3	Guthega Dam	Dec. 1951	Apr. 1955																								
4	Guthega Pressure Pipelines	Dec. 1951	Apr. 1955																								
5	Perisher Range Aqueduct	Oct. 1952	Mar. 1956																								
6	Munyang River Aqueduct	Feb. 1953	Mar. 1956																								
7	Rams Flat Aqueduct	Oct. 1954	Apr. 1956																								
8	Falls Creek Aqueduct	Nov. 1954	Apr. 1956																								
9	Eucumbene-Tumut Tunnel	Nov. 1954	July 1959																								
10	Tumut 1 Pressure Tunnel	Jan. 1955	Feb. 1959																								
11	Tumut 1 Underground Power Station	May 1955	Dec. 1959																								
12	Tumut Pound Dam	Nov. 1955	Mar. 1959																								
13	Eucumbene Dam	May 1956	May 1958																								
14	Tooma River Aqueduct	Jan. 1957	May 1961																								

371

Appendix B.1 (*cont.*)

No.	Name of the structure	Construction period From:	To:	1950–54					1955–59					1960–64					1965–69					1970–74				
				0	1	2	3	4	5	6	7	8	9	0	1	2	3	4	5	6	7	8	9	0	1	2	3	4
15	Burns Creek Aqueduct	Feb. 1957	May 1961								█	█	█	█	█												
16	Tumut 1 Tailwater Tunnel	Apr. 1957	Mar. 1959								█	█	█														
17	Deep Creek Aqueduct	Sep. 1957	May 1961								█	█	█	█	█												
18	Goodradigbee River Aqueduct	Nov. 1957	July 1960									█	█	█													
19	Tantangara Dam	June 1958	May 1960									█	█	█													
20	Murrumbidgee-Eucumbene Tunnel	June 1958	Feb. 1961									█	█	█	█												
21	Tooma-Tumut Tunnel	June 1958	Mar. 1961									█	█	█	█												
22	Tumut 2 Tailwater Tunnel	June 1958	Sep. 1961									█	█	█	█												
23	Tumut 2 Underground Power Station	June 1958	Jan. 1962									█	█	█	█	█											
24	Happy Jacks Dam	Sep. 1958	June 1959									█															
25	Tooma Dam	Oct. 1958	Mar. 1961									█	█	█	█												
26	Tumut 2 Pressure Tunnel	Nov. 1958	Sep. 1961									█	█	█	█												
27	Tumut 2 Dam	Mar. 1959	Sep. 1961										█	█	█												
28	Eight Mile Creek Aqueduct	Apr. 1959	June 1962										█	█	█	█											
29	Section Creek Aqueduct	Apr. 1959	June 1962										█	█	█	█											
30	Snow Ridge Aqueduct	Aug. 1961	Nov. 1961												█												
31	Eucumbene-Snowy Tunnel	Dec. 1961	Oct. 1965												█	█	█	█	█								
32	Murray 1 Pressure Tunnel	Mar. 1962	Jan. 1966													█	█	█	█								
33	Snowy-Geehi Tunnel	Mar. 1962	Feb. 1966													█	█	█	█								
34	Island Bend Dam	Aug. 1962	July 1965													█	█	█	█								
35	Geehi Dam	Oct. 1962	Feb. 1966													█	█	█	█								
36	Bourkes Gorge Intake	Nov. 1962	Aug. 1963													█	█										
37	Geehi River Aqueduct	Nov. 1962	June 1970													█	█	█	█	█	█	█	█				
38	Murray 1 Power Station	Dec. 1962	Aug. 1967													█	█	█	█	█	█						

39	Murray 1 Pressure Pipelines	Mar. 1963	Apr. 1967
40	Khancoban Dam	Oct. 1963	Mar. 1966
41	Gungarlin River Aqueduct	Oct. 1963	Oct. 1965
42	Burrungubugge River Aqueduct	Oct. 1963	Oct. 1965
43	Jindabyne-Island Bend Tunnel	July 1964	July 1968
44	Blowering Pressure Tunnel	Aug. 1964	July 1965
45	Bar Ridge Aqueduct	Oct. 1964	July 1965
46	Diggers Creek Aqueduct	Oct. 1964	July 1965
47	Blowering Dam	May 1965	Sep. 1968
48	Mowamba River Aqueduct	Aug. 1965	July 1967
49	Bourkes Gorge Aqueduct	Aug. 1965	Feb. 1967
50	Jindabyne Dam	Dec. 1965	Apr. 1967
51	Murray 2 Power Station	Feb. 1966	Sep. 1969
52	Murray 2 Dam	Mar. 1966	Dec. 1968
53	Murray 2 Pressure Tunnel	Mar. 1966	June 1969
54	Jindabyne Pumping Station	Mar. 1966	Feb. 1969
55	Jounama Dam	May 1966	Apr. 1968
56	Blowering Power Station	Oct. 1966	Oct. 1969
57	Middle Creek Aqueduct	Oct. 1966	Jun. 1967
58	Murray 2 Pressure Pipelines	Oct. 1966	Jan. 1969
59	Jindabyne Pressure Pipeline	Nov. 1966	Feb. 1969
60	Cascade Creek Aqueduct	Jan. 1967	May 1969
61	Talbingo Dam	Dec. 1967	Oct. 1970
62	Tumut 3 Power Station	Jan. 1968	Sep. 1973
63	Tumut 3 Headrace Channel	Apr. 1968	Mar. 1971
64	Tumut 3 Pressure Pipelines	Feb. 1969	July 1973
65	Tumut 3 Pipeline Inlet Structure	July 1969	May 1971

Source of data: Snowy Mountains Hydro-electric Authority (1993).

C Details of Diversion Scheme from the Clarence and Macleay River Basins

C.1 DETAILS OF DIVERSION SCHEMES FROM THE CLARENCE RIVER BASIN

Scheme reference	CLA-1	CLA-2	CLA-3	CLA-4	CLA-SUP-A	CLA-SUP-B	CLA-4 A	CLA-4 B	CLA-4 AB	CLA-5	CLA-6
Inland basin	Cond-amine	Cond-amine	Border rivers	Border rivers	—	—	Border rivers	Border rivers	Border rivers	Border rivers	Border rivers
Diversion method											
Pumping	×	—	—	×	×	×	×	×	×	×	×
Gravity	—	×	×	×	×	—	×	×	×	×	×
Annual diversion volume (10^6 m^3)	67	21	5.8	46	67	129	113	175	242	93	89
Tunnel/pipeline details											
Length (km)	29 (P)[b]	6.7 (T)[b]	25 (T)	2.1 (P) 67 (T)	4.9 (P) 7.8 (T)	15 (P)	2.1 (P) 67 (T)	2.1 (P) 67 (T)	2.1 (P) 67 (T)	4 (P) 22 (T)	0.8 (P) 27 (T)
Diameter (m)	1.9	2.7	2.7	1.6	1.9 2.7	2 × 1.9	2 × 1.8 2.8	3 × 1.8 3.3	3 × 2 3.7	2 × 1.6 3	2 × 1.6 2.9
Capacity (m^3 s^{-1})	4.3	4.9	6.1	2.9	4.3 6.9	8.3	7.2 7.5	11.3 11.8	15.5 16.1	6.0 6.4	5.7 6.2
Pumping capacity (m^3 s^{-1})	4.3	—	—	2.9	4.3	8.3	7.2	11.3	15.5	6.0	5.7
Total capital cost[a] ($M)	228	185	160	600	153	485	783	1225	1478	363	463
Total annual cost[a] ($M)	32	20	18	69	21	69	95	156	190	46	55
Cost[a] of water											
Capital cost ($/10^3 m^3)	3400	8900	27600	13000	2285	3760	6930	7000	6100	3900	5200
Annual cost ($/10^3 m^3)	480	950	3100	1500	315	535	840	890	785	495	620

Appendix C.1. (cont.)

Scheme reference	CLA-7	CLA-SUP-C	CLA-5 C	CLA-6C	CLA-7C	CLA-8	CLA-9	CLA-10	CLA-11	CLA-12	CLA-13	CLA-14
Inland basin	Border rivers	Border rivers	Border rivers	Border rivers	Border rivers	Border rivers	Border rivers	Border rivers	Border rivers	Gwydir	Gwydir	Gwydir
Diversion method												
Pumping	×	×	×	×	×	×	–	–	–	–	–	–
Gravity	×	–	×	×	×	×	×	×	×	×	×	×
Annual diversion volume (10^6 m^3)	89	666	759	755	755	89	21	21	13	17	17	17
Tunnel/pipeline details												
Length (km)	0.05 (P) 41 (T)	1.8 (P) 12 (T)	4.0 (P) 22 (T)	0.8 (P) 27 (T)	0.05 (P) 41 (T)	4.6 (P) 11 (T)	13 (T)	25 (T)	12 (T)	35 (T)	37 (T)	49 (T)
Diameter (m)	2 × 1.6 29	9 × 2 6.1	2 × 1.6 4.4	10 × 2 6.2	10 × 2 6.4	2 × 1.6 2.7	2.7	2.7	2.7	2.7	2.7	2.7
Capacity (m^3 s^{-1})	5.7 5.8	42.8 44.1	6 8	48.6 48.6	48.6 49.7	5.7 6.8	7.4	4.8	6.0	3.6	5.6	9.4
Pumping capacity (m^3 s^{-1})	5.7	42.8	49	49	49	5.7	–	–	–	–	–	–
Total capital cost[a] ($M)	500	740	1533	1548	1640	430	163	208	108	203	235	260
Total annual cost[a] ($M)	57	143	279	261	248	58	18	23	12	22	26	28
Cost[a] of water												
Capital cost ($/10^3 m^3)	5620	1110	2020	2050	2170	4830	7760	9900	8310	11 940	13 820	15 290
Annual cost ($/10^3 m^3)	640	215	370	345	330	650	860	1095	925	1295	1530	1650

Note: [a] The 1981 estimates of costs have been multiplied by a factor of 2.5 in order to represent the 2002 prices; [b] (P) stands for Pipeline and (T) for Tunnel.
Source: Clarence Valley Inter-Departmental Committee on Water Resources (1982).

C.2 DETAILS OF DIVERSION SCHEMES FROM THE MACLEAY RIVER BASIN

Scheme reference	MAC-1	MAC-2	MAC-3	MAC-4	MAC-5	MAC-SUP-A	MAC-SUP-B	MAC-SUP-C	MAC-SUP-D	MAC-SUP-E
Inland basin	Gwydir	Gwydir	Gwydir	Namoi	Namoi	—	—	—	Namoi	—
Diversion method										
Pumping	—	—	—	—	—	—	—	—	—	—
Gravity	×	×	×	×	×	×	×	×	×	×
Annual diversion volume (10^6 m^3)	15	33	33	33	22	23	30	29	10	17
Tunnel/pipeline details										
Length (km)	32 (T)[b]	34 (T)	46 (T)	63 (T)	28 (T)	24 (T)	A number of pipelines and tunnels involved	A number of pipelines and tunnels involved	3.9 (T)	15 (T)
Diameter (m)	2.7	2.7	2.7	2.7	2.7	2.7			2.7	2.7
Capacity (m^3 s^{-1})	6.0	7.1	5.4	4.0	5.0	3.1			7.7	13.8
Pumping capacity (m^3 s^{-1})	—	—	—	—	—	—	A number of pump stations	A number of pump stations	—	—
Total capital cost[a] ($M)	153	275	313	358	150	178	315	283	38	145
Total annual cost[a] ($M)	17	30	34	39	17	19	35	31	4	16
Cost[a] of water										
Capital cost ($/10^3 m3)	10 200	8330	9480	10 850	6820	7740	10 500	9760	3800	8530
Annual cost ($/10^3 m^3)	1130	910	1030	1180	770	825	1165	1070	400	940

Note: [a] The 1981 estimates of costs have been multiplied by a factor of 2.5 in order to represent the 2002 prices, [b] (T) stands for Tunnel.
Source: Water Resources Commission of N.S.W. (1981).

D Chronological Table of the Most Important Events in the Goldfields Pipeline Scheme, Western Australia

Year	Month	Day	Event
1892	Sep.	16	Discovery of gold in Coolgardie by A. Bailey and W. Ford
1893	June	10	Discovery of gold in Kalgoorlie by Patrick Hanna and two others
1894	July	1	Railway reached Southern Cross
	Mar.	23	Railway reached Coolgardie
1896	Mar.	23	John Forrest, Premier of Western Australia, announced that the Goldfields should be provided with adequate water supply from the coast
	July	17	Publication of C. Y. O'Connor's report about the Goldfields Pipeline Scheme
	Sept.	23	Royal Assent given for the Coolgardie Goldfields Water Supply Loan Bill
1897	Jan.	23	O'Connor left Albany for London to discuss his proposal with the experts
	Sep.	16	O'Connor returned to Albany after examination of his proposal in London
1898	Jan.	17	John Forrest informed O'Connor that the loan of £2 500 000 was approved
	Apr.	–	Works commenced on the construction of the Mundaring Weir
	Oct.	–	Coolgardie Goldfields Water Supply Construction Bill was passed
	Oct.	–	Contracts were signed for the supply of pipes by two Australian firms
1900	Jan.	–	Excavation of trenches started
	Mar.	–	Contract was signed for delivery and installation of 20 sets of pumps, boilers and accessories

Year	Month	Day	Event
1901	Jan.	–	Installation of the pumping machinery commenced
	Feb.	7	Forrest inspected the dam site before entering the Federal Parliament
	Feb.	17	The Select Committee presented its report to the Parliament
1902	Feb.	19	The Legislative Assembly appointed a Royal Commission
	Mar.	5	The Royal Commission interviewed the first witnesses
	Mar.	10	C. Y. O'Connor committed suicide
	Mar	–	Water first pumped from Pumping Station number one
	April	18	Water reached Northam
	Oct.	30	Water reached Southern Cross
	Dec.	22	Water reached Coolgardie
	Jan.	16	Water reached Kalgoorlie
1903	Jan.	22	Lady Forrest started pumping machinery at Mundaring Weir
	Jan.	24	John Forrest opened the scheme and praised C. Y. O'Connor

E Flooding of the Sahara Depressions

E.1 INTRODUCTION

The idea of flooding the Sahara depressions goes back to the second half of the nineteenth century. The aim was to divert waters of the Mediterranean Sea into the Sahara depressions in Tunisia and Algeria via a canal. The hope was that filling these depressions would alter the climate of the area. The history of the scheme is described in a comprehensive publication by Letolle and Bendjoudi (1997). They describe the geography, geology, hydrology and the socio-economic conditions of the project area. The following brief description is based on their work.

There is a vast depression in the Sahara running from the west of Gabès on the Mediterranean coast of Tunisia to the southeast of Biskra, with a general downward slope of east to west (Figure E.1). Chott[1] Melrhir and Chott Rharsa in the western part of this depression, with an area of about $8000\,km^2$, were proposed for flooding. It was believed that this depression was once part of the ancient Bay of Triton. Herodotus (485–425 BC) and other historians of antiquity described its shores as being highly rich and fertile.

Martins (1864) described that the last of the Chotts in the east is only $16\,km$ away from the sea. By breaking through the Gabès ridge, the basin of the Chotts could become an inland sea. Lavigne (1869) described his observations of the area as a journalist, and raised the possibility of creating an inland sea. Pomel (1872) was the first to say that an inland sea will not have any impact on the climate of the region. Over the period of 1 May 1872 to 1 June 1873, Captain François Elie Roudaire, a French army surveyor who was topographically surveying in the region of the Chotts, discovered that the western part of the Chott Melrhir was about $27\,m$ below sea level. He studied the writings and maps of the ancient historians and geographers and was convinced that the Chotts Basin was once connected to the Mediterranean Sea and was the Bay of Triton. The Bay dried out about the beginning of the Christian era due to formation of a ridge not very far from the Gulf of Gabès. Roudaire, who denied that he had any knowledge of the Lavigne (1869) article, became interested in creating an inland sea in the area.

Roudaire published his first paper in the *Bulletin de la Société de Géographie* in March 1874 and the second one entitled "Une Mer Intérieure en Algérie" for the general public in the *Revue des Deux Mondes* in May 1874. Following a background description and referring to the Bay of Triton, he proposed to excavate a canal connecting the Mediterranean Sea to the Chotts from Gabès. He provided some technical details and cost estimates, and justified his project on the following points:

- Evaporation from the vast lake surface would increase the humidity and the rainfall.
- It would be possible to navigate from Gabès to Biskra.

The article was well received by the general public and gained the support of Ferdinand de Lesseps[2] who was a member of the Academy of Sciences after 1869 when he completed the construction of the Suez Canal. de Lesseps supported Roudaire's proposal and gave him access to the prestigious tribune of the Academy of Science. Roudaire was financially supported by the government for an expedition, which was expected to include a geologist, a botanist and an archaeologist, to study the area. However, the project was opposed by a number of the Academy members. One of the arguments was that flooding of this depression, which has no outlet and high evaporation, would deposit enormous quantities of salt.

E.2 ROUDAIRE'S EXPEDITIONS

The first Roudaire expedition took place from December 1874 to April 1875. He was in charge of investigating the topography of the Chott Melrhir and identifying the area located below sea level. He was accompanied by a large number of people, including Henri Duveyrier who was in charge of all scientific observations including geology, botany and archaeology.

---

[1] The term chott is used for the dry (salt) lakes in the Sahara that stay dry in the summer, but receive some water in the winter.
[2] The French engineer who built the Suez Canal over the period of 1859 to 1869.

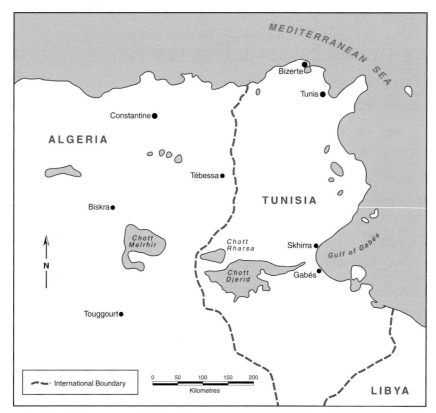

Figure E.1 Location map of the major Sahara depressions.

Roudaire and his assistants completed numerous measurements and found that the lowest point of the Chott Melrhir was 31 m below sea level. On their return to Paris, Duveyrier presented the results of the expedition, which included a topographic map of the proposed inland sea, to the French Geographical Society. Duveyrier became disinterested in the inland sea project and in 1892 committed suicide due to personal problems.

In August 1875, the second International Geographic Congress was held in Paris. The Roudaire project was strongly criticised by a number of participants. They believed that the Gabès ridge, with an elevation of about 30 m above sea level would prevent connection of the Mediterranean Sea and the depressions.

The second expedition took place in February 1876 and consisted of only three French participants. The objective was to explore the Tunisian side of the Chotts. Roudaire's (1877) report of the expedition consisted of the following sections:

(1) summary of the previous activities;
(2) activities undertaken in Tunisia;
(3) characteristics of the Chotts Basin and the Bay of Triton;
(4) description of the proposed canal;
(5) consequences of submerging the Chotts Basin, and
(6) assessment of the objections against the project.

The dispute over the inland sea continued during 1876 and 1877.

The third expedition was from November 1878 to May 1879. The objective was to undertake drilling between Gabès and Chotts to have better knowledge of the type of rocks required to be excavated. Roudaire was accompanied by de Lesseps at Gabès who returned to Paris shortly after because he was increasingly involved in the future Panama Canal. The drillings at the Gabès ridge showed the presence of the Cretaceous limestone between 8 and 10 m above sea level and under about 20 m of sand. Roudaire completed his second report by the end of October 1880 and published it in early 1881. However, the controversy about the project continued in scientific publications as well as in the newspapers.

E.3 COMMISSION OF INQUIRY

On 27 May 1882, the Government created a Commission of Inquiry headed by the Foreign Minister (Monsieur de Freycinet), to examine the inland sea proposal. The Commission consisted of three sub-commissions in charge of the (1) technical, (2) scientific,[3] and (3) diplomatic, military and commercial issues. Ferdinand de Lesseps was involved as

---

[3] This included the study of evaporation, water balance, and the impact of the inland sea on the atmosphere.

the Vice-president of the technical sub-commission. The Commission was composed of numerous experts, parliamentarians, military experts, and representatives of various government agencies. Discussions were very heated and exciting. Roudaire was constantly requested to attend numerous meetings and to answer various questions. The Commission estimated the cost of the project at 1.3 billion Francs (1882 prices) or about 520 million New Francs[4] in 2002 prices. Table E.1 lists some details of the project.

Although de Lesseps supported the project, the Commission, on 28 July 1882, recommended that, considering the high cost of creating the inland sea, which would be out of proportion with the expected results, the French Government should not encourage its undertaking. The Government published the totality of the background information and conclusions of the Commission in more than 500 pages in the *Official Journal*.

E.4 CONTINUATION OF THE INLAND SEA AFFAIR (1882–1936)

Ferdinand de Lesseps was furious about the decision of the Commission and requested concession of the land of the project area on 4 October 1882. Also, he announced in December 1882 creation of the "Société d'Etude de la Mer Intériure" with a capital of 200 000 Francs subscribed by himself and his friends in the Suez Company. Money supplied by the Society permitted Roudaire, who had no more access to public funds, to go to the area for the last time. He departed on 25 December 1882 to complete his drilling programme and topographic measurements.

In March 1883, de Lesseps again visited the area, with a party of distinguished engineers and contractors. His intention was to cross the desert lying between Gabès and the depressions of southern Tunisia, expecting to decide the fate of the project. On his return he again objected to the decision of the Commission and strongly recommended construction of the canal, which in his opinion would cost only 150 million Francs or about nine times less than the Commission's estimate. de Lesseps came under attack for not accepting the conclusions of the Commission. Finally he met with Monsieur de Freycinet, and to bring an end to the controversy, they decided not to recall the Commission.

All geologists who studied the area were convinced by 1884 that during the historic time no internal sea ever existed in the area, and the fundamental assumption of the project regarding the Bay of Triton was without foundation.

In spite of the friendship and support of de Lesseps, Roudaire gradually lost hope over the project, became ill and finally died on 14 January 1885. Nothing further happened[5] until the late 1920s and early 1930s when a number of publications by German authors and others appeared about the issue of the Bay of Triton. In 1936, a group of American scientists studied the issue in Tunisia without making any proposal.

E.5 DEVELOPMENTS FROM 1957 TO 1968

The discovery of oil occurred in both Libya and Algeria in 1956, and Tunisia in 1957. These were followed by independence of Algeria in 1962. In 1957, the "Association de Researches Technique pour l'Etude de la Mer Intériure Saharienne" with the acronym of ARTEMIS was founded (Ward, 1962, pp. 181–183). The aim was to further investigate the creation of an inland sea in the Sahara depressions. Because oil was discovered in both Algeria and Tunisia, ARTEMIS hoped that some of the revenues from oil might be used for the project. However, the outcome of the ARTEMIS investigations is unknown.

In 1958, a French Company was asked by the Tunisian Government to assess the flooding of the Chotts depressions

Table E.1. *Some technical features of the canal project*

Item	Detail
Length of the canal	224 km[(a)]
Slope of the canal	35 mm/km in the soft materials and 74 mm/km in the limestone
Cross section of the canal	714 m^2 in the soft materials and 459 m^2 in the limestone
Water depth in the canal	14 m
Excavation volume of soft materials	575×10^6 m^3
Excavation volume of limestone	27×10^6 m^3
Total volume of excavation	602×10^6 m^3

Note: [(a)] The 224 km length of the canal is due to the fact that the Chott Djerid and its continuation to the east are above sea level.
Source: Letolle and Bendjoudi (1997, Appendix 4b, pp. 195–206).

---

[4] One New Franc is equivalent to 100 old Francs and 520 million New Francs is about 80 million Euro.

[5] In 1905 Jules Verne published a book entitled *L'Invasion de la Mer* (Invasion of the Sea) about creation of an inland sea (details of the 1978 edition of the book is in the references). The summary of the story is that a surveying expedition attempts to find a canal route to link the Gulf of Gabès with the parts of Sahara lying below sea level. Work on the canal started but an earthquake breached the Gabès ridge and the sea rushed in, to accomplish in a few days what would have taken men many years (Verne, 1973, p. 209).

for the improvement of internal transport, creation of a micro-climate, and development of south Tunisia. The company reported on evaporation and characteristics of the canal in 1959. Because of the huge problems of excavating a canal at sea level, the company proposed excavation of a canal above sea level to be fed by pumping.

The Director of the Tunisian Atomic Energy Commission proposed the creation of an inland sea in 1968. He failed to provide technical details in his proposal. However, he argued that creation of the inland sea would prevent expansion of the Sahara and would create a fishing industry by producing 50 kg of fish/ha per year. Based on experiments in the United States and Russia, he advocated the use of atomic explosions to excavate the canal.

E.6 THE JOINT ALGERIA AND TUNISIA PROJECT (1983–85)

In November 1983, the governments of Algeria and Tunisia signed a protocol regarding the establishment of the "Société d'Etude Tuniso-Algérienne de la Mer Intérieure" with the acronym SETAMI. A group of experts from the two countries were charged with undertaking a feasibility study of the inland sea. The study was given in July 1984 to SWECO, a Swedish Engineering bureau that had studied a similar proposal for the Qattara Depression in Egypt over the period 1981–83. SWECO studied the meteorology, climatology, hydrology, technical issues, agronomy, and economics of the project.

The average annual evaporation rate of the area was estimated at about 2.7 m by the 1958–59 study undertaken by a French Company. SWECO used an energy balance method and estimated 2 m for the annual evaporation rate. Using a meteorological model developed at the University of Uppsala (Enger, 1984), SWECO showed that the increase of precipitation would be negligible over the major part of the region. However, an increase of 50–100 mm could be expected over a narrow strip on the southern flank of the Sahara Atlas near Biskra. These results were based on the assumption that the water level in the inland sea would be at sea level. With the water level at 15 m below sea level, a 20 percent reduction of the above estimate was predicted. SWECO also showed that:

(1) temperature changes would be negligible;
(2) the increased evaporation would prevent the cultivation of the palm trees; and
(3) the concentration of salt due to evaporation would result in stratification of the water and a return canal would be required for the discharge of the highly concentrated saline water.

SWECO simulated filling the depressions using two annual evaporation rates of 2 m and 2.7 m. Table E.2 shows some of the results.

The estimated capital cost of the project for three different canal capacities, with the interest rates of 4 percent and 10 percent, are shown in Table E.3. It indicates that the cost ranges from about US$11 to US$85 billion in 2002 prices. The operational costs for the three canal capacity options in 2002 prices would be US$11, US$32 and US$79 million per annum respectively.

The SWECO study concluded that the project:

(1) would have a negative impact on the agricultural production of the area;
(2) would not reduce the import of the foodstuff;
(3) would not create jobs except during the construction period; and
(4) would increase the deficit of the balance of payments for both Tunisia and Algeria.

SWECO pointed out that because of the high investment and operational costs of the project, which would be an important part of the GDP of both countries, with its hypothetical benefits, and its environmental and socio-economic impacts, the project was not profitable and could not be justified.

Table E.2. *Water level and corresponding stabilisation time for various canal capacities and evaporation rates*

Canal capacity ($m^3 s^{-1}$)	Water level and stabilisation time for the annual evaporation rate of			
	2 m		2.7 m	
	Water level (m)	Stabilisation time (years)	Water level (m)	Stabilisation time (years)
400	−5	40	−13	40
600	−2	20	−4	25
800	−2	12	−2	15
1500	−1	4	−1	6

Source: Letolle and Bendjoudi (1997, Figure 20).

Table E.3. *The estimated capital cost of the project for three canal capacity options*

Interest rate (%)	Canal capacity		
	Low (400 $m^3 s^{-1}$)	Medium (800 $m^3 s^{-1}$)	High (1500 $m^3 s^{-1}$)
4	US$11 billion	US$18 billion	US$44 billion
10	US$57 billion	US$41 billion	US$85 billion

Note: The 1985 SWECO cost estimates are converted to 2002 prices by multiplying them by a factor of 1.7.
Source: Letolle and Bendjoudi (1997, p. 125).

References

Enger, L. (1984). *A Three Dimensional Time-dependent Model for the Mesoscale: Some Last Results with a Preliminary Version.* Uppsala, Sweden: Department of Meteorology, Uppsala University. Report 80.

Lavigne, G. (1869). Le precement de Gabès. *La Revue Moderne* **55**: 322–355.

Letolle, R. and Bendjoudi, H. (1997). *Histoires d'une Mer au Sahara: Utopies et Politiques.* Paris: L'Harmattan.

Martins, C. (1864). Tableau physique du Sahara oriental de la province de Constantine. *Revue des Deux Mondes*, pp. 295–322.

Pomel, A. (1872). Le Sahara, observations de géologie et de géographie physique et biologique, avec des aperçus sur l'Atlas et le Soudan, et discussion de l'hypothèse de la mer saharienne á l'époque préhistorique. *Bull. Soc. Climatologique de l'Algérie* **8**: 133–265.

Roudaire, F. E. (1877). *Rapport à Monsieur le Minister de l'Instruction Publique sur la Mission des Chotts.* Paris: Challamel.

Verne, J. J. (1973). *Jules Verne: A Biography by Jean Jules-Verne.* Translated and adapted by Roger Greaves. London: Macdonald and Jane's.

Verne, J. (1978). *L'Invasion de la Mer.* Paris: Union General d'Edition.

Ward, E. (1962). *Sahara Story.* London: Robert Hale Limited.

F The Ord River Irrigation Scheme

F.1 INTRODUCTION

The Ord River Irrigation Scheme is located in the northern part of Western Australia and the Northern Territory (Figure F.1). The region experiences the tropical climate of the Kimberley region with a wet and a dry season. Average annual rainfall at Kununurra is 793 mm. Its mean daily maximum temperature ranges from 30.7°C in July to 38.9°C in November, while the mean daily minimum temperature varies from 15.4°C in July to 25.6°C in December (Australian Bureau of Meteorology website).[1]

Interest in the Kimberley region of Western Australia, including the Ord River Valley, began with Alexander[2] Forrest's report of his exploration of the Fitzroy and Ord River Valleys between 1875 and 1879. The potential of the Ord River Valley for irrigated agriculture was first recognised in the late 1930s (Department of National Resources, 1976; Young, 1979). In 1941 a party of engineers and scientists explored the area and established that the Ord River could be dammed at one of many sites and provide water to a large fertile plain. In the same year, a small experimental farm was established on the banks of the Ord River near the off-take of the present M1 Supply Channel (Figure F.1). In the next four years, the Western Australian Government sponsored more detailed investigations. These concluded that 22 000 ha of land could be irrigated, and a further 50 000 ha extending into the Northern Territory, warranted further examination. In late 1945, the experimental farm was abandoned and the Kimberley Research Station on the Ivanhoe Plain was established as a joint Federal–State venture. Its main aim was to determine whether irrigated agriculture could be successfully established in the area and to identify the most suitable crops and cropping techniques.

By 1958, investigations had shown that it was feasible to build a dam with a safe regulated yield to irrigate 72 000 ha of land in Western Australia and the Northern Territory. It had also been demonstrated that cotton, rice, and sugarcane would grow well under irrigation in the area. In May 1959, assistance was requested from the Federal Government for the construction of Stage 1 of the scheme to irrigate 12 000 ha of land. The request was approved in August 1959 and $10 million (about $105 million in 2002 prices) was provided to the Western Australian Government. Construction of the Scheme was undertaken in the following steps:

- **Step 1:** Commenced in 1960 and was completed in 1967 at a total capital cost of $20.8 million (about $180 million in 2002 prices). This included construction of the Kununurra Diversion Dam with a capacity of $97.4 \times 10^6 \, m^3$ and a height of 20 m, a pumping station, development of some 12 000 ha of land for irrigation and construction of the Kununurra Township.[3] The first five farms were allocated in August 1962 and farmers were able to use water in April 1963. By 1970–71, the total irrigated area increased to 4774 ha, of which 3632 ha were under cotton. Cotton yield increased to about 1082 kg per ha in 1970–71 from the initial 1963–64 average yield of 417 kg per ha (Department of National Resources, 1976, p. 47).

- **Step 2:** Involved construction of the Ord River Dam[4] at a cost of $22 million (about $187 million in 2002 prices) to form the major storage reservoir known as Lake Argyle with a capacity of $5.7 \times 10^9 \, m^3$, surface area of 740 km^2 at the spillway height. Construction of the dam commenced in early 1969, was officially opened on 30 June 1972, and overflowed in January 1974.

Over the years, more than 134 km of irrigation channels and 155 km of drains servicing 13 000 ha of land were developed with a net irrigable area of about 10 000 ha (Department of Resources Development, 1994, p. 12).

In the early years of production, cotton farmers made satisfactory profits because they were well subsidised. The level of subsidy was about 31.1 cents/kg in 1964–65 and they received a price of 60 cents/kg for their product. However, the subsidy declined rapidly to 3.8 cents/kg in 1971–72 while the price was 66.6 cents/kg (Department of National Resources, 1976, p. 47).

---

[1] http://www.bom.gov.au/climate/averages/tables/cw_002056.shtml (visited in August 2005).
[2] John Forrest's brother.
[3] It had a population of 337 in 1961, 4062 in 1991 and about 6100 in 2002.
[4] It has a height of 99 m, catchment area of 46 100 km^2, and annual rainfall of 550 mm.

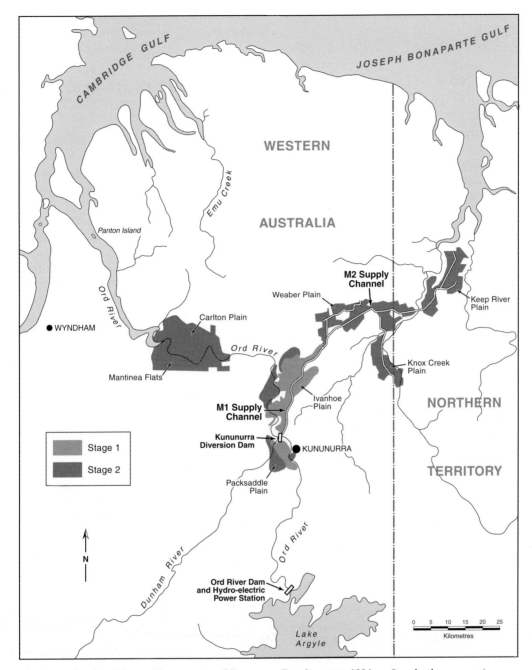

Figure F.1 Ord River Irrigation Scheme (Department of Resources Development, 1994, p. 8 and other sources).

Despite the availability of irrigation water and the agricultural experiments conducted at the Kimberley Research Station, the scheme faced numerous problems. At first it was thought that cotton would be the main crop, but insect pests (particularly *Heliothis* caterpillars) were a major problem and up to 50 applications of DDT were used during a single season. Eventually the cost of insecticide became equal to the value of the crop itself and in 1974 cotton production was abandoned (Smith, 1998, p. 171). Other problems encountered are that the soils in the irrigation area require heavy applications of nitrogen and phosphate fertilisers. Also, careful farm management is required, particularly in relation to the time of planting, irrigating, fertiliser application and harvesting to produce good yields. Rice, oil seeds and grain sorghum have been grown and some irrigated pastures for cattle fattening have been developed. A further problem was the remoteness of the area from the markets.[5] All these problems resulted in a very low

[5] Kununurra is located at about 3200 km from Perth, 4300 km from Sydney, 850 km from Darwin and 3500 km from Singapore.

level of agricultural activities in the Ord River Irrigation Area (ORIA) in the late 1970s.

F.2 HYDROLOGY AND WATER QUALITY OF THE ORD RIVER

The hydrology of the Ord River was investigated by Ruprecht and Rodgers (1999). They found that the Ord River has a catchment area of over 50 000 km^2 mostly in WA and partly in NT. The average annual rainfall of the catchment ranges from 780 mm in the north to 450 mm in the southern part of the catchment. Evaporation for Kununurra is about 3000 mm per annum, and is highest in October. The evaporation rate of Lake Argyle is approximately 2130 mm per year and is highest in October and January. Table F.1 shows the mean annual water balance components of the Lake.

The natural mean annual flow of the Ord River at the river mouth was about 4.5×10^9 m^3. Based on historical measurements, and modelled streamflow data, there is a 10 percent probability that the annual streamflow will be less than 1.1×10^9 m^3 per year, and a 10 percent probability that the streamflow will be over 8.2×10^9 m^3 per year.

There has been significant water resources development of the Ord River. With development of the first stage of the irrigation scheme, the average annual flow reduced to 3.2×10^9 m^3. This reduction is due to evaporation from Lake Argyle (see Table F.1) and water use for irrigation. With further irrigation development, the average annual flow of the Ord River is expected to reduce to about 2.3×10^9 m^3.

Major water quality issues of the Ord River are sedimentation of Lake Argyle, saltwater intrusion in the lower Ord River, and in more recent times the pesticide level in the Ord River downstream of the Kununurra Diversion Dam. The sediment load into Lake Argyle is estimated to be about 24×10^6 tonnes per year. However, because of the large storage volume of the Lake, the impact of this sediment load is minimal.

Dissolved salts in the upper catchment of the Ord River are low, averaging approximately 380 mg L^{-1} TDS. Intrusion of seawater into the lower end of the Ord River (mainly during high tide) is currently limited to about 25 km upstream of Panton Island (Figure F.1). Prior to the construction of the two dams, the saltwater/freshwater interface during the dry season extended further up the river than it does now.

Table F.1. *Annual water balance components of Lake Argyle*

Inputs (10^9 m^3)		Outputs (10^9 m^3)	
Inflow	3.94	Evaporation	1.75
Rainfall	0.65	Overflow	0.89
–	–	Releases	1.95
Total	4.59	Total	4.59

Source: Ruprecht and Rodgers (1999, p. 30).

F.3 ECONOMIC EVALUATION OF THE SCHEME

Davidson (1972) strongly opposed the Ord River Irrigation Scheme based on his economic analysis. His arguments were based on the fact that cotton had been produced on the Ord Scheme since 1964. Initially farmers obtained satisfactory returns because of the high level of subsidy paid by the Federal Government. The subsidy was limited to a total of $4 million per annum (about $38 million in 2002 prices) to be distributed among all Australian cotton growers. The subsidy was to be continued until 1969 and then phased out over a two-year period. During the 1960s cotton growing expanded rapidly on the Namoi Valley in northern NSW where yields were higher and costs lower than on the Ord Scheme. As the quantity of cotton grown in Australia expanded from 2500 tonnes in 1964 to 33 000 tonnes in 1969 (mainly in the Namoi Valley), the government subsidy per kg declined significantly. These reductions had a drastic effect on the profits of the Ord River farmers, resulting in significant decline in the area sown to cotton. Davidson's other arguments against the Scheme included:

- The Federal Government's financial support in 1968 for construction of the Ord River Dam was possibly made on political rather than economic grounds. It was made immediately before the Senate election in 1968 in which the Liberal Federal Government had every reason to expect a loss of votes in Western Australia.
- Experimental yields of other crops investigated including maize, sorghum and new variety of rice showed little promise of commercial success.
- There was no evidence that alternative crops would improve the economic position of the Ord River farmers.
- Suggestions that Ord Scheme might act as a food producing centre for the new mines was fallacious. These could be supplied with food at a lower cost from south Western Australia where cropping and processing costs were lower.
- After thirty years of experimental works on the Ord River Scheme and seven years of commercial cropping, techniques had not been developed which would enable farmers to produce crops without heavy subsidies.

In March 1992, the Kimberley Water Resources Development Advisory Board was appointed by the Western Australian Government to recommend actions to maximise the value to the community from the development of the Kimberley water resources. The Board commissioned Hassall & Associates to undertake a comprehensive study of the past, present and future economic valuation of the Scheme. Hassall & Associates submitted their two-volume report and its Executive Summary

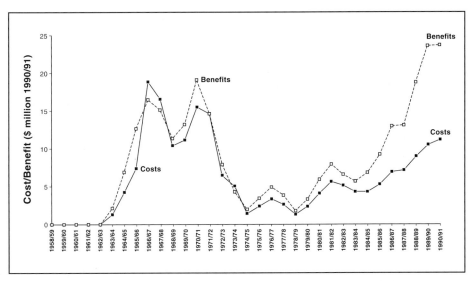

Figure F.2 Annual agricultural production costs and benefits of the Ord River Irrigation Area for the period of 1958–59 to 1990–91 (Hassall & Associates Pty Ltd, 1993, Figure 2).

in August 1993. Below is a brief description of their findings and conclusions.

In spite of agricultural failure in the 1970s, the performance and outlook of the Scheme improved significantly during the 1980s. This was partly attributed to the development of specialised field crops and horticultural industries, the strengthening market in Asia, and the tenacity of farmers. Developments in biological control of insect pests had also been successful in many areas, replacing the previous heavy reliance on insecticides.

In the 1990s, apart from an increase in total crop area, the range of cultivated crops extended to maize, sunflower, soybean, sorghum, peanuts, chickpeas, mangoes, banana and cucurbits.[6] Figure F.2 shows the annual agricultural production costs and benefits for the period of 1958–59 to 1990–91. It indicates that although costs and benefits up to 1978–79 were similar, from this year onward benefits clearly exceeded costs and by 1990–91 reached $25 million.

In 1990–91, the total crop area in the ORIA was estimated at 4407 ha. The major crops in terms of area were maize and chickpeas, and the main crops in terms of value were bananas and cucurbits. These latter crops can be grown during the Southern Hemisphere winter and therefore attracts high prices in the major Australian cities. Fruit was also exported to Asian markets such as Singapore and Hong Kong (Hassall & Associates Pty Ltd, 1993, p. 6).

While agricultural development was the primary objective of the Scheme, there have been other spin-offs in areas such as tourism and the mining industry. The Argyle Diamond Mine, developed in the 1970s, is the largest of these and a major user of the Ord River water.

The cost–benefit analysis undertaken by Hassall & Associates Pty Ltd (1993) for the period of 1958–59 to 1990–91 indicates that the Scheme incurred an overall loss of $497 million in 1990–91 values. The main factors which contributed to these results, were the considerable under-utilisation of water provided by the Scheme and the lack of success in crop production in the area until the 1980s. However, there was a sustained growth in total gross margins associated with agricultural production since 1980. As a result, the total annual benefits generated by agriculture exceeded significantly the annual costs of the Scheme in 1988 and in subsequent years.

Hassall & Associates Pty Ltd (1993) also analysed the cost–benefit of the Scheme for the period of 1992–2020 under the three following scenarios reflecting different economic and social conditions in Australia:

- **Continuity**, which assumed gradual development of global economy and structural evolution within Australia.
- **Growth**, which considered national and State commitment to integrating Australia more aggressively into the Asian economy.
- **Isolation**, which assumed that Australia is a passive and marginal participant in a dynamic and highly competitive world economy.

Table F.2 shows the 1992 area and the projected areas of agricultural activities within the ORIA for 2021 under these various development scenarios. In projecting the land use, a number of lower value field crops such as pumpkins, cucumber and zucchinis were omitted from the analysis.

An analysis of the economic costs and benefits associated with the future development of the ORIA was performed for each scenario. The results of the analysis for the future development (1990–91 to 2020–21) are presented in Table F.3. These results predict that the economic benefits from the expansion of the

---

[6] Crops such as melon, pumpkin, and cucumber.

Table F.2. *The 1992 and the projected areas (in ha) of the agricultural activities within the ORIA for 2021 under various scenarios*

Type of crop	Area in 1992	Development scenarios		
		Continuity	Growth	Isolation
Soybeans	400	845	345	1030
Maize	250	500	500	500
Chickpeas	400	1155	655	1470
Sorghum	500	3030	3000	1930
Sunflower	500	3030	3000	1930
Mango	120	1000	2000	1000
Banana	150	2000	2500	2000
Rockmelon	260	3000	3500	2250
Watermelon	260	3000	3500	2250
Lucerne	1000	5490	5500	4590
Sugarcane	0	30 200	34 010	3000
Total	3840	53 250	58 510	21 950

Source: Hassall & Associates Pty Ltd (1993, Table 1).

Table F.3. *The net present value ($ million in 1990–91 prices) for future development scenarios of the ORIA*

Discount rate	Continuity scenario	Growth scenario	Isolation scenario
0	2785	3415	1434
4	1119	1419	621
8	489	634	296

Note: The net present values equal the present value of all benefits over the project period, less the present value of all costs over the same period.
Source: Hassall & Associates Pty Ltd (1993, Table 3).

ORIA, would significantly exceed the costs for all future development scenarios.

In addition to the direct costs and benefits, construction of the Ord Scheme, and the subsequent economic activities that resulted from it, has had an impact on the other sectors of the economy of the Kimberley Region, the State of Western Australia and the nation as a whole. Hassall & Associates Pty Ltd (1993) analysed these impacts for five industry groups namely construction, agricultural production, tourism, agricultural processing and transport for two periods of 1958–59 to 1990–91 and 1991–92 to 2020–21. They concluded that significant expansion of irrigated agriculture in the Ord region over the next (from early 1990s) two to three decades was warranted in economic terms, and was likely to lead to a significant expansion in agricultural processing, transport, and tourism industries, with major and sustained economic benefits at the regional, State and national levels.

F.4 RECENT GROSS VALUES OF AGRICULTURAL PRODUCTION

The net cropped area of the ORIA was about 4000 ha over the period of 1989–90 to 1991–92 (Figure F.3). It increased to about 7000 ha in 1992–93, 10 000 ha in 1994–95, and 11 000 ha in 1996–97, which remained almost unchanged until 1999–2000. These lands were used to produce a diverse range of field, fodder and horticultural crops.

The value of crops in the ORIA has increased from $30 million in 1990–91 to $67.5 million in 1999–2000 (Figure F.4). The highest value crops in 1999–2000 were melons ($23.7 million), sugarcane ($20.3 million), and bananas ($5.2 million).

To assist development of the sugar industry, a sugar mill was constructed in 1995 by a joint venture between CSR Limited[7] and the Ord River District Cooperative, and was sold to a South Korean corporation (Cheil Jedang Corporation) in 2000.

F.5 HYDRO-POWER GENERATION

A 30 MW hydro-electric power station at the Ord River Dam, was constructed in 1995–96 by a private company (Pacific Hydro Limited), which also owns and operates the facilities. The pre-existing outlets were modified to accommodate the power station. Two 4.42 m diameter tunnels constructed at the time of the dam construction were connected to four steel intakes, each 2.2 m diameter. Two 15 MW generators are each driven by two 7.5 MW Francis turbines. The electricity is supplied to Kununurra, Wyndham and the Argyle Diamond Mine.[8]

F.6 STAGE 2 OF THE SCHEME

By 1994 some 13 000 ha of land was developed for irrigation, mainly on the Ivanhoe and Packsaddle Plains, and large tracts of additional irrigable land were also identified (Department of Resources Development, 1994). In 1994, the Western Australian Government decided to expand the project area by 64 000 hectares, consisting of about:

- 50 000 hectares mainly in the Upper Weaber Plain of WA (18 780 ha), Lower Weaber Plain of NT (4507 ha), Knox Creek Plain (WA & NT) and Lower Keep River Plain of NT (17 443 ha), and the Knox Creek Plain of WA and NT (9483 ha). These lands would be served by construction of

---

[7] http://www.csr.com.au (visited in August 2005).
[8] Further information regarding hydro-power generation is available at the Pacific Hydro website http://www.pacifichydro.com.au/projects.asp?articleZoneID=191#Article-275 (visited in August 2005).

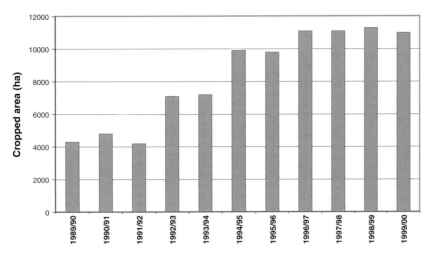

Figure F.3 Cropped area of the ORIA from 1989–90 to 1999–2000 (Department of Local Government and Regional Development, 2001, p. 6).

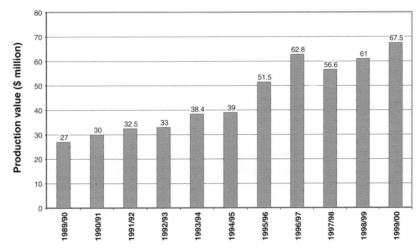

Figure F.4 Gross value of agricultural production of the ORIA from 1989–90 to 1999–2000 (Department of Local Government and Regional Development, 2001, p. 6).

the M2 Supply Channel beyond the present development. The net cropped area within the M2 Supply Channel area would be about 32 000 ha which is much less than the gross area of about 50 000 ha.

- 14 000 hectares, known as Riverside Developments in the Ivanhoe Plain and west of the Ord (1646 ha), Carlton Plain (9167 ha), and Mantinea Flats (3250 ha).

The above developments were possible because of the availability of substantial water resources stored in Lake Argyle.

On 19 April 1995, the Premier of WA and the Chief Minister of the NT signed a Memorandum of Understanding regarding the expansion of the Ord River Irrigation Scheme. The first part of the planned expansion was the development of the M2 Supply Channel area. Throughout 1995 and 1996 the Governments commissioned various preliminary studies in relation to the expansion of the ORIA, with a view to involve the private sector.

F.6.1 Wesfarmers/Marubeni Development Proposal

In 1997, a call for expressions of interest stimulated great interest from the private sector, and resulted in the submission of a number of detailed proposals from local and overseas organisations. The proponents for the development of the M2 Supply Channel area included a joint venture between Wesfarmers Sugar Company Pty Ltd (Wesfarmers) and Marubeni Corporation (Marubeni) and the Water Corporation of Western Australia as a co-proponent. In April 1998, the Governments of Western Australia and the Northern Territory granted an exclusive mandate to the Wesfarmers/Marubeni joint venture to investigate the feasibility of development of the M2 Supply Channel area. The Wesfarmers/Marubeni proposal consisted of the following components (Kinhill Pty Ltd, 2000):

- Sugarcane plantation on approximately 29 000 ha.
- Development of a sugar mill by Wesfarmers/Marubeni with a capacity of approximately 400 000 tpa of raw sugar and 160 000 tpa of molasses.
- Development of raw sugar and molasses storage and handling facilities at Wyndham Port by Wesfarmers/Marubeni.
- Sale of some 3000 ha of land to independent farmers on an unconditional basis with respect to the type of crops that may be grown.
- Construction of approximately 400 km of water supply channels and drains, as well as almost 150 km of flood protection levees.
- Establishment and management of over 40 000 ha of land for conservation purposes.

Numerous investigations into the development of Stage 2 of the Ord River Scheme were undertaken. These included:

- **Hydrology of the Ord River** (Ruprecht, and Rodgers, 1999) described in section F.2.
- **Groundwater investigation** (O'Boy et al., 2001) which demonstrated that management of groundwater recharge will be required to minimise problems associated with rising groundwater levels and perched watertables. High salt storage in some sub-soils makes it imperative that irrigation management maintain watertables at a safe depth to avoid soil salinity problems.
- **Interim water allocation plan** (Water and Rivers Commission, 1999), which considered allocation of 12×10^6 m^3 for the Argyle Diamond Mine, 300×10^6 m^3 for irrigation of the Stage 1, 740×10^6 m^3 for Stage 2,[9] and 195×10^6 m^3 for irrigation of the Carlton Plain and Mantinea Flats, making a total water allocation of 1247×10^6 m^3. These estimates, and particularly irrigation water requirements, are associated with various uncertainties due to crop type, crop water requirements, and irrigation water distribution efficiency. Apart from these water requirements, water would be needed to maintain ecological and geomorphological processes within the Ord River, hydro-power generation, and other activities (recreational, tourism and others). This interim water allocation plan was reviewed and amended by the Western Australian Department of Environment to take into account environmental flow requirements and the impacts of reduced dilution flows on the water quality (nutrients, pesticide and salt concentration) of the Ord River.
- **Environmental Review and Management Program/Draft Environmental Impact Statement** (Kinhill Pty Ltd, 2000). This report covers topics such as project objectives and background, the existing ORIA, description of the project, physical environment (climate, geology, soils, landform, erosion and others), surface and groundwater resources, terrestrial vegetation and flora, aquatic flora and fauna, environmental impacts and management, land use, air quality and noise, public health, community and social issues, specific aboriginal issues, and environmental framework and commitments.

F.6.2 Miriuwung and Gajerrong People and the Development Proposal

The land encompassing the Ord and Keep Rivers, including the project area, is of traditional and current significance to the Miriuwung and Gajerrong people. They lodged three native title claims over a significant portion of the East Kimberley in WA and NT in 1994 and 1995. Their claims cover the entire project area.

The early development of the ORIA occurred at a time when there was no legal requirement to formally consult the Aboriginal people and address their concerns. In recent years they have actively sought to have their traditional rights recognised, to participate in the economic development process and to share in its economic benefits.

The Western Australian Government undertook consultations with the Miriuwung and Gajerrong people from 1995 to 1998 (Kinhill Pty Ltd, 2000, Chapter 12). In 1998, the Federal Court of Australia determined that the Miriuwung and Gajerrong people hold native title right to an area of land, which was the subject of the first of the three native title claims. The area covers approximately 7900 km^2 and includes the Ord River, Lake Argyle and approximately 100 km^2 of the project area. Over the period of 1998–2001, Wesfarmers/Marubeni and Water Corporation were engaged in consultation with the Miriuwung and Gajerrong people regarding the settlement of native title issues of the project area through a negotiated approach. The Western Australian Government also offered a special package of benefits to the Miriuwung and Gajerrong people. The offer included granting title over about 161 000 ha of land, much of which has high heritage and cultural values. The Government also offered assistance to Aboriginal businesses that could provide services to the proposed agricultural developments. The assistance would include the provision of land on which facilities could be established, as well as assistance with business development plans, tendering and applications for financial support. This is in return for the surrender of native title rights over approximately 42 500 ha of

---

[9] This estimate is based on a net irrigable area of 32 000 ha crop (sugarcane), water demand of 22.1×10^3 m^3/ha, effective rainfall of 5.5×10^3 m^3/ha, on-farm and in-field efficiency of 90 percent and delivery system efficiency of 80 percent (Ruprecht and Rodgers, 1999, Table 5.3).

land required for the Ord Stage 2. The proposed development could not occur until all native title issues over the project area were resolved.

F.6.3 Withdrawal of the Wesfarmers/Marubeni Proposal

In a media release dated 12 December 2001, Wesfarmers Limited and Marubeni Corporation announced that they would not proceed with their proposed development of a raw sugar industry in Stage 2 of the Ord River Irrigation Scheme. This was after a three and a half year feasibility study undertaken by the joint venture at a cost of $3.8 million. The project cost, including all the required irrigation and other infrastructure, was estimated at $500 million. The Water Corporation of Western Australia was closely involved in the project and was expected to provide the irrigation infrastructure of the project. Withdrawal was for the following reasons:

- **Volatility of sugar price on the world market.** Following a period of relative stability, world sugar prices severely fluctuated throughout much of the 1990s.[10] There was little evidence to suggest that price volatility would not be significant in the medium term. Although the project proved cost competitive compared to many other world sugar exporters, there were concerns regarding the low cost position of the sugar industry in Brazil and the rate of growth of the industry. Brazil is the lowest cost sugar producer in the world, with about 30 percent of the world export market and has potential for considerable expansion.
- **Uncertainty about the amount of irrigation water.** Numerous uncertainties existed in relation to the volume of irrigation water supply (Water and Rivers Commission, 1999). Some of these issues are briefly described in section F.6.1 (dot point of Interim water allocation plan).
- **Unresolved land access issues.** Access to the land required for the project was subject to claims under the *Native Title Act*, and land rights legislation in the Northern Territory had not been resolved. Although from the outset, the Wesfarmers/Marubeni joint venture made it clear that they wanted to reach a negotiated agreement with the claimants rather than taking legal action, and that they had been in discussion with them since 1998, these issues remained unresolved (see section F.6.2).
- **Approval of the environmental issues.** Although, to a large extent, the environmental issues were investigated by both the Western Australian and the Northern Territory Governments (Environmental Protection Authority, 2000; Department of Lands, Planning and Environment, 2000) and the consultant (Kinhill Pty Ltd, 2000), approval was not granted by the respective governments. Moreover, construction of the hydro-power plant was undertaken without considering environmental issues. Therefore, environmental issues had to be negotiated with Pacific Hydro Limited.

On the same day that Wesfarmers/Marubeni withdrew their proposal, the Minister for State Development, Tourism and Small Business of Western Australia released a Media Statement announcing that the Western Australian and Northern Territory Governments continue to be committed to the project. Both Governments realise that there are important land access and heritage issues still to be resolved with the Miriuwung and Gajerrong people and will undertake to address these issues in a timely and fair manner. Much had been learned over the past three years and the joint venture agreed to provide the State with a significant amount of valuable technical information, which could assist with the establishment of a new project.

F.6.4 Future of the M2 Channel Irrigation Development Project

In April 2003 the Western Australian Government announced that it had:[11]

- Offered to the Miriuwung and Gajerrong people freehold title to a 50 000 ha area known as Yardungarl.
- Agreed to a Kimberley Land Council request that the Miriuwung and Gajerrong No. 1 claim consent determination and the Ord Stage 2 negotiations be combined. The aim was for a global settlement package to be negotiated with the Miriuwung and Gajerrong claimants.
- A package of non-native title benefits constituting native title compensation for extinguishment of native title.
- Native title and heritage clearances to allow the Ord Stage 2 projects to proceed.
- Appointed an inter-departmental team to coordinate the global settlement.

In August 2003, the Western Australian Government appointed a consultant to conduct the global negotiations. In September 2003, the Department of Industry and Resources, after completing a preliminary investigation of the cost of partially developing the M2 Supply area, in consultation with the Department of Agriculture and the Northern Territory Government, advertised for a consultant to develop a business case for the proposed development. The consultant's tasks included quantifying the Ord development costs and expected returns to an investor who decides to develop the area either on a staged or full development. The consultant report was

---

[10] The raw sugar price, which reached to about US$328 per tonnes in January 1995, dropped to US$120 in January 1999 and was about US$ 190 in September 2001 (Lichts, 2001, p. 99).

[11] http://www.doir.wa.gov.au/investment/F1543F681A8A41B7A9620 93479A2C906.asp (visited in August 2005).

completed in October 2004 and the Government was expected to make a decision about the project. However, no decision has been made so far (August 2005).

References

Davidson, B. R. (1972).[12] *The Northern Myth: A Study of the Physical and Economic Limits to Agricultural and Pastoral Development in Tropical Australia*. Third Edition. Melbourne: Melbourne University Press.

Department of Lands, Planning and Environment (2000). *Ord River Irrigation Scheme Stage 2: Biodiversity Assessment*. Darwin: Environment and Heritage Division, Department of Lands, Planning and Environment.

Department of Local Government and Regional Development (2001). *Kimberley Economic Perspective: An Update on the Economy of Western Australia's Kimberley Region*. Perth: DLGRD.

Department of National Resources (1976). *Ord Irrigation Project Western Australia: An Outline of its History, Resources and Progress*. Canberra: Australian Government Publishing Service.

Department of Resources Development (1994). *Ord River Irrigation Project: A Review of its Expansion Potential*. Perth: Government of Western Australia, Department of Resources Development.

Environmental Protection Authority (2000). *Ord River Irrigation Area Stage 2 (M2 Supply Channel), Kununurra: Part 1 – Biodiversity Implications*. Perth: EPA Bulletin 988.

Hassall & Associates Pty Ltd (1993). *The Ord River Irrigation Project, Past, Present, and Future: An Economic Evaluation*. Executive Summary. Report prepared for the Kimberley Water Resources Development Office.

Kinhill Pty Ltd (2000).[13] *Ord River Irrigation Area Stage 2, Proposed Development of the M2 Area: Environmental Review and Management Programme/Draft Environmental Impact Statement*. Victoria Park, Western Australia: Kinhill Pty Ltd.

Lichts, F. O. (2001). *F. O. Lichts World Sugar Statistics 2002*. Ratzeburg: F. O. Lichts GmbH.

O'Boy, C. A., Tickell, S. J., Yesertener, C., Commander, D. P., Jolly, P. and Laws, A. T. (2001). *Hydrogeology of the Ord River Irrigation Area, Western Australia and Northern Territory*. Hydrogeological Record Series, Report HG 7. East Perth: Water and Rivers Commission.

Ruprecht, J. K. and Rodgers, S. J. (1999). *Hydrology of the Ord River*. Water Resources Technical Series, Report No. WRT 24. East Perth: Water and Rivers Commission.

Smith, D. I. (1998). *Water in Australia: Resources and Management*. Melbourne: Oxford University Press.

Water and Rivers Commission (1999). *Draft Interim Water Allocation Plan, Ord River, Western Australia*. Water Resource Allocation and Planning Series No. WRAP 2. East Perth: Water and Rivers Commission, Policy and Planning Division.

Young, N. (1979). *Ord River Irrigation Area Review, 1978: A Joint Commonwealth and Western Australian Review*. Canberra: Australian Government Publishing Service.

---

[12] Earlier editions of this book were published in 1965 and 1966.

[13] This publication is also available at: http://www.lpe.nt.gov.au/enviro/EIAREG/ORD/download.htm (visited in August 2005).

G The West Kimberley Irrigation Scheme

G.1 INTRODUCTION

In May 1991, John Logan, a cotton grower from Narrabri[1] arrived in Broome to study prospects of developing land and water resources of the Fitzroy River region for large-scale irrigation production. Previously, he completed a topographic desktop evaluation of the area, and travelled to Indonesia in April 1991 to assess the likely future marketing opportunities for large-volume supply of cotton for export from the port of Broome (John Logan, personal communication, August 2003).

In 1992, the Hon Ernie Bridge, Minister for Water Resources of Western Australia, initiated a series of studies to investigate ways to take advantage of the Kimberley water resources. In 1993, the Kimberley Water Resources Development Office initiated studies to investigate development opportunities for the Fitzroy valley (*Water and the West Kimberley: A Community Newsletter, June 1996*).

Following is a brief description of the West Kimberley Irrigation Scheme mainly from the website of the Department of Industry and Resources.[2]

In February 1996, the Western Australian Government recognised the capacity for large-scale integrated agricultural development in the West Kimberley. Following calls for Expression of Interest, in mid-1997 the Government announced Western Agricultural Industries Pty Limited (WAI)[3] as the preferred proponent to develop the project. WAI was appointed by the WA Government to carry out feasibility studies into establishing an irrigated agricultural industry based on the groundwater resources of the Cretaceous Broome Sandstone[4] in the Canning Basin[5] and surface water resources of the Fitzroy River.

In April 1998, WAI and the then WA Premier signed a Memorandum of Understanding (MoU) which allowed WAI to undertake the feasibility studies for irrigation of about 20 000 ha of lands south of Broome from groundwater resources of the Canning Basin. WAI was also to consider the concept of using water taken from the Fitzroy River to irrigate an additional area of 155 000 ha.

The MoU required a strict environmental assessment process. The environmental impact assessments and plans for long-term environmental management strategies required examination of:

- Impacts of the proposed development on relevant ecosystems including wetlands, rivers and groundwater.
- Seasonal water regimes and sediment movement.
- Pest and weed management.
- Soil erosion, compaction, and maintenance of soil structure.
- Groundwater investigation to gain detailed understanding of the volumes and quality of renewable groundwater available from the La Grange Sub-basin of the Canning Basin, south of Broome (Figure G.1).

The WAI commenced feasibility studies into the viability of the project. Trials were conducted on Shamrock Gardens and other sites south of Broome. New transgenic cotton was trialled alongside conventional cotton plants and showed a marked decrease in the requirement for pest control. In conjunction with the cotton trial, WAI carried out tests to establish a water system based on sub-surface drip irrigation.

G.2 GROUNDWATER ALLOCATION AND STAKEHOLDERS CONCERNS

Groundwater resources were expected to be used for development of the irrigated cotton and other crops in the La Grange

---

[1] Currently, Chairman of the Western Agricultural Industries Pty Limited, Neutral Bay Sydney.
[2] http://www.doir.wa.gov.au/investment/html/wkimbis.htm (visited in June 2003)
[3] A joint Venture between Kimberley Agricultural Industries Pty Limited and Queensland Cotton Corporation Ltd.
[4] Broome Sandstone is wide spread in coastal areas of the Canning Basin. Its maximum recorded thickness is 320 m and has an average saturated thickness of about 130 m. It has an estimated potential yield of about $194 \times 10^6 \, m^3$ per year (Commander et al., 2002).
[5] For a description of the Canning Basin and its aquifer systems see Laws (1991), Ghassemi et al. (1992), and Ferguson et al. (1992).

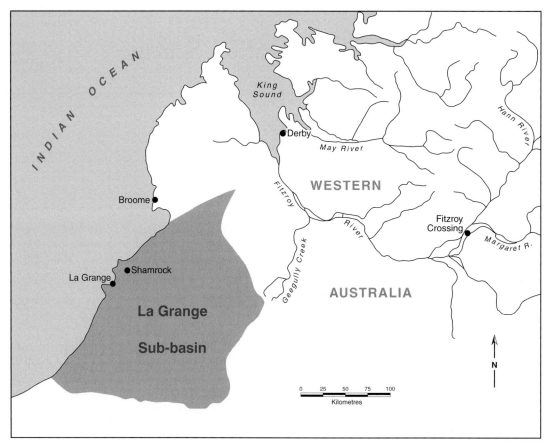

Figure G.1 Location of the La Grange Sub-basin of the Canning Basin (Yu, 2000, Map 2).

Sub-basin.[6] However, this raised a number of concerns from various stakeholders. Beckwith & Associates (1999) describe these concerns, some of which are summarised below:

- **Cotton farming** is viewed by many stakeholders as having a poor environmental record because of a historically heavy reliance on application of pesticides and fertilisers. These chemicals have the potential for contaminating groundwater resources. Moreover, aerial spraying of pesticides may affect waterbodies, residents, and fauna near areas of cotton farming.[7]
- **Sustainability of groundwater resources** to meet the long-term requirements of cotton farming as well as domestic supply, pastoral operations, and future land uses in the area.
- **Groundwater-dependent ecosystems**. The vegetation system in the area considered for irrigated cotton production is composed of acacia and low trees. The degree of dependence of these dominant vegetations on groundwater was unknown. Therefore a biodiversity study was required for assessing the flora and fauna values of those areas that would either be cleared for cotton production or are dependent on groundwater resources for their survival.
- **Coastal ecosystems** dependent on groundwater resources required careful assessment.

- **Native title claims**. The surface and groundwater resources of the Kimberley are part of the traditional country of Aboriginal people. Many of these resources are subject to native title claims.
- **Cultural values**. The Aboriginal cultural values, particularly those associated with water resources such as seasonal water holes, needed to be assessed and protected.
- **Impacts on local water users**. Stakeholders raised their concerns over how the WAI proposal would affect groundwater use by local station owners/operators, coastal tourism operators and the community near La Grange. Issues included: (1) the potential negative impact on local bores through increased salinity and/or drawdown effects; (2) the potential for pesticides and fertilisers to leach into

---

[6] Surface water was planned to be abstracted from the mouth of the Fitzroy River and conveyed by canal to the La Grange Sub-basin area. Then, this water was expected to be used to artificially recharge the aquifer (John Logan, WAI's Chairman, personal communication, June 2003).

[7] The whole basis for the cotton research in the area was to have a system that provides for a substantial reduction in the conventional use of pesticides. WAI devised a system whereby aerial spraying was not to be involved at all (Dr Gary Fitt, Chief Executive Officer, Australian Cotton CRC, personal communication, June 2003).

the groundwater, affecting local drinking water resources;[8] and (3) the availability and sustainability of groundwater for existing and future agricultural activities in the area.

- **Need for a precautionary approach.** Stakeholders advocated application of the precautionary principle when considering the environmental acceptability of the WAI proposal. Questions were raised about how the Water and Rivers Commission (WRC) would be able to stop or wind back the cotton industry once it had become established, if unexpected impacts occurred.

G.3 CULTURAL VALUES OF GROUNDWATER

Yu (2000) studied the cultural values of groundwater in the La Grange Sub-basin. This study was commissioned by the WRC in preparation of the Allocation Plan for the use of groundwater in order to ascertain the significance of groundwater to the Karajarri and other traditional owners of the affected area. The goals of the study were:

- To identify and document the Aboriginal cultural values of groundwater-dependent ecological and hydrological features within the study area.
- To provide an assessment of the significance of these environmental values with respect to the cultural values they possess.
- To identify any registered Aboriginal heritage sites in the study area that are linked to groundwater-dependent ecological or hydrological features.
- To involve the Karajarri and other Aboriginal groups with traditional lands in the study area, in the research process.
- To make specific recommendations regarding the avoidance of negative impacts on the groundwater-dependent cultural values within the study area.
- To ensure that the Aboriginal community and other groups involved, approve the resultant study report.

Major concerns of the traditional owners were (Yu, 2000):

- They were concerned about the effects of extracting too much groundwater. Based on their knowledge of groundwater systems, they feared that the underground *rivers* and *streams* will be irrevocably deprived of water and as a consequence their water resources will dry up or become salty.
- They were concerned that if the fresh groundwater supply to the springs and other water sources significantly reduces, their wetlands will become inundated by saline water.
- They were further concerned about the effects of large amounts of fresh groundwater being pumped everyday, believing that such activity would irreversibly change the cycle of groundwater flow and recharge of the aquifers, resulting in drawdown of the groundwater levels from the *jila*,[9] springs, soaks and others, thereby rendering them salty or dry.

In summary, from the traditional owners' point of view, the impacts on the traditional water sources caused by the extraction of groundwater were unacceptable.

G.4 COTTON RESEARCH

The history of cotton evaluation in the Broome and Derby areas is described in Yeates (2001, pp. 23–25). The first evaluation was conducted at Bidyadanga (La Grange) mission during the late 1800s. In 1922, cropping was attempted near Derby, which apparently failed due to drought, heat and insects. During 1993–96, a former cotton farmer from NSW initiated very small-scale trial evaluations near Broome at Kinkella Farm, Shamrock and Nita Downs stations and a further site was established at Dampier Downs.

Research was expanded with the WAI activities. WAI employed a full time research agronomist and support staff, and established a research site at Shamrock Station with an area of 15 ha in 1999. The key findings of this research on crop adaptation, crop water and nutrient requirements, variety assessments, pest management, and future research direction are described in Yeates (2001). The objective of WAI's West Kimberley Irrigation Project was to set a standard for World's Best Practice by developing and implementing an environmentally sustainable irrigated farming production system.

G.5 BENEFITS OF THE WAI PROPOSAL

The WAI proposed project was expected to be a significant investment in the West Kimberley. It was initially based on irrigating 20 000 ha of land and was expected to generate revenue of $80 million per annum and create 300 on-farm and support jobs. When regional multipliers were included, a gross value of $120 million was expected. If the project could expand to its full potential of 175 000 ha, the expected value of production would be in the order of $850 million per annum and was estimated to create 3000 on-farm and direct jobs.

WAI completed its preliminary feasibility studies into the viability and sustainability of a large-scale integrated agricultural industry based on genetically modified cotton using

---

[8] The objective of the research on irrigation was largely to ensure no leakage occurs below the root zone. This objective could be achieved through sub-surface drip irrigation and by carefully matching water application to daily crop requirements (Dr Ivan McLeod, Project Manager WAI, personal communication, August 2003).
[9] Permanent water holes, sometimes with visible surface water.

groundwater from the Canning Basin. This research determined that WAI's water efficient cotton production system could be used to produce high quality cotton over a sustained period. The Department of Agriculture collaborated with WAI in initial field trials with a range of crops assessed under an efficient sub-surface drip irrigation system, which was designed for the local environment. Field trials were suspended after the 1999 winter season when the Karajarri traditional owners refused access to land in support of their native title claim.

G.6 PROGRESS OF THE FEASIBILITY STUDY

On 22 November 2000, a variation to the MoU was executed by the then Premier of Western Australia, which resulted in a 3-year extension to allow feasibility studies to continue. The MoU variation set out agreed project milestones. Included in these were groundwater investigations and farm trials, which were central to development of a detailed blueprint for the project.

In March 2001, the Karajarri Traditional Lands Association, representing the Karajarri native title claimants, advised WAI that it opposed a proposed work program clearance for WAI's farm trials and groundwater investigations, which were required as part of the feasibility studies. It was understood that they wanted native title issues resolved first. Without the results of these investigations, WAI could not develop a blueprint of the proposed irrigation industry, and the WRC could not develop a water allocation plan for the area.

The MoU expired on 30 June 2003. However, in March 2003, WAI submitted a formal request for an extension of the MoU. On 28 October 2003, the WA Government extended the MoU until 30 June 2004. This was to allow the WAI to complete its investigations and feasibility studies; and resolve land access and land availability issues. WAI proposed to put an innovative package to the traditional owners of the area, and re-engage with the local community via an improved community consultation. The WA Government advised WAI that it strongly supports a negotiated native title outcome that ensures sustainable economic and social outcomes for the local communities and endorses WAI's proposal to re-engage with both the local indigenous and non-indigenous communities.

G.7 FAILURE OF THE PROPOSAL

On 18 November 2003, WAI applied for another extension of the MoU to carry on after June 2004. This was because of the complexity of the issues involved including the native title and securing a land use agreement with the traditional owners of the proposed area of development. Negotiations with traditional owners reached an impasse and finally, on 11 August 2004, the State Government announced that, in view of all the circumstances, it would not extend the MoU, which expired on 30 June 2004. Instead, the Government would engage in a new process with local communities that sought to identify what industries they wanted to see developed in their region.

References

Beckwith & Associates (1999). *La Grange Groundwater Allocation: A Kimberley Sub-Regional Allocation Plan*. Report prepared for the Water and Rivers Commission.

Commander, D. P., Laws, A. T. and Stone, R. (2002). Development of the Canning Basin groundwater resource. In *Proceedings of the 7th National Groundwater Conference: Balancing the Groundwater Budget*. 12–17 May 2002, Darwin, Northern Territory. CD-ROM.

Ferguson, J., Etminan, H. and Ghassemi, F. (1992). Salinity of deep formation water in the Canning Basin, Western Australia. *BMR Journal of Australian Geology & Geophysics* **13**: 93–105.

Ghassemi, F., Etminan, H. and Ferguson, J. (1992). A reconnaissance investigation of the major Palaeozoic aquifers in the Canning Basin, Western Australia, in relation to Zn–Pb mineralisation. *BMR Journal of Australian Geology & Geophysics* **13**: 37–57.

Laws, A. T. (1991). Outline of the groundwater resource potential of the Canning Basin Western Australia. In *Proceedings of the International Conference on Groundwater in Large Sedimentary Basins*. Perth, 9–13 July 1990. Australian Water Resources Council, Conference Series No. 20. Canberra: Australian Government Publishing Service, pp. 47–58.

Yeates, S. J. (2001). *Cotton Research and Development Issues in Northern Australia: A Review and Scoping Study*. Darwin, Northern Territory: Australian Cotton Cooperative Research Centre.

Yu, S. (2000). *Ngapa Kunangkul: Living Water, Report on the Aboriginal Cultural Values of Groundwater in the La Grange Sub-Basin*. Second Edition. Nedlands, Western Australia: Centre for Anthropological Research, University of Western Australia.

H Some Other Water Transfer Schemes in Australia

H.1 INTRODUCTION

Apart from case studies described in Chapters 4–10, numerous other inter-basin water transfer schemes are in operation in various parts of Australia. These include the Shoalhaven Diversion Scheme for Sydney water supply and hydro-power generation, the Thomson Diversion Scheme for Melbourne water supply, and a number of diversion schemes in Tasmania for hydro-power generation. These schemes are described in the following sections. Selected characteristics of some other schemes for water supply of Adelaide, Perth, Darwin and a number of small communities are presented in Table H.1 and their locations are shown in Figure H.1. Information regarding the following proposals is available in Osborne and Dunn (2004, pp. 100–105 and 110–112):

(1) transfer of water from Great Lake in Tasmania to Thomson Reservoir in Victoria;
(2) water transfer from Fly River in Papua New Guinea to Diamantina River in Queensland;
(3) building a national water grid; and
(4) digging a north to south trans-continental canal.

H.2 SHOALHAVEN DIVERSION SCHEME

H.2.1 Historic Background

Sydney's original water supply (1788–1826) came from the Tank Stream (Figure H.2). The stream, which wound its way through the colony before emptying into Sydney Harbour at Circular Quay, degraded into an open sewer and was abandoned in 1826.[1] Convicts then developed Busby's Bore, a 4 km tunnel leading from the Lachlan Swamps (now Centennial Park) and ending in the south-eastern corner of Hyde Park. By 1852, drought and increasing population led to a call for a more permanent water supply. A third water source, the Botany Swamps Scheme, began operations in late 1859 but within 20 years, this major source of freshwater supply was depleted. The Upper Nepean Scheme was Sydney's fourth source of water supply. Its construction commenced in 1880 and completed in 1888. The scheme diverted water from the Cataract, Cordeaux, Avon and Nepean rivers to Prospect Reservoir via 64 km of tunnels, canals and aqueducts known collectively as the Upper Canal. However, the Upper Nepean Scheme brought only temporary relief to Sydney's water supply.

The Federation Drought of 1901–02 brought Sydney close to a complete water famine. Following two Royal Commissions into Sydney's water supply, the authorities agreed that a dam be built on Cataract River. The successive building of Cataract, Cordeaux, Avon and Nepean dams between 1907 and 1935 greatly improved the Upper Nepean Scheme's capacity (Table H.2).

The potential of Warragamba was identified as early as 1845. However, construction of the dam only commenced in 1948 and was completed in 1960. It replaced the Upper Nepean Scheme as the primary metropolitan water source.

The plan to develop a water supply in the Shoalhaven Catchment[2] first arose during World War I. In 1968, the then Water Board consulted the Snowy Mountain Hydro-Electric Authority about the longer-term water needs of Sydney and the South Coast. There was concern that Warragamba Dam, which was opened only 8 years earlier, might prove inadequate to meet Sydney's water supply needs by the mid 1970s.[3] The advice was to proceed with the Shoalhaven Diversion Scheme. It was designed as a dual-purpose water transfer and hydro-electric power generation scheme. Construction began in 1971 under supervision of the Snowy Mountains Engineering Corporation. The Scheme was completed in 1977 at a cost of $128 million ($500 million in 2003 values). The Scheme transfers water from Shoalhaven River catchment to Wollondilly and Nepean rivers and plays a major role in Sydney and Wollongong's water supply, particularly during drought periods (Sydney Catchment Authority, 2002).

---

[1] http://www.sca.nsw.gov.au/dams/history.html (visited in September 2005).
[2] Description of land use and hydrology of the Shoalhaven catchment is available in Costin et al. (1984).
[3] http://www.sca.nsw.gov.au/dams/shoalhaven.html (visited in September 2005).

Table H.1. *Selected characteristics of other inter-basin water transfer projects in Australia*

No	Transfer From: River basin[a]	To: River basin[a]	State	Annual capacity ($10^6 \, m^3$)	Remark
1	Brisbane (I-43)	Condamine (IV-22)	QLD	4	Perserverance Creek diversion for Toowoomba water supply (transfer to MDB).
2	Manning (II-8)	Hunter (II-10)	NSW	20	Barnard River Scheme. Potential for diversion of $70 \times 10^6 \, m^3/yr$.
3	Macquarie (IV-21)	Hawkesbury (II-12)	NSW	14	Fish River water supply scheme (transfer out of MDB).
4	Glenelg (II-38)	Wimmera–Avon (IV-15)	VIC	76	Rocklands Dam supplies some of the water for the Wimmera Mallee stock and domestic requirements.
5	Moorabool (II-32)	Barwon (II-33)	VIC	6	Geelong water supply.
6	Otway Coast (II-35)	Lake Corangamite (II-34) and Hopkins River (II-36)	VIC	Not available	Diversion from Gellibrand River for water supply to Camperdown, Warnambool, and other towns.
7	Gawler (V-5)	Torrens (V-4)	SA	18	Warren, South Para, Little Para, and Barossa Dams. Adelaide water supply.
8	Onkaparinga (V-3)	Torrens (V-4)	SA	32	Mount Bold and Happy Valley Dams. Adelaide water supply.
9	Myponga (V-2)	Torrens (V-4)	SA	11	Myponga Dam. Adelaide water supply.
10	Murray (VI-14)	Swan Coast (VI-16)	WA	105	Water diverted from Serpentine, South Dandalup, and North Dandalup rivers for Perth water supply.
11	Adelaide (VIII-17)	Finniss (VIII-15)	NT	10	Manton Dam, Darwin water supply.

Note: [a] River basin numbers are in brackets following basin names.
Source: Australian Water Resources Council (1987, pp. 30–32).

H.2.2 Water Supply Component

The water supply component of the Scheme consists of four dams (Table H.3). Water pumped from the Scheme is primarily collected from the Tallowa Dam catchment area (Sydney Catchment Authority, 2002). Wingecarribee and Fitzroy Falls reservoirs have relatively small catchment areas totaling 71 km². During drought periods, water from the Shoalhaven can be fed into Warragamba and Upper Nepean dams[4] (Figure H.2). From Wingecarribee Reservoir water can be released into the Wingecarribee River, which flows into the Wollondilly River, feeding the main Sydney supply system via Warragamba Dam. Water can also be released from Wingecarribee Reservoir via Glenquarry Cut into Nepean River and Dam. The Scheme also supplies water to local communities and the NSW South Coast cities of Nowra and Port Kembla Wollongong.

The estimated annual diversion capacity of the Scheme is about $284 \times 10^6 \, m^3$ (Australian Water Resources Council, 1987, p. 30). However, the daily water supply from Shoalhaven varies widely depending on a number of parameters, including[5] rainfall, available water in Tallowa Dam, available room in Fitzroy Falls and Wingecarribee reservoirs to store water pumped from Tallowa Dam, Sydney Catchment Authority's Water Management Licence conditions which outline how the water can be transferred and how much can be diverted at different times of the year, and any flooding that may occur along Wingecarribee and Wollondilly rivers during high rainfall. For example in September 2005, volume of transferred water was $0.9 \times 10^6 \, m^3$ per day from 9 to 12 September. It increased to $1.5 \times 10^6 \, m^3$ per day on 13 September and $1.6 \times 10^6 \, m^3$ per day on 14 September. Then it dropped to $0.05 \times 10^6 \, m^3$ per day on

---

[4] http://www.sca.nsw.gov.au/dams/shoalscheme.html (visited in September 2005).

[5] http://www.sca.nsw.gov.au/dams/pumping.html (visited in September 2005).

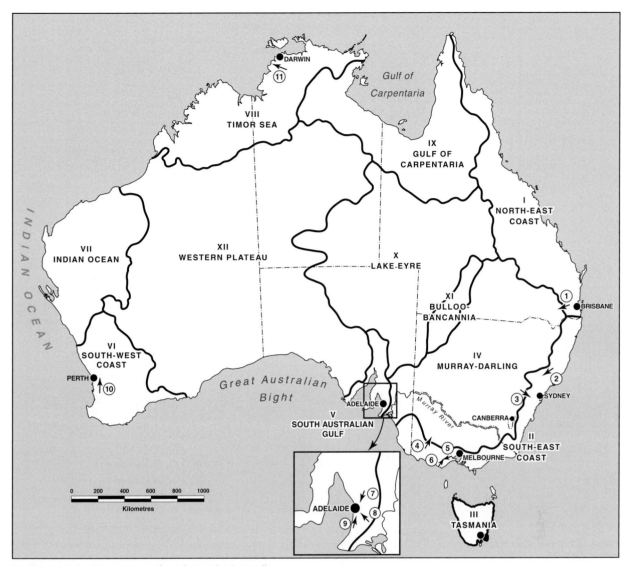

Figure H.1 Other inter-basin water transfer schemes in Australia.

15 and 16 September, and was expected to be nil from 17 to 21 September.[6]

Construction of the Welcome Reef Dam on the upper reaches of the Shoalhaven with a capacity of $2.7 \times 10^9 \, m^3$ and drainage area of $2700 \, km^2$, was supposed to begin in 2002 and be completed in 2005 at a cost of $1 billion. However, its construction was deferred indefinitely because of its environmental impacts. Instead, the NSW Government announced creation of the Welcome Reef Nature Reserve on 6000 ha of the area.[7]

Raising the height of the Tallowa Dam by 7 m is a key part of the NSW Government's Metropolitan Water Plan (DIPNR, 2004). It will allow additional water to be supplied to Sydney via Avon or Warragamba dams. Construction works on the dam are expected to commence in November 2006 and are due for completion in 2008.[8] New pipe networks will be built to transfer water instead of using the river systems. This will help natural ecosystems in the rivers and will reduce riverbank erosion.

The project is expected to provide Sydney metropolitan area with an additional $50–80 \times 10^6 \, m^3$ per year by 2010 and up to $110 \times 10^6 \, m^3$ per annum by 2020. This would cost $250 million for Stage 1, and if implemented $430 million for Stage 2 of the project. However, Shoalhaven residents are opposed to the project based on its environmental impacts and are strongly advocating the recycling of Sydney's water instead of transferring more water from Shoalhaven Catchment.[9]

---

[6] http://www.sca.nsw.gov.au/dams/pumping.html (visited in 16 September 2005).

[7] http://www.sca.nsw.gov.au/news/media/10.html (visited in September 2005).

[8] http://www.sca.nsw.gov.au/dams/shoaltran.html (visited in September 2005).

[9] http://www.smh.com.au/news/national/bitter-outpouring-at-carrs-water-plan/2005/07/24/1122143730141.html (visited in September 2005). In February 2006, under pressure from local residents, the NSW Government decided not to implement this option.

Figure H.2 Some of the main components of the early and present water supply schemes for Sydney (Sammut and Erskine, 1995, Figure 2).

Table H.2. *Selected characteristics of four large dams of the Upper Nepean Water Supply System and Warragamba Dam*

Name	Construction Commenced	Completed	Storage capacity ($10^6 \, m^3$)	Height (m)	Catchment area (km^2)	Average annual flow ($10^6 \, m^3$)
Cataract	1902	1907	94.3	49	130	70.3
Cordeaux	1918	1926	93.6	50	91	53.7
Avon	1921	1927	214.0	71	142	73.9
Nepean	1926	1935	81.4	75	319	106.7
Warragamba	1948	1960	2031.0	142	9051	950.0

Source: Sammut and Erskine (1995, Table 1), and the URL provided in the footnote.[10]

H.2.3 Hydro-Power Generation Component

Two hydro-power stations at Bendeela and Kangaroo Valley are parts of the Scheme and were completed in 1977. Bendeela Pumping and Power Station consists of two units of 40 MW producing 80 MW of electricity, and Kangaroo Valley Pumping and Power Station generates 160 MW of electricity through two units of 80 MW each (Eraring Energy, 2004, p. 15). Water is pumped up the system between Lake Yarrunga created by

[10] http://www.sca.nsw.gov.au/dams/facts.html (visited in September 2005).

Table H.3. *Major dams of the Shoalhaven Scheme in order of reservoir capacity*

Name	Height (m)	Length (m)	Capacity (10^6 m^3)	Catchment area (km^2)	Lake area (km^2)	Year of completion
Tallowa Dam	43	518	85.5	5750	9.3	1976
Wingecarribee Reservoir	19	1140	25.9	40	6.3	1974
Fitzroy Falls Reservoir	14	1530	23.5	31	5.2	–
Bendeela Pondage	15	2118	1.2	–	–	1972

Source: Sydney Catchment Authority (2002).

Tallowa Dam and Fitzroy Falls Reservoir using off-peak electricity. Some of the water is then released back down the system for generation of electricity during periods of peak demand. The power is fed into the statewide transmission grid and the Eraring Energy manages the system.[11]

H.2.4 Environmental Impacts

The Upper Nepean Scheme has progressively regulated the Upper Nepean River for water supply of Sydney and Wollongong with major hydrological impacts (Sammut and Erskine, 1995). Flow regulation has significantly changed the natural flow regime by substantially reducing the magnitude and frequency of floods. The four large dams (Cataract, Cordeaux, Avon and Nepean) have estimated or measured sediment trap efficiencies greater than 99 percent. This means that sediment loads in the downstream rivers are on average about one percent of their pre-dam values.

Implications of flow regulation are four-fold (Sammut and Erskine, 1995). First, during low flow periods, point source pollution has a dramatic impact on water quality because of the lack of dilution flows from the upper catchment. Second, the habitat of the Macquarie perch has been greatly impacted by extended periods of very low to zero flow. Rectification of this problem requires release of adequate environmental flows from reservoirs. Third, the reduced magnitude and frequency of floods caused by flow regulation has profound implications for channel morphology and distribution of riparian vegetation. Fourth, abrupt flow stoppage caused massive fish mortalities below Tallowa Dam when flow was stopped in the Shoalhaven River. This was due to concentration and isolation of fish immediately below dam and weirs where they were subject to increased predation and competition as well as poor water quality.

The Healthy Rivers Commission (1998) inquired the health of the Hawkesbury Nepean River System and identified six key challenges in order to meet future health of the system. These are:

(1) urban development and riverine corridors;
(2) sewage and storm water impacts on water quality;
(3) water sharing;
(4) extractive industries;
(5) sustainable agriculture; and
(6) river values and costs.

In order to improve the river health the Healthy Rivers Commission (1998) made a number of recommendations on:

(1) institutional arrangements;
(2) local government;
(3) water quality objectives;
(4) flow regime;
(5) riverine corridors;
(6) urban stormwater;
(7) sewage treatment and disposal;
(8) aquatic weeds;
(9) agriculture; and
(10) boating.

The Healthy Rivers Commission (1999) investigated the health of the Shoalhaven Catchment with respect to water quality, streamflow, aquatic plants and animals, riverside vegetation, bed and bank stability, and wetlands for each of its six divisions (Upper, Middle Western, Middle Forested, Northeastern, Downstream of Dams, and Estuary and Floodplain). The Commission concluded that more than half of the sub-catchments have high environmental stress. The main issues are stream bed and/or bank erosion, fish barriers, loss of riparian vegetation and gully erosion (Healthy Rivers Commission, 1999, pp. 78–80). Only a very few sub-catchments were assessed as being stressed due to pattern of water use and/or water quality.

H.2.5 Sydney Metropolitan Water Plan 2004

Sydney's current water use is unsustainable. Its dams[12] with a total storage capacity of 2.39×10^9 m^3 have an average annual yield of 600×10^6 m^3, while its water use was about 630×10^6 m^3 in 2000. It has been estimated that without taking any action and with Sydney's population growth of 40 000 per annum or one million over the period of 2004–29, water use will reach

---

[11] http://www.eraring-energy.com.au (visited in September 2005).
[12] Warragamba, Avon, Cataract, Cordeaux, Woronora, Nepean, Tallowa, Wingecarribee, Fitzroy Falls, Prospect and Blue Mountains. (http://www.sca.nsw.gov.au/dams/232.html visited in December 2005).

$800 \times 10^6 \, m^3$ by 2029 (DIPNR, 2004). Water is also required to protect the health of regulated river systems. In order to secure future water requirements of Sydney metropolitan area, the NSW Government has launched the Metropolitan Water Plan 2004, *Meeting the Challenges: Securing Sydney's Water Future* (DIPNR, 2004). This Plan has been developed to satisfy water requirements of Sydney's population over the next 25 years and to restore the health of the Hawkesbury–Nepean and other river systems. The Plan will be reviewed every 5 years to update it as new evidence become available. The following description of the Plan is based on DIPNR (2004) and other sources.

During the major 2002–05 drought, Sydney's reservoir capacities dropped to 42.6 percent in early October 2004 and to 38.8 percent in late October 2005. Mandatory water restrictions were imposed on 1 October 2003, and water consumption dropped by 10 percent or $63 \times 10^6 \, m^3$ over the period of October 2003–04. Sydney is faced with four significant factors in planning for its long-term future water management: population growth; drought; climate change; and river health. The aims of the Plan are to:

(1) minimise the risk of water shortages by diversifying sources of supply;
(2) ensure secure water supplies;
(3) protect and restore river health;
(4) adopt a partnership approach with the community;
(5) provide good quality and cost effective water supply services;
(6) foster innovation;
(7) increase the efficient use of water;
(8) match the grade of water to its end use;
(9) optimise the use of existing infrastructure;
(10) appropriately target future investment;
(11) make decisions adaptively; and
(12) ensure actions are acceptable to the public, affordable, feasible and sustainable.

The NSW Government spent $81 million[13] over the period of 1991–2004 to reduce water demand. As a result, almost 240 000 households have converted to water efficient products with a reported saving of $4.5 \times 10^6 \, m^3$ of water per annum; 4000 rainwater tanks have been installed, 200 businesses reduced their annual water use by $4.3 \times 10^6 \, m^3$, leakage detection and repair saved $15 \times 10^6 \, m^3$ in 2003–04, and reuse of recycled water reached $14 \times 10^6 \, m^3$ in 2004. These savings totalled about $40 \times 10^6 \, m^3$.

The Plan is philosophically opposed to the construction of costly and environmentally damaging new dams. Instead, in order to increase security of Sydney's water supply, it will undertake several measures including:

- **Modifying dam intake structure.** Modifying Avon, Warragamba and potentially Nepean dams at a cost of $106 million so that water at the bottom of the dams can be accessed.[14]

- **Water from Tallowa Dam.** Pumping more water ($50–110 \times 10^6 \, m^3$ per year) from Tallowa Dam into the Warragamba Dam (see Section H.2.2).

- **Groundwater use.** Extracting groundwater from Botany sand and Sydney and Hawkesbury sandstone aquifers after undertaking detailed investigations.

- **Water recycling.** Increasing use of recycled wastewater from $14 \times 10^6 \, m^3$ in 2004 to $22 \times 10^6 \, m^3$ by 2010 through several projects. A number of additional water recycling projects are under investigation. These include: (1) supplying $24 \times 10^6 \, m^3$ of recycled water through separate pipes for outdoor and toilet flushing in new land release areas; (2) using $32 \times 10^6 \, m^3$ of recycled water for farm irrigation; and (3) releasing up to $40 \times 10^6 \, m^3$ of high quality recycled water to river systems for environmental use. The cost of these and other water recycling projects is estimated at about $563 million.

- **Desalination of seawater.** The NSW Government allocated $4 million for planning and design of a desalination plant for Sydney using reverse osmosis. This plant is planned to assist Greater Sydney to overcome its water shortage if the 2002–05 drought prolongs in the following years. GHD-FICHTNER (2005a) investigated issues (infrastructure, environment, energy, cost and others) related to construction of desalination plants with capacities ranging from 100 000 to 500 000 m^3 per day[15] on the industrial land at Kurnell located at the south-east of Botany Bay. Water would be transferred to the Plant from the Tasman Sea via an intake tunnel. Another tunnel will return the concentrated saline water to the sea at a location removed from the intake point. The consultants estimated that the project would cost in the order of $470 million to $1.75 billion, and water could be delivered at about $1.44 per m^3, which is much higher than $1.11 per m^3 for Perth desalinised water (see section 9.7), because of the cost of about 110 MW electricity and other infrastructure requirements. GHD-FICHTNER (2005b) has assessed the environmental factors (impacts on marine and terrestrial ecology, indigenous and non-indigenous heritage, water quality, hydrology and drainage, air quality, and hazards and risks) associated with the desalination pilot plants and concluded that, by adopting the measures identified in their assessment, there would be no significant environmental impact as a result of undertaking the proposed works.[16] Construction of the plant was expected to commence in late 2006, taking

---

[13] This included more than $3 million in community education on water conservation.
[14] These bottom waters have poorer water quality.
[15] This is about 30 percent of the current annual Sydney Water Supply requirements.
[16] The desalination plant proposal has been roundly criticised on a variety of grounds. Opponents argue that the major desalination plant in Southern California, which is largely unused because of operational costs, exemplifies why desalination is a poor option.

26 months to be completed in early 2009. However, under strong pressure from environmentalists and other groups, on 8 February 2006, the NSW Government decided to halt the project indefinitely in favour of groundwater extraction from Leonay and Kangaloon aquifers. The project will be revived only if dam reserves fall to 30 percent of their capacities. Further information regarding this controversial option for Sydney water supply is available at URLs provided in the footnote.[17]

- **Reducing industrial demands.** Industry (excluding agriculture), businesses and government bodies use about 30 percent of water from the Sydney water system. Estimates indicate that savings of 10 to 30 percent can be achieved.
- **Reducing households demand.** The water conservation efforts of Sydney households have kept demand steady over the past 20 years. Sydney's householders used $387 \times 10^6 \, m^3$ of water in 2002–03, about $290 \, m^3$ per household. Through a partnership with the community, the Government intends to reduce household consumption.
- **Reducing agricultural demand.** The agricultural sector is the second highest consumer of water in the Greater Sydney area, using approximately $100 \times 10^6 \, m^3$ of river water per year. The Government objective is to increase agricultural water use efficiency by about 25 percent through the use of efficient irrigation methods and water metering. The government will also encourage use of recycled water for irrigation rather than the use of river water.[18]
- **Water pricing.** Historically, water price has been below its real value. The Government intends to change water price structure to encourage water saving by householders, businesses and other water users. For example, on 1 October 2005 a two tiered system was introduced for domestic users. The price of water for householders using up to $400 \, m^3$ per year has increased to $1.20 per m^3 and $1.48 per m^3 for consumption of more than $400 \, m^3$ per year.[19]

The rivers supplying Sydney have been dammed extensively and this has caused serious ecological impacts. The Hawkesbury–Nepean River is showing signs of substantial environmental stress, such as weed infestation, algal blooms, and deteriorating fish and oyster production. Poor river health will impact on the tourist, agriculture, fishing and recreation industries in the valley of Hawkesbury–Nepean. These industries are currently generating more than $2 billion each year. The Shoalhaven River also has substantial fishing, tourist and recreational industries. Moreover, the health of Sydney's river systems impacts on the ability of Sydney to secure a high quality water supply. Government and community have undertaken numerous measures[20] to improve the health of the regulated river systems.[21] The Government plans to implement new environmental flow regimes in the Hawkesbury–Nepean River, Woronora River and the Shoalhaven River. Ongoing monitoring will be used to examine if environmental releases are effective before works are built at successive dams.

H.3 THOMSON DIVERSION SCHEME

The early settlers in Melbourne depended on the Yarra River for their water supply, although some residents also collected rainwater (Gibbs, 1915). Melbourne's piped water supply commenced in 1857, with the completion of Yan Yean Reservoir (Figure H.3) when Melbourne and its suburbs had a population of about 100 000. Water from Goulburn Catchment (IV-5)[22] in the southern part of the Murray–Darling Basin was diverted to Yarra Catchment via Silver Creek–Wallaby Creek Aqueduct and stored in the Yan Yean Reservoir. The total cost of the works including street reticulation to the end of 1857 was £754 206 (Gibbs, 1915, p. 11) or about $47 million in 2003 values. The Aqueduct has an annual capacity of $13 \times 10^6 \, m^3$. In 1891, the newly formed Melbourne and Metropolitan Board of Works became responsible for Melbourne's water supply system. Melbourne water supply was progressively augmented.[23] Maroondah Dam was completed in 1927, followed by the small O'Shannassy Reservoir in 1928.[24] Silvan Reservoir was completed in 1932 to regulate the increased flow in the O'Shannassy Aqueduct from the Upper Yarra River and Coranderrk Creek diversions. When the Upper Yarra Dam with a capacity of $200 \times 10^6 \, m^3$ was completed in 1957, the total capacity of Melbourne's water supply was tripled to nearly $300 \times 10^6 \, m^3$. To meet growing demand in the western suburbs, the Greenvale Reservoir was completed in 1971. Following the 1967–68 drought, construction of the Cardinia Reservoir commenced in 1969 and completed in 1973, bringing Melbourne's total storage capacity to $610 \times 10^6 \, m^3$. The Sugarloaf Project, including a major pumping station and water treatment plant was completed in 1981, increasing Melbourne's total storage by $96 \times 10^6 \, m^3$.

---

[17] <http://www.sydneywater.com.au/EnsuringTheFuture/Desalination/index.cfm>, <http://www.nccnsw.org.au/water/projects/upload/Final%20submission%20Desal%20EPBC%20Kurnell.pdf>, and <http://www.tai.org.au/Publications_Files/Papers&Sub_Files/Desalination%20plant.pdf> (all three visited in November 2005).

[18] There are some problems with the use of recycled water in agriculture. In particular, this may increase the soil sodicity.

[19] http://www.sydneywater.com.au/Publications/_download.cfm?DownloadFile=FactSheets/UsageCharges.pdf (visited in November 2005).

[20] These include the release of $58.2 \times 10^6 \, m^3$ by the Sydney Catchment Authority in 2003–04 from its dams to Hawkesbury–Nepean, Woronora and Shoalhaven rivers; weed harvesting; reduction of nutrient discharge; implementation of the Sustaining the Catchments Regional Environmental Plan; and other measures (DIPNR, 2004, p. 20).

[21] These measures appear to have had limited success.

[22] For location see Australian Water Resources Council (1987).

[23] http://www.melbournewater.com.au/content/publications/fact_sheets/water/melbournes_water_supply_system.asp (visited in September 2005).

[24] The early history of Melbourne water supply through Yan Yean, Maroondah and O'Shannassy reservoirs and the Upper Yarra systems is available in Gibbs (1915).

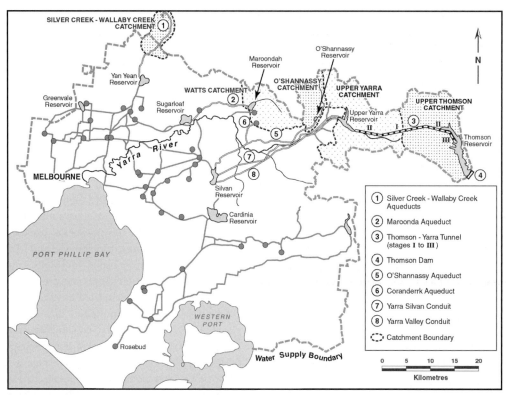

Figure H.3 Melbourne's Water Supply System (Melbourne Water, 2003; and Melbourne and Metropolitan Board of Works, 1975, Figure 1.1).

The Thomson River is a major tributary of the La Trobe River system in the Gippsland region of Victoria. The upper reaches of the Thomson River drains the high rainfall area and make it suitable for efficient water harvesting. The Thomson water diversion scheme consisted of three stages (Melbourne and Metropolitan Board of Works, 1975):

- **Stage I** consisted of construction of a 20 km tunnel with a diameter of 3.6 m under the Great Divide for diversion of the Upper Thomson River waters westward into the Yarra Catchment. Work on this stage commenced in 1969, completed in August 1974, and has been in operation since then. It had a diversion capacity of $41 \times 10^6 \, m^3$ per annum.
- **Stage II** was approved by the State Government in 1973 and consisted of the extension of the **Stage I** tunnel from the east by 5.6 km and from the west by 5.1 km. This stage, which was completed in 1977, doubled the area available for water harvesting and increased the diversion capacity of the project by $22.5 \times 10^6 \, m^3$ per annum.
- **Stage III** consisted of construction of a large dam on the Thomson River and linking its reservoir to **Stages I** and **II** tunnels by the **Stage III** tunnel with a length of 6 km. Construction works were completed in 1983 and became operational since 1984. Completion of this stage provided an additional $84.5 \times 10^6 \, m^3$ of water per annum for diversion to Melbourne water supply systems, with a total diversion capacity of $148 \times 10^6 \, m^3$ per annum. The 1975 total estimated cost of the Thomson project was about $237 million ($1.2 billion in 2003 values).

In 1991, the Melbourne and Metropolitan Board of Works merged with a number of smaller urban water authorities to form *Melbourne Water*. Currently *Melbourne Water* manages nine major reservoirs with a total capacity of about $1.8 \times 10^9 \, m^3$ (Table H.4). It also operates 1000 km of distribution mains, about 60 service reservoirs at 35 sites, more than 200 km of aqueducts, five major treatment plants and about 20 pump stations (Melbourne Water, 2003).

Approximately, 90 percent of Melbourne's water supply originates from uninhabited catchments covered with mountain ash forests[25] where 140 000 ha has been reserved for the primary purpose of harvesting water. These water supply catchments have restricted public access to prevent contamination of water supply.

H.3.1 Environmental Impacts

Melbourne and Metropolitan Board of Works (1975, Volume 2) provides an assessment of the environmental impacts of the Thomson Diversion Scheme. Some of the findings are:

- No unique or rare species will be threatened by construction and operation of the project.

---

[25] For descriptions and illustrations of flora and fauna of these catchments see Lindenmayer and Beaton (2000).

Table H.4. *Selected characteristics of Melbourne's major dams and reservoirs in order of reservoir capacity*

No.	Name of the reservoir	Catchment area (ha)	Type of dam[a]	Height (m)	Length (m)	Reservoir area (ha)	Operating since	Capacity ($10^6 \, m^3$)
1	Thomson	48 700	EF & RF	165	590	2230	1984	1068
2	Cardinia	2800	EF & RF	85	1542	1295	1973	287
3	Upper Yarra	33 670	EF & RF	90	610	750	1957	200
4	Sugarloaf	915	RF	89	1050	440	1981	96
5	Silvan	900	EF	40	650	333	1932	40
6	Yan Yean	2250	EF	10	963	560	1857	30
7	Greenvale	350	EF & RF	52	2500	174	1971	27
8	Maroondah	10 400	CG	41	291	200	1927	22
9	O'Shannassy	11 900	EF	34	226	27	1928	3
Total								1773

Note: [a] EF (Earthfill), RF (Rockfill), CG (Concrete Gravity).
Source: Melbourne Water (2003), and Melbourne Water website.[26]

- Local habitat will be destroyed in the area inundated by the reservoir.
- A migratory fish, the eel, would be prevented from using the upper portion of the Thomson River, but this will not threaten its existence.
- The impact of the dam on the downstream river ecosystems will be limited. Riparian releases could overcome most of the problems.
- There is no evidence that the biological balance in Lake Wellington at the lower end of the Thomson River will be changed by construction and operation of the project.
- There are no major adverse social implications and in the long-term the area will benefit from tourism.

The findings of the study have led to the general conclusion that no major environmental issues should arise as a result of construction and operation of the Thomson Dam and associated works. However, the report highlighted issues such as catchment management policies and practices, water quality control and resource utilisation to be considered for further investigations.

H.4 HYDRO-POWER GENERATION IN TASMANIA

Tasmania, with an area of 68 200 km², is the smallest State of Australia. It consists of a group of islands separated by Bass Strait from the mainland. Its highest mountain is Mt Ossa with an elevation of 1617 m (Figure H.4). It was settled in 1803 to secure British strategic interests against the French. Its population increased to 5400 in 1820, to 172 000 in 1901 and 477 000 in 2003 (see Table 3.1). Tasmania has a high average annual precipitation (rainfall and snow), ranging from about 500 mm on the eastern part of the State to more than 2500 mm on its western part (Figure 3.2 and the URL provided in the footnote).[27] The average annual pan evaporation is nearly 1500 mm in the northern region, while in the western, central and southern regions it is less than 750 mm. At 30 June 1998, Tasmania had 3.3 Mha of forested land partly with high conservation value and a further 72 000 ha of softwood plantation and 62 000 ha of hardwood plantation (Australian Bureau of Statistics, 1999, p. 236).

Tasmania has a network of rivers, and natural and man-made lakes (Figure H.4). South Esk River is the longest river (214 km) followed by Derwent (187 km) and Gordon (181 km). Major lakes include Lake Gordon, Lake Pedder and the Great Lake. While the area of Tasmania is about 1 percent of the total area of Australia, its average annual run-off ($45.6 \times 10^9 \, m^3$) is about 12 percent of the total average annual run-off on the country (Table 3.5).

Water resources of the State have been developed since the late nineteenth century for hydro-power generation in order to satisfy municipal and industrial demand.[28] The first hydro-electric power station was introduced in 1895 with construction of a 450 kW Duck Reach Hydro-Electric Station developed on the South Esk River near Launceston (see Figure H.5) by the local municipal authority (Australian Bureau of Statistics, 1978, p. 286), which had a population of about 20 000 at the time. The Hydro-Electric Power and Metallurgical Company began work on Tasmania's first significant hydro-power scheme at the Great Lake in 1911. However, before construction was completed, the Tasmanian Government bought the undertaking.

---

[26] http://www.melbournewater.com.au/content/publications/fact_sheets/water/water.asp (visited in September 2005).
[27] http://www.hydro.com.au/home/Energy/Tasmanian±Hydro±Electric±Schemes/ (visited in September 2005).
[28] Read (1986) provides a detailed history of hydro-power development and its administration in Tasmania.

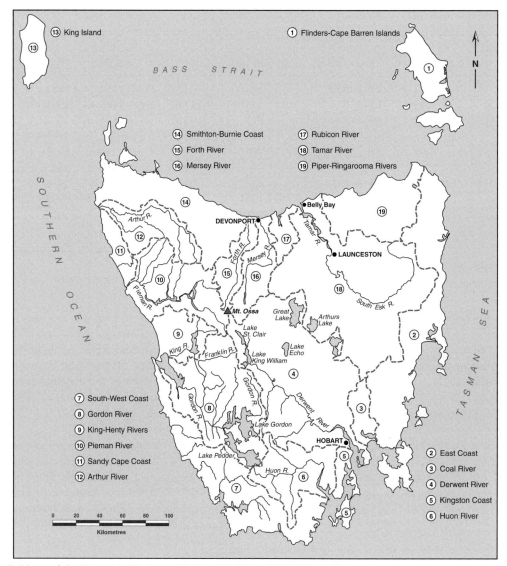

Figure H.4 Sub-divisions of the Tasmania Drainage Division III (Hale, 1979, Figure 1).

The Great Lake Scheme and Waddamana "A" Power Station was opened in 1916. In January 1930, the *Hydro-Electric Commission Act* 1929 was enacted and the Hydro-Electric Commission (HEC) was created to manage existing works and to control waters of the State. It had the sole right of generating, distributing and selling electricity throughout Tasmania. Until 1971, Tasmania was unique among Australian states in that its electric power system was based almost entirely on hydro-electric installation, but in 1971 a thermal oil-powered station commenced operating at Belly Bay[29] (Australian Bureau of Statistics, 1981, p. 250). In 2004, mainland Tasmania had 29 hydro-power stations with a total installed capacity of 2265 MW[30] (Hydro Tasmania, 2004, p. 116), one gas-powered station (240 MW) and one wind powered station (65 MW).

Information regarding hydro-power generation in Tasmania is available in Frost (1983) and Hydro Tasmania's website.[31] Major conflicts over HEC's proposal for construction of the Gordon-above-Olga Dam and the Gordon-below-Franklin Dam to generate hydro-electric power are described by Thompson (1981), Lowe (1984), and Smith and Handmer (1991). These conflicts led to registration of the area on the UNESCO's World Heritage List, a High Court challenge and finally halting the project by the Federal Government in 1983 because of its high conservation value. The cost–benefit analysis

---

[29] For location see Figure H.4.
[30] This is less than the 3756 MW installed hydro-power generating capacity of the Snowy Scheme (see Table 4.4).
[31] http://www.hydro.com.au/home/Energy/Tasmanian±Hydro± Electric±Schemes/Catchment±Areas/ (visited in September 2005). This URL and its links provide information about hydro-power generation in the following catchments: (1) Great Lake–South Esk with three power stations; (2) Derwent with 10 power stations; (3) Mersey–Forth with seven power stations; (4) Gordon–Pedder; (5) Pieman–Anthony; (6) King; and (7) Yolande.

APPENDIX H

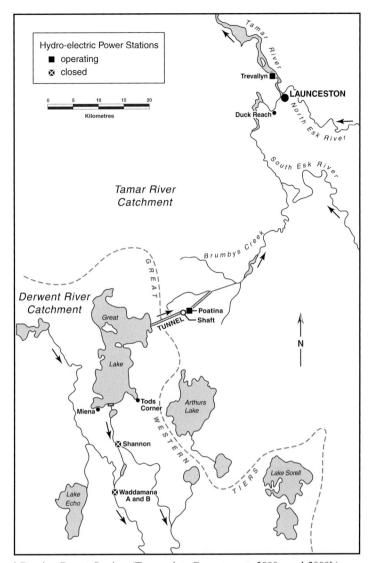

Figure H.5 The Great Lake and Poatina Power Station (Tasmanian Government, 2000a and 2000b).

of the halted project is available in Saddler *et al.* (1980), while Blackers (1994) provides an economic analysis of the King and Anthony[32] Schemes constructed in place of the halted project.

Most of Tasmania's hydro-electric power stations are located along the river systems, while inter-basin water transfer assisted a limited number of schemes. In the following sections only three schemes, which are summarised in Table H.5, are described. Apart from these schemes, the headwaters of the Henty River are diverted into the Pieman River hydro-electric scheme. Additionally, there are numerous small diversions. For example, three small tributary streams of the Franklin River are diverted into Lake King William (Locher, 1999).

H.4.1 The Great Lake Scheme

Development of the Great Lake for hydro-power generation dates back to May 1916 when the Waddamana "A" power Station (Figure H.5) with an installed capacity of 7 MW was commissioned. A new Miena dam in 1922 replaced the original low weir across the Shannon outlet of the Great Lake and resulted in availability of more water for power generation. The Shannon Power Station was constructed between 1924 and 1931 with a generating capacity of 10.5 MW.[33] The Waddamana "B" Power Station with a head of 344 m and an installed capacity of 48 MW commenced power generation in 1944 and its last turbine came into operation in 1949 (Australian Bureau of Statistics, 1978, p. 286).

---

[32] This last major hydro-electric power generation project was completed in 1994 (Hydro Tasmania, 2004, p. 4).
[33] Shannon website at the URL: http://www.hydro.com.au/home/ Energy/Tasmanian±Hydro±Electric±Schemes/ Catchment±Areas/Great±Lake±and±South±Esk±Catchment/ Waddamana±and±Shannon.htm (visited in September 2005).

Table H.5. *Inter-basin water transfer schemes for hydro-electric power generation in Tasmania*

Scheme	Transfer From: River basin[a]	To: River basin[a]	Annual capacity (10^6 m^3)	Remark
Great Lake	Derwent (4)	Tamar (18)	698	Great Lake and smaller catchments diverted to Poatina Hydro-electric power station.
Mersey–Forth	Mersey (16)	Forth (15)	830	Mersey–Forth hydro-electric power scheme.
Gordon River	Huon (6)	Gordon (8)	Not available	Upper catchment of Huon River (261 km^2) diverted for the Gordon Power Station via Lake Pedder and Lake Gordon.

Note: [a] Numbers in brackets refer to the catchment number shown in Figure H.4.
Source: Australian Water Resources Council (1987, p. 31).

In 1957 a decision was made to transfer water from Great Lake in Derwent River Catchment to the Tamar River Catchment (Figure H.5). This stopped water flowing south from the Great Lake and diverted it north to be used in a fall of 830 m down the face of the Great Western Tiers.[34] A 6 km tunnel was drilled under a ridge of the Great Western Tiers. From here, water flows down a large, high-pressure pipeline and a vertical shaft into Tasmania's first underground power station at Poatina. The generating capacity of the Scheme reached 250 MW by 1965, through five generators of 50 MW each. A sixth generator of 50 MW was commissioned in 1977, bringing the total installed capacity of the Station to 300 MW (Australian Bureau of Statistics, 1978, p. 287). This is the second largest power station in Tasmania after the Gordon Power Station (see Section H.4.3).

A variety of works was undertaken to increase the amount of water available for use at Poatina.[35] The storage capacity of the Great Lake was increased to 2.3×10^9 m^3 with the completion of a 22 m high rockfill dam at Miena (the third Miena Dam) in 1967. This dam was raised another 6 m in 1982 to provide a further 0.7×10^9 m^3 of storage capacity. Water leaving the Poatina Power Station flows through a 4.5 km tunnel to a canal and eventually into the South Esk and Tamar River, which feeds the Trevallyn Station near Launceston with an installed generating capacity of 80 MW. Construction of the Poatina Power Station led to the closure of three power stations in the south of the Great Lake. Shannon Power Station closed in 1964, followed by Waddamana "A" in 1965. Waddamana "B" was retained only for emergency and peak-load generation and was finally closed in 1994.

H.4.2 Mersey–Forth Hydro-Electric Scheme

The Mersey–Forth Hydro-electric Scheme[36] was started in 1963 and completed in 1973. It uses water from four rivers in the area: the Fisher, Mersey, Forth and Wilmot (Figure H.6). It comprises seven dams (Table H.6) and seven hydro-power stations. Rowallan and Parangana dams were built on the upper reaches of the Mersey River. Lake Rowallan is one of the main storages for the development. Water from Lake Rowallan flows through the Rowallan Power Station (11 MW) and down into Lake Parangana. A dam on the Fisher River raised the level of Lake Mackenzie. Water from the Lake Mackenzie flows to the Fisher Power Station and involves a drop of some 650 m. The Station has a capacity of 43 MW and discharges its water into Lake Parangana.

The water in Lake Parangana is transferred westward into the Forth River via a 6.8 km tunnel. After it leaves the tunnel, it passes down through the Lemonthyme Power Station with a capacity of 51 MW. A dam on the Wilmot River diverts water in an easterly direction into the valley of the Forth River. The water passes via tunnel and penstock through the Wilmot Power Station with a capacity of 30 MW on the edge of Lake Cethana. The combined flows of the Mersey, Fisher, Forth and Wilmot rivers are then used for power generation at three further power stations on the Forth River at Cethana (85 MW), Devils Gate (60 MW) and Paloona (28 MW). The combined generating capacity of the seven power stations is 308 MW.

---

[34] Poatina website at the URL: http://www.hydro.com.au/home/Energy/Tasmanian$\pm$Hydro$\pm$Electric$\pm$Schemes/Catchment$\pm$Areas/Great$\pm$Lake$\pm$and$\pm$South$\pm$Esk$\pm$Catchment/Poatina.htm (visited in September 2005).

[35] Poatina website at the URL: http://www.hydro.com.au/home/Energy/Tasmanian$\pm$Hydro$\pm$Electric$\pm$Schemes/Catchment$\pm$Areas/Great$\pm$Lake$\pm$and$\pm$South$\pm$Esk$\pm$Catchment/Poatina.htm (visited in September 2005).

[36] Description of the scheme provided here is based on the information available at the following URL: http://www.hydro.com.au/home/Energy/Tasmania$\pm$Hydro$\pm$Electric$\pm$Scheme/Catchment$\pm$Areas/Mersey$\pm$Forth/ (visited in September 2005).

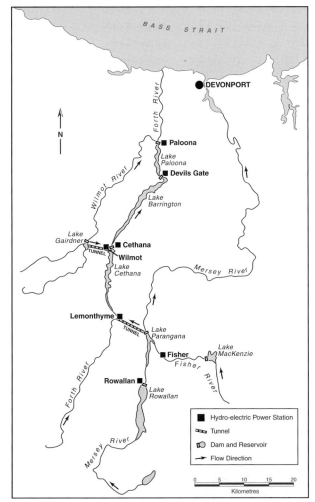

Figure H.6 The Mersey–Forth Hydro-electric Scheme (Tasmanian Government, 2000c, 2000d).

H.4.3 Gordon River Power Development Scheme

The Gordon River Power Development Scheme[37] has created the largest water storage in Australia, comprising of Lake Pedder and Lake Gordon[38] joined by the McPartlan Canal (Figure H.7) with a length of 2.6 km and design capacity of $71\,\mathrm{m}^3\,\mathrm{s}^{-1}$ (Faircloth, 1977, p. 3). The Scheme has combined part of the upper catchment of the Huon River[39] into the enlarged Lake Pedder. In 1972, following 12 years of escalating public protest and criticism, Lake Pedder[40] was flooded to generate electricity for industry (Australian Bureau of Statistics, 1999, p. 17). Lake Pedder was created by construction of the 38 m high Serpentine Dam in 1971 (Figure H.7 and Table H.7) and two other dams in 1973. These consisted of a 43 m high rockfill dam on the upper reaches of the Huon River at Scotts Peak[41] with a catchment area of $261\,\mathrm{km}^2$ (Table H.5), and the 17 m high Edgar Dam at the south-east corner of the Lake Pedder. The greatly enlarged Lake Pedder has a surface area of $242\,\mathrm{km}^2$ and a total storage capacity of $3.3 \times 10^9\,\mathrm{m}^3$. Lake Gordon was created in 1974 by construction of a 140 m high concrete dam across the Gordon River. Lake Gordon has a surface area of $272\,\mathrm{km}^2$ and storage capacity of $12.5 \times 10^9\,\mathrm{m}^3$.

Water from the two lakes is used in the underground Gordon Power Station situated near the Gordon Dam. Lake Pedder provides about 41 percent of the water used in the Gordon Power Station. Water from Lake Pedder flows through the McPartlan Canal into Lake Gordon. The underground Gordon Power Station is the largest in Tasmania. Water from Lake Gordon passes through a cylindrical intake at the base of an 80 m intake tower. It falls 140 m through a vertical intake shaft to three turbines. The first two generators were commissioned in 1978 and the third in 1988. Each turbine has an installed capacity of 144 MW generating a total of 432 MW.

H.4.4 Environmental Impacts

More than $1100\,\mathrm{km}^2$ of river valleys, wetlands and pre-existing lakes have been inundated through the development of hydro-electricity in Tasmania (Locher, 1999). Construction of the 107 dams and weirs affects 21 creeks, 25 rivers and seven estuaries. A total length of at least 1200 km of natural creeks and rivers are impacted by way of diversion and alteration of flow regimes. These hydrological changes have implications for sediment movement, stream channel instability, siltation, instream biota, fish passage, riparian habitats, and water quality (including turbidity).

The nature and severity of the impacts of the 29 hydro-power stations and 52 significant storages varies considerably depending on the surrounding land uses, water management values, and other features. For example:

- Issues on the west coast, in the King and Pieman-Anthony catchments (Figure H.4), predominantly involve the interactions of regulated flows and lake level fluctuations with mine wastes. For instance, the John Butters Power Scheme discharges into the King River which is highly impacted by decades of mine waste discharges from the Mount Lyell Copper Mine.

---

[37] Description of the Scheme provided here is mostly based on the information available at the following URL: http://www.hydro.com.au/home/Energy/Tasmania±Hydro±Electric±Scheme/Catchment±Areas/Gordon±Catchment/ (visited in September 2005).
[38] Information regarding geology, landforms, climate, hydrology, vegetation, habitat, forestry and cultural resources of the Lake Gordon–Lake Pedder catchment is available in Waterman (1980).
[39] Information regarding geology, landforms, climate, hydrology, vegetation, habitat, forestry and cultural resources of the Huon–Weld catchment is available in South West Tasmania Resources Survey (1979).
[40] Lake Pedder National Park was established in 1955.
[41] This represents the inner-basin water transfer component of the scheme.

Table H.6. *Selected characteristics of major dams of the Mersey–Forth Hydro-electric Scheme, in order of reservoir capacity*

No.	Name of the dam	Name of the lake	River	Year of completion	Type[a]	Height (m)	Reservoir capacity ($10^6 \, m^3$)	Surface area (km^2)
1	Devils Gate	Barrington	Forth	1969	VA	84	180	7
2	Rowallan	Rowallan	Mersey	1967	ER	43	131	9
3	Cethana	Cethana	Forth	1971	ER	110	109	4
4	Mackenzie	Mackenzie	Fisher	1972	ER	14	20	3
5	Paloona	Paloona	Forth	1971	ER	43	19	2
6	Parangana	Parangana	Mersey	1968	ER	53	15	1
7	Wilmot	Gairdner	Wilmot	1970	ER	34	9	1

Note: [a] VA (Concrete Arch); ER (Rockfill Embankment).
Source: National Registers of Large Dams.[42]

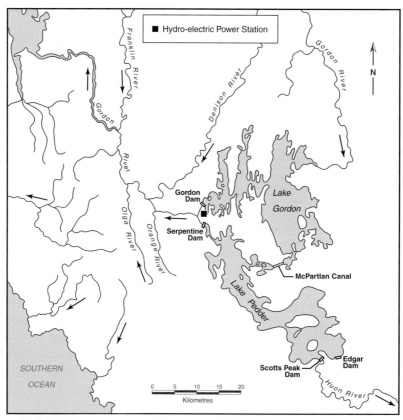

Figure H.7 Lake Gordon and Lake Pedder (Tasmanian Government, 2000d).

- The Mersey–Forth (Figure H.6), South Esk and Derwent schemes discharge into agricultural regions where availability of water for hydro-power versus irrigation as well as for environmental flow is an issue.
- The South Esk River discharges into the Trevallyn Power Station (Figure H.5). Here the interaction of power station operations with naturally occurring high sediment loads in the Tamar River is an issue for the local residents.
- The upper catchments of the Derwent and South Esk schemes (Figure H.4) are in Tasmania's Central Highlands, which is characterised by many lakes of differing sizes with highly valued recreational trout fisheries, as well as native fish species, a number of which are protected under threatened species legislation.
- Certain lakes in the upper catchments of the Derwent, South Esk, Gordon and Mersey schemes (e.g. Lakes Augusta, Pedder, and St Clair) are located in the

[42] http://www.ancold.org.au/dam_register.html (visited in September 2005).

Table H.7. *Selected characteristics of major dams of the Gordon River Scheme, in order of reservoir capacity*

No.	Name of the dam	Name of the lake	River	Year of completion	Type$^{(a)}$	Height (m)	Reservoir capacity (10^9 m^3)	Surface area (km^2)
1	Gordon	Gordon	Gordon	1974	VA	140	12.5	272
2	Serpentine	Pedder	Serpentine	1971	ER	38	3.3	242
3	Edgar	Pedder	Off stream	1973	TE	17	3.3	242
4	Scotts Peak	Pedder	Huon	1973	ER	43	3.3	242

Note: $^{(a)}$ VA (Concrete Arch); ER (Rockfill Embankment), TE (Earthfill Embankment).
Source: National Registers of Large Dams.[43]

Tasmanian Wilderness World Heritage Area. Therefore, protection of wilderness values, native species, and prevention of introduction of pest flora and fauna species are of high importance.

- Fish passage is an issue where power stations are located close to the estuarine regions. For example, Trevallyn on the South Esk River, Meadowbank on the Derwent River, and Paloona on the Forth River.

Over the past decades, particular events have called attention to environmental issues in isolated waterbodies (Locher, 1999). For example:

(1) the Lagoon of Islands experienced a major algal bloom in 1987 following construction of a diversion to increase the Lagoon's catchment area;
(2) the Pieman River experienced a significant fish kill in 1991 due to supersaturated dissolved oxygen levels in the discharge from the Reece Power Station; and
(3) the discharges of hydrogen sulphide following commissioning of the John Butters Power Station in 1992 occurred in the King River due to decomposition of organic materials deep in the newly formed lake.

These incidents were pivotal in the development of environmental policies and programs by Hydro Tasmania during the 1990s. These have improved its environmental performance with regard to waterway management. The Water Management Review program, for example, is a long-term program which assesses environmental and social issues associated with Hydro Tasmania's operations on a catchment-by-catchment basis. It makes a commitment to water management measures to address issues of concern and community values. Through this program, Hydro Tasmania is progressively considering environmental flow and other aquatic environmental issues throughout its operations by implementing a range of operational and capital works, including fish passage infrastructure, lake level operating rules, habitat enhancement programs, and development of recreational use guidelines.

The Mersey River has been seriously affected by construction of Rowallan and Parangana dams and water diversion to the Forth River (Figure H.6), with the average annual discharge at the Parangana Dam reduced to about 15 percent of the natural flow at this point. Hydro Tasmania undertook a major investigation in collaboration with all stakeholders, and now releases enough water from Lake Parangana so that the daily flow measured at Liena gauging station located about 10 km downstream of the dam to exceed 173 000 m^3 (equivalent to 2 m^3 s^{-1}). To assess the effectiveness of the flow release from Lake Parangana for environmental purposes, the Department of Primary Industries, in cooperation with Hydro Tasmania and the Inland Fisheries Service, have undertaken a comprehensive monitoring program since 1997 (DPIWE, 2005, p. 13). The aim of the monitoring has been to:

(1) determine the nature and extent of environmental benefits in the Mersey River from the releases from Lake Parangana; and
(2) monitoring changes in environmental conditions of the Mersey River due to releases from Lake Parangana.

The first stage of the program was undertaken during the period of 1997–2002 and involved an assessment of the response of biota (algae, macroinvertebrates and fish) to environmental releases. Stage 2 of this program is continuing with quantitative and qualitative assessment of macroinvertebrates and fish at different points along the river, and verifies the environmental benefit that has been brought about by this flow. The Water Management Plan developed under the Tasmanian Water Management Act for the Mersey River formalises the environmental flow release as well as addressing other community concerns in this river catchment (DPIWE, 2005).

Information regarding environmental flow assessments for 17 rivers in Tasmania including Derwent and its tributary Clyde, North Esk, and others are available at the URL provided in the footnote.[44] Estimation of the environmental flows have been based on a suite of methods, ranging from desktop studies in unstressed catchments to comprehensive studies of minimum flow requirements. New approaches that incorporate natural

---

[43] http://www.ancold.org.au/dam_register.html (visited in September 2005).
[44] http://www.dpiwe.tas.gov.au/inter.nsf/WebPages/JMUY-5F93LC?open (visited in November 2005).

variability in stream flow and the high flow water requirements are also being developed and used.[45]

References

Australian Bureau of Statistics (1978). *Tasmania Year Book 1978*. Hobart: ABS Tasmania Office.
Australian Bureau of Statistics (1981). *Tasmania Year Book 1981*. Hobart: ABS Tasmania Office.
Australian Bureau of Statistics (1999). *Tasmania Year Book 2000*. Hobart: ABS Regional Office.
Australian Water Resources Council (1987). *1985 Review of Australia's Water Resources and Water Use: Volume 1, Water Resources Data Set*. Canberra: Australian Government Publishing Service.
Blackers, A. (1994). Hydro-electricity in Tasmania revisited. *Australian Journal of Environmental Management* **1**: 110–120.
Costin, A. B., Greenway, M. A. and Wright, L. G. (1984). *Harvesting Water from Land*. Canberra: Centre for Resource and Environmental Studies. The Australian National University.
DPIWE (2005). *Mersey Water Management Plan*. Hobart: Department of Primary Industries, Water and Environment, Water Assessment and Planning Branch.
DIPNR (2004).[46] *Meeting the Challenges: Securing Sydney's Water Future*. Sydney: Department of Infrastructure, Planning and Natural Resources.
Eraring Energy (2004). *Eraring Energy: 2004 Annual Report*. Sydney: Eraring Energy.
Faircloth, P. (1977). *A Preliminary Review of Water Power Resources in the South West Study Area*. Sandy Bay, Tasmania: Steering Committee, South West Tasmania Resources Survey.
Frost, A. C. H. (1983). *Hydro-Electricity in Australia: Water 2000 Consultants Report No. 6*. Canberra: Australian Government Publishing Service.
GHD-FICHTNER (2005a).[47] *Planning for Desalination*. Sydney: GHD.
GHD-FICHTNER (2005b).[48] *Review of Environmental Factors for Desalination Pilot Testing*. Sydney: GHD.
Gibbs, G. A. (1915). *Water Supply Systems of the Melbourne and Metropolitan Board of Works*. Melbourne: D.W. Paterson.
Hale, G. E. A. ed. (1979). *Lower Gordon Region: Land Use, Resources and Special Features*. Hobart: Hydro-Electric Commission.
Healthy Rivers Commission (1998). *Independent Inquiry into the Hawkesbury–Nepean River System: Final Report*. Sydney: HRC of New South Wales.
Healthy Rivers Commission (1999). *Independent Inquiry into the Shoalhaven River Systems: Final Report*. Sydney: HRC of New South Wales.
Henry, F. J. J. (1939). *The Water Supply and Sewerage of Sydney*. Sydney: Halstead Press.
Hydro Tasmania (2004). *Hydro Tasmania: 2004 Annual Report*. Hobart: Hydro Tasmania.
Lindenmayer, D. and Beaton, E. (2000). *Life in the Tall Eucalypt Forests*. Sydney: New Holland Publishers.
Locher, H. (1999). Changing approaches to river management in the Tasmanian hydro electricity system. In *Proceedings of the Second Australian Stream Management Conference: The Challenge of Rehabilitating Australia's Streams, 8–11 February 1999, Adelaide, South Australia*. Clayton, Victoria: Cooperative Research Centre for Catchment Hydrology. pp. 395–400.
Lowe, D. (1984). *The Price of Power: The Politics Behind the Tasmanian Dams Case*. South Melbourne: Macmillan.
Melbourne, and Metropolitan Board of Works (1975). *Report on Environmental Study into Thomson Dam and Associated Works*. Volume 1, Proposed Works and Existing Environment: Volume 2, Environmental Implications and Conclusions. Melbourne: MMBW.
Melbourne Water (2003). *Our Precious Drinking Water*. Melbourne: Melbourne Water.
Osborne, M. and Dunn, C. (2004).[49] *Talking Water: An Australian Guidebook for the 21st Century*. Sydney: Farmhand Foundation.
Read, P. (1986). *The Organisation of Electricity Supply in Tasmania*. Hobart: University of Tasmania.
Saddler, H., Bennett, J., Reynolds, I. and Smith, B. (1980). *Public Choice in Tasmania: Aspects of the Lower Gordon Hydro-electric Development Proposal*. Canberra: Centre for Resource and Environmental Studies, the Australian National University.
Sammut, J. and Erskine, W. D. (1995). Hydrological impacts of flow regulation associated with the Upper Nepean Water Supply Scheme, NSW. *Australian Geographers* **26**(1): 71–86.
Smith, D. I. and Handmer, J. (1991). Water conflict and resolution: a case study of dams in southwest Tasmania. In Handmer, J. W., Dorcey, A. H. J., and Smith, D. I., eds. *Negotiating Water: Conflict Resolution in Australian Water Management*. Canberra: Centre for Resource and Environmental Studies, The Australian National University. pp. 47–62.
South West Tasmania Resources Survey (1979). *Huon–Weld Catchment*. Sandy Bay, Tasmania: Steering Committee, South West Tasmania Resources Survey.
Sydney Catchment Authority (2002). *Dams of Greater Sydney and Surrounds: Shoalhaven*. Sydney: SCA.
Tasmanian Government (2000a). *Tasmania South East, 1:250 000*. Hobart: Department of Primary Industries, Water and Environment.
Tasmanian Government (2000b). *Tasmania North East, 1:250 000*. Hobart: Department of Primary Industries, Water and Environment.
Tasmanian Government (2000c). *Tasmania North West, 1:250 000*. Hobart: Department of Primary Industries, Water and Environment.
Tasmanian Government (2000d). *Tasmania South West, 1:250 000*. Hobart: Department of Primary Industries, Water and Environment.
Tompson, P. (1981). *Power in Tasmania*. Hawthorn, Victoria: Australian Conservation Foundation.
Waterman, P. ed. (1980). *Lake Gordon–Lake Pedder Catchment*. Sandy Bay, Tasmania: Steering Committee, South West Tasmania Resources Survey.

---

[45] http://www.dpiwe.tas.gov.au/inter.nsf/WebPages/RPIO-4YD6S7?open (visited in November 2005).
[46] Available at the URL: http://www.dipnr.nsw.gov.au/waterplan/pdf/waterplanbroadband.pdf (visited in September 2005).
[47] Available at: http://www.sydneywater.com.au/EnsuringTheFuture/Desalination/index.cfm (visited in November 2005).
[48] Available at: http://www.sydneywater.com.au/EnsuringTheFuture/Desalination/index.cfm (visited in November 2005).
[49] Available at the URL http://www.farmhand.org.au/press.html (visited in October 2005).

I Selected Technical Features of the Central Valley Project in California

Table I.1. *Major divisions and units of the Central Valley Project*

No.	Divisions and Units	Websites[a]
1	American River Division, Folsom and Sly Park Units	http://www.usbr.gov/dataweb/html/folsom.html
2	American River Division, Auburn-Folsom South Unit	http://www.usbr.gov/dataweb/html/auburn.html
3	Delta Division	http://www.usbr.gov/dataweb/html/delta.html
4	Friant Division	http://www.usbr.gov/dataweb/html/friant.html
5	East Side Division, New Melones Unit	http://www.usbr.gov/dataweb/html/newmelones.html
6	Sacramento River Division, Sacramento Canals Unit	http://www.usbr.gov/dataweb/html/sacramento.html
7	San Felipe Division	http://www.usbr.gov/dataweb/html/sanfelipe.html
8	West San Joaquin Division, San Luis Unit	http://www.usbr.gov/dataweb/html/casanluis.html
9	Shasta/Trinity River Divisions	http://www.usbr.gov/dataweb/html/shasta.html

Note: [a] All websites visited in August 2005.

Table I.2. *Selected technical features of five major hydro-electric power plants of the Central Valley Project, in order of total capacity*

No.	Name	Number of generators	Capacity of each generator (kW)	Total capacity (kW)
1	Shasta Power Plant	7	2 × 143 750	619 850
			3 × 109 250	
			2 × 2300	
2	William R. Gianelli Generating Plant	8	53 000	424 000
3	New Melones Power Plant	2	150 000	300 000
4	Folsom Power Plant	3	76 176	228 530
5	Spring Creek Power Plant	2	86 250	172 500

Source: Autobee *et al.* (Unpublished, Appendix).

Table I.3. *Selected technical features of four major canals of the Central Valley Project, in order of length*

No.	Name	Type	Length (km)	Bottom width (m)	Maximum water depth (m)	Capacity (m^3 s^{-1})
1	Friant–Kern Canal	Concrete lined	243.3	11.0	4.7	113.3
2	Delta–Mendota Canal	Concrete lined	186.2	30.5	4.4	130.3
3	Tehama–Colusa Canal	Concrete lined	178.6	7.3	4.8	71.6
4	Contra Costa Canal	Concrete lined	76.7	7.3	2.0	9.9

Source: Autobee *et al.* (Unpublished, Appendix).

Table I.4. *Selected technical features of seven dams and reservoirs constructed by the U.S. Army Corps of Engineers in the Central Valley, in order of capacity*

No.	Name	Type	Year of completion	Height (m)	Capacity ($10^6 \, m^3$)	Surface area (km^2)
1	Pine Flat Dam and Lake	Concrete gravity	1954	131.1	1230	24.2
2	Isabella Dam and Lake	Earthfill	1953	56.4	700	56.7
3	New Hogan Dam and Lake	Earth and rockfill	1964	61.0	390	13.4
4	Buchanan Dam and H.V. Eastman Lake	Earth and rockfill	1975	62.5	185	7.2
5	Terminus Dam and Lake Kaweah	Earth and rockfill	1962	76.2	176	7.9
6	Hidden Dam and Hensly Lake	Earthfill	1974	46.7	111	6.4
7	Success Dam and Lake	Earthfill	1961	61.0	102	9.7

Source: Autobee *et al.* (Unpublished, Appendix).

Reference

Autobee, R., Simonds, W.J., Stene, E.A. and Whynot, W.E. (Unpublished). Central Valley Project: Historic Reclamation Projects Book. Denver, Colorado: U.S. Bureau of Reclamation.

J Selected Technical Features of the State Water Project in California

Table J.1. *Selected technical features of five major dams and reservoirs of the State Water Project, in order of capacity*

No.	Name	Type	Construction period	Height (m)	Capacity (10^6 m^3)	Surface area (km^2)
1	Oroville Dam and Lake	Earthfill	1962–68	234.7	4364	64.0
2	Sisk Dam and San Luis Reservoir	Earth and rockfill	1963–67	117.3	2502	50.7
3	Castaic Dam and Lake	Earthfill	1965–74	129.5	399	9.1
4	Pyramid Dam and Lake	Earth and rockfill	1969–73	121.9	211	5.3
5	Perris Dam and Lake	Earthfill	1970–74	39.0	162	9.4

Source: Department of Water Resources (1999, pp. 155 and 156).

Table J.2. *Selected technical features of five major pumping plants of the State Water Project, in order of total capacity*

No.	Name	Number of units	Normal static head (m)	Total capacity (m^3 s^{-1})	Total motor rating (HP)
1	Dos Amigos Pumping Plant	6	32.6–38.1	437.5	240 000
2	Castaic Pumping Plant	6	328.6	340.0	1 920 000
3	W. R. Gianelli Pumping Generating Plant	8	30.2–99.7	311.5	504 000
4	H. O. Banks Delta Pumping Plant	11	71.9–76.8	302.2	333 000
5	Thermalito Pumping Generating Plant	3	25.9–31.1	258.3	120 000

Source: Department of Water Resources (1999, p. 158).

Table J.3. *Length of channels, canals, pipelines, and tunnels of the State Water Project (in km)*

Facility	Channel	Canal	Pipeline	Tunnel	Total
North Bay Aqueduct	0.0	0.0	44.1	0.0	44.1
South Bay Aqueduct	0.0	13.5	52.9	2.6	69.0
California Aqueduct: Main Line	10.5	619.5	65.6	18.8	714.4
California Aqueduct: West Branch	14.8	14.6	10.3	11.6	51.3
California Aqueduct: Coastal Branch	0.0	24.1	157.6	4.3	186.0
Total	25.3	671.7	330.5	37.3	1064.8

Source: Department of Water Resources (1999, p. 159).

Reference

Department of Water Resources (1999). *California State Water Project Atlas*. Sacramento: DWR.

K Selected Characteristics of Some of the Completed or Proposed Inter-basin Water Transfer Projects in Australia, United States, Canada, China and India, in Chronological Order

Country	Name of project	Water transfer From	Water transfer To	Status	Year completed or proposed	Annual capacity (10^6 m^3)	Transfer method	Estimated cost of the project in 2002 values (US$ million)	Sectors benefiting from transfer	Reference
Australia	Goldfields Pipeline	Helena River	Kalgoorlie	Completed	1903	8.3 (initial) 28.8 (in 1997)	Pumping	200 (initial)	Municipal, mining and agriculture	Chapter 8
Australia	Morgan-Whyalla Pipelines	River Murray	Whyalla	Completed	1944 and 1966	75	Pumping	110	Municipal and industrial	Chapter 10
Australia	Mannum-Adelaide Pipeline	River Murray	Adelaide	Completed	1954	139	Pumping	NA	Municipal	Chapter 10
Australia	Swan Reach-Paskeville Pipeline	River Murray	Paskeville	Completed	1969	29	Pumping	NA	Municipal	Chapter 10
Australia	Tailem Bend-Keith Pipeline	River Murray	Keith	Completed	1969	11	Pumping	NA	Municipal	Chapter 10
Australia	Mareeba-Dimbulah	Barron River	Walsh River	Completed	1970	NA	Gravity	180	Irrigation	Chapter 10
Australia	Murray Bridge-Onkaparinga	River Murray	Onkaparinga River	Completed	1973	188	Pumping	110	Municipal	Chapter 10
Australia	Mersey-Forth Scheme	Mersey River	Forth River	Completed	1973	830	Gravity	NA	Hydro (308 MW)	Appendix H
Australia	Snowy Mountains Hydro-electric Scheme	Snowy River	Murray and Murrumbidgee rivers	Completed	1974	1140	Gravity	7000	Irrigation and hydro$^{(f)}$ (3756 MW)	Chapter 4
Australia	Shoalhaven Scheme	Shoalhaven River	Nepean and Wollondilly rivers	Completed	1977	284	Pumping	370	Municipal and hydro (240 MW)	Appendix H
Australia	Great Lake	Derwent River Catchment	Tamar River Catchment	Completed	1977	698	Gravity	NA	Hydro (380 MW)	Appendix H

(continued)

Appendix K (cont.)

Country	Name of project	Water transfer From	Water transfer To	Status	Year completed or proposed	Annual capacity (10^6 m^3)	Transfer method	Estimated cost of the project in 2002 values (US$ million)	Sectors benefiting from transfer	Reference
Australia	Thomson Scheme	Thomson River	Yarra River	Completed	1984	148	Gravity	890	Municipal	Appendix H
Australia	Gordon Scheme	Huon River, lakes Pedder and Gordon	Gordon River	Completed	1988	NA	Gravity	NA	Hydro (432 MW)	Appendix H
Australia	Flooding Lake Eyre	Sea	Lake Eyre	Proposed	1880s	NA	Gravity	NA	Climate and Rainfall	Chapter 7
Australia	Bradfield Scheme	Tully River	Thomson River	Proposed	1938	NA	Gravity	NA	Irrigation	Chapter 6
Australia	Great Boomerang	Queensland Coastal rivers	Inland Queensland rivers	Proposed	1941	NA	Gravity	NA	Irrigation and hydro	Chapter 7
Australia	Reid Scheme	Mitchell River	Diamantina River	Proposed	1946	NA	Gravity	NA	Irrigation	Chapter 6
Australia	Clarence Scheme (CLA-7C)	Clarence River	Border Rivers	Proposed	Pre 1980	755	Pumping and gravity	1230	Irrigation	Chapter 5
Australia	Newton Boyd Scheme	Clarence River	Border Rivers	Proposed	1985	1100	Gravity	2000	Irrigation and hydro (450 MW)	Chapter 5
Australia[a]	Water Research Foundation Scheme	Clarence River	Border Rivers and Gwydir River	Proposed	1985	2000	Gravity	2500	Irrigation and hydro (3000 MW)	Chapter 5
Australia	Kimberley Pipeline	Kimberley	Perth metropolitan	Proposed	1988	300	Pumping	9000[e]	Municipal	Chapter 9
United States	Los Angeles Aqueducts	Owens River	Los Angeles	Completed	1913 and 1970	594	Gravity	865[e]	Municipal and hydro	Chapter 11
United States	Strawberry Valley Project	Strawberry River	Diamond Fork River	Completed	1918	NA	Gravity	NA	Irrigation and hydro	Chapter 11
United States	Hetch Hetchy Aqueduct	Tuolumne River	San Francisco	Completed	1934	440	Gravity	1400[e]	Municipal	Chapter 11
United States	Mokelumne Aqueducts	Mokelumne River	San Francisco	Completed	1929, 1949 and 1963	NA	Gravity	NA	Municipal	Chapter 11
United States	Colorado River Aqueduct	Parker Dam	Los Angeles and San Diego	Completed	1941	1500	Pumping	NA	Municipal	Chapter 11
United States	Colorado-Big Thompson	Colorado River	Big Thompson River	Completed	1959	382	Gravity and pumping	NA	Irrigation, municipal and hydro	Chapter 11
United States	Central Valley Project	Sacramento River Catchment	San Joaquin River Valley	Completed	Late 1960s	8600	Gravity and pumping	NA	Irrigation, municipal, hydro,	Chapter 11

United States	San Juan-Chama Project	San Juan River, Colorado	Rio Chama, and Rio Grande, New Mexico	Completed	1970	137	Gravity	210$^{(e)}$	Irrigation and municipal	Chapter 11
United States	State Water Project	Sacramento River Catchment	San Joaquin River Valley, Central Coast and Southern California	Completed	1972	3000	Gravity and pumping	11 250$^{(e)}$	Municipal, irrigation, hydro, industry, flood control, fish and wildlife	Chapter 11
United States	Fryingpan-Arkansas Project	Fryingpan River	Arkansas River	Completed	1978	85	Gravity	NA	Irrigation, municipal and hydro	Chapter 11
United States	Central Utah Project	Strawberry and other tributaries of the Green River	Utah Lake Catchment	Under construction					Irrigation, municipal and hydro (200 MW)	NA
NA	Gravity	NA	Irrigation, municipal, industry, hydro, flood control, fish and wildlife	Chapter 11						
United States	Snake-Colorado Project	Snake River	Colorado River, South Pacific and Coastal Plain	Proposed	1963	3000	Gravity and pumping	NA	Irrigation, municipal and others	Chapter 11
United States	North American Water and Power Alliance (NAWAPA)	Rivers from Alaska and Western Canada	Two provinces in Canada, 22 states in USA and three states in Mexico	Proposed	1964	220 000 consisting of: 31 000 for Canada, 165 000 for USA, and 24 000 for Mexico	Gravity and pumping	600 000$^{(e)}$	Irrigation, municipal, industrial, hydro and others	Chapter 11
United States	Modified Snake-Colorado Project	Snake River and Columbia River	Great Basin, Snake River, South Pacific, Coastal Plain and Colorado River	Proposed	1965	6200	Gravity and pumping	NA	Irrigation, municipal and others	Chapter 11

(continued)

Appendix K (cont.)

Country	Name of project	Water transfer From	Water transfer To	Status	Year completed or proposed	Annual capacity (10^6 m^3)	Transfer method	Estimated cost of the project in 2002 values (US$ million)	Sectors benefiting from transfer	Reference
United States	Undersea Aqueduct Systems	North Coast and Pacific Rivers	Central Valley, South Pacific and Coastal Plain	Proposed	1967	13 600	Gravity and pumping	NA	Municipal, irrigation and others	Chapter 11
United States[b]	Beck Plan	Missouri River	Texas High Plains	Proposed	1967	12 300	Gravity and pumping	NA	Irrigation, municipal and others	Chapter 11
Canada	Long Lake Diversion	Albany River	Lake Superior	Completed	1939	1419	Gravity	NA	Hydro and transport of pulpwood	Chapter 12
Canada	Ogoki Diversion	Ogoki River	Lake Superior	Completed	1943	3564	Gravity	NA	Hydro	Chapter 12
Canada	Nechako-Kemano Diversion	Nechako River	Kemano River	Completed	1954	3630	Gravity	NA	Hydro (896 MW)	Chapter 12
Canada	Churchill Falls (Newfoundland)	Julian-Unkown, Naskaupi and Kanairiktok rivers	Churchill River	Completed	1971	16 600	Gravity	NA	Hydro	Chapter 12
Canada	Churchill-Nelson Diversion (Manitoba)	Churchill River	Nelson River	Completed	1976	24 440	Gravity	NA	Hydro (3932 MW)	Chapter 12
Canada[c]	La Grand Project	Caniapiscau, Eastmain, Opinaca and Sakami rivers	La Grande River	Completed	1996	53 900	Gravity	18 000	Hydro (15 550 MW)	Chapter 12
Canada	GRADN Canal	James Bay	Lake Huron, Lake Superior, Western Canada, Western USA and Mexico	Proposed	1965	36 000	Pumping and gravity	130 000[e] to 210 000[e]	Irrigation, municipal and hydro	Chapter 12
Canada	Kuiper Diversion Scheme	Mackenzie River Basin	Lake Winnipeg, Great Lakes and Southern USA	Proposed	1966	NA	Pumping and gravity	NA	Irrigation, municipal and hydro	Chapter 12
Canada	Central North American Water Project	Great Bear, Great Slave and other lakes	Lake Superior, Missouri River and Rio Grande	Proposed	1967	NA	Pumping and gravity	NA	Irrigation, municipal and hydro	Chapter 12

Country	Scheme	Water source	Destination	Status	Year	Distance (km)	Transfer method	Capacity (Mm³/year)	Purpose	Reference
Canada	Western States Water Augmentation Concept	Mackenzie River	Western and southern United States	Proposed	1968	47000	Pumping and gravity	400 000[e]	Irrigation, municipal and hydro	Chapter 12
Canada	Magnum Scheme	Finlay, Peace, Athabasca and Saskatchewan rivers	Missouri River in the United States	Proposed	1960s	31000	Gravity and pumping	NA	Irrigation, municipal and hydro	Chapter 12
China[d]	Wanjiazhai Water Transfer Project	Yellow River	Taiyum City	Completed	2005	1400	Pumping and gravity	NA	Municipal	Chapter 13
China	Eastern Route Scheme	Yangtze River	Yellow River Basin and Tianjin Municipality	Under construction	Since 2002	9000	Pumping and gravity	15000	Municipal, irrigation and the environment	Chapter 13
China	Middle Route Scheme	Hanjiang River (tributary of the Yangtze River)	Beijing and Tianjin municipality	Under construction	Since 2002	10000	Gravity	13000	Municipal, irrigation and the environment	Chapter 13
China	Western Route Scheme	Headwaters of the Yangtze River	Yellow River	Proposed	2002	17000	Pumping	37000	Municipal, irrigation and the environment	Chapter 13
India	Kurnool Cuddapah Canal	Krishna Basin	Pennar Basin	Completed	1863	2670	Gravity	NA	Irrigation and municipal	Chapter 14
India	Periyar Project	Periyar River catchment	Vaigai River catchment	Completed	1895	1290	Gravity	NA	Irrigation, municipal and hydro (140 MW)	Chapter 14
India	Parambikulam Aliyar Project	Five west and two east flowing rivers	Tamil Nadu and Kerala	Completed	1960s	NA	Pumping	NA	Irrigation, municipal and hydro (185 MW)	Chapter 14
India	Teluga Ganga Project	Krishna River in Andhra Pradesh	Chennai (Madras)	Completed	NA	2136	Gravity	NA	Municipal and irrigation	Chapter 14

Appendix K (*cont.*)

Country	Name of project	Water transfer From	Water transfer To	Status	Year completed or proposed	Annual capacity ($10^6 \, m^3$)	Transfer method	Estimated cost of the project in 2002 values (US$ million)	Sectors benefiting from transfer	Reference
India	Ravi-Beas-Sutlej Link	Ravi and Beas rivers	Sutlej River	Completed	NA	9360	Gravity	NA	Irrigation and municipal	Chapter 14
India	Ganga-Tamil Nadu Canal	Ganga River in the State of Bihar	Cauvery River in the State of Tamil Nadu	Proposed	1972	22 000	Pumping and gravity	33 000	Irrigation and municipal	Chapter 14
India	Himalayan and Garland Canals	(1) Ravi River; (2) Northern and Western Peninsular rivers	(1) Brahmaputra (2) Central, Eastern and Southern rivers	Proposed	1977	(1) 617 000 (2) 864 000	Pumping and gravity	15 500 000	Irrigation and municipal	Chapter 14
India	National River-Linking Project	(1) Himalayan rivers; (2) Peninsular rivers	(1) Ganga, Yamuna, Sabarmati, Mahanadi and others; (2) Peninsular rivers	Proposed	1980	200 000 to 300 000	Gravity and pumping	112 000 to 200 000	Irrigation and municipal	Chapter 14

Note: [a] For information about numerous other proposals regarding inland transfer of water from various coastal catchments of New South Wales see Chapter 5 and Appendix C; [b] For information about a number of other large-scale water diversion proposals developed in the 1960s see Table 11.6; [c] For information about other completed diversion projects in Canada see Table 12.9; [d] For information about other completed diversion projects in China see Table 13.6; [e] In 2004 values. [f] Hydro (Hydro-power).

Glossary

Allocation: An entitlement of water, specified in volumetric terms, often for a particular purpose over a specified period.

Anadromous: Any fish that spends a portion of its life cycle in freshwater and a portion in the sea.

Appropriation: The establishment of a water right through the intent and activities that demonstrate intent to capture, possess, and control water for beneficial use.

Aqueduct: A conduit (such as a pipeline) or artificial channel (such as concrete-lined or unlined canal) for transporting water.

Aquifer: An underground water-bearing layer of permeable rock, sand or gravel that is capable of yielding exploitable quantities of water to wells or springs.

Arable land: Land that is, or has the potential to be cultivated for crop production.

Assessment: An objective evaluation but not necessarily quantified or verified by experiment.

Australian Height Datum (AHD): The datum surface, which passes through mean sea level at 30 tide gauges around the Australian mainland coastline. It was adopted in May 1971 as the datum to which all vertical control for mapping is to be made.

Basin: A depression surrounded by elevated lands, which is drained by a river and its tributaries.

Basin of origin: The basin or watershed in which developed water originates.

Beneficial use: Diversion of a reasonable and appropriate amount of water used under reasonably efficient practices to accomplish without waste the purpose for which the appropriation is lawfully made. Beneficial use includes use of water for agriculture, municipal, mining, manufacturing, and aquaculture.

Benthic organisms: Organisms that live at the bottom of a waterbody or in the sediments.

Brackish water: Water with soluble salts content of greater than 1500 mg L^{-1} TDS but less than 5000 mg L^{-1} TDS.

Capital expenditure: Funds spent for permanent facilities and additions or improvements to plants or equipment.

Cap on diversion: A term that specifies the maximum volume of water that can be diverted from natural flows over a specified period.

Catchment: The entire area drained by a river and its tributaries.

Cloud seeding: A technique that adds substances (most commonly silver iodide) to clouds in order to alter their natural development and promote precipitation in a designated area.

Compact: A contract or agreement between two or more states and with the consent of the federal government. In the matter of water, compacts define the relative rights of two or more states on an interstate river to use the waters of that river.

Conjunctive use: A water management strategy that combines the use of surface and groundwater resources.

Consumptive use: The portion of water withdrawn that is used for crop production, industrial purposes, or consumed by humans or livestock.

Contingent valuation: A method, usually in the form of a survey questionnaire, of eliciting values for hypothetical goods or services. The method is used to estimate the value for certain environmental goods.

Cost–benefit analysis: A methodology for determining whether a project or activity generates a positive net benefit for society by evaluating all the costs and benefits over time.

Dam: A barrier built across a valley or river to store water.

Deflation: The removal of material from a beach or other land surface by wind action.

Depletion: Use of water in a manner that makes it no longer available.

Desalination: Removal of salt from water.

Developed water: Water that is collected using dams, reservoirs, conduits, tunnels, or other constructed facilities.

Discounting: Method by which future costs and benefits are converted into current values.

Discount rate: A numerical value that serves to convert a future value to a present value.

Divertible water resources: The average annual volume of water, which using current practice, could be removed from developed or potential surface water or groundwater resources on a sustainable basis at rates capable of serving urban, irrigation, industrial or extensive stock use. It does not include low yielding bores providing domestic or supplies by low yielding pumps such as windmills or surface water resources such as roof run-off or small farm dams.

Dryland salinity: Salinisation of land in non-irrigated areas caused by capillary rise from shallow watertables, which results in deposition of salt on the land surface.

Earthfill dam: An embankment type dam in which more than 50 percent of the total volume is formed of compacted fine-grained material.

Ecosystem: A dynamic complex of plants, animal and microorganism communities and their non-living environment, interacting as a functional unit.

Electrical conductivity (EC): The ability of a soil, water sample or solution to conduct electricity. It is proportional to the concentration of salts dissolved in solution. In Australia, it is measured in units of $\mu S\,cm^{-1}$ (microsiemens per centimetre). Approximately, $1\,\mu S\,cm^{-1}$ corresponds to about $0.6\,mg\,L^{-1}$ total dissolved solids.

Endemic: Native to a particular area and found nowhere else, or having originated in the region where it is now found.

Endocrine systems: A set of glands (hypothalamus, pituitary, thyroid, parathyroid, pancreas, adrenal, testicles and ovaries) that release hormones into the bloodstream.

Endocrine disruptors: Synthetic chemicals found in pesticides and industrial products such as DDT, polychlorinated biphenyls, and dioxins having the potential to disrupt normal function of endocrine systems in humans and wildlife. Health effects include cancers, infertility, birth defect, and behavioural problems.

Entitlement: A right to water, conferred by a state or federal government department.

Environmental impact: An alteration to the environment or any component of the environment, including economic, social or cultural aspects.

Eutrophication: Process by which waters become enriched with nutrients, primarily nitrogen and phosphorus, which stimulate the growth of algae.

Evaporation: The physical process for the change of water from a liquid to vapour form and its subsequent loss to atmosphere.

Evapotranspiration: The process of water vapour transfer into the atmosphere from vegetated land surface. It includes water evaporated from the soil surface and water transpired by plants.

Fallow: A phase when land is left bare before being actively cropped.

Fauna: The entire animals living within a given area or environment or during a stated period. In Roman mythology, the grandson of the God Saturn worshipped as God of the fields and of shepherds.

Floodplain: An area adjacent to a river or watercourse that is subject to flooding.

Flora: Aggregate of plants growing in a particular region or period. In Roman mythology, Goddess of flowers and springtime.

Freshwater: Water with soluble salts content of less than 1500 mg L^{-1} TDS.

Front Range: The portion of Colorado situated immediately east of the Rocky Mountains.

Groundwater: Sub-surface water contained in a saturated zone of soil or geological strata and capable of moving in response to gravity and hydraulic pressure gradient.

Gully: A small valley cut by running water.

Gully erosion: A form of erosion involving the formation of deep, steep-sided channels or gullies.

Habitat: The place where an animal or a plant normally lives and produces.

Headwaters: The upper parts of a river system.

Holistic: An approach that seeks to study the whole of a system, which may be "more than the sum of its parts". It extends to examine the linkages and interactions between parts of the system. This contrasts with a single-discipline, reductionist focus.

Hydrology: The science that treats the waters of the earth, their occurrence, circulation, distribution, their chemical, physical and biological properties, and their reaction with the environment, including the living things.

Incidence analysis: A method that disaggregates the overall impacts of the options according to the impact on individual community groups. The disaggregation is commonly undertaken in terms of the income grouping of those affected by a specific development. It is not an alternative to cost–benefit analysis but rather provides information on the distribution of benefits and costs.

Indigenous people: Native peoples distinguished by their close attachment and economic dependence upon ancestral lands.

Input–output analysis: A method of modelling an economy by specifying the relationship between inputs, intermediate outputs, outputs and final demands in an economy.

Instream water use: Water that is used, but not withdrawn from a surface water source for such purposes as hydro-electric power generation, navigation, fish propagation and recreation.

Inter-basin water transfer: Transfer of water from a river basin (or catchment) to another one by gravity or pumping, using facilities such as weirs, dams, canals, pipelines, and tunnels.

Internal rate of return: The discount rate in cost–benefit analysis at which the present value of benefits equal the present value of costs.

Karst: A region underlined by soluble carbonate rocks (limestone or dolomite) and characterised by distinctive features such as sinkholes, conduits, and caverns caused by dissolution.

Kilowatt hours (kWh): An electrical energy or work unit equal to 1000 Watts of power supplied to or taken from an electric circuit steadily for 1 hour.

Mitigation: An action designed to lessen or reduce adverse impacts due to a project's implementation or environmental problem.

Multi-criteria analysis: A group of methods for evaluation of problems, usually with multiple conflicting criteria. Each method has its own underlying principles, data requirements, and ways of incorporating preferences, data processing, generating solutions and presenting results.

Multiple objective programming: A valuable method in the assessment of options, which have several objectives which cannot be quantified in monetary terms. The method uses mathematical programming techniques to select projects based on explicit objectives.

Net present value: The sum of the discounted net benefits (benefits minus costs) in each time period, over the lifetime of the investment.

Overdraft: The condition of a groundwater basin in which the amount of water withdrawn by pumping exceeds the amount of water that recharges the basin.

Permanent crops: Crops such as cocoa, coffee and rubber that occupy the land for long periods and need not be replanted after each harvest.

Permafrost: Permanently frozen ground, found in Alpine, Arctic and Antarctic regions.

Physiographic region: A large area of the Earth's crust that has the following distinct characteristics: (1) it extends over a large contiguous area with similar relief features; (2) its landform has been shaped by a common set of geomorphic processes;

and (3) it possesses a common geological structure and history.

Piezometer: A tube inserted in and sealed into the ground but with the bottom open. The pressure of water in the soil at the bottom of the tube causes water to rise to a height in the tube, which is a measure of the hydraulic pressure at the bottom of the tube.

Piezometric surface: A surface defined by the level to which groundwater stands in piezometers.

Playa: The sandy, salty, or muddy floor of a desert basin with interior drainage, usually occupied by a shallow lake during the rainy season or after prolonged, heavy rains.

Pollution (of water): The alteration of the physical, chemical, or biological properties of water by the introduction of any substance into water that adversely affects any beneficial use of water.

Potable water: Water with soluble salts content of less than $500\,\text{mg}\,\text{L}^{-1}$ TDS.

Precipitation: The deposition of water in form of solid (snow) or liquid (rain) on the earth's surface from the atmosphere.

Present value: The value of returns in the future, converted into current dollars.

Prior appropriation doctrine: A legal concept in which the first person to appropriate water and apply it to beneficial use has the first right to use that amount of water from that source. Each successive appropriator may only take a share of the water remaining after all senior water rights are satisfied. This is the historical basis for Colorado water law and is sometimes known as the Colorado Doctrine or the principle of "first in time, first in rights".

Progradation: The outward building of a sedimentary deposit, such as the seaward advance of a delta or shoreline.

Pump-generating plant: A plant, which can either pump water or generate electricity, depending on the direction of water flow.

Recharge: The process by which water is added to a groundwater system either naturally or artificially.

Reclaimed water: Wastewater that has been treated so that it can be reused for most purposes.

Regulated river system: A river system, in which the natural flows are managed by major structures such as dams, weirs, distribution channels, inter-basin water transfer facilities and others.

Reliability: The high probability that a specified amount of water is available or can be delivered to or taken by a user in a specified period.

Reservoir: A man-made artificial lake, pond, tank, or basin into which water flows and is stored for future use.

Resettlement: Physical relocation of people whose home, land or common property resources are affected by a development, such as dam building.

Return flow: Water that reaches a surface water source after release from the point of use, thus becoming available for further use.

Reuse: The additional use of previously used water.

Reverse osmosis: A method to remove salts and other constituents from water by forcing water through membranes.

Rill: A shallow gutter or very small brook.

Rill erosion: A form of erosion involving formation of shallow gutters.

Riparian: Located on the banks of a stream or other body of water.

Riparian water rights doctrine: A legal concept in which owners of lands along the banks of a river/stream or water body have the right to reasonable use of the waters. The right is appurtenant to the land and does not depend on prior use. Riparian rights are not recognised in Colorado but are common in the eastern United States. They are used in California along with prior appropriation rights.

Risk–benefit analysis: A type of risk assessment whereby the risks or costs associated with an activity are compared to the potential benefits.

Rockfill dam: An embankment type dam in which more than 50 percent of the total volume is comprised of compacted or dumped cobbles, boulders, rock fragments, or quarried rock.

Run-off: That part of rainfall that flows off the land surface into the drainage system.

Safe yield: Rate of surface water diversion or groundwater extraction from a basin for consumptive use over an indefinite period of time that can be maintained without producing negative effects.

Salinisation: Accumulation of salts at soil surface or in the root zone of plants, usually due to capillary rise of saline water from a shallow watertable.

Saline water: Water with soluble salts content of greater than $5000\,\mathrm{mg\,L^{-1}}$ TDS.

Service area: The geographic area served by a water agency.

Sheet erosion: The removal of a fairly uniform layer of soil from the land surface by raindrop splash and/or run-off.

Siltation: Deposition of suspended sediments from water into channels, harbours, and reservoirs.

Surface water: Water that flows or lies on the ground surface.

TDS: See total dissolved solids.

Tillage: Mechanical disturbance of the soil by using various implements to alter the soil structure, usually done to create a seedbed, kill weeds or increase water entry.

Total dissolved solids (TDS): A quantitative measure, usually expressed in milligrams per litre ($\mathrm{mg\,L^{-1}}$) of the residual minerals dissolved in water that remain after evaporation.

Transboundary water: Water that crosses between, or is shared by nations or sub-national political units.

Transpiration: Release of water to the air by plants.

Tributary: A stream that flows into a larger stream or other body of water.

Third party effect: The impacts of water trade on others such as the environment, other consumptive water users or the community.

Unregulated river system: A river system which is not affected by major dams or other structures, but its flow regime could be far from natural due to changes in land use, diversion by pumping or by gravity, farm dams, levee systems, and others.

Wastewater: Used water from domestic (toilets, sinks, bathrooms, washing machines, and dishwashers) and industrial activities.

Water Conservancy District: A special unit of local government with authority to tax and incur bonded indebtedness in order to provide water supply services for persons or entities within its boundaries.

Water conservation: The efficient use of water on farms, in homes, or by industry by reducing water consumption, eliminating losses, and recycling wastewater.

Water quality: Description of the chemical, physical, and biological characteristics of water, usually in regard to its suitability for a particular purpose or use.

Water recycling: The treatment of wastewater to a level rending it suitable for a specific beneficial use.

Water right: A legally protected right, granted by law, to use a certain portion of the waters in a water supply system for beneficial purposes.

Watershed: The area or region drained by a river, stream, or reservoir.

Watertable: The upper surface of the saturated zone in an unconfined aquifer.

Water trading: Marketing arrangement that can include the permanent sale of a water right by the water right holder; a lease of the rights to use water from water rights holder; and the sale or lease of a contractual right to water supply.

Water withdrawal: Water removed from a groundwater reservoir or diverted from a surface water source for use.

Weir: A structure built across an open channel to raise the upstream water level or divert flow.

Wetland: The land area alongside rivers or fresh and saltwater bodies that is flooded all or part of the year.

Index

Aboriginal 38, 39, 64, 177, 178, 390, 394, 395
Acidification 19, 71, 74
Acid rain 305
Acid sulphate soils 72, 74, 123
Action Plan 76, 314
Adelaide 73, 76, 77, 134, 169, 174, 180, 397
Agreement 40, 75, 104, 123, 228, 251, 290, 341, 355
Agricultural 16, 17, 44, 65, 76, 81–3, 94, 122, 171, 183, 206, 224, 232, 242, 296, 304, 305, 386
Agricultural land 13, 15, 107, 126, 207, 228, 240, 247, 284, 325, 328
Agricultural production 96, 107, 110, 122, 216, 235, 264, 387, 388
Agriculture 10, 15, 59, 64, 68, 83, 107, 123, 125, 165, 169, 204, 209, 215, 223, 263, 275, 295, 298, 302, 312, 314, 319, 321, 335, 403
Airborne Laser Scanning 24
Alamo Canal 226, 227
Albany 151, 155, 365
Albany River 276
Alberta 211, 275, 290, 291
Albuquerque 251, 252
ALCAN See Aluminium Company of Canada
Algal 109, 187
Algal bloom 30, 74, 77, 239, 403, 411
Algeria 379
All-American Canal 217, 226–8, 243
Alpine 55, 56, 92, 103–5
Alternative 27, 238, 357
Aluminium Company of Canada 277, 279
Alva B. Adams Tunnel 245
Anadromous fish 231
Andhra Pradesh 331, 340
Aquaculture 72, 109, 110, 313
Aquatic 22, 32, 34, 72, 86, 101, 206, 273, 283, 313, 338
  ecosystem 33, 109, 209, 273
  species 32, 34, 240, 274
Arable land 14, 15, 19, 53, 62
Arbitration 43
Argyle Diamond Mine 387, 388
Arid 16, 54, 62, 75, 140, 148, 203, 250, 297, 312, 319, 320, 325
Arkansas River 247, 248
Arsenic 64, 326
Artificial recharge 238, 329
Asia 4–6, 169, 314, 387
Asian Development Bank 337
Assam 326
Assessment 23, 25–8, 36, 43, 45, 86, 111, 121, 122, 171, 335, 342

Athabasca River 286, 289
Atlantic Ocean 180, 261
Australia 34, 35, 38, 40, 51, 55, 61, 65, 67, 94, 97, 110, 268, 275, 346, 350, 353, 397
Australian National University 364, 366
Avon Dam 401
Avon River 397

Bailey, Arthur 151
Bangladesh 323, 336
Bank 312, 329, 363, 368
Banking 363
Barrier Ranges and Broken Hill Water Supply Syndicate 191
Barron River 184, 186, 187
Barron Water Resources Plan 187, 188
Bathurst 362, 364
Bay of Bengal 319
Bay of Triton 379–81
Beal, Jack 119
Beck Plan 211
Beijing 296, 297, 300, 301, 307–9, 312
Benefit 22, 23, 27, 35, 77, 83, 84, 102, 105, 110, 111, 116, 122, 123, 126, 147, 171, 237, 249, 276, 290, 292, 312–14, 387, 395, 405, 411
Benthic organism 29
Benthic species 109
BHP See Broken Hill Proprietary Company Limited
Bihar 326, 331
Biodiversity 19, 44, 71
Biota 32, 33
Biskra 379
Black Buck 338
Blainey, Geoffrey 157
Blowering Dam 96
Blue-green-alga 74
Bonneville Unit 255, 256
Boundary Waters Treaty 42, 273, 291, 292
Bowen River 189
Bowen Shire 189
Brackish 110, 151, 177, 215, 238, 253, 329
Bradfield, Bill 362
Bradfield, John, J.C. 125–8, 361
Bradfield Scheme 123, 125, 126, 128, 129, 131, 132, 149
Brahmaputra River 319, 323, 326, 327, 330, 331, 335, 336
Brazil 391
Bridge, Ernie 176, 393
Brine:
  discharge 238
  disposal 355, 358
Brisbane 361, 362

British 39, 321, 362, 365, 405
  Army engineers 321
  Columbia 271, 275, 290, 291
Broken Hill 157, 180, 189, 195, 196, 367
  Proprietary Company Limited 189, 190, 194
  Water Board 194
  Water Supply Act 192
  Water Supply and Sewerage Act 194
  Water Supply Company 192–4
  Water Supply Syndicate 191, 192
Broom 170, 393
Burdekin River 125, 126, 128, 129, 188, 189
Burrinjuck Dam 361

Cairns 184, 186–8
CALFED 235, 236
CALFED-Bay Delta Program 229–31, 351
California 30, 38, 202, 204, 205, 207–9, 215
  Aqueduct 224, 290
  Bay-Delta Authority 232
  Bay-Delta Program 240
  Development Company 226
  Irrigation Management Information System 235
  Seven Party Agreement 241
  Urban Water Conservation Council 236
  Water Code 229
  Water Plan 216, 229, 240
  Water Resources Development Bound Act 224
Canada 38, 209, 211, 261, 349, 351
Canada Water Act 272
Caniapiscau River 280
Canning Basin 393
Cap 76–8, 87
Capacity 240, 300, 309, 323, 384, 398
Captain Arthur Lawley 370
Captain Dasture 331
Captain François Elie Roudaire 379
Captain James Stirling 151
Cardinia Reservoir 403
Cataract Dam 361, 401
Cataract River 397
Cauvery River 331, 340
Cauvery Water Disputes Tribunal 340
C-BT Project See Colorado-Big Thompson Project
CeNAWP See Central North American Water Project
Central Arizona Project 99, 242
Central North American Water Project 284, 286
Central Utah Project 243, 253, 254, 256, 351
Central Utah Project Completion Act 255, 256

Central Utah Water Conservation District 255, 256
Central Valley 201, 208, 211, 233
Central Valley Project 216, 217, 222, 224, 231–3, 246, 348
Central Valley Project Improvement Act 223, 232, 351
Chaffy, George 226
Chambal River 341
Charles H. Boustead Tunnel 247
Chemicals 74, 266, 273, 275, 303, 313, 325, 394
Chenab River 331
Chennai 330, 331, 338
Cherrapunjee 321
Chhattisgarh 340
Chicago Diversion 281
Chicago River 281
Chiefly, Ben 92, 95, 362, 366, 368
Chiefly, Patrick 362
China 7, 9, 13, 19, 22, 270, 295, 319, 345, 349, 352, 357
Chott Melrhir 379
Chott Rharsa 379
Churchill Falls 270
Churchill-Nelson Diversion 279, 346, 349
Churchill River 270, 279, 285, 286, 289
Clarence River 107, 109–11, 113, 117, 121–3
Clean Water Act 206
Climate 54, 83, 92, 131, 139, 140, 215, 262, 270, 295, 297, 320, 379
Climate change 12, 13, 24, 25, 56, 81, 82, 86, 87, 166, 167, 208, 209, 216, 230, 239, 265, 270, 305, 316, 326, 330, 402
Cloud seeding 104, 239, 257, 355
Coachella Canal 217, 228, 236
Coachella Valley 226, 227
COAG See Council of the Australian Governments
Coffey, David, D. 117
Colonel Robert Bradford Marshall 223
Colorado 207, 244, 249
Colorado-Big Thompson Project 243, 244, 246, 247
Colorado River 99, 211, 216–8, 226, 228, 232, 247, 250, 253
   Aqueduct 217, 226, 229, 243, 348
   Basin 209, 210, 240, 285
   Basin Project Act 242
   Basin Salinity Control Act 242
   Basin Salinity Control Program 242
   Compact 42, 240, 355
   Salinity 242
Colorado Springs 248
Colorado Water Conservation Board 249, 250
Columbia River 211, 286
Commonwealth Bank 363
Communist Party 364
Community 76, 85, 86, 101, 102, 105, 107, 110, 121–3, 167, 168, 171, 236, 237, 257, 301, 314, 327, 338, 340, 343, 394, 403
Compensation 43, 212, 284, 314, 328
Condenser 151
Conflict 27, 34, 39, 40, 43, 85, 212, 213, 242, 276, 290, 314, 328, 343, 355, 406
Conjunctive:
   use 24, 256, 257, 329, 338, 354
   management 209, 229, 237, 238, 354
   water management program 238
Conservation 31, 73, 83, 102, 107, 109, 215, 220, 225, 234, 244, 249, 254, 256, 272, 284, 406
Consumptive use 186, 206

Contaminant 26, 74, 206, 217, 230, 266
Contamination 74, 109, 252, 265, 303, 325, 326
Continental Divide 201, 244, 245, 247, 251
Coolgardie 151, 153, 160, 369
Coolgardie Goldfields Water Supply:
   Construction Bill 157
   Loan Bill 153, 162, 164
Cooper Creek 126, 129, 134, 140, 141
Copi Hollow 197, 415
Cordeaux Dam 401
Cordeaux River 397
Corporatisation 101, 102
Cost 23, 26, 27, 61, 78, 81, 84, 96, 102–5, 107, 111, 113, 116, 119, 122, 123, 146, 155, 170, 178, 208, 210, 211, 217, 220, 227, 237, 274, 281, 284–6, 288, 290, 301, 307, 308, 311, 312, 314, 322, 325, 327, 329, 335, 336, 338, 341, 379, 387, 402
Cost-benefit 37, 116, 117, 121, 123, 149, 290, 338, 342, 387, 406
Cost effective 34, 113, 215, 236, 342, 357, 402
Cotton 384, 385, 393–5
Council of Great Lakes Governors 273
Council of the Australian Governments 39, 77, 78, 84, 85
Couston, James 158, 159
Crop 15, 19, 30, 31, 64, 117, 130, 131, 184, 186, 204, 209, 232, 264, 275, 305, 386
CSIRO 54, 62, 81, 86
CSR Limited 388
Cultivation 53, 298
Cultural 36, 394, 395
Curtin, John 363, 368
CUP See Central Utah Project
CUWCD See Central Utah Water Conservancy District
CVP See Central Valley Project
CVPIA See Central Valley Project Improvement Act
CWCB See Colorado Water Conservation Board

Dam 97, 118, 119, 128, 129, 154, 181, 223, 270, 292, 300, 306, 323, 326, 328, 336
Damanganga River 341
Danjiangkou Dam 310, 314
Darling Pipeline 195
Darling Range 153, 159, 167, 346, 369
Darling River 192, 194, 195
   Scheme 195
Darwin 397
Deccan Plateau 323
Decision-making 43, 44, 85, 273, 284
Deforestation 207, 298, 326, 336, 353
Degradation 19, 20, 22, 53, 71, 83, 87, 110, 231, 272
de Lesseps, Ferdinand 379–81
Delhi 330
Demand 284, 314
Depletion 44, 205, 313
Depression 145, 379, 381
Derby 395
Derwent River 405, 408
Desalination 27, 86, 123, 168, 174, 176, 177, 215, 221, 229, 235, 238, 253, 257, 312, 329, 355, 358, 402
Desert 295, 297
Devils Gate Power Station 408
Diamantina River 129, 135, 140, 141, 397
Diamond Gorge Dam 170, 177
Digital Elevation Model 140
Dilution Flow 390, 401

Disease 18, 28, 29, 152, 272, 314, 327
Displaced 301, 327, 328
Dispute 36, 40, 75, 96, 220, 240, 273, 281, 330, 355
Domestic 10, 20, 81–3, 110, 126, 141, 152, 154, 188, 300, 302
Dorman Long and Co. 362
Drainage 16, 28, 31, 71, 72, 109, 122, 242, 298, 306, 314
Dridan, J.R. 194
Drought 59, 60, 87, 104, 110, 125, 126, 129, 143, 190, 191, 195, 212, 215, 221, 230, 232, 238, 239, 244, 249, 252, 297, 304, 305, 307, 313, 326, 330, 361
Dryland 4, 31, 71, 83, 84, 116
Dust Bowl 148, 208
Duveyrier, Henri 379
Dysentery 152, 346

Earthquake 25, 219–21, 231, 311, 336, 352
East Bay 220, 221
   Municipal Utility District 220
Eastern Route Scheme 308, 309, 311, 313
EBMUD See East Bay Municipal Utility District
EC See Electrical conductivity
Ecology 77, 104, 313, 402
Ecological 29, 32–4, 43, 72, 78, 80, 110, 177, 196, 238, 283, 284, 289, 306, 312, 327, 337, 341
Economic 22, 23, 29, 35, 37, 43, 77, 79, 86, 87, 101, 105, 110, 111, 123, 147, 171, 210, 237, 239, 273, 283–5, 288, 291, 303–5, 313, 328
Economy 54, 65, 79, 81, 86, 215, 230, 262, 296, 312, 319
Ecosystem 11, 15, 19, 22, 23, 30–5, 40, 54, 59, 77, 78, 83, 84, 92, 109, 149, 165, 196, 209, 231, 240, 265, 274, 292, 313, 325, 327, 342, 393, 394, 399, 405
Edgar Dam 409
Education 237, 250, 257, 275, 329, 343, 344
Efficiency 20, 22, 77, 78, 81–6, 121, 123, 133, 149, 186, 213, 284, 322, 328
Electrical conductivity 187
Electricity 120, 126, 149, 223, 238, 245, 249, 277, 284, 292, 305, 402, 406
El Niño 55, 59, 141, 326
El Vado Dam 252
Endangered 73
   species 36, 256, 338, 341, 343
   Species Act 215, 223, 252, 255
Endocrine disruptors 11, 74, 206, 265, 275, 354
Energy 238, 239, 262, 274, 300
Environment 16, 19, 84, 87, 102, 104, 109, 147, 165, 217, 230, 233, 239, 249, 265, 284, 305, 306, 315
Environmental 10, 22–4, 30, 53, 85, 102, 103, 107, 110, 121–3, 207, 210, 213, 236, 238, 256, 273, 284, 288, 301, 305, 306, 312, 313, 329, 382
   assessment 28, 36, 102, 393
   degradation 31, 76, 107, 222, 306
   flow 23, 30–5, 78, 82, 83, 87, 102, 104, 107, 252, 401, 403, 411
   impact 28, 33, 37, 70, 101, 102, 137, 149, 166, 176, 240, 255, 289, 290, 300, 313, 327, 341, 342, 345, 357, 390, 399, 401, 402, 404, 409
   impact assessment 315, 333, 336, 340, 393
   management plan 333, 336
Environmentalists 78, 105, 212, 220, 223, 326, 333, 403

Erosion 5, 16, 26, 27, 29, 30, 44, 71, 72, 107, 109, 111, 146, 148, 272, 283, 284
Estuary 33, 109, 110, 409
Eucumbene River 91
Eungella Dam 188
Europe 362
Eutrophication 30, 74, 313
Evaporation 13, 25, 27, 55, 58, 71, 111, 125, 128, 129, 141, 142, 145, 149, 151, 180, 202, 204, 209, 242, 297, 299, 305, 321, 379, 382, 386, 405
Eyre, Edward John 139, 141

Fadden, Arthur 368
Fauna 74, 105, 176, 196, 327, 337, 394, 411
Feasibility 24, 25, 45, 94, 104, 111, 122, 123, 126, 128, 132, 172, 244, 253, 254, 332, 334, 336–8, 382, 393
Ferguson, Mephan 157, 158
Fertiliser 15, 16, 206, 237, 304, 325, 385, 394
Finlayson, James 158
Fish 29, 30, 33, 35, 77, 101, 109, 110, 122, 142, 211, 212, 223, 232, 249, 252–4, 271, 274, 282, 283, 289, 327, 328, 382, 403, 405, 409–11
Fisheries 19, 28, 29, 32, 137, 209, 254, 272, 284, 291, 313, 335, 336, 410
Fisher Power Station 408
Fisher River 408
Fishing 77, 211, 283
Fish kill 72, 110, 411
Fish mortality 401
Fitzroy River 170, 173, 177, 178, 384, 393
Flinders River 125, 126, 129, 135
Flood 13, 61, 72, 87, 91, 111, 121, 128, 140, 170, 188, 207, 210, 223, 224, 269, 275, 297, 300, 304–6, 314, 315, 326, 330, 335, 338, 387
Flooding 30, 139, 141, 221, 231, 304, 379
Floodplain 29, 31, 33, 72, 101, 102, 207, 298, 300
Flora 74, 105, 107, 196, 327, 337, 394, 411
Flow 25, 30, 34, 126, 129, 188, 240
Fly River 397
Folsom Dam 208
Food 15, 16, 19, 53, 77, 86, 110, 209, 305, 322, 335, 342, 386
Ford, William 151
Forest 71, 77, 107, 110, 131, 157, 167, 208, 209, 253, 263, 271, 282–4, 298, 300, 305, 327, 328, 336, 338, 340
Forestry 59, 83, 265
Forrest, Alexander 384
Forrest, John 153, 156–8, 160, 161, 364, 370
Forth River 408
Fountain Valley Conduit 248
France 366
Freeman, Ralph 362
Free Trade 17
Free Trade Agreement 288, 290, 291, 293
Fremantle 151, 157, 369, 370
Freshwater 8, 13, 19, 33, 137, 149, 177, 206, 224, 238, 239, 264, 274, 286, 289, 291, 299, 345
Friant Dam 208
Frigate Dam 281
Fryingpan-Arkansas Project 243, 247
FTA See Free Trade Agreement

GAB See Great Artesian Basin
Gabés 379–81
Ganga-Brahmaputra Delta 336, 353
Ganga River 319, 323, 325–7, 330, 331, 335, 336

Gardener, Henry 253
Garland Canals 331
Garrison River 289
GATT 291
GDP See Gross Domestic Product
Geography 201, 215, 261, 295, 319, 379
Geographic Information System 140, 141
Geology 24, 26, 139, 379
Geomorphology 29, 30
Georgina River 141
Germany 155, 157, 158
Gila River 285
Gilbert River 135
Gippsland 404
GIS See Geographic Information System
Glacier 304, 327
Glen Canyon Dam 204, 240, 253
Global warming 11, 13, 19, 58, 87, 92, 209, 292, 327, 343, 345, 357
Godavari Delta 321
Godavari River 323, 325, 330
Godavari Water Dispute Tribunal Award 340
Gold 109, 139, 151, 153, 223, 346, 364, 365, 370
Goldfields 151–3, 365, 369
  and Agricultural Areas Water Supply Scheme 162, 163, 170
  Pipeline Scheme 151, 154, 346, 369, 370
Gold rush 215, 218
Gordon Power Station 408, 409
Gordon River 405, 409
Gordon River Power Development Scheme 409
Granby Dam 245
GRAND Canal 286
Grand Canal 300, 307, 309
Grand Lake 244, 245
Gravity 25, 28, 111, 114, 119, 125, 126, 129, 135, 147, 174, 184, 218, 235, 275, 276, 309, 311, 332, 338
Great Artesian Basin 67, 75, 139, 141, 149
Great Basin 253, 254
Great Boomerang Scheme 147, 149
Great Dividing Range 51, 54, 86, 107, 118, 125, 129, 147, 184, 362
Great Indian Bustard 338
Great Lake 405, 408
Great Lake Scheme 407
Great Lakes 202, 210, 261, 264, 271, 273, 276, 285–7, 292
  Basin 292
  Basin Diversions 281, 349
  Charter 273
Great Plains 201, 261
Great Salt Lake 253
Green, Francis 361
Greenhouse 28, 101, 103, 174, 177, 178, 272, 292
Green Mountain Dam 245
Green River 254
Greenvale Reservoir 403
Gross Domestic Product 65, 262, 296, 307, 314, 320, 336
Groundwater 8, 11, 12, 16, 22, 26, 39, 66, 68, 69, 71, 72, 74, 75, 78, 83, 84, 86, 141, 149, 151, 168, 187, 204, 207, 216, 225, 238, 264, 298, 301, 303–6, 315, 322, 325, 329, 335, 342, 390, 393, 402, 403
Gujarat 325
Gulf of Carpentaria 184
Gulf of Gabés 379
Gulf of Mexico 201
Guthega Pondage 96
Guthega Project 96

Habitat 30, 32, 101, 104, 205, 212, 223, 232, 239, 253, 271, 272, 274, 283, 337, 405, 409
Haihe River 299, 301, 302, 304, 307
Hanjiang River 310
Hannan, Patrick 151, 160
Harper, N.W. 161
Harvey O. Banks Pumping Plant 224
Health 16, 29, 32, 83, 84, 102, 109, 110, 171, 209, 217, 230, 237, 272, 274, 275, 283, 315, 338, 344, 354
Heavy metals 74, 267, 275, 326, 354
Hebei 308, 309
Heilong River 299
Helena River 153, 346, 369
Henty River 407
Herbert River 125, 126, 128
Herbicide 19, 206, 325
Herodotus 379
Heron Reservoir 217
Hetch Hetchy Aqueduct 217, 219, 222, 348, 351, 357
High Court 96, 363, 368, 406
High Line Canal 254
Himalaya 295, 323, 331
Himalayan rivers 332
Himalayan Rivers Development Component 331, 335
Hodgson, T.C. 154, 159
Holistic 32
Holistic methods 33
Hong Kong 387
Hoover Dam 99, 240, 242
Hoskins, G.Y.C. 157
Huaihe River 299, 301, 304, 305, 307
Huanghe (Yellow) River 307
Huang-Huai-Hai Plain 295, 301, 310
Hudson Bay 261, 279, 286, 289
Hudson, William 92, 95, 366
Huon River 409
Hyderabad 330
Hydrocarbon 74
Hydro-electric 27, 28, 36, 66, 94, 98, 111, 118, 123, 126, 147, 149, 176, 186, 218, 228, 248, 276, 281, 284, 286, 292, 300
  Commission 406
  Commission Act 406
Hydrogeology 26, 205
Hydrology 12, 13, 25, 91, 140, 223, 264, 379, 386, 390
Hydro-power 28, 105, 107, 111, 122, 170, 187, 246, 270, 279, 300, 301, 305, 321, 324, 330, 335, 338, 388, 390, 397, 400, 405
Hydro Tasmania 411

Idriess, Ion 147, 149, 367
IID See Imperial Irrigation District
IJC See International Joint Commission
Illinois-Michigan Canal 281
Illinois River 281
Impact 22, 26–8, 36, 77, 86, 87, 102, 105, 109–11, 121, 123, 131, 205, 209, 238, 272, 273, 281, 284, 292, 301, 313, 338, 350, 382
Imperial Dam 191, 242
Imperial Irrigation District 227, 228, 236
Imperial Valley 226–8
India 7, 9, 19, 22, 38, 42, 319, 350, 352, 353
Indian Peninsula 91
Indian Tribe 255
Indian Wolf 338
Indigenous 23, 27, 36, 38, 39, 84, 187, 279, 283, 284, 292, 301, 343, 396, 402
Indira Gandhi Canal 319, 331

Indonesia 393
Indus River 319, 323
Indus Waters Treaty 42
Industrial 20, 66, 70, 111, 114, 132, 223, 232, 246, 248, 305
Industry 10, 82, 107, 110, 147, 149, 165, 171, 302, 403
Infrastructure 54, 72, 78, 83, 111, 188, 212, 215, 230, 237, 239, 247, 250, 275, 301, 307, 312, 314, 402
Inland sea 379–82
Inner Mongolia 300, 312
Insecticide 385, 387
Insect pest 384, 387
Institution 18, 41, 273, 306, 329, 342
Institutional 84, 215, 314, 315
Integrated 16, 18, 43, 45, 75, 78, 84, 122, 123, 230, 231, 240, 273, 306, 329
  catchment management 282, 292, 357
  Water Supply Scheme 167, 168
Integration 22, 214
International Joint Commission 273, 291, 292
International Water Management Institute 342, 344
Inter-State Water Dispute Act 330
Investigation 24, 26, 87, 92, 94, 100, 122, 123, 126, 128, 131, 155, 169, 181, 187, 244, 247, 251, 253, 332, 411
Ireland 369
Irrigated 14, 64, 83, 101, 186
  land 15, 64, 83, 122, 204, 227, 246, 257, 322
Irrigation 10, 14, 16, 17, 26, 28, 29, 36, 66, 68, 70, 77, 82–5, 94, 103, 107, 111, 114, 116, 126, 147, 189, 204, 207, 210, 213, 247, 248, 267, 270, 274, 275, 298, 300, 306, 312, 314, 319, 321, 325, 330, 335, 338, 342, 386
Ivanhoe Plain 384, 388
IWMI See International Water Management Institute
IWSS See Integrated Water Supply Scheme

James Bay 276, 280, 286
James, David 190
Jensen Unit 256
Jiangsu 308, 309
Jindabyne Dam 91, 104
Jurisdiction 38, 40, 41, 84, 273, 284, 292, 314, 355

Kalgoorlie 151, 153, 154, 160, 170, 369
Karnataka 324
Karstic 301
Kemano Power Plant 279
Kenny Dam 277
Kenogami River 276, 279
Kerala 330, 340, 341
Kerala-Tamil Nadu Agreement 356
Kern County 238
Kimberley 168, 169, 171, 174, 176, 384, 393
  Aqueduct Scheme 174
  Pipeline Scheme 165, 166, 169, 171
  Research Station 384
  Water Resources Development Office 393
King George V 365
Kosciuszko National Park 104, 105
Krishna River 323, 331
Kuiper Diversion Scheme 286
Kununurra 384, 386, 388
  Diversion Dam 384, 386
Kurnool Cuddapah Canal 330
Kwinana Desalination Plant 168
Kwinana Power Station 178

Labor Party 362–4, 368
La Grande (James Bay) Diversion 280, 349
La Grande River 270, 280
La Grange sub-basin 393, 395
Lake Argyle 174, 384, 386, 389, 390
Lake Bonneville 253
Lake Cethana 408
Lake Eucumbene 97, 104
Lake Erie 281
Lake Eyre 51, 55, 61, 75, 131, 139, 143, 146, 365
Lake Gordon 405, 409
Lake Granby 245
Lake Havasu 229
Lake Huron 281, 286, 287
Lake Jindabyne 104
Lake King William 407
Lake Mackenzie 408
Lake Manitoba 289
Lake Mathews 229
Lake Mead 204, 242
Lake Michigan 281, 287
Lake Nipigon 276
Lake Ontario 273
Lake Parangana 408, 411
Lake Pedder 405, 409
Lake Powell 204, 253, 257
Lake Rowallan 408
Lake Superior 264, 276, 281, 285, 287
Lake Torrens 131, 145, 146
Lake Wellington 405
Lake Williston 289
Lake Winnipeg 270, 279, 284–6
Land 19, 70, 86, 87, 126, 226
  clearing 31, 71
  cover 263, 297
  degradation 7, 10, 17, 53, 83
  subsidence 11, 205, 222, 238, 301, 311, 313
  use 23, 25, 64, 203, 239, 387, 394, 409
LANDSAT 143
Lang, T.A. 95
Large-scale 19, 34, 35, 71, 87, 107, 211, 263, 284, 290, 292, 293, 312, 321, 326, 330, 393
La Trobe River 404
Launceston 405, 408
Lead 190, 193
Legislation 37–40, 71, 75, 95, 153, 275, 292, 306, 363, 410
Legislative 159, 215, 230, 291
Lemmon, Nelson 92, 366
Lemonthyme Power Station 408
Leopard 338
Levee 230, 231
Liaohe River 299, 301, 304, 305
Liberal Party 368, 369
Libya 381
Living Murray 78
Loess Plateau 295, 300
Logan, John 393
London 153, 155, 156, 158, 365, 366
Long Lake Diversion 276, 281, 349
Long-term 16, 19, 25, 27, 36, 54, 79, 82, 83, 85–7, 92, 96, 110, 122, 123, 165, 172, 228, 237, 250, 297, 305, 314, 402
Los Alamos 252
Los Angeles 217, 226, 227, 229, 236, 237
  Aqueducts 217
  River 217, 218
Loss 26, 110, 111, 128, 129, 132, 168, 207, 209, 227, 269, 304
Loss of habitat 107, 298, 338
Loss of wildlife 231

Luni River 335
Lyons, Joseph 368

Mackenzie River 264, 275, 286, 289
Macleay River 113, 114, 123, 361
Macquarie perch 401
Madhya Pradesh 341
Magnum Diversion Scheme 289
Mahanadi River 330, 335
Maher, John 153, 160
Malaria 327
Malnutrition 16
Management 18, 19, 22, 23, 26, 32, 34, 35, 39, 74, 78, 84, 94, 101, 102, 122, 194, 213, 223, 230, 240, 253, 267, 272, 284, 291, 292, 306, 312, 329
Manitoba 279, 284
  - Hydro 279, 280
Manning River 113–15, 123
Mannum-Adelaide Pipeline 183
Mao Zedong 308
Mapleton Canal 254
Mareeba-Dimbulah 180
  Irrigation Scheme 184, 187, 347
  Irrigation Area 185, 186
Market 17, 37, 38, 43, 64, 82, 101, 110, 153, 176, 232, 284, 296, 312, 314, 385, 387
Master Plan 86, 87
McKell, William 366
MDBC See Murray-Darling Commission
MDIA See Mareeba-Dimbulah Irrigation Area
Mediterranean Sea 144, 379, 380
Mekong River 19, 301, 314
  Commission 42
Melbourne 82, 94, 121, 368, 403
Melbourne Water 404
Memorandum of Understanding 236, 389, 393, 396
Menindee:
  Lakes 180, 192, 196
  Lakes Storage Scheme 195
  Water Conservation Act 195
Menzies, Robert 95, 96, 364, 368
Merigan, E.L. 95
Mersey-Forth Hydro-electric Scheme 408
Mersey River 408, 411
Metropolitan Water District 218, 227–9
Mexico 209, 218, 227, 240, 242, 286
Microclimate 28
Middle Route Scheme 308, 310
Mine 74, 189, 267
Mining 20, 74, 82, 109, 141, 149, 165, 171, 188, 190, 194, 267
Missi Falls 279
Mississippi River 204, 208, 281
Missouri River 284
Mitchell River 135, 184
Model 12, 13, 25, 26, 30, 44, 56, 79, 84, 143, 315, 327, 382, 386
Modelling 25, 32, 37, 43, 82, 143, 209
Mokelumne Aqueducts 217, 220, 348
Mokelumne River 220
Mongolia 297
Monitoring 26, 29, 84, 121, 267, 303, 315, 411
Mono Lake 218
Monsoon 13, 140, 297, 304, 320, 326, 342
Monterey Agreement 225, 232
Morgan 73, 181, 195
Morgan-Whyalla Pipelines 181
Mormons 253
MoU See Memorandum of Understanding
Mount Lofty Ranges 180

INDEX

Mt Charlotte 154
Mt Elbert 248, 249
Mt Kosciuzsko 51, 93
Muir, John 219, 351
Mumbai 330, 338, 341
Mundaring 157
   Reservoir 369
   Weir 151, 154, 160, 167, 370
Municipal 206, 223, 232, 236, 242, 246–8
Murray-Darling Basin 59, 66, 70–4, 76–9, 81, 86, 120, 121, 403
   Agreement 39, 75, 87, 102, 180, 196
   Commission 39, 75, 76, 78, 196
   Ministerial Council 75, 77, 78
Murrumbidgee River 91, 93, 94, 97, 101–4, 117, 361
Multidisciplinary 41
MWD See Metropolitan Water District

NAFTA See North American Free Trade Agreements
Nambe Falls Dam 252
Namoi Catchment 44
Namoi Valley 386
Narmada 323
National Environmental Policy Act 255, 351
National Indravati Tiger Reserve Sanctuary 340
National Parks and Wildlife Service 107
National Perspective Plan 331
National River-Linking Project 330, 331, 335, 336, 338, 341, 342, 352, 357
National Water:
   Initiative 39, 40, 84–7
   Commission 84, 85, 87
   Development Agency 332–5, 338, 340, 341
   Grid 331
   Pipeline Scheme 173, 174
   Policy 329, 341, 343
   Quality Assessment Program 206
   Sector Prospective Plan 342
   Saving Office 306
Native 110, 284, 292, 357, 411
   title claim 390, 394, 396
Natural resources 36, 41, 76, 215, 237, 305, 314, 328
NAWAPA See North American Water and Power Alliance
NCWCD See Northern Colorado Water Conservancy District
Nebraska 245
Nechako-Kemano Diversion 277, 349
Negotiation 28, 75, 211, 228, 232, 240, 284, 290, 308, 356, 396
Nelson River 270, 279, 286
Nepal 335, 336
Nepean Dam 366, 401
Nepean River 397
New Mexico 240, 242, 243, 250
   State Water Plan 252
Newton Boyd Scheme 117
New Zealand 366, 369
NGOs 27, 34, 336, 342
Niagara:
   Falls 276
   River 276
   River Water Diversion Treaty 276
Nile Waters Treaty 42
Nolan's Stephens Creek Water Supply Company 192
Northam 153, 154, 159

North America 4, 9, 13, 15, 65, 208, 209, 261, 273, 353, 362
North American Free Trade Agreements 290, 291, 293
North American Water and Power Alliance 211, 284–6, 290
North China Plain 295, 301, 304, 305, 308, 309, 313–15
North Dakota 289
Northern Colorado Water Conservancy District 247
Northern Territory 384, 389
North Thompson Project 290
North Thompson River 290
Nutrient 16, 19, 29, 30, 62, 74, 101, 107, 110, 187, 206, 237, 272, 289, 303
NWC See National Water Commission
NWDA See National Water Development Agency
NWI See National Water Initiative

O'Connor, C.Y. 154, 155, 157–9, 365, 369
Office of Water Use Efficiency 235, 236
Ogoki Diversion 276, 281, 349
Oil 381
Olympic Dam 141
Ontario 271, 291
Ord River 169, 174, 384, 386, 390
   Dam 170, 384
   Irrigation Area 386, 387, 389
   Irrigation Scheme 169, 384, 386
ORIA See Ord River Irrigation Area
Orissa 340
O'Shannassy Reservoir 403
O'Shaughnessy Dam 219, 220
Overdraft 216, 237, 301, 307, 325
Owens River 218, 239
OWUE See Office of Water Use Efficiency
Oyster 72, 110, 403

Pacific Hydro Limited 388, 391
Pacific Ocean 147, 149, 201, 231, 261, 279
Pakistan 42
Palaeoclimatic 209
Palmer, Charles 159
Paloona Power Station 408
Pamba River 341
Panama Canal 380
Papua New Guinea 397
Parambikulam Aliyar Project 330, 356
Parangana Dam 408
Pardee Dam 220
Parker Dam 229, 242
Parsons, Ralph, M. 211
Pathogen 74, 206, 265, 326
Peace River 270, 289
Pearl 169, 367
Pearl River 299, 301
Peninsular rivers 332
Peninsular Rivers Development Component 331, 334, 337, 338, 341
Periyar Project 330
Perth 82, 151, 153, 157, 165, 167, 169, 171, 177, 364, 369, 397
Pest 28, 393
Pesticide 16, 19, 74, 206, 267, 325, 386, 394
Planning 23, 27, 29, 231, 235, 239, 247, 284, 292, 329, 335
Poatina Power Station 408
Policy 19, 34, 38–40, 44, 54, 60, 64, 78, 81, 82, 87, 107, 178, 213, 215, 237, 253, 273, 274, 288, 290, 292, 296, 306, 314, 322, 329, 342

Pollution 18, 19, 110, 231, 265, 274, 303–5, 312, 313, 327, 338, 342, 401
Pool, James 190
Population 6, 9, 14, 16, 19, 22, 28, 29, 51, 52, 54, 75, 82, 86, 87, 95, 96, 101, 105, 107, 110, 123, 151, 165, 191, 202, 208, 212, 215, 220, 224, 229, 244, 250, 252, 261, 264, 296, 298, 301, 302, 305, 307, 319, 329, 335, 345, 402
Port Augusta 144, 145, 147, 149, 181
Power Plant 223, 239
Power station 24, 97, 99, 101, 189
Precipitation 8, 12, 13, 55, 91, 104, 125, 143, 204, 213, 215, 216, 240, 243, 250, 262, 271, 297, 299, 305, 307, 320, 323, 382, 405
   enhancement 104, 229, 239
President:
   Kennedy 252
   Richard Nixon 223
   Roosevelt 223
Price 70, 98, 116, 117, 119, 123, 126, 153, 161, 176, 209, 284, 315, 341
Pricing 85, 269, 274
Prior appropriation 38
Prospect Reservoir 397
Public consultation 75
Public participation 27, 292
Pueblo Dam 247
Pueblo Reservoir 248
Pumping 25, 27, 111, 113, 116, 123, 126, 149, 153, 172, 187, 213, 276, 285, 332, 338, 382
   station 97, 99, 111, 133, 153, 154, 159, 170, 172, 183, 197, 309, 415

Quebec 271, 280, 284
Queensland 137, 180, 184, 188, 361, 366
Queen Victoria 365

Rainfall 12, 25, 31, 55, 58, 65, 82, 91, 107, 125, 127, 131, 140, 189, 297, 304, 320, 326, 379, 386, 405
Rajasthan 321, 325, 331, 341
Rao, K.L. 331
Rason, C.H. 160
Rasp, Charles 190
Rat Hole Tank 191
Rat River 279
Ravi-Beas-Sutlej links 331
Recycling 22, 85, 269
Recreation 252, 254, 270, 403
Red Sea 145, 149
Reform 16, 39, 74, 81, 85
Reid, L.B.S. 135
Reid Scheme 135, 137
Reservoir 10, 204, 216, 223, 252, 253, 283, 285–7, 300, 309, 321, 323, 330, 338
Resettlement 36, 39, 301, 314, 334, 338, 340, 356
Reuse 27, 85, 86, 315
Reverse osmoses 178, 238, 402
Rio Chama 250, 251
Rio Grande 247, 250, 252, 285
   Compact 251
Riparian rights 38, 39, 234
Risk 37, 54, 71, 110, 305, 315, 336, 352, 354, 402
River Basin Commission 315
River Basin Coordinating Council 315
River Murray 76, 87, 91, 93, 94, 96, 101, 103, 104, 117, 123, 180, 194
   Commission 39

River Murray (cont.)
  Pipelines 180, 347, 350
  Water Agreement 39
Rocky Mountains 201, 202, 204, 261, 262
Rocky Mountain National Park 244, 245
Rockwood, Charles 226
Rowallan Dam 408
Rowallan Power Station 408
Ruedi Dam 247
Run-off 9, 13, 25, 30, 31, 39, 65, 82, 87, 91, 107, 110, 125, 126, 130, 141, 143, 209, 239, 240, 299, 305, 313, 321, 323, 405

Sabarmati River 335, 341
Sacramento River 221–3, 231, 237, 239
Sacramento-San Joaquin Delta 208, 222, 224, 228
Sahara 144, 379
Saint John River 265
Sakami River 281
Saline water 13, 110, 206, 215, 231, 253, 299, 382, 395
Salinisation 5, 10, 16, 26, 31, 71, 72, 77, 87, 101, 299, 313, 314, 327, 338
Salinity 59, 71, 73, 103, 105, 122, 132, 216, 239, 313, 322, 338
Salmon 223, 231
Salmonides 29
Salt 146, 242, 379, 382
Saltwater 205, 231, 272, 313, 386
San Diego 226, 228, 229, 240
  County Water Authority 228
San Francisco 217–19
  Bay 222, 224, 231
San Joaquin River 222, 237
San Joaquin Valley 220, 224, 234
San Juan-Chama Project 243, 247, 250
San Juan River 250–2
San Luis Unit 224
Santa Barbara 238
Santa Clara 238
Santa Fe 252
Saskatchewan 211
Saskatchewan-Nelson Basin 289
Saskatchewan River 286, 289, 290
Scenario 44, 54, 79, 81–6, 166, 172, 216, 342, 387
Scullin, James 362
SDCWA See San Diego County Water Authority
Seawater 11, 123, 145, 168, 174, 177, 237, 238, 325, 329, 386
Sediment 24, 26, 29, 30, 101, 170, 187, 188, 209, 240, 298, 300, 301, 313, 325, 336, 401, 409
Sedimentation 26, 29, 107, 110, 123, 300, 338, 386
Seismic 25, 311
Semi-arid 16, 54, 203, 240, 250, 320
Serpentine Dam 409
Settlement 37, 105, 131, 209, 222, 250, 298, 338
Shadow Mountain Lake 245
Shandong 308, 309
Shanghai 296, 301
Shannon Power Station 407, 408
Shasta Dam 208, 223, 231, 232
Shasta Lake 290
Shoalhaven Diversion Scheme 346, 350, 397
Shoalhaven River 111, 401, 403
Sierra Club 219, 255, 351, 357
Sierra Mountains 218, 220
Sierra Nevada 201, 218, 239
Siltation 194, 327, 328

Silver 139, 190, 239
Silverton 191
Simulation 25, 26, 56, 143, 327
Singapore 387
Snow 12, 13, 55, 91, 92, 104, 263, 269, 292, 297, 405
Snowfall 202, 209, 239
Snowmelt 91
Snowy Hydro Corporation Act 101
Snowy Hydro Ltd 101, 102, 104
Snowy Mountains 91, 94, 95, 104, 366
  Engineering Corporation 100, 397
  Hydro-electric Agreement 41, 96, 356
  Hydro-electric Authority 92, 95, 364, 366, 397
  Hydro-electric Power Act 95, 356, 369
  Hydro-electric Scheme 91, 346, 350, 363
Snowy-Murray Diversion 96
Snowy River 91, 94, 101–3, 105
Snowy Tumut Diversion 96
Snowy Water Inquiry 102, 104
SNWTP See South to North Water Transfer Project
Social 29, 35, 37, 43, 77, 87, 101, 110, 123, 171, 210, 283, 284, 288, 300, 301, 314, 333
Socio-economic 16, 334, 379, 382
SOI See Southern Oscillation Index
Soil 26, 31, 53, 61, 63, 74, 83, 132, 135, 149, 184, 223, 229, 295, 311
  conservation 26, 31, 101, 208, 306
  degradation 5, 7
  erosion 28, 30, 109, 208, 300, 393
  loss 30
  moisture 12, 271, 327
Songhua River 301, 305
Souris River 289
South Australia 181, 192–4
Southern Cross 151, 160, 369
Southern Indian Lake 279
Southern Oscillation Index 55, 61
South Esk River 405, 408, 410
South Platte River 244, 245, 247
South to North Water Transfer Project 307, 312–14, 349, 352
Spanish Fork Diversion Dam 254
Spanish Fork Power Plant 254
Spanish Fork River 253
Spawning 29, 30, 35, 104, 122, 283
Spencer Gulf 182
Spooner, William 96
Stakeholders 22, 26–8, 34, 43, 44, 78, 85, 214, 237, 284, 292, 329, 393, 411
State Council 308
State Water Project 216, 217, 224, 228, 231, 232, 249, 348
State Water Resources Control Board 227, 235
Statewide Water Supply Initiative 244, 249, 250
Stephens Creek 191, 192
  Reservoir 192–5
Stirling Dam 167
St Lawrence River 264, 265, 272, 273, 275, 276, 281, 285
Stockdale Company 193
Stockdale, Harry 192
Strategic Plan 229, 246, 300
Strategic Water Resources Policy 75
Strategy 74, 165, 177, 178, 181, 253, 282, 291, 306, 341
Strawberry:
  Aqueduct 256
  Dam 254
  Reservoir 253

River 256
Valley Project 243, 253, 254
Strike 362, 364
Strzelecki, Paul Edmund 93
Sturt, Charles 189
Suez Canal 379
Sugar price 391
Supreme Court 281, 330, 333
Surface water 26, 65, 67, 69, 77, 86, 168, 204, 216, 264, 299, 301, 323, 325, 335
Sustainability 18, 19, 40, 43, 77, 205, 234, 284, 395
Sustainable 16, 19, 53, 77, 84, 102, 122, 165, 169, 188, 230, 274, 305, 315, 325, 329, 342, 357, 395, 402
Swan Coastal Plain 167
Swan Reach-Paskeville Pipeline 183
Swan River 151, 369
SWECO 382
SWP See State Water Project
SWRCD See State Water Resources Control Board
SWSI See Statewide Water Supply Initiative
Sydney 82, 94, 125, 362, 366, 367, 397, 401
  Harbour Bridge 361

Tahtsa Lake 279
Tailem Bend-Keith Pipeline 183
Talbot, J.S. 160
Tallowa Dam 398, 399, 401, 402
Tamar River 408, 410
Tamil Nadu 330, 331, 341
Tantangara Dam 97
Tasmania 174, 397, 405
Tasmanian Wilderness World Heritage 410
TDS See Total Dissolved Solids
Telugu Ganga Project 331
Temperature 55, 56, 82, 101, 104, 127, 141, 189, 209, 226, 239, 263, 270, 297, 305, 314, 326
Texas 209, 251
Texas Water Plan 211
Thailand 301
Thermo-electric 206
Thomson Dam 405
Thomson Diversion Scheme 347, 350, 397, 403, 404
Thomson River 125, 126, 129, 404, 405
Three Gorges Dam 36, 311, 313, 352
Tianjin 296, 307–9
Tibet 311, 323
Tibet Plateau 295, 300, 323
Tidal 170, 176, 313
Tide 286
Tinaroo Falls Dam 184–7
Topography 24, 149, 295, 320
Total Dissolved Solids 66, 67, 181, 242, 386
Tourism 107, 110, 123, 169, 171, 215, 335, 387, 388, 394, 405
Trade 17, 84, 85, 274, 290, 291
Treaties 40, 41
Tribal 328, 329
Trinity River 240
Tully River 125, 126
Tumut Pound Dam 96
Tumut River 91, 96, 97, 103
Tunisia 379–81
Tuolumne River 219, 220
Turbidity 74, 109, 183, 272, 283, 298
Tuross River 114, 115, 123
Turquoise Lake 248
Twin Lakes 248
Typhoid 152, 191, 346, 367

INDEX

Umberumberka
  Creek 193
  Creek Dam 193, 194
  Reservoir 194
  Water Trust 193
Uncertainty 37, 44, 177, 336, 391
UNEP 5, 7
UNESCO 406
United States 9, 37, 38, 51, 131, 157, 201, 261, 270, 273, 275, 281, 284, 286, 287, 290–2, 347, 351
  Reclamation Service 244, 253
Upper Colorado River Basin Compact 251
Upper Nepean Scheme 397
Urban 81, 82, 84, 86, 107, 114, 131, 206, 210, 223, 224, 275, 305
U.S. Army Corp of Engineers 208, 223, 288, 348
USBR See U.S. Bureau of Reclamation
U.S. Bureau of Reclamation 99, 208, 227, 235, 243, 244, 246, 247, 252, 254, 255
USRS See United States Reclamation Service
Utah 242, 243, 253
Utah Lake 253
Utah State Water Plan 257
Uttar Pradesh 326

Vegetation 15, 26, 28, 29, 31, 33, 71, 82, 83, 101, 109, 140, 188, 283, 401
Vembanad Lake 341
Vernal Unit 256

WAI See Western Agricultural Industries Pty Limited
Walsh River 136, 184
Warragamba Dam 397, 402
Wasatch Divide 253
Waste 206, 227, 267, 325
Wastewater 24, 27, 85, 86, 107, 109, 167, 168, 215, 236, 237, 239, 257, 266, 273–5, 292, 303, 304, 309, 312–15, 357, 402
Water:
  Allocation 30, 34, 37, 38, 45, 75, 82, 312, 329, 343, 390
  Balance 26, 143, 216, 313, 327, 334, 335
  Bank 35, 213, 232
  Conservation 165, 184, 208, 213, 218, 221, 229, 230, 232, 234, 235, 253, 257, 273, 274, 292, 306, 309, 329, 343, 354, 403
  consumption 107, 217, 218, 267, 291, 308, 314, 402
  demand 27, 274, 275, 305, 342, 402
  export 284, 290
  loss 11, 27, 275
  management 235, 237, 273, 402, 409
  market 22, 37, 84, 235, 354
  plan 213, 252
  planning 229, 230, 249, 250
  pollution 11, 206, 306, 309, 313, 315
  pricing 75, 81, 84, 87, 257, 292, 306, 307, 309, 312, 322, 343, 403
  quality 11, 19, 23, 25, 26, 29, 34, 44, 71, 73, 74, 77, 81, 101, 109, 110, 122, 123, 165, 187, 206, 209, 216, 223, 230, 238, 254, 264, 265, 272, 273, 282, 298, 303, 309, 315, 325, 329, 335, 386, 390, 401, 402, 405, 409
  recycling 221, 229, 232, 235–7, 253, 274, 399, 402
  requirement 25, 26, 55, 79, 82, 85, 86, 123, 155, 186, 210, 224, 247, 257, 292, 312, 340, 390, 402
  resources 9, 13, 19, 22, 23, 29, 33, 34, 44, 58, 65, 70, 75, 81, 84, 86, 87, 94, 110, 123, 125, 139, 141, 165, 169, 204, 209–11, 250, 252, 253, 264, 272, 292, 293, 299–301, 305–7, 314, 323, 326, 329, 386, 405
  rights 38–40, 81, 213, 218, 232, 234, 235, 239, 253, 306, 312
  saving 16, 104, 226, 228, 235, 236, 292, 306, 307, 312, 315, 403
  supply 36, 67, 73, 76, 79, 86, 92, 107, 109, 111, 114, 126, 141, 151, 165, 187, 191, 196, 212, 215, 224, 239, 249, 252, 264, 267, 270, 273, 300, 305, 306, 321, 325, 330, 361, 397, 402–4
  trading 168, 229, 232, 234, 257, 354, 358
  use 13, 16, 18, 20, 23, 64, 67, 68, 77, 79, 81–4, 107, 168, 207, 216, 227, 237, 247, 250, 257, 269, 274, 301, 306, 312, 325, 341
  use efficiency 18, 24, 27, 165, 178, 207, 235, 236, 253, 274, 306, 314, 322, 329, 341, 343, 354, 357, 386, 401, 403
  users 212, 215, 232, 244, 247, 250, 267, 275, 312, 314
  withdrawal 9, 10, 16, 165, 206, 267, 292, 302, 307, 313, 345
Water Corporation 162, 164, 168, 177
Water Dispute Tribunal 330
Waterfall 147
Waterlogging 10, 16, 77, 122, 322, 325, 327, 338
Water Quality Assessment Authority 326
Water Quality Review Committee 326
Water Research Foundation of Australia 120
WCD See World Commission on Dams
Weed 29, 101, 109, 403
Welcome Reef Dam 399
Welcome Reef Nature Reserve 399
Welland Canal 281
Wesfarmers/Marubeni joint venture 389, 391
Western Agricultural Industries Pty Limited 393, 395, 396
Western States Water Augmentation Concept 286
West Bengal 326
Western Australia 153, 165, 177, 364, 369, 384, 393
Western Route Scheme 308, 311
West Kimberley Irrigation Scheme 393
Wetland 33–5, 75, 77, 83, 110, 167, 196, 205, 223, 231, 271, 313, 327, 328, 393, 395, 401, 409
White, G.B. 121
White Leeds Dam 191
White Scheme 121
Whyalla 181
Wildlife 28, 29, 75, 176, 212, 223, 224, 232, 252–4, 271, 274, 284, 327, 336, 341
Willow Creek 252
  Dam 245
Wilmot Power Station 408
Wilmot River 408
Wollondilly River 397
Woomera 182
World Bank 22, 36, 312, 314, 336, 337
World Commission on Dams 9, 22, 23, 356
World Heritage List 406
World Trade Organisation 17
World War I 362, 366, 367, 397
World War II 5, 40, 137, 147, 182, 194, 245, 264, 276, 363, 364, 366
World Wildlife Fund 312
Woronora Dam 366
WTO See World Trade Organisation
Wyoming 242

Xinjiang 300, 304

Yamuna River 330
Yangtze Delta 300
Yangtze Plain 295
Yangtze River 295, 299, 301, 304, 307–9, 311, 313
Yarra Dam 403
Yarragadee aquifer 168
Yarra River 403
Yellow (Huanghe) River 19, 299–301, 304, 305, 307, 309, 311, 313
Yield 17, 25, 26, 29, 30, 82, 83, 105, 116, 117, 165, 169, 286, 298, 313, 315, 325, 385, 386
Yosemite National Park 219, 220
Yukon River 286
Yuma Desalting Plant 242

Zhou Enlai 308
Zinc 190, 193